Lecture Notes in Artificial Intelligence 7364

Subseries of Lecture Notes in Computer Science

Bernhard Gramlich Dale Miller
Uli Sattler (Eds.)

Automated Reasoning

6th International Joint Conference, IJCAR 2012
Manchester, UK, June 26-29, 2012
Proceedings

Springer

Series Editors

Randy Goebel, University of Alberta, Edmonton, Canada
Jörg Siekmann, University of Saarland, Saarbrücken, Germany
Wolfgang Wahlster, DFKI and University of Saarland, Saarbrücken, Germany

Volume Editors

Bernhard Gramlich
Technische Universität Wien, Fakultät für Informatik
Favoritenstr. 9, E185/2, 1040 Wien, Austria
E-mail: gramlich@logic.at

Dale Miller
INRIA Saclay and Laboratoire d'Informatique, École Polytechnique
Route de Saclay, 91128 Palaiseau Cedex, France
E-mail: dale.miller@inria.fr

Uli Sattler
The University of Manchester, School of Computer Science
Oxford Road, Manchester, M13 9PL, UK
E-mail: sattler@cs.man.ac.uk

ISSN 0302-9743 e-ISSN 1611-3349
ISBN 978-3-642-31364-6 e-ISBN 978-3-642-31365-3
DOI 10.1007/978-3-642-31365-3
Springer Heidelberg Dordrecht London New York

Library of Congress Control Number: 2012940337

CR Subject Classification (1998): I.2.3-4, F.4.1, D.2.4, F.3, F.4, I.2, G.1

LNCS Sublibrary: SL 7 – Artificial Intelligence

Typesetting: Camera-ready by author, data conversion by Scientific Publishing Services, Chennai, India

Printed on acid-free paper

Springer is part of Springer Science+Business Media (www.springer.com)

Preface

This volume contains the proceedings of the 6th International Joint Conference on Automated Reasoning (IJCAR 2012) held during June 26–29, 2012, in Manchester, UK. This year's meeting was a merging of several leading events in automated reasoning, namely, CADE (International Conference on Automated Deduction), FroCoS (International Symposium on Frontiers of Combining Systems), FTP (International Workshop on First-Order Theorem Proving), and TABLEAUX (International Conference on Automated Reasoning with Analytic Tableaux and Related Methods). During the meeting there were four different systems competitions and, during the two days following the meeting, there were 15 workshops. IJCAR 2012 was part of the Alan Turing Year 2012 and colocated with The Alan Turing Centenary Conference held June 22–25, 2012. Previous editions of IJCAR took place in Siena, Italy (2001), Cork, Ireland (2004), Seattle, USA (2006), Sydney, Australia (2008) and Edinburgh, UK (2010).

The call for papers invited authors to submit either full papers (of 15 pages) or system descriptions (of seven pages). We received a total of 116 submissions and eventually accepted 32 full papers and nine system descriptions. Each submission was reviewed by at least three Program Committee members and their selected reviewers.

We are pleased that Nikolaj Bjørner, Yuri Matiyasevich, Robert Nieuwenhuis, and Nicole Schweikardt accepted to give invited talks during the technical part of the program. We are also honored that Peter Andrews, Martin Davis, and John Alan Robinson, three pioneers in automated reasoning, accepted to give evening talks. Another highlight was the Herbrand Award ceremony, where CADE Inc. honored Melvin Fitting for his exceptional contributions to the field of automated deduction.

We wish to thank the Program Committee members and their reviewers for their efforts in helping to evaluate the submissions. They have generously shared their knowledge of the field and provided the authors with helpful feedback: it has been a pleasure to work with them. The EasyChair conference management system was a great help in dealing with all aspects of putting together our program and the proceedings.

We also wish to thank the sponsors of this meeting: the *Artificial Intelligence Journal*, Microsoft Research, and the University of Manchester.

The local organization of the conference as well as the organization of its satellite events and competitions are challenging, time-consuming tasks, and we are extremely thankful to everybody who volunteered to contribute to these, in particular but not restricted to Vicki Chamberlin, Birte Glimm, Konstantin Korovin, Ruth Maddocks, Rina Srabonian, Geoff Sutcliffe, and Andrei Voronkov.

Finally, we would like to thank all authors for submitting their work to IJCAR 2012: this resulted in what we believe was an exciting technical program.

May 2012

Bernhard Gramlich
Dale Miller
Uli Sattler

Conference Organization

Conference Chairs

Konstantin Korovin	University of Manchester, UK
Andrei Voronkov	University of Manchester, UK

Program Committee Chairs

Bernhard Gramlich	TU Wien, Austria
Dale Miller	INRIA Saclay - Île-de-France, France
Uli Sattler	University of Manchester, UK

Satellite Events Chair

Birte Glimm	Ulm University, Germany

Competitions Chair

Geoff Sutcliffe	University of Miami, USA

Program Committee

Takahito Aoto	RIEC, Tohoku University, Japan
Franz Baader	TU Dresden, Germany
Peter Baumgartner	NICTA, ANU, Australia
Maria Paola Bonacina	Università degli Studi di Verona, Italy
Torben Braüner	Roskilde University, Denmark
Hans De Nivelle	University of Wrocław, Poland
Michael Fink	TU Wien, Austria
Jacques Fleuriot	University of Edinburgh, UK
Silvio Ghilardi	Università degli Studi di Milano, Italy
Jürgen Giesl	RWTH Aachen, Germany
Bernhard Gramlich	TU Wien, Austria
Reiner Hähnle	Chalmers University of Technology, Sweden
Florent Jacquemard	ENS de Cachan, France
Deepak Kapur	University of New Mexico, Albuquerque, USA
Yevgeny Kazakov	The University of Oxford, UK
Hélène Kirchner	INRIA Rocquencourt, France
Konstantin Korovin	University of Manchester, UK
Martin Lange	Universität Kassel, Germany

Stéphane Lengrand	LIX, Ecole Polytechnique, France
Carsten Lutz	Universität Bremen, Germany
Christopher Lynch	Clarkson University, Potsdam, USA
Christoph Lüth	DFKI and Universität Bremen, Germany
George Metcalfe	Universität Bern, Switzerland
Dale Miller	INRIA Saclay - Île-de-France, France
Aleksandar Nanevski	IMDEA-Software, Madrid, Spain
Tobias Nipkow	TU München, Germany
Albert Oliveras	Technical University of Catalonia, Barcelona, Spain
Nicolas Peltier	LIG/IMAG, Grenoble, France
Frank Pfenning	Carnegie Mellon University, USA
Grigore Rosu	University of Illinois at Urbana-Champaign, USA
Michael Rusinowitch	Loria-INRIA-Lorraine, Nancy, France
Uli Sattler	University of Manchester, UK
Viorica Sofronie-Stokkermans	MPI für Informatik, Saarbrücken, Germany
Georg Struth	University of Sheffield, UK
Aaron Stump	The University of Iowa, USA
Geoff Sutcliffe	University of Miami, USA
René Thiemann	University of Innsbruck, Austria
Cesare Tinelli	The University of Iowa, USA
Alwen Tiu	ANU, Canberra, Australia
Bow-Yaw Wang	Academia Sinica, Taipei, Taiwan
Christoph Weidenbach	MPI für Informatik, Saarbrücken, Germany
Michael Zakharyaschev	Birkbeck College, London, UK
Hans Zantema	Eindhoven University of Technology, The Netherlands

Additional Reviewers

Alberti, Francesco
Aravantinos, Vincent
Badban, Bahareh
Baelde, David
Bersani, Marcello
Blanchette, Jasmin Christian
Bobot, François
Boy de La Tour, Thierry
Brauner, Paul
Brockschmidt, Marc
Brotherston, James
Bruttomesso, Roberto
Bubel, Richard
Bundy, Alan
Cerami, Marco
Chaudhuri, Kaustuv
Ciobaca, Stefan
Cuenca Grau, Bernardo
Dao-Tran, Minh
De Moura, Leonardo
Denney, Ewen
Eades, Harley
Echenim, Mnacho
Faella, Marco
Felgenhauer, Bertram
Fontaine, Pascal
Frehse, Goran
Friedmann, Oliver
Fu, Peng
Fuchs, Alexander

Fuhs, Carsten
Galmiche, Didier
Gherardi, Guido
Gimenez, Stéphane
Gnaedig, Isabelle
Goncalves, Rafael
Goncharov, Sergey
Goré, Rajeev
Gorin, Daniel
Gregoire, Benjamin
Guelev, Dimitar
Guenot, Nicolas
Göller, Stefan
Haarslev, Volker
Haemmerlé, Rémy
Harrison, John
Heijltjes, Willem
Hermant, Olivier
Houtmann, Clement
Hoyrup, Mathieu
Huang, Guan-Shieng
Hutter, Dieter
Ji, Ran
Jovanovic, Dejan
Jung, Jean Christoph
Kaminski, Mark
Kesner, Delia
King, Tim
Knapp, Alexander
Komendantskaya, Ekaterina
Koshimura, Miyuki
Krennwallner, Thomas
Kuhtz, Lars
Kuznets, Roman
Köpf, Boris
Lange, Christoph
Larchey-Wendling, Dominique
Loup, Ulrich
Lozes, Etienne
Ma, Feifei
McKinley, Richard
Mclaughlin, Sean
Meikle, Laura
Michaliszyn, Jakub
Middeldorp, Aart
Morawska, Barbara
Moreau, Pierre-Etienne
Moser, Georg
Mossakowski, Till
Mousavi, Mohammad Reza
Nahon, Fabrice
Navarro Perez, Juan Antonio
Neurauter, Friedrich
Noguera, Carles
Nonnengart, Andreas
Obua, Steven
Otop, Jan
Otto, Carsten
Papapanagiotou, Petros
Payet, Etienne
Peñaloza, Rafael
Popescu, Andrei
Qi, Guilin
Quesada, Luis
Rabe, Florian
Ranise, Silvio
Redl, Christoph
Ringeissen, Christophe
Rodriguez-Carbonell, Enric
Ruemmer, Philipp
Sano, Katsuhiko
Sasaki, Katsumi
Sato, Haruhiko
Schmidt, Renate A.
Schmidt-Schauß, Manfred
Schneider, Thomas
Schröder, Lutz
Schuppan, Viktor
Schwind, Camilla
Schüller, Peter
Scott, Phil
Serbanuta, Traian
Seylan, Inanc
Simancik, Frantisek
Simpson, Alex
Stefanescu, Andrei
Sticksel, Christoph
Stratulat, Sorin
Ströder, Thomas
Stuckey, Peter

Studer, Thomas
Suda, Martin
Tacchella, Armando
Tamura, Naoyuki
Tsarkov, Dmitry
Turhan, Anni-Yasmin
Urban, Josef
Van Delft, Bart
Viganò, Luca
Vigneron, Laurent
Villaret, Mateu
Waldmann, Uwe
Wansing, Heinrich
Widmann, Florian
Wiedijk, Freek
Willemse, Tim
Wilson, Sean
Xiao, Guohui
Zanardo, Alberto
Zeilberger, Noam

Table of Contents

Invited Talks

Full Papers and System Descriptions

Taking Satisfiability to the Next Level with Z3
(Abstract)

Nikolaj Bjørner

Microsoft Research
nbjorner@microsoft.com

Several applications from program analysis, design and testing rely critically on solving SMT problems. Many applications build on top of SMT solvers in sophisticated ways by carefully crafting the solver interaction. We illustrate partial correctness checking as an SMT problem and we introduce a procedure for model finding of recursive Horn clauses with arithmetic.

The Satisfiability Modulo Theories [1, 2] (SMT) solver Z3 [3], from Microsoft Research, is a core of several advanced program analysis, testing and model-based development tools. These tools rely on components using logic for describing states and transformations between system states. Consequently, they require a logic inference engine for reasoning about the state transformations. Z3 is particularly appealing because it combines specialized solvers for domains that are of relevance for computation and it integrates crucial innovations in automated deduction. It is tempting to build custom ad-hoc solvers for each application, but extending and scaling these require a high investment and will inevitably miss advances from automated deduction. New applications introduce new challenges for Z3 and provide inspiration for improving automated deduction techniques. It is not uncommon that when improvements to Z3 are made based on one application, other applications benefit as well.

A different dimension for advancing SMT solving is by raising the bar for interfacing with SMT. We describe recent progress in the context of Z3 on supporting property checking of recursively defined predicates. This includes symbolic software verification of safety properties as an SMT problem. While symbolic software model checkers have been developed for more than a decade now, comparisons and evaluation was only recently introduced [4] for tools capable of analyzing C programs. In spite of a high overlap of methodology in the underlying deduction problems, currently separate techniques are developed for other programming languages. We will illustrate SMT as a basis for symbolic software model checking of safety properties. This perspective is not limited to safety properties; besides making the case promoted here, [5] shows how to also formulate termination by including second-order predicates.

1 Some Applications of Z3

Figure 1 shows a number of Microsoft tools using Z3. Most can be tried online at http://rise4fun.com. The program verification systems F⋆ [6] that integrates a rich refinement type system, VCC [7] that was used to verify large

B. Gramlich, D. Miller, and U. Sattler (Eds.): IJCAR 2012, LNAI 7364, pp. 1–8, 2012.

portions of the Viridian Hyper-V product, HAVOC [8] that has been used to check Windows kernel code and Dafny [9] use the Boogie[1] verification condition generator. Boogie converts programs with contracts into SMT formulas. The SLAyer [10] system, that works on Windows device drivers, uses separation logic when checking memory safety of systems code. The Symbolic Difference tool SymDiff [11], that is used actively for compiler verification and validation, uses procedure summaries when checking for equivalence of procedures from two different programs. We have in a rough manner categorized these systems as *property driven*, as they share a common trait of taking contracts (pre/post conditions and invariant annotations) and refining these according to the control-flow graph. A different set of tools takes advantage of symbolic simulation of program executions. We say they are *execution guided*, by either over or under-approximations of executions. These include the SLAM/SDV [12] static driver verifier that ships with Windows Server editions as part of the Driver Development Kit, the Terminator tool [13], Poirot [14], Yogi [15] that is now also being added to the SDV distribution, Pex and SAGE [16] that use dynamic symbolic execution. Pex is used by thousands of .NET developers to enhance unit-testing, and SAGE is used heavily internally at Microsoft to find security vulnerabilities in media readers. The right column of the figure contains *model-based* tools. They rely on a model to either test [17], develop [18, 19] or synthesize software [20]. There are several other remarkable systems, including [21–25], that integrate and develop creative algorithms around Z3.

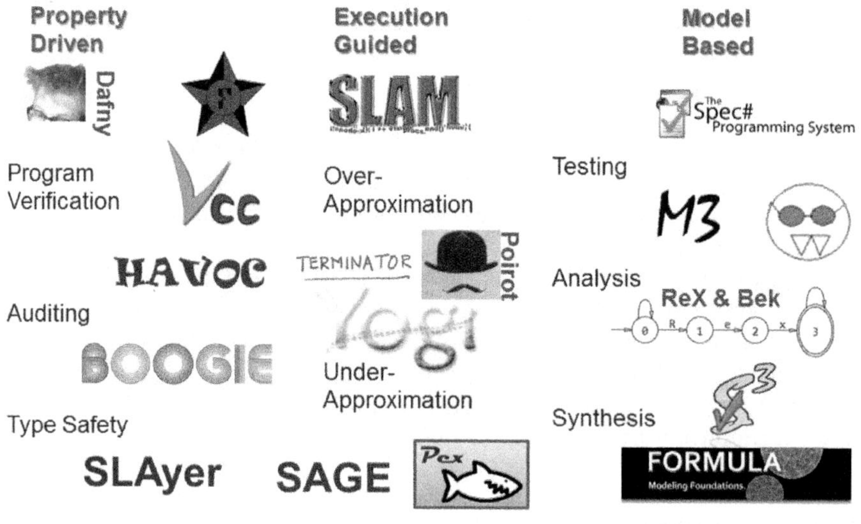

Fig. 1. Some Microsoft Systems using Z3

[1] `http://research.microsoft.com/~leino/papers/krml178.pdf`

2 Solving Recursive Predicates

This section recalls connections with program properties as recursive predicates and invariants as solutions to the predicates [26, 5].

2.1 Solving Recursive Predicates Is an SMT Problem

The set of reachable states of a transition system is a recursively defined predicate. The set of reachable configurations of a procedural program is also a recursively defined predicate. Partial correctness checking relies on establishing that the least fixed-point of the recursively defined predicates is contained in the specification property. A way to specify recursive predicates is by using Horn clauses in the style of pure Prolog. The central tool for establishing partial correctness of programs is to find inductive invariants that imply the specification property. Inductive invariants are interpretations of predicates that satisfy the Horn clauses. Correctness of a specification property amounts to checking that it is implied by the least fixed-point of the recursive predicates. Equivalently, there is an inductive invariant that implies the specification property. In other words, checking partial program correctness is simply put a first-order Satisfiability Modulo Theories problem: it suffices to find an interpretation (inductive invariant) for the Horn clauses from the recursive relation and for the implication to the specification property.

Nevertheless, SMT solvers have been mainly targeted at large quantifier-free problems and for checking unsatisfiability of quantified formulas, and mostly used as stepping stones for finding models of (recursive) Horn clauses. The model finding problem for such clauses requires finding (or at least establishing that there are) interpretations for formulas with quantifiers and theories. In particular, arithmetic is prolific in some applications extracted from software analysis, which makes the model finding problem highly intractable. SMT solvers today are not directly targeted at formulas from symbolic model checking applications. Many symbolic software model checking tools are still able to successfully use SMT solving. They integrate with SMT solvers using abstraction refinement loops based on interpolants, predicate abstraction and other techniques that have been developed with program verification in mind. The number of techniques developed for automatically checking specification properties is abundant. Regardless, in a nutshell they solve an SMT problem.

2.2 From Programs and Properties to SMT

Let us illustrate how partial correctness properties of programs correspond to SMT. Consider a two-process version of Lamport's Bakery algorithm.

initially $y_1 := y_2 := 0;$

$$P_1 :: \left[\begin{array}{l}\textbf{loop forever do}\\ \left[\begin{array}{l}\ell_0 : y_1 := y_2 + 1;\\ \ell_1 : \textbf{await } y_2 = 0 \vee y_1 \leq y_2;\\ \ell_2 : \textbf{critical};\\ \ell_3 : y_1 := 0;\end{array}\right]\end{array}\right] \;||\; P_2 :: \left[\begin{array}{l}\textbf{loop forever do}\\ \left[\begin{array}{l}\ell_0 : y_2 := y_1 + 1;\\ \ell_1 : \textbf{await } y_1 = 0 \vee y_2 \leq y_1;\\ \ell_2 : \textbf{critical};\\ \ell_3 : y_2 := 0;\end{array}\right]\end{array}\right]$$

The safety property of interest is *mutual exclusion*: processes P_1 and P_2 cannot simultaneously execute **critical**. We can capture the set of reachable states of this program and the safety property as a set of Horn clauses that are satisfiable if and only if the safety property holds.

The relation R encodes the reachable configurations for the control locations for P_1 and P_2 and the two variables y_1, y_2. The first clause states that the state where both processes are at ℓ_0 and $y_1 = y_2 = 0$ is reachable. The next two clauses encode a step by either P_1 or P_2. The transition relation T encodes each of the steps. For example, if $y_1 \leq y_2 \vee y_2 = 0$, then P_1 can move from ℓ_1 to ℓ_2. We exploit that P_1 and P_2 are symmetric by swapping y_1 and y_2 so we can reuse the transition relation. Capitalized variables are implicitly universally quantified in the clauses.

$$
\begin{aligned}
&R(\ell_0, \ell_0, 0, 0).\\
R(L', M, Y_1', Y_2) &\leftarrow R(L, M, Y_1, Y_2) \wedge T(L, L', Y_1, Y_2, Y_1').\\
R(L, M', Y_1, Y_2') &\leftarrow R(L, M, Y_1, Y_2) \wedge T(M, M', Y_2, Y_1, Y_2').\\
&T(\ell_0, \ell_1, Y_1, Y_2, Y_2 + 1).\\
T(\ell_1, \ell_2, Y_1, Y_2, Y_1) &\leftarrow Y_1 \leq Y_2 \vee Y_2 = 0.\\
&T(\ell_2, \ell_3, Y_1, Y_2, Y_1).\\
&T(\ell_3, \ell_0, Y_1, Y_2, 0).\\
\mathit{false} &\leftarrow R(\ell_2, \ell_2, Y_1, Y_2).
\end{aligned}
$$

The interpretation $R(L, M, Y_1, Y_2) := S(L, M, Y_1, Y_2) \wedge S(M, L, Y_2, Y_1)$ where

$$
S(L, M, Y_1, Y_2) := \begin{array}{l}
((L = \ell_1 \vee L = \ell_2 \vee L = \ell_3) \rightarrow Y_1 > 0)\\
\wedge\ (L = \ell_0 \rightarrow Y_1 = 0)\\
\wedge\ ((L = \ell_2 \wedge (M = \ell_3 \vee M = \ell_4)) \rightarrow Y_2 < Y_1)\\
\wedge\ ((L = \ell_2 \wedge M = \ell_2) \rightarrow Y_1 \neq Y_2)
\end{array}
$$

satisfies the Horn clauses.

To illustrate recursive procedures we use a (contrived) program P that balances the values of variables y_1, y_2. The corresponding Horn clauses are listed on the right. They are satisfiable if and only if y_1, y_2 are unchanged after P returns .

$$
\begin{array}{l}
P(\mathsf{var}\ y_1, \mathsf{var}\ y_2) ::\\
\left[\begin{array}{l}
\textbf{if } \star \textbf{ then}\\
\left[\begin{array}{l}
\ell_0 : y_1 := y_1 + 1;\\
\ell_1 : P(y_2, y_1);\\
\ell_2 : P(y_1, y_2);\\
\ell_3 : y_1 := y_1 - 1;
\end{array}\right]
\end{array}\right]
\end{array}
\qquad
\begin{aligned}
&P(Y_1, Y_2, Y_1, Y_2).\\
P(Y_1, Y_2, Y_5 - 1, Y_6) &\leftarrow \left[\begin{array}{l} P(Y_2, Y_1 + 1, Y_3, Y_4)\\ \wedge\ P(Y_4, Y_3, Y_5, Y_6)\end{array}\right].\\
\mathit{false} &\leftarrow \left[\begin{array}{l} P(Y_1, Y_2, Y_3, Y_4)\\ \wedge\ ((Y_1 \neq Y_3) \vee (Y_2 \neq Y_4))\end{array}\right].
\end{aligned}
$$

The relevant inductive property is $P(Y_1, Y_2, Y_3, Y_4) := Y_1 = Y_3 \wedge Y_2 = Y_4$.

2.3 Generalized Property Directed Reachability

The IC3 algorithm [27] is a new algorithm that has so far been used successfully in the context of hardware model checking. IC3 is most often referred to as PDR, which is a shorthand for Property Directed Reachability. Z3 contains an implementation of PDR, where the algorithm is generalized to handle both programs with procedure calls and constraints using arithmetic [28]. The PDR algorithm has striking analogies with Conflict Directed Clause Learning [29], a basis of modern DPLL-based SAT solvers. PDR simultaneously strengthens an abstraction of reachable states and prunes a search for counter-examples. It maintains the original transition relations, and only requires creating abstractions of reachable states as conjunctions of state properties (as conflict clauses). A clever use of induction, reminiscent of subsumption and clause minimization, allows strengthening state properties.

We will not spell out the entire formal apparatus of our generalization of PDR here, but we now give an intuition of the PDR generalization using Bakery. The Horn clauses that are used to define R correspond to a predicate transformer $\mathcal{F}$ (likely better known as the strongest post-condition):

$$\begin{aligned} & \Theta \vee \\ \mathcal{F}(R)(L, M, Y_1, Y_2) := \; & (\exists L_0, Y_0 \,.\, T(L_0, L, Y_0, Y_2, Y_1) \wedge R(L_0, M, Y_0, Y_2)) \vee \\ & (\exists M_0, Y_0 \,.\, T(M_0, M, Y_0, Y_1, Y_2) \wedge R(L, M_0, Y_1, Y_0)) \end{aligned}$$

where $\Theta := L = M = \ell_0 \wedge Y_1 = Y_2 = 0$. The safety property of interest is $S := \neg(L = \ell_2 \wedge M = \ell_2)$.

The algorithm maintains over-approximations of reachable states $R_0, \ldots, R_N$ satisfying, for every $0 \leq i < N$ the implications in (1), and also $R_0 := \Theta$. In our notation, each arrow stands for one implication. The implications in (1) are therefore $R_i \rightarrow S$, $R_i \rightarrow R_{i+1}$ and $\mathcal{F}(R_i) \rightarrow R_{i+1}$. Initially, $N := 0$ and $R_1 := R_2 := \ldots := \mathit{true}$. The algorithm increments N every time it can establish that $R_N \rightarrow S$. It concludes that S is invariant if there is an i, such that $R_{i+1} \rightarrow R_i$ (together with the invariant (1), this implies that R_i is an inductive invariant that implies S). Since $R_0 \rightarrow S$ because $R_0 = \Theta = (L = M = \ell_0 \wedge Y_1 = Y_2 = 0)$ and $S = \neg(L = M = \ell_2)$ we can set $N := 1$. The next action is to check if $R_N \wedge \neg S$ is satisfiable. It is, and one possible model is denoted by $\mathcal{M} := L = M = \ell_2 \wedge Y_1 = Y_2 = 0$. This state violates the safety property, but is it reachable from the current unfolding? It would be if $\mathcal{F}(R_0) \wedge \mathcal{M}$ were satisfiable. It is not satisfiable. There are several unsatisfiable cores explaining unsatisfiability. One is $L = \ell_0 \vee M = \ell_0$, which we can conjoin to R_1. In other words, we update $R_1 := R_1 \wedge (L = \ell_0 \vee M = \ell_0)$ and $N := 2$. After a few steps, yada-yada-yada, we arrive to the point where $N = 4$, $R_4 = \mathit{true}$ and we examine whether $R_4 \wedge \neg S$ is satisfiable. Again, a possible assignment is $\mathcal{M} := L = M = \ell_2 \wedge Y_1 = Y_2 = 0$. The formula $\mathcal{F}(R_3) \wedge \mathcal{M}$ is also satisfiable, and extends to a model $\mathcal{M}_3$ that assigns values to the existentially quantified variables that are used in the recursive invocation of R. For example $\mathcal{M}_3 := L = \ell_1 \wedge M = \ell_2 \wedge Y_1 = Y_2 = 0$.

$$\begin{array}{ccccc} S & & R_{i+1} & & \\ & \nwarrow \quad \nearrow & & \nwarrow & \\ & R_i & & & \mathcal{F}(R_i) \end{array} \qquad (1)$$

We can repeat this backwards propagation a few steps to drive the counter-example candidate to the initial state. The attempt is illustrated on the right.

During propagation, $\mathcal{F}(R_2) \wedge \mathcal{M}_3$ is satisfiable and there is a model of the arguments to the recursive invocation of R, $\mathcal{M}_2 := L = M = \ell_1 \wedge Y_1 = Y_2 = 0$. One more time: $\mathcal{F}(R_1) \wedge \mathcal{M}_2$ is satisfiable and a model for the recursive arguments is $\mathcal{M}_1 := L = \ell_0 \wedge M = \ell_1 \wedge Y_1 = -1 \wedge Y_2 = 0$.

$$
\begin{array}{c}
L = M = \ell_2 \wedge Y_1 = Y_2 = 0 \models \mathcal{F}(R_3) \wedge \neg S \\
\uparrow \\
L = \ell_1 \wedge M = \ell_2 \wedge Y_1 = Y_2 = 0 \models \mathcal{F}(R_2) \\
\uparrow \\
L = \ell_1 \wedge M = \ell_1 \wedge Y_1 = Y_2 = 0 \models \mathcal{F}(R_1) \\
\vdots \\
L = \ell_0 \wedge M = \ell_1 \wedge Y_1 = -1 \wedge Y_2 = 0 \models \neg\mathcal{F}(R_0)
\end{array}
$$

We are almost done, but the final push to $\mathcal{F}(R_0) \wedge \mathcal{M}_1$ fails because R_0 implies $Y_1 = 0$. We could add the clause $Y_1 \neq -1$ to R_1 to rule out this spurious counter-example, but we can do better.

The approach is to identify the strongest conflict lemma that contradicts $\mathcal{M}_1$ to prune the maximal number of spurious counter-examples. In this case, the strongest conflict lemma is $Y_1 \geq 0$. Z3 uses Farkas lemma. Recall that Farkas lemma justifies so-called *theory conflicts* for real linear arithmetic; it establishes that a conjunction of linear inequalities is infeasible iff there is a linear combination that adds up to an infeasible constraint $1 \leq 0$. Z3 uses Farkas lemma to find strongest conflict lemma during proof search. The theory conflicts use literals that come from either $\mathcal{M}_1$ or $\mathcal{F}(R_0)$. For example $Y_1 \leq -1 \wedge Y_1 \geq 0$ is a theory conflict, where $Y_1 \leq -1$ is an inequality of $\mathcal{M}_1$. The other inequality in the theory conflict comes from $\mathcal{F}(R_0)$. The basic idea is that we could have weakened $\mathcal{M}_1$ to be the negation $Y_1 < 0$ and still use this theory conflict. In general, Farkas lemma ensures that there are suitable coefficients to the literals from $\mathcal{F}(R_0)$, such that the linear combination multiplied by these coefficients is a conflict lemma to $\mathcal{M}_1$.

It takes generalized PDR a few iterations to converge on an inductive invariant for Bakery. We show [28] that this abstraction refinement technique is a decision procedure for timed push-down systems, and in spite of the general problem being highly intractable, this method alone also suffices for several applications from software verification. It is still very easy to produce examples where this method diverges. The quest is on for extending this approach with other methods that can produce sufficient inductive invariants on a broader range of applications.

There is an intimate connection with our generalization of PDR with interpolation [30]. Every time PDR increments N, the labeling of predicates are interpolants for a bounded DAG unfolding of the Horn clauses [31, 32] (predicates are given fresh names based on their depth in the DAG). Our method interleaves the generation of the DAG interpolant with satisfiability search and leverages the partial results (conflict clauses) to prune the search space. Current work includes extending the method to synthesize interpolants for uninterpreted functions, and by reduction also the theory of arrays and algebraic data-types.

3 Conclusions

Several applications from program analysis, design and testing rely critically on solving SMT problems. Many applications build on top of SMT solvers in sophisticated ways by carefully crafting the solver interaction. We illustrated partial correctness checking as an SMT problem and we introduced a procedure for model finding of recursive Horn clauses with arithmetic.

Acknowledgments. Z3 is developed by Leonardo de Moura, Christoph Wintersteiger and the author. The perspective of symbolic software analysis as SMT owes to discussions with Leonardo de Moura, Andrey Rybalchenko, Ken McMillan, Tony Hoare, Josh Berdine, Ethan Jackson and Karthick Jayaraman. The generalization of PDR is joint work with Kryštof Hoder.

References

1. Barrett, C.W., Sebastiani, R., Seshia, S.A., Tinelli, C.: Satisfiability Modulo Theories. In: Biere, A., Heule, M., van Maaren, H., Walsh, T. (eds.) Handbook of Satisfiability. Frontiers in Artificial Intelligence and Applications, vol. 185, pp. 825–885. IOS Press (2009)
2. de Moura, L., Bjørner, N.: Satisfiability Modulo Theories: Introduction & Applications. Comm. ACM (2011)
3. de Moura, L., Bjørner, N.: Z3: An Efficient SMT Solver. In: Ramakrishnan, C.R., Rehof, J. (eds.) TACAS 2008. LNCS, vol. 4963, pp. 337–340. Springer, Heidelberg (2008)
4. Beyer, D.: Competition on Software Verification - (SV-COMP). In: [33], pp. 504–524
5. Grebenshchikov, S., Lopes, N.P., Popeea, C., Rybalchenko, A.: Synthesizing software verifiers from proof rules. In: PLDI (2012)
6. Swamy, N., Chen, J., Fournet, C., Strub, P.Y., Bhargavan, K., Yang, J.: Secure distributed programming with value-dependent types. In: Chakravarty, M.M.T., Hu, Z., Danvy, O. (eds.) ICFP, pp. 266–278. ACM (2011)
7. Cohen, E., Dahlweid, M., Hillebrand, M., Leinenbach, D., Moskal, M., Santen, T., Schulte, W., Tobies, S.: VCC: A Practical System for Verifying Concurrent C. In: Berghofer, S., Nipkow, T., Urban, C., Wenzel, M. (eds.) TPHOLs 2009. LNCS, vol. 5674, pp. 23–42. Springer, Heidelberg (2009)
8. Condit, J., Hackett, B., Lahiri, S.K., Qadeer, S.: Unifying type checking and property checking for low-level code. In: Shao, Z., Pierce, B.C. (eds.) POPL, pp. 302–314. ACM (2009)
9. Rustan, K., Leino, M.: Developing Verified Programs with Dafny. In: Joshi, R., Müller, P., Podelski, A. (eds.) VSTTE 2012. LNCS, vol. 7152, p. 82. Springer, Heidelberg (2012)
10. Berdine, J., Cook, B., Ishtiaq, S.: SLAyer: Memory Safety for Systems-Level Code. In: Gopalakrishnan, G., Qadeer, S. (eds.) CAV 2011. LNCS, vol. 6806, pp. 178–183. Springer, Heidelberg (2011)
11. Lahiri, S., Hawblitzel, C., Kawaguchi, M., Rebelo, H.: SymDiff: A language-agnostic semantic diff tool for imperative programs, (Tool description) (2012)
12. Ball, T., Rajamani, S.K.: The SLAM project: debugging system software via static analysis. SIGPLAN Not. 37, 1–3 (2002)
13. Cook, B., Podelski, A., Rybalchenko, A.: Terminator: Beyond Safety. In: Ball, T., Jones, R.B. (eds.) CAV 2006. LNCS, vol. 4144, pp. 415–418. Springer, Heidelberg (2006)
14. Lal, A., Qadeer, S., Lahiri, S.: Corral: A Solver for Reachability Modulo Theories (2012)

15. Nori, A.V., Rajamani, S.K., Tetali, S., Thakur, A.V.: The YOGI Project: Software Property Checking via Static Analysis and Testing. In: Kowalewski, S., Philippou, A. (eds.) TACAS 2009. LNCS, vol. 5505, pp. 178–181. Springer, Heidelberg (2009)
16. Godefroid, P., de Halleux, J., Nori, A.V., Rajamani, S.K., Schulte, W., Tillmann, N., Levin, M.Y.: Automating Software Testing Using Program Analysis. IEEE Software 25, 30–37 (2008)
17. Grieskamp, W., Kicillof, N., MacDonald, D., Nandan, A., Stobie, K., Wurden, F.L.: Model-Based Quality Assurance of Windows Protocol Documentation. In: ICST, pp. 502–506. IEEE Computer Society (2008)
18. Jackson, E.K., Kang, E., Dahlweid, M., Seifert, D., Santen, T.: Components, platforms and possibilities: towards generic automation for MDA. In: EMSOFT, pp. 39–48 (2010)
19. Veanes, M., Bjørner, N.: Symbolic Automata: The Toolkit. In: [33], pp. 472–477
20. Srivastava, S., Gulwani, S., Foster, J.S.: From program verification to program synthesis. In: Hermenegildo, M.V., Palsberg, J. (eds.) POPL, pp. 313–326. ACM (2010)
21. Yang, J., Hawblitzel, C.: Safe to the last instruction: automated verification of a type-safe operating system. In: Zorn, B.G., Aiken, A. (eds.) PLDI, pp. 99–110. ACM (2010)
22. Yang, J., Yessenov, K., Solar-Lezama, A.: A language for automatically enforcing privacy policies. In: [34], pp. 85–96
23. Köksal, A.S., Kuncak, V., Suter, P.: Constraints as control. In: [34], pp. 151–164
24. Chugh, R., Rondon, P.M., Jhala, R.: Nested refinements: a logic for duck typing. In: [34], pp. 231–244
25. Madhusudan, P., Qiu, X., Stefanescu, A.: Recursive proofs for inductive tree data-structures. In: [34], pp. 123–136
26. Bjørner, N., Browne, A., Manna, Z.: Automatic Generation of Invariants and Assertions. In: Montanari, U., Rossi, F. (eds.) CP 1995. LNCS, vol. 976, pp. 589–623. Springer, Heidelberg (1995)
27. Bradley, A.R.: SAT-Based Model Checking without Unrolling. In: Jhala, R., Schmidt, D. (eds.) VMCAI 2011. LNCS, vol. 6538, pp. 70–87. Springer, Heidelberg (2011)
28. Hoder, K., Bjørner, N.: Generalized Property Directed Reachability. In: SAT (2012)
29. Silva, J.P.M., Sakallah, K.A.: GRASP: A Search Algorithm for Propositional Satisfiability. IEEE Trans. Computers 48, 506–521 (1999)
30. McMillan, K.L.: Interpolants from Z3 proofs. In: FMCAD (2011)
31. Gupta, A., Popeea, C., Rybalchenko, A.: Solving Recursion-Free Horn Clauses over LI+UIF. In: Yang, H. (ed.) APLAS 2011. LNCS, vol. 7078, pp. 188–203. Springer, Heidelberg (2011)
32. Albarghouthi, A., Gurfinkel, A., Chechik, M.: From under-approximations to over-approximations and back. In: [33], pp. 157–172
33. Flanagan, C., König, B. (eds.): TACAS 2012. LNCS, vol. 7214. Springer, Heidelberg (2012)
34. Field, J., Hicks, M. (eds.): Proceedings of the 39th ACM SIGPLAN-SIGACT Symposium on Principles of Programming Languages, POPL 2012, Philadelphia, Pennsylvania, USA, January 22-28. ACM (2012)

Enlarging the Scope of Applicability of Successful Techniques for Automated Reasoning in Mathematics

Yuri Matiyasevich

Steklov Institute of Mathematics, St. Petersburg, Russia

In mathematics sometimes methods from one area can be fruitfully applied for getting results in another area, occasionally looking very remote from the other area. A well-known example is given by analytic geometry that enables us, besides proving "elementary" geometrical theorems, to establish otherwise untractable results like unsolvability of the problems of angle trisection and doubling the cube by compass and straightedge and to reduce calculation of the kissing numbers of spheres to verification of a first-order formula about real numbers (and that could be done, in principle, by Tarski algorithm).

In automated reasoning in mathematics we witness spectacular achievements in some narrow areas such as integration in closed form or proving combinatorial identities. The author's suggestion is to try to enlarge the scope of applicability of such successful techniques by proper (mathematical) reductions of other problems to the required forms. This will be illustrated on the expressive power of the language of binomial coefficients [1,2].

References

1. Margenstern, M., Matiyasevich, Y.: A binomial representation of the $3x+1$ problem. Acta Arith. 91(4), 367–378 (1999)
2. Matiyasevich, Y.: Some arithmetical restatements of the four color conjecture. Theor. Comput. Sci. 257(1-2), 167–183 (2001)

B. Gramlich, D. Miller, and U. Sattler (Eds.): IJCAR 2012, LNAI 7364, p. 9, 2012.

SAT and SMT Are Still Resolution: Questions and Challenges

Robert Nieuwenhuis*

Abstract. The aim of this invited talk is to discuss strengths, limitations and challenges around one of the simplest yet most powerful practical automated deduction formalisms, namely propositional SAT and its extensions. We will see some of the reasons why *CDCL SAT solvers* are so effective for finding solutions to so diverse real-world problems, using a single fully automatic push-button strategy, and, by extending them to SAT Modulo Theories (SMT), also to optimization problems and problems with complex (e.g., arithmetic) constraints for which a full encoding into SAT would be too large and/or inefficient. We will give some examples of trade-offs regarding full SAT encodings vs SMT theory solvers, and discuss why SAT and even SMT are just binary resolution strategies, the consequences of this fact, and possible ways to overcome it. Many aspects of the discussion carry over to first-order logic and beyond.

SAT. In spite of its simplicity, SAT has become very important for practical applications, especially in the multi-billion industry of electronic design automation (EDA), and, in general, hardware and software verification. Research on SAT has been pushed by huge industrial needs and resources and, as a result, modern *Conflict-Driven Clause Learning* (CDCL) SAT solvers also work impressively well on real-world problems from *many* other sources, using a *single, fully automatic, push-button* strategy (see the handbook [BHvMW09] for all details and further references on SAT and SAT encodings). Hence, modeling and using SAT has essentially become a *declarative* task. On the negative side, propositional logic is a low-level language and hence modeling and encoding *tools* are required.

Example: As a running example, consider the simple well-known NP-complete problem of *vertex cover*: given a graph, find a subset of at most K of its vertices containing for each edge (v, v') at least one of v and v'. For a graph with vertices $v_1, \ldots, v_n$, this problem can be encoded into SAT using n propositional variables x_i meaning "vertex v_i is in the cover", having for each edge (v_i, v_j) a propositional clause $x_i \vee x_j$, and adding more clauses for expressing that at most K of the variables $x_1, \ldots, x_n$ can be true, i.e., adding a SAT encoding for the *cardinality constraint* $x_1 + \ldots + x_n \leq K$, which can be, for example, a *cardinality network* [ANORC11], using $O(n \log^2 K)$ auxiliary variables and $O(n \log^2 K)$ clauses.

---

* Technical Univ. of Catalonia (UPC), Barcelona, Spain. Partially supported by Spanish Min. of Science &Innovation, SweetLogics project TIN2010-21062-C02-01.

B. Gramlich, D. Miller, and U. Sattler (Eds.): IJCAR 2012, LNAI 7364, pp. 10–13, 2012.

SMT. Even though a lot of work has been done on defining good SAT encodings for cardinality and other complex (not only arithmetic) constraints, such encodings may become too large and/or inefficient. *SAT Modulo Theories* (SMT) was developed as an answer to this situation (see [NOT06, BSST09]). The idea is to encode only part of the constraints into SAT and considering the remaining constraints as a background *theory*. Similarly to the filtering algorithms in Constraint Programming, during the CDCL SAT solving process a *theory solver* uses efficient *specialized algorithms* to detect additional propagations (and inconsistencies) with respect to this theory.

Example (cont.): For the cardinality constraint $x_1 + \ldots + x_n \leq K$ of our vertex cover example, each time the CDCL SAT solver reaches a partial model with K true variables $y_1 \ldots y_K$, the theory solver can propagate, setting all other x_i to false. If $\neg x_i$ then becomes part of a conflict, conflict analysis will use the clause $\neg y_1 \vee \ldots \vee \neg y_K \vee \neg x_i$ provided by the theory solver as the *reason* explaining the propagation of $\neg x_i$. This is similar to the *lazy clause generation* approach of [OSC09].

Trade-Offs. Many problems have numerous large (arithmetic and other) constraints. For example, in certain industrial scheduling contexts, one has to express that resources' capacities are never exceeded. If $t_{i,h}$ denotes that task i is active at hour h, and task i uses a_i units of a certain resource, e.g., trucks, of which K units are available, then for each h and resource there will be a pseudo-Boolean constraint of the form $a_1 \cdot t_{1,h} + \ldots + a_n \cdot t_{n,h} \leq K$ (if there are n tasks). If the number of available trucks is not an important bottleneck, the propagations caused by the "truck constraints" will cause relatively few conflicts, few reason clauses will have to be generated, and then the SMT approach behaves very efficiently. But if due to the limited number of trucks the problem is (close to) unsatisfiable, the theory solver may end up enumerating reasons that amount to the full SAT encoding, and moreover a very naive one! It would probably be better not to handle the trucks resource only with SMT, and use, for certain hours h, a more compact propositional encoding with auxiliary variables instead.

Example (cont.): If a cardinality constraint $x_1 + \ldots + x_n \leq K$ like the one of our vertex cover example is a bottle neck in a problem, causing it to be (close to) unsatisfiability, then the theory solver is likely to enumerate most, if not all, exponentially many $\binom{n}{K+1}$ reasons of the form $\neg y_1 \vee \ldots \vee \neg y_K \vee \neg x_i$. Indeed, on an unsatisfiable input problem with two cardinality constraints $x_1 + \ldots + x_n \leq K$ and $x_1 + \ldots + x_n > K$, the SMT approach would need to generate *all* these reasons. In this case one could have used a cardinality network encoding.

Of course it is hard to know in advance which constraints are bottle necks and which ones are not. A very nice first answer to this challenge can be found in recent work by Abío and Stuckey [AS12], showing how to generate the efficient compact SAT encoding *on demand:* initially a constraint is handled in SMT mode, but a cardinality network is built piece by piece only for those variables that trigger many conflicts. In the extreme case this can cause the full cardinality network to be generated. Indeed, this hybrid approach frequently behaves much better than any of its two ingredients, and is never importantly worse, both for cardinality constraints and for pseudo-Boolean ones. Interesting challenges remain, such as: how to do this for other types of constraints? is there a general, i.e., not constraint-specific method for this? can the CDCL SAT solver further improve with special heuristics for splitting on the auxiliary variables?

Resolution. It is well-known that CDCL SAT solvers, when given an unsatisfiable problem, can generate a trace from which one can reconstruct a binary resolution refutation [ZM03]. It is easy to see that CDCL SAT indeed amounts to a model-search-driven resolution strategy, and, as a consequence, the runtime of the SAT solver is lower bounded by the size of the resolution refutation. SMT solvers do not escape from this limitation: the trace produces a resolution refutation from the initial clauses *and* the set of all generated reasons.

The challenge is of course to overcome this limitation of CDCL and SMT solvers, for example, by (roughly) introducing definitions of new variables, thus obtaining the power of the *extended resolution* rule. Some first steps in this direction have been taken [AKS10, Hua10], but in terms of practical performance this challenge remains wide open.

Optimization. In many practical applications in fact one deals with optimization problems of the form $(S, cost)$, where S is a clause set over Boolean variables $x_1 \dots x_n$, with an cost function $cost\colon \mathbb{B}^n \rightarrow \mathbb{R}$, and the aim is to find a model A of S such that $cost(A)$ is minimized.

Example (cont.): The problem of finding a *minimum* vertex cover, i.e., finding the smallest possible K, is a classical optimization problem. In a SAT encoding using a cardinality network for the constraint $x_1 + \ldots + x_n \leq K$, one can start with a large K and, due to the characteristics of the cardinality network, each time a solution is found, by adding unit clauses one can progressively strengthen the constraint lowering the K one by one (see e.g., [ANORC11]).

A more general approach is branch and bound. Following the ideas of [NO06], the cost function is dealt with as an SMT theory that becomes progressively stronger: each time a solution is found, the theory is strengthened to allow only lower cost solutions from then on. Not surprisingly, the closer the problem comes to unsatisfiability, the harder it gets.

By the same line of reasoning as before, it turns out that such branch-and-bound solvers can generate resolution-like independently verifiable optimality proof certificates, even in the presence of theory-based (i.e., cost-based) lower bounding, propagation and backjumping [LNORC11]. Again, in this context the resolution-related limitations apply.

Many challenges remain in the context of optimization: better lower bounding procedures for improved propagation, and more knowledge about variable selection and polarity choice heuristics in the context of branch and bound.

Acknowledgement. The author wishes to thank Ignasi Abío for many fruitful discussions around these topics.

References

[AKS10] Audemard, G., Katsirelos, G., Simon, L.: A restriction of extended resolution for clause learning sat solvers. In: Proceedings of the Twenty-Fourth AAAI Conference on Artificial Intelligence. AAAI (2010)

[ANORC11] Asín, R., Nieuwenhuis, R., Oliveras, A., Rodríguez-Carbonell, E.: Cardinality networks: a theoretical and empirical study. Constraints 16(2), 195–221 (2011)

[AS12] Abío, I., Stuckey, P.J.: Conflict-directed lazy decomposition (submitted, 2012)

[BHvMW09] Biere, A., Heule, M.J.H., van Maaren, H., Walsh, T. (eds.): Handbook of Satisfiability. Frontiers in Artificial Intelligence and Applications, vol. 185. IOS Press (February 2009)

[BSST09] Barrett, C., Sebastiani, R., Seshia, S.A., Tinelli, C.: Satisfiability Modulo Theories. In: Biere, et al. (eds.) [BHvMW09], vol. 185, ch. 26, pp. 825–885 (February 2009)

[Hua10] Huang, J.: Extended clause learning. Artif. Intell. 174(15), 1277–1284 (2010)

[LNORC11] Larrosa, J., Nieuwenhuis, R., Oliveras, A., Rodríguez-Carbonell, E.: A framework for certified boolean branch-and-bound optimization. J. Autom. Reasoning 46(1), 81–102 (2011)

[NO06] Nieuwenhuis, R., Oliveras, A.: On SAT Modulo Theories and Optimization Problems. In: Biere, A., Gomes, C.P. (eds.) SAT 2006. LNCS, vol. 4121, pp. 156–169. Springer, Heidelberg (2006)

[NOT06] Nieuwenhuis, R., Oliveras, A., Tinelli, C.: Solving SAT and SAT Modulo Theories: from an Abstract Davis-Putnam-Logemann-Loveland Procedure to DPLL(T). Journal of the ACM 53(6), 937–977 (2006)

[OSC09] Ohrimenko, O., Stuckey, P.J., Codish, M.: Propagation via lazy clause generation. Constraints 14(3), 357–391 (2009)

[ZM03] Zhang, L., Malik, S.: Validating SAT Solvers Using an Independent Resolution-Based Checker: Practical Implementations and Other Applications. In: 2003 Conference on Design, Automation and Test in Europe Conference, DATE 2003, pp. 10880–10885. IEEE Computer Society (2003)

Unification Modulo Synchronous Distributivity

Siva Anantharaman[1], Serdar Erbatur[2], Christopher Lynch[3], Paliath Narendran[2], and Michael Rusinowitch[4]

[1] LIFO - Université d'Orléans, France
siva@univ-orleans.fr
[2] University at Albany–SUNY, USA
{se,dran}@cs.albany.edu
[3] Clarkson University, USA
clynch@clarkson.edu
[4] Loria-INRIA Lorraine, Nancy, France
rusi@loria.fr

Abstract. Unification modulo the theory defined by a single equation which specifies that a binary operator distributes synchronously over another binary operator is shown to be undecidable. It is the simplest known theory, to our knowledge, for which unification is undecidable: it has only one defining axiom and moreover, every congruence class is finite (so the matching problem is decidable).

Keywords: Equational unification, Intercell Turing machine, Decidability.

1 Preliminaries

It is well known that unification plays a very major role in all formal deduction mechanisms. Syntactic unification – also known as unification modulo the empty theory – is known to be decidable from around 1930, and optimized algorithms for it are well-known as well [2]. Semantic (or equational) unification is an extension of syntactic unification, to meet the situation where terms in the underlying signature are bound by some given equational theory. Several such theories are of great practical interest, in particular the theories of commutativity, associativity, associativity-commutativity; and decision procedures for unification modulo these theories are well-known from around 1970-1980 [3]. Another equational theory of practical interest is distributivity, which specifies that a binary operator distributes over another binary operator – a typical example being that of multiplication over addition on integers. Unification modulo such a distributivity is known to be decidable [11,13]. Note that the distributivity of multiplication over addition on integers is 'asynchronous' when used two-sided, in the sense that it then works 'argument-wise' below addition. There are other instances of distributivity for which a different theory is needed; for instance, if B stands for the division operation on nonzero rational numbers and $*$ for multiplication, then the following property is satisfied:

B. Gramlich, D. Miller, and U. Sattler (Eds.): IJCAR 2012, LNAI 7364, pp. 14–29, 2012.

$$\mathcal{E}: \qquad B(u,x) * B(v,y) \;=\; B(u*v, x*y)$$

In contrast with the example mentioned earlier, here the binary operator B distributes over $*$ synchronously, i.e., in parallel on its arguments. Note that the property $\mathcal{E}$ is also satisfied by the RSA-based implementation of the *blind signature scheme* for cryptography [4] (B stands in this case for the product of an integer m with a random number r raised to a given key e, and $*$ is the usual product on integers). Yet another model for $\mathcal{E}$, of practical interest, is the '*Exchange Law for concurrent processes*' as defined in [5]. The equation $\mathcal{E}$ can be turned easily into a terminating rewrite rule, oriented either way; it forms a convergent rewrite system in both cases. The theory defined by this equation $\mathcal{E}$ will also be referred to as $\mathcal{E}$ in the sequel.

Our objective in this paper is two-fold. We first present (in Section 2) a sound and complete set of inference rules for the $\mathcal{E}$-unification problem. A dependency graph is associated in a natural manner with any $\mathcal{E}$-unification problem $\mathcal{P}$, given in a 'standard form' (see definition below); and it is shown that the problem admits a solution if and only if the dependency graph remains bounded under the inferences. We then show that the $\mathcal{E}$-unification problem is undecidable in general, by reduction from the boundedness problem for deterministic Intercell Turing Machines (ITM), which is known to be undecidable [10]; this is done in Section 5. Such a reduction is rendered possible by suitably encoding the relations between the nodes and the paths on the dependency graph of $\mathcal{P}$ as string rewrite relations (string equations), which can be subsequently interpreted as the transition rules of an ITM[1]; the technical developments needed for this are presented in Sections 3 and 4.

2 A Set of Inference Rules for Elementary $\mathcal{E}$-Unification

Our signature consists of a (countably infinite) set of variables $\mathcal{X}$ and the two binary symbols B and '$*$'; the variables of $\mathcal{X}$ will be denoted by lower or upper case letters from u or U, to z or Z, with or without suffixes and primes. Note that B and '$*$' are cancellative: by this, we mean that for all terms s_1, t_1, s_2, t_2, $B(s_1,t_1) =_{\mathcal{E}} B(s_2,t_2)$ if and only if $s_1 =_{\mathcal{E}} s_2$ and $t_1 =_{\mathcal{E}} t_2$; similarly for '$*$'. (One easy way to show this is to use $\mathcal{E}$ as a rewrite rule $B(u*v, x*y) \to B(u,x) * B(v,y)$.)

Without loss of generality, the equations of the given unification problem $\mathcal{P}$ are assumed to be in a *standard form,* i.e., in one of the following forms:

$$X =^? V, \; X =^? B(V,Y), \; X =^? V * Y$$

where X, Y, V are variables. A set of equations is said to be in *dag-solved form* (or *d-solved form*) if and only if they can be arranged as a list

$$x_1 =^? t_1, \; \ldots, \; x_n =^? t_n$$

[1] The reader can see that our undecidability proof is influenced by the techniques in [10] and [7].

where (a) each left-hand side x_i is a distinct variable, and (b) $\forall 1 \leq i \leq j \leq n$: x_i does not occur in t_j ([8]). The following relations on the variables of $\mathcal{P}$ will be needed in the sequel:

- $U \succ_{r_*} V$ iff there is an equation $U =^? Y * V$
- $U \succ_{l_*} V$ iff there is an equation $U =^? V * Y$
- $U \succ_{r_B} V$ iff there is an equation $U =^? B(Y, V)$
- $U \succ_{l_B} V$ iff there is an equation $U =^? B(V, Y)$
- $U \succ_{*} V$ iff $U \succ_{r_*} V$ or $U \succ_{l_*} V$
- $U \succ_{B} V$ iff $U \succ_{r_B} V$ or $U \succ_{l_B} V$

The following transformation (inference) rules for $\mathcal{E}$-unification are given, where $\mathcal{EQ}$ stands for a set of equations in the problem $\mathcal{P}$, the symbol $\uplus$ stands for disjoint set union, and $\cup$ is usual set union.

(1) *Variable Elimination:*

$$\frac{\{X =^? V\} \uplus \mathcal{EQ}}{\{X =^? V\} \cup [V/X](\mathcal{EQ})} \quad \text{if } X \text{ occurs in } \mathcal{EQ}$$

(2) *Cancellation on B:*

$$\frac{\mathcal{EQ} \uplus \{X =^? B(V, Y),\ X =^? B(W, T)\}}{\mathcal{EQ} \cup \{X =^? B(W, T),\ V =^? W,\ Y =^? T\}}$$

(3) *Cancellation on '$*$':*

$$\frac{\mathcal{EQ} \uplus \{X =^? V * Y,\ X =^? W * T\}}{\mathcal{EQ} \cup \{X =^? W * T,\ V =^? W,\ Y =^? T\}}$$

(4) *Splitting:*

$$\frac{\mathcal{EQ} \uplus \{X =^? B(V, Y),\ X =^? W * Z\}}{\mathcal{EQ} \cup \{X =^? W * Z,\ W =^? B(V_0, Y_0),\ Z =^? B(V_1, Y_1),\ V =^? V_0 * V_1, Y =^? Y_0 * Y_1\}}$$

(5) *Occur-Check:*

$$\frac{\mathcal{EQ}}{FAIL} \quad \text{if } X\ (\succ_* \cup \succ_B)^+\ X \text{ for some } X$$

An outline of the algorithm is as follows: As long as the rules are applicable, rule (5) ("Occur-Check"), and rule (1) ("Variable Elimination"), are to be applied most eagerly; the cancellation rules (2) and (3) come next. The splitting rule (4) is applied with the lowest priority, i.e., only when no other rule is applicable.

The variable X in the specification of the splitting rule is referred to as a *peak*. In other words, a peak is any variable Z such that $Z \succ_{l_*} U$, $Z \succ_{r_*} V$, $Z \succ_{l_B} X$ and $Z \succ_{r_B} Y$ for some variables U, V, X, Y. Note that rule (4) introduces fresh variables; it also moves some variables from the right side to the left. This may give rise to further applications of (2) and (3). Furthermore, splitting may not terminate. For instance, $U =^? B(Y, X)$ and $U =^? Y * Z$ will cause an infinite loop, by using rule (4) forever. It is easy to conclude that there will be no solution in such a situation. The proof of correctness for this algorithm is similar to the one in Tiden-Arnborg [13].

We define a relation $\Rightarrow$ between sets of equations $\mathcal{S}$ and $\mathcal{S}'$ as follows: $\mathcal{S} \Rightarrow \mathcal{S}'$ if and only if $\mathcal{S}'$ can be obtained from $\mathcal{S}$ by applying one of the rules (1) – (5).

Lemma 1. *Rules* (1) – (5) *are sound and complete for unification modulo* $\mathcal{E}$.

Proof. The soundness of rules (1), (2) and (3) is easily seen[2]. Now, if there is an occur-check cycle of any length for a variable X, then clearly there is no solution for $\mathcal{EQ}$; and it is obvious that rule (5) catches such cycles in the problem if they exist. Hence rule (5) is also sound. We now show, explicitly, that the splitting rule (4) is sound.

Let $\mathcal{S} = \mathcal{EQ} \uplus \{X =^? B(V,Y),\ X =^? W * Z)\}$ where $\mathcal{EQ}$ is a set of equations and V, W, X, Y, Z are variables. And let $\mathcal{S}'$ be

$$\mathcal{EQ} \cup \{X =^? W * Z,\ W =^? B(V_0, Y_0),\ Z =^? B(V_1, Y_1),\ V =^? V_0 * V_1,\ Y =^? Y_0 * Y_1\}$$

where V_0, V_1, Y_0, Y_1 are new variables not occurring in $Var(\mathcal{S})$, the set of variables that occur in $\mathcal{S}$. Thus $\mathcal{S} \Rightarrow \mathcal{S}'$ by rule (4). We have then:

Claim (i) Any unifier of $\mathcal{S}'$ is a unifier of $\mathcal{S}$: Indeed, suppose θ is a unifier of $\mathcal{S}'$. It is easy to check then that $\theta(X) =_{\mathcal{E}} \theta(B(V,Y))$.

Claim (ii) Let σ be a unifier of $\mathcal{S}$. Then there is a substitution σ' such that σ' is an extension of σ and σ' is a unifier of $\mathcal{S}'$: For proving this we reason on terms in normal form under the convergent rewrite system:

$$B(u * v,\ x * y) \ \rightarrow\ B(u,x) * B(v * y).$$

Since σ is a unifier of $\mathcal{S}$, the normal forms of $\sigma(V)$ and $\sigma(Y)$ must be product terms, i.e., terms of the form $s_0 * s_1$ and $t_0 * t_1$ respectively. Then $\sigma(W) = B(s_0, t_0)$ and $\sigma(Z) = B(s_1, t_1)$. Thus $\sigma \circ \{V_0 := s_0, V_1 := s_1, Y_0 := t_0, Y_1 := t_1\}$ is a unifier of $\mathcal{S}'$; this proves (ii).

To show completeness, first note that if the algorithm terminates on $\mathcal{S}$ without failure, the resulting system is in d-solved form. On the other hand, if the algorithm does not terminate, then it has to be because there is infinite splitting — i.e., the splitting rule (4) is applied infinitely often.

Claim (iii) If there is infinite splitting there is no unifier: Assume the contrary. Let θ be a normalized unifier of $\mathcal{S}$. If there is infinite splitting, then there is an infinite sequence of variables $V_i = V_{i_1} \succ V_{i_2} \succ \ldots$ where $V_i \in Dom(\theta)$. From what was shown above, there must also be an infinite sequence of unifiers $\theta = \theta_1, \theta_2, \ldots$ where each unifier is an extension of the previous one. But this leads to a contradiction, since if γ is a unifier, $V, V' \in Dom(\gamma)$ and $V \succ V'$, then $|\gamma(V)| > |\gamma(V')|$. □

A sufficient condition for nonunifiability, an $\mathcal{E}$-unification problem to be unsatisfiable can be formulated as cycle checking, on suitably defined relations on the variables of $\mathcal{P}$ over some of the models for the equation $\mathcal{E}$; see [1].

3 From $\mathcal{E}$-Unification Problems to Thue Systems

Before we give the reduction proving the undecidability of this unification problem, we need a few preliminaries.

---

[2] A general proof for soundness of these rules can be found in [12].

As explained in the previous section, the set of variables in a problem could get larger since fresh variables may be created when splitting occurs. If a variable X is split, then we add the equation $X = X_0 * X_1$ to the problem. In general, new variables may be split further and starting from a variable X we may obtain a variable X_β where β is a string of 0s and 1s. We shall agree that the general discipline for creating new variables is specified as: $X_\beta = X_{\beta 0} * X_{\beta 1}$, where $\beta \in \{0,1\}^*$. Note that if $\beta = \lambda$, the empty string, then $X_\beta = X$, an original variable in the problem. For a set of variables V, we define $\overline{V} = \{X_\beta \mid X \in V,\ \beta \in \{0,1\}^*\}$ to denote the set of all variables which may originate from V through splitting. In the next section we define our dependency graph notation and describe splitting and variable elimination in a graph setting.

3.1 The Dependency Graph

It is common to represent the problems by dependency graphs induced by the relations ($\succ_{l_*}$ etc.) among the variables. Each node corresponds to a variable and each directed edge is labeled w.r.t. the relation among the variables in the nodes. Interpretation of the unification problems through the relations among variables in a graph setting was used in [13]. Here we have four types of edges in the dependency graph: l_*, r_*, l_B and r_B. If two variables X and Y are related through $\succ_{l_*}$, then nodes induced by X and Y are connected by a directed edge labeled as l_*; similarly for the other relations. For instance, for the problem given as: $U =^? V,\ U =^? U_0 * U_1,\ U =^? B(X,Y)$, the dependency graph is given in Figure 1; note that V does not appear as a node on this graph.

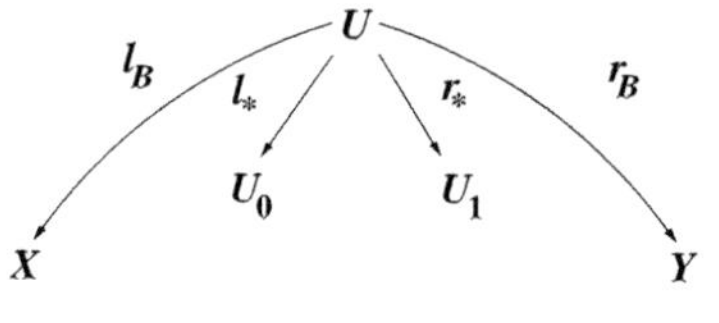

Fig. 1. Graph for: $U =^? V,\ U =^? U_0 * U_1,\ U =^? B(X,Y)$

Let S be the initial set of equations of the problem, and let G_0 be its dependency graph. Dependency graphs are not stable; they get updated each time an inference rule applies; thus we get a sequence of graphs $G_0, G_1, G_2, \ldots$. Recall, in particular that rule (1) is applied eagerly, e.g., when there is an equation of the form $U =^? V$. The consequence of its application is that U then merges into V, more precisely V replaces U in the problem as well as on the dependency graph. The problem just considered thus becomes:

$$U =^? V,\ V =^? U_0 * U_1,\ V =^? B(X,Y).$$

Its dependency graph is obtained from the one in Figure 1, by changing the label of the node U to V. It is important to note that the variable U has not been deleted from the problem, which still contains the equation $U =^? V$; the only change is that V now represents U on the graph. In intuitive terms, we shall say: on applying rule (1), two (or more) nodes merge on the dependency graph, and two (or more) paths merge by merging their end nodes; and any variable of the problem has a unique representative node on the dependency graph, up to variable equality. Alternatively, one could label the nodes of the graph with the equivalence classes of the variables of the problem, the equivalence being defined up to variable equality.

To suit our purposes in the sequel, we agree to shorten the labels of the edges of the dependency graph as follows: We replace l_* by 0 and r_* by 1; and we also replace l_B by L and r_B by R.

The splitting rule (4) is the only rule that adds nodes to the dependency graph, and new edges joining these new nodes. When a variable is split, 0- and 1-edges are added; the other equations introduced by rule (4) cause L- and R-edges to be added too. Thus, for the problem just mentioned above, (after having applied rule (1)) we apply splitting, and the problem thus derived is:

$$U =^? V,\ V =^? U_0{*}U_1,\ X =^? X_0{*}X_1,\ Y =^? Y_0{*}Y_1,\ U_0 =^? B(X_0, Y_0),\ U_1 =^? B(X_1, Y_1)$$

The dependency graph for the problem thus derived is given in Figure 2, using the short labels for its edges. Note that the edges from (U or) V to X and Y have been dropped out from the earlier graph (as well as the equations to which they corresponded).

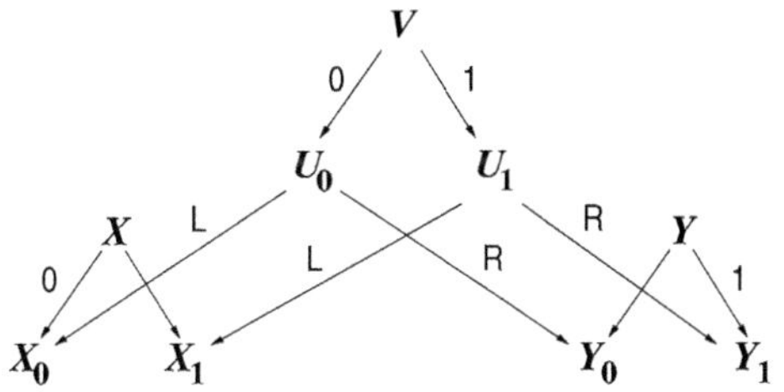

Fig. 2. Graph for the problem of Figure 1 after applying rules (1) and (4)

It is in general necessary to use other rules again after applying rule (4). For instance, consider the problem given as: $\{X =^? B(Y, Z),\ X =^? X_0 * X_1,\ X_0 =^? B(U, V)\}$. Variables Y and Z split and we obtain $X_0 =^? B(Y_0, Z_0)$ as one of the resulting equations. Therefore it is now necessary to apply first rule (2), followed by rule (1) to the equations $Y_0 =^? U$ and $Z_0 =^? V$ thereby derived.

For the purpose of proving the undecidability of $\mathcal{E}$-unification, we slightly modify our view of the dependency graph representation. Mainly, we don't explicitly delete any nodes or edges from the graph – all we do is to merge nodes. This leads to a more general vision of the dependency graph, that could be said to be the *relation graph*. Since nodes are merged, we assume that a node may have several labels. (See Figure 3.) Thus there is an *onto* function ϕ defined from $Var(\mathcal{S})$ to $V(G)$, the set of vertices of the graph. That is, each variable points to exactly one node in the graph but a node can be pointed to by more than

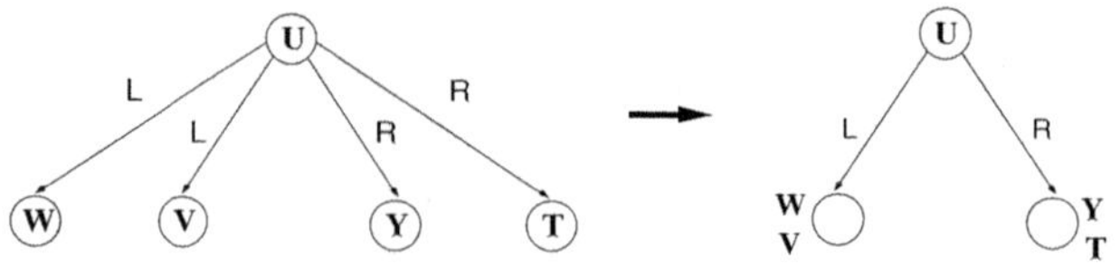

Fig. 3. Applying the cancellation rule (2)

one variable. Now using ϕ define a relation as a tuple (X, μ, Y) where X and Y are variables and μ is either L or R. In other words, (X, μ, Y) holds if and only if $\phi(X)$ and $\phi(Y)$ are connected with an edge labelled as μ.

Given a problem $\mathcal{P}$, let $G = G_\infty$ be the set of *persistent* nodes and edges (i.e., those not dropped out) in the sequence $G_0 \cup G_1 \ldots$, of graphs, updated along the inferences. We now characterize this graph in terms of string relations (equalities) over $\overline{Var(\mathcal{S})} \cup \{L, R\}$. If X and Y are two distinct variables such that $\phi(X) = \phi(Y)$, then we write $X =_G Y$. If there is a directed path $\Gamma \in \{L, R\}^*$ from $\phi(X)$ to $\phi(Y)$ on G, then we express this 'path relation' between X and Y as $\Gamma^{\mathrm{rev}} X =_G Y$, where Γ^{rev} is the reverse string of Γ. For paths of 0- and 1-edges, we define a similar relation: if there is a directed path $\beta \in \{0, 1\}^*$ from $\phi(X)$ to $\phi(Y)$ on G, we write $X_\beta =_G Y$.

We denote the length of any string $\Pi \in \{L, R\}^*$ (resp. $\beta \in \{0, 1\}^*$) as $|\Pi|$ (resp. as $|\beta|$). We say that a variable $X \in \mathcal{X}$ "exists" (in G) if there is a node in G with X as one of its labels.

Lemma 2. *(i) Let $\nu \in \{L, R\}$, $U, Y \in \mathcal{X}$ such that $\nu U =_G Y$. If U_0 exists (and also U_1, i.e., U splits), then $\nu U_0 =_G Y_0$ and $\nu U_1 =_G Y_1$.*

(ii) Let $\Pi \in \{L, R\}^+$, $U, Y \in \mathcal{X}$ such that $\Pi U =_G Y$. If U_0 exists (and also U_1, i.e., U splits), then $\Pi U_0 =_G Y_0$ and $\Pi U_1 =_G Y_1$.

(iii) Let $\Pi \in \{L, R\}^+$, $\beta \in \{0, 1\}^*$ such that $\Pi X =_G Y_\beta$. If X_0 exists (and also X_1, i.e., X splits), then $\Pi X_0 =_G Y_{\beta 0}$ and $\Pi X_1 =_G Y_{\beta 1}$.

Proof. Assertion (i) follows from seeing G as the relation graph on $\overline{Var(\mathcal{S})}$, in the sense defined above. Assertions (ii) and (iii) are proved by induction on the lengths of Π and β; see [1] for the details. □

Lemma 3. *Let $\Pi \in \{L, R\}^+$, $\alpha, \beta \in \{0, 1\}^*$ and X, Y such that $\Pi X =_G Y_\beta$. If X_α exists then $\Pi X_\alpha =_G Y_{\beta\alpha}$.*

Proof. By induction on the length of α, and Lemma 2 □

3.2 Thue Systems Associated with $\mathcal{E}$-Unification Problems

We henceforth speak of any $\mathcal{E}$-unification problem as a set of equations in standard form, often denoted as $\mathcal{S}$. With any given $\mathcal{E}$-unification problem $\mathcal{S}$, we shall associate a Thue system (i.e., string rewrite system), and subsequently relate the Thue congruence thus obtained, to the path relations on the dependency graph of $\mathcal{S}$, as defined in the previous section.

Let $V = Var(\mathcal{S})$ be the set of variables of the given problem $\mathcal{S}$. The alphabet over which the Thue system is defined is $\Sigma = V \cup \{L, R\} \cup \{0, 1\}$. We obtain string equations from an $\mathcal{E}$–unification problem as follows. For equations of the form $X =^? B(Y, Z)$ we create string equations $LX = Y$ and $RX = Z$. For $X =^? U * Y$, we form $X0 = U$ and $X1 = Y$. (Notice the connection between these and the binary relations defined in Section 3 - e.g. if $X \succ_{l_B} Y$, then $LX = Y$.) Let S_{Th} denote the set of string equations (the Thue system) thus associated with $\mathcal{S}$. Every such string equation is either of the form $\mu X = Y$ for $\mu \in \{L, R\}$, or of the form $X\nu = Y$ for $\nu \in \{0, 1\}$, with $X, Y \in V$. There is a close connection between the congruence on strings over Σ, modulo these string equations, denoted by $=_{S_{Th}}$, and the congruence in the graph context, denoted by $=_G$, that was introduced in the previous section, on the dependency graph G of the problem $\mathcal{S}$.

The next couple of results show the relation between $=_{S_{Th}}$ and $=_G$; it is assumed in their statements that $X, Y \in Var(\mathcal{S})$, $\Pi \in \{L, R\}^*$, $\alpha, \beta \in \{0, 1\}^*$:

Proposition 1. *For every* X, Y, Π, α, β, $\Pi X \alpha =_{S_{Th}} Y\beta$ *if and only if there exists* $\alpha', \beta', \gamma \in \{0, 1\}^*$ *such that* $\alpha = \alpha'\gamma$, $\beta = \beta'\gamma$ *and* $\Pi X_{\alpha'} =_G Y_{\beta'}$.

Corollary 1. *For every* X, Y, Π, β, $\Pi X =_{S_{Th}} Y\beta$ *if and only if* $\Pi X =_G Y_\beta$.

The "if" part of the above Proposition is easy; its "only if" part, as well as the above Corollary, are proved by induction on the number of derivation steps needed to deduce that $\Pi X\alpha =_{S_{Th}} Y\beta$; for the details, see [1].

Let $\mathcal{S}$ be an $\mathcal{E}$-unification problem and S_{Th} its associated Thue system. We now relate the path relation on the graph of $\mathcal{S}$ and the Thue congruence associated with $\mathcal{S}$. For a variable $X \in Var(\mathcal{S})$, we define its extent[3] $ext(X)$ as follows:

$$ext(X) = \{\Pi \in \{L, R\}^* \mid \exists Y \in Var(\mathcal{S}) \wedge \beta \in \{0, 1\}^* \text{ such that } S_{Th} \vdash \Pi X = Y\beta\}$$

The finiteness of $ext(X)$ for every X is closely connected to the unifiability of the problem. If $ext(X)$ is infinite for X, it obviously means that X splits infinitely many times and vice versa[4]. The following result is given without proof since it is (now) obvious:

Proposition 2. *An* $\mathcal{E}$*-unification problem* $\mathcal{S}$ *is solvable if and only if no failure rule applies and* $ext(X)$ *is finite for every* X *in* $Var(\mathcal{S})$.

We define this as a general concept for Thue systems. Let T be a Thue system with the alphabet Σ. Let Δ be a nonempty subset of Σ. T is said to have *finite* Δ*-span* if and only if $\forall q \in \Delta$, $ext(q)$ is finite where

$$ext(q) = \{\Pi \in (\Sigma \setminus \Delta)^* \mid \exists q' \in \Delta \wedge \beta \in (\Sigma \setminus \Delta)^* \text{ such that } \Pi q =_T q'\beta\}$$

[3] This follows the definition of *extent* by Jahama and Kfoury [7].

[4] Note that if $ext(X)$ is finite when X splits infinitely, then there exist a longest path $\Pi \in \{L, R\}^*$ with length k. But this is a contradiction since one can find another path Π' s.t. $|\Pi'| > k$ since X splits infinitely.

4 Thue Systems and Intercell Turing Machines

We give a review of relevant literature, mainly based on the notation used in [10]. An Intercell Turing Machine (ITM) is defined as a triple $M = \langle Q, \Sigma, \delta \rangle$, where Q is a set of states, Σ is a finite tape alphabet, and δ is a transition relation defined as $\delta \subseteq Q \times D \times \Sigma \times \Sigma \times Q$. Here D points to the direction of the move of the tape head (assumed placed between two tape cells) and is one of $\{-1, +1\}$. An instantaneous description (ID) of M is defined as a quadruple $\langle w_1, q, m, w_2 \rangle$ where q is the current state of the machine, $w_1 w_2$ is the string over Σ that forms the current tape content, m is an integer, and the header is between the cells $m-1$ and m, and it also separates w_1 and w_2. A move of M, from one ID to another, is denoted as a relation $\vdash_M$ formally defined as follows, where $s, t \in \Sigma$, $w_1, w_2 \in \Sigma^*$ and $q_1, q_2 \in Q$, and q_1 is the current state:

- **left-move**: For $\langle q_1, -1, s, t, q_2 \rangle \in \delta$, $\langle w_1 s, q_1, m, w_2 \rangle \vdash_M \langle w_1, q_2, m-1, t w_2 \rangle$
- **right-move**: For $\langle q_1, +1, s, t, q_2 \rangle \in \delta$, $\langle w_1, q_1, m, s w_2 \rangle \vdash_M \langle w_1 t, q_2, m{+}1, w_2 \rangle$

An ITM is said to be *deterministic* if and only if:

(i) the set of states Q splits as left-move and right-move states Q_l and Q_r so that $\delta \subseteq (Q_l \times \{-1\} \times \Sigma \times \Sigma \times Q) \cup (Q_r \times \{+1\} \times \Sigma \times \Sigma \times Q)$; and
(ii) δ is partial function from $Q \times D \times \Sigma$ to $\Sigma \times Q$.

This implies that there is at most one possible move from any given ID of a deterministic ITM: a left-move or a right-move.

The symmetric closure of an ITM $M = \langle Q, \Sigma, \delta \rangle$ is defined as the ITM $M_s = \langle Q, \Sigma, \delta_s \rangle$ where:

$$\delta_s = \delta \cup \{\langle q_1, -x, a, b, q_2 \rangle \mid \langle q_2, x, b, a, q_1 \rangle \in \delta\}$$

An ITM M is said to be *symmetric* iff $M = M_s$ holds. An ITM M is said to be *bounded* iff there exists a positive integer n such that for any arbitrary ID I of M, the number of different IDs reachable by M from I is at most n.

The following results on the boundedness problem are shown in [10] by using the ideas from [6].

Lemma 4. *It is undecidable to check whether a deterministic ITM is bounded.*

Lemma 5. *A deterministic ITM M is bounded if and only if its symmetric closure M_s is bounded.*

Corollary 2. *Given a deterministic ITM M, it is undecidable to check whether its symmetric closure M_s is bounded.*

Let $M = \langle Q, \Sigma, \delta \rangle$ be a deterministic ITM with tape alphabet $\Sigma = \{0, 1\}$. We shall use L, R to represent tape symbols $0, 1$ respectively, to the left of the tape head. Under such a vision, any transition of M can be expressed as a string rewrite rule of the form:

$$q_1 a \sim b q_2$$

where $a \in \{0, 1\}$, $b \in \{L, R\}$, and $\sim \in \{\leftarrow, \rightarrow\}$. For instance, $q_2 t \leftarrow L q_1$ represents the left-move $\langle q_1, -1, 0, t, q_2 \rangle$; and $q_1 s \rightarrow R q_2$ represents the right-move $\langle q_1, +1, s, 1, q_2 \rangle$ of the ITM.

Let R_M be the string rewrite system consisting of all these rules and let S_M be the Thue system obtained by symmetrizing the rewrite rules, i.e., making them bidirectional.

The extent of a state q in the ITM M is defined in terms of the system S_M: $ext(q) = \{\Pi \in \{L, R\}^* \mid \exists q' \in Q \wedge \beta \in \{0,1\}^* \text{ such that } S_M \vdash \Pi q = q'\beta\}$

Lemma 6. *Let $M = \langle Q, \Sigma, \delta\rangle$ be a deterministic ITM. Then M is bounded if and only if $ext(q)$ is finite for every state q in M.*

Proof. If $ext(q)$ is infinite for some state q in M, then clearly M is unbounded. On the other hand, suppose M is unbounded, and assume that $ext(q)$ is finite for every $q \in Q$. Let k be the length of the longest string that appears in any $ext(q)$. (Recall that Q is finite.) Since M is unbounded, there are configurations C and C' such that $C = \langle \Pi, q, m, w\beta\rangle$ and $C' = \langle \Pi\mathcal{B}, q', m+p, \beta\rangle$, such that C' is reachable from C, with $p = |w| = |\mathcal{B}| > k$, and the head never moves left past the mth cell, nor right past the $(m+p)^{th}$ cell. Then the configurations $\langle \epsilon, q, m, w\rangle$ and $\langle \mathcal{B}, q', m+p, \epsilon\rangle$ are reachable from each other as well by definition, and thus $qw \leftrightarrow^*_{S_M} \mathcal{B}q'$. Since $|\mathcal{B}| > k$ this contradicts the assumption that k is the length of the longest string in any $ext(q)$. □

Corollary 3. *Let $M = \langle Q, \Sigma, \delta\rangle$ be a deterministic ITM. Then M is bounded if and only if S_M has a finite Q-span.*

5 The Undecidability of Elementary $\mathcal{E}$-Unification

The undecidability result is by a reduction of the boundedness problem for deterministic ITMs. We shall proceed as follows: for each transition t of M we add two new (dummy) states along with their transitions — the reason for this will be clear later — but making sure that the resulting system, denoted M', is still deterministic, and furthermore, M' is bounded if and only if M is bounded. We then symmetrize M' to obtain M'_s. Thus M'_s is bounded if and only if M' is bounded (by Lemma 5) if and only if M is bounded. We shall finally show how to construct an $\mathcal{E}$-unification problem such that the problem is solvable if and only if M'_s is bounded.

For any deterministic ITM $M = \langle Q, \Sigma, \delta\rangle$ with tape alphabet $\Sigma = \{0, 1\}$, we construct another deterministic ITM M' based on M. For that, we first introduce the following notation: for $a \in \{0,1\}$, we set $1-a = 1$ if $a = 0$, and $1-a = 0$ if $a = 1$. Analogously, for $b \in \{L, R\}$ we let $\bar{b} = R$ if $b = L$ and $\bar{b} = L$ if $b = R$. Recall that in Section 4 a move of M was specified as $q_1 a \sim bq_2$ where $\sim$ is either $\leftarrow$ or $\rightarrow$. Let us consider first the (rightward) move $q_1 a \rightarrow bq_2$; we then add a left-move state and a right-move state, denoted as w' and w respectively, for each transition of M along with the following transitions:

$$
\begin{aligned}
q_1\,(1-a) &\leftarrow bw' \\
wa &\rightarrow \bar{b}q_2 \\
w\,(1-a) &\rightarrow \bar{b}w' \\
w\,(1-a) &\leftarrow \bar{b}w'
\end{aligned}
$$

The construction is the same for the leftward move $q_1 a \leftarrow bq_2$: the same set of states and transitions get added, i.e., the direction of the move does not affect the modifications. In all other cases, different state pairs are added to M' for different transitions of M; one could thus adopt the notation w_t and w'_t corresponding to every transition t in M. So the extension M' is defined as the ITM $\langle Q', \Sigma, \delta' \rangle$, where:

$$Q' = Q \cup \{ w_t, w'_t \mid \text{ for each corresponding transition } t \text{ of } M \}$$

and δ' consists of δ plus the new transitions induced by the extra moves above. M' is also deterministic because every state in Q' has only one possible move. Note that M and M' have the same tape alphabet $\Sigma = \{0, 1\}$. (See illustrative Figure 4 for the right-move $q_1 0 \rightarrow Lq_2$ of M, and the corresponding part in the extended ITM M'.)

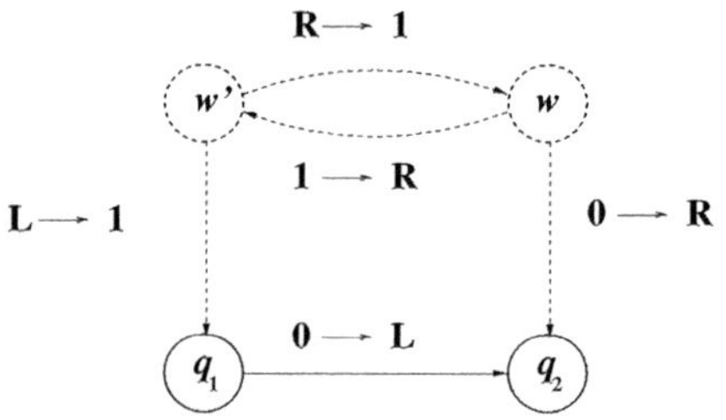

Fig. 4. Extension M' of a deterministic ITM M. Edges/Nodes not in M are dashed.

Lemma 7. *M is bounded if and only if M' is bounded.*

Proof. The "if" part is trivial, due to the fact that M' includes all transitions of M. For the "only if" part, we need to show that any ID involving w or w' can reach finitely many different IDs. Note that the move $w\,(1-a) \leftrightarrow \bar{b}w'$ will cause only one different ID for both types of IDs. The moves back and forth between IDs corresponding to $w\,(1-a)$ and $\bar{b}w'$ don't affect the number of different IDs reachable from them. In the remaining transitions w reaches an ID with q_2 and w' to an ID with q_1. Then we are done since we assume M is bounded and q_1 and q_2 are from M. □

Both M and M' are deterministic but not symmetric. For our purpose we symmetrize M'. This is done by first finding symmetric closure M'_s of M' along with the transition set δ'_s as defined in Section 4. Thus we get $M'_s = \langle Q', \Sigma, \delta'_s \rangle$. And we deduce the following, from Lemma 5.

Lemma 8. *M is bounded if and only if M'_s is bounded.*

Let $S_{M'_s}$ be the Thue system for the symmetric ITM M'_s. For each transition $q_1 a \sim bq_2$ of M, $S_{M'_s}$ will contain string equalities of the following forms:

$$\begin{aligned} q_1 a &\leftrightarrow b q_2 \\ q_1 (1-a) &\leftrightarrow b w' \\ w a &\leftrightarrow \bar{b} q_2 \\ w (1-a) &\leftrightarrow \bar{b} w' \end{aligned}$$

Obviously $S_{M'_s}$ is an extension of S_M, the Thue system of M. But note that $S_{M'}$ and $S_{M'_s}$ are the same.

We now show how each string equality can be simulated using unification problems. Let $\mathcal{S}$ be the set of equations that we create. We first look at original transitions in M. Note that depending on what a and b are in a transition t we have four possible types of moves. Then, we construct M'_s as described above. The variables of the unification problem are exactly the *states* of M'_s. Therefore for each possible pair (a, b) that t involves a unique equation modulo $\mathcal{E}$ with variables q_1, q_2, w and w' is constructed as shown below:

<u>Case 1:</u> If the transition of M'_s is $q_1 0 \leftrightarrow Lq_2$, then we (effectively) add the following equation to $\mathcal{S}$:

$$B(q_1, w) =^?_{\mathcal{E}} q_2 * w' \tag{1}$$

Since we assume that $\mathcal{S}$ is in standard form in our unification procedure, we transform this into standard form by using another new variable. Hence we create the equations: $u =^?_{\mathcal{E}} B(q_1, w)$, $u =^?_{\mathcal{E}} q_2 * w'$, where u is fresh. Note that if we apply the splitting rule to (1), we get:

$$u = q_2 * w',\ q_1 = q_{10} * q_{11},\ w = w_0 * w_1,\ q_2 = B(q_{10}, w_0),\ w' = B(q_{11}, w_1) \tag{2}$$

Therefore we see that the string equation (and hence the move of M) indeed corresponds to one of the relations among variables in (2). Note that $q_1 0 = q_{10}$ and $Lq_2 = q_{10}$. This is the motivation behind the reduction. In addition there are three other similar relations which can be observed in (2).

$$\begin{aligned} q_1\, 1 &= L\, w' \\ w\, 0 &= R\, q_2 \\ w\, 1 &= R w' \end{aligned}$$

The motivation for the construction of M' and M'_s should now be clear. In fact, the equalities above are included in $S_{M'}$. Recall that in Section 3.2 we defined the Thue system for a $\mathcal{E}$-unification problem $\mathcal{S}$ and denoted it as S_{Th}. Hence the congruence induced by $S_{M'}$ is subsumed by S_{Th}, or, $\leftrightarrow^*_{S_{M'}} \subset \leftrightarrow^*_{S_{Th}}$.

<u>Case 2:</u> If the move of M'_s is $q_1 1 \leftrightarrow Lq_2$, then we add:

$$u =^?_{\mathcal{E}} B(q_1, w),\ u =^?_{\mathcal{E}} w' * q_2$$

<u>Case 3:</u> If the move of M'_s is $q_1 1 \leftrightarrow Rq_2$, then add:

$$u =^?_{\mathcal{E}} B(w, q_1),\ u =^?_{\mathcal{E}} w' * q_2$$

Case 4: If the move of M'_s is $q_1 0 \leftrightarrow Rq_2$, then add:

$$u =^?_{\mathcal{E}} B(w, q_1),\ u =^?_{\mathcal{E}} q_2 * w'$$

We proceed now to show the following in several steps: S_{Th} has finite $Var(\mathcal{S})$-span if and only if $S_{M'}$ has finite Q'-span; the "if" part is easy to prove:

Lemma 9. *If S_{Th} has finite $Var(\mathcal{S})$-span then $S_{M'}$ has finite Q'-span.*

Proof. Trivial since the congruence induced by $S_{M'}$ is a subset of the congruence induced by S_{Th} and $Q' \subseteq Var(\mathcal{S}) = Q' \cup \{u_t \mid$ for each transition t in $M'_s\}$. □

We next prove the converse of Lemma 9, i.e., that finite Q'-span of $S_{M'}$ implies finite $Var(\mathcal{S})$-span of S_{Th}. For that we consider the string rewrite systems which are obtained by orienting equations in S_{Th} and $S_{M'}$ according to the order defined below. We then apply Knuth-Bendix completion to those systems as specified in [9]. It was shown there that those final string rewrite systems are possibly infinite and *lex-confluent*[5]. Hence for any Thue system T there exists a lex-confluent system equivalent to it.

Let Σ' be the alphabet of $S_{M'}$. Note that $\Sigma' = \Sigma \cup Q'$. Let $\mathcal{U}$ be the set of variables u_t that are added to ensure that the unification problem $\mathcal{S}$ is in standard form. Then the alphabet of S_{Th} is $\Sigma'' = \Sigma' \cup \mathcal{U}$. Equations in $S_{M'}$ can be oriented with the help of a length+lexicographical ordering on strings in Σ'^* induced by a total ordering $\succ$ on Σ'. One such ordering can be defined as $x > y$ if and only if:

(1) Either $|x| > |y|$;
(2) Or $|x| = |y|$, $x = ax'$, $y = by'$ where $a, b \in \Sigma'$; and, either $a \succ b$ or ($a = b$ and $x' > y'$).

with the following assumptions on the symbol ordering $\succ$:
Symbols in Σ are ordered as $L \succ R \succ 0 \succ 1$
Symbols in Q are ordered as $q_1 \succ q_2 \succ \ldots$
For any $X \in \Sigma$ and $Y \in Q$, we assume $X \succ Y$.

Let $C^*_{M'}$ be the resulting confluent system. A total length+lexicographic ordering on Σ'' can be defined similarly with an additional assumption such that the variables u_t are greater than the symbols in Q'. Thus every equation in S_{Th} will be oriented in such a way that the variables u_t are on the left. We denote by $\overrightarrow{S_{Th}}$ this new rewrite system.

Lemma 10. *$C^*_{M'} \cup \overrightarrow{S_{Th}}$ is confluent.*

Proof. It is not hard to see the forms of the rules in both subsystems. $\overrightarrow{S_{Th}}$ has rules of the form $\Pi u_t \rightarrow q'$ or $u_t\beta \rightarrow q$ for u_t and $\Pi \in \{L, R\}$ and $\beta \in \{0, 1\}$. On the other hand, $C^*_{M'}$ (possibly infinite) includes rules like $\Pi q \rightarrow q'\beta$ with Π and β defined as before. We already know that $C^*_{M'}$ is confluent, i.e., any critical pair in $C^*_{M'}$ is joinable. Note that left-hand sides in $\overrightarrow{S_{Th}}$ do overlap and thus give

[5] This is abbreviation for length+lexicographic confluence, which was introduced in [9].

rise to critical pairs. But these critical pairs are joinable by the rules of $C^*_{M'}$. For instance suppose $\overrightarrow{S_{Th}}$ has $Lu_t \to q$ and $u_t0 \to q'$. Here Lu_t0 is an overlap which gives rise to the critical pair $(Lq', q0)$. But we have $Lq' \to q0$ already in $C^*_{M'}$ by definition. □

To prove the converse of Lemma 9, we assume that $S_{M'}$ has finite Q'-span in the sense we explained earlier; we begin by showing that the state variables have finite extent in both S_{Th} and $S_{M'}$.

Lemma 11. *Let* $\Pi \in \{L, R\}^*$ *and* $\beta \in \{0,1\}^*$ *and* $q_1, q_2 \in Q'$ *be two states in* M', *where* $q_1, q_2 \in Var(\mathcal{S})$. *Then* $\Pi q_1 \leftrightarrow^*_{S_{M'}} q_2\beta$ *if and only if* $\Pi q_1 \leftrightarrow^*_{S_{Th}} q_2\beta$.

Proof. The "only if" part is obvious since $\leftrightarrow^*_{S_{M'}}$ is subsumed by $\leftrightarrow^*_{S_{Th}}$. Conversely, suppose $\Pi q_1 \leftrightarrow^*_{S_{M'}} q_2\beta$ holds. In this case note that the only applicable rules in $C^*_{M'} \cup \overrightarrow{S_{Th}}$ are the rules in $C^*_{M'}$. By Lemma 10 $C^*_{M'} \cup \overrightarrow{S_{Th}}$ is confluent and then by assumption, Πq_1 and $q_2\beta$ will rewrite to the same term w.r.t. $S_{M'}$. Then the rules involving the non-state variables u_t do not affect the derivation and hence $\overrightarrow{S_{Th}}$ does not affect the rewrite steps. As a result Πq_1 and $q_2\beta$ will have the same rewrite proof w.r.t. S_{Th}. □

The next couple of results are easy consequences of the above lemmas:

Lemma 12. *Let* $\Pi \in \{L, R\}^*$, $\beta \in \{0,1\}^*$, $q \in Q'$, $u_t \in \mathcal{U}$. *Then* $\Pi q \leftrightarrow^*_{S_{Th}} u_t\beta$ *if and only if there exist* $b \in \{0,1\}$, $q' \in Q'$, $\beta' \in \{0,1\}^*$ *and a rule* $u_tb \to q'$ *in* $\overrightarrow{S_{Th}}$ *such that: (i)* $\beta = b\beta'$, *and (ii)* $\Pi q \leftrightarrow^*_{S_{M'}} q'\beta'$

Proof. The "if" part follows from Lemma 11, thanks to conditions (i) and (ii). For the "only if" part, the assumption implies the existence of $b \in \{0,1\}$, such that $\beta = b\beta'$. By the construction of $\overrightarrow{S_{Th}}$, there exists a state q' and a rule $u_tb \to q'$; it is not hard then to show that $\Pi q \leftrightarrow^*_{S_{M'}} q'\beta'$. □

Corollary 4. *Let* $q \in Q'$ *be a state of* M'. *If* $ext(q)$ *is finite with respect to* $S_{M'}$ *then it is also finite with respect to* S_{Th}.

It follows then that the new variables u_t in $Var(\mathcal{S})$ have finite extent in S_{Th}, under the assumption that $S_{M'}$ has finite Q'-span:

Lemma 13. *Let* $\Pi \in \{L, R\}^*$ *and* $\beta \in \{0,1\}^*$, *and* $u_t \in Var(\mathcal{S})$. *If* $ext(u_t)$ *is infinite then there is a state* $q \in Q'$ *such that* $ext(q)$ *is infinite.*

Proof. Note that $\overrightarrow{S_{Th}}$ has rules of the form $u_tb \to q'$ with $b \in \{0,1\}$. Thus the result follows from the definition of *ext*. □

Corollary 5. *If* $S_{M'}$ *has finite* Q'*-span, then* S_{Th} *has finite* $Var(\mathcal{S})$*-span.*

Lemma 14. $S_{M'}$ *has finite* Q'*-span if and only if* S_{Th} *has finite* $Var(\mathcal{S})$*-span.*

Proof. The "only if" assertion is the preceding Corollary; and the "if" assertion is Lemma 9. □

Lemma 15. *M'_s is bounded if and only if $\mathcal{S}$ is unifiable.*

Proof. In Sections 3.1 and 3.2 we showed that the unification problem $\mathcal{S}$ is solvable if and only if S_{Th} has finite $Var(\mathcal{S})$-span. By Lemma 14 S_{Th} has finite $Var(\mathcal{S})$-span if and only if $S_{M'}$ has finite Q'-span. Finally the result follows since M' is bounded iff $S_{M'}$ has finite Q'-span by Corollary 3. □

From the lemmas established in this section, we finally get our main result:

Theorem 1. *Unifiability modulo $\mathcal{E}$ is undecidable.*

6 Conclusion

The equational theory $\mathcal{E}$ studied in this paper is defined by a single equation, which is orientable either way to give a convergent term rewrite system, and for which every congruence class is finite. It is surprising that the unification problem could be undecidable for such a "weak" theory.

Since elementary unification modulo $\mathcal{E}$ is undecidable, so are unification with free constants and general unification. Matching modulo $\mathcal{E}$ appears actually to be tractable (decidable in polynomial time); this could be of some interest.

References

1. Anantharaman, S., Erbatur, S., Lynch, C., Narendran, P., Rusinowitch, M.: Unification modulo Synchronous Distributivity. Technical Report SUNYA-CS-12-01, Dept. of Computer Science, University at Albany—SUNY (2012), `http://www.cs.albany.edu/~ncstrl/treports/Data/README.html`
2. Baader, F., Snyder, W.: Unification Theory. In: Handbook of Automated Reasoning, pp. 440–526. Elsevier Sc. Publishers B.V. (2001)
3. Baader, F., Nipkow, T.: Rewriting and all that. Cambridge University Press, New York (1998)
4. Chaum, D.: Security without Identification: Transaction System to Make Big Brother Obsolete. Communications of the ACM 28(2), 1030–1044 (1985)
5. Hoare, C.A.R., Hussain, A., Möller, B., O'Hearn, P.W., Petersen, R.L., Struth, G.: On Locality and the Exchange Law for Concurrent Processes. In: Katoen, J.-P., König, B. (eds.) CONCUR 2011. LNCS, vol. 6901, pp. 250–264. Springer, Heidelberg (2011)
6. Hooper, P.K.: The Undecidability of the Turing Machine Immortality Problem. J. Symbolic Logic 31(2), 219–234 (1966)
7. Jahama, S., Kfoury, A.J.: A General Theory of Semi-Unification. Technical Report 1993-018, Dept. of Computer Science, Boston University (December 1993)
8. Jouannaud, J.-P., Kirchner, C.: Solving equations in abstract algebras: a rule-based survey of unification. In: Computational Logic: Essays in Honor of Alan Robinson, pp. 360–394. MIT Press, Boston (1991)
9. Kapur, D., Narendran, P.: The Knuth-Bendix Completion Procedure and Thue Systems. SIAM Journal on Computing 14(4), 1052–1072 (1985)
10. Kfoury, A.J., Tiuryn, J., Urzyczyn, P.: The Undecidability of the Semi-Unification Problem. Information and Computation 102(1), 83–101 (1993)

11. Schmidt-Schauß, M.: A Decision Algorithm for Distributive Unification. Theoretical Comp. Science 208(1-2), 111–148 (1998)
12. Snyder, W.: A Proof Theory for General Unification, pp. 25–26. Birkhäuser, Basel (1991)
13. Tiden, E., Arnborg, S.: Unification Problems with One-sided Distributivity. Journal of Symbolic Computation 3(1–2), 183–202 (1987)

SAT Encoding of Unification in $\mathcal{ELH}_{R^+}$ w.r.t. Cycle-Restricted Ontologies*

Franz Baader, Stefan Borgwardt, and Barbara Morawska

Theoretical Computer Science, TU Dresden, Germany
{baader,stefborg,morawska}@tcs.inf.tu-dresden.de

Abstract. Unification in Description Logics has been proposed as an inference service that can, for example, be used to detect redundancies in ontologies. For the Description Logic $\mathcal{EL}$, which is used to define several large biomedical ontologies, unification is NP-complete. An NP unification algorithm for $\mathcal{EL}$ based on a translation into propositional satisfiability (SAT) has recently been presented. In this paper, we extend this SAT encoding in two directions: on the one hand, we add general concept inclusion axioms, and on the other hand, we add role hierarchies ($\mathcal{H}$) and transitive roles (R^+). For the translation to be complete, however, the ontology needs to satisfy a certain cycle restriction. The SAT translation depends on a new rewriting-based characterization of subsumption w.r.t. $\mathcal{ELH}_{R^+}$-ontologies.

1 Introduction

The Description Logic (DL) $\mathcal{EL}$, which offers the constructors conjunction ($\sqcap$), existential restriction ($\exists r.C$), and the top concept ($\top$), has recently drawn considerable attention since, on the one hand, important inference problems such as the subsumption problem are polynomial in $\mathcal{EL}$, even in the presence of general concept inclusion axioms (GCIs) [11,4]. On the other hand, though quite inexpressive, $\mathcal{EL}$ can be used to define biomedical ontologies, such as the large medical ontology SNOMED CT.[1]

Unification in DLs has been proposed in [8] as a novel inference service that can, for instance, be used to detect redundancies in ontologies. For example, assume that one developer of a medical ontology defines the concept of a *patient with severe injury of the frontal lobe* as

$$\exists \mathsf{finding}.(\mathsf{Frontal\_lobe\_injury} \sqcap \exists \mathsf{severity}.\mathsf{Severe}), \tag{1}$$

whereas another one represents it as

$$\exists \mathsf{finding}.(\mathsf{Severe\_injury} \sqcap \exists \mathsf{finding\_site}.\exists \mathsf{part\_of}.\mathsf{Frontal\_lobe}). \tag{2}$$

These two concept descriptions are not equivalent, but they are nevertheless meant to represent the same concept. They can obviously be made equivalent by

* Supported by DFG under grant BA 1122/14-1.
[1] See http://www.ihtsdo.org/snomed-ct/

B. Gramlich, D. Miller, and U. Sattler (Eds.): IJCAR 2012, LNAI 7364, pp. 30–44, 2012.

treating the concept names Frontal_lobe_injury and Severe_injury as variables, and substituting the first one by Injury $\sqcap$ $\exists$finding_site.$\exists$part_of.Frontal_lobe and the second one by Injury $\sqcap$ $\exists$severity.Severe. In this case, we say that the descriptions are unifiable, and call the substitution that makes them equivalent a *unifier*.

To motivate our interest in unification w.r.t. GCIs, role hierarchies, and transitive roles, assume that the developers use the descriptions (3) and (4) instead of (1) and (2):

$$\begin{aligned} &\exists \mathsf{finding}.\exists \mathsf{finding\_site}.\exists \mathsf{part\_of}.\mathsf{Brain} \sqcap \\ &\exists \mathsf{finding}.(\mathsf{Frontal\_lobe\_injury} \sqcap \exists \mathsf{severity}.\mathsf{Severe}) \end{aligned} \quad (3)$$

$$\begin{aligned} &\exists \mathsf{status}.\mathsf{Emergency} \sqcap \\ &\exists \mathsf{finding}.(\mathsf{Severe\_injury} \sqcap \exists \mathsf{finding\_site}.\exists \mathsf{part\_of}.\mathsf{Frontal\_lobe}) \end{aligned} \quad (4)$$

The descriptions (3) and (4) are not unifiable without additional background knowledge, but they are unifiable, with the same unifier as above, if the GCIs

$$\exists \mathsf{finding}.\exists \mathsf{severity}.\mathsf{Severe} \sqsubseteq \exists \mathsf{status}.\mathsf{Emergency},$$
$$\mathsf{Frontal\_lobe} \sqsubseteq \exists \mathsf{proper\_part\_of}.\mathsf{Brain}$$

are present in a background ontology and this ontology additionally states that part_of is transitive and proper_part_of is a subrole of part_of.

Most of the previous results on unification in DLs did not consider such additional background knowledge. In [8] it was shown that, for the DL $\mathcal{FL}_0$, which differs from $\mathcal{EL}$ by offering value restrictions ($\forall r.C$) in place of existential restrictions, deciding unifiability is an ExpTime-complete problem. In [5], we were able to show that unification in $\mathcal{EL}$ is of considerably lower complexity: the decision problem is NP-complete. The original unification algorithm for $\mathcal{EL}$ introduced in [5] was a brutal "guess and then test" NP-algorithm, but we have since then also developed more practical algorithms. On the one hand, in [7] we describe a goal-oriented unification algorithm for $\mathcal{EL}$, in which nondeterministic decisions are only made if they are triggered by "unsolved parts" of the unification problem. On the other hand, in [6], we present an algorithm that is based on a reduction to satisfiability in propositional logic (SAT). In [7] it was also shown that the approaches for unification of $\mathcal{EL}$-concept descriptions (without any background ontology) can easily be extended to the case of an acyclic TBox as background ontology without really changing the algorithms or increasing their complexity. Basically, by viewing defined concepts as variables, an acyclic TBox can be turned into a unification problem that has as its unique unifier the substitution that replaces the defined concepts by unfolded versions of their definitions.

For GCIs, this simple trick is not possible, and thus handling them requires the development of new algorithms. In [1,2] we describe two such new algorithms: one that extends the brute-force "guess and then test" NP-algorithm from [5] and a more practical one that extends the goal-oriented algorithm from [7]. Both algorithms are based on a new characterization of subsumption w.r.t. GCIs in $\mathcal{EL}$, which we prove using a Gentzen-style proof calculus for subsumption. Unfortunately, these algorithms are complete only for cycle-restricted TBoxes, i.e., finite sets of GCIs that satisfy a certain restriction on cycles, which, however, does not prevent all cycles. For example, the cyclic GCI $\exists \mathsf{child}.\mathsf{Human} \sqsubseteq \mathsf{Human}$ satisfies this restriction, whereas the cyclic GCI $\mathsf{Human} \sqsubseteq \exists \mathsf{parent}.\mathsf{Human}$ does not.

In the present paper, we still cannot get rid of cycle-restrictedness of the ontology, but extend the results of [2] in two other directions: (i) we add transitive roles (indicated by the subscript R^+ in the name of the DL) and role hierarchies (indicated by adding the letter $\mathcal{H}$ to the name of the DL) to the language, which are important for medical ontologies [17,15]; (ii) we provide an algorithm that is based on a translation into SAT, and thus allows us to employ highly optimized state-of-the-art SAT solvers [10] for implementing the unification algorithm. In order to obtain the SAT translation, using the characterization of subsumption from [2] is not sufficient, however. We had to develop a new rewriting-based characterization of subsumption.

In the next section, we introduce the DLs considered in this paper and the important inference problem subsumption. In Section 3 we define unification for these DLs and recall some of the existing results for unification in $\mathcal{EL}$. In particular, we introduce in this section the notion of cycle-restrictedness, which is required for the results on unification w.r.t. GCIs to hold. In Section 4 we then derive rewriting-based characterizations of subsumption. Section 5 contains the main result of this paper, which is a reduction of unification in $\mathcal{ELH}_{R^+}$ w.r.t. cycle-restricted ontologies to propositional satisfiability. The proof of correctness of this reduction strongly depends on the characterization of subsumption shown in the previous section.

2 The Description Logics $\mathcal{EL}$, $\mathcal{EL}^+$, and $\mathcal{ELH}_{R^+}$

The expressiveness of a DL is determined both by the formalism for describing concepts (the concept description language) and the terminological formalism, which can be used to state additional constraints on the interpretation of concepts and roles in a so-called ontology.

Syntax and Semantics

The concept description language considered in this paper is called $\mathcal{EL}$. Starting with a finite set N_C of *concept names* and a finite set N_R of *role names*, $\mathcal{EL}$*-concept descriptions* are built from concept names using the constructors *conjunction* ($C \sqcap D$), *existential restriction* ($\exists r.C$ for every $r \in N_R$), and *top* ($\top$).

Since in this paper we only consider $\mathcal{EL}$-concept descriptions, we will sometimes dispense with the prefix $\mathcal{EL}$.

On the semantic side, concept descriptions are interpreted as sets. To be more precise, an *interpretation* $\mathcal{I} = (\Delta^{\mathcal{I}}, \cdot^{\mathcal{I}})$ consists of a non-empty domain $\Delta^{\mathcal{I}}$ and an interpretation function $\cdot^{\mathcal{I}}$ that maps concept names to subsets of $\Delta^{\mathcal{I}}$ and role names to binary relations over $\Delta^{\mathcal{I}}$. This function is extended to concept descriptions as shown in the semantics column of Table 1.

Table 1. Syntax and semantics of $\mathcal{EL}$

Name	Syntax	Semantics
concept name	A	$A^\mathcal{I} \subseteq \Delta^\mathcal{I}$
role name	r	$r^\mathcal{I} \subseteq \Delta^\mathcal{I} \times \Delta^\mathcal{I}$
top	$\top$	$\top^\mathcal{I} = \Delta^\mathcal{I}$
conjunction	$C \sqcap D$	$(C \sqcap D)^\mathcal{I} = C^\mathcal{I} \cap D^\mathcal{I}$
existential restriction	$\exists r.C$	$(\exists r.C)^\mathcal{I} = \{x \mid \exists y : (x,y) \in r^\mathcal{I} \wedge y \in C^\mathcal{I}\}$
general concept inclusion	$C \sqsubseteq D$	$C^\mathcal{I} \subseteq D^\mathcal{I}$
role inclusion	$r_1 \circ \cdots \circ r_n \sqsubseteq s$	$r_1^\mathcal{I} \circ \cdots \circ r_n^\mathcal{I} \subseteq s^\mathcal{I}$

Ontologies

A *general concept inclusion (GCI)* is of the form $C \sqsubseteq D$ for concept descriptions C, D, and a *role inclusion* is of the form $r_1 \circ \cdots \circ r_n \sqsubseteq s$ for role names $r_1, \ldots, r_n, s$. Both are called *axioms*. Role inclusions of the form $r \circ r \sqsubseteq r$ are called *transitivity axioms* and of the form $r \sqsubseteq s$ *role hierarchy axioms*. An interpretation $\mathcal{I}$ *satisfies* such an axiom if the corresponding condition in the semantics column of Table 1 holds, where $\circ$ in this column stands for composition of binary relations.

An $\mathcal{EL}^+$*-ontology* is a finite set of axioms. It is an $\mathcal{ELH}_{R^+}$*-ontology* if all its role inclusions are transitivity or role hierarchy axioms, and an $\mathcal{EL}$*-ontology* if it contains only GCIs. An interpretation is a *model* of an ontology if it satisfies all its axioms.

Subsumption, Equivalence, and Role Hierarchy

A concept description C is *subsumed* by a concept description D w.r.t. an ontology $\mathcal{O}$ (written $C \sqsubseteq_\mathcal{O} D$) if every model of $\mathcal{O}$ satisfies the GCI $C \sqsubseteq D$. We say that C is *equivalent* to D w.r.t. $\mathcal{O}$ ($C \equiv_\mathcal{O} D$) if $C \sqsubseteq_\mathcal{O} D$ and $D \sqsubseteq_\mathcal{O} C$. If $\mathcal{O}$ is empty, we also write $C \sqsubseteq D$ and $C \equiv D$ instead of $C \sqsubseteq_\mathcal{O} D$ and $C \equiv_\mathcal{O} D$, respectively. As shown in [11,4], subsumption w.r.t. $\mathcal{EL}^+$-ontologies (and thus also w.r.t. $\mathcal{ELH}_{R^+}$- and $\mathcal{EL}$-ontologies) is decidable in polynomial time.

Since conjunction is interpreted as intersection, the concept descriptions $(C \sqcap D) \sqcap E$ and $C \sqcap (D \sqcap E)$ are always equivalent. Thus, we dispense with parentheses and write nested conjunctions in flat form $C_1 \sqcap \cdots \sqcap C_n$. Nested existential restrictions $\exists r_1.\exists r_2.\ldots\exists r_n.C$ will sometimes also be written as $\exists r_1 r_2 \ldots r_n.C$, where $r_1 r_2 \ldots r_n$ is viewed as a word over the alphabet of role names, i.e., an element of N_R^*.

The *role hierarchy* induced by $\mathcal{O}$ is a binary relation $\trianglelefteq_\mathcal{O}$ on N_R, which is defined as the reflexive-transitive closure of the relation $\{(r,s) \mid r \sqsubseteq s \in \mathcal{O}\}$. Using elementary reachability algorithms, the role hierarchy can be computed in polynomial time in the size of $\mathcal{O}$. It is easy to see that $r \trianglelefteq_\mathcal{O} s$ implies that $r^\mathcal{I} \subseteq s^\mathcal{I}$ for all models $\mathcal{I}$ of $\mathcal{O}$.

3 Unification

In order to define unification, we first introduce the notion of a substitution operating on concept descriptions. For this purpose, we partition the set N_C of concepts names into a set N_v of concept variables (which may be replaced by substitutions) and a set N_c of concept constants (which must not be replaced by substitutions). A *substitution* σ maps every variable to an $\mathcal{EL}$-concept description. It can be extended from variables to $\mathcal{EL}$-concept descriptions as follows:

- $\sigma(A) := A$ for all $A \in N_c \cup \{\top\}$,
- $\sigma(C \sqcap D) := \sigma(C) \sqcap \sigma(D)$ and $\sigma(\exists r.C) := \exists r.\sigma(C)$.

A concept description C is *ground* if it does not contain variables, and a substitution is *ground* if all concept descriptions in its range are ground. Obviously, a ground concept description is not modified by applying a substitution, and if we apply a ground substitution to any concept description, then we obtain a ground description. An ontology is *ground* if it does not contain variables.

Definition 1. *Let $\mathcal{O}$ be a ground ontology. A* unification problem *w.r.t. $\mathcal{O}$ is a finite set $\Gamma = \{C_1 \sqsubseteq^? D_1, \ldots, C_n \sqsubseteq^? D_n\}$ of subsumptions between $\mathcal{EL}$-concept descriptions. A substitution σ is a* unifier *of Γ w.r.t. $\mathcal{O}$ if σ* solves *all the subsumptions in Γ w.r.t. $\mathcal{O}$, i.e., if $\sigma(C_1) \sqsubseteq_{\mathcal{O}} \sigma(D_1), \ldots, \sigma(C_n) \sqsubseteq_{\mathcal{O}} \sigma(D_n)$. We say that Γ is* unifiable *w.r.t. $\mathcal{O}$ if it has a unifier w.r.t. $\mathcal{O}$.*

We call Γ w.r.t. $\mathcal{O}$ an $\mathcal{EL}$-, $\mathcal{EL}^+$-, or $\mathcal{ELH}_{R^+}$-unification problem depending on whether and what kind of role inclusions are contained in $\mathcal{O}$.

Three remarks regarding the definition of unification problems are in order. First, note that some of the previous papers on unification in DLs used equivalences $C \equiv^? D$ instead of subsumptions $C \sqsubseteq^? D$. This difference is, however, irrelevant since $C \equiv^? D$ can be seen as a shorthand for the two subsumptions $C \sqsubseteq^? D$ and $D \sqsubseteq^? C$, and $C \sqsubseteq^? D$ has the same unifiers as $C \sqcap D \equiv^? C$.

Second, note that—as in [2]—we have restricted the background ontology $\mathcal{O}$ to be ground. This is not without loss of generality. In fact, if $\mathcal{O}$ contained variables, then we would need to apply the substitution also to its axioms, and instead of requiring $\sigma(C_i) \sqsubseteq_{\mathcal{O}} \sigma(D_i)$ we would thus need to require $\sigma(C_i) \sqsubseteq_{\sigma(\mathcal{O})} \sigma(D_i)$, which would change the nature of the problem considerably. The treatment of unification w.r.t. acyclic TBoxes in [7] actually considers a more general setting, where some of the primitive concepts occurring in the TBox may be variables. The restriction to ground general TBoxes is, however, appropriate for the application scenario sketched in the introduction. In this scenario, there is a fixed background ontology, which is extended with definitions of new concepts by several knowledge engineers. Unification w.r.t. the background ontology is used to check whether some of these new definitions actually are redundant, i.e., define the same intuitive concept. Here, some of the primitive concepts newly introduced by one knowledge engineer may be further defined by another one, but we assume that the knowledge engineers use the vocabulary from the background ontology unchanged, i.e., they define *new* concepts rather than adding definitions for concepts that already occur in the background ontology. An instance of this scenario can, e.g., be found in [12], where different extensions of SNOMED CT are checked for overlaps, albeit not by using unification, but by simply testing for equivalence.

Third, though arbitrary substitutions σ are used in the definition of a unifier, it is actually sufficient to consider ground substitutions such that all concept descriptions $\sigma(X)$ in the range of σ contain only concept and role names occurring in Γ or $\mathcal{O}$. It is an easy consequence of well-known results from unification theory [9] that Γ has a unifier w.r.t. $\mathcal{O}$ iff it has such a ground unifier.

Relationship to Equational Unification

Unification was originally not introduced for Description Logics, but for equational theories [9]. In [7] it was shown that unification in $\mathcal{EL}$ (w.r.t. the empty ontology) is the same as unification in the equational theory *SLmO* of semilattices with monotone operators [16]. As argued in [2], unification in $\mathcal{EL}$ w.r.t. a ground $\mathcal{EL}$-ontology corresponds to unification in *SLmO* extended with a finite set of ground identities. In contrast to GCIs, role inclusions add non-ground identities to *SLmO* (see [16] and [3] for details).

This unification-theoretic point of view sheds some light on our decision to restrict unification w.r.t. general TBoxes to the case of general TBoxes that are ground. In fact, if we lifted this restriction, then we would end up with a generalization of rigid E-unification [14,13], in which the theory *SLmO* extended with the identities expressing role inclusions is used as a background theory. To the best of our knowledge, such variants of rigid E-unification have not been considered in the literature, and are probably quite hard to solve.

Flat Ontologies and Unification Problems

To simplify the technical development, it is convenient to normalize the TBox and the unification problem appropriately. To introduce this normal form, we need the notion of an atom.

An *atom* is a concept name or an existential restriction. Obviously, every $\mathcal{EL}$-concept description C is a finite conjunction of atoms, where $\top$ is considered to be the empty conjunction. We call the atoms in this conjunction the *top-level atoms* of C. An atom is called *flat* if it is a concept name or an existential restriction of the form $\exists r.A$ for a concept name A.

The GCI $C \sqsubseteq D$ or subsumption $C \sqsubseteq^? D$ is called *flat* if C is a conjunction of $n \geq 0$ flat atoms and D is a flat atom. The ontology $\mathcal{O}$ (unification problem Γ) is called *flat* if all the GCIs in $\mathcal{O}$ (subsumptions in Γ) are flat. Given a ground ontology $\mathcal{O}$ and a unification problem Γ, we can compute in polynomial time (see [3]) a flat ontology $\mathcal{O}'$ and a flat unification problem Γ' such that

- Γ has a unifier w.r.t. $\mathcal{O}$ iff Γ' has a unifier w.r.t. $\mathcal{O}'$;
- the type of the unification problem ($\mathcal{EL}$, $\mathcal{EL}^+$, or $\mathcal{ELH}_{R^+}$) is preserved.

For this reason, we will assume in the following that all ontologies and unification problems are flat.

Cycle-Restricted Ontologies

The decidability and complexity results for unification w.r.t. $\mathcal{EL}$-ontologies in [2], and also the corresponding ones in the present paper, only hold if the ontologies satisfy a restriction that prohibits certain cyclic subsumptions.

Definition 2. *The $\mathcal{EL}^+$-ontology $\mathcal{O}$ is called* cycle-restricted *iff there is no nonempty word $w \in N_R^+$ and $\mathcal{EL}$-concept description C such that $C \sqsubseteq_\mathcal{O} \exists w.C$.*

Note that cycle-restrictedness is not a syntactic condition on the form of the axioms in $\mathcal{O}$, but a semantic one on what follows from $\mathcal{O}$. Nevertheless, for $\mathcal{ELH}_{R^+}$-ontologies, this condition can be decided in polynomial time [3]. Basically, one first shows that the $\mathcal{ELH}_{R^+}$-ontology $\mathcal{O}$ is cycle-restricted iff $A \not\sqsubseteq_\mathcal{O} \exists w.A$ holds for all nonempty words $w \in N_R^+$ and all $A \in N_C \cup \{\top\}$. Then, one shows that $A \sqsubseteq_\mathcal{O} \exists w.A$ for some $w \in N_R^+$ and $A \in N_C \cup \{\top\}$ implies that there are $n \geq 1$ role names $r_1, \ldots, r_n$ and $A_1, \ldots, A_n \in N_C \cup \{\top\}$ such that

$$(*) \quad A \sqsubseteq_\mathcal{O} \exists r_1.A_1, A_1 \sqsubseteq_\mathcal{O} \exists r_2.A_2, \ldots, A_{n-1} \sqsubseteq_\mathcal{O} \exists r_n.A_n \text{ and } A_n = A.$$

Using the polynomial-time subsumption algorithm for $\mathcal{ELH}_{R^+}$, we can build a graph whose nodes are the elements of $N_C \cup \{\top\}$ and where there is an edge from A to B with label r iff $A \sqsubseteq_\mathcal{O} \exists r.B$. Then we can use standard reachability algorithms to check whether this graph contains a cycle of the form $(*)$. The restriction to $\mathcal{ELH}_{R^+}$ stems from the fact that the proof of correctness of this algorithm is based on Lemma 7 below, which we cannot show for $\mathcal{EL}^+$.

The main reason why we need cycle-restrictedness of $\mathcal{O}$ is that it ensures that a substitution always induces a strict partial order on the variables.[2] To be more precise, assume that γ is a substitution. For $X, Y \in N_v$ we define

$$X >_\gamma Y \text{ iff } \gamma(X) \sqsubseteq_\mathcal{O} \exists w.\gamma(Y) \text{ for some } w \in N_R^+. \tag{5}$$

Transitivity of $>_\gamma$ is an easy consequence of transitivity of subsumption, and cycle-restrictedness of $\mathcal{O}$ yields irreflexivity of $>_\gamma$.

Lemma 3. *If $\mathcal{O}$ is a cycle-restricted $\mathcal{EL}^+$-ontology, then $>_\gamma$ is a strict partial order on N_v.*

4 Subsumption w.r.t. $\mathcal{EL}^+$- and $\mathcal{ELH}_{R^+}$-Ontologies

Subsumption w.r.t. $\mathcal{EL}^+$-ontologies can be decided in polynomial time [4]. For the purpose of deciding unification, however, we do not simply want a decision procedure for subsumption, but are more interested in a characterization of subsumption that helps us to find unifiers. The characterization of subsumption derived here is based on a rewrite relation that uses axioms as rewrite rules from right to left.

[2] Why we need this order will become clear in Section 5.

Proving Subsumption by Rewriting

Throughout this subsection, we assume that $\mathcal{O}$ is a flat $\mathcal{EL}^+$-ontology. Intuitively, an axiom of the form $A_1 \sqcap \ldots \sqcap A_n \sqsubseteq B \in \mathcal{O}$ is used to replace B by $A_1 \sqcap \ldots \sqcap A_n$ and an axiom of the form $r_1 \circ \ldots \circ r_n \sqsubseteq s \in \mathcal{O}$ to replace $\exists s.C$ by $\exists r_1 \ldots r_n.C$. In order to deal with associativity, commutativity, and idempotency of conjunction, it is convenient to represent concept descriptions as sets of atoms rather than as conjunctions of atoms.

Given an $\mathcal{EL}$-concept description C, the *description set* $\mathsf{s}(C)$ associated with C is defined by induction:

- $\mathsf{s}(A) := \{A\}$ for $A \in N_C$ and $\mathsf{s}(\top) := \emptyset$;
- $\mathsf{s}(C \sqcap D) := \mathsf{s}(C) \cup \mathsf{s}(D)$ and $\mathsf{s}(\exists r.C) := \{\exists r.\mathsf{s}(C)\}$.

For example, if $C = A \sqcap \exists r.(A \sqcap \exists r.\top)$, then $\mathsf{s}(C) = \{A, \exists r.\{A, \exists r.\emptyset\}\}$. We call *set positions* the positions in $\mathsf{s}(C)$ at which there is a set. In our example, we have three set positions, corresponding to the sets $\{A, \exists r.\{A, \exists r.\emptyset\}\}$, $\{A, \exists r.\emptyset\}$, and $\emptyset$. The set position that corresponds to the whole set $\mathsf{s}(C)$ is called the *root position.*

Our *rewrite rules* are of the form $N \leftarrow M$, where N, M are description sets. Such a rule *applies* at a set position p in $\mathsf{s}(C)$ if the corresponding set $\mathsf{s}(C)|_p$ contains M, and its *application* replaces $\mathsf{s}(C)|_p$ by $(\mathsf{s}(C)|_p \setminus M) \cup N$ (see [3] for a more formal definition of set positions and of the application of rewrite rules).

Given a flat $\mathcal{EL}^+$-ontology $\mathcal{O}$, the corresponding rewrite system $R(\mathcal{O})$ consists of the following rules:

- *Concept inclusion* ($\mathbf{R}_c$): For every $C \sqsubseteq D \in \mathcal{O}$, $R(\mathcal{O})$ contains the rule
$$\mathsf{s}(C) \leftarrow \mathsf{s}(D).$$
- *Role inclusion* ($\mathbf{R}_r$): For every $r_1 \circ \cdots \circ r_n \sqsubseteq s \in \mathcal{O}$ and every $\mathcal{EL}$-concept description C, $R(\mathcal{O})$ contains the rule
$$\mathsf{s}(\exists r_1 \ldots r_n.C) \leftarrow \mathsf{s}(\exists s.C).$$
- *Monotonicity* ($\mathbf{R}_m$): For every atom D, $R(\mathcal{O})$ contains the rule
$$\mathsf{s}(D) \leftarrow \emptyset.$$

Definition 4. *Let N, M be description sets. We write $N \leftarrow_{\mathcal{O}} M$ if N can be obtained from M by the application of a rule in $R(\mathcal{O})$. The relation $\xleftarrow{*}_{\mathcal{O}}$ is defined to be the reflexive, transitive closure of $\leftarrow_{\mathcal{O}}$, i.e., $N \xleftarrow{*}_{\mathcal{O}} M$ iff there is a chain*

$$N = M_\ell \leftarrow_{\mathcal{O}} M_{\ell-1} \leftarrow_{\mathcal{O}} \ldots \leftarrow_{\mathcal{O}} M_0 = M$$

of $\ell \geq 0$ rule applications. We call such a chain a derivation *of N from M w.r.t. $\mathcal{O}$. A rewriting step in such a derivation is called a* root step *if it applies a rule of the form ($\boldsymbol{R}_c$) at the root position. We write $N \xleftarrow{(n)}_{\mathcal{O}} M$ to express that there is a derivation of N from M w.r.t. $\mathcal{O}$ that uses* at most n root steps.

For example, if $\mathcal{O}$ contains the axioms $\top \sqsubseteq \exists r.B$ and $s \sqsubseteq r$, then the following is a derivation w.r.t. $\mathcal{O}$:

$$\{A, \exists s.\{A\}\} \leftarrow_{\mathcal{O}} \{A, \exists r.\{A\}\} \leftarrow_{\mathcal{O}} \{A, \exists r.\{A, \exists r.\{B\}\}\} \leftarrow_{\mathcal{O}} \{A, \exists r.\{A, \exists r.\emptyset\}\}$$

This is a derivation without a root step, which first applies a rule of the form $(\mathbf{R}_m)$, then one of the form $(\mathbf{R}_c)$ (not at the root position), and finally one of the form $(\mathbf{R}_r)$. This shows $\mathsf{s}(A \sqcap \exists s.A) \overset{(0)}{\longleftarrow}_{\mathcal{O}} \mathsf{s}(A \sqcap \exists r.(A \sqcap \exists r.\top))$.

The following theorem states that subsumption w.r.t. $\mathcal{O}$ corresponds to the existence of a derivation w.r.t. $\mathcal{O}$ whose root steps are bounded by the number of GCIs in $\mathcal{O}$ (see [3] for a proof of this result).

Theorem 5. *Let $\mathcal{O}$ be a flat $\mathcal{EL}^+$-ontology containing n GCIs and C, D be two $\mathcal{EL}$-concept descriptions. Then $C \sqsubseteq_{\mathcal{O}} D$ iff $\mathsf{s}(C) \overset{(n)}{\longleftarrow}_{\mathcal{O}} \mathsf{s}(D)$.*

A Structural Characterization of Subsumption in $\mathcal{ELH}_{R^+}$

Our translation of unification problems into propositional satisfiability problems depends on a structural characterization of subsumption, which we can unfortunately only show for $\mathcal{ELH}_{R^+}$ ontologies. Throughout this subsection, we assume that $\mathcal{O}$ is a flat $\mathcal{ELH}_{R^+}$-ontology. We say that r is *transitive* if the transitivity axiom $r \circ r \sqsubseteq r$ belongs to $\mathcal{O}$.

Definition 6. *Let C, D be atoms. We say that C is* structurally subsumed *by D w.r.t. $\mathcal{O}$ ($C \sqsubseteq^{\mathsf{s}}_{\mathcal{O}} D$) iff*

- *$C = D$ is a concept name,*
- *$C = \exists r.C'$, $D = \exists s.D'$, $C' \sqsubseteq_{\mathcal{O}} D'$, and $r \trianglelefteq_{\mathcal{O}} s$, or*
- *$C = \exists r.C'$, $D = \exists s.D'$, and $C' \sqsubseteq_{\mathcal{O}} \exists t.D'$ for a transitive role t with $r \trianglelefteq_{\mathcal{O}} t \trianglelefteq_{\mathcal{O}} s$.*

On the one hand, structural subsumption is a stronger property than $C \sqsubseteq_{\mathcal{O}} D$ since it requires the atoms C and D to have "compatible" top-level structures. On the other hand, it is weaker than subsumption w.r.t. the empty ontology, i.e., whenever $C \sqsubseteq D$ holds for two atoms C and D, then $C \sqsubseteq^{\mathsf{s}}_{\mathcal{O}} D$, but not necessarily vice versa. If $\mathcal{O} = \emptyset$, then the three relations $\sqsubseteq$, $\sqsubseteq^{\mathsf{s}}_{\mathcal{O}}$, $\sqsubseteq_{\mathcal{O}}$ coincide on atoms. Like $\sqsubseteq$ and $\sqsubseteq_{\mathcal{O}}$, $\sqsubseteq^{\mathsf{s}}_{\mathcal{O}}$ is reflexive, transitive, and closed under applying existential restrictions (see [3] for proofs of the results mentioned in this paragraph).

Using the connection between subsumption and rewriting stated in Theorem 5, we can now prove a characterization of subsumption in the presence of an $\mathcal{ELH}_{R^+}$-ontology $\mathcal{O}$ that expresses subsumption in terms of structural subsumptions and derivations w.r.t. $\leftarrow_{\mathcal{O}}$. Recall that all $\mathcal{EL}$-concept descriptions are conjunctions of atoms, that $C \sqsubseteq_{\mathcal{O}} D_1 \sqcap \cdots \sqcap D_m$ iff $C \sqsubseteq_{\mathcal{O}} D_j$ for all $j \in \{1, \ldots, m\}$, and $C \sqsubseteq_{\mathcal{O}} D$ iff there is an ℓ such that $\mathsf{s}(C) \overset{(\ell)}{\longleftarrow}_{\mathcal{O}} \mathsf{s}(D)$.

Lemma 7. *Let $\mathcal{O}$ be a flat $\mathcal{ELH}_{R^+}$-ontology, $C_1, \ldots, C_n$, D be atoms, and $\ell \geq 0$. Then $\mathsf{s}(C_1 \sqcap \cdots \sqcap C_n) \xleftarrow{(\ell)}_{\mathcal{O}} \mathsf{s}(D)$ iff there is*

1. *an index $i \in \{1, \ldots, n\}$ such that $C_i \sqsubseteq^{\mathsf{s}}_{\mathcal{O}} D$; or*
2. *a GCI $A_1 \sqcap \cdots \sqcap A_k \sqsubseteq B$ in $\mathcal{T}$ such that*
   a) for every $p \in \{1, \ldots, k\}$ we have $\mathsf{s}(C_1 \sqcap \cdots \sqcap C_n) \xleftarrow{(\ell-1)}_{\mathcal{O}} \mathsf{s}(A_p)$,
   b) $\mathsf{s}(C_1 \sqcap \cdots \sqcap C_n) \xleftarrow{(\ell)}_{\mathcal{O}} \mathsf{s}(B)$, and
   c) $B \sqsubseteq^{\mathsf{s}}_{\mathcal{O}} D$.

A detailed proof of this lemma is given in [3]. Here, we only want to point out that this proof makes extensive use of the transitivity of $\sqsubseteq^{\mathsf{s}}_{\mathcal{O}}$, and that this is the main reason why we cannot deal with general $\mathcal{EL}^+$-ontologies. In fact, while it is not hard to extend the definition of structural subsumption to more general kinds of ontologies, it is currently not clear to us how to do this such that the resulting relation is transitive; and without transitivity of structural subsumption, we cannot show a characterization analogous to the one in Lemma 7.

5 Reduction of Unification w.r.t. Cycle-Restricted $\mathcal{ELH}_{R^+}$-Ontologies to SAT

The main idea underlying the NP-membership results in [5] and [2] is to show that any $\mathcal{EL}$-unification problem that is unifiable w.r.t. the empty ontology and w.r.t. a cycle-restricted $\mathcal{EL}$-ontology, respectively, has a so-called local unifier. Here, we generalize the notion of a local unifier to the case of unification w.r.t. cycle-restricted $\mathcal{ELH}_{R^+}$-ontologies, but then go a significant step further. Instead of using an algorithm that "blindly" generates all local substitutions and then checks whether they are unifiers, we reduce the search for a local unifier to a propositional satisfiability problem.

Local Unifiers

Let Γ be a flat unification problem and $\mathcal{O}$ be a flat, cycle-restricted $\mathcal{ELH}_{R^+}$-ontology. We denote by At the set of atoms occurring as subdescriptions in subsumptions in Γ or axioms in $\mathcal{O}$ and define

$$\mathsf{At}_{\mathsf{tr}} := \mathsf{At} \cup \{\exists t.D' \mid \exists s.D' \in \mathsf{At},\ t \trianglelefteq_{\mathcal{O}} s,\ t \text{ transitive}\}.$$

Furthermore, we define the set of *non-variable atoms* by $\mathsf{At}_{\mathsf{nv}} := \mathsf{At}_{\mathsf{tr}} \setminus N_v$. Though the elements of $\mathsf{At}_{\mathsf{nv}}$ cannot be variables, they may contain variables if they are of the form $\exists r.X$ for some role r and a variable X. We call a function S that associates every variable $X \in N_v$ with a set $S_X \subseteq \mathsf{At}_{\mathsf{nv}}$ an *assignment*. Such an assignment induces the following relation $>_S$ on N_v: $>_S$ is the transitive closure of

$$\{(X, Y) \in N_v \times N_v \mid Y \text{ occurs in an element of } S_X\}.$$

We call the assignment S *acyclic* if $>_S$ is irreflexive (and thus a strict partial order). Any acyclic assignment S induces a unique substitution σ_S, which can be defined by induction along $>_S$:

– If X is a minimal element of N_v w.r.t. $>_S$, then we set $\sigma_S(X) := \bigsqcap_{D \in S_X} D$.
– Assume that $\sigma(Y)$ is already defined for all Y such that $X >_S Y$. Then we define $\sigma_S(X) := \bigsqcap_{D \in S_X} \sigma_S(D)$.

We call a substitution σ *local* if it is of this form, i.e., if there is an acyclic assignment S such that $\sigma = \sigma_S$. Since N_v and $\mathsf{At}_{\mathsf{nv}}$ are finite, there are only finitely many local substitutions. Thus, if we know that any solvable unification problem has a local unifier, then we can enumerate (or guess, in a nondeterministic machine) all local substitutions and then check whether any of them is a unifier. Thus, in general many substitutions will be generated that only in the subsequent check turn out not to be unifiers. In contrast, our SAT reduction will ensure that only unifiers are generated.

The Reduction

Here, we reduce unification w.r.t. cycle-restricted $\mathcal{ELH}_{R^+}$-ontologies to the satisfiability problem for propositional logic, which is NP-complete. This shows that this unification problem is in NP. But more importantly, it immediately allows us to apply highly optimized SAT solvers for solving such unification problems.

As before, we assume that Γ is a flat unification problem and $\mathcal{O}$ is a flat, cycle-restricted $\mathcal{ELH}_{R^+}$-ontology. Let $\mathcal{T}$ be the subset of $\mathcal{O}$ that consists of the GCIs in $\mathcal{O}$. We define the set

$$\text{Left} := \mathsf{At} \cup \{C_1 \sqcap \cdots \sqcap C_n \mid C_1 \sqcap \cdots \sqcap C_n \sqsubseteq^? D \in \Gamma \text{ for some } D \in \mathsf{At}\}$$

that contains all atoms of Γ and $\mathcal{O}$ and all left-hand sides of subsumptions from Γ. For $L \in \text{Left}$ and $C \in \mathsf{At}$, we write "$C \in L$" if C is a top-level atom of L.

The propositional variables we use for the reduction are of the form $[L \sqsubseteq D]^i$ for $L \in \text{Left}$, $D \in \mathsf{At}_{\mathsf{tr}}$, and $i \in \{0, \ldots, |\mathcal{T}|\}$. The intuition underlying these variables is that every satisfying propositional valuation induces an acyclic assignment S such that the following holds for the corresponding substitution σ_S: $[L \sqsubseteq D]^i$ is evaluated to true by the assignment iff $\mathsf{s}(\sigma_S(L))$ can be derived from $\mathsf{s}(\sigma_S(D))$ using at most i root steps, i.e., $\mathsf{s}(\sigma_S(L)) \xleftarrow{(i)}_{\mathcal{O}} \mathsf{s}(\sigma_S(D))$.

Additionally, we use the propositional variables $[X > Y]$ for $X, Y \in N_v$ to express the strict partial order $>_S$ induced by the acyclic assignment S.

The auxiliary function Dec is defined as follows for $C \in \mathsf{At}$, $D \in \mathsf{At}_{\mathsf{tr}}$:

$$\text{Dec}(C \sqsubseteq D) = \begin{cases} \mathbf{1} & \text{if } C = D \\ [C \sqsubseteq D]^{|\mathcal{T}|} & \text{if } C \text{ and } D \text{ are ground} \\ \text{Trans}(C \sqsubseteq D) & \text{if } C = \exists r.C',\ D = \exists s.D', \text{ and } r \trianglelefteq_{\mathcal{O}} s\ , \\ [C \sqsubseteq D]^{|\mathcal{T}|} & \text{if } C \text{ is a variable} \\ \mathbf{0} & \text{otherwise} \end{cases}$$

$$\text{Trans}(C \sqsubseteq D) = [C' \sqsubseteq D']^{|\mathcal{T}|} \vee \bigvee_{\substack{t \text{ transitive} \\ r \trianglelefteq_{\mathcal{O}} t \trianglelefteq_{\mathcal{O}} s}} [C' \sqsubseteq \exists t.D']^{|\mathcal{T}|}.$$

Note that $C' \in \mathsf{At}$ and $D', \exists t.D' \in \mathsf{At}_{\mathsf{tr}}$ by the definition of $\mathsf{At}_{\mathsf{tr}}$ and since Γ and $\mathcal{O}$ are flat. Here, **0** and **1** are Boolean constants representing the truth values 0 (*false*) and 1 (*true*), respectively.

The unification problem will be reduced to satisfiability of the following set of propositional formulae. For simplicity, we do not use only clauses here. However, our formulae can be transformed into clausal form by introducing polynomially many auxiliary propositional variables and clauses.

Definition 8. *Let Γ be a flat unification problem and $\mathcal{O}$ a flat, cycle-restricted $\mathcal{ELH}_{R^+}$-ontology. The set $C(\Gamma, \mathcal{O})$ contains the following propositional formulae:*

(I) Translation of the subsumptions of Γ. *For every $L \sqsubseteq^? D$ in Γ, we introduce a clause asserting that this subsumption must hold:*

$$\rightarrow [L \sqsubseteq D]^{|\mathcal{T}|}.$$

(II) Translation of the relevant properties of subsumption.

1) For all ground atoms $C \in \mathsf{At}$, $D \in \mathsf{At}_{\mathsf{tr}}$ and $i \in \{0, \ldots, |\mathcal{T}|\}$ such that $C \not\sqsubseteq_{\mathcal{O}} D$, we introduce a clause preventing this subsumption:

$$[C \sqsubseteq D]^i \rightarrow \quad .$$

2) For every variable Y, $B \in \mathsf{At}_{\mathsf{nv}}$, $i, j \in \{0, \ldots, |\mathcal{T}|\}$, and $L \in$ Left, we introduce the clause

$$[L \sqsubseteq Y]^i \wedge [Y \sqsubseteq B]^j \rightarrow [L \sqsubseteq B]^{\min\{|\mathcal{T}|, i+j\}}.$$

3) For every $L \in$ Left $\setminus N_v$ and $D \in \mathsf{At}_{\mathsf{tr}}$, we introduce the following formulae, depending on L and D:

a) If D is a ground atom and L is not a ground atom, we introduce

$$[L \sqsubseteq D]^i \rightarrow \bigvee_{C \in L} \mathrm{Dec}(C \sqsubseteq D) \vee \bigvee_{\substack{A_1 \sqcap \cdots \sqcap A_k \sqsubseteq B \in \mathcal{O} \\ B \sqsubseteq_{\mathcal{O}} D}} ([L \sqsubseteq A_1]^{i-1} \wedge \cdots \wedge [L \sqsubseteq A_k]^{i-1})$$

for all $i \in \{1, \ldots, |\mathcal{T}|\}$ and

$$[L \sqsubseteq D]^0 \rightarrow \bigvee_{C \in L} \mathrm{Dec}(C \sqsubseteq D).$$

b) If D is a non-variable, non-ground atom, we introduce

$$[L \sqsubseteq D]^i \rightarrow \bigvee_{C \in L} \mathrm{Dec}(C \sqsubseteq D) \vee \bigvee_{A \text{ atom of } \mathcal{O}} ([L \sqsubseteq A]^i \wedge \mathrm{Dec}(A \sqsubseteq D))$$

for all $i \in \{1, \ldots, |\mathcal{T}|\}$ and

$$[L \sqsubseteq D]^0 \rightarrow \bigvee_{C \in L} \mathrm{Dec}(C \sqsubseteq D).$$

(III) Translation of the relevant properties of $>$.

1) Transitivity and irreflexivity of $>$ is expressed by the clauses

$$[X > X] \rightarrow \quad \text{and } [X > Y] \wedge [Y > Z] \rightarrow [X > Z]$$

for all $X, Y, Z \in N_v$.

2) The connection between $>$ and $\sqsubseteq$ is expressed using the clause

$$[X \sqsubseteq \exists r.Y]^i \rightarrow [X > Y]$$

for every $X, Y \in N_v$, $\exists r.Y \in \mathsf{At}_{\mathsf{tr}}$, and $i \in \{0, \ldots, |\mathcal{T}|\}$.

It is easy to see that the set $C(\Gamma, \mathcal{O})$ can be constructed in time polynomial in the size of Γ and $\mathcal{O}$. In particular, subsumptions $B \sqsubseteq_{\mathcal{O}} D$ between ground atoms B, D can be checked in polynomial time in the size of $\mathcal{O}$ [4].

There are several differences between $C(\Gamma, \mathcal{O})$ and the clauses constructed in [6] to solve unification in $\mathcal{EL}$ w.r.t. the empty ontology. The propositional variables employed in [6] are of the form $[C \not\sqsubseteq D]$ for atoms C, D of Γ, i.e., they stand for non-subsumption rather than subsumption. The use of single atoms C instead of whole left-hand sides L also leads to a different encoding of the subsumptions from Γ in part (I). The clauses in (III) are identical up to negation of the variables $[X \sqsubseteq \exists r.Y]^i$. But most importantly, in [6] the properties of subsumption expressed in (II) need only deal with subsumption w.r.t. the empty ontology, whereas here we have to take a cycle-restricted $\mathcal{ELH}_{R^+}$-ontology into account. We do this by expressing the characterization of subsumption given in Lemma 7. This is also the reason why the propositional variables $[L \sqsubseteq D]^i$ have an additional index i: in fact, in Lemma 7 we refer to the number of root steps in the derivation that shows the subsumption, and this needs to be modeled in our SAT reduction.

Theorem 9. *The unification problem Γ is solvable w.r.t. $\mathcal{O}$ iff $C(\Gamma, \mathcal{O})$ is satisfiable.*

Since $C(\Gamma, \mathcal{O})$ can be constructed in polynomial time and SAT is in NP, this shows that unification w.r.t. cycle-restricted $\mathcal{ELH}_{R^+}$-ontologies is in NP. NP-hardness follows from the known NP-hardness of $\mathcal{EL}$-unification w.r.t. the empty ontology [5].

Corollary 10. *Unification w.r.t. cycle-restricted $\mathcal{ELH}_{R^+}$-ontologies is an NP-complete problem.*

To prove Theorem 9, we must show soundness and completeness of the reduction.

Soundness of the Reduction. Let τ be a valuation of the propositional variables that satisfies $C(\Gamma, \mathcal{O})$. We must show that then Γ has a unifier w.r.t. $\mathcal{O}$. To this purpose, we use τ to define an assignment S by

$$S_X := \{D \in \mathsf{At}_{\mathsf{nv}} \mid \exists i \in \{0, \ldots, |\mathcal{T}|\} : \tau([X \sqsubseteq D]^i) = 1\}.$$

Using the clauses in (III), it is not hard to show [3] that $X >_S Y$ implies $\tau([X > Y]) = 1$. Due to the irreflexivity clause in (III), this yields that the assignment S is acyclic. Thus, it induces a substitution σ_S. A proof of the following lemma can be found in [3].

Lemma 11. *If* $\tau([L \sqsubseteq D]^i) = 1$ *for* $L \in \text{Left}$, $D \in \mathsf{At}_{\mathsf{tr}}$, *and* $i \in \{0, \ldots, |\mathcal{T}|\}$, *then* $\sigma_S(L) \sqsubseteq_\mathcal{O} \sigma_S(D)$.

Because of the clauses in (I), this lemma immediately implies that σ_S is a unifier of Γ w.r.t. $\mathcal{O}$.

Completeness of the Reduction. Given a unifier γ of Γ w.r.t. $\mathcal{O}$, we can define a valuation τ that satisfies $C(\Gamma, \mathcal{O})$ as follows.

Let $L \in \text{Left}$ and $D \in \mathsf{At}_{\mathsf{tr}}$ and $i \in \{0, \ldots, |\mathcal{T}|\}$. We set $\tau([L \sqsubseteq D]^i) := 1$ iff $\mathsf{s}(\gamma(L)) \xleftarrow{(i)}_\mathcal{O} \mathsf{s}(\gamma(D))$. According to Theorem 5, we thus have $\tau([L \sqsubseteq D]^i) = 0$ for all $i \in \{0, \ldots, |\mathcal{T}|\}$ iff $\gamma(L) \not\sqsubseteq_\mathcal{O} \gamma(D)$. Otherwise, there is an $i \in \{0, \ldots, |\mathcal{T}|\}$ such that $\tau([L \sqsubseteq D]^j) = 1$ for all $j \geq i$, and $\tau([L \sqsubseteq D]^j) = 0$ for all $j < i$.

To define the valuation of the remaining propositional variables $[X > Y]$ with $X, Y \in N_v$, we set $\tau([X > Y]) = 1$ iff $X >_\gamma Y$, where $>_\gamma$ is defined as in (5), i.e., $X >_\gamma Y$ iff $\gamma(X) \sqsubseteq_\mathcal{O} \exists w.\gamma(Y)$ for some $w \in N_R^+$.

The following lemma, whose proof can be found in [3], shows completeness of our reduction using Lemma 7.

Lemma 12. *The valuation* τ *satisfies* $C(\Gamma, \mathcal{O})$.

Note that cycle-restrictedness of $\mathcal{O}$ is needed in order to satisfy the irreflexivity clause $[X > X] \rightarrow$ (see Lemma 3). We cannot dispense with this clause since it is needed in the proof of soundness to obtain acyclicity of the assignment S constructed there. In fact, only because S is acyclic can we define the substitution σ_S, which is then shown to be a unifier.

6 Conclusions

We have shown that unification w.r.t. cycle-restricted $\mathcal{ELH}_{R^+}$-ontologies can be reduced to propositional satisfiability. This improves on the results in [1,2] in two respects. First, it allows us to deal also with ontologies that contain transitivity and role hierarchy axioms, which are important for medical ontologies. Second, the SAT reduction can easily be implemented and enables us to make use of highly optimized SAT solvers, whereas the goal-oriented algorithm in [1], while having the potential of becoming quite efficient, requires a high amount of additional optimization work. The main topic for future research is to investigate whether we can get rid of cycle-restrictedness.

References

1. Baader, F., Borgwardt, S., Morawska, B.: Unification in the description logic $\mathcal{EL}$ w.r.t. cycle-restricted TBoxes. LTCS-Report 11-05, Theoretical Computer Science, TU Dresden (2011), http://lat.inf.tu-dresden.de/research/reports.html

2. Baader, F., Borgwardt, S., Morawska, B.: Extending unification in $\mathcal{EL}$ towards general TBoxes. In: Proc. of the 13th Int. Conf. on Principles of Knowledge Representation and Reasoning. AAAI Press (2012) (short paper)
3. Baader, F., Borgwardt, S., Morawska, B.: SAT encoding of unification in $\mathcal{ELH}_{\mathcal{R}+}$ w.r.t. cycle-restricted ontologies. LTCS-Report 12-02, Theoretical Computer Science, TU Dresden (2012), `http://lat.inf.tu-dresden.de/research/reports.html`
4. Baader, F., Brandt, S., Lutz, C.: Pushing the $\mathcal{EL}$ envelope. In: Kaelbling, L.P., Saffiotti, A. (eds.) Proc. of the 19th Int. Joint Conf. on Artificial Intelligence, pp. 364–369. Morgan Kaufmann, Los Altos (2005)
5. Baader, F., Morawska, B.: Unification in the Description Logic $\mathcal{EL}$. In: Treinen, R. (ed.) RTA 2009. LNCS, vol. 5595, pp. 350–364. Springer, Heidelberg (2009)
6. Baader, F., Morawska, B.: SAT Encoding of Unification in $\mathcal{EL}$. In: Fermüller, C.G., Voronkov, A. (eds.) LPAR-17. LNCS, vol. 6397, pp. 97–111. Springer, Heidelberg (2010)
7. Baader, F., Morawska, B.: Unification in the description logic $\mathcal{EL}$. Logical Methods in Computer Science 6(3) (2010)
8. Baader, F., Narendran, P.: Unification of concept terms in description logics. J. of Symbolic Computation 31(3), 277–305 (2001)
9. Baader, F., Snyder, W.: Unification theory. In: Robinson, J.A., Voronkov, A. (eds.) Handbook of Automated Reasoning, pp. 445–532. The MIT Press (2001)
10. Biere, A., Heule, M., van Maaren, H., Walsh, T. (eds.): Handbook of Satisfiability. IOS Press (2009)
11. Brandt, S.: Polynomial time reasoning in a description logic with existential restrictions, GCI axioms, and—what else? In: de Mántaras, R.L., Saitta, L. (eds.) Proc. of the 16th Eur. Conf. on Artificial Intelligence. pp. 298–302 (2004)
12. Campbell, J.R., Lopez Osornio, A., de Quiros, F., Luna, D., Reynoso, G.: Semantic interoperability and SNOMED CT: A case study in clinical problem lists. In: Kuhn, K., Warren, J., Leong, T.Y. (eds.) Proc. of the 12th World Congress on Health (Medical) Informatics, pp. 2401–2402. IOS Press (2007)
13. Degtyarev, A., Voronkov, A.: The undecidability of simultaneous rigid E-unification. Theor. Comput. Sci. 166(1&2), 291–300 (1996)
14. Gallier, J.H., Narendran, P., Plaisted, D.A., Snyder, W.: Rigid E-unification: NP-completeness and applications to equational matings. Inf. Comput. 87(1/2), 129–195 (1990)
15. Seidenberg, J., Rector, A.L.: Representing Transitive Propagation in OWL. In: Embley, D.W., Olivé, A., Ram, S. (eds.) ER 2006. LNCS, vol. 4215, pp. 255–266. Springer, Heidelberg (2006)
16. Sofronie-Stokkermans, V.: Locality and subsumption testing in $\mathcal{EL}$ and some of its extensions. In: Proc. Advances in Modal Logic (2008)
17. Suntisrivaraporn, B., Baader, F., Schulz, S., Spackman, K.: Replacing SEP-Triplets in SNOMED CT Using Tractable Description Logic Operators. In: Bellazzi, R., Abu-Hanna, A., Hunter, J. (eds.) AIME 2007. LNCS (LNAI), vol. 4594, pp. 287–291. Springer, Heidelberg (2007)

UEL: Unification Solver for the Description Logic $\mathcal{EL}$ – System Description

Franz Baader, Julian Mendez, and Barbara Morawska

Theoretical Computer Science, TU Dresden, Germany
{baader,mendez,morawska}@tcs.inf.tu-dresden.de

Abstract. UEL is a system that computes unifiers for unification problems formulated in the description logic $\mathcal{EL}$. $\mathcal{EL}$ is a description logic with restricted expressivity, but which is still expressive enough for the formal representation of biomedical ontologies, such as the large medical ontology SNOMED CT. We propose to use UEL as a tool to detect redundancies in such ontologies by computing unifiers of two formal concepts suspected of expressing the same concept of the application domain. UEL can be used as a plug-in of the popular ontology editor Protégé, or as a standalone unification application.

1 Motivation

The description logic (DL) $\mathcal{EL}$, which offers the concept constructors conjunction ($\sqcap$), existential restriction ($\exists r.C$), and the top concept ($\top$), has recently drawn considerable attention since, on the one hand, important inference problems such as the subsumption problem are polynomial in $\mathcal{EL}$ [1,8,2]. On the other hand, though quite inexpressive, $\mathcal{EL}$ can be used to define biomedical ontologies, such as the large medical ontology SNOMED CT.[1]

Unification in DLs has been proposed in [6] as a novel inference service that can, for instance, be used to detect redundancies in ontologies. For example, assume that one developer of a medical ontology defines the concept of a *patient with severe head injury* as

$$\mathsf{Patient} \sqcap \exists \mathsf{finding}.(\mathsf{Head\_injury} \sqcap \exists \mathsf{severity}.\mathsf{Severe}), \tag{1}$$

whereas another one represents it as

$$\mathsf{Patient} \sqcap \exists \mathsf{finding}.(\mathsf{Severe\_injury} \sqcap \exists \mathsf{finding\_site}.\mathsf{Head}). \tag{2}$$

These two concept descriptions are not equivalent, but they are nevertheless meant to represent the same concept. They can obviously be made equivalent by treating the concept names $\mathsf{Head\_injury}$ and $\mathsf{Severe\_injury}$ as variables, and substituting the first one by $\mathsf{Injury} \sqcap \exists \mathsf{finding\_site}.\mathsf{Head}$ and the second one by $\mathsf{Injury} \sqcap \exists \mathsf{severity}.\mathsf{Severe}$. In this case, we say that the descriptions are unifiable,

[1] See http://www.ihtsdo.org/snomed-ct/

B. Gramlich, D. Miller, and U. Sattler (Eds.): IJCAR 2012, LNAI 7364, pp. 45–51, 2012.

Table 1. Syntax and semantics of $\mathcal{EL}$

Name	Syntax	Semantics
concept name	A	$A^{\mathcal{I}} \subseteq \Delta^{\mathcal{I}}$
role name	r	$r^{\mathcal{I}} \subseteq \Delta^{\mathcal{I}} \times \Delta^{\mathcal{I}}$
top	$\top$	$\top^{\mathcal{I}} = \Delta^{\mathcal{I}}$
conjunction	$C \sqcap D$	$(C \sqcap D)^{\mathcal{I}} = C^{\mathcal{I}} \cap D^{\mathcal{I}}$
existential restriction	$\exists r.C$	$(\exists r.C)^{\mathcal{I}} = \{x \mid \exists y : (x, y) \in r^{\mathcal{I}} \wedge y \in C^{\mathcal{I}}\}$
concept definition	$A \equiv C$	$A^{\mathcal{I}} = C^{\mathcal{I}}$

and call the substitution that makes them equivalent a *unifier*. Intuitively, such a unifier proposes definitions for the concept names that are used as variables: in our example, we know that, if we define Head_injury as Injury ⊓ ∃finding_site.Head and Severe_injury as Injury ⊓ ∃severity.Severe, then the two concept descriptions (1) and (2) are equivalent w.r.t. these definitions. Of course, this example was constructed such that the unifier actually provides sensible definitions for the concept names used as variables. In general, the existence of a unifier only says that there is a structural similarity between the two concepts. The developer that uses unification as a tool for finding redundancies in an ontology or between two different ontologies needs to inspect the unifier(s) to see whether the definitions it suggests really make sense.

In [3] it was shown that unification in $\mathcal{EL}$ is an NP-complete problem. Basically, this problem is in NP since every solvable unification problem has a "local" unifier, i.e., one built from parts of the unification problem. The NP algorithm introduced in [3] is a brutal "guess and then test" algorithm, which guesses a local substitution and then checks whether it is a unifier. In [5], a more practical $\mathcal{EL}$-unification algorithm was introduced, which tries to transform the given unification problems into a solved form, and makes nondeterministic decisions only if triggered by the problem. While having the potential of becoming quite efficient, this algorithm still requires a high amount of additional optimization work before it can be used in practice. Our system UEL[2] is based on a third kind of algorithm, which encodes the unification problem into a set of propositional clauses [4], and then solves it using an existing highly optimized SAT solver.

2 $\mathcal{EL}$ and Unification in $\mathcal{EL}$

In order to explain what UEL actually computes, we need to recall the relevant definitions and results for $\mathcal{EL}$ and unification in $\mathcal{EL}$ (see [7,1,5] for details).

Starting with a finite set N_C of *concept names* and a finite set N_R of *role names*, $\mathcal{EL}$*-concept descriptions* are built from concept names using the constructors *conjunction* ($C \sqcap D$), *existential restriction* ($\exists r.C$ for every $r \in N_R$), and *top* ($\top$). On the semantic side, concept descriptions are interpreted as sets.

[2] Version 1.0.0 of this system, as described in this paper, is available for download at `http://sourceforge.net/projects/uel/files/uel/1.0.0/`

To be more precise, an *interpretation* $\mathcal{I} = (\Delta^{\mathcal{I}}, \cdot^{\mathcal{I}})$ consists of a non-empty domain $\Delta^{\mathcal{I}}$ and an interpretation function $\cdot^{\mathcal{I}}$ that maps concept names to subsets of $\Delta^{\mathcal{I}}$ and role names to binary relations over $\Delta^{\mathcal{I}}$. This function is extended to concept descriptions as shown in the semantics column of Table 1.

A *concept definition* is of the form $A \equiv C$ for a concept name A and a concept description C. A *TBox* $\mathcal{T}$ is a finite set of concept definitions such that no concept name occurs more than once on the left-hand side of a definition in $\mathcal{T}$. The TBox $\mathcal{T}$ is called *acyclic* if there are no cyclic dependencies between its concept definitions. Given a TBox $\mathcal{T}$, we call a concept name A a *defined concept* if it occurs as the left-side of a concept definition $A \equiv C$ in $\mathcal{T}$. All other concept names are called *primitive concepts.* An interpretation is a *model* of a TBox $\mathcal{T}$ if $A^{\mathcal{I}} = C^{\mathcal{I}}$ holds for all definitions $A \equiv C$ in $\mathcal{T}$.

Subsumption asks whether a given concept description C is a subconcept of another concept description D: C is *subsumed* by D w.r.t. $\mathcal{T}$ ($C \sqsubseteq_{\mathcal{T}} D$) if every model of $\mathcal{T}$ satisfies $C^{\mathcal{I}} \subseteq D^{\mathcal{I}}$. We say that C is *equivalent* to D w.r.t. $\mathcal{T}$ ($C \equiv_{\mathcal{T}} D$) if $C \sqsubseteq_{\mathcal{T}} D$ and $D \sqsubseteq_{\mathcal{T}} C$. For the empty TBox, we write $C \sqsubseteq D$ and $C \equiv D$ instead of $C \sqsubseteq_{\emptyset} D$ and $C \equiv_{\emptyset} D$, and simply talk about subsumption and equivalence (without saying "w.r.t. $\emptyset$").

In order to define unification, we partition the set N_C of concept names into a set N_v of concept variables (which may be replaced by substitutions) and a set N_c of concept constants (which must not be replaced by substitutions). Intuitively, N_v are the concept names that have possibly been given another name or been specified in more detail in another concept description describing the same notion. A *substitution* σ maps every variable to a concept description. It can be extended to concept descriptions in the usual way. Unification in $\mathcal{EL}$ was first considered w.r.t. the empty TBox [3]. In this setting, an *$\mathcal{EL}$-unification problem* is a finite set $\Gamma = \{C_1 \equiv^? D_1, \ldots, C_n \equiv^? D_n\}$ of equations. A substitution σ is a *unifier* of Γ if σ *solves* all the equations in Γ, i.e., if $\sigma(C_1) \equiv \sigma(D_1), \ldots, \sigma(C_n) \equiv \sigma(D_n)$. We say that Γ is *solvable* if it has a unifier.

As mentioned before, the main reason for solvability of unification in $\mathcal{EL}$ to be in NP is that any solvable unification problem has a local unifier. Basically, any unification problem Γ determines a polynomial number of so-called *non-variable atoms*, which are concept constants or existential restrictions of the form $\exists r.A$ for a role name r and a concept constant or variable A. An *assignment* S maps every concept variable X to a subset S_X of the set $\mathsf{At}_{\mathsf{nv}}$ of non-variable atoms of Γ. Such an assignment induces the following relation $>_S$ on N_v: $>_S$ is the transitive closure of $\{(X, Y) \in N_v \times N_v \mid Y \text{ occurs in an element of } S_X\}$. We call the assignment S *acyclic* if $>_S$ is irreflexive (and thus a strict partial order). Any acyclic assignment S induces a unique substitution σ_S, which can be defined by induction along $>_S$:

- If X is a minimal element of N_v w.r.t. $>_S$, then we define $\sigma_S(X) := \bigsqcap_{D \in S_X} D$.
- Assume that $\sigma(Y)$ is already defined for all Y such that $X >_S Y$. Then we define $\sigma_S(X) := \bigsqcap_{D \in S_X} \sigma_S(D)$.

We call a substitution σ *local* if it is of this form, i.e., if there is an acyclic assignment S such that $\sigma = \sigma_S$. In [3] it is shown that any solvable unification problem has a local unifier. Consequently, one can enumerate (or guess, in a nondeterministic machine) all acyclic assignments and then check whether any of them induces a substitution that is a unifier. Using this brute-force approach, in general many local substitutions will be generated that only in the subsequent check turn out not to be unifiers.

In contrast, the SAT reduction introduced in [4] ensures that only assignments that induce unifiers are generated. The set of propositional clauses $C(\Gamma)$ generated by the reduction contains two kinds of propositional letters: $[A \not\sqsubseteq B]$ for $A, B \in \mathrm{At}_{\mathsf{nv}}$ and $[X > Y]$ for concept variables X, Y. Intuitively, setting $[A \not\sqsubseteq B] = 1$ means that the local substitution σ_S induced by the corresponding assignment S satisfies $\sigma_S(A) \not\sqsubseteq \sigma_S(B)$, and setting $[X > Y] = 1$ means that $X >_S Y$. The clauses in $C(\Gamma)$ are such that Γ has a unifier iff $C(\Gamma)$ is satisfiable. In particular, any propositional valuation τ satisfying $C(\Gamma)$ defines an assignment S^τ with $S^\tau_X := \{A \mid \tau([X \not\sqsubseteq A]) = 0, A \in \mathrm{At}_{\mathsf{nv}}\}$, which induces a local unifier of Γ. Conversely, any local unifier of Γ can be obtained in this way. Thus, by generating all propositional valuations satisfying $C(\Gamma)$ we can generate all local unifiers of Γ.

In [5], *unification w.r.t. an acyclic TBox* $\mathcal{T}$ was introduced. In this setting, the concept variables are a subset of the primitive concepts of $\mathcal{T}$, and substitutions are applied both to the concept descriptions in the unification problem and to the right-hand sides of the definitions in $\mathcal{T}$. To deal with such unification problems, one does not need to develop a new algorithm. In fact, by viewing the defined concepts of $\mathcal{T}$ as variables, one can turn $\mathcal{T}$ into a unification problem, which one simply adds to the given unification problem Γ. As shown in [5], there is a 1–1-correspondence between the unifiers of Γ w.r.t. $\mathcal{T}$ and the unifiers of this extended unification problem.

3 Things Not Mentioned in the Theoretical Papers

When implementing UEL, we had to deal with several issues that are abstracted away in the theoretical papers describing unification algorithms for $\mathcal{EL}$.

Primitive Definitions. In addition to concept definitions, as introduced above, biomedical ontologies often contain so-called *primitive definitions* $A \sqsubseteq C$ where A is a concept name and C is a concept description. Models $\mathcal{I}$ of $A \sqsubseteq C$ need to satisfy $A^\mathcal{I} \subseteq C^\mathcal{I}$. Thus, primitive definitions formulate necessary conditions for concept membership, but these conditions are not sufficient. SNOMED CT contains about 350,000 primitive definitions and only 40,000 concept definitions.

By using a trick first introduced by Nebel [9], primitive definitions $A \sqsubseteq C$ can be turned into concept definitions $A \equiv C \sqcap A\_UNDEF$, where A_UNDEF is a new concept name that stands for the undefined part of the definition of A. In the resulting acyclic TBox, these new concept names are primitive concepts, and thus can be declared to be variables. In this case, a unifier σ suggests how to complete the definition of A by providing the concept description $\sigma(A\_UNDEF)$.

Unifiers as Acyclic TBoxes. Given an acyclic assignment S computed by the SAT reduction, our system UEL actually does not produce the corresponding local unifier σ_S as output, but rather the acyclic TBox $\mathcal{T}_S := \{X \equiv \bigsqcap_{D \in S_X} D \mid X \in N_v\}$. This TBox solves the input unification problem Γ w.r.t. $\mathcal{T}$ in the sense that $C \equiv_{\mathcal{T} \cup \mathcal{T}_S} D$ holds for all equations $C \equiv^? D$ in Γ. This is actually what the developer that employs unification wants to know: how must the concept variables be defined such that the concept descriptions in the equations become equivalent? Another advantage of this representation of the output is that the size of S and thus of $\mathcal{T}_S$ is polynomial in the size of the input Γ and $\mathcal{T}$, while the size of the concept descriptions $\sigma_S(X)$ may be exponential in this size. In the following, we will also call the TBoxes $\mathcal{T}_S$ unifiers.

Internal Variables. The unification algorithms for $\mathcal{EL}$ actually assume that the unification problem is first transformed into a so-called flat form. This form can easily be generated by introducing auxiliary variables. These new variables have system-generated names, which do not make sense to the user. Thus, they should not show up in the output acyclic TBox $\mathcal{T}_S$. By replacing such auxiliary defined concepts in $\mathcal{T}_S$ by their definitions as long as auxiliary names occur, we can transform $\mathcal{T}_S$ into an acyclic TBox that satisfies this requirement, actually without causing an exponential blow-up of the size of the TBox.

Reachable sub-TBox. As mentioned above, acyclic TBoxes are treated by viewing them as part of the unification problem. For very large TBoxes like SNOMED CT, adding the whole TBox to the unification problem is neither viable nor necessary. In fact, it is sufficient to add the reachable part of the TBox, i.e., the definitions onto which the concept descriptions in the unification problem depend. This reachable part is usually rather small, even for very large TBoxes.

Enumeration of All Local Unifiers. Depending on how many concept names are turned into variables, a unification problem can have many local unifiers. If the SAT solver has provided a satisfying propositional valuation, we can add a clause to the SAT problem that prevents the re-computation of this unifier, and call the SAT solver with this new SAT instance. While computing a single unifier is usually quite fast, computing all of them can take much longer. Thus, we enable the user to compute and then inspect one unifier at a time. If this unifier makes sense, i.e., suggests reasonable definitions for the variables, then the user can stop. Otherwise, by pressing a button, the computation of the next local unifier can be initiated. For this to work well, it is important that "good" unifiers are computed first. For the moment, we have interpreted "good" as meaning small, i.e., we want to compute those unifiers first that are generated by acyclic assignments for which the sets S_X are small. It has turned out that the SAT reduction sketched above actually leads to computing unifiers in the opposite order, at least if we use a SAT solver that tries to minimize the number of propositional letters that are set to 1. In fact, setting a letter of the form $[X \not\sqsubseteq A]$ for $X \in N_v$ and $A \in \mathsf{At}_{\mathsf{nv}}$ to 0 rather than 1 adds A to S_X. This problem can be overcome by using propositional letters $[A \sqsubseteq B]$ with the obvious meaning, and basically replacing $[A \not\sqsubseteq B]$ in the SAT reduction by $\neg[A \sqsubseteq B]$.

4 The System UEL and How to Use It

UEL was implemented in Java 1.6 and is compatible with Java 1.7. It uses the OWL API 3.2.4[3] to read ontologies. It has a visual interface that can be used as a Protégé 4.1 plug-in, or as a standalone application. The unification problem generated by the user through this interface is translated into a propositional formula in conjunctive normal form using the DIMACS CNF format,[4] which is the most popular format used by SAT solvers. As SAT solver, we currently use SAT4J,[5] which is implemented in Java. This configuration is, however, parametrized and can be easy changed to any SAT solver that accepts DIMACS CNF input and returns the computed satisfying propositional valuation.

After opening UEL's visual interface, the first step is to open one or two ontologies. The second option enables unification of concepts defined in different ontologies. The user can then choose two concepts to be unified.[6] This is done by choosing two concept names that occur on the left-hand sides of concept definitions or primitive definitions. UEL then computes the subontologies reachable from these concept names, and turns the primitive definitions in these subontologies into concept definitions.

After choosing the concepts to be unified, pressing the button  opens a dialog window in which the user is presented with the primitive concepts contained in these subontologies (including the ones with ending _*UNDEF*). The user can then decide which of these primitive concepts should be viewed as variables in the unification problem

Once the user has chosen the variables, UEL computes the unification problem defined this way, and transforms it into a clause set in DIMACS CNF format. It also opens a dialog window with control buttons. By pressing the button , the user triggers the computation of the first unifier (or later, of the next one). Each computed unifier is shown (as an acyclic TBox) in the dialog window. The button  can be used to go back to the previously computed unifier. The button  can be used to trigger the computation of all (remaining) unifiers, and the button  allows to jump back to the first unifier. Unifiers already computed are stored, and thus need not be recomputed during navigation. Each unifier (i.e., the acyclic TBox representing it) can be saved using the RDF/OWL or the KRSS format by pressing the button . The format for saving is determined by the file ending typed by the user (.krss or .owl).

The user can use the button  to retrieve internal details about the computation process. These details include the unification problem created internally by UEL, the number of all concept variables (user chosen and internal variables), the number of propositional letters, and the number of propositional clauses that are checked for satisfiability by the SAT solver.

---

[3] `http://owlapi.sourceforge.net`

[4] `http://www.satcompetition.org/2004/format-solvers2004.html`

[5] `http://www.sat4j.org`

[6] Note that a finite set of equations $\{C_1 \equiv^? D_1, \ldots, C_n \equiv^? D_n\}$ can always be encoded into the single equation $\{\exists r_1.C_1 \sqcap \ldots \sqcap \exists r_n.C_n \equiv^? \exists r_1.D_1 \sqcap \cdots \sqcap \exists r_n.D_n\}$, where $r_1, \ldots, r_n$ are pairwise distinct role names.

5 An Example

We consider a modified version of our example in the first section, where the TBox gives (1) as definition for the concept name Patient_with_severe_head_injury and (2) as definition for the concept name Patient_with_severe_injury_at_head. In addition, the TBox contains two primitive definitions, saying that Head_injury and Severe_injury are subconcepts of Injury. We load this TBox into UEL and choose Patient_with_severe_head_injury and Patient_with_severe_injury_at_head as the concepts to be unified. The system then offers us the primitive concepts Patient, Severe, Head as well as Head_injury_UNDEF, Severe_injury_UNDEF as possible variables, of which we choose only the latter two.

The SAT translation generates a SAT problem consisting of 3976 clauses and containing 320 different propositional letters. The first unifier computed by UEL is the substitution

$$\{\mathsf{Head\_injury\_UNDEF} \mapsto \exists \mathsf{finding\_site.Head},$$
$$\mathsf{Severe\_injury\_UNDEF} \mapsto \exists \mathsf{severity.Severe}\}.$$

This unifier thus completes the primitive definitions of the concepts Head_injury and Severe_injury to concept definitions Head_injury $\equiv$ Injury $\sqcap$ finding_site.Head and Severe_injury $\equiv$ Injury $\sqcap$ $\exists$severity.Severe.

However, the unification problem has 127 additional local unifiers. Some of them are similar to the first one, but contain "redundant" conjuncts. Others do not make much sense in the application (e.g., ones where Patient occurs in the images of the variables). Computing all 128 local unifiers at once (after pressing the button ⏩) takes less than 1 second.

References

1. Baader, F.: Terminological cycles in a description logic with existential restrictions. In: Proc. IJCAI 2003 (2003)
2. Baader, F., Brandt, S., Lutz, C.: Pushing the $\mathcal{EL}$ envelope. In: Proc. IJCAI 2005 (2005)
3. Baader, F., Morawska, B.: Unification in the Description Logic $\mathcal{EL}$. In: Treinen, R. (ed.) RTA 2009. LNCS, vol. 5595, pp. 350–364. Springer, Heidelberg (2009)
4. Baader, F., Morawska, B.: SAT Encoding of Unification in $\mathcal{EL}$. In: Fermüller, C.G., Voronkov, A. (eds.) LPAR-17. LNCS, vol. 6397, pp. 97–111. Springer, Heidelberg (2010)
5. Baader, F., Morawska, B.: Unification in the description logic $\mathcal{EL}$. Logical Methods in Computer Science 6(3) (2010)
6. Baader, F., Narendran, P.: Unification of concept terms in description logics. J. of Symbolic Computation 31(3), 277–305 (2001)
7. Baader, F., Nutt, W.: Basic Description Logics. In: The Description Logic Handbook. Cambridge University Press (2003)
8. Brandt, S.: Polynomial time reasoning in a description logic with existential restrictions, GCI axioms, and—what else? In: Proc. ECAI 2004 (2004)
9. Nebel, B.: Reasoning and Revision in Hybrid Representation Systems. LNCS, vol. 422. Springer, Heidelberg (1990)

Effective Finite-Valued Semantics for Labelled Calculi

Matthias Baaz[1], Ori Lahav[2,⋆], and Anna Zamansky[1]

[1] Vienna University of Technology
[2] Tel Aviv University
{baaz,annaz}@logic.at, orilahav@post.tau.ac.il

Abstract. We provide a systematic and modular method to define non-deterministic finite-valued semantics for a natural and very general family of *canonical labelled calculi*, of which many previously studied sequent and labelled calculi are particular instances. This semantics is *effective*, in the sense that it naturally leads to a decision procedure for these calculi. It is then applied to provide simple *decidable* semantic criteria for crucial syntactic properties of these calculi, namely (strong) analyticity and cut-admissibility.

1 Introduction

There are two contrary aims in logic: the first is to find calculi that characterize a given semantics, the second is to find semantics for a logic that is only given as a formal calculus. Roughly speaking, the former aim has been reached for all (ordinary) finite-valued logics (including, of course, classical logic), as well as for non-deterministic finite-valued logics ([2, 3]). As for the latter, there is no known systematic method of constructing for a given general calculus, a corresponding "well-behaved" semantics. By "well-behaved" here we mean that it is *effective* in the sense of naturally inducing a decision procedure for its underlying logic. Moreover, it is desirable that such semantics can be applied to provide simple semantic characterization of important syntactic properties of the corresponding calculi, which are hard to establish by other means. Analyticity and cut-admissibility are just a few cases in point.

In [6] and [4] two families of *labelled sequent calculi* have been studied in this context.[1] [6] considers labelled calculi with generalized forms of cuts and identity

⋆ Supported by The Israel Science Foundation (grant no. 280-10) and by FWF START Y544-N23.

[1] A remark is in order here on the relationship between the labelled calculi studied here and the general framework of labelled deductive systems (LDS) from [8]. Both frameworks consider consequence relations between labelled formulas. Methodologically, however, they have different aims: [8] constructs a system for a given logic defined in semantic terms, while we define a semantics for a given labelled system. Moreover, in LDS *anything* is allowed to serve as labels, while we assume a finite set of labels. In this sense, our labelled calculi are a particular instance of LDS.

B. Gramlich, D. Miller, and U. Sattler (Eds.): IJCAR 2012, LNAI 7364, pp. 52–66, 2012.

axioms and a restricted form of logical rules, and provides some necessary and sufficient conditions for such calculi to have a characteristic finite-valued matrix. In [4] labelled calculi with a less restrictive form of logical rules (but a more restrictive form of cuts and axioms) are considered. The calculi of [4], satisfying a certain coherence condition, have a semantic characterization using a natural generalization of the usual finite-valued matrix called *non-deterministic matrices* ([2, 3]). The semantics provided in [6, 4] for these families of labelled calculi is well-behaved in the sense defined above, that is the question of whether a sequent s follows in some (non-deterministic) matrix from a finite set of sequents $\mathcal{S}$, can be reduced to considering legal *partial* valuations, defined on the subformulas of $\mathcal{S} \cup \{s\}$. This naturally induces a decision procedure for such logics.

In this paper we show that the class of labelled calculi that have a finite-valued well-behaved semantics is substantially larger than all the families of calculi considered in the literature in this context. We start by defining a general class of fully-structural and propositional labelled calculi, called *canonical labelled calculi*, of which the labelled calculi of [6, 4] are particular examples. In addition to the weakening rule, canonical labelled calculi have rules of two forms: primitive rules and introduction rules. The former operate on labels and do not mention any connective. The generalized cuts and axioms of [6] are specific instances of such rules. As for the latter, each such rule introduces one logical connective of the language. To provide semantics for such calculi in a systematic and modular way, we generalize the notion of non-deterministic matrices to *partial non-deterministic matrices* (PNmatrices), by allowing empty sets of options in the truth tables of logical connectives. Although applicable to a much wider range of calculi, the semantic framework of finite PNmatrices shares a crucial property with both standard and non-deterministic matrices: any calculus that has a characteristic PNmatrix is decidable. Moreover, as opposed to the results in [6, 4], *no* conditions are required for a canonical labelled calculi to have a characteristic PNmatrix: *all* such calculi have one, and so *all* of them are decidable. We then apply PNmatrices to provide simple *decidable* characterizations of two crucial syntactic properties: strong analyticity and strong cut-admissibility.

Due to lack of space, most proofs are omitted, and will appear in an extended version.

2 Preliminaries

In what follows $\mathcal{L}$ is a propositional language, and $\pounds$ is a finite non-empty set of labels. We assume that $p_1, p_2, \ldots$ are the atomic formulas of $\mathcal{L}$. We denote by $Frm_{\mathcal{L}}$ the set of all wffs of $\mathcal{L}$. We usually use φ, ψ as metavariables for formulas, Γ, Δ for finite sets of formulas, l for labels, and L for sets of labels.

Definition 1. *A* labelled formula *is an expression of the form $l : \psi$, where $l \in \pounds$ and $\psi \in Frm_{\mathcal{L}}$. A* sequent *is a finite set of labelled formulas. An n-*clause *is a sequent consisting of atomic formulas from $\{p_1, \ldots, p_n\}$. Given a set $L \subseteq \pounds$, we write $(L : \psi)$ instead of (the sequent) $\{l : \psi \mid l \in L\}$.*

Given a labelled formula γ, we denote by $frm[\gamma]$ the (ordinary) formula appearing in γ, and by $sub[\gamma]$ the set of subformulas of the formula $frm[\gamma]$. frm and sub are extended to sets of labelled formulas and to sets of sets of labelled formulas in the obvious way.

Remark 1. The usual (two-sided) sequent notation $\psi_1, \ldots, \psi_n \Rightarrow \varphi_1, \ldots, \varphi_m$ can be interpreted as $\{f : \psi_1, \ldots, f : \psi_n, t : \varphi_1, \ldots, t : \varphi_m\}$, i.e. a sequent in the sense of Definition 1 for $\pounds = \{t, f\}$.

Definition 2. *An* $\mathcal{L}$-substitution *is a function* $\sigma : Frm_{\mathcal{L}} \to Frm_{\mathcal{L}}$*, which satisfies* $\sigma(\diamond(\psi_1, \ldots, \psi_n)) = \diamond(\sigma(\psi_1), \ldots, \sigma(\psi_n))$ *for every* n*-ary connective* $\diamond$ *of* $\mathcal{L}$*. A substitution is extended to labelled formulas, sequents, etc. in the obvious way.*

3 Canonical Labelled Calculi

In this section we define the family of *canonical labelled calculi*. This is a general family of labelled calculi, which includes many natural subclasses of previously studied systems. These include the system **LK** for classical logic, the canonical sequent calculi of [2], the signed calculi of [4] and the labelled calculi of [6].[2]

All canonical labelled calculi have in common the *weakening* rule. In addition, they include rules of two types: *primitive rules* and *introduction rules*. Each rule of the latter type introduces exactly one logical connective, while rules of the former type operate on labels and do no mention any logical connectives. Next we provide precise definitions.

Definition 3 (Weakening). *The* weakening *rule allows to infer* $s \cup s'$ *from* s *for every two sequents* s *and* s'*.*

Definition 4 (Primitive Rules). *A* primitive rule for $\pounds$ *is an expression of the form* $\{L_1, \ldots, L_n\}/L$ *where* $n \geq 0$ *and* $L_1, \ldots, L_n, L \subseteq \pounds$*. An* application of a primitive rule $\{L_1, \ldots, L_n\}/L$ *is any inference step of the following form:*

$$\frac{(L_1 : \psi) \cup s_1 \quad \ldots \quad (L_n : \psi) \cup s_n}{(L : \psi) \cup s_1 \cup \ldots \cup s_n}$$

where ψ *is a formula, and* s_i *is a sequent for every* $1 \leq i \leq n$*.*

Example 1. Suppose $\pounds = \{a, b, c\}$ and consider the primitive rule $\{\{a\}, \{b\}\}/\{b, c\}$. This rule allows to infer $(\{b, c\} : \psi) \cup s_1 \cup s_2$ from $\{a : \psi\} \cup s_1$ and $\{b : \psi\} \cup s_2$ for every two sequents s_1, s_2 and a formula ψ.

[2] The family of canonical labelled calculi also includes the systems dealt with in [9]. [9] extends the results of [2] by considering also "semi-canonical calculi", which are obtained from (two-sided) canonical calculi by discarding either the cut rule, the identity axioms or both of them. Clearly, these systems are particular instances of canonical labelled calculi, defined in this paper.

Definition 5. *A primitive rule for* $\pounds$ *of the form* $\emptyset/L$ *is called a* canonical axiom. *Its applications provide all axioms of the form* $(L:\psi)$.

Example 2. Axiom schemas of two-sided sequent calculi usually have the form $\psi \Rightarrow \psi$. Using the notation from Remark 1, it can be presented as the canonical axiom $\emptyset/\{t,f\}$.

Definition 6. *A primitive rule for* $\pounds$ *of the form* $\{L_1,\dots,L_n\}/\emptyset$ *is called a* canonical cut. *Its applications allow to infer* $s_1 \cup \ldots \cup s_n$ *from the sequents* $(L_i:\psi)\cup s_i$ *for every* $1 \leq i \leq n$ *(the formula* ψ *is called the* cut-formula*).*

Example 3. Applications of the cut rule for two-sided sequent calculi are usually presented by the following schema:

$$\frac{\Gamma_1 \Rightarrow \psi, \Delta_1 \quad \Gamma_2, \psi \Rightarrow \Delta_2}{\Gamma_1, \Gamma_2 \Rightarrow \Delta_1, \Delta_2}$$

Using the notation from Remark 1, the corresponding canonical cut has the form $\{\{t\},\{f\}\}/\emptyset$.

Definition 7 (Introduction Rules). *A* canonical introduction rule *for an n-ary connective* $\diamond$ *of* $\mathcal{L}$ *and* $\pounds$ *is an expression of the form* $\mathcal{S}/L : \diamond(p_1,\dots,p_n)$, *where* $\mathcal{S}$ *is a finite set of n-clauses (see Definition 1) (called* premises*), and* L *is a non-empty subset of* $\pounds$. *An* application *of a canonical introduction rule* $\{c_1,\dots,c_m\}/L : \diamond(p_1,\dots,p_n)$ *is any inference step of the following form:*[3]

$$\frac{\sigma(c_1)\cup s_1 \quad \ldots \quad \sigma(c_m)\cup s_m}{(L:\sigma(\diamond(p_1,\dots,p_n)))\cup s_1 \cup \ldots \cup s_m}$$

where σ *is an* $\mathcal{L}$*-substitution, and* s_i *is a sequent for every* $1 \leq i \leq m$.

Example 4. The introduction rules for the classical conjunction in **LK** are usually presented as follows:

$$\frac{\Gamma, \psi, \varphi \Rightarrow \Delta}{\Gamma, \psi \wedge \varphi \Rightarrow \Delta} \qquad \frac{\Gamma_1 \Rightarrow \Delta_1, \psi \quad \Gamma_2 \Rightarrow \Delta_2, \varphi}{\Gamma_1, \Gamma_2 \Rightarrow \Delta_1, \Delta_2, \psi \wedge \varphi}$$

Using the notation from Remark 1, the corresponding canonical rules are:

$$r_1 = \{\{f:p_1, f:p_2\}\}/\{f\} : p_1 \wedge p_2 \qquad r_2 = \{\{t:p_1\},\{t:p_2\}\}/\{t\} : p_1 \wedge p_2$$

Their applications have the forms:

$$\frac{\{f:\psi, f:\varphi\}\cup s}{\{f:\psi\wedge\varphi\}\cup s} \qquad \frac{\{t:\psi\}\cup s_1 \quad \{t:\varphi\}\cup s_2}{\{t:\psi\wedge\varphi\}\cup s_1 \cup s_2}$$

---

[3] Note the full separation between a rule and its application: $p_1,\dots,p_n$ appearing in the rule serve as schematic variables, which are replaced by actual formulas of the language in the application.

Definition 8 (Canonical Labelled Calculi). *A* canonical labelled calculus **G** *for* $\mathcal{L}$ *and* $\pounds$ *includes the weakening rule, a finite set of primitive rules for* $\pounds$*, and a finite set of introduction rules for the connectives of* $\mathcal{L}$ *and* $\pounds$*. We say that a sequent s* is derivable *in a canonical labelled calculus* **G** *from a set of sequents* $\mathcal{S}$ *(and denote it by* $\mathcal{S} \vdash_{\mathbf{G}} s$*) if there exists a derivation in* **G** *of s from* $\mathcal{S}$*.*

Notation: Given a canonical labelled calculus **G** for $\mathcal{L}$ and $\pounds$, we denote by $\mathsf{P}_{\mathbf{G}}$ the set of primitive rules of **G**. In addition, for every connective $\diamond$ of $\mathcal{L}$, we denote by $\mathsf{R}^{\diamond}_{\mathbf{G}}$ the set of canonical introduction rules for $\diamond$ of **G**.

Example 5. The standard sequent system **LK** can be represented as a canonical labelled calculus for the language of classical logic and $\{t, f\}$ (see Remark 1 and Examples 2 to 4).

Henceforth, to improve readability, we usually omit the parentheses from the set appearing before the "/" symbol in primitive rules and canonical introduction rules.

Example 6. For $\pounds = \{a, b, c\}$, the canonical labelled calculus $\mathbf{G}_{abc}$ includes the primitive rules $\emptyset/\{a, b\}, \emptyset/\{b, c\}, \emptyset/\{a, c\}$, and $\{a, b, c\}/\emptyset$. It also has the following canonical introduction rules for a ternary connective $\circ$:

$$\{a : p_1, c : p_2\}, \{a : p_3, b : p_2\}/\{a, c\} : \circ(p_1, p_2, p_3)$$

$$\{c : p_2\}, \{a : p_3, b : p_3\}, \{c : p_1\}/\{b, c\} : \circ(p_1, p_2, p_3)$$

Their applications are of the forms:

$$\frac{\{a : \psi_1, c : \psi_2\} \cup s_1 \quad \{a : \psi_3, b : \psi_2\} \cup s_2}{(\{a, c\} : \circ(\psi_1, \psi_2, \psi_3)) \cup s_1 \cup s_2}$$

$$\frac{\{c : \psi_2\} \cup s_1 \quad \{a : \psi_3, b : \psi_3\} \cup s_2 \quad \{c : \psi_1\} \cup s_3}{(\{b, c\} : \circ(\psi_1, \psi_2, \psi_3)) \cup s_1 \cup s_2 \cup s_3}$$

Note that the canonical labelled calculi studied here are substantially more general than the signed calculi of [4] and the labelled calculi of [6], as the primitive rules of both of these families of calculi include only canonical cuts and axioms. Moreover, in the latter only introduction rules which introduce a singleton are allowed, which is not the case for the calculus in Example 6. In the former, all systems have $\emptyset/\pounds$ as their only axiom, and the set of cuts is always assumed to be $\{\{l_1\}, \{l_2\}/\emptyset \mid l_1 \neq l_2\}$ (again leaving the calculus in Example 6 out of scope).

4 Partial Non-deterministic Matrices

Non-deterministic matrices (Nmatrices) are a natural generalization of the notion of a standard many-valued matrix. These are structures, in which the truth value of a complex formula is chosen non-deterministically out of a *non-empty* set of options (determined by the truth values of its subformulas). For further discussion on Nmatrices we refer the reader to [2, 3]. In this paper we introduce a further generalization of the concept of an Nmatrix, in which this set of options is allowed to be *empty*. Intuitively, empty sets of options in a truth table corresponds to forbidding some combinations of truth values. As we shall see, this will allow us to characterize a wider class of calculi than that obtained by applying usual Nmatrices. However, as shown in the sequel, the property of *effectiveness* is preserved in PNmatrices, and like finite-valued matrices and Nmatrices, (calculi characterized by) finite-valued PNmatrices are decidable.

4.1 Introducing PNmatrices

Definition 9. *A* partial non-deterministic matrix *(*PNmatrix *for short)* $\mathcal{M}$ *for* $\mathcal{L}$ *and* $\pounds$ *consists of: (i) set* $\mathcal{V}_\mathcal{M}$ *of truth values, (ii) a function* $\mathcal{D}_\mathcal{M} : \pounds \to P(\mathcal{V}_\mathcal{M})$ *assigning a set of (designated) truth values to the labels of* $\pounds$*, and (iii) a function* $\diamond_\mathcal{M} : {\mathcal{V}_\mathcal{M}}^n \to P(\mathcal{V}_\mathcal{M})$ *for every* n*-ary connective* $\diamond$ *of* $\mathcal{L}$*. We say that* $\mathcal{M}$ *is* finite *if so is* $\mathcal{V}_\mathcal{M}$*.*

Definition 10. *Let* $\mathcal{M}$ *be a PNmatrix for* $\mathcal{L}$ *and* $\pounds$*.*

1. *An* $\mathcal{M}$*-legal* $\mathcal{L}$*-valuation is a function* $v : Frm_\mathcal{L} \to \mathcal{V}_\mathcal{M}$ *satisfying the condition* $v(\diamond(\psi_1, \ldots, \psi_n)) \in \diamond_\mathcal{M}(v(\psi_1), \ldots, v(\psi_n))$ *for every compound formula* $\diamond(\psi_1, \ldots, \psi_n) \in Frm_\mathcal{L}$*.*
2. *Let* v *be an* $\mathcal{M}$*-legal* $\mathcal{L}$*-valuation. A sequent* s *is* true *in* v *for* $\mathcal{M}$ *(denoted by* $v \models_\mathcal{M} s$*) if* $v(\psi) \in \mathcal{D}_\mathcal{M}(l)$ *for some* $l : \psi \in s$*. A set* $\mathcal{S}$ *of sequents is* true *in* v *for* $\mathcal{M}$ *(denoted by* $v \models_\mathcal{M} \mathcal{S}$*) if* $v \models_\mathcal{M} s$ *for every* $s \in \mathcal{S}$*.*
3. *Given a set of sequents* $\mathcal{S}$ *and a single sequent* s*,* $\mathcal{S} \vdash_\mathcal{M} s$ *if for every* $\mathcal{M}$*-legal* $\mathcal{L}$*-valuation* v*,* $v \models_\mathcal{M} s$ *whenever* $v \models_\mathcal{M} \mathcal{S}$*.*

We now define a special subclass of PNmatrices, in which no empty sets of truth values are allowed in the truth tables of logical connectives. This corresponds to the case of Nmatrices from [2–4].

Definition 11. *We say that a PNmatrix* $\mathcal{M}$ *for* $\mathcal{L}$ *and* $\pounds$ *is* proper *if* $\mathcal{V}_\mathcal{M}$ *is non-empty and* $\diamond_\mathcal{M}(x_1, \ldots, x_n)$ *is non-empty for every* n*-ary connective* $\diamond$ *of* $\mathcal{L}$ *and* $x_1, \ldots, x_n \in \mathcal{V}_\mathcal{M}$*.*

Remark 2. Nmatrices in their original formulation can be viewed as proper PNmatrices for $\mathcal{L}$ and $\pounds$, where $\pounds$ is a singleton. In this case $\mathcal{D}_\mathcal{M}$ is practically a set of designated truth values. This is useful to define consequence relations between sets of formulas and formulas in the following way: $T \vdash_\mathcal{M} \psi$ if whenever the formulas of T are "true in v for $\mathcal{M}$" (that is $v(\varphi) \in \mathcal{D}_\mathcal{M}$ for every $\varphi \in T$), also ψ is "true in v for $\mathcal{M}$" ($v(\psi) \in \mathcal{D}_\mathcal{M}$). However, in this paper we study consequence relations of a different type, namely relations between a set of labelled sequents and a labelled sequent. We need, therefore, a notion of "being true for $\mathcal{M}$" for every $l \in \pounds$. This is achieved by taking $\mathcal{D}_\mathcal{M}$ to be a function from $\pounds$ to $P(\mathcal{V}_\mathcal{M})$. Finally, note that for simplicity of presentation, unlike in previous works, we allow the set of designated truth values (for every $l \in \pounds$) to be empty or to include all truth values in $\mathcal{V}_\mathcal{M}$.

Example 7. Let $\pounds = \{a, b\}$ and suppose that $\mathcal{L}$ contains one unary connective $\star$. The PNmatrices $\mathcal{M}_1$ and $\mathcal{M}_2$ are defined as follows: $\mathcal{V}_{\mathcal{M}_1} = \mathcal{V}_{\mathcal{M}_2} = \{t, f\}$, $\mathcal{D}_{\mathcal{M}_1}(a) = \mathcal{D}_{\mathcal{M}_2}(a) = \{t\}$ and $\mathcal{D}_{\mathcal{M}_1}(b) = \mathcal{D}_{\mathcal{M}_2}(b) = \{f\}$. The respective truth tables for $\star$ are defined as follows:

x	$\star_{\mathcal{M}_1}(x)$
t	$\{f\}$
f	$\{t, f\}$

x	$\star_{\mathcal{M}_2}(x)$
t	$\emptyset$
f	$\{t, f\}$

While both $\mathcal{M}_1$ and $\mathcal{M}_2$ are (finite) PNmatrices, only $\mathcal{M}_1$ is proper. Note that in this case we have $\{a : p_1\} \vdash_{\mathcal{M}_2} \emptyset$, simply because there is no $\mathcal{M}_2$-legal $\mathcal{L}$-valuation that assigns t to p_1.

Finally, we extend the notion of *simple refinements* of Nmatrices ([3]) to the context of PNmatrices:

Definition 12. *Let $\mathcal{M}$ and $\mathcal{N}$ be PNmatrices for $\mathcal{L}$ and $\pounds$. We say that $\mathcal{N}$ is a* simple refinement *of $\mathcal{M}$, denoted by $\mathcal{N} \subseteq \mathcal{M}$, if $\mathcal{V}_{\mathcal{N}} \subseteq \mathcal{V}_{\mathcal{M}}$, $\mathcal{D}_{\mathcal{N}}(l) = \mathcal{D}_{\mathcal{M}}(l) \cap \mathcal{V}_{\mathcal{N}}$ for every $l \in \pounds$, and $\diamond_{\mathcal{N}}(x_1, \ldots, x_n) \subseteq \diamond_{\mathcal{M}}(x_1, \ldots, x_n)$ for every n-ary connective $\diamond$ of $\mathcal{L}$ and $x_1, \ldots, x_n \in \mathcal{V}_{\mathcal{N}}$.*

Proposition 1. *Let $\mathcal{M}$ and $\mathcal{N}$ be PNmatrices for $\mathcal{L}$ and $\pounds$, such that $\mathcal{N} \subseteq \mathcal{M}$. Then:* (1) *Every $\mathcal{N}$-legal $\mathcal{L}$-valuation is also $\mathcal{M}$-legal; and* (2) $\vdash_{\mathcal{M}} \subseteq \vdash_{\mathcal{N}}$.

4.2 Decidability

For a denotational semantics to be useful, it should be *effective*: the question of whether some conclusion follows from a finite set of assumptions, should be decidable by considering some *computable* set of partial valuations defined on some *finite* set of "relevant" formulas. Usually, the "relevant" formulas are taken as all subformulas occurring in the conclusion and the assumptions. Next, we show that the semantics induced by PNmatrices is effective in this sense.

Definition 13. *Let $\mathcal{M}$ be a PNmatrix for $\mathcal{L}$ and $\pounds$, and let $\mathcal{F} \subseteq Frm_{\mathcal{L}}$ closed under subformulas. An $\mathcal{M}$-legal $\mathcal{F}$-valuation is a function $v : \mathcal{F} \to \mathcal{V}_{\mathcal{M}}$ satisfying $v(\diamond(\psi_1, \ldots, \psi_n)) \in \diamond_{\mathcal{M}}(v(\psi_1), \ldots, v(\psi_n))$ for every formula $\diamond(\psi_1, \ldots, \psi_n) \in \mathcal{F}$. $\models_{\mathcal{M}}$ is defined for $\mathcal{F}$-valuations exactly as for $\mathcal{L}$-valuations. We say that an $\mathcal{M}$-legal $\mathcal{F}$-valuation is* extendable *in $\mathcal{M}$ if it can be extended to an $\mathcal{M}$-legal $\mathcal{L}$-valuation.*

In proper PNmatrices, all partial valuations are extendable:

Proposition 2. *Let $\mathcal{M}$ be a* proper *PNmatrix for $\mathcal{L}$ and $\pounds$, and let $\mathcal{F} \subseteq Frm_{\mathcal{L}}$ closed under subformulas. Then any $\mathcal{M}$-legal $\mathcal{F}$-valuation is* extendable *in $\mathcal{M}$.*

Proof. The proof goes exactly like the one for Nmatrices in [1]. Note that the non-emptiness of $\mathcal{V}_{\mathcal{M}}$ is needed in order to extend the empty valuation. Clearly, the different definition of $\mathcal{D}_{\mathcal{M}}$ is immaterial here. □

However, this is not the case for arbitrary PNmatrices:

Example 8. Consider the PNmatrix $\mathcal{M}_2$ from Example 7. Let v be the $\mathcal{M}_2$-legal $\{p_1\}$-valuation defined by $v(p_1) = t$. Obviously, there is no $\mathcal{M}_2$-legal $\mathcal{L}$-valuation that extends v (as there is no way to assign a truth value to $\star p_1$). Thus v is not extendable in $\mathcal{M}_2$.

Theorem 1. *Let $\mathcal{M}$ be a PNmatrix for $\mathcal{L}$ and $\pounds$ and $\mathcal{F} \subseteq Frm_{\mathcal{L}}$ closed under subformulas. An $\mathcal{M}$-legal $\mathcal{F}$-valuation v is extendable in $\mathcal{M}$ iff v is $\mathcal{N}$-legal for some proper PNmatrix $\mathcal{N} \subseteq \mathcal{M}$.*

Corollary 1. *Given a finite PNmatrix $\mathcal{M}$ for $\mathcal{L}$ and $\pounds$, a finite $\mathcal{F} \subseteq Frm_{\mathcal{L}}$ closed under subformulas. and a function $v : \mathcal{F} \to \mathcal{V}_{\mathcal{M}}$, it is decidable whether v is an $\mathcal{M}$-legal $\mathcal{F}$-valuation which is extendable in $\mathcal{M}$.*

Proof. Checking whether v is $\mathcal{M}$-legal is straightforward. To verify that it is extendable in $\mathcal{M}$, we go over all (finite) proper PNmatrices $\mathcal{N} \subseteq \mathcal{M}$ (there is a finite number of them since $\mathcal{M}$ is finite), and check whether v is $\mathcal{N}$-legal for some such $\mathcal{N}$. We return a positive answer iff we have found some $\mathcal{N} \subseteq \mathcal{M}$ such that v is $\mathcal{N}$-legal. The correctness is guaranteed by Theorem 1. □

Corollary 2. *Given a finite PNmatrix $\mathcal{M}$ for $\mathcal{L}$ and $\pounds$, a finite set $\mathcal{S}$ of sequents, and a sequent s, it is decidable whether $\mathcal{S} \vdash_{\mathcal{M}} s$ or not.*

In the literature of Nmatrices (see e.g. [1]) effectiveness is usually identified with the property given in Proposition 2.[4] In this case Corollary 1 trivially holds: to check that v is an extendable $\mathcal{M}$-legal $\mathcal{F}$-valuation, it suffices to check that it is $\mathcal{M}$-legal, as extendability is a priori guaranteed. However, the results above show that this property is not a necessary condition for decidability. To guarantee the latter, instead of requiring that *all* partial valuations are extendable, it is sufficient to have an algorithm that establishes which of them are.

4.3 Minimality

In the next section, we show that the framework of PNmatrices provides a semantic way of characterizing canonical labelled calculi. A natural question in this context is how one can obtain *minimal* such characterizations. Next we provide lower bounds on the number of truth values that are needed to characterize $\vdash_{\mathcal{M}}$ of some PNmatrix $\mathcal{M}$ satisfying a separability condition defined below. Moreover, we provide a method to extract from a given (separable) PNmatrix an equivalent PNmatrix with the *minimal* number of truth values.

Definition 14. *Let $\mathcal{M}$ be a PNmatrix for $\mathcal{L}$ and $\pounds$.*

1. *A truth value $x \in \mathcal{V}_{\mathcal{M}}$ is called* useful *in $\mathcal{M}$ if $x \in \mathcal{V}_{\mathcal{N}}$ for some proper PNmatrix $\mathcal{N} \subseteq \mathcal{M}$.*

[4] This property is sometimes called (semantic) *analyticity*. Note that in this paper the term 'analyticity' refers to a *proof-theoretic* property (see Definition 20).

2. *The PNmatrix* $R[\mathcal{M}]$ *is the simple refinement of* $\mathcal{M}$*, defined as follows:* $\mathcal{V}_{R[\mathcal{M}]}$ *consists of all truth values in* $\mathcal{V}_{\mathcal{M}}$ *which are useful in* $\mathcal{M}$*; for every* $l \in \pounds$, $\mathcal{D}_{R[\mathcal{M}]}(l) = \mathcal{D}_{\mathcal{M}}(l) \cap \mathcal{V}_{R[\mathcal{M}]}$*; and for every n-ary connective* $\diamond$ *of* $\mathcal{L}$ *and* $x_1, \ldots, x_n \in \mathcal{V}_{R[\mathcal{M}]}$, $\diamond_{R[\mathcal{M}]}(x_1, \ldots, x_n) = \diamond_{\mathcal{M}}(x_1, \ldots, x_n) \cap \mathcal{V}_{R[\mathcal{M}]}$.

Proposition 3. *Let* $\mathcal{M}$ *be a PNmatrix for* $\mathcal{L}$ *and* $\pounds$*, and let* v *be an* $\mathcal{M}$*-legal* $\mathcal{L}$*-valuation. Then:* (1) *For every formula* ψ, $v(\psi)$ *is useful in* $\mathcal{M}$*; and* (2) *Every* $\mathcal{M}$*-legal* $\mathcal{L}$*-valuation is also* $R[\mathcal{M}]$*-legal.*

Corollary 3. $\vdash_{\mathcal{M}} = \vdash_{R[\mathcal{M}]}$ *for every PNmatrix* $\mathcal{M}$.

Proof. One direction follows from Proposition 1, simply because $R[\mathcal{M}]$ is a simple refinement of $\mathcal{M}$ by definition. The converse is easily established using Proposition 3. We leave the details to the reader. □

Definition 15. *Let* $\mathcal{M}$ *be a PNmatrix for* $\mathcal{L}$ *and* $\pounds$*. We say that two truth values* $x_1, x_2 \in \mathcal{V}_{\mathcal{M}}$ *are* separable *in* $\mathcal{M}$ *for* $l \in \pounds$ *if* $x_1 \in \mathcal{D}_{\mathcal{M}}(l) \Leftrightarrow x_2 \notin \mathcal{D}_{\mathcal{M}}(l)$ *holds.* $\mathcal{M}$ *is called* separable *if every pair of truth values in* $\mathcal{V}_{\mathcal{M}}$ *are separable in* $\mathcal{M}$ *for some* $l \in \pounds$.

We are now ready to obtain a *lower bound* on the number of truth values needed to characterize $\vdash_{\mathcal{M}}$ for a given separable PNmatrix $\mathcal{M}$:

Theorem 2. *Let* $\mathcal{M}$ *be a separable PNmatrix for* $\mathcal{L}$ *and* $\pounds$*. If* $\vdash_{\mathcal{M}} = \vdash_{\mathcal{N}}$ *for some PNmatrix* $\mathcal{N}$ *for* $\mathcal{L}$ *and* $\pounds$*, then* $\mathcal{N}$ *contains at least* $|\mathcal{V}_{R[\mathcal{M}]}|$ *truth values.*

Remark 3. As done for usual matrices, it is also possible to define $\vdash_F$, the consequence relation induced by *a family of proper PNmatrices* to be $\bigcap_{\mathcal{N} \in F} \vdash_{\mathcal{N}}$. A PNmatrix can then be thought of as a succinct presentation of a family of proper PNmatrices in the following sense. The consequence relation induced by a PNmatrix $\mathcal{M}$ can be shown to be equivalent to the relation induced by the family of all the proper PNmatrices $\mathcal{N}$, such that $\mathcal{N} \subseteq \mathcal{M}$. Conversely, for every family of proper PNmatrices it is possible to construct an equivalent PNmatrix.

5 Finite PNmatrices for Canonical Labelled Systems

Definition 16. *We say that a PNmatrix* $\mathcal{M}$ *(for* $\mathcal{L}$ *and* $\pounds$*) is* characteristic *for a canonical labelled calculus* $\mathbf{G}$ *(for* $\mathcal{L}$ *and* $\pounds$*) if* $\vdash_{\mathcal{M}} = \vdash_{\mathbf{G}}$.

Next we provide a systematic way to obtain a characteristic PNmatrix $\mathcal{M}_{\mathbf{G}}$ for every canonical labelled calculus $\mathbf{G}$. The intuitive idea is as follows: the primitive rules of $\mathbf{G}$ determine the set of the truth values of $\mathcal{M}_{\mathbf{G}}$, while the introduction rules for the logical connectives dictate their corresponding truth tables. The semantics based on PNmatrices is thus *modular*: each such rule corresponds to a certain semantic condition, and the semantics of a system is obtained by joining the semantic effects of each of its derivation rules.

Definition 17. *Let $r = \{L_1, \ldots, L_n\}/L_0$ be a primitive rule for $\pounds$. Define:*

$$r^* = \{L \subseteq \pounds \mid L_i \cap L = \emptyset \text{ for some } 1 \leq i \leq n \text{ or } L_0 \cap L \neq \emptyset\}$$

Example 9. For an axiom $r = \emptyset/L_0$, we have $r^* = \{L \subseteq \pounds \mid L_0 \cap L \neq \emptyset\}$. For a cut $r = \{L_1, \ldots, L_n\}/\emptyset$, $r^* = \{L \subseteq \pounds \mid L_i \cap L = \emptyset \text{ for some } 1 \leq i \leq n\}$. In particular, continuing Examples 2 and 3 (for $\pounds = \{t, f\}$), $r^* = \{\{t\}, \{f\}, \{t, f\}\}$ for the classical axiom, and $r^* = \{\emptyset, \{t\}, \{f\}\}$ for the classical cut.

Definition 18. *Let $\diamond$ be an n-ary connective, and let $r = \mathcal{S}/L_0 : \diamond(p_1, \ldots, p_n)$ be a canonical introduction rule for $\diamond$ and $\pounds$. For every $L_1, \ldots, L_n \subseteq \pounds$, define:*

$$r^*[L_1, \ldots, L_n] = \begin{cases} \{L \subseteq \pounds \mid L_0 \cap L \neq \emptyset\} & \forall s \in \mathcal{S}.((L_1 : p_1) \cup \ldots \cup (L_n : p_n)) \cap s \neq \emptyset \\ P(\pounds) & \text{otherwise} \end{cases}$$

Example 10. Let $\pounds = \{t, f\}$. Recall the usual introduction rules for conjunction from Example 4. By Definition 18:

$$r_1^*[L_1, L_2] = \begin{cases} \{\{f\}, \{t, f\}\} & f \in L_1 \cup L_2 \\ P(\{t, f\}) & \text{otherwise} \end{cases}$$

$$r_2^*[L_1, L_2] = \begin{cases} \{\{t\}, \{t, f\}\} & t \in L_1 \cap L_2 \\ P(\{t, f\}) & \text{otherwise} \end{cases}$$

Definition 19 (The PNmatrix $\mathcal{M}_{\mathbf{G}}$). *Let $\mathbf{G}$ be a canonical labelled calculus for $\mathcal{L}$ and $\pounds$. The PNmatrix $\mathcal{M}_{\mathbf{G}}$ (for $\mathcal{L}$ and $\pounds$) is defined by:*

1. $\mathcal{V}_{\mathcal{M}_{\mathbf{G}}} = \{L \subseteq \pounds \mid L \in r^* \text{ for every } r \in \mathsf{P}_{\mathbf{G}}\}$.
2. *For every* $l \in \pounds$, $\mathcal{D}_{\mathcal{M}_{\mathbf{G}}}(l) = \{L \in \mathcal{V}_{\mathcal{M}_{\mathbf{G}}} \mid l \in L\}$.
3. *For every n-ary connective $\diamond$ of $\mathcal{L}$ and $L_1, \ldots, L_n \in \mathcal{V}_{\mathcal{M}_{\mathbf{G}}}$:*

$$\diamond_{\mathcal{M}_{\mathbf{G}}}(L_1, \ldots, L_n) = \{L \in \mathcal{V}_{\mathcal{M}_{\mathbf{G}}} \mid L \in r^*[L_1, \ldots, L_n] \text{ for every } r \in \mathsf{R}^{\diamond}_{\mathbf{G}}\}$$

Example 11. Let $\pounds = \{t, f\}$ and consider the calculus $\mathbf{G}_\wedge$ whose primitive rules include only the classical axiom, and the classical cut (see Examples 2 and 3), and whose only introduction rules are the two usual rules for conjunction (see Example 4). By Example 9 and the construction above, $\mathcal{V}_{\mathcal{M}_{\mathbf{G}_\wedge}} = \{\{t\}, \{f\}\}$, $\mathcal{D}_{\mathcal{M}_{\mathbf{G}_\wedge}}(t) = \{t\}$, and $\mathcal{D}_{\mathcal{M}_{\mathbf{G}_\wedge}}(f) = \{f\}$. Using Example 10, we obtain the following interpretation of $\wedge$:

$\wedge_{\mathcal{M}_{\mathbf{G}_\wedge}}$	$\{t\}$	$\{f\}$
$\{t\}$	$\{t\}$	$\{f\}$
$\{f\}$	$\{f\}$	$\{f\}$

Example 12. Let $\pounds = \{a, b, c\}$, and assume that $\mathcal{L}$ contains only a unary connective $\star$. Let us start with the calculus $\mathbf{G}_0$, the primitive rules of which include the canonical axiom $\emptyset/\{a, b, c\}$ and the canonical cuts $\{a\}, \{c\}/\emptyset$ and $\{a\}, \{b\}/\emptyset$, while $\mathbf{G}_0$ has no introduction rules. Here we have $\mathcal{V}_{\mathcal{M}_{\mathbf{G}_0}} = \{\{a\}, \{b\}, \{c\}, \{b, c\}\}$, $\mathcal{D}_{\mathcal{M}_{\mathbf{G}}}(a) = \{\{a\}\}$, $\mathcal{D}_{\mathcal{M}_{\mathbf{G}}}(b) = \{\{b\}, \{b, c\}\}$ and $\mathcal{D}_{\mathcal{M}_{\mathbf{G}}}(c) = \{\{c\}, \{b, c\}\}$. $\star_{\mathcal{M}_{\mathbf{G}_0}}$ is

given in the table below (it is completely non-deterministic). One can now obtain a calculus $\mathbf{G}_1$ by adding the rule $\{a : p_1\}/\{b, c\} : \star p_1$. This leads to a refinement of the truth table, described below. Finally, one can obtain the calculus $\mathbf{G}_2$ by adding $\{b : p_1\}/\{a\} : \star p_1$, resulting in another refinement of truth table, also described below.

x	$\star_{\mathcal{M}_{\mathbf{G}_0}}(x)$	$\star_{\mathcal{M}_{\mathbf{G}_1}}(x)$	$\star_{\mathcal{M}_{\mathbf{G}_2}}(x)$
$\{a\}$	$\{\{a\},\{b\},\{c\},\{b,c\}\}$	$\{\{b\},\{b,c\}\}$	$\{\{b\},\{b,c\}\}$
$\{b\}$	$\{\{a\},\{b\},\{c\},\{b,c\}\}$	$\{\{a\},\{b\},\{c\},\{b,c\}\}$	$\{\{a\}\}$
$\{c\}$	$\{\{a\},\{b\},\{c\},\{b,c\}\}$	$\{\{a\},\{b\},\{c\},\{b,c\}\}$	$\{\{a\},\{b\},\{c\},\{b,c\}\}$
$\{b,c\}$	$\{\{a\},\{b\},\{c\},\{b,c\}\}$	$\{\{a\},\{b\},\{c\},\{b,c\}\}$	$\{\{a\}\}$

Theorem 3 (Soundness and completeness). *For every canonical labelled calculus* $\mathbf{G}$*,* $\mathcal{M}_{\mathbf{G}}$ *is a* characteristic PNmatrix *for* $\mathbf{G}$.

Corollary 4 (Decidability). *Given a canonical labelled calculus* $\mathbf{G}$*, a finite set* $\mathcal{S}$ *of sequents, and a sequent* s*, it is decidable whether* $\mathcal{S} \vdash_{\mathbf{G}} s$ *or not.*

Corollary 5. *The question whether a given canonical labelled calculus* $\mathbf{G}$ *is consistent (i.e.* $\not\vdash_{\mathbf{G}} \emptyset$*) is decidable.*

$\mathcal{M}_{\mathbf{G}}$ provides a semantic characterization for $\mathbf{G}$, however it may not be a minimal one (in terms of the number of truth values). For a minimal semantic representation, we should consider the equivalent PNmatrix $R[\mathcal{M}_{\mathbf{G}}]$:

Corollary 6 (Minimality). *For every canonical labelled calculus* $\mathbf{G}$*,* $R[\mathcal{M}_{\mathbf{G}}]$ *is a minimal (in terms of number of truth values) characteristic PNmatrix for* $\mathbf{G}$.

Proof. The claim follows by Theorem 2 from the fact that $\mathcal{M}_{\mathbf{G}}$ is separable for every system $\mathbf{G}$. □

6 Proof-Theoretic Applications

In this section we apply the semantic framework of PNmatrices to provide *decidable* semantic criteria for syntactic properties of canonical labelled calculi that are usually hard to generally characterize by other means. We focus on the notions of *analyticity* and *cut-admissibility*, extended to the context of reasoning with assumptions.

6.1 Strong Analyticity

Strong analyticity is a crucial property of a useful (propositional) calculus, as it implies its consistency and decidability. Intuitively, a calculus is *strongly analytic* if whenever a sequent s is provable in it from a set of assumptions $\mathcal{S}$, then s can be proven using only the formulas available within $\mathcal{S}$ and s.

Definition 20. *A canonical labelled calculus* $\mathbf{G}$ *is* strongly analytic *if whenever* $\mathcal{S} \vdash_{\mathbf{G}} s$*, there exists a derivation in* $\mathbf{G}$ *of* s *from* $\mathcal{S}$ *consisting solely of (sequents consisting of) formulas from* $sub[\mathcal{S} \cup \{s\}]$*.*

Below we provide a *decidable* semantic characterization of strong analyticity of canonical labelled calculi:

Theorem 4 (Characterization of Strong Analyticity). *Let* $\mathbf{G}$ *be a canonical labelled calculus for* $\mathcal{L}$ *and* $\pounds$*. Suppose that* $\mathbf{G}$ *does not include the (trivial) primitive rule* $\emptyset/\emptyset$*. Then,* $\mathbf{G}$ *is strongly analytic iff* $\mathcal{M}_{\mathbf{G}}$ *is proper.*

Corollary 7. *The question whether a given canonical labelled calculus is strongly analytic is decidable.*

6.2 Strong Cut-Admissibility

As the property of strong analyticity is sometimes difficult to establish, it is traditional in proof theory to investigate the property of *cut-admissibility*, which means that whenever s is provable in $\mathbf{G}$, it has a cut-free derivation in $\mathbf{G}$. In this paper we investigate a stronger notion of this property, defined as follows for labelled calculi:

Definition 21. *A labelled calculus* $\mathbf{G}$ *enjoys* strong cut-admissibility *if whenever* $\mathcal{S} \vdash_{\mathbf{G}} s$*, there exists a derivation in* $\mathbf{G}$ *of* s *from* $\mathcal{S}$ *in which only formulas from* $frm[\mathcal{S}]$ *serve as cut-formulas.*

Due to the special form of primitive and introduction rules of canonical calculi (which, except for canonical cuts, enjoy the subformula property), the above property guarantees strong analyticity:

Proposition 4. *Let* $\mathbf{G}$ *be a canonical labelled calculus. If* $\mathbf{G}$ *enjoys strong cut-admissibility, then* $\mathbf{G}$ *is strongly analytic.*

Although for two-sided canonical sequent calculi the notions of strong analyticity and strong cut-admissibility coincide (see [3]), this is not the case for general labelled calculi, for which the converse of Proposition 4 does not necessarily hold, as shown by the following example:

Example 13. Let $\pounds = \{a, b, c\}$, and assume that $\mathcal{L}$ contains only a unary connective $\star$. Let $\mathbf{G}$ be the canonical labelled calculus $\mathbf{G}$ for $\mathcal{L}$ and $\pounds$, the primitive rules of which include only the canonical cuts $\{a\}, \{b\}/\emptyset$, $\{a\}, \{c\}/\emptyset$, and $\{b\}, \{c\}/\emptyset$, and its only introduction rules are $\{a : p_1\}/\{a, b\} : \star p_1$ and $\{a : p_1\}/\{b, c\} : \star p_1$. To see that this system is strongly analytic, by Theorem 4, it suffices to construct $\mathcal{M}_{\mathbf{G}}$ and check that it is proper. The construction proceeds as follows: $\mathcal{V}_{\mathcal{M}_{\mathbf{G}}} = \{\emptyset, \{a\}, \{b\}, \{c\}\}$, $\mathcal{D}_{\mathcal{M}_{\mathbf{G}}}(l) = \{l\}$ for $l \in \{a, b, c\}$, and the truth table for $\star$ is the following:

x	$\star_{\mathcal{M}_{\mathbf{G}}}(x)$
$\emptyset$	$\{\emptyset, \{a\}, \{b\}, \{c\}\}$
$\{a\}$	$\{\{b\}\}$
$\{b\}$	$\{\emptyset, \{a\}, \{b\}, \{c\}\}$
$\{c\}$	$\{\emptyset, \{a\}, \{b\}, \{c\}\}$

This is a proper PNmatrix, and so **G** is strongly analytic. However, it impossible to derive the sequent $\{b : \star p_1\}$ from the singleton set $\{\{a : p_1\}\}$ using only p_1 as a cut-formula. This is possible by applying the two introduction rules of **G** and then using the cut $\{a\}, \{c\}/\emptyset$ (with $\star p_1$ as the cut-formula). Thus although this system is strongly analytic, it does not enjoy strong cut-admissibility.

The intuitive explanation is that non-eliminable applications of canonical cuts (like the one in the above example) are not harmful for strong analyticity because they enjoy the subformula property. Thus, the equivalence between strong analyticity and cut-admissibility can be restored if we enforce the following condition:

Definition 22. *A canonical labelled calculus* **G** *for* $\mathcal{L}$ *and* $\pounds$ *is* cut-saturated *if for every canonical cut* $\{L_1, \ldots, L_n\}/\emptyset$ *of* **G** *and* $l \in \pounds$, **G** *contains the primitive rule* $\{L_1, \ldots, L_n\}/\{l\}$.

Proposition 5. *For every canonical labelled calculus* **G**, *there is an equivalent cut-saturated canonical labelled calculus* **G**$'$ *(i.e.* $\vdash_{\mathbf{G}} = \vdash_{\mathbf{G}'}$*).*

Example 14. Revisiting the system from Example 13, we observe that **G** is not cut-saturated. To obtain a cut-saturated equivalent system **G**$'$, we add (among others) the three primitive rules: $r_1 = \{\{a\}, \{b\}\}/\{c\}$, $r_2 = \{\{a\}, \{c\}\}/\{b\}$, and $r_3 = \{\{b\}, \{c\}\}/\{a\}$. Note that the addition of these rules does not affect the set of truth values, i.e., $\mathcal{V}_{\mathcal{M}_{\mathbf{G}}} = \mathcal{V}_{\mathcal{M}_{\mathbf{G}'}}$. However, we can now derive $\{b : \star p_1\}$ from $\{\{a : p_1\}\}$ without any cuts by the two introduction rules and the new rule r_2. Moreover, by Theorem 5 below, **G**$'$ *does* enjoy strong cut-admissibility.

We are now ready to provide a decidable semantic characterization of strong cut-admissibility.

Theorem 5. *Let* **G** *be a cut-saturated canonical labelled calculus for* $\mathcal{L}$ *and* $\pounds$. *Suppose that* **G** *does not include the (trivial) primitive rule* $\emptyset/\emptyset$. *Then the following statements concerning* **G** *are equivalent: (i)* $\mathcal{M}_{\mathbf{G}}$ *is proper, (ii)* **G** *is strongly analytic, and (iii)* **G** *enjoys strong cut-admissibility.*

7 Conclusions and Further Research

Establishing proof-theoretical properties of syntactic calculi is in many cases a complex and error-prone task. For instance, proving that a calculus admits cut-elimination is often carried out using heavy syntactic arguments and many case-distinctions, leaving room for mistakes and omissions. This leads to the need of *automatizing* the process of reasoning about calculi. However, a faithful formalization is an elusive goal, as such important properties as cut-admissibility, analyticity and decidability, as well as the dependencies between them are little understood for the general case. We believe that the *abstract* view on labelled calculi taken in this paper is a substantial step towards finding the right level of abstraction for reasoning about these properties. Moreover, the simple and decidable semantic characterizations of these properties for canonical labelled calculi are a key to their faithful axiomatization in this context. To provide these characterizations, we have introduced *PNmatrices*, a generalization of Nmatrices, in which empty entries in logical truth tables are allowed, while still preserving the *effectiveness* of the semantics. A characteristic PNmatrix $\mathcal{M}_{\mathbf{G}}$ has been constructed for every canonical labelled calculus, which in turn implies its decidability. If in addition $\mathcal{M}_{\mathbf{G}}$ has no empty entries (i.e, is proper) — which is *decidable*, $\mathbf{G}$ is strongly analytic. For cut-saturated canonical calculi, the latter is also equivalent to strong cut-admissibility.

The results of this paper extend the theory of canonical sequent calculi of [2], as well as of the labelled calculi of [6] and signed calculi of [4], all of which are particular instances of canonical labelled calculi defined in this paper. Moreover, the semantics obtained for these families of calculi in the above mentioned papers, coincide with the PNmatrices semantics obtained for them here. It is particularly interesting to note that [6] provides a list of conditions, under which a labelled calculus has a characteristic finite-valued logic. These conditions include (i) reducibility of cuts (which can be shown to be equivalent to the criterion of coherence of [4]), which entails that $\mathcal{M}_{\mathbf{G}}$ is proper, and (ii) eliminability of compound axioms,[5] which entails that $\mathcal{M}_{\mathbf{G}}$ is completely deterministic (in other words, it can be identified with an ordinary finite-valued matrix). We conclude that, as shown in this paper, none of the conditions required in any of the mentioned papers [2, 6, 4] from a "well-behaved" calculus are necessary when moving to the more general semantic framework of PNmatrices, where *any* canonical labelled calculus has an effective finite-valued semantics.

An immediate direction for further research is investigating the applications of the theory of canonical labelled calculi developed here. One possibility is exploiting this theory for sequent calculi, whose rules are more complex than the canonical ones, but which can be reformulated in terms of canonical *labelled* calculi. This applies, e.g., to the large family of sequent calculi for paraconsistent logics given in [5]. For example, consider the (two-sided) Gentzen-type system $\mathbf{G_K}$ of [5] over the language $\mathcal{L}_C = \{\wedge, \vee, \supset, \circ, \neg\}$, obtained from **LK** by discarding the left rule for negation and adding the following schemas for the unary connective $\circ$:

$$(\circ \Rightarrow)\ \frac{\Gamma_1 \Rightarrow \psi, \Delta_1 \qquad \Gamma_2 \Rightarrow \neg\psi, \Delta_2}{\Gamma_1, \Gamma_2, \circ\psi \Rightarrow \Delta_1, \Delta_2} \qquad (\Rightarrow \circ)\ \frac{\Gamma, \psi, \neg\psi \Rightarrow \Delta}{\Gamma \Rightarrow \circ\psi, \Delta}$$

Clearly, these schemas cannot be formulated as canonical rules in the sense of [2] (since they use $\neg\psi$ as a principal formula). However, we can reformulate $\mathbf{G_K}$ in terms of canonical labelled calculi by using the set of labels $\mathcal{L}_4 = \{t^+, t^-, f^+, f^-\}$, where t and f denote the side on which the formula occurs, and $+$ and $-$ determine whether its occurrence is positive or negative (i.e. preceded with negation). Now each (two-sided) rule of $\mathbf{G_K}$ can be translated into

[5] This property intuitively means that compound axioms can be reduced to atomic ones. It is called 'axiom-expansion' in [7].

a labelled *canonical* rule over $\mathcal{L}_4$. For instance, $(\circ \Rightarrow)$ and $(\Rightarrow \circ)$ above are translated into $\{t^+ : p_1\}, \{t^- : p_1\}/\{f^+\} : \circ p_1$ and $\{f^+ : p_1, f^- : p_1\}/\{t^+\} : \circ p_1$ respectively. Adding further rules, it can be shown that for each (non-canonical) two-sided calculus $\mathbf{G}$ from [5], an equivalent labelled *canonical* calculus $\mathbf{G}'$ can be constructed (this automatically implies the decidability of the calculi from [5]). Detailed analysis of such situations is left for future work. Another direction is generalizing the results of this paper to more complex classes of labelled calculi, e.g., like those defined in [10] for inquisitive logic. Extending the results to the first-order case is another future goal. Finally, it would be interesting to explore the relation between the systems studied in this paper and the resolution proof systems of [11].

Acknowledgement. The research leading to these results has received funding from the European Community's Seventh Framework Programme (FP7/2007-2013) under grant agreement no. 252314.

References

1. A.A.: Multi-valued Semantics: Why and How. Studia Logica 92, 163–182 (2009)
2. Avron, A., Lev, I.: Non-deterministic Multiple-valued Structures. Journal of Logic and Computation 15 (2005)
3. Avron, A., Zamansky, A.: Non-deterministic Semantics for Logical Systems. Handbook of Philosophical Logic 16, 227–304 (2011)
4. Avron, A., Zamansky, A.: Canonical Signed Calculi, Non-deterministic Matrices and Cut-Elimination. In: Artemov, S., Nerode, A. (eds.) LFCS 2009. LNCS, vol. 5407, pp. 31–45. Springer, Heidelberg (2008)
5. Avron, A., Konikowska, B., Zamansky, A.: Modular Construction of Cut-Free Sequent Calculi for Paraconsistent Logics. To Appear in Proceedings of Logic in Computer Science (LICS 2012) (2012)
6. Baaz, M., Fermüller, C.G., Salzer, G., Zach, R.: Labelled Calculi and Finite-valued Logics. Studia Logica 61, 7–33 (1998)
7. Ciabattoni, A., Terui, K.: Towards a semantic characterization of cut elimination. Studia Logica 82(1), 95–119 (2006)
8. Gabbay, D.M.: Labelled Deductive Systems, Volume 1. Oxford Logic Guides, vol. 33. Clarendon Press/Oxford Science Publications, Oxford (1996)
9. Lahav, O.: Non-deterministic Matrices for Semi-canonical Deduction Systems. To Appear in Proceedings of IEEE 42nd International Symposium on Multiple-Valued Logic (ISMVL 2012) (2012)
10. Sano, K.: Sound and Complete Tree-Sequent Calculus for Inquisitive Logic. In: Ono, H., Kanazawa, M., de Queiroz, R. (eds.) WoLLIC 2009. LNCS, vol. 5514, pp. 365–378. Springer, Heidelberg (2009)
11. Stachniak, Z.: Resolution Proof Systems: An Algebraic Theory. Kluwer Academic Publishers (1996)

A Simplex-Based Extension of Fourier-Motzkin for Solving Linear Integer Arithmetic*

François Bobot[1], Sylvain Conchon[1], Evelyne Contejean[1], Mohamed Iguernelala[1], Assia Mahboubi[2], Alain Mebsout[1], and Guillaume Melquiond[2]

[1] LRI, Université Paris Sud, CNRS, Orsay F-91405
[2] INRIA Saclay–Île-de-France, Orsay, F-91893

Abstract. This paper describes a novel decision procedure for quantifier-free linear integer arithmetic. Standard techniques usually relax the initial problem to the rational domain and then proceed either by projection (*e.g. Omega-Test*) or by branching/cutting methods (*branch-and-bound, branch-and-cut, Gomory cuts*). Our approach tries to bridge the gap between the two techniques: it interleaves an exhaustive search for a model with bounds inference. These bounds are computed provided an oracle capable of finding constant positive linear combinations of affine forms. We also show how to design an efficient oracle based on the Simplex procedure. Our algorithm is proved sound, complete, and terminating and is implemented in the ALT-ERGO theorem prover. Experimental results are promising and show that our approach is competitive with state-of-the-art SMT solvers.

1 Introduction

Linear arithmetic is ubiquitous in many domains ranging from software and hardware verification, linear programming, compiler optimization to planning and scheduling. Decision procedures for the quantifier-free linear fragment over integers (QF-LIA) are widely studied in Satisfiability Modulo Theories. Most of the procedures used by state-of-the-art SMT solvers are extensions of either the Simplex algorithm or the Fourier-Motzkin method. Both techniques first relax the initial problem to the rational domain and then proceed by branching/cutting methods or by projection.

Given a conjunction $\bigwedge_i \sum_j a_{i,j}\, x_j + b_i \leq 0$ of constraints over rationals, the Simplex algorithm [16] finds an instantiation of the variables x_j satisfying these constraints or a contradiction if they are unsatisfiable. Three well-known extensions of this method to decision procedures over integers are *branch-and-bound*, *Gomory's cutting-planes*, and *branch-and-cut* [16]. Intuitively, these extensions prune non-integer solutions from the search space until they find an integer assignment or a contradiction. The Simplex algorithm is exponential in the worst case but behaves rather well in practice. On the other hand, the complexity of

* Work financially supported by the French ANR project ANR-08-005 DeCert.

B. Gramlich, D. Miller, and U. Sattler (Eds.): IJCAR 2012, LNAI 7364, pp. 67–81, 2012.

QF-LIA is NP-complete [16] and known algorithms are not as efficient in practice as the Simplex on the rational case.

By contrast, the idea behind the Fourier-Motzkin [16] algorithm is to perform successive variable eliminations in a breadth-first manner generating additional constraints with fewer variables. The original system is satisfiable in the rationals if a fixpoint is reached without deriving a trivially inconsistent inequality $c \leq 0$ where c is a positive rational. In the opposite case we conclude that the system is unsatisfiable. The *Omega-Test* [15] extends this algorithm to a decision procedure over integers by performing additional projection-based checks when the constraints are satisfiable in the rationals. These methods do not scale in practice because they introduce a (double) exponential number of inequalities, which saturates the memory.

In this paper, we present a novel decision procedure for conjunctions of quantifier-free linear integer arithmetic constraints. Our approach is not an instance of any of the above techniques. Roughly speaking, it interleaves an exhaustive search for a model with bounds inference. New bounds are computed by solving auxiliary linear optimization problems using the Simplex algorithm. Intuitively, each auxiliary problem simulates a run of the Fourier-Motzkin algorithm that would eliminate all the variables at once. In order to facilitate the reading of this article, we summarize hereafter the main ideas of our contribution.

After recalling some useful notations and mathematical background, we characterize in Section 2 when the solution set described by a conjunction of constraints can be effectively bounded along some direction. If there is no such bound, we prove that the solution set contains infinitely many integer solutions.

This characterization is based on finding constant positive linear combinations of affine forms. In Section 3, we first show that Fourier-Motzkin is a suitable algorithm to compute such combinations. Then, we explain how to cast this problem into a linear optimization problem, which can hence be solved by an efficient Simplex-based algorithm.

In Section 4, we show how to build a decision procedure for QF-LIA. The procedure uses the algorithms of Section 3 to find bounds on the solution set. If there are none, then the procedure stops since there are infinitely many integer solutions. Otherwise it performs a case-split analysis along the bounded direction and calls itself recursively to solve the simpler subproblems.

We have implemented our framework in the ALT-ERGO theorem prover [4]. In Section 5, we measure its performances on a subset of the QF-LIA benchmark and compare its performances with some state-of-the-art SMT solvers. Section 6 presents future and related works.

2 Preliminary Results

2.1 Background and Notations

In all what follows, if $m, n \in \mathbb{N}$, then $[|m, n|]$ denotes the integer interval bounded by m and n. We denote matrices by upper case letters like A and column vectors by lower case letters like x. We denote A^t the transpose of the matrix A and

Ax denotes the matrix product of the matrix A by the vector x. If A is a $m \times n$ matrix, $a_{i,j}$ denotes the element of A at position (i, j), a_i denotes the i-th row vector of A (of size n), and A_j the j-th column vector of A (of size m). If x is a vector, x_i denotes its i-th coordinate. If x and y are n-vectors of the same ordered vector space, $x \geq y$ denotes the conjunction of constraints $\forall i \in [|1, n|], x_i \geq y_i$, with similar notations for $\leq$ and the associated strict orders. For instance, if x is a vector, $x > 0$ denotes the conjunction of constraints $\forall i \in [|1, n|], x_i > 0$. We equip $\mathbb{Q}^n$ with the usual scalar product associated with its canonical basis. We measure distances using the *supremum* norm $\|\cdot\|_\infty$. $B_\infty(x, r)$ denotes the closed ball centered in x and of radius r for that norm.

We recall that affine maps $\psi : \mathbb{Q}^n \to \mathbb{Q}^m$ are the maps of shape $\psi = \phi + t_c$, with $\phi : \mathbb{Q}^n \to \mathbb{Q}^m$ a linear map and t_c the translation of direction $c \in \mathbb{Q}^m$. An affine map $\psi : \mathbb{Q}^n \to \mathbb{Q}$ is called an affine form on $\mathbb{Q}^n$. For instance, a constant map $\psi_c : \mathbb{Q}^n \to \mathbb{Q}^m$, with value c, is an affine map since it is the sum of the zero linear map and of the translation of direction $c \in \mathbb{Q}^m$.

Definition 1 (Positive linear combination of affine forms). *Let $(\psi_i)_{i \in [|1,k|]}$ be a family of affine forms on $\mathbb{Q}^n$. An affine form ψ on $\mathbb{Q}^n$ is a* positive linear combination *of the $(\psi_i)_{i \in [|1,k|]}$ if there exists $(\lambda_i)_{i \in [|1,k|]}$ a family of nonnegative scalars such that*

$$\psi = \sum_{i=1}^{k} \lambda_i \psi_i \quad \text{and} \quad \sum_{i=1}^{k} \lambda_i > 0$$

We recall the original formulation of Farkas lemma [16,9] on rationals:

Theorem 1 (Farkas' lemma). *Given a matrix $A \in \mathbb{Q}^{m \times n}$, and c a vector in $\mathbb{Q}^n$, then*

$$\exists x \in \mathbb{Q}^m, x \geq 0 \ \wedge \ Ax = c \quad \Leftrightarrow \quad \forall y \in \mathbb{Q}^n, y^t A \geq 0 \Rightarrow y^t c \geq 0$$

In the sequel, we use the following equivalent formulation:

Theorem 2 (Theorem of alternatives). *Let A be a matrix in $\mathbb{Q}^{m \times n}$ and b a vector in $\mathbb{Q}^m$. The system $Ax + b \leq 0$ has no solution if and only if there exists $\lambda \in \mathbb{Q}^m$ such that $\lambda \geq 0$ and $A^t \lambda = 0$ and $b^t \lambda > 0$.*

2.2 Convex Polytopes with an Infinite Number of Integer Points

We consider a closed convex subset $K \subset \mathbb{Q}^n$ defined by a linear system of constraints:

$$K := \{x \in \mathbb{Q}^n \mid Ax + b \leq 0\}$$

where $A \in \mathbb{Q}^{m \times n}$, $b \in \mathbb{Q}^m$. By definition, K is the convex polytope of the (rational) solutions of the linear system $Ax + b \leq 0$. We want to determine whether $K \cap \mathbb{Z}^n$ is empty or not, or in other words, whether the system $Ax + b \leq 0$ has integer solutions. For $i \in [|1, m|]$, we denote by L_i the following affine forms:

$$L_i : \left| \begin{array}{l} \mathbb{Q}^n \longrightarrow \mathbb{Q} \\ x \longmapsto (Ax + b)_i \end{array} \right.$$

Theorem 3. *If there is no constant positive linear combination of the linear forms* (L_i)*, then for all* $R \in \mathbb{Q}^+$*,* K *contains a ball* $B_\infty(w, R)$ *with* $w \in \mathbb{Q}^n$*.*

Proof. Let $R \in \mathbb{Q}^+$. We define $\gamma \in \mathbb{Q}^m$ such that for every $i \in [|1, m|]$:

$$\gamma_i := R\|a_i\|_1 = R\sum_j |a_{i,j}|$$

and we consider the convex $K' := \{x \in \mathbb{Q}^n \mid Ax + \gamma + b \leq 0\}$. Suppose for contradiction that K' is empty. Hence by Theorem 2, there exists $\lambda \in \mathbb{Q}^m$ such that $A^t\lambda = 0$. So, $\lambda^t(Ax + b) = \sum_i \lambda_i L_i$ is constant, which contradicts the hypothesis.

Therefore K' is not empty and contains a vector w such that $Aw + b + \gamma \leq 0$. We now prove that $B_\infty(w, R) \subseteq K$. Let u be a vector such that $\|u\|_\infty \leq R$. By triangular inequality we have

$$\forall i \in [|1, \ldots, m|] \quad (Au)_i \leq |(Au)_i| \leq \|a_i\|_1\|u\|_\infty \leq R\|a_i\|_1 = \gamma_i$$

hence

$$A(w + u) + b = Aw + b + Au \leq Aw + b + \gamma \leq 0$$

which proves that $w + u$ belongs to the convex K. □

Corollary 1. *If there is no constant positive linear combination of the* (L_i) *then* $K \cap \mathbb{Z}^n$ *contains infinitely many points, for* $n > 0$*.*

Proof. For any $N \in \mathbb{N}$ and any $x \in \mathbb{Q}^n$, the ball $B_\infty(x, N)$ contains at least $(2N)^n$ points with integer coordinates. □

Lemma 1. *If* $\sum_i \lambda_i L_i$ *is a positive linear combination of the* (L_i) *equal to a constant* c*, then*

- *if* c *is positive, then* K *should be empty;*
- *otherwise for every* k *such that* $\lambda_k \neq 0$*, and for any* $x \in K$*,* $L_k(x)$ *is bounded by* $\frac{c}{\lambda_k} \leq L_k(x) \leq 0$

Proof. For any $x \in K$, we have $\lambda_i L_i(x) \leq 0$ by definition of K and non-negativeness of λ_i. Hence $\sum_i \lambda_i L_i(x) \leq 0$, which concludes the first case.

Since $\sum_i \lambda_i L_i = c$, then for k such that $\lambda_k \neq 0$ and for any $x \in K$, we have $c - \lambda_k L_k(x) = \sum_{i \neq k} \lambda_i L_i(x) \leq 0$, which concludes the second case. □

Note that, if the constant c is zero, then all the inequalities $L_k \leq 0$ associated to a nonzero λ_k are in fact equalities.

2.3 Intersection with an Affine Subspace

In addition to K, we now consider another convex $K' \subset \mathbb{Q}^n$ defined by ℓ equations:

$$K' := \{x \in \mathbb{Q}^n \mid A'x + b' = 0\}$$

where $A' \in \mathbb{Z}^{\ell \times n}$, $b' \in \mathbb{Z}^{\ell}$, and study the intersection $K \cap K' \cap \mathbb{Z}^n$. We prove a sufficient condition for this intersection to contain an infinite number of points when $K \cap \mathbb{Z}^n$ is known to be infinite.

Let $(e_1, \ldots, e_n)$ be the canonical basis of $\mathbb{Q}^n$. We suppose that there exists $i_1, \ldots, i_j$ such that K is invariant by any translation of direction e_{i_k} for $k \in [|1, j|]$. Hence if we pose $E := \langle e_{i_1}, \ldots, e_{i_j} \rangle$ the vector space generated by these vectors, K is invariant by any translation of direction $e \in E$. We denote $\pi : \mathbb{Q}^n \to \mathbb{Q}^{n-j}$ the orthogonal projection along E (on the orthogonal complement $E^{\perp}$ of E). Note that since we consider vectors as column matrices of their coordinates on the canonical basis, computing the projection $\pi(x)$ of a vector x boils down to annihilating the coordinates $i_1, \ldots, i_j$ of x.

Theorem 4. *Assume that there are no constant positive linear combinations of the (L_i) and that $K' \cap \mathbb{Z}^n$ contains at least one point. Then if $\pi(K) \subseteq \pi(K')$, $K \cap K' \cap \mathbb{Z}^n$ contains infinitely many points.*

See `http://hal.inria.fr/hal-00687640` for the detailed proof.

3 Constant Positive Linear Combinations of Affine Forms

In this section, we are interested in computing constant positive linear combinations of affine forms. More precisely, we intend to build an oracle which takes as input a set of affine forms (L_i) (or equivalently, a matrix A and a vector b) and meets the following specifications:

1. if there is no constant positive linear combination of the (L_i), it says so;
2. otherwise, it returns such a combination $\sum_i \lambda_i L_i$.

We first present a method based on the Fourier-Motzkin procedure. Then, we describe an efficient implementation based on the Simplex algorithm and prove its soundness, completeness, and termination.

3.1 The Fourier-Motzkin Procedure

Let $K := \{x \in \mathbb{Q}^n \mid Ax + b \leq 0\}$ be a closed convex where $A \in \mathbb{Q}^{m \times n}$ and $b \in \mathbb{Q}^m$, and C the set of the affine forms $L_i : x \mapsto (Ax + b)_i$. Fourier-Motzkin can be seen as an algorithm that attempts to compute constant positive linear combinations $\sum \lambda_i L_i$ in order to decide whether K is empty or not. For that purpose, it eliminates iteratively all the variables from the set of affine forms. More precisely, the iteration k of the procedure consists in

1. choosing a variable x_k to eliminate,
2. partitioning the current set C_k of affine forms into a set $C^0_{x_k}$ not containing x_k and a set $C^+_{x_k}$ (resp. $C^-_{x_k}$) where x_k has positive (resp. negative) coefficients,
3. computing the set C_{k+1} of new affine forms:
$$C_{k+1} := C^0_{x_k} \;\cup\; \Pi(C^+_{x_k} \times C^-_{x_k})$$
where Π calculates a positive combination $L_{i,j} = \alpha_{i,j} L_i + \beta_{i,j} L_j$ not containing x_k for each $L_i \in C^+_{x_k}$ and $L_j \in C^-_{x_k}$.

Notice that if either $C^+_{x_k}$ or $C^-_{x_k}$ is empty, then Π returns an empty set. The iterative process terminates when all the variables are eliminated and it returns a (possibly empty) set C_f of constant affine forms. We know that K is empty if there exists $c \in C_f$ such that $c > 0$. Moreover, given a constant $c \in C_f$, it is easy to retrieve a positive linear combination $\sum_i \lambda_i L_i = c$. For that, we recursively unfold the definitional equalities $L_{i,j} = \alpha_{i,j} L_i + \beta_{i,j} L_j$ computed by Π.

Example 1. Consider the following set of affine forms:

$$C_1 : \left\{ \begin{array}{lll} L_1 = 2x + y, & L_2 = -2x + 3y - 5, & L_3 = x + z + 1, \\ L_4 = x + 5y + z, & L_5 = -x - 4y + 3, & L_6 = 3x - 2y + 2 \end{array} \right\}$$

Eliminating z from C_1 is immediate since it only appears positively:

$$C_2 : \left\{ \begin{array}{lll} L_1 = 2x + y, & L_2 = -2x + 3y - 5, & L_5 = -x - 4y + 3, \\ L_6 = 3x - 2y + 2 & & \end{array} \right\}$$

We eliminate the variable x and compute the set C_3 below using the combinations: $L_7 = L_1 + L_2, \quad L_8 = L_1 + 2L_5, \quad L_9 = 2L_6 + 3L_2, \quad L_{10} = L_6 + 3L_5$

$$C_3 : \left\{ \begin{array}{lll} L_7 = 4y - 5, & L_8 = -7y + 6, & L_9 = 5y - 11, \\ L_{10} = -14y + 11 & & \end{array} \right\}$$

Finally, the variable y is in turn eliminated thanks to the following combinations: $L_{11} = 7L_7 + 4L_8, \quad L_{12} = 7L_7 + 2L_{10}, \quad L_{13} = 7L_9 + 5L_8, \quad L_{14} = 14L_9 + 5L_{10}$

The iterative process terminates and returns the set

$$C_4 : \{ L_{11} = -11, \quad L_{12} = -13, \quad L_{13} = -47 \quad L_{14} = -99 \}$$

Moreover, unfolding the equalities introduced by Π yields

$$\left\{ \begin{array}{lllll} -11 = L_{11} &=& 7L_7 + 4L_8 &= \cdots =& 11L_1 + 7L_2 + 8L_5 \\ -13 = L_{12} &=& 7L_7 + 2L_{10} &= \cdots =& 7L_1 + 7L_2 + 6L_5 + 2L_6 \\ -47 = L_{13} &=& 7L_9 + 5L_8 &= \cdots =& 5L_1 + 21L_2 + 10L_5 + 14L_6 \\ -99 = L_{14} &=& 14L_9 + 5L_{10} &= \cdots =& 42L_2 + 15L_5 + 33L_6 \end{array} \right.$$

A constant positive linear combination $c = \sum \lambda_i L_i$ can now be used in conjunction with Lemma 1 to refine the bounds on the initial set of affine forms. Since for any vector $x \in K$, and for any j, $L_j(x) \leq 0$, we have $c = \sum \lambda_i L_i(x) \leq \lambda_j L_j(x)$, and we obtain a lower bound $\frac{c}{\lambda_j}$ on L_j as soon as $\lambda_j \neq 0$.

Example 2. Using the linear combination $11L_1 + 7L_2 + 8L_5 = -11$, we can make the deductions $-1 \leq L_1$, $-\frac{11}{7} \leq L_2$ and $-\frac{11}{8} \leq L_3$ in the rationals. Furthermore, these deductions are refined as follows in the integers: $-1 \leq L_1$, $\lceil -\frac{11}{7} \rceil = -1 \leq L_2$ and $\lceil -\frac{11}{8} \rceil = -1 \leq L_3$.

3.2 Computing the Linear Combinations Using a Simplex

While the Fourier-Motzkin algorithm can be used to compute all the relevant constant positive linear combinations of affine forms, it does not scale in practice. In the following, we describe an efficient Simplex-based alternative and show its soundness, completeness, and termination. As opposed to the Fourier-Motzkin algorithm, this new approach will only attempt to compute one particular constant positive linear combination.

Let $K := \{x \in \mathbb{Q}^n \mid Ax + b \leq 0\}$ be a closed convex where $A \in \mathbb{Q}^{m\times n}$ and $b \in \mathbb{Q}^m$, and C the set of the affine forms $L_i : x \mapsto (Ax+b)_i$ of the form $\sum_{j=1}^{n} a_{i,j}\, x_j + b_i$. Consider the combination $\sum \lambda_i\, L_i$ of the affine forms. This sum unfolds as follows:

$$\lambda_1\,(\sum_{j=1}^{n} a_{1,j}\, x_j + b_1) \quad + \quad \cdots \quad + \quad \lambda_m\,(\sum_{j=1}^{n} a_{m,j}\, x_j + b_m)$$

and factorizing the x_i gives:

$$x_1\,(\sum_{i=1}^{m} a_{i,1}\,\lambda_i) \quad + \quad \cdots \quad + \quad x_n\,(\sum_{i=1}^{m} a_{i,n}\,\lambda_i) \quad + \quad \sum_{i=1}^{m} b_i\,\lambda_i$$

Since we are only interested in computing constant positive linear combinations, we require that for every k, $\sum_{i=1}^{m} a_{i,k}\,\lambda_i = 0$, which eliminates the variable x_k. Moreover, we look for the combinations that maximize the value of $\sum_{i=1}^{m} b_i\,\lambda_i$, since this will improve efficiency, as described in Section 4.2. More precisely, we compute such a constant positive linear combination by solving the following problem in the rationals:

$$\begin{array}{ll} \textit{maximize} & \sum_{i=1}^{m} b_i\,\lambda_i \\ \textit{subject to} & A^t\lambda = 0 \quad \wedge \quad \sum_{i=1}^{m}\lambda_i > 0 \quad \wedge \quad \bigwedge_{i=1}^{m}\lambda_i \geq 0 \end{array}$$

This problem reminds of the dual Simplex input, but here we have equalities $A^t\lambda = 0$ instead of the usual inequalities and an extra constraint $\sum \lambda_i > 0$.

In order to solve the above problem, we first introduce a slack variable s and a positive parameter ε to transform the strict inequality $\sum_{i=1}^{m}\lambda_i > 0$ into $\sum_{i=1}^{m}\lambda_i - \varepsilon = s \wedge s \geq 0$, following Lemma 1 of [8]. Then we solve the system of equalities in $\mathbb{Q}$ modulo the constraints $\bigwedge_{i=1}^{m}\lambda_i \geq 0 \wedge s \geq 0$. This returns unsat if this system is inconsistent in $\mathbb{Q}$ modulo the non-negativeness constraints, or a matrix of the form

$$\begin{pmatrix} & & \lambda_1 & \lambda_2 & \cdots & s \\ \lambda_{b_1} & \mapsto & c_{1.1} & c_{1.2} & \cdots & c_{1.m+1} \\ \lambda_{b_2} & \mapsto & c_{2.1} & c_{2.2} & \cdots & c_{2.m+1} \\ \vdots & & & & & \\ \lambda_{b_{n+1}} & \mapsto & c_{n+1.1} & c_{n+1.2} & \cdots & c_{n+1.m+1} \end{pmatrix}$$

Finally, we initialize the Simplex algorithm with this matrix and try to maximize the objective function. The Simplex returns either unsat if the given system has no solution, or unbound if the objective function has no upper bound, or a maximum m and a valuation ν for the vector λ.

If the Simplex algorithm returns unsat, then the oracle answers that there is no constant positive linear combination. If it returns unbound, the oracle just returns a positive constant. Otherwise, the Simplex algorithm necessarily returns a solution with a non-positive maximum for the objective function. Indeed, if the maximum were to be positive, one could multiply coordinate-wise any solution λ by a constant larger than 1 and obtain another solution with a larger objective value. The oracle then returns the corresponding linear combination. Note that as soon as the Simplex exploration discovers a positive value for the objective function, the answer will eventually be unbound so it can exit immediately.

3.3 Soundness, Completeness, and Termination

On top of the Simplex algorithm we only add some substitutions, so the termination of this oracle follows directly from the one of the Simplex algorithm.

Let us justify that the introduction of the parameter ε does affect neither the soundness nor the completeness of the oracle. Let us denote by $P_{>0}$ the original problem and by $P_{\geq\varepsilon}$ the problem we actually send to the Simplex algorithm. First, remember that for both problems, the answer is either unsat or unbound or a solution with a non-positive evaluation of the objective function: indeed, if ν is a solution, so is $\alpha\nu$ for any scalar α with the constraint $\alpha > 0$ for $P_{>0}$, and $\alpha > 1$ for $P_{\geq\varepsilon}$, and the value of the objective function is multiplied accordingly. Moreover, any solution of $P_{\geq\varepsilon}$ is obviously a solution of $P_{>0}$. Let us now proceed with the proof by case analysis.

1. If $P_{>0}$ is unsat, so is $P_{\geq\varepsilon}$ by inclusion of solutions.
2. If $P_{\geq\varepsilon}$ is unsat, let us assume by contradiction that $P_{>0}$ is not unsat, hence has a solution ν. Then $\frac{\varepsilon}{\sum \nu_i}\nu$ is a solution of $P_{\geq\varepsilon}$, a contradiction.
3. If $P_{\geq\varepsilon}$ is unbound, so is $P_{>0}$ by inclusion of solutions.
4. If $P_{>0}$ is unbound, we show that $P_{\geq\varepsilon}$ is also unbound. Let M be an arbitrary large value. By hypothesis on $P_{>0}$, there is a solution ν of $P_{>0}$ with an evaluation of the objective function greater than M. Then, if $\sum \nu_i \geq \varepsilon$, ν is a solution of $P_{\geq\varepsilon}$ with an evaluation of the objective function greater than M. Otherwise, $\frac{\varepsilon}{\sum \nu_i}\nu$ is a solution of $P_{\geq\varepsilon}$, with an evaluation of the objective function greater than $\frac{\varepsilon}{\sum \nu_i}M$, hence greater that M since $\varepsilon > \sum \nu_i$.
5. If $P_{>0}$ (resp. $P_{\geq\varepsilon}$) has a solution with a non-positive evaluation of the objective function, so has $P_{\geq\varepsilon}$ (resp. $P_{>0}$), since the other cases are impossible, as shown above.

4 The Decision Procedure

Let us now build a decision procedure for QF-LIA based upon an oracle that follows the interface described at the beginning of Section 3 and the theorems presented in Section 2.

4.1 The Algorithm

Let $A \in \mathbb{Q}^{m \times n}$ and $b \in \mathbb{Q}^m$. The procedure shall decide whether the system $Ax + b \leq 0$ has a solution $x \in \mathbb{Z}^n$. Let $L = (L_i)_{i \in [|1,m|]}$ be the associated family of affine forms. Figure 1 sketches the algorithm. It takes as input both the system L of inequalities and an additional argument Eq representing affine relations between variables, *e.g.* a set of equalities, or a substitution, or an echelon matrix, etc. This last argument is initially empty. The result of the decision procedure is stored in the *sols* variable. It is a finite set of integer solutions, possibly empty, or an indeterminate infinite set of integer solutions.

```
1  global sols ← ∅
2  procedure lia(L = (L_i), Eq)
3      remove trivial inequalities c ≤ 0 with c constant from L
4      if some c was positive then return
5      if L = ∅ then
6          sols ← sols ∪ check_1(Eq)
7          return
8      call oracle(L)
9      if there is no constant positive linear combination then
10         sols ← sols ∪ check_∞(Eq)
11         return
12     let Σ λ_i L_i = c the constant positive linear combination found by the oracle
13     if c > 0 then return
14     choose k such that λ_k ≠ 0, and μ > 0 such that μL_k has integer coefficients only
15     for all v from ⌈μ c/λ_k⌉ to 0 do
16         create a substitution σ from μL_k(x_1, ..., x_n) = v
17         if there is no possible substitution then continue to next iteration
18         remove L_k from L
19         apply σ to L
20         call lia(L, Eq ∪ {σ})
21     return
```

Fig. 1. Algorithm for the decision procedure

The *check* functions at lines 6 and 10 compute the integer solutions of a system of equations, but the special shape of the systems they deal with allows important optimizations that will be detailed below.

The algorithm is recursive. Recursive calls are performed on smaller and smaller systems L until complete resolution. Branching is caused by the loop on line 15. The results are merged along the various branches at lines 6 and 10. One can also consider that there are implicit statements $sols \leftarrow sols \cup \emptyset$ at lines 4, 13, and 17. Notice that the algorithm performs computations only when going from the root to the leaves of the call tree. For the sake of clarity, we have described a simple version of the algorithm. An actual implementation would likely be more complicated. For instance, it would exit as soon as a branch finds an infinity of solutions or even a single solution if one is interested only in satisfiability. It could also use *splitting on demand* [3] at line 15.

From lines 3 to 7, the algorithm deals with degenerate systems that contain no inequalities or only trivial inequalities. It then calls the oracle on L. If it answers that no suitable combination of the affine forms exists, Corollary 1 can be applied. There are infinitely many solutions with integer coordinates, assuming Eq imposes no restriction (hence the call to $check_\infty$). The decision procedure is done in this branch.

Otherwise, the oracle returns a constant positive linear combination $\sum_i \lambda_i L_i$ equal to c. In that case, Lemma 1 can be applied. If c is positive, the procedure is done too: the system has no solution.

Otherwise, we have $\frac{c}{\lambda_k} \leq L_k \leq 0$ for all k such that $\lambda_k \neq 0$. The decision procedure chooses a value for k. Since the coefficients of μL_k are in $\mathbb{Z}$, for any point $x \in K \cap \mathbb{Z}^n$, $\mu L_k(x)$ is an integer between $\mu\frac{c}{\lambda_k}$ and 0. For each integer $v \in [\mu\frac{c}{\lambda_k}, 0]$, the decision procedure considers the equality $\mu L_k = v$ from which it infers a substitution if possible, applies the result to all the other affine forms $(L_i)_{i \neq k}$ and removes L_k from the system while updating Eq with the substitution. The decision procedure is then called recursively. If no solution is found after a complete exploration of all the possible integer values in $[\mu\frac{c}{\lambda_k}, 0]$, then the procedure returns at line 21 without updating *sols*. This is what happens if L has rational solutions but no integer solutions.

Note that, if the constant c is zero, then several equalities might appear at once. An optimized procedure should therefore compute a substitution taking all of them into account, rather than one after the other, as is done in Figure 1.

We now give more details on the computations performed at the leaves of the call tree by the *check* functions. The choice of these functions depends on the substitution scheme at line 16. We describe here two possible scenarios.

Integer Substitution with Slack Variables. Let us first consider the case where the substitution introduces slack variables. The variables $x_1, \ldots, x_n$ of L_k are expressed as affine combinations of new variables $x_{n+1}, \ldots, x_{n+\ell}$, such that the integer solutions of $L_k(x_1, \ldots, x_n) = v$ are completely parameterized by these new variables. Removing L_k from L and applying the substitution produces a system equisatisfiable to L and solutions to the original system can be trivially computed thanks to Eq.

A way to obtain the substitution is the Generalized GCD test [2] with the approach given by Pugh in the Omega-Test [15]. Note that the substitution may have only one solution, *e.g.* $2x = 6$. The substitution may also not exist, *e.g.* $5x = 2$, in which case, the exit case described line 17 applies.

In that case, functions *check* are implemented as follows. For $check_1$, the solutions are constrained purely by Eq. More precisely, the set of integer solutions is parameterized by the set of variables that are never the target of substitution in Eq. Moreover, the Eq system has been built only from adding successively (cf. line 20) new substitutions not featuring the variables previously substituted (cf. line 19), so it is never inconsistent. Therefore, only two situations are possible when Eq is passed to $check_1$: either Eq involves all the variables of the system, in which case there is exactly one solution, or it does not (some variables have not been substituted) and there are infinitely many integer solutions. In this implementation Eq is always a system with integer coefficients, and function $check_\infty$ is trivial for integer substitutions. Indeed, it just returns an indeterminate infinite set, whatever the value of Eq, since some variables have to be unsubstituted at that point.

In some context, *e.g.* an SMT solver, one might need more information than what this indeterminate set seems to carry. If one needs some explicit witnesses, the proof of Theorem 3 explains how to effectively compute them from L. Witnesses for the original system can then be deduced from Eq. If one needs to know which equalities are implied by the system, then Eq describes them entirely. Indeed, the decision procedure will not exit at line 11 if some constant positive linear combination still exists in L.

Gaussian Elimination. Let us now consider the case of a substitution performed by a simple Gaussian elimination on rational numbers; it does not introduce any slack variables but the coefficients involved in the substitutions are rationals, possibly non integers. Function $check_1$ now has to test whether the set of equalities Eq admits some solution and to return them. In this scenario, there can be either zero solution, or one, or an infinite number of them. The implementation of function $check_\infty$ can in that case take benefit of Theorem 4. The hypotheses of the theorem are actually verified thanks to the Gaussian elimination, and the vector space E of Theorem 4 is generated by the vectors of the canonical basis associated with the variables already substituted. Since these variables have been eliminated from L, the set of solutions of L is obviously invariant by translation along these coordinates. By construction of Eq the hypothesis of inclusion of the respective projections also holds. Therefore if the system of equalities Eq admits at least one integer solution, then there are an infinite number of solutions for the problem considered in the current branch. Otherwise there is no solution for this branch.

Note that, whichever of these substitution schemes is used, the solver for linear integer arithmetic embedded in the *check* functions has to deal with equations only and is therefore simple.

Producing Explanations. An important feature when developing a decision procedure for SMT is to provide the most precise explanations that improve the backtrack level when branching. In our setting, the explaination of each inconsistency or lower bound $\frac{c}{\lambda_k}$ inferred by the oracle is the explanations of the inequalities $L_i \leq 0$ such that $\lambda_i \neq 0$. The explanations of the inequalities that have not participated in the inference process are thus discarded.

4.2 Soundness, Completeness, and Termination

Termination is obvious, assuming the oracle is itself terminating. Indeed, at each recursive call, one affine form at least is removed from the system. Note that it has to be effectively removed from the system; otherwise the oracle may just return the linear combination that bounds this form again, hence causing the procedure to enter an infinite recursion.

Soundness depends on the completeness of the oracle: if the oracle does not find any constant positive linear combination, there should be none. Theorems of Section 2.2 then cover all the possible cases. Completeness of the decision procedure comes from termination and soundness.

While the oracle can return any constant positive linear combination, for efficiency reasons, it should strive to find a positive constant if possible, and zero if not. Indeed, this ensures that the algorithm will not branch too early.

5 Experimental Results

We implemented the decision procedure with the Simplex-based oracle in a modified version[1] of ALT-ERGO [4]. Equalities are handled using a rewriting system that relies on substitutions with integer slack variables [14]. Inequalities are added to a dictionary associating affine forms with integer interval domains. The case-split analysis is implemented as a recursive function with non-chronological backtracking and uses a heuristic that privileges affine forms with smaller intervals. In the current implementation, we do not use a traditional Simplex to cut down the search space.

In this section, we benchmark our implementation and compare its performances with some leading state-of-the-art SMT solvers including MATHSAT5 v5.1.3 [11], Z3 v3.2 [5] and YICES2 v2.0-prototype [6]. We could not include the MISTRAL solver of [7] because it was not possible to obtain it. The test suite contains 1070 instances taken from the QF-LIA category of SMT-LIB[2]. This includes the following families:

- CAV-2009: randomly-generated instances used in [7]. Most of them are satisfiable. They are reported very hard for modern SMT solvers,
- SLACKS: reformulation of CAV-2009 instances used in [12] that introduces slack variables to bound all variables,
- CUT-LEMMAS: crafted instances encoding the validity of cutting planes in $\mathbb{Z}$,
- PRIME-CONE: crafted instances used in [12] that encode a tight n-dimensional cone around the point whose coordinates are the first n prime numbers.
- PIDGEONS (*sic*): crafted instances encoding the pigeonhole principle. They are reported hard for any solver not using cutting planes [12],
- PB2010: industrial instances coming from the PB competition (2010),
- MIPLIB2003: instances generated from some optimization problems in [1].

We have selected these families because they are known to be well-suited for stressing the integer-reasoning part of solvers. Moreover, contrarily to some other families, they do not require solvers to be especially efficient for other tasks:

[1] A prototype is available at `http://alt-ergo.lri.fr/ijcar2012/`

[2] The SMT-LIB library: `http://www.smtlib.org`

preprocessing, *if-then-else* handling, SAT solving, theory propagation. In fact, a large part of the QF-LIA benchmark, *e.g.* NEC-SMT, does not even require fast LIA solvers [7].

All measures were obtained on a 64-bit machine with a quad-core Intel Xeon processor at 3.2 GHz and 24 GB of memory. Provers were given a time limit of 600 seconds and a memory limit of 2 GB for each test. The results of our experiments are reported in Figure 2. The first two columns show the families we considered and the number of their instances. For each prover, we report both the number of solved instances within the required time for every family and the time needed for solving them (not counting timeouts). The last rows summarize the total number of solved instances and the accumulated time for each prover.

SMT SOLVERS		ALT-ERGO		MATHSAT5		MATHSAT5+CFP		YICES 2		Z3	
families	**#inst.**	**solved**	**time**	**solved**	**time**	**solved**	**time**	**solved**	**time**	**solved**	**time**
CAV-2009	591	590	**253**	588	4857	589	4544	386	11664	590	5195
SLACKS	233	233	**67**	166	3551	155	6545	142	6102	187	9897
CUT-LEMMAS	93	93	**216**	62	3424	59	2775	92	1892	67	3247
PRIME-CONE	37	37	**0.4**	37	1	37	2.2	37	2.3	37	14
PIDGEONS	19	19	2	19	0.16	19	0.16	19	**0.01**	19	0.28
PB2010	81	23	390	38	743	34	1540	25	8.3	64	1831
MIPLIB2003	16	2	34.7	12	432	12	501	11	145.4	12	**241**
total	1070	997	**963.1**	922	13008	905	15907	712	19814	976	20425
total QF-LIA[3]	5882	4410	68003	5597	**47635**	5524	50481	3220	71324	5597	54503

Fig. 2. Experimental results. Underlined values are for tools that have proved the most instances. Bolded results are for tools that have proved both the most instances and the fastest.

Although the first two families were reported very hard for modern SMT solvers, our approach only requires 320 seconds to solve almost all the instances. Thus, it significantly outperforms the other solvers' approaches. This observation also applies for the third and the fourth families. From the results of the sixth and the seventh families, we notice that our technique does not perform well on large difference-logic-like problems compared to MATHSAT5 and Z3's. We think this is partly due to our naive implementation of the Simplex algorithm which computes on dense matrices while sparse matrices would be better suited for these problems. We plan to implement advanced techniques in the very near future such as the revised Simplex method [16] to overcome this issue.

The last row of Table 2 shows the results for the whole QF-LIA benchmark. There are two reasons for the poor results. First, ALT-ERGO has yet to be tuned for parts other than the LIA solver. Second, some families, *e.g.* BOFILL, contain

[3] The time limit is 180 seconds for the tests of the complete benchmark.

large intervals that need splitting, and the decision procedure does not deal efficiently with them. This will possibly require a combination of our approach with other established techniques for integers.

6 Conclusion and Future Works

We have presented a new decision procedure for quantifier-free linear integer arithmetic that combines a model search mechanism with bounds inference. These bounds are discovered thanks to a Simplex-based oracle that computes constant positive linear combinations of affine forms. We proved the soundness, the completeness and the termination of our method and implemented it in the ALT-ERGO SMT solver.

Designing efficient decision procedures for QF-LIA has been an active research topic in the SMT community over the last decade. An efficient integration of the Simplex algorithm in the DPLL(T) framework has been proposed in [8]. This integration rests on a preprocessing step that enables fast backtracking and efficient theory propagation. The contribution of [7] is seen as a generalization of *branch-and-bound*. Using the notion of *the defining constraints* of a vertex, it derives additional inequalities that prune higher dimensional subspaces not containing integer solutions. In our setting, the Simplex algorithm is instead used on auxiliary problems to refine the search space by bounds inference.

The approach described in [10] focuses on combining several existing techniques using heuristics and *layering* to take advantage of each of them. We believe that the ideas we described in this paper can naturally be used to enhance this combination approach. Yet another different contribution described in [12] consists in extending the inference rules of the CDCL procedure with linear arithmetic reasoning. This tight integration naturally takes advantage of the good CDCL properties: model search, dynamic variable reordering, propagation, conflicts explanation, and backjumping. The extension of our framework with an efficient conflict learning mechanism as done in [12] or in [13] for the rationals would greatly improve our decision procedure.

As reflected by our contribution, the Simplex we use does not directly work on the initial problem nor on its dual. Therefore, fast incrementality and backtracking techniques developed for Simplex-based approaches are not suitable for our setting. To alleviate this issue, we have used memoization techniques to reuse previously computed results at the expense of a larger memory footprint. In the near future, we plan to better integrate with DPLL(T) by extending our method with a conflict resolution technique, a cleverer case-split analysis, and an efficient theory propagation. We also believe that a combination, *à la* MATHSAT, of state-of-the-art techniques with ours would be beneficial. Furthermore, the use of advanced data-structures and algorithms such as sparse matrices and the revised Simplex would greatly enhance our implementation.

References

1. Achterberg, T., Koch, T., Martin, A.: MIPLIB 2003. Operations Research Letters 34(4), 361–372 (2006)
2. Banerjee, U.: Dependence Analysis for Supercomputing. Kluwer Academic Publishers, Norwell (1988)
3. Barrett, C., Nieuwenhuis, R., Oliveras, A., Tinelli, C.: Splitting on Demand in SAT Modulo Theories. In: Hermann, M., Voronkov, A. (eds.) LPAR 2006. LNCS (LNAI), vol. 4246, pp. 512–526. Springer, Heidelberg (2006)
4. Bobot, F., Conchon, S., Contejean, E., Iguernelala, M., Lescuyer, S., Mebsout, A.: The Alt-Ergo Automated Theorem Prover, `http://alt-ergo.lri.fr`
5. de Moura, L., Bjørner, N.: Z3, an efficient SMT solver, `http://research.microsoft.com/projects/z3`
6. de Moura, L., Dutertre, B.: Yices: An SMT Solver, `http://yices.csl.sri.com`
7. Dillig, I., Dillig, T., Aiken, A.: Cuts from Proofs: A Complete and Practical Technique for Solving Linear Inequalities over Integers. In: Bouajjani, A., Maler, O. (eds.) CAV 2009. LNCS, vol. 5643, pp. 233–247. Springer, Heidelberg (2009)
8. Dutertre, B., de Moura, L.: A Fast Linear-Arithmetic Solver for DPLL(T). In: Ball, T., Jones, R.B. (eds.) CAV 2006. LNCS, vol. 4144, pp. 81–94. Springer, Heidelberg (2006)
9. Farkas, G.: Über die theorie der einfachen ungleichungen. Journal für die Reine und Angewandte Mathematik 124, 1–27 (1902)
10. Griggio, A.: A practical approach to satisability modulo linear integer arithmetic. Journal on Satisfiability, Boolean Modeling and Computation 8, 1–27 (2012)
11. Griggio, A., Schaafsma, B., Cimatti, A., Sebastiani, R.: MathSAT 5: An SMT Solver for Formal Verification, `http://mathsat.fbk.eu//`
12. Jovanović, D., de Moura, L.: Cutting to the Chase Solving Linear Integer Arithmetic. In: Bjørner, N., Sofronie-Stokkermans, V. (eds.) CADE 2011. LNCS, vol. 6803, pp. 338–353. Springer, Heidelberg (2011)
13. Korovin, K., Voronkov, A.: Solving Systems of Linear Inequalities by Bound Propagation. In: Bjørner, N., Sofronie-Stokkermans, V. (eds.) CADE 2011. LNCS, vol. 6803, pp. 369–383. Springer, Heidelberg (2011)
14. Kroening, D., Strichman, O.: Decision Procedures: An Algorithmic Point of View, 1st edn. Springer Publishing Company, Incorporated (2008)
15. Pugh, W.: The Omega test: a fast and practical integer programming algorithm for dependence analysis. In: Proceedings of the 1991 ACM/IEEE Conference on Supercomputing 1991, pp. 4–13. ACM, New York (1991)
16. Schrijver, A.: Theory of linear and integer programming. Wiley-Interscience series in discrete mathematics and optimization. John Wiley & Sons (1998)

How Fuzzy Is My Fuzzy Description Logic?*

Stefan Borgwardt, Felix Distel, and Rafael Peñaloza

Theoretical Computer Science, TU Dresden, Germany
{stefborg,felix,penaloza}@tcs.inf.tu-dresden.de

Abstract. Fuzzy Description Logics (DLs) with t-norm semantics have been studied as a means for representing and reasoning with vague knowledge. Recent work has shown that even fairly inexpressive fuzzy DLs become undecidable for a wide variety of t-norms. We complement those results by providing a class of t-norms and an expressive fuzzy DL for which ontology consistency is linearly reducible to crisp reasoning, and thus has its same complexity. Surprisingly, in these same logics crisp models are insufficient for deciding fuzzy subsumption.

1 Introduction

Description logics (DLs) [1] are a family of logic-based knowledge representation formalisms, which can be used to represent the knowledge of an application domain in a formal way. In particular, they have been successfully used for the representation of medical knowledge in large-scale ontologies like SNOMED CT[1] and GALEN.[2] However, in their standard form DLs are not suited for dealing with imprecise or vague knowledge. For example, in the medical domain a high body temperature is often a symptom for a disease. When trying to represent this knowledge, it is not possible to give a precise characterization of the concept HighTemperature: one cannot define a point where a temperature becomes high. However, 37°C should belong "less" to this concept than, say 39°C.

Fuzzy variants of description logics have been proposed as a formalism for modeling this kind of imprecise knowledge, by providing a *degree of membership* of individuals to concepts—typically a number from the interval $[0, 1]$. One could thus express that 36°C and 39°C belong to HighTemperature with degrees 0.7 and 0.9, respectively. A more thorough description of the use of fuzzy semantics in medical applications can be found in [20].

A great variety of fuzzy DLs can be found in the literature (for two relevant surveys see [18,12]). In fact, fuzzy DLs have several degrees of freedom for defining their expressiveness. In addition to the choice of concept constructors (e.g. conjunction $\sqcap$ or existential restriction $\exists$), and the type of axioms allowed (like acyclic concept definitions or general concept inclusions), which define the

* Partially supported by the DFG under grant BA 1122/17-1 and in the Collaborative Research Center 912 "Highly Adaptive Energy-Efficient Computing".

1 http://www.ihtsdo.org/snomed-ct/

2 http://www.opengalen.org/

B. Gramlich, D. Miller, and U. Sattler (Eds.): IJCAR 2012, LNAI 7364, pp. 82–96, 2012.

underlying logical language, one must also decide how to interpret the different constructors, through a choice of functions over the domain of fuzzy values $[0,1]$. As in mathematical fuzzy logic [13], these functions are typically determined by a continuous *t-norm* that interprets conjunction.

Research in fuzzy DLs has focused on three specific t-norms, namely the Gödel, Łukasiewicz, and product t-norms. However, there are uncountably many continuous t-norms, each with different properties. For example, under the product t-norm semantics, existential restrictions ($\exists$) and value restrictions ($\forall$) are not interdefinable, while under the Łukasiewicz t-norm they are. Even after fixing the t-norm, one can still choose whether to interpret negation by the involutive negation operator, or using the residual negation, which need not be involutive. An additional level of liberty comes from selecting the class of models over which reasoning is considered: either all models, or so-called witnessed models only [14].

The majority of the reasoning algorithms available have been developed for the Gödel semantics, either by a reduction to crisp reasoning [6], or by a simple adaptation of the known algorithms for crisp DLs [23,24,25,27]. However, methods capable of dealing with other t-norms have also been explored [7,8,9,26,22]. Usually, these algorithms reason w.r.t. witnessed models.[3]

Very recently, it was shown that the tableaux-based algorithms for logics with semantics based on t-norms other than the Gödel t-norm and allowing general concept inclusions were incorrect [2,5]. This raised doubts about the decidability of the reasoning problems in these logics, and eventually led to a plethora of undecidability results for fuzzy DLs [2,3,4,11]. These undecidability results were then extended to a wide variety of fuzzy DLs in [10]. In fact, it has been shown that for a large class of t-norms ontology consistency easily becomes undecidable. More precisely, for every t-norm that "starts" with the Łukasiewicz t-norm, consistency of *crisp* ontologies is undecidable for any fuzzy DL that can express conjunction, existential restrictions and the residual negation.

In this paper we counterbalance these undecidability results by considering continuous t-norms *not* starting with the Łukasiewicz t-norm—in particular, the Gödel and product t-norms are of this kind. We show that consistency of *fuzzy* ontologies is again decidable, even for the very expressive DL $\mathcal{SHOI}$, which allows for nominals and transitive and inverse roles, if negation is interpreted using residual negation. Moreover, for any of these t-norms, an ontology is consistent w.r.t. fuzzy semantics iff it is consistent w.r.t. to *crisp* semantics. Thus, ontology consistency in fuzzy $\mathcal{SHOI}$ is ExpTime-complete for every t-norm not starting with the Łukasiewicz t-norm; for all other t-norms, or if the involutive negation is used, this problem is undecidable [10].

To some extent, the fact that fuzzy ontology consistency can be reduced to crisp reasoning is not very surprising, since fuzzy logics are not, nor should they be considered to be, a formalism for dealing with inconsistencies. Yet, it shines a negative light on the capacity of fuzzy DLs for dealing with imprecise knowledge: the decidable fuzzy DLs considered in this paper are not fuzzy, but mere syntactic extensions of classical DLs. However, there are other DL reasoning problems for

[3] In fact, witnessed models were introduced in [14] to correct the algorithm from [27].

which this is not true: we show that crisp reasoning is insufficient for deciding subsumption or instance checking. Thus, even for the logics considered in this paper, where satisfiability is "crisp", reasoning in general is fuzzy.

In the next section, we introduce some basic notions from t-norms and fuzzy description logics. Section 3 shows some properties of t-norms that do not start with the Łukasiewicz t-norm. In Sections 4 and 5 we prove that consistency and satisfiability w.r.t. these t-norms are essentially crisp reasoning problems. In the end we provide an example that shows that crisp reasoning is insufficient for deciding subsumption or instance checking. Specifically, we provide a subsumption relation that holds in every crisp and finite model, but does not hold in general.

2 Preliminaries

We first recall the basic notions of t-norms and mathematical fuzzy logic [17,13], which we then use to define the semantics of fuzzy DLs.

2.1 Mathematical Fuzzy Logic

Mathematical fuzzy logic generalizes classical logic by replacing *true* and *false* by a larger set of truth values. Here, we use the real interval $[0, 1]$ as truth values and generalize propositional conjunction $\wedge$ by a *t-norm*: an associative, commutative, and monotone binary operator on $[0, 1]$ that has 1 as its unit element. Classical implication is then generalized by the *residuum* $\Rightarrow$ of the t-norm, if it exists. The residuum is a binary operator on $[0, 1]$ that satisfies $x \otimes y \leq z$ iff $y \leq x \Rightarrow z$ for all $x, y, z \in [0, 1]$. A consequence of this definition is that, for all $x, y \in [0, 1]$,

- $1 \Rightarrow x = x$ and
- $x \leq y$ iff $x \Rightarrow y = 1$.

A t-norm is called *continuous* if it is continuous as a function from $[0, 1]^2$ to $[0, 1]$. In this paper, we consider only continuous t-norms and often call them simply t-norms. Any continuous t-norm $\otimes$ has a unique residuum $\Rightarrow$ given by $x \Rightarrow y = \sup\{z \in [0, 1] \mid x \otimes z \leq y\}$. Based on the residuum, one can define a unary *residual negation* by $\ominus x = x \Rightarrow 0$. To generalize disjunction, the *t-conorm* $\oplus$ defined as $x \oplus y = 1 - ((1 - x) \otimes (1 - y))$ can be used. Notice that 0 is the unit of the t-conorm, and hence

$$x \oplus y = 0 \text{ iff } x = 0 \text{ and } y = 0. \tag{1}$$

Three important continuous t-norms, together with their t-conorms and residua, are depicted in Table 1. These are *fundamental* in the sense that every continuous t-norm can be constructed from these three as follows.

Definition 1 (ordinal sum). *Let I be a set and for each $i \in I$ let $\otimes_i$ be a continuous t-norm and $a_i, b_i \in [0, 1]$ such that $a_i < b_i$ and the intervals (a_i, b_i) are pairwise disjoint. The* ordinal sum *of the t-norms $\otimes_i$ is the t-norm $\otimes$ with*

$$x \otimes y = \begin{cases} a_i + (b_i - a_i)\left(\frac{x-a_i}{b_i-a_i} \otimes_i \frac{y-a_i}{b_i-a_i}\right) & \text{if } x, y \in [a_i, b_i] \text{ for some } i \in I, \\ \min\{x, y\} & \text{otherwise.} \end{cases}$$

Table 1. The three fundamental continuous t-norms

Name	t-norm ($x \otimes y$)	t-conorm ($x \oplus y$)	residuum ($x \Rightarrow y$)
Gödel	$\min\{x, y\}$	$\max\{x, y\}$	$\begin{cases} 1 & \text{if } x \leq y \\ y & \text{otherwise} \end{cases}$
product	$x \cdot y$	$x + y - x \cdot y$	$\begin{cases} 1 & \text{if } x \leq y \\ y/x & \text{otherwise} \end{cases}$
Łukasiewicz	$\max\{x + y - 1, 0\}$	$\min\{x + y, 1\}$	$\min\{1 - x + y, 1\}$

The ordinal sum of a class of continuous t-norms is itself a continuous t-norm, and its residuum is given by

$$x \Rightarrow y = \begin{cases} 1 & \text{if } x \leq y, \\ a_i + (b_i - a_i)\left(\frac{x-a_i}{b_i-a_i} \Rightarrow_i \frac{y-a_i}{b_i-a_i}\right) & \text{if } a_i \leq y < x \leq b_i \text{ for some } i \in I, \\ y & \text{otherwise,} \end{cases}$$

where $\Rightarrow_i$ is the residuum of $\otimes_i$, for each $i \in I$. Intuitively, this means that the t-norm $\otimes$ and its residuum "behave like" $\otimes_i$ and its residuum in each of the intervals $[a_i, b_i]$, and like the Gödel t-norm and residuum everywhere else.

Theorem 2 ([21]). *Every continuous t-norm is isomorphic to the ordinal sum of copies of the Łukasiewicz and product t-norms.*

Motivated by this representation as an ordinal sum, we say that a continuous t-norm $\otimes$ *starts with the Łukasiewicz t-norm* if in its representation as ordinal sum there is an $i \in I$ such that $a_i = 0$ and $\otimes_i$ is isomorphic to the Łukasiewicz t-norm.

An element $x \in (0, 1)$ is called a *zero divisor* for $\otimes$ if there is a $z \in (0, 1)$ such that $x \otimes z = 0$. Of the three fundamental continuous t-norms, only the Łukasiewicz t-norm has zero divisors. In fact, every element in the interval $(0, 1)$ is a zero divisor for this t-norm. A continuous t-norm can only have zero divisors if it starts with the Łukasiewicz t-norm.

Lemma 3 ([17]). *A continuous t-norm has zero divisors iff it starts with the Łukasiewicz t-norm.*

2.2 The Fuzzy Description Logic $\otimes$-$\mathcal{SHOI}$

A fuzzy description logic usually inherits its syntax from the underlying crisp description logic. In this paper, we consider the constructors of $\mathcal{SHOI}$ with the addition of $\rightarrow$, which in the crisp case can be expressed by $\sqcup$ and $\neg$.

Definition 4 (syntax). *Let* $\mathsf{N_C}$, $\mathsf{N_R}$, *and* $\mathsf{N_I}$, *be disjoint sets of* concept, role, *and* individual names, *respectively, and* $\mathsf{N_R^+} \subseteq \mathsf{N_R}$ *be a set of* transitive role names. *The set of* (complex) roles *is* $\mathsf{N_R} \cup \{r^- \mid r \in \mathsf{N_R}\}$. *The set of* (complex) concepts *is defined by the following syntax rule:*

$$C ::= A \mid \top \mid \bot \mid \{a\} \mid \neg C \mid C \sqcap C \mid C \sqcup C \mid C \rightarrow C \mid \exists s.C \mid \forall s.C,$$

where A is a concept name, a is an individual name, and s is a complex role.

The *inverse* of a complex role s (denoted by $\overline{s}$) is s^- if $s \in \mathsf{N_R}$ and r if $s = r^-$. A role s is *transitive* if either s or $\overline{s}$ belongs to $\mathsf{N_R^+}$.

Let now $\otimes$ be a continuous t-norm. As a generalization of $\mathcal{SHOI}$, where concepts are interpreted by subsets of a domain, in the fuzzy DL $\otimes$-$\mathcal{SHOI}$ they are interpreted by *fuzzy sets*, which are functions specifying the membership degree of each domain element to the concept. The interpretation of the constructors is based on the t-norm $\otimes$ and the induced operators $\oplus$, $\Rightarrow$, and $\ominus$.

Definition 5 (semantics). *An* interpretation *is a pair* $\mathcal{I} = (\Delta^\mathcal{I}, \cdot^\mathcal{I})$, *where the* domain $\Delta^\mathcal{I}$ *is a non-empty set and* $\cdot^\mathcal{I}$ *is a function that assigns to every concept name* A *a function* $A^\mathcal{I}\colon \Delta^\mathcal{I} \to [0,1]$, *to every individual name* a *an element* $a^\mathcal{I} \in \Delta^\mathcal{I}$, *and to every role name* r *a function* $r^\mathcal{I}\colon \Delta^\mathcal{I} \times \Delta^\mathcal{I} \to [0,1]$ *such that* $r^\mathcal{I}(x,y) \otimes r^\mathcal{I}(y,z) \leq r^\mathcal{I}(x,z)$ *holds for all* $x,y,z \in \Delta^\mathcal{I}$ *if* $r \in \mathsf{N_R^+}$. *The function* $\cdot^\mathcal{I}$ *is extended to complex roles and concepts as follows for every* $x,y \in \Delta^\mathcal{I}$,

- $(r^-)^\mathcal{I}(x,y) = r^\mathcal{I}(y,x)$,
- $\top^\mathcal{I}(x) = 1, \quad \bot^\mathcal{I}(x) = 0$,
- $\{a\}^\mathcal{I}(x) = 1$ *if* $a^\mathcal{I} = x$ *and* 0 *otherwise*,
- $(\neg C)^\mathcal{I}(x) = \ominus C^\mathcal{I}(x)$,
- $(C_1 \sqcap C_2)^\mathcal{I}(x) = C_1^\mathcal{I}(x) \otimes C_2^\mathcal{I}(x)$,
- $(C_1 \sqcup C_2)^\mathcal{I}(x) = C_1^\mathcal{I}(x) \oplus C_2^\mathcal{I}(x)$,
- $(C_1 \to C_2)^\mathcal{I}(x) = C_1^\mathcal{I}(x) \Rightarrow C_2^\mathcal{I}(x)$,
- $(\exists s.C)^\mathcal{I}(x) = \sup_{z \in \Delta^\mathcal{I}} s^\mathcal{I}(x,z) \otimes C^\mathcal{I}(z)$, *and*
- $(\forall s.C)^\mathcal{I}(x) = \inf_{z \in \Delta^\mathcal{I}} s^\mathcal{I}(x,z) \Rightarrow C^\mathcal{I}(z)$.

An interpretation $\mathcal{I}$ *is called* finite *if its domain* $\Delta^\mathcal{I}$ *is finite, and* crisp *if* $A^\mathcal{I}(x), r^\mathcal{I}(x,y) \in \{0,1\}$ *for all* $A \in \mathsf{N_C}$, $r \in \mathsf{N_R}$, *and* $x,y \in \Delta^\mathcal{I}$.

Knowledge is encoded using DL axioms, which restrict the class of interpretations that are considered. The fuzzy DL $\otimes$-$\mathcal{SHOI}$ extends the axioms of $\mathcal{SHOI}$ by specifying a degree to which the restrictions should hold.

Definition 6 (axioms). *An* axiom *is either an* assertion *of the form* $\langle a{:}C, \ell\rangle$ *or* $\langle (a,b){:}s, \ell\rangle$, *a* general concept inclusion (GCI) *of the form* $\langle C \sqsubseteq D, \ell\rangle$, *or a* role inclusion *of the form* $\langle s \sqsubseteq t, \ell\rangle$, *where* C *and* D *are concepts,* $a,b \in \mathsf{N_I}$, s,t *are complex roles, and* $\ell \in (0,1]$. *An axiom is called* crisp *if* $\ell = 1$.

An interpretation $\mathcal{I}$ satisfies *an assertion* $\langle a{:}C, \ell\rangle$ *if* $C^\mathcal{I}(a^\mathcal{I}) \geq \ell$ *and an assertion* $\langle (a,b){:}s, \ell\rangle$ *if* $s^\mathcal{I}(a^\mathcal{I}, b^\mathcal{I}) \geq \ell$. *It* satisfies *the GCI* $\langle C \sqsubseteq D, \ell\rangle$ *if* $C^\mathcal{I}(x) \Rightarrow D^\mathcal{I}(x) \geq \ell$ *holds for all* $x \in \Delta^\mathcal{I}$. *It* satisfies *a role inclusion* $\langle s \sqsubseteq t, \ell\rangle$ *if* $s^\mathcal{I}(x,y) \Rightarrow t^\mathcal{I}(x,y) \geq \ell$ *holds for all* $x,y \in \Delta^\mathcal{I}$.

An ontology $(\mathcal{A}, \mathcal{T}, \mathcal{R})$ *consists of a finite set* $\mathcal{A}$ *of assertions* (ABox), *a finite set* $\mathcal{T}$ *of GCIs* (TBox), *and a finite set* $\mathcal{R}$ *of role inclusions* (RBox). *It is* crisp *if every axiom in* $\mathcal{A}$, $\mathcal{T}$, *and* $\mathcal{R}$ *is crisp. An interpretation* $\mathcal{I}$ *is a* model *of this ontology if it satisfies all its axioms.*

The combination of axioms in an ontology may entail some knowledge of the domain that is not explicitly represented. Reasoning can then be used to make this knowledge explicit. We consider the standard reasoning problems of crisp $\mathcal{SHOI}$, extended with a degree to which they hold.

Definition 7 (reasoning problems). *Let $\mathcal{O}$ be an ontology, C, D be concepts, a an individual, and $\ell \in [0,1]$. $\mathcal{O}$ is called* consistent *if it has a model.*

C is ℓ-satisfiable *w.r.t. $\mathcal{O}$ if there is a model $\mathcal{I}$ of $\mathcal{O}$ and $x \in \Delta^{\mathcal{I}}$ such that $C^{\mathcal{I}}(x) \geq \ell$. C is* ℓ-subsumed *by D w.r.t. $\mathcal{O}$ with $\ell \in [0,1]$ if every model of $\mathcal{O}$ satisfies the GCI $\langle C \sqsubseteq D,\ \ell\rangle$. The individual a is an* ℓ-instance *of C w.r.t. $\mathcal{O}$ if every model of $\mathcal{O}$ satisfies the assertion $\langle a{:}C,\ \ell\rangle$.*

The best satisfiability (subsumption, instance) degree *of C (C and D, a and C) w.r.t. $\mathcal{O}$ is the supremum of all $\ell \in [0,1]$ such that C is ℓ-satisfiable (C is ℓ-subsumed by D, a is an ℓ-instance of C) w.r.t. $\mathcal{O}$.*

Recall that the semantics of the quantifiers require the computation of a supremum or infimum of the membership degrees of a possibly infinite set of elements of the domain. As is standard in the fuzzy DL community, we restrict reasoning to a special kind of models, called witnessed models [14]. For example, consider the axiom $\langle \top \sqsubseteq \exists r.\top,\ 1\rangle$. There are models where an individual has infinitely many r-successors with role degree smaller than 1, as long as the supremum of the role degrees is 1. Witnessed models prevent these situations and ensure that there actually exists an r-successor with degree 1.

Definition 8 (witnessed). *An interpretation $\mathcal{I}$ is called* witnessed *if for every $x \in \Delta^{\mathcal{I}}$, every role s and every concept C there are $y_1, y_2 \in \Delta^{\mathcal{I}}$ such that*

$$(\exists s.C)^{\mathcal{I}}(x) = s^{\mathcal{I}}(x, y_1) \otimes C^{\mathcal{I}}(y_1), \qquad (\forall s.C)^{\mathcal{I}}(x) = s^{\mathcal{I}}(x, y_2) \Rightarrow C^{\mathcal{I}}(y_2).$$

We will show that, if the t-norm $\otimes$ has no zero divisors, then consistency w.r.t. witnessed models in $\otimes$-$\mathcal{SHOI}$ is effectively the same problem as consistency in crisp $\mathcal{SHOI}$. Moreover, the precise values appearing in the axioms in the ontology are then irrelevant. The same is not true, however, for subsumption or instance checking. To obtain these results, we exploit some properties those t-norms.

3 Properties of T-Norms without Zero Divisors

By Lemma 3, continuous t-norms without zero divisors are exactly those that do not start with the Łukasiewicz t-norm. In particular, this includes the two other basic continuous t-norms, the Gödel and product t-norms.

Proposition 9. *For any t-norm $\otimes$ without zero divisors and every $x \in [0,1]$,*

1. *$x \Rightarrow y = 0$ iff $x > 0$ and $y = 0$, and*
2. *$\ominus x = 0$ iff $x > 0$.*

Proof. We prove the *if*-direction of the first claim. Assume $x > 0$ and $y = 0$. Then $x \Rightarrow y = x \Rightarrow 0 = \sup\{z \mid z \otimes x = 0\}$. Since $\otimes$ has no zero divisors, $z \otimes x > 0$ for all $z > 0$. Therefore $\{z \mid z \otimes x = 0\} = \{0\}$ and thus $x \Rightarrow y = 0$. The *only if*-direction holds for all t-norms [17]. The second statement follows from the first one since $\ominus x = x \Rightarrow 0$. □

The main result of this paper is based on the function $\mathbb{1}$ that maps fuzzy truth values to crisp truth values by defining, for all $x \in [0,1]$,

$$\mathbb{1}(x) = \begin{cases} 1 & \text{if } x > 0 \\ 0 & \text{if } x = 0. \end{cases}$$

For a t-norm without zero divisors it follows from Proposition 9 that $\mathbb{1}(x) = \ominus\ominus x$ for all $x \in [0,1]$. This function is compatible with negation, the t-norm, the corresponding t-conorm, implication and suprema. It is also compatible with minima, provided that they exist.

Lemma 10. *Let $\otimes$ be a t-norm without zero divisors. For all $x, y \in [0,1]$ and all non-empty sets $X \subseteq [0,1]$ it holds that*

1. $\mathbb{1}(\ominus x) = \ominus\mathbb{1}(x)$,
2. $\mathbb{1}(x \otimes y) = \mathbb{1}(x) \otimes \mathbb{1}(y)$,
3. $\mathbb{1}(x \oplus y) = \mathbb{1}(x) \oplus \mathbb{1}(y)$,
4. $\mathbb{1}(x \Rightarrow y) = \mathbb{1}(x) \Rightarrow \mathbb{1}(y)$,
5. $\mathbb{1}(\sup\{x \mid x \in X\}) = \sup\{\mathbb{1}(x) \mid x \in X\}$, *and*
6. *if* $\min\{x \mid x \in X\}$ *exists then* $\mathbb{1}(\min\{x \mid x \in X\}) = \min\{\mathbb{1}(x) \mid x \in X\}$.

Proof. It holds that $\mathbb{1}(\ominus x) = \ominus\ominus\ominus x = \ominus\mathbb{1}(x)$ which proves 1. Since $\otimes$ does not have zero divisors it holds that $x \otimes y = 0$ iff $x = 0$ or $y = 0$. This yields $\mathbb{1}(x \otimes y) = 0$ iff $\mathbb{1}(x) = 0$ or $\mathbb{1}(y) = 0$. Because there are no zero divisors, this shows that

$$\mathbb{1}(x \otimes y) = 0 \text{ iff } \mathbb{1}(x) \otimes \mathbb{1}(y) = 0. \tag{2}$$

Both $\mathbb{1}(x \otimes y)$ and $\mathbb{1}(x) \otimes \mathbb{1}(y)$ can only have the values 0 or 1. Hence, (2) is sufficient to prove the second statement. Following similar arguments we obtain from (1) that $\mathbb{1}(x \oplus y) = 0$ holds iff $\mathbb{1}(x) \oplus \mathbb{1}(y) = 0$. This suffices to prove 3. We use Proposition 9 to prove 4:

$$\mathbb{1}(x \Rightarrow y) = \begin{cases} 1 & \text{iff } x = 0 \text{ or } y > 0 \\ 0 & \text{iff } x > 0 \text{ and } y = 0 \end{cases} = \begin{cases} 1 & \text{iff } \mathbb{1}(x) = 0 \text{ or } \mathbb{1}(y) = 1 \\ 0 & \text{iff } \mathbb{1}(x) = 1 \text{ and } \mathbb{1}(y) = 0 \end{cases}$$
$$= \mathbb{1}(x) \Rightarrow \mathbb{1}(y).$$

To prove 5, observe that $\sup X = 0$ iff $X = \{0\}$, which yields

$$\mathbb{1}\big(\sup X\big) = 0 \Leftrightarrow \sup X = 0 \Leftrightarrow X = \{0\}$$
$$\Leftrightarrow \{\mathbb{1}(x) \mid x \in X\} = \{0\} \Leftrightarrow \sup\{\mathbb{1}(x) \mid x \in X\} = 0.$$

Assume now that $\min X = x_{\min}$ exists. Then we have

$$\mathbb{1}(\min X) = 0 \Leftrightarrow x_{\min} = 0 \Leftrightarrow 0 \in \{\mathbb{1}(x) \mid x \in X\} \Leftrightarrow \min\{\mathbb{1}(x) \mid x \in X\} = 0.$$

This shows that $\mathbb{1}(\min X) = 0$ iff $\min\{\mathbb{1}(x) \mid x \in X\} = 0$, which proves 6. □

Notice that in general $\mathbb{1}$ is not compatible with the infimum. Consider for example the set $X = \{\frac{1}{n} \mid n \in \mathbb{N}\}$. Then $\inf X = 0$ and hence $\mathbb{1}(\inf X) = 0$, but $\inf\{\mathbb{1}(\frac{1}{n}) \mid n \in \mathbb{N}\} = \inf\{1\} = 1$. This is the main reason why we consider witnessed models only. In fact, the construction provided in the next section does not work for general model reasoning.

4 The Crisp Model Property

The existing undecidability results for Fuzzy DLs all rely heavily on the fact that one can design ontologies that allow only models with infinitely many truth values. We shall see that for t-norms without zero divisors one cannot construct such an ontology in $\otimes$-$\mathcal{SHOI}$. It is even true that all consistent $\otimes$-$\mathcal{SHOI}$-ontologies have a crisp (and finite) model.

Definition 11. *A fuzzy DL $\mathcal{L}$ has the* crisp model property *if every consistent $\mathcal{L}$-ontology has a crisp model.*

For the rest of this paper we assume that $\otimes$ is a continuous t-norm that does not have zero divisors. These t-norms share the useful properties described in Section 3. In particular, Lemma 10 allows us to construct a crisp interpretation from a fuzzy interpretation by simply applying the function $\mathbb{1}$.

Let $\mathcal{I}$ be a witnessed fuzzy interpretation for the concept names $\mathsf{N_C}$ and role names $\mathsf{N_R}$. We construct the interpretation $\mathcal{J}$ over the domain $\Delta^{\mathcal{J}} := \Delta^{\mathcal{I}}$ by defining, for all concept names $A \in \mathsf{N_C}$, all role names $r \in \mathsf{N_R}$, and all $x, y \in \Delta^{\mathcal{I}}$,

$$A^{\mathcal{J}}(x) = \mathbb{1}\left(A^{\mathcal{I}}(x)\right) \text{ and } r^{\mathcal{J}}(x,y) = \mathbb{1}\left(r^{\mathcal{I}}(x,y)\right).$$

To show that $\mathcal{J}$ is a valid interpretation, we first verify the transitivity condition for all $r \in \mathsf{N_R^+}$ and all $x, y, z \in \Delta^{\mathcal{J}}$. From Lemma 10, we obtain

$$r^{\mathcal{J}}(x,y) \otimes r^{\mathcal{J}}(y,z) = \mathbb{1}\big(r^{\mathcal{I}}(x,y)\big) \otimes \mathbb{1}\big(r^{\mathcal{I}}(y,z)\big) = \mathbb{1}\left(r^{\mathcal{I}}(x,y) \otimes r^{\mathcal{I}}(y,z)\right).$$

Since $\mathcal{I}$ satisfies the transitivity condition and $\mathbb{1}$ is monotonic, we have

$$\mathbb{1}\left(r^{\mathcal{I}}(x,y) \otimes r^{\mathcal{I}}(y,z)\right) \le \mathbb{1}\big(r^{\mathcal{I}}(x,z)\big) = r^{\mathcal{J}}(x,z),$$

and thus $r^{\mathcal{J}}(x,y) \otimes r^{\mathcal{J}}(y,z) \le r^{\mathcal{J}}(x,z)$.

Lemma 12. *For all complex roles s and $x, y \in \Delta^{\mathcal{I}}$, $s^{\mathcal{J}}(x,y) = \mathbb{1}(s^{\mathcal{I}}(x,y))$.*

Proof. If s is a role name, this follows directly from the definition of $\mathcal{J}$. If $s = r^-$ for some $r \in \mathsf{N_R}$, then $s^{\mathcal{J}}(x,y) = r^{\mathcal{J}}(y,x) = \mathbb{1}(r^{\mathcal{I}}(y,x)) = \mathbb{1}(s^{\mathcal{I}}(x,y))$.

In a similar way, the interpretation $\mathcal{J}$ preserves the compatibility of $\mathbb{1}$ to the different constructors.

Lemma 13. *For all complex concepts C and $x \in \Delta^{\mathcal{I}}$, $C^{\mathcal{J}}(x) = \mathbb{1}\left(C^{\mathcal{I}}(x)\right)$.*

Proof. We use induction over the structure of C. The claim holds trivially for $C = \bot$ and $C = \top$. For $C = A \in \mathsf{N_C}$ it follows immediately from the definition of $\mathcal{J}$. It also holds for $C = \{a\}$, $a \in \mathsf{N_I}$, because $\{a\}^{\mathcal{I}}(x)$ can only take the values 0 or 1 for all $x \in \Delta^{\mathcal{I}}$.

Assume now that the concepts D and E satisfy $D^{\mathcal{J}}(x) = \mathbb{1}(D^{\mathcal{I}}(x))$ and $E^{\mathcal{J}}(x) = \mathbb{1}(E^{\mathcal{I}}(x))$ for all $x \in \Delta^{\mathcal{I}}$. In the case where $C = D \sqcap E$, Lemma 10 yields that for all $x \in \Delta^{\mathcal{I}}$

$$\begin{aligned} C^{\mathcal{J}}(x) = D^{\mathcal{J}}(x) \otimes E^{\mathcal{J}}(x) &= \mathbb{1}\big(D^{\mathcal{I}}(x)\big) \otimes \mathbb{1}\big(E^{\mathcal{I}}(x)\big) \\ &= \mathbb{1}\big(D^{\mathcal{I}}(x) \otimes E^{\mathcal{I}}(x)\big) = \mathbb{1}\big(C^{\mathcal{I}}(x)\big). \end{aligned}$$

Likewise, the compatibility of $\mathbb{1}$ with the t-conorm, the residuum, and the negation entails the result for the cases $C = D \sqcup E$, $C = D \to E$, and $C = \neg D$.

For $C = \exists s.D$, where s is a complex role and D is a concept description satisfying $D^{\mathcal{J}}(x) = \mathbb{1}(D^{\mathcal{I}}(x))$ for all $x \in \Delta^{\mathcal{I}}$, we obtain

$$\mathbb{1}\big(C^{\mathcal{I}}(x)\big) = \mathbb{1}\big((\exists s.D)^{\mathcal{I}}(x)\big) = \mathbb{1}\big(\sup_{y\in\Delta^{\mathcal{I}}} \{s^{\mathcal{I}}(x,y) \otimes D^{\mathcal{I}}(y)\}\big)$$
$$= \sup_{y\in\Delta^{\mathcal{I}}} \big\{\mathbb{1}\big(s^{\mathcal{I}}(x,y)\big) \otimes \mathbb{1}\big(D^{\mathcal{I}}(y)\big)\big\} \tag{3}$$

because $\mathbb{1}$ is compatible with the supremum and the t-norm. Lemma 12 yields

$$\sup_{y\in\Delta^{\mathcal{I}}} \big\{\mathbb{1}(r^{\mathcal{I}}(x,y)) \otimes \mathbb{1}(D^{\mathcal{I}}(y))\big\} = \sup_{y\in\Delta^{\mathcal{I}}} \big\{r^{\mathcal{J}}(x,y) \otimes D^{\mathcal{J}}(y)\big\} = (\exists r.D)^{\mathcal{J}}(x). \tag{4}$$

Equations (3) and (4) prove $\mathbb{1}(C^{\mathcal{I}}(x)) = C^{\mathcal{J}}(x)$ for the case where $C = \exists r.D$. If $C = \forall r.D$, we have

$$\mathbb{1}\big(C^{\mathcal{I}}(x)\big) = \mathbb{1}\big(\inf_{y\in\Delta^{\mathcal{I}}} \{r^{\mathcal{I}}(x,y) \Rightarrow D^{\mathcal{I}}(y)\}\big). \tag{5}$$

Since $\mathcal{I}$ is witnessed, there must exist some $y_0 \in \Delta^{\mathcal{I}}$ such that

$$r^{\mathcal{I}}(x,y_0) \Rightarrow D^{\mathcal{I}}(y_0) = \inf_{y\in\Delta^{\mathcal{I}}} \big\{r^{\mathcal{I}}(x,y) \Rightarrow D^{\mathcal{I}}(y)\big\};$$

that is, $\min_{y\in\Delta^{\mathcal{I}}} \big\{r^{\mathcal{I}}(x,y) \Rightarrow D^{\mathcal{I}}(y)\big\}$ exists. Thus, Part 6. of Lemma 10 is applicable and $\mathbb{1}(C^{\mathcal{I}}(x)) = C^{\mathcal{J}}(x)$ follows in analogy to the case for existential restrictions. □

With the help of this lemma we can show that the crisp interpretation $\mathcal{J}$ satisfies all the axioms that are satisfied by $\mathcal{I}$.

Lemma 14. *Let $\mathcal{O} = (\mathcal{A}, \mathcal{T}, \mathcal{R})$ be a $\otimes$-$\mathcal{SHOI}$-ontology. If $\mathcal{I}$ is a witnessed model of $\mathcal{O}$, then $\mathcal{J}$ is also a witnessed model of $\mathcal{O}$.*

Proof. We prove that $\mathcal{J}$ satisfies all assertions, GCIs, and role inclusions from $\mathcal{O}$. Let $\langle a{:}C,\ \ell\rangle$, $\ell \in (0,1]$, be a concept assertion from $\mathcal{A}$. Since the assertion is satisfied by $\mathcal{I}$, $C^{\mathcal{I}}(a^{\mathcal{I}}) \geq \ell > 0$ holds. Lemma 13 yields $C^{\mathcal{J}}(a^{\mathcal{J}}) = 1 \geq \ell$. The same argument can be used for role assertions.

Let now $\langle C \sqsubseteq D,\ \ell\rangle$ be a GCI from $\mathcal{T}$. Let x be an element $x \in \Delta^{\mathcal{I}}$. As the GCI is satisfied by $\mathcal{I}$, we get $C^{\mathcal{I}}(x) \Rightarrow D^{\mathcal{I}}(x) \geq \ell > 0$. By Lemmata 10 and 13, we obtain

$$C^{\mathcal{J}}(x) \Rightarrow D^{\mathcal{J}}(x) = \mathbb{1}(C^{\mathcal{I}}(x)) \Rightarrow \mathbb{1}(D^{\mathcal{I}}(x)) = \mathbb{1}(C^{\mathcal{I}}(x) \Rightarrow D^{\mathcal{I}}(x)) = 1 \geq \ell,$$

and thus $\mathcal{J}$ satisfies the GCI $\langle C \sqsubseteq D,\ \ell\rangle$. A similar argument, using Lemma 12 instead of Lemma 13, shows that $\mathcal{J}$ satisfies all role inclusions in $\mathcal{R}$. □

The previous results show that by applying $\mathbb{1}$ to the truth degrees we obtain a crisp model $\mathcal{J}$ from any fuzzy model $\mathcal{I}$ of a $\otimes$-$\mathcal{SHOI}$-ontology $\mathcal{O}$.

Theorem 15. *$\otimes$-$\mathcal{SHOI}$ has the crisp model property if $\otimes$ has no zero divisors.*

In the next section we will use this result to show that ontology consistency and concept satisfiability can be decided in exponential time.

5 Consistency and Satisfiability

For a given $\otimes$-$\mathcal{SHOI}$-ontology $\mathcal{O}$, we define crisp($\mathcal{O}$) to be the crisp $\mathcal{SHOI}$-ontology that is obtained from $\mathcal{O}$ by replacing all the truth values appearing in the axioms by 1. For example, for the ontology

$$\mathcal{O} = \{\langle a{:}C,\ 0.2\rangle, \langle (a,b){:}r,\ 0.8\rangle, \langle C \sqsubseteq D,\ 0.5\rangle, \langle r \sqsubseteq s,\ 0.1\rangle\}$$

we obtain

$$\text{crisp}(\mathcal{O}) = \{\langle a{:}C,\ 1\rangle, \langle (a,b){:}r,\ 1\rangle, \langle C \sqsubseteq D,\ 1\rangle, \langle r \sqsubseteq s,\ 1\rangle\}.$$

Lemma 16. *Let $\mathcal{O}$ be a $\otimes$-$\mathcal{SHOI}$-ontology and $\mathcal{I}$ be a crisp interpretation. Then $\mathcal{I}$ is a model of $\mathcal{O}$ iff it is a model of* crisp($\mathcal{O}$).

Proof. Assume that crisp($\mathcal{O}$) has a model $\mathcal{I}$. Let $\langle C \sqsubseteq D,\ \ell\rangle$, $\ell > 0$, be an axiom from $\mathcal{O}$. Since $\mathcal{I}$ is a model of crisp($\mathcal{O}$), it must satisfy $\langle C \sqsubseteq D,\ 1\rangle$; that is, $C^{\mathcal{I}}(x) \Rightarrow D^{\mathcal{I}}(x) \geq 1 \geq \ell$ holds for all $x \in \Delta^{\mathcal{I}}$. Thus $\mathcal{I}$ satisfies $\langle C \sqsubseteq D,\ \ell\rangle$. The proof that $\mathcal{I}$ satisfies assertions and role inclusions is analogous. Hence $\mathcal{I}$ is also a model of $\mathcal{O}$.

For the other direction, assume that $\mathcal{I}$ satisfies $\langle C \sqsubseteq D,\ \ell\rangle$. As $\mathcal{I}$ is a crisp interpretation it holds that $C^{\mathcal{I}}(x) \Rightarrow D^{\mathcal{I}}(x) \in \{0,1\}$ for all $x \in \Delta^{\mathcal{I}}$. Together with $C^{\mathcal{I}}(x) \Rightarrow D^{\mathcal{I}}(x) \geq \ell > 0$ we obtain $C^{\mathcal{I}}(x) \Rightarrow D^{\mathcal{I}}(x) = 1$. Thus, $\mathcal{I}$ satisfies the GCI $\langle C \sqsubseteq D,\ 1\rangle$. The same argument can be used for role inclusions and assertions. Thus, $\mathcal{I}$ is also a model of crisp($\mathcal{O}$). □

In particular, a $\otimes$-$\mathcal{SHOI}$-ontology $\mathcal{O}$ has a crisp model iff crisp($\mathcal{O}$) has a crisp model. Together with Theorem 15, this shows that a $\otimes$-$\mathcal{SHOI}$-ontology $\mathcal{O}$ is consistent iff crisp($\mathcal{O}$) has a crisp model. Therefore, one can use reasoning in crisp $\mathcal{SHOI}$ to decide consistency of $\otimes$-$\mathcal{SHOI}$-ontologies. Reasoning in crisp $\mathcal{SHOI}$ is known to be ExpTime-complete [15].

Corollary 17. *Deciding consistency in $\otimes$-$\mathcal{SHOI}$ is* ExpTime-*complete.*

Similar arguments show that satisfiability is decidable in $\otimes$-$\mathcal{SHOI}$. Since any concept is 0-satisfiable, we can assume in the following that the concept C is ℓ-satisfiable w.r.t. an ontology $\mathcal{O}$ with $\ell > 0$. Then there is a model $\mathcal{I}$ of $\mathcal{O}$ satisfying $C^{\mathcal{I}}(x) \geq \ell > 0$. Thus, the model $\mathcal{J}$ of $\mathcal{O}$ constructed in Section 4 also satisfies $C^{\mathcal{J}}(x) = 1 \geq \ell$. This shows that if C is ℓ-satisfiable w.r.t. $\mathcal{O}$ for some $\ell > 0$, it is also 1-satisfiable w.r.t. $\mathcal{O}$, and in particular 1-satisfiable w.r.t. crisp($\mathcal{O}$). Clearly, the implication in the other direction also holds.

Lemma 18. *Deciding ℓ-satisfiability in $\otimes$-$\mathcal{SHOI}$ is* ExpTime-*complete. Furthermore, the best satisfiability degree of a concept C w.r.t. $\mathcal{O}$ is either* 0 *or* 1 *and can be computed in exponential time.*

Lemma 16 and Corollary 15 still hold when we restrict the semantics to the slightly less expressive logics $\otimes$-$\mathcal{SHO}$, which does not allow for inverse roles, or $\otimes$-$\mathcal{SI}$ which does not allow for nominals and role hierarchies. The crisp DLs $\mathcal{SHO}$ and $\mathcal{SI}$ are known to have the finite model property [16,19], and $\otimes$-$\mathcal{SI}$ and $\otimes$-$\mathcal{SHO}$ inherit the finite model property from their crisp ancestors.

Theorem 19. *The logics* $\otimes$-$\mathcal{SHO}$ *and* $\otimes$-$\mathcal{SI}$ *and their sublogics have the finite model property.*

This theorem contradicts a recent result stating that the sublogic Π-$\mathcal{ALC}$ of $\otimes$-$\mathcal{SHOI}$, where $\otimes$ is the product t-norm, does not have the finite model property [5, Theorem 3.8]. As a matter of fact, the proof from [5] is based on the erroneous claim that every model $\mathcal{I}$ of the assertion $\langle a{:}A,\ 0.5\rangle$ must be such that $A^{\mathcal{I}}(a^{\mathcal{I}}) = 0.5$. The case of an interpretation with $A^{\mathcal{I}}(a^{\mathcal{I}}) = 1$, which also satisfies this assertion, is not considered in the induction argument.

6 Subsumption and Instance Checking

We now show that, despite the crisp model property, ℓ-subsumption of concepts w.r.t. $\otimes$-$\mathcal{SHOI}$-ontologies cannot be decided using crisp reasoning. Moreover, this holds even if the ontology is restricted to be crisp itself.

Consider first the ontology $\mathcal{O}_1$ containing only the GCI $\langle \top \sqsubseteq A,\ \ell\rangle$ for some $\ell \in (0,1)$. Since $\ell > 0$, for every crisp model $\mathcal{I}$ of $\mathcal{O}_1$ and $x \in \Delta^{\mathcal{I}}$, $A^{\mathcal{I}}(x) = 1$ holds. Thus, $\top$ is 1-subsumed by A w.r.t. $\mathcal{O}_1$ when reasoning is restricted to crisp models. However, the interpretation $\mathcal{I}_1 = (\{x\}, \cdot^{\mathcal{I}}_1)$, where $A^{\mathcal{I}_1}(x) = \ell$, is also a model of $\mathcal{O}_1$, but violates the axiom $\langle \top \sqsubseteq A,\ 1\rangle$. In fact, the best subsumption degree of $\top$ and A w.r.t. $\mathcal{O}_1$ is ℓ, which is smaller than 1. Notice that this example only assumes that the logic can express concept names, the top concept, and fuzzy GCIs. Moreover, it is irrelevant which t-norm $\otimes$ was chosen for the semantics.

Proposition 20. *For every fuzzy DL* $\otimes$-$\mathcal{L}$ *that allows the top constructor and fuzzy GCIs,* ℓ*-subsumption cannot be decided over crisp models only.*

If the logic uses a t-norm $\otimes$ without zero divisors and is able to express the residual negation, then this proposition holds even if the ontology is crisp. Take for instance the ontology $\mathcal{O}_2$ containing the axiom $\langle \top \sqsubseteq \neg\neg A,\ 1\rangle$. As before, it is easy to see that every crisp model of $\mathcal{O}_2$ also satisfies $\langle \top \sqsubseteq A,\ 1\rangle$. On the other hand, the best subsumption degree of $\top$ and A w.r.t. $\mathcal{O}_2$ is 0.

To show this, we construct a model $\mathcal{I}_2$ of $\mathcal{O}_2$ that violates $\langle \top \sqsubseteq A,\ \ell\rangle$ for every $\ell > 0$. The interpretation $\mathcal{I}_2 = (\mathbb{N}, \cdot^{\mathcal{I}_2})$ is given by $A^{\mathcal{I}_2}(i) = 1/i$ for every $i \geq 1$. $\mathcal{I}_2$ is indeed a model of $\mathcal{O}_2$ since $A^{\mathcal{I}_2}(i) > 0$ and hence $(\neg\neg A)^{\mathcal{I}_2}(i) = 1$ for every $i \geq 1$. However, for every $\ell > 0$ there is an $i \in \mathbb{N}$ such that $0 < 1/i < \ell$ and hence $\mathcal{I}_2$ violates the axiom $\langle \top \sqsubseteq A,\ \ell\rangle$. Thus, the best subsumption degree of $\top$ and A w.r.t. $\mathcal{O}_2$ is 0.

Proposition 21. *Let* $\otimes$ *be a t-norm without zero divisors and* $\otimes$-$\mathcal{L}$ *be a fuzzy DL with residual negation. Then* ℓ*-subsumption cannot be decided over crisp models only. This holds even for* ℓ*-subsumption w.r.t. crisp ontologies.*

In the special case where $\otimes$ is the product t-norm, the problem is more pronounced, since reasoning cannot be restricted to *finite* models either, as we show next. Consider the ontology

$$\mathcal{O} = \{\langle \top \sqsubseteq \neg\neg A,\ 1\rangle,\quad \langle \top \sqsubseteq \exists r.\top,\ 1\rangle,\quad \langle \exists r.A \sqsubseteq A \sqcap A,\ 1\rangle\}.$$

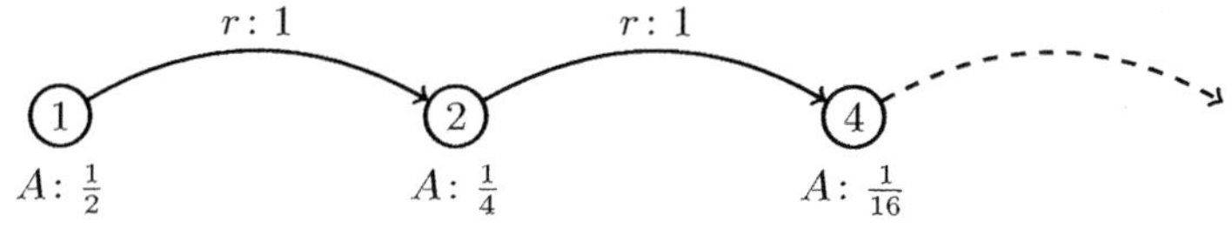

Fig. 1. A model where $\langle \top \sqsubseteq A,\ \ell\rangle$ does not hold for any $\ell > 0$.

We show that every finite model of $\mathcal{O}$ also satisfies the GCI $\langle \top \sqsubseteq A,\ 1\rangle$, but the best subsumption degree of $\top$ and A w.r.t. $\mathcal{O}$ is 0.

Let first $\mathcal{I}$ be a model of $\mathcal{O}$ that violates $\langle \top \sqsubseteq A,\ 1\rangle$. We show that $\mathcal{I}$ must be infinite. To do this, we show by induction that for every $n \geq 1$ there exist $x_1, \ldots, x_n \in \Delta^{\mathcal{I}}$ such that $1 > A^{\mathcal{I}}(x_1) > \ldots > A^{\mathcal{I}}(x_n) > 0$; since $A^{\mathcal{I}}(x_i) \neq A^{\mathcal{I}}(x_j)$ for every $i \neq j$, this implies that $\Delta^{\mathcal{I}}$ must contain infinitely many individuals.

For the induction base, since $\mathcal{I}$ violates $\langle \top \sqsubseteq A,\ 1\rangle$, there must be an $x \in \Delta^{\mathcal{I}}$ such that $A^{\mathcal{I}}(x) < 1$. As $\mathcal{I}$ satisfies the first axiom of $\mathcal{O}$, it also follows that $A^{\mathcal{I}}(x) > 0$. Thus, if we set $x_1 = x$, then the claim holds for $n = 1$. Suppose now that it holds for $n \geq 1$, we show that it also holds for $n + 1$. Since $\mathcal{I}$ is a witnessed model of $\mathcal{O}$, the second axiom implies that there must exist a $y \in \Delta^{\mathcal{I}}$ such that $r^{\mathcal{I}}(x_n, y) = r^{\mathcal{I}}(x_n, y) \otimes \top(y) = 1$. The third axiom then implies that

$$\begin{aligned} A^{\mathcal{I}}(x_n) > \left(A^{\mathcal{I}}(x_n)\right)^2 &\geq (\exists r.A)^{\mathcal{I}}(x_n) \\ &\geq r^{\mathcal{I}}(x_n, y) \otimes A^{\mathcal{I}}(y) = A^{\mathcal{I}}(y). \end{aligned}$$

Since $\mathcal{I}$ satisfies the first axiom, it additionally holds that $A^{\mathcal{I}}(y) > 0$. Thus, setting $x_{n+1} = y$ yields the result.

It remains only to show that the best subsumption degree of $\top$ and A w.r.t. $\mathcal{O}$ is 0. We build a model $\mathcal{I}_0$ of $\mathcal{O}$ that violates $\langle \top \sqsubseteq A,\ \ell\rangle$ for every $\ell > 0$. Let $\mathcal{I}_0 = (\{2^i \mid i \geq 0\}, \cdot^{\mathcal{I}_0})$ be given by $A^{\mathcal{I}_0}(x) = 2^{-x}$, and

$$r^{\mathcal{I}_0}(x, y) = \begin{cases} 1 & y = 2x \\ 0 & y \neq 2x \end{cases}$$

for all $x, y \in \Delta^{\mathcal{I}_0}$ (cf. Figure 1).

We verify that $\mathcal{I}_0$ is a model of $\mathcal{O}$. First, since $2^{-i} > 0$ for every $i \geq 0$, it follows that $A^{\mathcal{I}_0}(x) > 0$ for all $x \in \Delta^{\mathcal{I}}$. Thus, $\mathcal{I}_0$ satisfies the first axiom of $\mathcal{O}$. For every $x \in \Delta^{\mathcal{I}}$ it also holds that

$$\begin{aligned} (\exists r.\top)^{\mathcal{I}_0}(x) &= r^{\mathcal{I}_0}(x, 2x) = 1 \qquad \text{and} \\ (\exists r.A)^{\mathcal{I}_0}(x) &= r^{\mathcal{I}_0}(x, 2x) \otimes A^{\mathcal{I}_0}(2x) \\ &= 2^{-2x} = 2^{-x} \cdot 2^{-x} = A^{\mathcal{I}_0}(x) \otimes A^{\mathcal{I}_0}(x), \end{aligned}$$

satisfying the remaining two axioms of the ontology. The fact that this model is witnessed is a trivial consequence of the fact that every individual of the domain has exactly one r-successor with degree different from 0.

This all means that $\top$ is *not* ℓ-subsumed by A w.r.t. $\mathcal{O}$ for any $\ell > 0$, but $\top$ is subsumed by A with degree 1 in every finite model of $\mathcal{O}$. Notice that all the axioms in $\mathcal{O}$ are crisp. We thus have the following result.

Proposition 22. *Let $\otimes$ be the product t-norm and $\otimes$-$\mathcal{L}$ be a fuzzy DL with conjunction, existential restriction, and residual negation. Then ℓ-subsumption cannot be decided over finite models only. This holds even for ℓ-subsumption w.r.t. crisp ontologies.*

Notice that the three ontologies $\mathcal{O}_1, \mathcal{O}_2$, and $\mathcal{O}$ presented in this section contain only GCIs. In this case it follows that a concept C is ℓ-subsumed by D iff any individual a is an ℓ-instance of the concept $C \to D$. Likewise, the best subsumption degree of C and D is equivalent to the best instance degree of a and $C \to D$. Thus, if the fuzzy DL allows for the constructor $\to$, then Propositions 20, 21, and 22 also hold for ℓ-instance checking, i.e. ℓ-instances cannot be checked by a reduction to crisp reasoning. This is true even if the ontology is crisp. Moreover, under product t-norm semantics, finite models are insufficient for instance checking w.r.t. crisp ontologies.

7 Conclusions

We have shown that for every t-norm $\otimes$ that does not have zero divisors, consistency of $\otimes$-$\mathcal{SHOI}$ ontologies is ExpTime-complete. Indeed, to decide this problem it suffices to test consistency of the crisp version of the ontology. For all other t-norms—those having zero divisors—it was previously shown that consistency becomes undecidable already for a fairly inexpressive DL, allowing only for conjunction, existential restrictions and residual negation.

It is worth pointing out that the correctness of our reduction to crisp reasoning strongly depends on the fact that $\otimes$-$\mathcal{SHOI}$ ontologies, as presented in this paper, cannot express upper bounds for the membership degrees. If one extends this logic to allow for these upper bounds, either by the introduction of the involutive negation $1 - x$ or by axioms of the form $\langle \alpha \leq \ell \rangle$, then ontology consistency becomes undecidable for every t-norm except the Gödel t-norm.

In crisp DLs, ontology consistency is the "main" decision problem in the sense that all other standard problems—like concept satisfiability, subsumption and instance checking—are polynomially reducible to it. In crisp DLs, a is a (1-)instance of C w.r.t. an ontology $\mathcal{O}$ iff the ontology obtained by adding the assertion $\langle a{:}\neg C,\ 1 \rangle$ to $\mathcal{O}$ is inconsistent. However, for any t-norm without zero divisors, this last axiom only states that $a^{\mathcal{I}}(C) = 0$ must hold in every model, which is much stronger than the required condition $a^{\mathcal{I}}(C) < 1$. Indeed, despite $\otimes$-$\mathcal{SHOI}$ having the crisp model property, crisp reasoning is insufficient for deciding subsumption and instance checking. Moreover, under the product t-norm semantics, finite models cannot decide these problems, even for those sublogics of $\otimes$-$\mathcal{SHOI}$ that have the finite model property.

These results leave open the decidability status of subsumption and instance checking in fuzzy DLs. This is one of the main problems we intend to examine in future work. In this respect it is worth to point out that, so far, all the existing decision procedures for fuzzy DLs depend on crisp- or finite-model reasoning. This suggests that if, e.g. subsumption turns out to be decidable in these logics, a different kind of decision procedure would have to be developed.

References

1. Baader, F., Calvanese, D., McGuinness, D., Nardi, D., Patel-Schneider, P.F.: The Description Logic Handbook: Theory, Implementation, and Applications. Cambridge University Press (2003)
2. Baader, F., Peñaloza, R.: Are fuzzy description logics with general concept inclusion axioms decidable? In: Proc. of the 2011 IEEE Int. Conf. on Fuzzy Systems (FUZZ-IEEE 2011), pp. 1735–1742. IEEE Press (2011)
3. Baader, F., Peñaloza, R.: GCIs make reasoning in fuzzy DLs with the product t-norm undecidable. In: Rosati, R., Rudolph, S., Zakharyaschev, M. (eds.) Proc. of the 24th Int. Workshop on Description Logics (DL 2011), Barcelona, Spain. CEUR Workshop Proceedings, vol. 745 (2011)
4. Baader, F., Peñaloza, R.: On the Undecidability of Fuzzy Description Logics with GCIs and Product T-norm. In: Tinelli, C., Sofronie-Stokkermans, V. (eds.) FroCos 2011. LNCS (LNAI), vol. 6989, pp. 55–70. Springer, Heidelberg (2011)
5. Bobillo, F., Bou, F., Straccia, U.: On the failure of the finite model property in some fuzzy description logics. Fuzzy Sets and Systems 172(23), 1–12 (2011)
6. Bobillo, F., Delgado, M., Gómez-Romero, J., Straccia, U.: Fuzzy description logics under Gödel semantics. International Journal of Approximate Reasoning 50(3), 494–514 (2009)
7. Bobillo, F., Straccia, U.: A fuzzy description logic with product t-norm. In: Proc. of the 2007 IEEE Int. Conf. on Fuzzy Systems FUZZ-IEEE 2007, pp. 1–6. IEEE Press (2007)
8. Bobillo, F., Straccia, U.: On qualified cardinality restrictions in fuzzy description logics under Łukasiewicz semantics. In: Proc. of the 12th Int. Conf. on Information Processing and Management of Uncertainty in Knowledge-Based Systems (IPMU 2008), pp. 1008–1015 (2008)
9. Bobillo, F., Straccia, U.: Fuzzy description logics with general t-norms and datatypes. Fuzzy Sets and Systems 160(23), 3382–3402 (2009)
10. Borgwardt, S., Peñaloza, R.: Undecidability of fuzzy description logics. In: Proc. of the 13th Int. Conf. on Principles of Knowledge Representation and Reasoning (KR 2012), Rome, Italy. AAAI Press (to appear, 2012)
11. Cerami, M., Straccia, U.: On the undecidability of fuzzy description logics with GCIs with Łukasiewicz t-norm. Technical report, Computing Research Repository (2011), arXiv:1107.4212v3 [cs.LO]
12. García-Cerdaña, Á., Armengol, E., Esteva, F.: Fuzzy description logics and t-norm based fuzzy logics. International Journal of Approximate Reasoning 51, 632–655 (2010)
13. Hájek, P.: Metamathematics of Fuzzy Logic (Trends in Logic). Springer (2001)
14. Hájek, P.: Making fuzzy description logic more general. Fuzzy Sets and Systems 154(1), 1–15 (2005)

15. Hladik, J.: A tableau system for the description logic $\mathcal{SHIO}$. In: Proceedings of the Doctoral Programme of IJCAR 2004. CEUR Worksop Proceedings, vol. 106, pp. 21–25 (2004)
16. Horrocks, I., Sattler, U., Tobies, S.: A PSpace-algorithm for deciding $\mathcal{ALCNI}_{R+}$-satisfiability. LTCS-Report 98-08, RWTH Aachen, Germany (1998)
17. Klement, E.P., Mesiar, R., Pap, E.: Triangular Norms. Springer (2000)
18. Lukasiewicz, T., Straccia, U.: Managing uncertainty and vagueness in description logics for the semantic web. Journal of Web Semantics 6(4), 291–308 (2008)
19. Lutz, C., Areces, C., Horrocks, I., Sattler, U.: Keys, nominals, and concrete domains. Journal of Artificial Intelligence Research 23, 667–726 (2004)
20. Molitor, R., Tresp, C.B.: Extending Description Logics to Vague Knowledge in Medicine. In: Szczepaniak, P., Lisboa, P.J.G., Tsumoto, S. (eds.) Fuzzy Systems in Medicine. STUDFUZZ, vol. 41, pp. 617–635. Springer (2000)
21. Mostert, P.S., Shields, A.L.: On the structure of semigroups on a compact manifold with boundary. Annals of Mathematics 65, 117–143 (1957)
22. Stoilos, G., Stamou, G.B.: A framework for reasoning with expressive continuous fuzzy description logics. In: Grau, B.C., Horrocks, I., Motik, B., Sattler, U. (eds.) Proc. of the 22nd Int. Workshop on Description Logics (DL 2009). CEUR Workshop Proceedings, vol. 477 (2009)
23. Stoilos, G., Stamou, G.B., Tzouvaras, V., Pan, J.Z., Horrocks, I.: The fuzzy description logic f-$\mathcal{SHIN}$. In: Proc. of the 1st Int. Workshop on Uncertainty Reasoning for the Semantic Web (URSW 2005), pp. 67–76 (2005)
24. Stoilos, G., Straccia, U., Stamou, G.B., Pan, J.Z.: General concept inclusions in fuzzy description logics. In: Proc. of the 17th Eur. Conf. on Artificial Intelligence (ECAI 2006). Frontiers in Artificial Intelligence and Applications, vol. 141, pp. 457–461. IOS Press (2006)
25. Straccia, U.: Reasoning within fuzzy description logics. Journal of Artificial Intelligence Research 14, 137–166 (2001)
26. Straccia, U., Bobillo, F.: Mixed integer programming, general concept inclusions and fuzzy description logics. In: Proc. of the 5th EUSFLAT Conf (EUSFLAT 2007), pp. 213–220. Universitas Ostraviensis (2007)
27. Tresp, C.B., Molitor, R.: A description logic for vague knowledge. In: Proc. of the 13th Eur. Conf. on Artificial Intelligence (ECAI 1998), Brighton, UK, pp. 361–365. J. Wiley and Sons (1998)

Truthful Monadic Abstractions*

Taus Brock-Nannestad and Carsten Schürmann

IT University of Copenhagen
{tbro,carsten}@itu.dk

Abstract. In intuitionistic sequent calculi, detecting that a sequent is unprovable is often used to direct proof search. This is for instance seen in backward chaining, where an unprovable subgoal means that the proof search must backtrack. In undecidable logics, however, proof search may continue indefinitely, finding neither a proof nor a disproof of a given subgoal.

In this paper we characterize a family of truth-preserving abstractions from intuitionistic first-order logic to the monadic fragment of classical first-order logic. Because they are truthful, these abstractions can be used to disprove sequents in intuitionistic first-order logic.

1 Introduction

Two common methods of proof search in intuitionistic sequent calculi are forward and backward chaining. In forward chaining, the hypotheses are used to derive new facts. The proof succeeds if the goal is derived, and fails if this process saturates (i.e. no new facts can be derived) without deriving the goal.

Backward chaining proceeds bottom-up from the goal, continually splitting subgoals into other subgoals or closing subgoals if these are an immediate consequence of what is in the context. Backtracking is employed if a subgoal can be neither split nor immediately satisfied. In the case of backward chaining, the proof search succeeds if all subgoals are satisfied, and it fails if backtracking returns to the original goal without finding a proof.

In undecidable logics such as intuitionistic first-order logic, both of these approaches may fail to terminate. Forward chaining may fail to saturate, continually adding new facts to the context, and backward chaining may have an unbounded number of points at which backtracking may occur. Consider, for instance, an attempt to prove that 5 is an even number. Let the predicate $D(x, y)$ — signifying that y is twice the value of x — be given by the following context

$$\Gamma = D(z, z), \forall x.\forall y.D(x, y) \supset D(s(x), s(s(y)))$$

The statement that 5 is even is equivalent to the following sequent

$$\Gamma \Longrightarrow \exists x.D(x, s(s(s(s(s(z))))))$$

* This work supported in part by DemTech grant 10-092309 from the Danish Council for Strategic Research, Programme Commission on Strategic Growth Technologies.

B. Gramlich, D. Miller, and U. Sattler (Eds.): IJCAR 2012, LNAI 7364, pp. 97–110, 2012.

Clearly, attempting to prove the above sequent cannot succeed, but a naive forward chaining proof attempt may fail to terminate, as any number of distinct hypotheses $D(s(z), s(s(z))), D(s(s(z)), s(s(s(s(z))))), \ldots$ may be generated by the left implication rule. Thus, the proof search never saturates, and we cannot conclude that the above sequent is not derivable.

Of course, the above sequent fails to be derivable even in classical logic, hence one might try to feed it to a classical first-order theorem prover, such as Spass [13] or Vampire [11], or a countermodel generator, such as for example Paradox [2] hoping to generate a counterexample. This technique will work for the above simple example, but in general there is no guarantee that the tools will terminate either, as we have reduced one undecidable problem to another.

If we insist on having decidability, we must let go of either *soundness*, in which case whatever proofs we find are not necessarily correct, or *completeness*, in which case we may not be able to find a proof even if a proof exists. Thus, if we primarily care about disproving sequents, having an unsound but complete and decidable procedure is sufficient.

In this paper, we define a family of *truthful* (or TI [3]) abstractions i.e. functions that preserve provability. Thus, each abstraction is a function α that satisfies the truthfulness condition:

$$\Gamma \Longrightarrow A \quad \text{entails} \quad \alpha(\Gamma) \longrightarrow \alpha(A)$$

where the sequent on the left is in the *base* logic, and the sequent on the right is in the *abstraction* logic. By choosing a logic in which derivability is decidable, we get the aforementioned decision procedure.

In our case, our abstractions map into the monadic fragment of classical first-order logic which has the finite model property, and is thus decidable [4]. In the monadic fragment, only unary predicates and functions may appear, but all other connectives are allowed. Thus, the defining feature of the abstraction is how it acts on atoms. This mapping can be intuitively understood as mapping arbitrary terms — which may be seen as being tree-shaped — onto paths through these trees. Thus, the following formula

$$P(f(g(a), b), c) \supset P(f(c, b), g(a))$$

will have the following abstractions:

$$P_1(f_1(g_1(a))) \supset P_1(f_1(c))$$

$$P_1(f_2(b)) \supset P_1(f_2(b))$$

$$P_2(c) \supset P_2(g_1(a))$$

Here, the subscripts indicate which argument became part of the path.

This paper is organized as follows. We reiterate the definition of intuitionistic first-order logic, our choice of base logic, in Section 2. In Section 3 we define the

$$\frac{P \in \Gamma}{\Gamma \Longrightarrow P}\,\text{init} \qquad \frac{}{\Gamma \Longrightarrow \top}\,\top\text{R} \qquad \frac{}{\Gamma, \bot \Longrightarrow C}\,\bot\text{L}$$

$$\frac{\Gamma, A \wedge B, A, B \Longrightarrow C}{\Gamma, A \wedge B \Longrightarrow C}\,\wedge\text{L} \qquad \frac{\Gamma \Longrightarrow A \quad \Gamma \Longrightarrow B}{\Gamma \Longrightarrow A \wedge B}\,\wedge\text{R}$$

$$\frac{\Gamma \Longrightarrow A_i}{\Gamma \Longrightarrow A_1 \vee A_2}\,\vee\text{R}_i \qquad \frac{\Gamma, A \vee B, A \Longrightarrow C \quad \Gamma, A \vee B, B \Longrightarrow C}{\Gamma, A \vee B \Longrightarrow C}\,\vee\text{L}$$

$$\frac{\Gamma, A \supset B \Longrightarrow A \quad \Gamma, A \supset B, B \Longrightarrow C}{\Gamma, A \supset B \Longrightarrow C}\,\supset\text{L} \qquad \frac{\Gamma, A \Longrightarrow B}{\Gamma \Longrightarrow A \supset B}\,\supset\text{R}$$

$$\frac{\Gamma, \exists x.A, [a/x]A \Longrightarrow C}{\Gamma, \exists x.A \Longrightarrow C}\,\exists\text{L}^a \qquad \frac{\Gamma \Longrightarrow [t/x]A}{\Gamma \Longrightarrow \exists x.A}\,\exists\text{R}$$

$$\frac{\Gamma, \forall x.A, [t/x]A \Longrightarrow C}{\Gamma, \forall x.A \Longrightarrow C}\,\forall\text{L} \qquad \frac{\Gamma \Longrightarrow [a/x]A}{\Gamma \Longrightarrow \forall x.A}\,\forall\text{R}^a$$

Fig. 1. Base Logic: First-order intuitionistic logic

abstraction formally. Section 4 describes a few examples, illustrating the applicability of the decision procedure. In Section 5 we sketch briefly an implementation of truthful abstractions and report on our experimental results on the intuitionistic fragment ILTP[10] of TPTP[12], using Spass as a decision procedure for the monadic fragment of classical logic. We assess results, describe related work and conclude in Section 6.

2 Base Logic

In this paper we focus on intuitionistic first-order logic as a base logic, which is of particular interest to us because the result seems to scale to proof search problems in type theories, which we plan to study in future work. In general, the idea of truthful monadic abstractions is applicable in other settings as well.

In particular, we work with a sequent calculus formulation of first-order intuitionistic logic (FOL). The syntactic categories of terms and formulas are defined as follows:

$$t ::= x \mid c \mid f(t_1, \ldots, t_n)$$
$$A, B ::= P(t_1, \ldots, t_n) \mid A \wedge B \mid A \vee B \mid A \supset B \mid \top \mid \bot \mid \forall x.A \mid \exists x.A$$

Here, x is a variable and c a constant. We let F denote the set of function symbols f and assume there is a function $\Sigma : F \to \mathbb{N}$ that records the arity of each function symbol. The connectives and the inference rules depicted in Figure 1 are standard. The superscripted variables on the $\forall$R and $\exists$L rules indicate that these rules are subject to the eigenvariable condition.

3 Truthfulness

There are many different decidable fragments of classical first-order logic. Most of these fragments impose restrictions on the shape of the formulas that may occur e.g. by restricting the quantifier prefixes, number of variables or guardedness of the formulas. For our purposes, we require that there is no restriction on the shape of the formulas, hence these fragments are *a priori* not suitable as targets for our abstraction.

The truthful abstraction that we present in this section maps sequents from the base logic into the *abstraction logic*, the monadic fragment of classical first-order logic, which we will not define in any more detail than to say that all predicate and function symbol are required to be unary. Base and abstraction logic formulas both share the same connectives, and as we will see in this section, we will define the truthful abstraction as a composition of three conceptually simpler abstractions through two intermediate logics.

1. Base logic: Intuitionistic first-order logic
2. Intermediate logic 1: Intuitionistic first-order logic with monadic predicates
3. Intermediate logic 2: Monadic intuitionistic first-order logic
4. Abstraction logic: Monadic classical first-order logic

Since the composition of truthful abstractions is again a truthful abstraction, we are able to compose the three and obtain the desired truthful abstraction from base to abstraction logic. The compositionality of truthful abstractions bears another interesting opportunity: Any logic that can be mapped truthfully into the base logic gives rise to a truthful abstraction into monadic first-order logic.

In the remainder of this section, we discuss each of the three abstractions in turn, and establish the desired completeness result.

3.1 From Predicates to Monadic Predicates

We describe the mapping from the base logic into the intermediate logic 1, i.e. from intuitionistic first-order formulas to intuitionistic first-order logic with monadic predicates. An obvious candidate for such a mapping is a homomorphism that replaces predicates with monadic predicates, dropping all but one argument. Notice, that the rules init, $\exists$R, and $\forall$L from Figure 1 pose a particular challenge in the proof of truthfulness, because abstraction and substitution must commute in a suitable fashion for the proof to go through.

Let π be a function from the set of predicates to $\mathbb{N}$ such that $\pi(P)$ points to a valid index, i.e. it is at most equal to the arity of P. We then define a map, α, as follows:

$$\begin{aligned}
\alpha(\top) &= \top \\
\alpha(\bot) &= \bot \\
\alpha(A \odot B) &= \alpha(A) \odot \alpha(B) \\
\alpha(Qx.A) &= Qx.\alpha(A) \\
\alpha(P(t_1, \ldots, t_n)) &= P(t_i) \text{ where } i = \pi(P)
\end{aligned}$$

where $\odot$ is $\wedge$,$\vee$ or $\supset$ and Q is either $\forall$ or $\exists$.

Since α is only defined on formulas and not terms, this definition behaves nicely with regard to substitution, as the following lemma shows.

Lemma 1 (Substitutivity). *For any term t, formula A, and α defined as above, the equality $\alpha([t/x]A) = [t/x]\alpha(A)$ holds.*

Proof. By straightforward structural induction. □

Proving that α is a truthful abstraction is now easy.

Theorem 1 (Truthfulness). *If $\Gamma \Longrightarrow A$ then $\alpha(\Gamma) \Longrightarrow \alpha(A)$.*

Proof. By structural induction on the derivation of $\Gamma \Longrightarrow A$. Most cases are immediate. We show here a few of the more interesting cases.

Case: $\mathcal{D} :: \dfrac{}{\Gamma, P(t_1, \ldots, t_n) \Longrightarrow P(t_1, \ldots, t_n)}\text{init}$

$\alpha(P(t_1, \ldots, t_n)) = P(t_i)$ for some i	by definition of α.
$\alpha(\Gamma), P(t_i) \Longrightarrow P(t_i)$	by init.

Case:

$$\dfrac{\begin{array}{c}\mathcal{D}\\ \Gamma \Longrightarrow [t/x]A\end{array}}{\Gamma \Longrightarrow \exists x.A}\exists\text{R}$$

$\alpha(\Gamma) \Longrightarrow \alpha([t/x]A)$	by i.h. on $\mathcal{D}$.
$\alpha(\Gamma) \Longrightarrow [t/x]\alpha(A)$	by Lemma 1
$\alpha(\Gamma) \Longrightarrow \exists x.\alpha(A)$	by $\exists$R.
$\alpha(\Gamma) \Longrightarrow \alpha(\exists x.A)$	by definition of α.

Case:

$$\dfrac{\begin{array}{c}\mathcal{D}\\ \Gamma, \forall x.A, [t/x]A \Longrightarrow C\end{array}}{\Gamma, \forall x.A \Longrightarrow C}\forall\text{L}$$

$\alpha(\Gamma), \alpha(\forall x.A), \alpha([t/x]A) \Longrightarrow \alpha(C)$	by i.h. on $\mathcal{D}$.
$\alpha(\Gamma), \forall x.\alpha(A), [t/x]\alpha(A) \Longrightarrow \alpha(C)$	by definition of α and Lemma 1.
$\alpha(\Gamma), \forall x.\alpha(A) \Longrightarrow \alpha(C)$	by $\forall$L.
$\alpha(\Gamma), \alpha(\forall x.A) \Longrightarrow \alpha(C)$	by definition of α.

□

With this abstraction in place, we may now assume (without loss of generality) that all predicates are unary until the end of this section.

3.2 From Terms to Monadic Terms

We now tend to the most challenging part of the abstraction, a mapping from intermediate logic 1 to intermediate logic 2 that maps intuitionistic first-order logic with monadic predicates into monadic intuitionistic first-order logic, replacing n-ary function symbols of the term language into unary function symbols. Here again, the abstraction and substitution applications from rules init, $\exists$R, and $\forall$L must commute. This is more complex than before, because the abstraction defined here will abstract terms as well.

Let μ be any endofunction on the set of terms satisfying $\mu(x) = x$ for all variables x. Its homomorphic extension to formulas and contexts is given by the following definition

$$
\begin{array}{ccc}
\mu(P(t)) = P(\mu(t)) & \mu(\top) = \top & \mu(\bot) = \bot \\
\mu(A \wedge B) = \mu(A) \wedge \mu(B) & \mu(A \vee B) = \mu(A) \vee \mu(B) & \mu(A \supset B) = \mu(A) \supset \mu(B) \\
\mu(\forall x.A) = \forall x.\mu(A) & \mu(\exists x.A) = \exists x.\mu(A) & \\
\mu(\cdot) = \cdot & \mu(\Gamma, A) = \mu(\Gamma), \mu(A) &
\end{array}
$$

where μ must map arbitrary terms t to monadic terms $\mu(t)$, which refer only to function symbols of arity 1. The intermediate logic 2 under the image of μ must admit the following three rules of inference that we obtain from Figure 1. The other inference rules of the intermediate logic 1 remain unchanged.

$$
\frac{}{\Gamma, P(\mu(t)) \longrightarrow P(\mu(t))}\ \text{init}^{\mu}
$$

$$
\frac{\Gamma \longrightarrow [\mu(t)/x]A}{\Gamma \longrightarrow \exists x.A}\ \exists\text{R}^{\mu}
\qquad
\frac{\Gamma, \forall x.A, [\mu(t)/x]A \longrightarrow C}{\Gamma, \forall x.A \longrightarrow C}\ \forall\text{L}^{\mu}
$$

Theorem 2. *Let μ be a function defined as above. The following properties are then equivalent:*

1. *(Truthfulness) If $\Gamma \Longrightarrow A$ is provable, then so is $\mu(\Gamma) \longrightarrow \mu(A)$.*
2. *(Substitutivity on terms) For all terms t, t':*

$$\mu([t/x]t') = [\mu(t)/x]\mu(t').$$

3. *(Substitutivity on formulas) For all terms t and formulas A:*

$$\mu([t/x]A) = [\mu(t)/x]\mu(A)$$

Proof. (2)$\Rightarrow$(3) follows from a straightforward structural induction on A.

(1)$\Rightarrow$(2) Let t and t' be given. The sequent

$$P(t), \forall x.P(x) \supset Q(t') \Longrightarrow Q([t/x]t')$$

is easily seen to be derivable, as the following derivation shows:

$$\dfrac{\dfrac{\overline{P(t) \Longrightarrow P(t)}\ \text{init} \qquad \overline{Q([t/x]t') \Longrightarrow Q([t/x]t')}\ \text{init}}{P(t), P(t) \supset Q([t/x]t') \Longrightarrow Q([t/x]t')}\ \supset\text{L}}{P(t), \forall x.P(x) \supset Q(t') \Longrightarrow Q([t/x]t')}\ \forall\text{L}$$

hence by assumption the sequent

$$\mu(P(t), \forall x.P(x) \supset Q(t')) \longrightarrow \mu(Q([t/x]t'))$$

which is equal to

$$P(\mu(t)), \forall x.P(x) \supset Q(\mu(t')) \longrightarrow Q(\mu([t/x]t'))$$

is likewise derivable. It may be shown (e.g. by means of *focusing*[1,6]) that the following must then be a proof (with occurrences of contraction elided)

$$\dfrac{\dfrac{\overline{P(\mu(t)) \longrightarrow P(\mu(s))}\ \text{init}^{\mu} \qquad \overline{Q([\mu(s)/x]\mu(t')) \longrightarrow Q(\mu([t/x]t'))}\ \text{init}^{\mu}}{P(\mu(t)), P(\mu(s)) \supset Q([\mu(s)/x]\mu(t')) \longrightarrow Q(\mu([t/x]t'))}\ \supset\text{L}^{\mu}}{P(\mu(t)), \forall x.P(x) \supset Q(\mu(t')) \longrightarrow Q(\mu([t/x]t'))}\ \forall\text{L}^{\mu}$$

for some term $\mu(s)$. Substituting the equality $\mu(t) = \mu(s)$ from the left init rule into the equality $[\mu(s)/x]\mu(t') = \mu([t/x]t')$ from the right init rule yields the desired equality $\mu([t/x]t') = [\mu(t)/x]\mu(t')$.

(3)⇒(1) By structural induction on the derivation of $\Gamma \Longrightarrow A$. All cases are straightforward. The substitutivity property is used in the ∃R and ∀L cases.

$$\dfrac{\begin{array}{c}\mathcal{D}\\ \Gamma \Longrightarrow [t/x]A\end{array}}{\Gamma \Longrightarrow \exists x.A}\ \exists\text{R}$$

Applying the induction hypothesis to $\mathcal{D}$, we get a derivation of $\mu(\Gamma) \longrightarrow \mu([t/x]A)$. By assumption, this is equal to a derivation $\mu(\Gamma) \longrightarrow [\mu(t)/x]\mu(A)$ to which we may apply the $\exists\text{R}^{\mu}$ rule to get a derivation of $\mu(\Gamma) \longrightarrow \mu(\exists x.A)$. The case for ∀L is similar. □

Definition 1 (Monadic Terms). *Monadic terms are built using the following syntax:*

$$m ::= x \mid c \mid f(m)$$

where, again, x represents a variable and c a constant.

A crucial part of the definition of our abstraction is the concept of *monadic subterm.*

Definition 2 (Monadic Subterm). *The judgment $m \prec t$ (read: m is a monadic subterm of t) is defined by the following inference rules:*

$$\frac{}{x \prec x} \qquad \frac{}{c \prec c} \qquad \frac{m \prec t_i}{f_i(m) \prec f(t_1, \ldots, t_n)}$$

where x is a variable, c is a constant and $n \geq 1$.

Loosely speaking, a monadic term m is a monadic subterm of a term t, written $m \prec t$ if m encodes a maximal path within the syntax tree of t, i.e. a directed path from the root to a leaf. For example, the term $f(g(a, b), h(c))$ has three monadic subterms: $f_1(g_1(a))$, $f_1(g_2(b))$ and $f_2(h_1(c))$.

We are particularly interested in a specific class of monadic subterms, those given by the following definition:

Definition 3. *A function μ is* regular *if*

1. *there exists a function $\sigma : F \to \mathbb{N}$ such that $\sigma(f) \leq \Sigma(f)$ for all functions f.*
2. *For all functions f of arity n and terms $t_1, \ldots, t_n$ the following equation holds*

$$\mu(f(t_1, \ldots, t_n)) = f_i(\mu(t_i))$$

where $i = \sigma(f)$.

The following property is straightforward to prove:

Lemma 2. *For any regular function μ and term t, we have $\mu(t) \prec t$.*

Note that the monadic subterms under a regular function μ are much fewer than compared to monadic subterms in general. For instance, the term $f(f(a, b), c)$ has only two regular monadic subterms: $f_1(f_1(a))$ and $f_2(c)$. The monadic subterm $f_1(f_2(b))$ cannot be generated using a regular function. This might seem overly restrictive, but as we shall see below, regular functions are the only ones that are of interest when proving truthfulness.

The above lemma has a partial converse, as the following theorem shows:

Theorem 3. *Let μ be an endofunction on the set of terms. The following properties are then equivalent:*

1. *For all terms t and t'*

$$\mu(t) \prec t \qquad \text{and} \qquad \mu([t/x]t') = [\mu(t)/x]\mu(t')$$

2. *The function μ is regular.*

Proof. (1)⇒(2) We will show that μ satisfies the defining equations of a regular function. In particular, we show that for any function f of arity n and any terms $t_1, \ldots, t_n$, the following equality holds:

$$\mu(f(t_1, \ldots, t_n)) = f_i(\mu(t_i))$$

for some value i that only depends on f.

Assume that f has arity n. Let $x_1, \ldots, x_n$ be freshly chosen variables that do not appear in the terms $t_1, \ldots, t_n$. Then

$$\mu(f(x_1,\ldots,x_n)) \prec f(x_1,\ldots,x_n)$$

hence $\mu(f(x_1,\ldots,x_n))$ must be of the form $f_i(x_i)$ for some i. To see that i cannot depend on the choice of variables $x_1,\ldots,x_n$, we note that if we had chosen instead the variables $y_1,\ldots,y_n$, we would have the following string of equations

$$\begin{aligned}\mu(f(y_1,\ldots,y_n)) &= \mu([y_1/x_1]\cdots[y_n/x_n]f(x_1,\ldots,x_n))\\ &= [\mu(y_1)/x_1]\cdots[\mu(y_n)/x_n]\mu(f(x_1,\ldots,x_n))\\ &= [y_1/x_1]\cdots[y_n/x_n]f_i(x_i)\\ &= f_i(y_i)\end{aligned}$$

where i was the index we had determined previously. Thus, the value of this i can only depend on the function f, hence we may define a function $\sigma : F \to \mathbb{N}$ by $\sigma(f) = i$. Now, given the terms $t_1,\ldots,t_n$, we have the following equalities

$$\begin{aligned}\mu(f(t_1,\ldots,t_n)) &= \mu([t_1/x_1]\cdots[t_n/x_n]f(x_1,\ldots,x_n))\\ &= [\mu(t_1)/x_1]\cdots[\mu(t_n)/x_n]\mu(f(x_1,\ldots,x_n))\\ &= [\mu(t_1)/x_1]\cdots[\mu(t_n)/x_n]f_i(x_i)\\ &= f_i(\mu(t_i))\end{aligned}$$

where $i = \sigma(f)$, hence μ satisfies the definition of a regular function.

(2)⇒(1) Assuming μ is regular, we now need to show that it satisfies the substitution property

$$\mu([t/x]t') = [\mu(t)/x]\mu(t')$$

for all terms t, t'. We show this by induction on the structure of t'. In the case where x does not occur in t', the result is immediate, as both sides become equal to $\mu(t')$. If $t' = x$, then both sides equal $\mu(t)$. Now, assume $t' = f(t_1,\ldots,t_n)$ and that $\sigma(f) = i$. Then

$$\begin{aligned}\mu([t/x]t') &= \mu([t/x]f(t_1,\ldots,t_n))\\ &= \mu(f([t/x]t_1,\ldots,[t/x]t_n))\\ &= f_i(\mu([t/x]t_i))\\ &= f_i([\mu(t)/x]\mu(t_i)) && \text{by the induction hypothesis}\\ &= [\mu(t)/x]f_i(\mu(t_i))\\ &= [\mu(t)/x]\mu(f(t_1,\ldots,t_n))\\ &= [\mu(t)/x]\mu(t').\end{aligned}$$

This completes the proof. □

With the above theorems and lemmas, we may now prove the main theorem:

Theorem 4. *Let μ be any function satisfying $\mu(t) \prec t$ for all terms t and let α be defined as in Section 3.1. The following properties are then equivalent:*

1. $\alpha(\Gamma) \Longrightarrow \alpha(A)$ *implies* $\mu(\alpha(\Gamma)) \longrightarrow \mu(\alpha(A))$ *in the monadic fragment.*
2. μ *is regular.*

Proof. If μ is regular, then the substitutivity property follows from Theorem 3, and hence truthfulness by Theorem 2. That the derivation is in the monadic fragment follows from the fact that the image of a regular function is a subset of the set of monadic terms.

Conversely, if μ is truthful, then μ satisfies the substitutivity property by Theorem 2, and is thus regular by Theorem 3. □

As a consequence of this theorem we get that, assuming the reasonable requirement that $\mu(t) \prec t$ for all terms t, it is exactly the regular functions that induce truthful abstractions.

Note that because we only need to consider regular functions, there is no need to keep track of the indices on the abstracted predicates and functions, as these can always be recovered using the σ function defined above.

3.3 From Intuitionistic to Classical Logic

The third abstraction mapping the intermediate logic 2 into the abstraction logic is trivial. In fact, if monadic intuitionistic first-order logic were decidable, we would already be done, but it is not [8]. Monadic classical logic, on the other hand is. We chose therefore the obvious complete embedding of intuitionistic logic into classical logic as the third and final abstraction. This abstraction is obviously truthful.

4 Examples

First, we will consider the example given in the introduction. Recall that the predicate $D(x, y)$ was given by the following axioms

$$\Gamma = D(z, z), \forall x.\forall y.D(x, y) \supset D(s(x), s(s(y)))$$

There are two abstractions of the sequent

$$\Gamma \Longrightarrow \exists x.D(x, s(s(s(s(s(z))))))$$

corresponding to the choice of either $\pi(D) = 1$ or $\pi(D) = 2$. In the first case, we get the following abstracted sequent

$$D_1(z), \forall x.\forall y.D_1(x) \supset D_1(s(x)) \longrightarrow \exists x.D_1(x)$$

which is immediately provable by instantiating x by z. The second abstraction is

$$D_2(z), \forall x.\forall y.D_2(y) \supset D_2(s(s(y))) \longrightarrow \exists x.D_2(s(s(s(s(s(z))))))$$

and in this case, a counterexample is found immediately by Spass. It therefore follows that the original sequent is not provable.

As a second example, we consider the following: Let Γ be the following context:

$$\begin{aligned}
&N(z),\\
&\forall x.N(x) \supset N(s(x)),\\
&\forall x.N(x) \supset S(z,x,x),\\
&\forall x_1,x_2,x_3.N(x_1) \wedge N(x_2) \wedge N(x_3) \wedge S(x_1,x_2,x_3) \supset S(s(x_1),x_2,s(x_3)).
\end{aligned}$$

The two predicates N and S specify the Peano numerals and triples satisfying $x_1 + x_2 = x_3$ respectively.

A basic lemma about sums is that zero is a right identity for addition, i.e. that the following sequent is provable

$$\Gamma \Longrightarrow \forall x.N(x) \supset S(x,z,x)$$

In order to prove this sequent, we would need induction that is not available to us in plain first-order logic. There are three different abstractions of the above sequent, based on which argument of the predicate S is preserved. As N is already a unary predicate, it is unchanged by these abstractions, hence we elide it in the following:

$$\begin{aligned}
&\forall x.N(x) \supset S(z), \forall x_1,x_2,x_3.N(x_1) \wedge N(x_2) \wedge N(x_3) \wedge S(x_1) \supset S(s(x_1))\\
&\qquad\qquad\qquad\qquad\qquad\qquad\qquad\qquad\longrightarrow \forall x.N(x) \supset S(x)\\
&\forall x.N(x) \supset S(x), \forall x_1,x_2,x_3.N(x_1) \wedge N(x_2) \wedge N(x_3) \wedge S(x_2) \supset S(x_2)\\
&\qquad\qquad\qquad\qquad\qquad\qquad\qquad\qquad\longrightarrow \forall x.N(x) \supset S(z)\\
&\forall x.N(x) \supset S(x), \forall x_1,x_2,x_3.N(x_1) \wedge N(x_2) \wedge N(x_3) \wedge S(x_3) \supset S(s(x_3))\\
&\qquad\qquad\qquad\qquad\qquad\qquad\qquad\qquad\longrightarrow \forall x.N(x) \supset S(x)
\end{aligned}$$

Of these three abstracted sequents, the first sequent (and only this sequent) is unprovable, hence the unabstracted sequent is not provable, as one would expect.

5 Experimental Results

To test our abstraction, we applied it to a selection of problems from the ILTP[10] problem library. This is a rather coarse test, as it corresponds to only using the abstraction to disprove the original goal, and not the subgoals that are encountered during proof search. We feel, however, that even this relatively limited test shows the efficacy of this way of disproving sequents.

The abstraction was implemented as a transformation for the `tptp2X` tool, which is part of the TPTP[12] problem library. By doing this, we were able to leverage the pre-existing methods for parsing and printing problems in the TPTP format.

We applied the abstraction to a selection of ILTP problems that were flagged as Theorem, Non-Theorem or Unsolved. The resulting monadic problems were then tested with the Spass 3.5 theorem prover running on a laptop with a 2.66GHz processor and 4GB memory. A time limit of 10 seconds was set for the theorem prover. The results are summarized in Table 1.

Table 1. Experimental results

Status	Problems	Abstractions	Proved	Disproved	Timed out
Theorem	167	7706	167	0	0
Non-Theorem	56	110	50	6	0
Unsolved	78	14855	5	59	14

A problem is tallied in the Disproved column if at least one abstraction was disproved, and in the Timeout column if there were attempts that timed out and no attempts that found a disproof. By testing our implementation on problems that are known to be theorems, we have an empirical verification that our implementation is indeed a truthful abstraction.

Table 2. Distribution of abstraction results

Problem	Disproved	Proved	Timed out
KRS173	2	0	0
SWV016	0	288	0
SWV018	0	288	0
SYN322	2	0	0
SYN330	0	2	0
SYN344	1	1	0
SYN419	1022	0	2
SYN420	973	0	51
SYN421	997	0	27
SYN422	999	0	25
SYN423	379	0	645
SYN424	201	0	823
SYN425	963	0	61
SYN426	164	0	860
SYN427	244	0	780
SYN428	614	0	410
SYN429	753	0	271
SYN513	0	32	0
SYN514	32	0	0
SYN515	32	0	0
SYN516	32	0	0
SYN517	32	0	0
SYN518	3	24	5
SYN519	0	26	6
SYN520	0	22	10
SYN521	32	0	0
SYN540	8	24	0
SYN541	32	0	0
SYN544	181	32	43
SYN545	207	0	49
SYN546	178	0	78
SYN547	256	0	0

To give an accurate view of the usefulness of the abstraction, we have only included the results where we were able to test all abstractions of a given problem. This means that we have left out several problems from the NLP problem set, for which the number of abstractions in some cases exceeded 100,000. This should not be interpreted to mean that our implementation cannot be used for these problems, but rather that we thought it more important to give a complete set of results for the chosen problems. We also left out problems where there was no conjecture to prove or disprove.

It may seem surprising that it was not possible to disprove 50 of the 56 non-theorems tested. This is in part because 48 of these problems were already within the monadic fragment. Also, most of these problems are provable in classical logic, and can therefore not be disproved by our abstractions.

For each problem, finding a single abstraction without proof is enough to disprove the entire problem. To give a better view of how the three possible outcomes are distributed, we have collected a representative sample of the problems from the Unsolved problem set in Table 2. This table shows for each problem how many proof attempts resulted in the three possible outcomes.

6 Conclusion and Future Work

In this paper we have presented a family of truthful abstractions from intuitionistic first-order logic to monadic classical first-order logic. We have characterized the shape of a possible truthful abstractions and validated the usefulness of the technique experimentally.

Although these abstractions were motivated using proof search in intuitionistic logic, they apply equally to classical first-order logic, hence there might be some benefit in applying these abstractions to proof search in classical first-order logic as well.

One drawback of using the abstractions described in this paper is that a single sequent may have a large number of possible abstractions. Thus, to disprove a sequent it may be necessary to attempt disproofs of many more abstracted sequents in the hope of finding a single disproof. Clearly this approach can be parallelized in the trivial way of simply running all of these proof searches in parallel, but there might be gains to be had from either detecting abstracted sequents that are trivially true or combining several abstractions into a single abstraction.

Related to our work is the work on the Scott system [5] that combines the tableau method as model generation with automated theorem proving. The tableau method not only detects unsatisfiability of the negated conjecture but also generates models for it. This is similar to the use of model generating systems during refutation proofs. Thus, certain classes of false conjectures can be detected by generating counter-models. However, the relationship between these classes and the class characterized by the procedure presented in this paper is unclear yet and is left for future work.

Our work can be seen as an application of *Unsound Theorem Proving* [7] to intuitionistic first-order logic. In this paper we have only explored a small subset of all the decidable fragments of first-order logic. There are many more such fragments, and the overall methodology may apply to these fragments as well.

Another point for future work would be to test this approach more thoroughly on the ILTP[10] and TPTP[12] problem libraries to see what proportion of unprovable problems may be refuted by using our decision procedure.

Finally, using the compositionality of truthful abstractions, it might be interesting to extend this approach to "larger" source logic, for instance the meta-logic of the Twelf theorem prover[9].

References

1. Andreoli, J.: Logic programming with focusing proofs in linear logic. Journal of Logic and Computation 2(3), 297 (1992)
2. Claessen, K., Sorensson, N.: New techniques that improve mace-style finite model finding. In: Baumgartner, P., Fermueller, C. (eds.) Proceedings of the CADE-19 Workshop: Model Computation - Principles, Algorithms, Applications, Miami, USA (2003)
3. Giunchiglia, F., Walsh, T.: A theory of abstraction. Artif. Intell. 57(2-3), 323–389 (1992)
4. Gurevich, Y.: The decision problem for the logic of predicates and of operations. Algebra and Logic 8, 160–174 (1969), doi:10.1007/BF02306690
5. Hodgson, K., Slaney, J.: Tptp, casc and the development of a semantically guided theorem prover. AI Commun. 15, 135–146 (2002)
6. Liang, C., Miller, D.: Focusing and Polarization in Intuitionistic Logic. In: Duparc, J., Henzinger, T.A. (eds.) CSL 2007. LNCS, vol. 4646, pp. 451–465. Springer, Heidelberg (2007), doi: 10.1007/978-3-540-74915-8
7. Lynch, C.: Unsound Theorem Proving. In: Marcinkowski, J., Tarlecki, A. (eds.) CSL 2004. LNCS, vol. 3210, pp. 473–487. Springer, Heidelberg (2004), doi: 10.1007/978-3-540-30124-0_36
8. Okee, J.: A semantical proof of the undecidability of the monadic intuitionistic predicate calculus of the first order. Notre Dame J. Formal Logic 16, 552–554 (1975)
9. Pfenning, F., Schürmann, C.: System Description: Twelf - A Meta-Logical Framework for Deductive Systems. In: Ganzinger, H. (ed.) CADE-16. LNCS (LNAI), vol. 1632, pp. 202–206. Springer, Heidelberg (1999)
10. Raths, T., Otten, J., Kreitz, C.: The iltp problem library for intuitionistic logic. Journal of Automated Reasoning 38, 261–271 (2007)
11. Riazanov, A., Voronkov, A.: The design and implementation of vampire. AI Commun. 15(2-3), 91–110 (2002)
12. Sutcliffe, G.: The tptp problem library and associated infrastructure. Journal of Automated Reasoning 43, 337–362 (2009)
13. Weidenbach, C., Dimova, D., Fietzke, A., Kumar, R., Suda, M., Wischnewski, P.: SPASS Version 3.5. In: Schmidt, R.A. (ed.) CADE 2009. LNCS, vol. 5663, pp. 140–145. Springer, Heidelberg (2009)

Satallax: An Automatic Higher-Order Prover

Chad E. Brown

Saarland University, Saarbrücken, Germany

Abstract. Satallax is an automatic higher-order theorem prover that generates propositional clauses encoding (ground) tableau rules and uses MiniSat to test for unsatisfiability. We describe the implementation, focusing on flags that control search and examples that illustrate how the search proceeds.

Keywords: higher-order logic, simple type theory, higher-order theorem proving.

1 Introduction

Satallax is an automatic theorem prover for classical higher-order logic with extensionality and choice. The search proceeds by generating propositional clauses that simulate tableau rules. Once the set of propositional clauses is unsatisfiable, the original higher-order problem is solved. An abstract description of the search procedure is given in [6]. The corresponding tableau calculus is proven sound and complete relative to Henkin models in [1], and the search procedure is proven sound and complete in [6].

In this system description we discuss the implementation of the search procedure. A number of flags can be used to guide search. We discuss the most important of these flags and give example problems from TPTP v5.3.0 [10] to illustrate how these flags affect the behavior of Satallax. (From now on, we use TPTP to refer to TPTP v5.3.0.)

Satallax won the THF division of the CASC-23 competition at CADE-23 in 2011 [11]. Out of 300 problems with a 5 minute time limit, Satallax 2.1 solved 246. LEO-II 1.2.8 [3] came in second, solving 208 problems. Among the 300 problems, there were 15 problems that only Satallax could solve. Most of these 15 problems were related to a choice operator. Since Satallax is the only system that directly supports reasoning with choice operators, it clearly has an advantage on such problems. By contrast, there were 18 problems LEO could solve but no other system could. Many of these involved first-order equational reasoning (e.g., group theory problems).

The first versions of Satallax (1.0-1.4) were coded in Steel-Bank Common Lisp during 2009-2010. Starting with version 2.0 in 2010, Satallax has been implemented in Objective Caml with the exception of some code implementing a foreign function interface to MiniSat functions. MiniSat [8] is implemented in C++. The latest version of Satallax is Satallax 2.3 (approximately 13,000 lines Objective Caml code and 100 lines of C++) which uses MiniSat 2.2.0 (approximately 2,000 lines of C++). Satallax is available at `satallax.com`.

B. Gramlich, D. Miller, and U. Sattler (Eds.): IJCAR 2012, LNAI 7364, pp. 111–117, 2012.

2 Preliminaries

We will assume familiarity with simple type theory and only briefly review to make the notation clear. A more detailed presentation can be found in [1]. Simple types σ, τ are either base types (o, ι, α, β) or function types $\sigma\tau$ (for functions from σ to τ). The type o is the type of propositions. Terms (s, t) are either variables (x, y, z, ...), logical constants ($\bot$, $\rightarrow$, $\forall_\sigma$, $=_\sigma$ and ε_σ), applications st or λ-abstractions $\lambda x.s$. Variables have a corresponding type and we only consider well-typed terms. A term of type o is called a formula. We write s^x_t for the capture-avoiding substitution of t for x in s.

We use notation $\forall x_\sigma.s$ or $\forall x.s$ (where x has type σ) for $\forall_\sigma(\lambda x.s)$. We use infix notation $s \rightarrow t$ and $s =_\sigma t$ (or $s = t$) for $(\rightarrow s)t$ and $(=_\sigma s)t$, respectively. We write $\neg s$ for $s \rightarrow \bot$. We write stu for $(st)u$ except $\neg st$ means $\neg(st)$. Since the THF problems in the TPTP problem library make use of logical connectives such as $\vee$ and $\wedge$, we also use notation $s \vee t$ for $\neg s \rightarrow t$, $s \wedge t$ for $\neg(s \rightarrow \neg t)$, $s \leftrightarrow t$ for $s =_o t$, $\exists x.s$ for $\neg\forall x.\neg s$ and $\exists! x.s$ for $\exists x.s \wedge \forall y.s^x_y \rightarrow x = y$. The scope of the λ, $\forall$, $\exists$ and $\exists!$ binders is as far to the right as is consistent with parentheses. We also use the usual notation for quantifying over several variables. For example, $\forall xy.s$ means $\forall x.\forall y.s$ and $\exists xy.s$ means $\exists x.\exists y.s$.

A β-redex is of the form $(\lambda x.s)t$ and this redex reduces to s^x_t. An η-redex is of the form $(\lambda x.sx)$ where x is not free in s and this redex reduces to s. We also reduce terms $\neg\neg s$ to s. All typed terms have a normal form. Satallax normalizes eagerly.

A branch A is a finite set of normal formulas. Given a theorem proving problem, we take all the axioms of the problem and combine them with the negation of the conjecture (if a conjecture is given) to form a branch A. The goal is then to prove A is (Henkin-)unsatisfiable.

3 Basic Search Procedure and Implementation

We briefly describe the link between higher-order formulas and propositional literals. The general technique of using a propositional abstraction is standard and is used by SMT solvers (e.g., see [7]). Let Atom be a set of propositional atoms. A *literal* is an atom a or a negated atom $\overline{a}$. Let $\overline{\overline{a}}$ be a. Let $\lfloor - \rfloor$ be a function from formulas to propositional literals such that $\lfloor \neg s \rfloor$ is $\overline{\lfloor s \rfloor}$. (We assume if $\lfloor s \rfloor = \lfloor t \rfloor$, then s and t are equivalent up to renaming and normalization.) A clause is a finite set of literals. We write a clause $\{l_1, \ldots, l_n\}$ as $l_1 \sqcup \cdots \sqcup l_n$. (We use $\sqcup$ instead of $\vee$ to distinguish the propositional clause level from the higher-order formula level.)

A *quasi-state* Σ is determined by sets of passive and active formulas, sets of passive and active terms (to be used as instantiations) and a set of propositional clauses. An active formula or term is one that must still be processed, while a passive formula or term is one that has already been processed and can now only contribute when processing a new active formula or term.[1] A *state* is a

[1] A reviewer pointed out that in some of the literature on superposition-based theorem proving, the terms "active" and "passive" are used in the opposite way. We keep the current terminology to be consistent with [6].

quasi-state satisfying a finite number of conditions that can be found in [6]. The idea of the conditions can be easily summarized as follows: For every (instance of) a tableau rule that can be formed using passive formulas and passive terms, there are corresponding propositional clauses in the state. Also, every literal l in a clause is either $\lfloor s \rfloor$ for some active or passive formula s or is $\overline{\lfloor s \rfloor}$ for some passive formula s.

Given a branch A to refute, we can start from any initial state for A. A state is *initial for* A if for every formula s in A, s is either active or passive in the state and the unit clause $\lfloor s \rfloor$ is a clause in the state.

On an abstract level, a state is transformed into a successor state by processing an active formula (making the formula passive), by processing an active term (making the term passive), or by the generation of a new active term (to use as an instantiation). The successor state may have new active formulas, new active terms and new clauses. In reality, the situation is a bit more complicated. First, there must be an enumeration scheme that creates new active terms for higher-order quantifiers (if the original problem contains higher-order quantifiers). Also, there are two tableau rules with more than one principal formula: mating and confrontation.

We describe the mating rule. Suppose we are processing an active formula $ps_1 \cdots s_n$ where p is a variable. In order to process $ps_1 \cdots s_n$ we should, for each passive formula $\neg pt_1 \cdots t_n$ (a *mate*), make each disequation $s_i \neq t_i$ an active formula in the new state and add a clause

$$\overline{\lfloor ps_1 \cdots s_n \rfloor} \sqcup \lfloor pt_1 \cdots t_n \rfloor \sqcup \overline{\lfloor s_1 = t_1 \rfloor} \sqcup \cdots \sqcup \overline{\lfloor s_n = t_n \rfloor}.$$

The way the implementation actually handles this case is to create a command for each pair of mates. When the command is executed, the disequations are added as active formulas and the corresponding clause is added to the new state. The confrontation rule is similar to the mating rule, but operates on an equation $s =_\alpha t$ and a disequation $u \neq_\alpha v$ at a base type α (other than o).

The particular behavior of the search depends on 33 boolean flags and 79 integer flags. We will describe a few of these flags in Section 4 and give examples illustrating how they affect search.

Active and passive terms of a state are used as instantiations for quantifiers. In the implementation, the initial state always starts with two passive terms $\bot$ and $\neg\bot$ which act as instantiations for type o. New active terms s and t of a base type α (other than o) appear during the search when a disequation $s \neq_\alpha t$ is processed. If no active term of a base type α appears, then eventually a default element must be inserted as an active term. This default element will either be a new variable of type α, the term $\varepsilon_\alpha(\lambda x.\bot)$, or some term of type α that has appeared during the search already. For the sake of completeness, new active terms of function types must be enumerated during the search. There are two different enumeration processes for such terms. Under some flag settings other active terms are inserted into the state. Since some logical constants (e.g., $=_\sigma$) depend on a type σ, there is also an enumeration process for generating types.

A *command* is one of the following:

1. Process a formula s. Unless s is already passive (meaning it has already been processed), make s passive and add new active formulas, active terms, proposition clauses, and commands.
2. Process a term t_σ as an instantiation. Unless t is already passive, make t passive and for each passive formula $\forall_\sigma s$ add the normal form u of st as a new active formula, add the command of processing u, and add the clause $\overline{\lfloor \forall_\sigma s \rfloor} \sqcup \lfloor u \rfloor$.
3. Apply an instance of the mating rule.
4. Apply an instance of the confrontation rule.
5. Create a default element of a base type α.
6. Work on enumerating a new type. Once a type σ has been generated by the type enumeration process, we can imitate (in the sense of higher-order unification) logical constants $=_\sigma$, $\forall_\sigma$ and ε_σ when enumerating instantiation terms.
7. Work on enumerating a term of a given type σ with local variables $x_1, \ldots, x_n$.
8. Given a term $t_{\sigma_1 \cdots \sigma_n \alpha}$, work on enumerating a term of type α of the form $ts_1 \cdots s_n$ with local variables $x_1, \ldots, x_n$.
9. Use iterative deepening to enumerate all closed terms of a type up to a certain depth. This is an alternative to the previous enumeration commands which we will not discuss further.
10. Filter out a passive formula s if the set of clauses implies $\overline{\lfloor s \rfloor}$. We will not discuss filtering further.

A collection of commands is put into a priority queue. The purpose of many of the integer flags is to determine the priority of new commands as they are generated. Search proceeds by taking one of the highest priority commands and processing it. The search ends successfully when the set of propositional clauses is propositionally unsatisfiable.

Example 1. We discuss the simple problem SYO357^5 from the TPTP in detail to illustrate the search procedure. The source for this problem is [2]. The conjecture is $(\forall P_{\alpha o}.(a \vee \neg a) \wedge Pu \rightarrow (b \vee \neg b) \wedge Pv) \rightarrow \forall Q_{\alpha o}.Qu \rightarrow Qv$. Let s^1 be the negation of this conjecture. We start with an initial state with a single active formula s^1 and a single clause $\lfloor s^1 \rfloor$. We process this formula, making it passive, adding two new active formulas s^2: $\forall P_{\alpha o}.(a \vee \neg a) \wedge Pu \rightarrow (b \vee \neg b) \wedge Pv$ and s^3: $\neg \forall Q_{\alpha o}.Qu \rightarrow Qv$ and new clauses $\overline{\lfloor s^1 \rfloor} \sqcup \lfloor s^2 \rfloor$ and $\overline{\lfloor s^1 \rfloor} \sqcup \lfloor s^3 \rfloor$. We process s^2, but since there are no passive terms of type αo this adds no new active formulas or clauses. Since this is the first time a universal quantifier over type αo has been encountered, we add the command for enumerating a term of type αo. We process s^3, using Q as a fresh variable, adding the active formula s^4: $\neg(Qu \rightarrow Qv)$ and clause $\overline{\lfloor s^3 \rfloor} \sqcup \lfloor s^4 \rfloor$. We process s^4 adding active formulas Qu and $\neg Qv$ and clauses $\overline{\lfloor s^4 \rfloor} \sqcup \lfloor Qu \rfloor$ and $\overline{\lfloor s^4 \rfloor} \sqcup \overline{\lfloor Qv \rfloor}$. At this point, one thing Satallax will do is to process Qu and then $\neg Qv$ which adds the command for mating these two formulas. This line of actions does not contribute to the solution. Instead, we return to the command for enumerating a term of type αo. The command is executed by choosing a fresh variable x of type α and adding a command for enumerating a term s of type o with x free. We execute this new command by finding all possible heads for a term of the form $\lambda x.-$. In the language of higher-order unification, these heads are either the result of a projection or of an imitation. No projection is possible (because α is not o). Possible imitations are $Q_{\alpha o}$ and logical constants of the form $\varepsilon_{\sigma_1 \cdots \sigma_n o}$. The instantiation we want has Q at the head. One new command is to enumerate a term of type o with Q at the head and with x (possibly) free. To do this we only need to enumerate a term of type α with x free (to use as the argument of Q). We obtain such a term by projecting the local variable x and obtain the closed term $\lambda x.Qx$. In normal form, the term is Q. We add this as a new active term Q and immediately process this Q. We use this term Q as an instantiation for the passive universally quantified formula s^2 giving the new active formula s^5: $(a \vee \neg a) \wedge Qu \rightarrow (b \vee \neg b) \wedge Qv$ and new clause $\overline{\lfloor s^2 \rfloor} \sqcup \lfloor s^5 \rfloor$. The rest of the search is straightforward. We process s^5 and then the formulas that arise from continuing to process the resulting formulas. This yields the following clauses which (combined with the clauses above) are propositionally unsatisfiable.

$$\begin{array}{l}
\overline{\lfloor s^5 \rfloor} \sqcup \overline{\lfloor (a \vee \neg a) \wedge Qu \rfloor} \sqcup \lfloor (b \vee \neg b) \wedge Qv \rfloor \\
\lfloor (a \vee \neg a) \wedge Qu \rfloor \sqcup \overline{\lfloor a \vee \neg a \rfloor} \sqcup \overline{\lfloor Qu \rfloor} \\
\lfloor a \vee \neg a \rfloor \sqcup \overline{\lfloor a \rfloor} \\
\lfloor a \vee \neg a \rfloor \sqcup \lfloor a \rfloor \\
\overline{\lfloor (b \vee \neg b) \wedge Qv \rfloor} \sqcup \lfloor b \vee \neg b \rfloor \\
\overline{\lfloor (b \vee \neg b) \wedge Qv \rfloor} \sqcup \lfloor Qv \rfloor
\end{array}$$

4 Flags

We now consider a few of the most important flags that affect search.

Some flags affect what happens before the search begins. If the boolean flag LEIBEQ_TO_PRIMEQ is true, then subterms of the form $\forall P_{\sigma o}.Ps \rightarrow Pt$ or $\forall P.\neg Ps \rightarrow \neg Pt$ (where P is free in neither s nor t) are rewritten to $s =_\sigma t$. Also, subterms of the forms $\forall R_{\sigma\sigma o}.(\forall x.Rxx) \rightarrow Rst$ or $\forall R_{\sigma\sigma o}.\neg Rst \rightarrow \neg\forall x.Rxx$ (where R is free in neither s nor t) are rewritten to $s = t$. This is often a good idea because dealing with equalities is usually easier than dealing with higher-order quantifiers. Two particular examples from the TPTP in which this is good idea are SEV288^5 $(\lambda x_\alpha.\lambda y_\alpha.\forall q_\alpha.qx \rightarrow qy) = (\lambda x.\lambda y.x = y)$ and SEV121^5 $(\lambda x_\iota.\lambda y.x = y) = (\lambda x.\lambda y.\forall p_{\iota\iota o}.(\forall z.pzz) \rightarrow pxy)$. In both cases, the problem becomes trivial after rewriting the quantified formulas into equations. An example in which this is a bad idea is SYO357^5 (Example 1) because the conjecture becomes $(\forall P_{\alpha o}.(a \vee \neg a) \wedge Pu \rightarrow (b \vee \neg b) \wedge Pv) \rightarrow u = v$. The instantiation needed for P is $(\lambda z_\alpha.u = z)$ which is more complicated than the instantiation Q used in Example 1.

Another flag that controls preprocessing is SPLIT_GLOBAL_DISJUNCTIONS. If this flag is true, then the initial branch is split into several branches each of which is refuted independently.

Example 2. The formula $(\forall xy.x = y \rightarrow (\phi x \leftrightarrow \psi x)) \rightarrow ((\exists! x.\phi x) \leftrightarrow \exists! x.\psi x)$ is the conjecture of SEU550^2 from the TPTP. This can be split into two independent branches to refute, each of which is refuted in the same way.

An example for which setting SPLIT_GLOBAL_DISJUNCTIONS to true is a bad idea is SYO181^5 (a propositional encoding of McCarthy's Mutilated Checkerboard problem [9]) because the preprocessing would split it into over 2^{271} independent subgoals. On the other hand, if SPLIT_GLOBAL_DISJUNCTIONS is false, then SYO181^5 is easy to solve.

If the flag INITIAL_SUBTERMS_AS_INSTANTIATIONS is true, then we seed the initial state with active terms for each subterm of the initial branch. If the flag INSTANTIATE_WITH_FUNC_DISEQN_SIDES is true, then each time a functional disequation $s \neq_{\sigma\tau} t$ is processed the terms s and t are added as active terms.

One of the most successful additions to the basic search procedure is the use of higher-order clauses and pattern unification to find instantiations. If the flag ENABLE_PATTERN_CLAUSES is set to true, then processing universally quantified formulas may generate higher-order clauses with meta-variables. For example, an assumption $\forall xy.x \subseteq y \to y \subseteq x \to x =_\iota y$ will generate a clause of the form $?X \not\subseteq ?Y | ?Y \not\subseteq ?X | ?X =_\iota ?Y$. Afterwards, whenever a ground term $s \neq_\iota t$ is processed, the propositional clause

$$\overline{\lfloor \forall xy.x \subseteq y \to y \subseteq x \to x = y \rfloor} \sqcup \overline{\lfloor s \subseteq t \rfloor} \sqcup \overline{\lfloor t \subseteq s \rfloor} \sqcup \lfloor s = t \rfloor$$

is added and $s \not\subseteq t$ and $t \not\subseteq s$ are added as active formulas. An example is SEU506^2 in the TPTP.

If the boolean flag TREAT_CONJECTURE_AS_SPECIAL is true, then the conjecture (and subformulas of the conjecture) are processed before the other formulas. The integer flag AXIOM_DELAY determines how long the other formulas are delayed. The integer flag RELEVANCE_DELAY delays formulas longer if they do not have variables in common with the conjecture.

Each time a new propositional clause is sent to MiniSat, it will compute a satisfying assignment (if the set of clauses is still satisfiable). If the integer flag NOT_IN_PROP_MODEL_DELAY is non-zero, then Satallax will use the current propositional satisfying assignment to direct the search. In particular, Satallax will delay processing active formulas until the corresponding propositional literal is true in the current satisfying assignment.[2]

There are many more flags we will not discuss here. There are 279 predefined modes (collections of flag settings) in Satallax 2.3. Given a five minute timeout, the default strategy schedule (sequence of modes with a timeout) contains 37 modes. The TPTP (v5.3.0) contains 2924 THF (higher-order) problems. Of these, 343 are known to be satisfiable. Of the remaining 2581 THF problems, the strategy schedule can prove 1817 (70%) – including the examples above.

5 Conclusion and Future Work

In terms of CASC, Satallax has already proven to be a successful prover. However, there is much room for improvement. One possibility would be to integrate

[2] The idea of using the satisfying assignment to direct the search was suggested by Koen Claessen at CADE-23.

Satallax with an SMT solver [7,4]. Another possibility would be to solve for set variables using the techniques described in [5]. Also, integrating Satallax with an interactive proof assistant (e.g., Coq) would provide new ground upon which to judge its effectiveness.

Acknowledgements. Thanks to Gert Smolka for his support.

References

1. Backes, J., Brown, C.E.: Analytic Tableaux for Higher-Order Logic with Choice. Journal of Automated Reasoning 47(4), 451–479 (2011)
2. Benzmüller, C.: Equality and Extensionality in Automated Higher-Order Theorem Proving. PhD thesis, Universität des Saarlandes (1999)
3. Benzmüller, C., Paulson, L.C., Theiss, F., Fietzke, A.: LEO-II - A Cooperative Automatic Theorem Prover for Classical Higher-Order Logic (System Description). In: Armando, A., Baumgartner, P., Dowek, G. (eds.) IJCAR 2008. LNCS (LNAI), vol. 5195, pp. 162–170. Springer, Heidelberg (2008)
4. Blanchette, J.C., Böhme, S., Paulson, L.C.: Extending Sledgehammer with SMT Solvers. In: Bjørner, N., Sofronie-Stokkermans, V. (eds.) CADE 2011. LNCS (LNAI), vol. 6803, pp. 116–130. Springer, Heidelberg (2011)
5. Brown, C.E.: Solving for Set Variables in Higher-Order Theorem Proving. In: Voronkov, A. (ed.) CADE 2002. LNCS (LNAI), vol. 2392, pp. 408–422. Springer, Heidelberg (2002)
6. Brown, C.E.: Reducing Higher-Order Theorem Proving to a Sequence of SAT Problems. In: Bjørner, N., Sofronie-Stokkermans, V. (eds.) CADE 2011. LNCS (LNAI), vol. 6803, pp. 147–161. Springer, Heidelberg (2011)
7. De Moura, L., Bjørner, N.: Satisfiability modulo theories: introduction and applications. Commununications of the ACM 54(9), 69–77 (2011)
8. Eén, N., Sörensson, N.: An Extensible SAT-solver. In: Giunchiglia, E., Tacchella, A. (eds.) SAT 2003. LNCS, vol. 2919, pp. 502–518. Springer, Heidelberg (2004)
9. McCarthy, J.: A Tough Nut for Proof Procedures. Stanford Artificial Intelligence Memo No. 16 (July 1964)
10. Sutcliffe, G.: The TPTP Problem Library and Associated Infrastructure: The FOF and CNF Parts, v3.5.0. Journal of Automated Reasoning 43(4), 337–362 (2009)
11. Sutcliffe, G.: The CADE-23 Automated Theorem Proving System Competition - CASC-23. AI Communications 25(1), 49–63 (2012)

From Strong Amalgamability to Modularity of Quantifier-Free Interpolation

Roberto Bruttomesso[1], Silvio Ghilardi[1], and Silvio Ranise[2]

[1] Università degli Studi di Milano, Milan, Italy
[2] FBK (Fondazione Bruno Kessler), Trento, Italy

Abstract. The use of interpolants in verification is gaining more and more importance. Since theories used in applications are usually obtained as (disjoint) combinations of simpler theories, it is important to modularly re-use interpolation algorithms for the component theories. We show that a sufficient and necessary condition to do this for quantifier-free interpolation is that the component theories have the 'strong (sub-)amalgamation' property. Then, we provide an equivalent syntactic characterization, identify a sufficient condition, and design a combined quantifier-free interpolation algorithm handling both convex and non-convex theories, that subsumes and extends most existing work on combined interpolation.

1 Introduction

Algorithms for computing interpolants are more and more used in verification, e.g., in the abstraction-refinement phase of software model checking [16]. Of particular importance in practice are those algorithms that are capable of computing *quantifier-free* interpolants in presence of some background theory. Since theories commonly used in verification are obtained as combinations of simpler theories, methods to modularly combine available quantifier-free interpolation algorithms are desirable. This paper studies the modularity of quantifier-free interpolation.

Our starting point is the well-known fact [1] that quantifier-free interpolation (for universal theories) is equivalent to the model-theoretic property of *amalgamability*. Intuitively, a theory has the amalgamation property if any two structures $\mathcal{M}_1, \mathcal{M}_2$ in its class of models sharing a common sub-model $\mathcal{M}_0$ can be regarded as sub-structures of a larger model $\mathcal{M}$, called the amalgamated model. Unfortunately, this property is not sufficient to derive a modularity result for quantifier-free interpolation. As shown in this paper, a stronger notion is needed, called *strong amalgamability* [19], that has been thoroughly analyzed in universal algebra and category theory [21,28]. A theory has the strong amalgamation property if in the amalgamated model $\mathcal{M}$, elements from the supports of $\mathcal{M}_1, \mathcal{M}_2$ not belonging to the support of $\mathcal{M}_0$ *cannot be identified.* An example of an amalgamable but not strongly amalgamable theory is the theory of fields: let $\mathcal{M}_0$ be a real field and $\mathcal{M}_1, \mathcal{M}_2$ be two copies of the complex numbers, the imaginary unit in $\mathcal{M}_1$ must be identified with the imaginary unit of

B. Gramlich, D. Miller, and U. Sattler (Eds.): IJCAR 2012, LNAI 7364, pp. 118–133, 2012.

$\mathcal{M}_2$ (or with its opposite) in any amalgamating field $\mathcal{M}$ since the polynomial x^2+1 cannot have more than two roots (more examples will be discussed below, many examples are also supplied in the catalogue of [21]). We show that *strong amalgamability is precisely what is needed for the modularity of quantifier-free interpolation*, in the following sense (here, for simplicity, we assume that theories are universal although in the paper we generalize to arbitrary ones): (a) if T_1 and T_2 are signature disjoint, both stably infinite and strongly amalgamable, then $T_1 \cup T_2$ is also strongly amalgamable and hence quantifier-free interpolating and (b) a theory T is strongly amalgamable iff the disjoint union of T with the theory $\mathcal{EUF}$ of equality with uninterpreted symbols has quantifier-free interpolation (Section 3). The first two requirements of (a) are those for the correctness of the Nelson-Oppen method [26] whose importance for combined satisfiability problems is well-known.

Since the proof of (a) is non-constructive, the result does not provide an algorithm to compute quantifier-free interpolants in combinations of theories. To overcome this problem, we reformulate the notion of *equality interpolating* theory T in terms of the capability of computing some terms that are equal to the variables occurring in disjunctions of equalities entailed (modulo T) by pairs of quantifier-free formulae and show that equality interpolation is *equivalent* to strong amalgamation (Section 4). To put equality interpolation to productive work, we show that *universal* theories admitting elimination of quantifiers are equality interpolating (Section 4.1). This implies that the theories of recursively defined data structures [27], Integer Difference Logic, Unit-Two-Variable-Per-Inequality, and Integer Linear Arithmetic with division-by-n [5] are all equality interpolating. Our notion of equality interpolation is a strict generalization of the one in [32] so that all the theories that are equality interpolating in the sense of [32] are also so according to our definition, e.g., the theory of LISP structures [26] and Linear Arithmetic over the Reals (Section 4.2). Finally, we describe a combination algorithm for the generation of quantifier-free interpolants from finite sets of quantifier-free formulae in unions of signature disjoint, stably infinite, and equality interpolating theories (Section 5). The algorithm uses as sub-modules the interpolation algorithms of the component theories and is based on a sequence of syntactic manipulations organized in groups of syntactic transformations modelled after a non-deterministic version of the Nelson-Oppen combination schema (see, e.g., [31]). For proofs and additional information on related topics, see the Appendixes of the Technical Report [8].

2 Formal Preliminaries

We assume the usual syntactic and semantic notions of first-order logic (see, e.g., [12]). The equality symbol "=" is included in all signatures considered below. For clarity, we shall use "$\equiv$" in the meta-theory to express the syntactic identity between two symbols or two strings of symbols. Notations like $E(\underline{x})$ means that the expression (term, literal, formula, etc.) E contains free variables only from the tuple $\underline{x}$. A 'tuple of variables' is a list of variables without repetitions and a 'tuple of terms' is a list of terms (possibly with repetitions). Finally, whenever we use a notation like $E(\underline{x}, \underline{y})$ we implicitly assume not only that both the $\underline{x}$ and the $\underline{y}$ are pairwise distinct, but also that $\underline{x}$ and $\underline{y}$ are disjoint. A formula is *universal* (*existential*) iff it is obtained from a quantifier-free formula by prefixing it with a string of universal (existential, resp.) quantifiers.

Theories, Elimination of Quantifiers, and Interpolation. A *theory* T is a pair (Σ, Ax_T), where Σ is a signature and Ax_T is a set of Σ-sentences, called the *axioms* of T (we shall sometimes write directly T for Ax_T). The *models* of T are those Σ-structures in which all the sentences in Ax_T are true. A Σ-formula ϕ is *T-satisfiable* if there exists a model $\mathcal{M}$ of T such that ϕ is true in $\mathcal{M}$ under a suitable assignment a to the free variables of ϕ (in symbols, $(\mathcal{M}, \mathsf{a}) \models \phi$); it is *$T$-valid* (in symbols, $T \vdash \varphi$) if its negation is T-unsatisfiable or, equivalently, φ is provable from the axioms of T in a complete calculus for first-order logic. A theory $T = (\Sigma, Ax_T)$ is *universal* iff there is a theory $T' = (\Sigma, Ax_{T'})$ such that all sentences in $Ax_{T'}$ are universal and the sets of T-valid and T'-valid sentences coincide. A formula φ_1 *T-entails* a formula φ_2 if $\varphi_1 \to \varphi_2$ is *T-valid* (in symbols, $\varphi_1 \vdash_T \varphi_2$ or simply $\varphi_1 \vdash \varphi_2$ when T is clear from the context). The *satisfiability modulo the theory T ($SMT(T)$) problem* amounts to establishing the T-satisfiability of quantifier-free Σ-formulae.

A theory T admits *quantifier-elimination* iff for every formula $\phi(\underline{x})$ there is a quantifier-free formula $\phi'(\underline{x})$ such that $T \vdash \phi \leftrightarrow \phi'$. A theory T *admits quantifier-free interpolation* (or, equivalently, *has quantifier-free interpolants*) iff for every pair of quantifier-free formulae ϕ, ψ such that $\psi \wedge \phi$ is T-unsatisfiable, there exists a quantifier-free formula θ, called an *interpolant*, such that: (i) ψ T-entails θ, (ii) $\theta \wedge \phi$ is T-unsatisfiable, and (iii) only the variables occurring in both ψ and ϕ occur in θ. A theory admitting quantifier elimination also admits quantifier-free interpolation. A more general notion of quantifier-free interpolation property, involving also free function symbols, is analyzed in an Appendix of the extended version [8].

Embeddings, Sub-structures, and Combinations of Theories. The support of a structure $\mathcal{M}$ is denoted with $|\mathcal{M}|$. An embedding is a homomorphism that preserves and reflects relations and operations (see, e.g., [10]). Formally, a *Σ-embedding* (or, simply, an embedding) between two Σ-structures $\mathcal{M}$ and $\mathcal{N}$ is any mapping $\mu : |\mathcal{M}| \longrightarrow |\mathcal{N}|$ satisfying the following three conditions: (a) it is a injective function; (b) it is an algebraic homomorphism, that is for every n-ary function symbol f and for every $a_1, \ldots, a_n \in |\mathcal{M}|$, we have $f^{\mathcal{N}}(\mu(a_1), \ldots, \mu(a_n)) = \mu(f^{\mathcal{M}}(a_1, \ldots, a_n))$; (c) it preserves and reflects interpreted predicates, i.e. for every n-ary predicate symbol P, we have $(a_1, \ldots, a_n) \in P^{\mathcal{M}}$ iff $(\mu(a_1), \ldots, \mu(a_n)) \in P^{\mathcal{N}}$. If $|\mathcal{M}| \subseteq |\mathcal{N}|$ and the embedding $\mu : \mathcal{M} \longrightarrow \mathcal{N}$ is just the identity inclusion $|\mathcal{M}| \subseteq |\mathcal{N}|$, we say that $\mathcal{M}$ is a *substructure* of $\mathcal{N}$ or that $\mathcal{N}$ is a *superstructure* of $\mathcal{M}$. As it is well-known, the truth of a universal (resp. existential) sentence is preserved through substructures (resp. superstructures).

A theory T is *stably infinite* iff every T-satisfiable quantifier-free formula (from the signature of T) is satisfiable in an infinite model of T. By compactness, it is possible to show that T is stably infinite iff every model of T embeds into an infinite one (see [14]). A theory T is *convex* iff for every conjunction of literals δ, if $\delta \vdash_T \bigvee_{i=1}^{n} x_i = y_i$ then $\delta \vdash_T x_i = y_i$ holds for some $i \in \{1, ..., n\}$.

Let T_i be a stably-infinite theory over the signature Σ_i such that the $SMT(T_i)$ problem is decidable for $i = 1, 2$ and Σ_1 and Σ_2 are disjoint (i.e. the only shared symbol is equality). Under these assumptions, the Nelson-Oppen combination method [26] tells us that the SMT problem for the combination $T_1 \cup T_2$ of the theories T_1 and T_2 (i.e. the union of their axioms) is decidable.

3 Strong Amalgamation and Quantifier-Free Interpolation

We first generalize the notions of amalgamability and strong amalgamability to arbitrary theories.

Definition 1. *A theory T has the* sub-amalgamation property *iff whenever we are given models $\mathcal{M}_1$ and $\mathcal{M}_2$ of T and a common substructure $\mathcal{A}$ of them, there exists a further model $\mathcal{M}$ of T endowed with embeddings $\mu_1 : \mathcal{M}_1 \longrightarrow \mathcal{M}$ and $\mu_2 : \mathcal{M}_2 \longrightarrow \mathcal{M}$ whose restrictions to $|\mathcal{A}|$ coincide.*[1]

A theory T has the strong sub-amalgamation property *if the embeddings μ_1, μ_2 satisfy the following additional condition: if for some m_1, m_2 we have $\mu_1(m_1) = \mu_2(m_2)$, then there exists an element a in $|\mathcal{A}|$ such that $m_1 = a = m_2$.*

If the theory T is universal, any substructure of a model of T is also a model of T and we can assume that the substructure $\mathcal{A}$ in the definition above is also a model of T. In this sense, Definition 1 introduces generalizations of the standard notions of amalgamability and strong amalgamability for universal theories (see, e.g., [21] for a survey). The result of [1] relating universal theories and quantifier-free interpolation can be easily extended.

Theorem 1. *A theory T has the sub-amalgamation property iff it has quantifier-free interpolants.*

A theory admitting quantifier elimination has the sub-amalgamation property: this follows, e.g., from Theorem 1 above. On the other hand, quantifier elimination is not sufficient to guarantee the strong sub-amalgamation property. In fact, from Theorem 3 below and the counterexample given in [4], it follows that Presburger arithmetic does not have the strong sub-amalgamation property, even if we add congruences modulo n to the language. However, in Section 4, we shall see that it is sufficient to enrich the signature of Presburger Arithmetic with (integer) division-by-n (for every $n \geq 1$) to have strong amalgamability.

---

[1] For the results of this paper to be correct, the notion of structure (and of course that of substructure) should encompass the case of structures with empty domains. Readers feeling uncomfortable with empty domains can assume that signatures always contain an individual constant.

Examples. For any signature Σ, let $\mathcal{EUF}(\Sigma)$ be the pure equality theory over Σ. It is easy to see that $\mathcal{EUF}(\Sigma)$ is universal and has the strong amalgamation property by building a model $\mathcal{M}$ of $\mathcal{EUF}(\Sigma)$ from two models $\mathcal{M}_1$ and $\mathcal{M}_2$ sharing a substructure $\mathcal{M}_0$ as follows. Without loss of generality, assume that $|\mathcal{M}_0| = |\mathcal{M}_1| \cap |\mathcal{M}_2|$; let $|\mathcal{M}|$ be $|\mathcal{M}_1| \cup |\mathcal{M}_2|$ and arbitrarily extend the interpretation of the function and predicate symbols to make them total on $|\mathcal{M}|$.

Let us now consider two variants $\mathcal{AX}_{\text{ext}}$ and $\mathcal{AX}_{\texttt{diff}}$ of the theory of arrays considered in [7,9]. The signatures of $\mathcal{AX}_{\text{ext}}$ and $\mathcal{AX}_{\texttt{diff}}$ contain the sort symbols ARRAY, ELEM, and INDEX, and the function symbols $rd : \texttt{ARRAY} \times \texttt{INDEX} \longrightarrow \texttt{ELEM}$ and $wr : \texttt{ARRAY} \times \texttt{INDEX} \times \texttt{ELEM} \longrightarrow \texttt{ARRAY}$. The signature of $\mathcal{AX}_{\texttt{diff}}$ also contains the function symbol $\texttt{diff} : \texttt{ARRAY} \times \texttt{ARRAY} \longrightarrow \texttt{INDEX}$. The set $\mathcal{AX}_{\text{ext}}$ of axioms contains the following three sentences:

$$\forall y, i, j, e.\ i \neq j \Rightarrow rd(wr(y,i,e), j) = rd(y,j), \qquad \forall y, i, e.\ rd(wr(y,i,e), i) = e,$$
$$\forall x, y.\ x \neq y \Rightarrow (\exists i.\ rd(x,i) \neq rd(y,i))$$

whereas the set of axioms for $\mathcal{AX}_{\texttt{diff}}$ is obtained from that of $\mathcal{AX}_{\text{ext}}$ by replacing the third axiom with its Skolemization:

$$\forall x, y.\ \ x \neq y \Rightarrow rd(x, \texttt{diff}(x,y)) \neq rd(y, \texttt{diff}(x,y)) \ .$$

In [9], it is shown that $\mathcal{AX}_{\texttt{diff}}$ has the strong sub-amalgamation property while $\mathcal{AX}_{\text{ext}}$ does not. However $\mathcal{AX}_{\text{ext}}$ (which is *not* universal) enjoys the following property (this is the standard notion of amalgamability from the literature): given two models $\mathcal{M}_1$ and $\mathcal{M}_2$ of $\mathcal{AX}_{\text{ext}}$ sharing a substructure $\mathcal{M}_0$ *which is also a model of* $\mathcal{AX}_{\text{ext}}$, there is a model $\mathcal{M}$ of $\mathcal{AX}_{\text{ext}}$ endowed with embeddings from $\mathcal{M}_1, \mathcal{M}_2$ agreeing on the support of $\mathcal{M}_0$.

The application of Theorem 1 to $\mathcal{EUF}(\Sigma)$, $\mathcal{AX}_{\texttt{diff}}$, and $\mathcal{AX}_{\text{ext}}$ allows us to derive in a uniform way results about quantifier-free interpolation that are available in the literature: that $\mathcal{EUF}(\Sigma)$ (see, e.g., [13, 24]) and $\mathcal{AX}_{\texttt{diff}}$ [7, 9] have quantifier-free interpolants, and that $\mathcal{AX}_{\text{ext}}$ does not [20].

3.1 Modularity of Quantifier-Free Interpolation

Given the importance of combining theories in SMT solving, the next step is to establish whether sub-amalgamation is a modular property. Unfortunately, this is not the case since the combination of two theories admitting quantifier-free interpolation may not admit quantifier-free interpolation. For example, the union of the theory $\mathcal{EUF}(\Sigma)$ and Presburger arithmetic does not admit quantifier-free interpolation [4]. Fortunately, strong sub-amalgamation is modular when combining stably infinite theories.

Theorem 2. *Let T_1 and T_2 be two stably infinite theories over disjoint signatures Σ_1 and Σ_2. If both T_1 and T_2 have the strong sub-amalgamation property, then so does $T_1 \cup T_2$.*

Theorems 1 and 2 obviously imply that strong sub-amalgamation is sufficient for the modularity of quantifier-free interpolation for stably infinite theories.

Corollary 1. *Let T_1 and T_2 be two stably infinite theories over disjoint signatures Σ_1 and Σ_2. If both T_1 and T_2 have the strong sub-amalgamation property, then $T_1 \cup T_2$ admits quantifier-free interpolation.*

We can also show that strong sub-amalgamation is necessary as explained by the following result.

Theorem 3. *Let T be a theory admitting quantifier-free interpolation and Σ be a signature disjoint from the signature of T and containing at least a unary predicate symbol. Then, $T \cup \mathcal{EUF}(\Sigma)$ admits quantifier-free interpolation iff T has the strong sub-amalgamation property.*

Although Corollary 1 is already useful to establish whether combinations of theories admit quantifier-free interpolants, proving the strong sub-amalgamability property can be complex. In the next section, we study an alternative ("syntactic") characterization of strong sub-amalgamability that can be more easily applied to commonly used theories.

4 Equality Interpolation and Strong Amalgamation

There is a tight relationship between the strong sub-amalgamation property of a theory T and the fact that disjunctions of equalities among variables are entailed by T. To state this precisely, we need to introduce some preliminary notions. Given two finite tuples $\underline{t} \equiv t_1, \ldots, t_n$ and $\underline{v} \equiv v_1, \ldots, v_m$ of terms,

$$\text{the notation } \underline{t} \cap \underline{v} \neq \emptyset \text{ stands for the formula } \bigvee_{i=1}^{n} \bigvee_{j=1}^{m} (t_i = v_j).$$

We use $\underline{t}_1\underline{t}_2$ to denote the juxtaposition of the two tuples $\underline{t}_1$ and $\underline{t}_2$ of terms. So, for example, $\underline{t}_1\underline{t}_2 \cap \underline{v} \neq \emptyset$ is equivalent to $(\underline{t}_1 \cap \underline{v} \neq \emptyset) \vee (\underline{t}_2 \cap \underline{v} \neq \emptyset)$.

Definition 2. *A theory T is* equality interpolating *iff it has the quantifier-free interpolation property and satisfies the following condition:*

– *for every quintuple $\underline{x}, \underline{y}_1, \underline{z}_1, \underline{y}_2, \underline{z}_2$ of tuples of variables and pair of quantifier-free formulae $\delta_1(\underline{x}, \underline{z}_1, \underline{y}_1)$ and $\delta_2(\underline{x}, \underline{z}_2, \underline{y}_2)$ such that*

$$\delta_1(\underline{x}, \underline{z}_1, \underline{y}_1) \wedge \delta_2(\underline{x}, \underline{z}_2, \underline{y}_2) \vdash_T \underline{y}_1 \cap \underline{y}_2 \neq \emptyset \tag{1}$$

there exists a tuple $\underline{v}(\underline{x})$ of terms such that

$$\delta_1(\underline{x}, \underline{z}_1, \underline{y}_1) \wedge \delta_2(\underline{x}, \underline{z}_2, \underline{y}_2) \vdash_T \underline{y}_1\underline{y}_2 \cap \underline{v} \neq \emptyset \ . \tag{2}$$

We are now in the position to formally state the equivalence between strong sub-amalgamation and equality interpolating property.

Theorem 4. *A theory T has the strong sub-amalgamation property iff it is equality interpolating.*

4.1 Equality Interpolation at Work

We now illustrate some interesting applications of Theorem 4 so that, by using Corollary 1, we can establish when combinations of theories admit quantifier-free interpolation. To ease the application of Theorem 4, we first study the relationship between quantifier-elimination and equality interpolation for universal theories.

Theorem 5. *A universal theory admitting quantifier elimination is equality interpolating.*

Interestingly, the proof of this theorem (see [8]) is constructive and shows how an available quantifier elimination algorithm (for a universal theory) can be used *to find the terms* $\underline{v}$ *satisfying condition* (2) *of Definition 2*; this is key to the combined interpolation algorithm presented in Section 5 below.
Examples. The theory $\mathcal{RDS}$ of *recursive data structures* [27] consists of two unary function symbols *car* and *cdr* and a binary function symbol *cons*, and it is axiomatized by the following infinite set of sentences:

$$\forall x, y.car(cons(x, y)) = x, \quad \forall x, y.cdr(cons(x, y)) = y, \qquad \text{(CCC)}$$
$$\forall x, y.cons(car(x), cdr(x)) = x, \quad \forall x.x \neq t(x)$$

where t is a term obtained by finitely many applications of *car* and *cdr* to the variable x (e.g., $car(x) \neq x$, $cdr(cdr(x)) \neq x$, $cdr(car(x)) \neq x$, and so on). Clearly, $\mathcal{RDS}$ is universal; the fact that it admits elimination of quantifiers is known since an old work by Mal'cev [17].

Following [12], we define the theory $\mathcal{IDL}$ of *integer difference logic* to be the theory whose signature contains the constant symbol 0, the unary function symbols *succ* and *pred*, and the binary predicate symbol $<$, and which is axiomatized by adding to the irreflexivity, transitivity and linearity axioms for $<$ the following set of sentences:

$$\forall x.succ(pred(x)) = x, \qquad \forall x.pred(succ(x)) = x,$$
$$\forall x, y.x < succ(y) \leftrightarrow (x < y \vee x = y), \quad \forall x, y.pred(x) < y \leftrightarrow (x < y \vee x = y).$$

$\mathcal{IDL}$ is universal and the fact that admits elimination of quantifiers can be shown by adapting the procedure for a similar theory of natural numbers with successor and ordering in [12]. The key observation is that the atoms of $\mathcal{IDL}$ are equivalent to formulae of the form $i \bowtie f^n(j)$ (for $n \in \mathbb{Z}$, $\bowtie \in \{=, <\}$) where i, j are variables or the constant 0, $f^0(j)$ is j, $f^k(j)$ abbreviates $succ(succ^{k-1}(j))$ when $k > 0$ or $pred(pred^{k-1}(j))$ when $k < 0$. (Usually, $i \bowtie f^n(j)$ is written as $i - j \bowtie n$ or as $i \bowtie j + n$ from which the name of "integer difference logic.")

The theory $\mathcal{LAI}$ of Linear Arithmetic over the Integers contains the binary predicate symbol $<$, the constant symbols 0 and 1, the unary function symbol $-$, the binary function symbol $+$ and the unary function symbols $div[n]$ (integer division by n, for $n > 1$). The term $x\ div[n]$ (which is new with respect to the language of Presburger arithmetic) represents the unique q such that $x = qn + r$ for some $r = 0, \ldots, n-1$. As axioms, we take a set of sentences such that all true sentences in the standard model of the integers can be derived. This can be achieved for instance by adding to the axioms for totally ordered Abelian groups the following sentences (below $x\ rem[n]$ abbreviates $x - n(x\ div[n])$, moreover kt denotes the sum $t + \cdots + t$ having k addends all equal to the term t and k stands for $k1$):

$$0 < 1, \quad \forall y. \neg(0 < y \wedge y < 1), \quad \text{and} \quad \forall x. x\ rem[n] = 0 \vee \cdots \vee x\ rem[n] = n - 1\,.$$

$\mathcal{LAI}$ can be seen as a variant of Presburger Arithmetic obtained by adding the functions $div[n]$ instead of the 'congruence modulo n' relations (for $n = 1, 2, 3, ...$), which are needed to have quantifier elimination (see, e.g., [12]). For the application of Theorem 5, the problem with adding the 'congruence modulo n' is that the resulting theory is not universal. Instead, $\mathcal{LAI}$ is universal and the fact that admits elimination of quantifiers can be derived [8] by adapting existing quantifier-elimination procedures (e.g., the one in [12]) and observing that x is congruent to y modulo n can be defined as $x\ rem[n] = y\ rem[n]$.

By Theorem 5, $\mathcal{RDS}$, $\mathcal{IDL}$, and $\mathcal{LAI}$ are equality interpolating. In [8], the theory $\mathcal{UTVPI}$ of Unit-Two-Variable-Per-Inequality (see, e.g., [11]) is shown to be also equality interpolating via Theorem 5.

4.2 A Comparison with the Notion of Equality Interpolation in [32]

We now show that the notion of equality interpolating theories proposed here reduces to that of [32] when considering convex theories.

Proposition 1. *A convex theory T admitting quantifier-free interpolation is equality interpolating iff for every pair y_1, y_2 of variables and for every pair of conjunctions of literals $\delta_1(\underline{x}, \underline{z}_1, y_1), \delta_2(\underline{x}, \underline{z}_2, y_2)$ such that*

$$\delta_1(\underline{x}, \underline{z}_1, y_1) \wedge \delta_2(\underline{x}, \underline{z}_2, y_2) \vdash_T y_1 = y_2 \tag{3}$$

there exists a term $v(\underline{x})$ such that

$$\delta_1(\underline{x}, \underline{z}_1, y_1) \wedge \delta_2(\underline{x}, \underline{z}_2, y_2) \vdash_T y_1 = v \wedge y_2 = v. \tag{4}$$

The implication (3) $\Rightarrow$ (4) is exactly the definition of equality interpolation in [32]. In the following, a convex quantifier-free interpolating theory satisfying (3) $\Rightarrow$ (4) will be called *YMc equality interpolating*. By Proposition 1, an YMc equality interpolating (convex) theory is also equality interpolating according to Definition 2. For example, the theory $\mathcal{LST}$ of list structures [26] contains the function symbols of $\mathcal{RDS}$, a unary predicate symbol *atom*, and it is axiomatized by the axioms of $\mathcal{RDS}$ labelled (CCC) and the sentences:

$$\forall x, y. \neg atom(cons(x, y)), \quad \forall x. \neg atom(x) \to cons(car(x), cdr(x)) = x\,.$$

$\mathcal{LST}$ is a (universal) convex theory [26] that was shown to be YMc equality interpolating in [32]. By Proposition 1, we conclude that $\mathcal{LST}$ is equality interpolating in the sense of Definition 2. In [32], also Linear Arithmetic over the Reals ($\mathcal{LAR}$) is shown to be YMc equality interpolating (the convexity of $\mathcal{LAR}$ is well-known from linear algebra). By Proposition 1, $\mathcal{LAR}$ is equality interpolating in the sense of Definition 2. The same result can be obtained from Theorem 5 above by identifying a set of universal axioms for the theory and showing that they admit quantifier elimination. For the axioms to be universal, it is essential to include *multiplication by rational coefficients* in the signature of the theory, i.e. the unary function symbols $q * \_$ for every $q \in \mathbb{Q}$. If this is not the case, the theory is not sub-amalgamable and thus not equality interpolating: to see this, consider the embedding of the substructure $\mathbb{Z}$ into two copies of the reals. A direct counterexample to (3) $\Rightarrow$ (4) of Proposition 1 can be obtained by taking $\delta_i(x, y_i) \equiv y_i + y_i = x$ for $i = 1, 2$ so that $v(x) \equiv \frac{1}{2} * x$ in (4) and the function symbol $\frac{1}{2} * \_$ is required.

For *non-convex* theories, the notion of equality interpolation in this paper is strictly more general than the one proposed in the extended version of [32]. Such a notion, to be called *YM equality interpolating* below, requires quantifier-free interpolation and the following condition:

– for every tuples $\underline{x}$, $\underline{z}_1$, $\underline{z}_2$ of variables, further tuples $\underline{y}_1 = y_{11}, \ldots, y_{1n}$, $\underline{y}_2 = y_{21}, \ldots, y_{2n}$ of variables, and pairs $\delta_1(\underline{x}, \underline{z}_1, \underline{y}_1), \delta_2(\underline{x}, \underline{z}_2, \underline{y}_2)$ of conjunctions of literals,

$$\text{if } \delta_1(\underline{x}, \underline{z}_1, \underline{y}_1) \wedge \delta_2(\underline{x}, \underline{z}_2, \underline{y}_2) \vdash_T \bigvee_{i=1}^{n} (y_{1i} = y_{2i}) \text{ holds,}$$

then there exists a tuple $\underline{v}(\underline{x}) = v_1, \ldots, v_n$ of terms such that

$$\delta_1(\underline{x}, \underline{z}_1, \underline{y}_1) \wedge \delta_2(\underline{x}, \underline{z}_2, \underline{y}_2) \vdash_T \bigvee_{i=1}^{n} (y_{1i} = v_i \wedge v_i = y_{2i}).$$

We show that the notion of YM equality interpolation implies that of equality interpolation proposed in this paper. Indeed, if a convex theory is YMc equality interpolating, then it is also YM equality interpolating. Since $\mathcal{EUF}(\Sigma)$ is convex and YMc equality interpolating (as shown in [32]), it is YM equality interpolating. By Theorems 3 and 4 (and the combination result of [32]), if a theory T is YM equality interpolating, it is also equality interpolating in the sense of Definition 2. The converse does not hold, i.e. our notion is *strictly weaker* than YM equality interpolation. To prove this, we define a (non-convex) theory T_{cex} that has the strong sub-amalgamation property but is not YM equality interpolating. Let the signature of T_{cex} contain three propositional letters p_1, p_2 and p_3, three constant symbols c_1, c_2, and c_3, and a unary predicate Q. T_{cex} is axiomatized by the following sentences: exactly one among p_1, p_2 and p_3 holds, c_1, c_2, and c_3 are distinct, $Q(x)$ holds for no more than one x, and $p_i \to Q(c_i)$ for $i = 1, 2, 3$. It is easy to see that T_{cex} is stably infinite and has the strong sub-amalgamation property (T_{cex} is non-convex since $Q(x) \wedge y_1 = c_1 \wedge y_2 = c_2 \wedge y_3 = c_3$ implies the disjunction $x = y_1 \vee x = y_2 \vee x = y_3$ without implying any single disjunct). Now, notice that $Q(x) \wedge Q(y) \vdash_{T_{cex}} x = y$. According to the definition of the YM equality interpolating property (see above), there should be a *single* ground term v such that $Q(x) \wedge Q(y) \vdash_{T_{cex}} x = v \wedge y = v$. This cannot be the case since we must choose among one of the three constants c_1, c_2, c_3 to find such a term v and none of these choices fits our purposes. Hence, T_{cex} is not YM equality interpolating although it has the strong sub-amalgamation property and hence it is equality interpolating according to Definition 2.

To conclude the comparison with [32], since the notion of equality interpolation of this paper is *strictly weaker* than that of YM equality interpolation, the scope of applicability of our result about the modularity of theories admitting quantifier-free interpolation (i.e. Corollary 1 above) is *broader* than the one in the extended version of [32].

5 An Interpolation Algorithm for Combinations of Theories

Although the notion of equality interpolation together with Corollary 1 allow us to establish the quantifier-free interpolation property for all those theories obtained by combining a theory axiomatizing a container data structure (such as $\mathcal{EUF}$, $\mathcal{RDS}$, $\mathcal{LST}$, or $\mathcal{AX}_{\mathtt{diff}}$) with relevant fragments of Arithmetics (such as $\mathcal{LAR}$, $\mathcal{IDL}$, $\mathcal{UTVPI}$, or $\mathcal{LAI}$), just knowing that quantifier-free interpolants exist may not be sufficient. It would be desirable to compute interpolants for combinations of theories by modularly reusing the available interpolation algorithms for the component theories. This is the subject of this section.

To simplify the technical development, we work with ground formulae over signatures expanded with free constants instead of quantifier free formulae as done in the previous sections. We use the letters $A, B, \ldots$ to denote finite sets of ground formulae; the logical reading of a set of formulae is the conjunction of its elements. For a signature Σ and set A of formulae, Σ^A denotes the signature Σ expanded with the free constants occurring in A. Let A and B be two finite sets of ground formulae in the signatures Σ^A and Σ^B, respectively, and $\Sigma^C := \Sigma^A \cap \Sigma^B$. Given a term, a literal, or a formula φ we call it:

- *AB-common* iff it is defined over Σ^C;
- *A-local* (resp. *B-local*) if it is defined over Σ^A (resp. Σ^B);
- *A-strict* (resp. *B-strict*) iff it is A-local (resp. B-local) but not AB-common;
- *AB-mixed* if it contains symbols in both $(\Sigma^A \setminus \Sigma^C)$ and $(\Sigma^B \setminus \Sigma^C)$;
- *AB-pure* if it does not contain symbols in both $(\Sigma^A \setminus \Sigma^C)$ and $(\Sigma^B \setminus \Sigma^C)$.

(Sometimes in the literature about interpolation, "A-local" and "B-local" are used to denote what we call here "A-strict" and "B-strict").

5.1 Interpolating Metarules

Our combined interpolation method is based on the abstract framework introduced in [7, 9] (to which, the interested reader is pointed for more details) and used also in [6] that is based on 'metarules.' A metarule applies (bottom-up) to a pair A, B of finite sets of ground formulae[2] producing an equisatisfiable pair of sets of formulae. Each metarule comes with a proviso for its applicability and an instruction for the computation of the interpolant. As an example, consider the metarule (Define0):

[2] In [6, 7, 9], metarules manipulate pairs of finite sets of literals instead of ground formulae; the difference is immaterial.

Table 1. Interpolating Metarules (taken from [7,9]): each rule has a proviso *Prov.* and an instruction *Instr.* for recursively computing the new interpolant ϕ' from the old one(s) $\phi, \phi_1, \ldots, \phi_k$. Metarules are applied *bottom-up* and interpolants are computed *top-down*. Notation $\phi(t/a)$ is used for substitution.

Close1	Close2	Propagate1	Propagate2
$\dfrac{}{A \mid B}$	$\dfrac{}{A \mid B}$	$\dfrac{A \mid B \cup \{\psi\}}{A \mid B}$	$\dfrac{A \cup \{\psi\} \mid B}{A \mid B}$
Prov.: A is unsat. *Instr.*: $\phi' \equiv \bot$.	*Prov.*: B is unsat. *Instr.*: $\phi' \equiv \top$.	*Prov.*: $A \vdash \psi$ and ψ is AB-common *Instr.*: $\phi' \equiv \phi \wedge \psi$.	*Prov.*: $B \vdash \psi$ and ψ is AB-common *Instr.*: $\phi' \equiv \psi \rightarrow \phi$.

Define0	Define1	Define2
$\dfrac{A \cup \{a = t\} \mid B \cup \{a = t\}}{A \mid B}$	$\dfrac{A \cup \{a = t\} \mid B}{A \mid B}$	$\dfrac{A \mid B \cup \{a = t\}}{A \mid B}$
Prov.: t is AB-common, a fresh. *Instr.*: $\phi' \equiv \phi(t/a)$.	*Prov.*: t is A-local and a is fresh. *Instr.*: $\phi' \equiv \phi$.	*Prov.*: t is B-local and a is fresh. *Instr.*: $\phi' \equiv \phi$.

Disjunction1	Disjunction2
$\dfrac{\cdots \quad A \cup \{\psi_k\} \mid B \quad \cdots}{A \mid B}$	$\dfrac{\cdots \quad A \mid B \cup \{\psi_k\} \quad \cdots}{A \mid B}$
Prov.: $\bigvee_{k=1}^{n} \psi_k$ is A-local and $A \vdash \bigvee_{k=1}^{n} \psi_k$. *Instr.*: $\phi' \equiv \bigvee_{k=1}^{n} \phi_k$.	*Prov.*: $\bigvee_{k=1}^{n} \psi_k$ is B-local and $B \vdash \bigvee_{k=1}^{n} \psi_k$. *Instr.*: $\phi' \equiv \bigwedge_{k=1}^{n} \phi_k$.

Redplus1	Redplus2	Redminus1	Redminus2
$\dfrac{A \cup \{\psi\} \mid B}{A \mid B}$	$\dfrac{A \mid B \cup \{\psi\}}{A \mid B}$	$\dfrac{A \mid B}{A \cup \{\psi\} \mid B}$	$\dfrac{A \mid B}{A \mid B \cup \{\psi\}}$
Prov.: $A \vdash \psi$ and ψ is A-local. *Instr.*: $\phi' \equiv \phi$.	*Prov.*: $B \vdash \psi$ and ψ is B-local. *Instr.*: $\phi' \equiv \phi$.	*Prov.*: $A \vdash \psi$ and ψ is A-local. *Instr.*: $\phi' \equiv \phi$.	*Prov.*: $B \vdash \psi$ and ψ is B-local. *Instr.*: $\phi' \equiv \phi$.

ConstElim1	ConstElim2	ConstElim0
$\dfrac{A \mid B}{A \cup \{a = t\} \mid B}$	$\dfrac{A \mid B}{A \mid B \cup \{b = t\}}$	$\dfrac{A \mid B}{A \cup \{c = t\} \mid B \cup \{c = t\}}$
Prov.: a is A-strict and does not occur in A, t. *Instr.*: $\phi' \equiv \phi$.	*Prov.*: b is B-strict and does not occur in B, t. *Instr.*: $\phi' \equiv \phi$.	*Prov.*: c, t are AB-common, c does not occur in A, B, t. *Instr.*: $\phi' \equiv \phi$.

$$\frac{A \cup \{a = t\} \mid B \cup \{a = t\}}{A \mid B} \qquad \begin{array}{l} \textit{Proviso}: t \text{ is } AB\text{-common, } a \text{ is fresh} \\ \textit{Instruction}: \phi' \equiv \phi(t/a). \end{array}$$

It is not difficult to see that the $A \cup B$ is equisatisfiable to $A \cup B \cup \{a = t\}$ since a is a fresh constant that has been introduced to re-name the AB-common term

t according to the proviso of (Define0). The instruction attached to (Define0) allows for the computation of the interpolant ϕ' by eliminating the fresh constant a from the recursively known interpolant ϕ.

The idea is to build an *interpolating metarules refutation* for a given unsatisfiable $A_0 \cup B_0$, i.e. a labeled tree having the following properties: (i) nodes are labeled by pairs of finite sets of ground formulae; (ii) the root is labeled by A_0, B_0; (iii) the leaves are labeled by a pair $\tilde{A}, \tilde{B}$ such that $\bot \in \tilde{A} \cup \tilde{B}$; (iv) each non-leaf node is the conclusion of a metarule and its successors are the premises of that metarule (the complete list of metarules is in Table 1). Once an interpolating metarules refutation has been built, it is possible to recursively compute the interpolant by using (top-down) the instructions attached to the metarules in the tree:

Proposition 2 ([7,9]). *If there exists an interpolating metarules refutation for A_0, B_0 then there is a quantifier-free interpolant for A_0, B_0 (i.e., there exists a quantifier-free AB-common sentence ϕ such that $A_0 \vdash \phi$ and $B_0 \wedge \phi \vdash \bot$). The interpolant ϕ is recursively computed by applying the relevant interpolating instructions of the metarules.*

The idea to design the combination algorithm is the following. We design transformations instructions that can be non-deterministically applied to a pair A_0, B_0. Each of the transformation instructions is *justified by metarules*, in the sense that it is just a special sequence of applications of metarules. The instructions are such that, whenever they are applied exhaustively to a pair such that $A_0 \cup B_0$ is unsatisfiable, they produce a tree which is an interpolating metarules refutation for A_0, B_0 from which an interpolant can be extracted according to Proposition 2.

5.2 A Quantifier-Free Interpolating Algorithm

Let T_i be a stably-infinite and equality interpolating theory over the signature Σ_i such that the $SMT(T_i)$ problem is decidable and $\Sigma_1 \cap \Sigma_2 = \emptyset$ (for $i = 1, 2$). We assume the availability of algorithms for T_1 and T_2 that are able not only to compute quantifier-free interpolants but also the tuples $\underline{v}$ of terms in Definition 2 for equality interpolation. Since the $SMT(T_i)$ problem is decidable for $i = 1, 2$, it is always possible to build an equality interpolating algorithm by enumeration; in practice, better algorithms can be designed (see [32] for $\mathcal{EUF}$, $\mathcal{LST}$, $\mathcal{LAR}$ and [8] for the possibility to use quantifier elimination to this aim).

Let $\Sigma := \Sigma_1 \cup \Sigma_2, T := T_1 \cup T_2$, and A_0, B_0 be a T-unsatisfiable pair of finite sets of ground formulae over the signature $\Sigma^{A_0 \cup B_0}$. Like in the Nelson-Oppen combination method, we have a pre-processing step in which we purify A_0 and B_0 so as to eliminate from them the literals which are neither Σ_1- nor Σ_2-literals. To do this, it is sufficient to repeatedly apply the technique of "renaming terms by constants" described below. Take a term t (occurring in a literal from A_0 or from B_0), add the equality $a = t$ for a fresh constant a and replace all the occurrences of t by a. The transformation can be justified by the following sequence of metarules: Define1, Define2, Redplus1, Redplus2, Redminus1, Redminus2. For example, in the case of the renaming of some term t in A_0, the metarule Define1 is used to add the explicit definition $a = t$ to A_0, the metarule Redplus1 to add the formula

$\phi(a/t)$ for each $\phi \in A_0$, and the metarule Redminus1 to remove from A_0 all the formulæ ϕ in which t occurs (except $a = t$).

Because of purification, from now on, *we assume to manipulate pairs* A, B *of sets of ground formulæ where literals built up of only* Σ_1*- or of only* Σ_2*-symbols occur* (besides free constants): this invariant will be in fact maintained during the execution of our algorithm. Given such a pair A, B, we denote by A_1 and A_2 the subsets of Σ_1^A- and Σ_2^A-formulae belonging to A; the sub-sets B_1 and B_2 of B are defined similarly. Notice that *it is false* that $A \equiv A_1 \cup A_2$ and $B \equiv B_1 \cup B_2$, since quantifier-free formulae can mix Σ_1- and Σ_2-symbols even if the literals they are built from do not.

Before presenting our interpolation algorithm for the combination of theories, we need to import a technique, called *Term Sharing*, from [7]. Suppose that A contains a literal $a = t$, where the term t is AB-common and the free constant a is A-strict (a symmetric technique applies to B instead of A). Then it is possible to "make a AB-common" in the following way. First, introduce a fresh AB-common constant c with the explicit definition $c = t$ (to be inserted both in A and in B, as justified by metarule (Define0)); then replace the literal $a = t$ by $a = c$ and replace a by c everywhere else in A; finally, delete $a = c$ too. The result is a pair (A, B) where basically nothing has changed but a has been renamed to an AB-common constant c (the transformation can be easily justified by a suitable subset of the metarules). Intuitively, the reason why Term Sharing works is because, in the end, the new constant will have to be replaced with the AB-common term t, so the interpolant is not affected by the renaming of a to c.

An *A-relevant atom* is either an atomic formula occurring in A or it is an A-local equality between free constants; an *A-assignment* is a Boolean assignment α to relevant A-atoms satisfying A, seen as a set of propositional formulæ (relevant B-atoms and B-assignments are defined similarly). Below, we use the notation α to denote both the assignment α and the set of literals satisfied by α.

We are now in the position to present the collection of transformations that should be applied non-deterministically and exhaustively to a pair of purified sets of ground formulæ (all the transformations below can be justified by metarules, the justification is straightforward and left to the reader). In the following, let $i \in \{1, 2\}$ and $X \in \{A, B\}$.

Terminate$_i$: if $A_i \cup B_i$ is T_i-unsatisfiable and $\bot \notin A \cup B$, use the interpolation algorithm for T_i to find a ground AB-common θ such that $A_i \vdash_{T_i} \theta$ and $\theta \wedge B_i \vdash_{T_i} \bot$; then add θ and $\bot$ to B.

Decide$_X$: if there is no X-assignment α such that $\alpha \subseteq X$, pick one of them (if there are none, add $\bot$ to X); then update X to $X \cup \alpha$.

Share$_i$: let $\underline{a} = a_1, \ldots, a_n$ be the tuple of the current A-strict free constants and $\underline{b} = b_1, \ldots, b_m$ be the tuple of the current B-strict free constants. Suppose that $A_i \cup B_i$ is T_i-satisfiable, but $A_i \cup B_i \cup \{\underline{a} \cap \underline{b} = \emptyset\}$ is T_i-unsatisfiable. Since T_i is equality interpolating, there must exist AB-common Σ_i-ground terms $\underline{v} \equiv v_1, \ldots, v_p$ such that

$$A_i \cup B_i \vdash_{T_i} (\underline{a} \cap \underline{v} \neq \emptyset) \vee (\underline{b} \cap \underline{v} \neq \emptyset).$$

Thus the union of $A_i \cup \{\underline{a} \cap \underline{v} = \emptyset\}$ and of $B_i \cup \{\underline{b} \cap \underline{v} = \emptyset\}$ is not T_i-satisfiable and invoking the available interpolation algorithm for T_i, we can compute a ground AB-common Σ_i-formula θ such that $A \vdash_{T_i} \theta \vee \underline{a} \cap \underline{v} \neq \emptyset$ and $\theta \wedge B \vdash_{T_i} \underline{b} \cap \underline{v} \neq \emptyset$. We choose among $n * p + m * p$ alternatives in order to non-deterministically update A, B. For the first $n * p$ alternatives, we add some $a_i = v_j$ (for $1 \leq i \leq n$, $1 \leq j \leq p$) to A. For the last $m * p$ alternatives, we add θ to A and some $\{\theta, b_i = v_j\}$ to B (for $1 \leq i \leq m$, $1 \leq j \leq p$). Term sharing is finally applied to the updated pair in order to decrease the number of the A-strict or B-strict free constants.

Let $\mathsf{CI}(T_1, T_2)$ be the procedure that, once run on an unsatisfiable pair A_0, B_0, first purifies it, then non-deterministically and exhaustively applies the transformation rules above, and finally extracts an interpolant by using the instructions associated to the metarules.

Theorem 6. *Let T_1 and T_2 be two signature disjoint, stably-infinite, and equality interpolating theories having decidable SMT problems. Then, $\mathsf{CI}(T_1, T_2)$ is a quantifier-free interpolation algorithm for the combined theory $T_1 \cup T_2$.*

Algorithm $\mathsf{CI}(T_1, T_2)$ paves the way to reuse quantifier-free interpolation algorithms for both conjunctions (see, e.g., [29]) or arbitrary Boolean combinations of literals (see, e.g., [11]). In particular, the capability of reusing interpolation algorithms that can efficiently handle the Boolean structure of formulae seems to be key to enlarge the scope of applicability of verification methods based on interpolants [23]. Indeed, one major issue to address to make $\mathsf{CI}(T_1, T_2)$ practically usable is to eliminate the non-determinism. We believe this is possible by adapting the Delayed Theory Combination approach [3].

6 Conclusion and Related Work

The results of this paper cover several results for the quantifier-free interpolation of combinations of theories that are known from the literature, e.g., $\mathcal{EUF}$ and $\mathcal{LST}$ [32], $\mathcal{EUF}$ and $\mathcal{LAR}$ [11,25,29], $\mathcal{EUF}$ and $\mathcal{LAI}$ [5], $\mathcal{LST}$ with $\mathcal{LAR}$ [32], and $\mathcal{AX}_{\mathtt{diff}}$ with $\mathcal{IDL}$ [6]. To the best of our knowledge, the quantifier-free interpolation of the following combinations are new: (a) $\mathcal{RDS}$ with $\mathcal{LAR}$, $\mathcal{IDL}$, $\mathcal{UTVPI}$, $\mathcal{LAI}$, and $\mathcal{AX}_{\mathtt{diff}}$, (b) $\mathcal{LST}$ with $\mathcal{IDL}$, $\mathcal{UTVPI}$, $\mathcal{LAI}$, and $\mathcal{AX}_{\mathtt{diff}}$, and (c) $\mathcal{AX}_{\mathtt{diff}}$ with $\mathcal{LAR}$, $\mathcal{UTVPI}$, and $\mathcal{LAI}$.

In Section 4.2, we have extensively discussed the closely related work of [32], where the authors illustrate a method to derive interpolants in a Nelson-Oppen combination procedure, provided that the component theories satisfy certain hypotheses. The combination method in [15] has been designed to be efficiently incorporated in state-of-the-art SMT solvers but is complete only for convex theories. An interpolating theorem prover is described in [25], where a sequent-like calculus is used to derive interpolants from proofs in propositional logic, equality with uninterpreted functions, linear rational arithmetic, and their combinations. The "split" prover in [18] applies a sequent calculus for the synthesis of interpolants along the lines of that in [25] and is tuned for predicate abstraction. The "split" prover can handle combinations of theories involving that of arrays without extensionality and fragments of Linear Arithmetic. The CSIsat [2] permits the computation of quantifier-free interpolants over a combination of $\mathcal{EUF}$ and $\mathcal{LAR}$ refining the combination method in [32] as suggested in [29]. A version of MathSAT [11] features interpolation capabilities for $\mathcal{EUF}, \mathcal{LAR}, \mathcal{IDL}, \mathcal{UTVPI}$ and $\mathcal{EUF}+\mathcal{LAR}$ by extending Delayed Theory Combination [3]. Theorem 6 is the key to combine the strength of these tools and to widen the scope of applicability of available interpolation algorithms to richer combinations of theories. Methods [5, 20, 22, 23] for the computation of quantified interpolants in the combination of the theory of arrays and Presburger Arithmetic have been proposed. Our work focus on quantifier-free interpolants by identifying suitable variants of the component theories (e.g., $\mathcal{AX}_{\mathtt{diff}}$ instead of $\mathcal{AX}_{\text{ext}}$ and $\mathcal{LAI}$ instead of Presburger Arithmetic). Orthogonal to our approach is the work in [30] where interpolation algorithm are developed for extensions of convex theories admitting quantifier-free interpolation.

The framework proposed in this paper allows us to give a uniform and coherent view of many results available in the literature and we hope that it will be the starting point for new developments.

References

1. Bacsich, P.D.: Amalgamation properties and interpolation theorems for equational theories. Algebra Universalis 5, 45–55 (1975)
2. Beyer, D., Zufferey, D., Majumdar, R.: cSIsat: Interpolation for LA+EUF. In: Gupta, A., Malik, S. (eds.) CAV 2008. LNCS, vol. 5123, pp. 304–308. Springer, Heidelberg (2008)
3. Bozzano, M., Bruttomesso, R., Cimatti, A., Junttila, T., Ranise, S., van Rossum, P., Sebastiani, R.: Efficient Satisfiability Modulo Theories via Delayed Theory Combination. In: Etessami, K., Rajamani, S.K. (eds.) CAV 2005. LNCS, vol. 3576, pp. 335–349. Springer, Heidelberg (2005)
4. Brillout, A., Kroening, D., Rümmer, P., Wahl, T.: An Interpolating Sequent Calculus for Quantifier-Free Presburger Arithmetic. In: Giesl, J., Hähnle, R. (eds.) IJCAR 2010. LNCS, vol. 6173, pp. 384–399. Springer, Heidelberg (2010)
5. Brillout, A., Kroening, D., Rümmer, P., Wahl, T.: Beyond Quantifier-Free Interpolation in Extensions of Presburger Arithmetic. In: Jhala, R., Schmidt, D. (eds.) VMCAI 2011. LNCS, vol. 6538, pp. 88–102. Springer, Heidelberg (2011)
6. Bruttomesso, R., Ghilardi, S., Ranise, S.: A Combination of Rewriting and Constraint Solving for the Quantifier-Free Interpolation of Arrays with Integer Difference Constraints. In: Tinelli, C., Sofronie-Stokkermans, V. (eds.) FroCos 2011. LNCS, vol. 6989, pp. 103–118. Springer, Heidelberg (2011)
7. Bruttomesso, R., Ghilardi, S., Ranise, S.: Rewriting-based Quantifier-free Interpolation for a Theory of Arrays. In: RTA (2011)
8. Bruttomesso, R., Ghilardi, S., Ranise, S.: From Strong Amalgamation to Modularity of Quantifier-Free Interpolation. Technical Report RI 337-12, Dip. Scienze dell'Informazione, Univ. di Milano (2012)
9. Bruttomesso, R., Ghilardi, S., Ranise, S.: Quantifier-Free Interpolation of a Theory of Arrays. Logical Methods in Computer Science (to appear, 2012)
10. Chang, C., Keisler, J.H.: Model Theory, 3rd edn. North-Holland, Amsterdam (1990)

11. Cimatti, A., Griggio, A., Sebastiani, R.: Efficient Interpolant Generation in Satisfiability Modulo Theories. In: Ramakrishnan, C.R., Rehof, J. (eds.) TACAS 2008. LNCS, vol. 4963, pp. 397–412. Springer, Heidelberg (2008)
12. Enderton, H.B.: A Mathematical Introduction to Logic. Academic Press, New York (1972)
13. Fuchs, A., Goel, A., Grundy, J., Krstić, S., Tinelli, C.: Ground Interpolation for the Theory of Equality. In: Kowalewski, S., Philippou, A. (eds.) TACAS 2009. LNCS, vol. 5505, pp. 413–427. Springer, Heidelberg (2009)
14. Ghilardi, S.: Model theoretic methods in combined constraint satisfiability. Journal of Automated Reasoning 33(3-4), 221–249 (2004)
15. Goel, A., Krstić, S., Tinelli, C.: Ground Interpolation for Combined Theories. In: Schmidt, R.A. (ed.) CADE 2009. LNCS, vol. 5663, pp. 183–198. Springer, Heidelberg (2009)
16. Henzinger, T., McMillan, K.L., Jhala, R., Majumdar, R.: Abstractions from Proofs. In: POPL (2004)
17. Mal'cev, A.I.: Axiomatizable classes of locally free algebras of certain types. Sibirsk. Mat. Ž. 3, 729–743 (1962)
18. Jhala, R., McMillan, K.L.: A Practical and Complete Approach to Predicate Refinement. In: Hermanns, H. (ed.) TACAS 2006. LNCS, vol. 3920, pp. 459–473. Springer, Heidelberg (2006)
19. Jónsson, B.: Universal relational systems. Math. Scand. 4, 193–208 (1956)
20. Kapur, D., Majumdar, R., Zarba, C.: Interpolation for Data Structures. In: SIGSOFT 2006/FSE-14, pp. 105–116 (2006)
21. Kiss, E.W., Márki, L., Pröhle, P., Tholen, W.: Categorical algebraic properties. A compendium on amalgamation, congruence extension, epimorphisms, residual smallness, and injectivity. Studia Sci. Math. Hungar. 18(1), 79–140 (1982)
22. Kovács, L., Voronkov, A.: Finding Loop Invariants for Programs over Arrays Using a Theorem Prover. In: Chechik, M., Wirsing, M. (eds.) FASE 2009. LNCS, vol. 5503, pp. 470–485. Springer, Heidelberg (2009)
23. McMillan, K.: Interpolants from Z3 proofs. In: Proc. of FMCAD (2011)
24. McMillan, K.L.: An Interpolating Theorem Prover. In: Jensen, K., Podelski, A. (eds.) TACAS 2004. LNCS, vol. 2988, pp. 16–30. Springer, Heidelberg (2004)
25. McMillan, K.L.: An Interpolating Theorem Prover. Theor. Comput. Sci. 345(1), 101–121 (2005)
26. Nelson, G., Oppen, D.C.: Simplification by Cooperating Decision Procedures. ACM Transactions on Programming Languages and Systems 1(2), 245–257 (1979)
27. Oppen, D.C.: Reasoning about recursively defined data structures. Journal of the ACM 27, 403–411 (1980)
28. Ringel, C.M.: The intersection property of amalgamations. J. Pure Appl. Algebra 2, 341–342 (1972)
29. Rybalchenko, A., Sofronie-Stokkermans, V.: Constraint Solving for Interpolation. J. of Symbolic Logic 45(11), 1212–1233 (2010)
30. Sofronie-Stokkermans, V.: Interpolation in Local Theory Extensions. In: Furbach, U., Shankar, N. (eds.) IJCAR 2006. LNCS (LNAI), vol. 4130, pp. 235–250. Springer, Heidelberg (2006)
31. Tinelli, C., Harandi, M.T.: A new correctness proof of the Nelson-Oppen combination procedure. In: Proc. FroCoS 1996, pp. 103–119 (1996)
32. Yorsh, G., Musuvathi, M.: A Combination Method for Generating Interpolants. In: Nieuwenhuis, R. (ed.) CADE 2005. LNCS (LNAI), vol. 3632, pp. 353–368. Springer, Heidelberg (2005); Extended version available as Technical Report MSR-TR-2004-108, Microsoft Research (October 2004)

SPARQL Query Containment under RDFS Entailment Regime

Melisachew Wudage Chekol[1], Jérôme Euzenat[1], Pierre Genevès[2], and Nabil Layaïda[1]

[1] INRIA and LIG
[2] CNRS
{melisachew.chekol,firstname.lastname}@inria.fr

Abstract. The problem of SPARQL query containment is defined as determining if the result of one query is included in the result of another for any RDF graph. Query containment is important in many areas, including information integration, query optimization, and reasoning about Entity-Relationship diagrams. We encode this problem into an expressive logic called μ-calculus: where RDF graphs become transition systems, queries and schema axioms become formulas. Thus, the containment problem is reduced to formula satisfiability test. Beyond the logic's expressive power, satisfiability solvers are available for it. Hence, this study allows to exploit these advantages.

1 Introduction

SPARQL is a W3C recommended query language for RDF. The language is being extended with different entailment regimes and regular path expressions[1]. The semantics of SPARQL relies on the definition of basic graph pattern matching that is built on top of RDF simple entailment [11]. However, it may be desirable to use SPARQL to query triples entailed from subclass, subproperty, range, domain, and other relations which can be represented using RDF schema. The SPARQL specification defines the results of queries based on RDF simple entailment. The specification also presents a general parametrized definition of graph pattern matching that can be expanded to other entailments beyond simple entailment. Query answering under the RDFS entailment regime can be achieved via: (1) materialization (computing the deductive closure of the queried graph) [10], (2) rewriting the queries using the schema, and (3) hybrid (combining materialization and query rewriting). We use a technique based on the approaches (1) and (2) to study the problem of SPARQL query containment under the RDFS entailment regime.

Query containment is defined as determining if the result of one query is included in the result of another one for any RDF graph. It has been a central point of research due to its vital role in query optimization, information integration

[1] SPARQL1.1, working draft http://www.w3.org/TR/sparql11-query/

B. Gramlich, D. Miller, and U. Sattler (Eds.): IJCAR 2012, LNAI 7364, pp. 134–148, 2012.

and reasoning about Entity-Relationship diagrams [14]. In [5], a double exponential upper bound is proved for containment of union of conjunctive queries (UCQs) under expressive description logic constraints. Beyond UCQs, containment of (two-way) regular path queries (2RPQs) have been studied extensively [7,3]. These languages are used to query graph databases and containment has been shown to be PSPACE-complete and EXPTIME-hard under the presence of functionality constraints [7]. On the other hand, the containment of conjunctive 2RPQs is EXPSPACE-complete, this bound jumps to 2EXPTIME when considered under expressive description logic (DL) constraints [6]. In fact, this problem has already been implicitly addressed in [5] when $\mathcal{DLR}$ (DLs with n-ary relations) constraints are used. More recently, Path SPARQL (PSPARQL [1]) query containment has been studied in [8] where a double exponential upper bound is established. In this work, we consider the same approach as [8] and prove that containment of PSPARQL queries under RDF schema axioms has a double exponential upper bound. However, it is exponential if the query on the right hand side has a tree structure (cf. for example, [5]). Further, paths are being included in the new version of SPARQL (called SPARQL1.1), thus this work can be used to test containment of path SPARQL queries under the RDFS entailment regime.

To study containment, we apply an approach which has already been successfully applied for XPath [9]. SPARQL is interpreted over graphs, hence we encode it in a graph logic, specifically the alternation-free fragment of the μ-calculus [15] with converse and nominals [18] interpreted over labeled transition systems. We show that this logic is powerful enough to deal with query containment for union of conjunctive SPARQL queries under the RDFS entailment regime. Furthermore, this logic admits exponential time decision procedures that is implemented in practice [18,19,9]. Hence, our approach opens a way to take advantage of these implementations. We introduce a translation of RDF graphs into transition systems and SPARQL queries and RDF schema into μ-calculus formulae. Then, we show how query containment in SPARQL under RDFS entailment can be reduced to unsatisfiability in the μ-calculus.

In summary, the contribution of this work is fourfold: (1) we formulate the problem of query containment under the RDFS entailment regime in three different ways, (2) since paths are included in the new version of SPARQL, this work can be used to determine containment of path queries (under RDF schema as well), (3) we show how to extend the schema language to the description logic $\mathcal{SH}$ (short for, role transitivity $\mathcal{S}$ and role hierarchy $\mathcal{H}$), and (4) we prove a double exponential upper bound for containment.

2 Preliminaries

This section introduces the foundations of RDF(S), SPARQL, and μ-calculus.

2.1 RDF(S)

RDF is a language used to express structured information on the Web as graphs. We present a compact formalization of RDF [11]. Let U, B, and L be three disjoint infinite sets denoting the set of URIs (identifying a resource), blank nodes (denoting an unidentified resource) and literals (a character string or some other type of data) respectively. We abbreviate any union of these sets as for instance, UBL $= U \cup B \cup L$. A triple of the form $(s, p, o) \in \text{UB} \times U \times \text{UBL}$ is called an *RDF triple*. s is the *subject*, p is the *predicate*, and o is the *object* of the triple. Each triple can be thought of as an edge between the subject and the object labelled by the predicate, hence a set of RDF triples is often referred to as an *RDF graph*. RDF has a model theoretic semantics [11].

Example 1 (RDF Graph). Consider the following RDF graph (all identifiers correspond to URIs and _:b is a blank node):

$$G = \{(john, childOf, mary), (childOf, \texttt{sp}, ancestor), (\_{:}\mathrm{b}, hasFather, john), \\ (ancestor, \texttt{dom}, Person), (ancestor, \texttt{range}, Person)\}$$

RDF Schema (RDFS) may be considered as a simple ontology language expressing subsumption relations between classes or properties [11]. Technically, this is an RDF vocabulary used for expressing axioms constraining the interpretation of graphs. The RDFS vocabulary and its semantics are given in [11]. There, inference rules (shown in Table 1) are given which allow to deduce or infer new triples using the schema and RDF graph.

Table 1. RDFS inference Rules

Subclass (sc)	Subproperty (sp)	Typing (dom, range)
$\dfrac{(a, \texttt{sc}, b)\ (b, \texttt{sc}, c)}{(a, \texttt{sc}, c)}$ (1)	$\dfrac{(a, \texttt{sp}, b)\ (b, \texttt{sp}, c)}{(a, \texttt{sp}, c)}$ (3)	$\dfrac{(a, \texttt{dom}, b)\ (x, a, y)}{(x, \texttt{type}, b)}$ (5)
$\dfrac{(a, \texttt{sc}, b)\ (x, \texttt{type}, a)}{(x, \texttt{type}, b)}$ (2)	$\dfrac{(a, \texttt{sp}, b)\ (x, a, y)}{(x, b, y)}$ (4)	$\dfrac{(a, \texttt{range}, b)\ (x, a, y)}{(y, \texttt{type}, b)}$ (6)

Example 2. Using the inference rules, we can infer the triples {(john,type,Person), (mary,type,Person), (john, ancestor,mary)}. Hence, the deductive closure of graph G in Example 1 contains:

$$cl(G) = \{(john, childOf, mary), (childOf, \texttt{sp}, ancestor), (\_{:}\mathrm{b}, hasFather, john), \\ (john, \texttt{type}, Person), (mary, \texttt{type}, Person), (john, ancestor, mary), \\ (ancestor, \texttt{dom}, Person)\}$$

2.2 SPARQL

SPARQL is a W3C recommended query language for RDF [17]. PSPARQL (Path SPARQL) extends SPARQL with regular expression patterns [1]. The only difference between the syntax of SPARQL and PSPARQL is on triple patterns. In this study, we refer to both SPARQL and PSPARQL queries as SPARQL unless explicitly stated. Triple patterns in PSPARQL contain regular expressions in property positions instead of only URIs or variables as it is the case in SPARQL.

Queries are formed based on the notion of query patterns defined inductively from triple patterns: a tuple $t \in \text{UBV} \times e \times \text{UBLV}$, with V a set of variables disjoint from UBL and e a regular expression pattern defined over U and V, is called a triple pattern. Triple patterns grouped together using connectives AND and UNION[2] form *graph patterns* (a.k.a query patterns). A set of triple patterns is called basic graph pattern.

Definition 1. *A SPARQL query pattern q is inductively defined as follows:*

$$q = \ t \in \text{UBV} \times e \times \text{UBLV} \mid q_1 \textit{ AND } q_2 \mid q_1 \textit{ UNION } q_2$$
$$e = \ uri \mid x \mid e \mid e' \mid e \cdot e' \mid e^+ \mid e^*$$

Definition 2. *A SPARQL SELECT query is defined as $q(\overrightarrow{w})$ where $\overrightarrow{w}$ is a tuple of variables in V that are called distinguished variables, and q is a query pattern.*

Example 3 (SPARQL queries). Consider the following queries $q(?x)$ and $q'(?x)$– refer to Table 1 for vocabulary terms– on the graph of Example 1 and 2:

```
SELECT ?x WHERE {  ?x  type Person .  }

SELECT ?x WHERE {{ ?x  ?p  ?y .  ?p  sp*.dom.sc* Person . }
      UNION {?y  ?p  ?x .  ?p  sp*.range.sc* Person . } }
```

Definition 3 (SPARQL under RDFS entailment semantics). *Given an RDF graph G and a basic graph pattern P, a partial mapping function ρ is a solution for G and P under RDFS-entailment, $\rho \in [\![P]\!]_G$, if:*

- *the domain of ρ is exactly the set of variable in P, i.e., $dom(\rho) = V(P)$,*
- *terms in the range of ρ occur in G,*
- *If P′, obtained from P by replacing blank nodes with either URIs, blank nodes, or RDF literals is such that: the RDF graph $sk(\rho(P'))$ is RDFS-entailed by $sk(G)$. The function $sk(.)$ replaces blank nodes with fresh URIs (URIs that are neither in the queried graph nor in the query).*

Since SPARQL's entailment regimes only change the evaluation of basic graph patterns, the evaluation of query patterns can be defined in the standard way [17,16]. The evaluation of query patterns over an RDF graph G is defined inductively:

$$[\![q_1 \text{ AND } q_2]\!]_G = [\![q_1]\!]_G \bowtie [\![q_2]\!]_G$$
$$[\![q_1 \text{ UNION } q_2]\!]_G = [\![q_1]\!]_G \cup [\![q_2]\!]_G \qquad [\![q(\overrightarrow{w})]\!]_G = \pi_{\overrightarrow{w}}([\![q]\!]_G)$$

The projection operator $\pi_{\overrightarrow{w}}$ selects only those part of the mappings relevant to variables in $\overrightarrow{w}$. For detailed discussions we refer the reader to [10,1].

Example 4. The answers to query q and q' (under simple entailment semantics) of Example 3 on graphs G of Example 1 and $cl(G)$ of Example 2 are: $[\![q]\!]_G = \emptyset$ but $[\![q]\!]_{cl(G)} = \{john, mary\}$ and $[\![q']\!]_G = \{john, mary\}$. Thus, $[\![q]\!]_{cl(G)} = [\![q']\!]_G$. Clearly, $[\![q]\!]_G \subseteq [\![q']\!]_G$. Note also that, q when evaluated over G under the RDFS entailment is equivalent to q' evaluated under simple entailment semantics.

[2] We do not consider OPTIONAL and FILTER query patterns as containment over full SPARQL (equally expressive as relational algebra [2]) is undecidable.

Beyond these particular examples, the goal of query containment is to determine whether this holds for any graph.

Definition 4 (Containment). *Given an RDFS schema $\mathcal{S}$ and queries q and q' with the same arity, q is contained in q' under the RDFS entailment regime, denoted $q \sqsubseteq^{\mathcal{S}}_{\text{rdfs}} q'$, iff for any graph G satisfying the schema $\mathcal{S}$, $[\![q]\!]_G \subseteq [\![q']\!]_G$.*

The evaluation of SPARQL queries (also under the RDFS entailment regime) is proved to be PSPACE-complete. However, the evaluation problem is NP-complete for the fragment containing only AND and UNION query patterns [16,1,10].

To determine containment, SPARQL queries are encoded as μ-calculus formulas, next we present a brief introductory about this logic.

2.3 μ-Calculus

The modal μ-calculus [15] is an expressive logic which adds recursive features to modal logic using fixpoint operators. The syntax of the μ-calculus is composed of countable sets of *atomic propositions* AP, a set of *nominals* Nom, a set of *variables* Var, a set of *programs* Prog for navigating in graphs. A μ-calculus formula, φ, can be defined inductively as follows:

$$\varphi ::= \top \mid \bot \mid p \mid X \mid \neg\varphi \mid \varphi \vee \psi \mid \varphi \wedge \psi \mid \langle a \rangle \varphi \mid [a]\varphi \mid \mu X \varphi \mid \nu X \varphi$$

where $p \in AP, X \in Var$ and $a \in Prog$ is either an atomic program or its converse $\bar{a}$. The greatest and least fixpoint operators (ν and μ), respectively introduce general and finite recursion in graphs [15].

The semantics of the μ-calculus is given over a transition system, $K = (S, R, L)$ where S is a non-empty set of nodes, $R : Prog \to 2^{S \times S}$ is the transition function, and $L : AP \to 2^S$ assigns a set of nodes to each atomic proposition or nominal where it holds, such that $L(p)$ is a *singleton* for each nominal p. For converse programs, R can be extended as $R(\bar{a}) = \{(s', s) \mid (s, s') \in R(a)\}$. Besides, a valuation function $V : \text{Var} \to 2^S$ is used to assign a set of nodes to each variable. For a valuation V, variable X, and a set of nodes $S' \subseteq S$, $V[X/S']$ is the valuation that is obtained from V by assigning S' to X. The semantics of a formula, in terms of a transition system K (a.k.a. Kripke structure) and a valuation function, is represented by $[\![\varphi]\!]^K_V$. The semantics of basic μ-calculus formulae is defined as follows:

$$[\![p]\!]^K_V = L(p), p \in AP \cup Nom,\ L(p) \text{ is singleton for } p \in Nom$$

$$[\![X]\!]^K_V = V(X), X \in Var \quad [\![\neg\varphi]\!]^K_V = S \backslash [\![\varphi]\!]^K_V \quad [\![\top]\!]^K_V = S$$

$$[\![\varphi \wedge \psi]\!]^K_V = [\![\varphi]\!]^K_V \cap [\![\psi]\!]^K_V, \quad [\![\varphi \vee \psi]\!]^K_V = [\![\varphi]\!]^K_V \cup [\![\psi]\!]^K_V$$

$$[\![\langle a \rangle \varphi]\!]^K_V = \{s \in S | \exists s' \in S.(s, s') \in R(a) \ \wedge s' \in [\![\varphi]\!]^K_V\}$$

$$[\![[a]\varphi]\!]^K_V = \{s \in S | \forall s' \in S.(s, s') \in R(a) \Rightarrow s' \in [\![\varphi]\!]^K_V\}$$

$$[\![\mu X \varphi]\!]^K_V = \bigcap \{S' \subseteq S | [\![\varphi]\!]^K_{V[X/S']} \subseteq S'\}$$

$$[\![\nu X \varphi]\!]^K_V = \bigcup \{S' \subseteq S | S' \subseteq [\![\varphi]\!]^K_{V[X/S']}\}$$

3 RDF Graphs as Transition Systems

μ-calculus formulas are interpreted over labeled transition systems. Thus, we propose an encoding of an RDF graph as a transition system in which nodes correspond to RDF entities and RDF triples. Edges relate entities to the triples they occur in. Different edges are used for distinguishing the functions (subject, object, predicate). Expressing predicates as nodes, instead of atomic programs, makes it possible to deal with full RDF expressiveness in which a predicate may also be the subject or object of a statement.

Definition 5 (Transition system associated to an RDF graph [8]). *Given an RDF graph, $G \subseteq UB \times U \times UBL$, the transition system associated to G, $\sigma(G) = (S, R, L)$ over $AP = \text{UBL} \cup \{s', s''\}$, is such that:*

- $S = S' \cup S''$ *with S' and S'' the smallest sets such that* $\forall u \in U_G, \exists n_u \in S'$, $\forall b \in B_G, \exists n_b \in S'$, *and* $\forall t \in G, \exists n_t \in S''$,
- $\forall t = (s, p, o) \in G$, $\langle n_s, n_t \rangle \in R(s)$, $\langle n_t, n_p \rangle \in R(p)$, *and* $\langle n_t, n_o \rangle \in R(o)$,
- $L : \text{AP} \to 2^S$; $\forall u \in U_G, L(u) = \{n_u\}$, $\forall b \in B_G, L(b) = S'$, $L(s') = S'$, $\forall l \in L_G, L(l) = \{n_l\}$ *and* $L(s'') = S''$,
- $\forall n_t, n_{t'} \in S''$, $\langle n_t, n_{t'} \rangle \in R(d)$.

The program d is introduced to render each triple accessible to the others and thus facilitate the encoding of queries. The function σ associates what we call a *restricted transition system* to any RDF graph. Formally, we say that a transition system K is a *restricted transition system* iff there exists an RDF graph G such that $K = \sigma(G)$.

A restricted transition system is thus a bipartite graph composed of two sets of nodes: S', those corresponding to RDF entities, and S'', those corresponding to RDF triples. For example, Figure 1 shows the restricted transition system associated with the graph of Example 1. When checking for query containment, we consider the following restrictions: (i) the set of programs is fixed: $Prog = \{s, p, o, d, \bar{s}, \bar{p}, \bar{o}, \bar{d}\}$, and (ii) a model must be a restricted transition system. The latter constraint can be expressed in the μ-calculus as follows:

Proposition 1 (RDF restriction on transition systems [8]). *A formula φ is satisfied by some restricted transition system if and only if $\varphi \wedge \varphi_r$ is satisfiable by some transition system, i.e.* $\exists K_r [\![\varphi]\!]^{K_r} \neq \emptyset \iff \exists K [\![\varphi \wedge \varphi_r]\!]^{K} \neq \emptyset$, *where:*

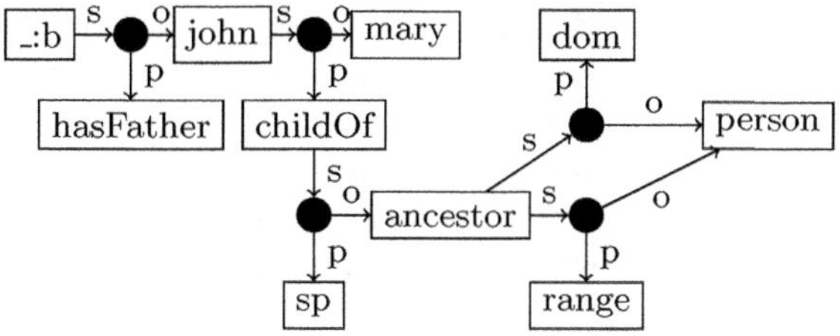

Fig. 1. Transition system encoding the RDF graph of Example 1. Nodes in S'' are black anonymous nodes; nodes in S' are the other nodes (d-transitions are not displayed).

$$\varphi_r = \nu X.\theta \wedge \kappa \ \wedge (\neg\langle d\rangle\top \vee \langle d\rangle X)$$

in which $\theta = \langle\bar{s}\rangle s' \wedge \langle p\rangle s' \wedge \langle o\rangle s' \wedge \neg\langle s\rangle\top \wedge \neg\langle\bar{p}\rangle\top \wedge \neg\langle\bar{o}\rangle\top$ *and* $\kappa = [\bar{s}]\xi \wedge [p]\xi \wedge [o]\xi$ *with*

$$\xi = (\neg\langle\bar{s}\rangle\top \wedge \neg\langle o\rangle\top \wedge \neg\langle p\rangle\top \wedge \neg\langle d\rangle\top \wedge \neg\langle\bar{d}\rangle\top \wedge \neg\langle s\rangle s' \wedge \neg\langle\bar{o}\rangle s' \wedge \neg\langle\bar{p}\rangle s').$$

The formula φ_r ensures that θ and κ hold in every node reachable by a d edge, i.e. in every s'' node. The formula θ forces each s'' node to have a subject, predicate and object. The formula κ navigates from a s'' node to every reachable s' node, and forces the latter not to be directly connected to other subject, predicate or object nodes.

If a μ-calculus formula ψ appears under the scope of a least μ or greatest ν fixed point operator over all the programs $\{s, p, o, d, \bar{s}, \bar{p}, \bar{o}, \bar{d}\}$ as, $\mu X.\psi \vee \langle s\rangle X \vee \langle p\rangle X \vee \cdots$ or $\nu X.\psi \wedge [s]X \wedge [p]X \wedge \cdots$ then, for the sake of legibility, we denote the recursion components of the respective formulae as $mu(X)$ for the μ recursion part and $nu(X)$ for the ν recursion part.

4 Encoding SPARQL Queries

In this section, we show how to encode queries as μ-calculus formulas. Then, in the next section, we use this encoding to test query containment under the RDFS entailment regime. Before discussing the encoding procedure, we briefly assess the issue of blank nodes. Blank nodes are existential variables that denote the existence of unnamed resources. Their definition matches the definition of non-distinguished variables in a query. Thus, blank nodes in the queries can be considered as non-distinguished variables. As a result, every occurrence of a blank node in the query is replaced by a fresh variable.

Queries are translated into μ-calculus formulas. The principle of the translation is that each triple pattern is associated with a sub-formula stating the existence of the triple somewhere in the graph. Hence, they are quantified by μ (least fixed point) so as to put them out of the context of a state. In this translation, variables are replaced by nominals which will be satisfied when they are at the corresponding position in such triple relations. A function called $\mathcal{A}$ is used to encode queries inductively on the structure of query patterns. AND and UNION are translated into boolean connectives $\wedge$ and $\vee$, respectively. When encoding $q \sqsubseteq q'$, we call q left-hand side query and q' right-hand side query. Cyclic dependencies among the non-distinguished variables in the query on the right-hand side create problems in the encoding process: because variables in cycles cannot be simply encoded using atomic propositions (APs) or $\top$. As APs can be true in several nodes in the transition system (resulting in the loss of connectedness). Thus, we provide separate encodings for q and q'.

Encoding left-hand side query: to encode the left-hand side query, one proceeds by encoding the distinguished or non-distinguished variables and constants using nominals. Basically, the variables and constants are frozen (i.e., equivalent to obtaining a canonical instance of the query). Afterwards, a recursive function $\mathcal{A}$ is used to inductively construct a formula. Regular expression patterns that appear in the query are encoded using the function $\mathcal{R}$. It takes two arguments (the predicate which is a regular expression pattern and the object of a triple).

$$\mathcal{A}((x,e,z)) = \mu X.(\langle \bar{s} \rangle x \wedge \mathcal{R}(e,z)) \vee mu(X)$$

$$\mathcal{A}(q_1 \text{ AND } q_2) = \mathcal{A}(q_1) \wedge \mathcal{A}(q_2) \qquad \mathcal{A}(q_1 \text{ UNION } q_2) = \mathcal{A}(q_1) \vee \mathcal{A}(q_2)$$

$$\mathcal{R}(uri,y) = \langle p \rangle uri \wedge \langle o \rangle y \qquad \mathcal{R}(x,y) = \langle p \rangle x \wedge \langle o \rangle y$$

$$\mathcal{R}(e \mid e',y) = (\mathcal{R}(e,y) \vee \mathcal{R}(e',y)) \qquad \mathcal{R}(e \cdot e',y) = \mathcal{R}(e, \langle s \rangle \mathcal{R}(e',y))$$

$$\mathcal{R}(e^+,y) = \mu X.\mathcal{R}(e,y) \vee \mathcal{R}(e, \langle s \rangle X) \qquad \mathcal{R}(e^*,y) = \mathcal{R}(e^+,y) \vee \langle \bar{s} \rangle y$$

In order to encode the right-hand side query, we need the notion of cyclic queries.

Definition 6 (Cyclic Query). *A SPARQL query is referred to as cyclic if a transition graph induced from the query patterns is cyclic. The transition graph*[3] *is constructed in the same way as done in Definition 5.*

Example 5. Consider $q(x) = (x, a \cdot e, y)$ AND (y, b, z) AND (z, c^*, r) AND (r, d, y) which is cyclic, as shown graphically,

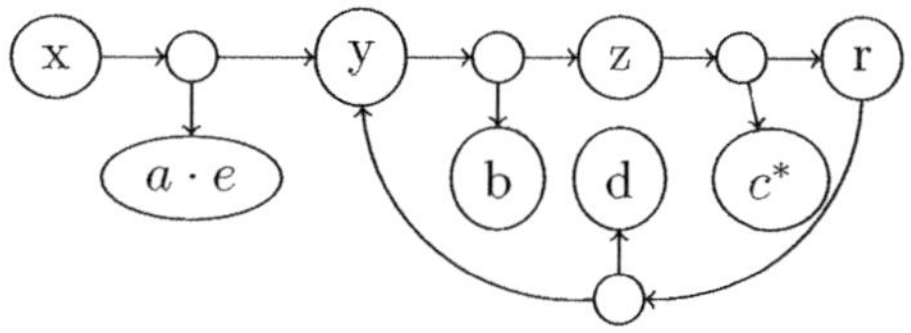

Encoding right-hand side query: the distinguished variables and constants are encoded as nominals whereas the non-distinguished variables are encoded as:

- if a non-distinguished variable appears only once, then it is encoded as $\top$.
- if a non-distinguished variable appears multiple times, then one performs the subsequent steps:
  1. for each $t_i \in q$, $t(t_i) = n_i$, i.e., introduce a nominal for each triple,
  2. for each $z \in t_i = (x_i, e_i, y_i) \in q$, a set of mappings containing formula assignments are generated as:

$$m_i = \{ z \mapsto \psi \mid \begin{cases} \psi = \varphi(s, e_i) \text{ if } subject(z) \wedge e_i \notin var(q) \\ \psi = \langle s \rangle t(t_i) \text{ if } subject(z) \wedge e_i \in var(q) \\ \psi = \varphi(o, e_i) \text{ if } object(z) \wedge e_i \notin var(q) \\ \psi = \langle \bar{o} \rangle t(t_i) \text{ if } object(z) \wedge e_i \in var(q) \\ \psi = \langle \bar{p} \rangle t(t_i) \text{ if } predicate(z) \wedge e_i \in var(q) \end{cases} \}$$

[3] The transition graph is similar to the tuple-graph used in [5] to detect dependency among variables.

s and o denote subject and object of a triple and φ is defined as:

$$\begin{aligned} \varphi(s,a) &= \langle s\rangle\langle p\rangle a & \varphi(o,a) &= \langle\bar{o}\rangle\langle p\rangle a \\ \varphi(s,a\cdot b) &= \varphi(s,a) & \varphi(o,a.b) &= \varphi(o,b) \\ \varphi(s,a \mid b) &= \big(\varphi(s,a)\vee\varphi(s,b)\big) & \varphi(o,a\mid b) &= \big(\varphi(o,a)\vee\varphi(o,b)\big) \\ \varphi(s,a^+) &= \varphi(s,a) & \varphi(o,a^+) &= \varphi(o,a) \\ \varphi(s,a^*) &= \varphi(s,a) & \varphi(o,a^*) &= \varphi(o,a) \end{aligned}$$

Note that there is an exponential number of m_i's in terms of the number of non-distinguished variables. More precisely, there are at most $\mathcal{O}(k^n)$ mappings, where n is the number of triples in which non-distinguished variables appear and k is the number of non-distinguished variables.

– finally function $\mathcal{A}$ works inductively on the query structure using m to generate the formula. As for the left-hand side query, $\mathcal{R}$ is used to produce the encodings of regular expressions.

$$\mathcal{A}(q,m) = \bigvee_{i=1}^{|m|}\mathcal{A}(q,m_i) \qquad d(m,x) = \begin{cases} \psi & \text{if } (x\mapsto\psi)\in m \\ \top & \text{if } \mathit{unique}(x) \\ x & \text{otherwise} \end{cases}$$

$$\begin{aligned} \mathcal{A}\big((x,e,z),m\big) &= \mu X.\big(\langle\bar{s}\rangle d(m,x)\wedge\mathcal{R}(d(m,e),d(m,e))\big)\vee mu(X) \\ \mathcal{A}(q_1 \text{ AND } q_2,m) &= \mathcal{A}(q_1,m)\wedge\mathcal{A}(q_2,m) \\ \mathcal{A}(q_1 \text{ UNION } q_2,m) &= \mathcal{A}(q_1,m)\vee\mathcal{A}(q_2,m) \end{aligned}$$

Example 6 (Encoding queries). Consider the encoding of $q \sqsubseteq q'$, where

$$q(x,z) = (x,(c\mid d)\cdot(a\mid b),z) \qquad q'(x,z) = (x,(c\mid d),y) \text{ AND } (y,a\mid b,z)$$

– The encoding of q is obtained by freezing the query and recursively constructing the formula using $\mathcal{A}$.

$$\begin{aligned} \mathcal{A}(q) &= \mu X.\langle\bar{s}\rangle x\wedge\mathcal{R}\big((c\mid d)\cdot(a\mid b),z\big)\vee mu(X) \\ &= \mu X.\langle\bar{s}\rangle x\wedge(\langle p\rangle c\vee\langle p\rangle d)\wedge\langle o\rangle\langle s\rangle(((\langle p\rangle a\vee\langle p\rangle b)\wedge\langle o\rangle z)\vee mu(X) \end{aligned}$$

– The encoding of q' is as follows:
  - the constants and distinguished variables are encoded as nominals,
  - $y\in var(q')$ is encoded as $\varphi(o,(c\mid d))$, since y is an object of the triple $(x,(c\mid d),y)$. Hence, $m_1 = \{y\mapsto(\langle\bar{o}\rangle\langle p\rangle c\vee\langle\bar{o}\rangle\langle p\rangle d)\}$. On the other hand, y can also be encoded as $\varphi(s,(a\mid b))$, since y is a subject of the triple $(y,a\mid b,z)$. Thus, we get $m_2 = \{y\mapsto(\langle s\rangle\langle p\rangle a\vee\langle s\rangle\langle p\rangle b)\}$.
  - finally, we use $\mathcal{A}$ to encode q' recursively, $\mathcal{A}(q',m) = \mathcal{A}(q',m_1)\vee\mathcal{A}(q',m_2)$

$$\begin{aligned} =\ &\big(\mu X.\langle\bar{s}\rangle x\wedge(\langle p\rangle c\vee\langle p\rangle d)\wedge\langle o\rangle(\langle\bar{o}\rangle\langle p\rangle c\vee\langle\bar{o}\rangle\langle p\rangle d)\vee mu(X) \\ &\quad\wedge\mu Y.\langle\bar{s}\rangle(\langle\bar{o}\rangle\langle p\rangle c\vee\langle\bar{o}\rangle\langle p\rangle d)\wedge(\langle p\rangle a\vee\langle p\rangle b)\wedge\langle o\rangle z\vee mu(Y)\big)\vee \\ &\big(\mu X.\langle\bar{s}\rangle x\wedge(\langle p\rangle c\vee\langle p\rangle d)\wedge\langle o\rangle(\langle s\rangle\langle p\rangle a\vee\langle s\rangle\langle p\rangle b)\vee mu(X) \\ &\quad\wedge\mu Y.\langle\bar{s}\rangle(\langle s\rangle\langle p\rangle a\vee\langle s\rangle\langle p\rangle b)\wedge(\langle p\rangle a\vee\langle p\rangle b)\wedge\langle o\rangle z\vee mu(Y)\big) \end{aligned}$$

Example 7 (Containment test). We show containment of the following queries: select all descendants and ancestors (q) whose names are "john" and (q') who share the same name.

$q(x, y) = (x, name, \text{``}john\text{''})$ AND $(x, ancestor^*, z)$ AND $(z, name, \text{``}john\text{''})$

$q'(x, y) = (x, name, y)$ AND $(x, ancestor^*, z)$ AND $(z, name, y)$

We proceed by first obtaining their encodings. Consider the encoding of $q \sqsubseteq q'$, we encode triple patterns using θ and $m = \{y \mapsto \langle\bar{o}\rangle name\}$.

$$\begin{aligned}\mathcal{A}(q) = & \big(\mu X.\theta(x, name, \text{``}john\text{''}) \vee mu(X)\big)\wedge \\ & \big(\mu X.\theta(x, ancestor^*, z) \vee mu(X)\big)\wedge \\ & \big(\mu X.\theta(z, name, \text{``}john\text{''}) \vee mu(X)\big) \\ \neg\mathcal{A}(q', m) = & \big(\nu X.\neg\theta(x, name, \langle\bar{o}\rangle name) \wedge nu(X)\big)\vee \\ & \big(\nu X.\neg\theta(x, ancestor^*, z) \wedge nu(X)\big)\vee \\ & \big(\nu X.\neg\theta(z, name, \langle\bar{o}\rangle name) \wedge nu(X)\big)\end{aligned}$$

The formula $\mathcal{A}(q) \wedge \neg\mathcal{A}(q', m)$ is unsatisfiable because $\mathcal{A}(q)$ demands its model to satisfy the encoding of each triple pattern somewhere in the transition system. On the contrary, the formula $\neg\mathcal{A}(q', m)$ requests this model to satisfy the negation of the encoding of the triples in the entire transition system. Hence, this leads to a contradiction and no such model exists for the formula. Therefore, $q \sqsubseteq q'$. On the other hand, it can be verified similarly to arrive at $q' \not\sqsubseteq q$.

5 Query Containment under RDFS Entailment

In the following, we propose three approaches to determine query containment under the RDFS entailment regime: encoding the RDFS semantics, query rewriting, and encoding the schema approaches.

5.1 Encoding the RDFS Semantics Approach

When queries are evaluated under the RDFS entailment regime, the queried graph is materialized or saturated using RDFS inference rules (or simply rules) and the schema. Henceforth, implicit or inferred triples are considered when computing the result of the query. Since no specific graphs are considered when dealing with containment, we encode schema and rules. In addition, blank nodes that appear in the schema graph are skolemized, i.e., replaced by fresh constants that do not appear neither in the queries nor schema.

Definition 7. *The encoding of an RDF schema graph $S = \{t_1, \cdots, t_n\}$ is produced by encoding each schema triple $t_i = (x, y, z) \in S$ such that:*

$$\Phi_S = \bigwedge_{i=1 \wedge t_i \in S}^{n} \big(\mu X.(\langle\bar{s}\rangle x \wedge \langle p\rangle y \wedge \langle o\rangle z) \vee mu(X)\big)$$

x, y, and z are atomic propositions corresponding to triple elements.

Definition 8 (Encoding inference rules). *The μ-calculus encoding of RDFS inference rules of Table 1 is the disjunction of formulas (1) to (6) such that:*

$$
\begin{aligned}
&(1)\ \ \nu X.\big(\theta(x,\mathtt{sc},\theta(y,\mathtt{sc},z)) \Rightarrow \theta(x,\mathtt{sc},z)\big) \wedge nu(X)\\
&(2)\ \ \nu X.\big(\theta(x,\mathtt{type},\theta(a,\mathtt{sc},b)) \Rightarrow \theta(x,\mathtt{type},b)\big) \wedge nu(X)\\
&(3)\ \ \nu X.\big(\theta(x,\mathtt{sp},\theta(y,\mathtt{sp},z)) \Rightarrow \theta(x,\mathtt{sp},z)\big) \wedge nu(X)\\
&(4)\ \ \nu X.\big(\theta(x,\theta(a,\mathtt{sp},b),y) \Rightarrow \theta(x,b,y)\big) \wedge nu(X)\\
&(5)\ \ \nu X.\big(\theta(x,\theta(a,\mathtt{dom},b),y) \Rightarrow \theta(x,\mathtt{type},b)\big) \wedge nu(X)\\
&(6)\ \ \nu X.\big(\theta'(x,\theta(a,\mathtt{range},b),y) \Rightarrow \theta(y,\mathtt{type},b)\big) \wedge nu(X)
\end{aligned}
$$

$$\theta(x,y,z) = x \wedge \langle s\rangle(\langle p\rangle y \wedge \langle o\rangle z) \qquad \theta'(x,y,z) = z \wedge \langle \bar{o}\rangle\big(\langle p\rangle(y \wedge \langle \bar{s}\rangle x\big)$$

We denote this formula by Φ_R.

So far, we have produced the encoding of SPARQL queries $\mathcal{A}(q)$ and $\mathcal{A}(q,m)$, RDFS inference rules Φ_R, and schema triples (axioms) Φ_S. In the following, we reduce query containment to unsatisfiability in μ-calculus and prove the correctness of this reduction.

Lemma 1. *Given an RDF schema S and a graph G, $G \models S \Leftrightarrow \Phi_S$ is satisfiable.*

Lemma 2. *For any SPARQL query q, q is satisfiable iff $\mathcal{A}(q)$ and $\mathcal{A}(q,m)$ are satisfiable.*

Proof. (sketch) We prove for $\mathcal{A}(q,m)$, the proof for $\mathcal{A}(q)$ is immediate.
($\Rightarrow$) a model obtained from an instance of q can be converted into a transition system that satisfies $\mathcal{A}(q,m)$.
($\Leftarrow$) any formula corresponding to a query encoding is satisfiable. However, each satisfying model may not be a restricted transition system. Thus, we use $\mathcal{A}(q,m) \wedge \varphi_r$ (Proposition 1), to guarantee that satisfying models are restricted transition systems. As such, it can be shown that a model of the formula $\mathcal{A}(q,m) \wedge \varphi_r$ can be turned into a graph G that satisfies q.

For the sake of legibility, we denote $\Phi_R \wedge \Phi_S \wedge A(q) \wedge \neg A(q',m) \wedge \varphi_r$ by $\Phi(S,q,q')$.

Theorem 1 (Soundness and Completeness). *Given SPARQL queries q and q' and a schema S, $\Phi(S,q,q')$ is unsatisfiable if and only if $q \sqsubseteq^S_{\mathrm{rdfs}} q'$.*

Proof. ($\Rightarrow$) we prove the contrapositive, $q \not\sqsubseteq^S_{\mathrm{rdfs}} q' \Rightarrow \Phi(S,q,q')$ is satisfiable. Assume there exists a graph G that entails the schema graph S, also assume that there exists a tuple $\vec{a} \in [\![q]\!]_G$ and $\vec{a} \notin [\![q']\!]_G$. We construct a restricted transition system K from G. Using Lemma 1, it is obvious that Φ_S is satisfiable in K. Besides, $[\![\varphi_r]\!]^K \neq \emptyset$ (cf. Proposition 1). Now let us use $\vec{a}$ to instantiate the distinguished variables in q and q'. Using the encodings of the instantiated queries and from Lemma 2, one deduces that $[\![\mathcal{A}(q)]\!]^K \neq \emptyset$ and $[\![\mathcal{A}(q',m)]\!]^K = \emptyset$. The later is not satisfiable in K because the nominals corresponding to the constants are not satisfied. Consequently, $[\![\neg\mathcal{A}(q',m)]\!]^K \neq \emptyset$ and $\mathcal{A}(q)\wedge\neg\mathcal{A}(q',m)$ is satisfiable. Therefore, we arrive at $\Phi(S,q,q')$ is satisfiable.

($\Leftarrow$) we show that if $\Phi(S,q,q')$ is satisfiable, then $q \not\sqsubseteq_{rdfs} q'$. Consider a restricted transition system model K for $\Phi(S,q,q')$. We construct an RDF graph G from K. From Lemma 1, it follows that $G \models S$. Thus, it remains to verify that $[\![q]\!]_G \not\subseteq [\![q']\!]_G$. To do so, we start from the assumption, $[\![\mathcal{A}(q)\wedge\neg\mathcal{A}(q',m)]\!]^K \neq \emptyset$. Subsequently, $[\![\mathcal{A}(q)]\!] \neq \emptyset$ and $[\![\mathcal{A}(q',m)]\!]^K = \emptyset$ because G contains all those triples that satisfy q and not q'. Besides, if q' contains a cycle, the constraints expressed by $\neg\mathcal{A}(q',m)$ are satisfied due to the ability, in a μ-calculus extended with nominals and converse, to express a formula that is satisfied in cyclic models. Therefore, $q \not\sqsubseteq^S_{\mathrm{rdfs}} q'$.

5.2 Query Rewriting Approach

SPARQL query containment under RDFS entailment regime can be determined by rewriting queries using the RDFS inference rules (shown in Table 1) and then reducing the encoding of the rewriting to unsatisfiability test. The rewriting is done using PSPARQL as explained in the following definition.

Definition 9 (SPARQL to PSPARQL). *Given a SPARQL query q, a rewriting function τ produces its PSPARQL equivalent as follows:*

$$\tau((s,sc,o)) = (s,sc^+,o) \quad \tau((s,sp,o)) = (s,sp^+,o)$$
$$\tau((s,p,o)) = (s,x,o) \text{ AND } (x,sp^*,p) \text{ such that } p \notin \{sc,sp,type\}$$
$$\tau((s,type,o)) = (s,type.sc^*,o) \text{ UNION } (s,x,y) \text{ AND } (x,\mathtt{sp}^*.\mathtt{dom.sc}^*,o)$$
$$\text{UNION } (y,x,s) \text{ AND } (x,\mathtt{sp}^*.\mathtt{range.sc}^*,o)$$
$$\tau((s,x,o)) = (s,x,o) \text{ when } x \text{ is a variable}$$
$$\tau(q_1 \text{ AND } q_2) = \tau(q_1) \text{ AND } \tau(q_2) \quad \tau(q_1 \text{ UNION } q_2) = \tau(q_1) \text{ UNION } \tau(q_2)$$

Definition 10 (Containment under RDFS entailment). *Given an RDF schema S, queries q and q', and a rewriting function τ. q is contained in q' under RDFS entailment, denoted $q \sqsubseteq^S_{\mathrm{rdfs}} q'$, if and only if $\tau(q) \sqsubseteq^S \tau(q')$.*

Theorem 2 (Soundness and Completeness). *Given an RDF schema S and SPARQL queries q and q', $q \sqsubseteq^S_{\mathrm{rdfs}} q' \Leftrightarrow \Phi_S \wedge \mathcal{A}(\tau(q)) \wedge \neg\mathcal{A}(\tau(q'),m) \wedge \varphi_r$ is unsatisfiable.*

Proof. The proof of this theorem follows from that of Theorem 1.

5.3 Encoding the Schema Approach

In this approach, in order to determine query containment under the RDFS entailment regime, we encode the schema triples (axioms) as formulae. As a consequence, the encoding of the axioms constrains a model satisfying the formula. We consider *subclass*, *subproperty*, *domain*, *range*, and *transitivity* (Tr(sc) or Tr(sp)) schema axioms.

Definition 11. *Given a set of axioms $s_1, s_2, ..., s_n$ of a schema $\mathcal{S}$, the μ-calculus encoding of $\mathcal{S}$ is: $\eta(\mathcal{S}) = \eta(s_1) \wedge \eta(s_2) \wedge ... \wedge \eta(s_n)$.*
We use a function η to translate each s_i into an equivalent μ-calculus formula:

$$\eta((C_1, \texttt{sc}, C_2)) = \nu X.\ (C_1 \Rightarrow C_2) \wedge nu(X)$$
$$\eta((R_1, \texttt{sp}, R_2)) = \nu X.\ (R_1 \Rightarrow R_2) \wedge nu(X)$$
$$\eta((R, \texttt{dom}, C)) = \nu X.\ \big(\langle s\rangle(\langle p\rangle R \Rightarrow \langle p\rangle \texttt{type} \wedge \langle o\rangle C)\big) \wedge nu(X)$$
$$\eta((R, \texttt{range}, C)) = \nu X.\ \big(\langle \bar{o}\rangle\langle p\rangle R \Rightarrow \langle s\rangle(\langle p\rangle \texttt{type} \wedge \langle o\rangle C)\big) \wedge nu(X)$$
$$\eta(Tr(sc)) = \nu X.\big(\theta(x, \texttt{sc}, \theta(y, \texttt{sc}, z)) \Rightarrow \theta(x, \texttt{sc}, z)\big) \wedge nu(X)$$
$$\eta(Tr(sp)) = \nu X.\big(\theta(x, \texttt{sp}, \theta(y, \texttt{sp}, z)) \Rightarrow \theta(x, \texttt{sp}, z)\big) \wedge nu(X)$$

In the following, for legibility, we denote $\Phi(\mathcal{S}, q, q') = \eta(\mathcal{S}) \wedge \mathcal{A}(q) \wedge \neg\mathcal{A}(q', m) \wedge \varphi_r$.

Theorem 3 (Soundness and Completeness). *Given queries q, q', and a set of RDF schema axioms $\mathcal{S}$, $q \sqsubseteq^{\mathcal{S}}_{\text{rdfs}} q'$ if and only if $\Phi(\mathcal{S}, q, q')$ is unsatisfiable.*

Proof. (sketch) *Soundness:* $\Phi(\mathcal{S}, q, q')$ unsatisifiable implies that $q \sqsubseteq^{\mathcal{S}}_{\text{rdfs}} q'$. We show the contrapositive, if $q \not\sqsubseteq^{\mathcal{S}}_{\text{rdfs}} q'$, then $\Phi(\mathcal{S}, q, q')$ is satisfiable, holds. One can verify that every model G of $\mathcal{S}$ in which there is at least one triple satisfying q but not q' can be turned into a transition system model for $\Phi(\mathcal{S}, q, q')$.
Completeness: $\Phi(\mathcal{S}, q, q')$ satisfiable implies $q_1 \not\sqsubseteq^{\mathcal{S}}_{\text{rdfs}} q_2$. Assume that there exists a restricted transition system K that satisfies $\Phi(\mathcal{S}, q, q')$. This entails that, $[\![\varphi_r]\!]^K \neq \emptyset$ (cf. Proposition 1). Now, from $K = (S, R, L)$ we need to construct an RDF graph G that is model of $\mathcal{S}$ such that $q \not\sqsubseteq^{\mathcal{S}}_{\text{rdfs}} q'$ holds:

- for every RDFS concept C in the schema, $\{(s, \texttt{type}, C) \mid \forall s', s'' \in S \wedge t \in S'.(s', t) \in R(s) \wedge (t, s'') \in R(p) \wedge (t, s) \in R(o) \wedge s \in [\![C]\!]^K\}$.
- for each RDFS property P in the schema, $\{(s, P, s') \in G \mid \forall t \in [\![P]\!]^K \wedge t' \in S.(s, t') \in R(s) \wedge (t', t) \in R(p) \wedge (t', s') \in R(o)\}$,
- add every schema axiom to G and for each triple $t_i \in q$, add t_i to G.

Since every RDF graph entails its schema graph, we obtain that G is a model of $\mathcal{S}$. Thus, it remains to show that $[\![q]\!]_G \not\subseteq [\![q']\!]_G$. From our assumption, one anticipates $[\![\mathcal{A}(q) \wedge \neg\mathcal{A}(q')]\!]^K \neq \emptyset$ which implies $[\![\mathcal{A}(q)]\!]^K \neq \emptyset$ and $[\![\mathcal{A}(q', m)]\!]^K = \emptyset$. Note here that, if a formula φ is satisfiable in a restricted transition system K, then $[\![\varphi]\!]^K = S$. Further, it is holds that $[\![q]\!]_G \neq \emptyset$ and $[\![q']\!]_G = \emptyset$ because G contains all those triples that satisfy q and not q'. Therefore, we get $[\![q]\!]_G \not\subseteq [\![q']\!]_G$. Since cycles in queries can be expressed by a formula in a μ-calculus extended with nominals and inverse, the constraints expressed by $\neg\mathcal{A}(q', m)$ are satisfied in a transition system containing cycles.

5.4 Complexity

Due to duplication in the encoding of the right hand side query q', the size of $|\mathcal{A}(q', m)|$ is exponential in terms of the non-distinguished variables that appear in cycles in the query. Thus, we obtain a 2ExpTime upper bound for containment independent of the approaches. That is, the complexity bound applies to all the approaches. As pointed out in [5], the problem is solvable in ExpTime if there is no cycle in the query on the right hand side. In this case, this complexity is a lower bound due to the complexity of satisfiability in μ-calculus.

Proposition 2. *SPARQL query containment under the RDFS entailment can be solved in a time of $2^{\mathcal{O}(n)}$, where n is the size of the encoding.*

All the three approaches have the same complexity bound, the difference lies on their extensibility. While encoding the RDFS semantics (§5.2) and query rewriting (§5.1) approaches are tied to the schema language which makes it harder for easy extension, the schema encoding approach (§5.3) can be extended to use a more expressive schema language than RDFS. For instance, we can extend the schema language to $\mathcal{SH}$ where a concept C can be a bottom concept ($\top$), an atomic concept A, or a complex concept $\neg C$ or $C \sqcap D$. A role r is an atomic role. An $\mathcal{SH}$ TBox consists of concept inclusion, role inclusion and role transitivity axioms [13]. Role inclusion and transitivity axioms can be encoded in the same way as it is done in Definition 11. The encoding of concept inclusion axioms is slightly different, thus, we extend η as follows:

$$\eta((C, \mathsf{sc}, D)) = \nu X. \ (\omega(C) \Rightarrow \omega(D)) \wedge nu(X)$$
$$\omega(\bot) = \bot \qquad \omega(A) = A \qquad \omega(\neg C) = \neg\omega(C) \qquad \omega(C \sqcap D) = \omega(C) \wedge \omega(D)$$

We can expand the proof of Theorem 1, to prove the correctness of this reduction. And thus, retaining the double exponential upper bound. Beyond this, we can even extend $\mathcal{SH}$ to the fragments of $\mathcal{SROIQ}$ [12]. More specifically, the fragments without number restrictions. The expressiveness of the schema language is limited as such due to the expressive power of the logic used for the encoding: μ-calculus with nominals and converse becomes undecidable when extended with graded modalities [4].

6 Conclusion

In this work, we have presented a translation of RDF graphs into labeled transition systems over which μ-calculus formulas are interpreted. We also have provided functions to produce the encodings of queries, inference rules and schema as formulas. Henceforth, query containment under RDFS entailment is reduced to formula satisfiability test in the μ-calculus. We introduced three approaches to achieve this, namely (1) encoding the RDFS semantics, (2) query rewriting, and (3) encoding the schema. Unlike (1) and (2), the third approach can be extended for a more expressive schema language as shown in §5.4, while maintaining a double exponential upper bound complexity. The power of the logic and our encoding allows for taking advantage of more expressive schema language. For instance, a good candidate could be the description logic $\mathcal{SROIQ}$ [12] underlying OWL 2.

In the future, we plan to investigate the optimality of the upper bound considering a more expressive schema language than RDF schema. Additionally, we plan to study containment of path queries with counting quantifiers (SPARQL 1.1 property hierarchies) using a fragment of μ-calculus called graded μ-calculus [4].

References

1. Alkhateeb, F., Baget, J.-F., Euzenat, J.: Extending SPARQL with regular expression patterns (for querying RDF). J. Web Semantics 7(2), 57–73 (2009)
2. Angles, R., Gutierrez, C.: The Expressive Power of SPARQL. In: Sheth, A.P., Staab, S., Dean, M., Paolucci, M., Maynard, D., Finin, T., Thirunarayan, K. (eds.) ISWC 2008. LNCS, vol. 5318, pp. 114–129. Springer, Heidelberg (2008)
3. Barceló, P., Hurtado, C., Libkin, L., Wood, P.: Expressive languages for path queries over graph-structured data. In: PODS 2010, pp. 3–14. ACM (2010)
4. Bonatti, P.A., Lutz, C., Murano, A., Vardi, M.Y.: The Complexity of Enriched μ-Calculi. In: Bugliesi, M., Preneel, B., Sassone, V., Wegener, I. (eds.) ICALP 2006. LNCS, vol. 4052, pp. 540–551. Springer, Heidelberg (2006)
5. Calvanese, D., De Giacomo, G., Lenzerini, M.: Conjunctive Query Containment and Answering under Description Logics Constraints. ACM Trans. on Computational Logic 9(3), 22.1–22.31 (2008)
6. Calvanese, D., Ortiz, M., Simkus, M.: Containment of regular path queries under description logic constraints. In: Proc. of the 22nd Int. Joint Conf. on Artificial Intelligence, IJCAI 2011 (2011)
7. Calvanese, D., Rosati, R.: Answering Recursive Queries under Keys and Foreign Keys is Undecidable. In: Proc. of the 10th Int. Workshop on Knowledge Representation meets Databases (KRDB 2003), vol. 79, pp. 3–14 (2003)
8. Chekol, M.W., Euzenat, J., Genevès, P., Layaïda, N.: PSPARQL query containment. In: DBPL 2011 (August 2011)
9. Genevès, P., Layaïda, N., Schmitt, A.: Efficient Static Analysis of XML Paths and Types. In: PLDI 2007, pp. 342–351. ACM, New York (2007)
10. Glimm, B.: Using SPARQL with RDFS and OWL entailment. In: Polleres, A., d'Amato, C., Arenas, M., Handschuh, S., Kroner, P., Ossowski, S., Patel-Schneider, P. (eds.) Reasoning Web 2011. LNCS, vol. 6848, pp. 137–201. Springer, Heidelberg (2011)
11. Hayes, P.: RDF Semantics. W3C Recommendation (2004)
12. Horrocks, I., Kutz, O., Sattler, U.: The even more irresistible SROIQ. In: Proc. of KR 2006, pp. 57–67 (2006)
13. Horrocks, I., Sattler, U., Tobies, S.: Practical Reasoning for Expressive Description Logics. In: Ganzinger, H., McAllester, D., Voronkov, A. (eds.) LPAR 1999. LNCS, vol. 1705, pp. 161–180. Springer, Heidelberg (1999)
14. Ioannidis, Y.E.: Query Optimization. ACM Comput. Surv. 28(1), 121–123 (1996)
15. Kozen, D.: Results on the propositional μ-calculus. Theor. Comp. Sci. 27, 333–354 (1983)
16. Pérez, J., Arenas, M., Gutierrez, C.: Semantics and complexity of SPARQL. ACM Transactions on Database Systems (TODS) 34(3), 16 (2009)
17. Prud'hommeaux, E., Seaborne, A.: SPARQL Query Language for RDF. W3C Rec (2008)
18. Tanabe, Y., Takahashi, K., Hagiya, M.: A Decision Procedure for Alternation-Free Modal μ-calculi. In: Advances in Modal Logic, pp. 341–362 (2008)
19. Tanabe, Y., Takahashi, K., Yamamoto, M., Tozawa, A., Hagiya, M.: A Decision Procedure for the Alternation-Free Two-Way Modal μ-Calculus. In: Beckert, B. (ed.) TABLEAUX 2005. LNCS (LNAI), vol. 3702, pp. 277–291. Springer, Heidelberg (2005)

Automated Verification of Recursive Programs with Pointers

Frank de Boer[1,2], Marcello Bonsangue[1], and Jurriaan Rot[1]

[1] Leiden Institute of Advanced Computer Science (LIACS), Niels Bohrweg 1, 2333 CA Leiden, Netherlands
{marcello,frb}@liacs.nl
[2] Centrum Wiskunde en Informatica (CWI), Science Park 123, 1098 XG Amsterdam, Netherlands
frb@cwi.nl

Abstract. We present a fully automated method for the verification of annotated recursive programs with dynamic pointer structures. Assertions are expressed in a dialect of dynamic logic extended with nominals and tailored to heap structures, in which one can express complex reachability properties. Verification conditions are generated using a novel calculus for computing the strongest postcondition of statements manipulating the heap, such as dynamic allocation and field-assignment. Further, we introduce a new decidable tableaux-based method and its prototype implementation to automatically check these verification conditions.

1 Introduction

Programs with pointers give rise to so-called heap structures which store the dynamically allocated program variables. A heap can be viewed as a labelled transition system (or Kripke structure) where the transitions, labelled with field names, between the states model the navigation structure.

In this paper we present a formal method for validating annotated recursive programs with pointers. We consider programs for manipulating pointer structures, annotated with assertions in a dialect of propositional dynamic logic [8] to describe the dynamically evolving heap structures. The modalities of this dialect contain regular expressions over the field names and tests on the program variables. The program variables themselves are represented in the logic as a certain kind of propositional variables, so-called nominals, which hold in exactly one state. Assertions in this dialect are used to specify the pre- and postconditions of (recursive) procedures and invariants of while statements, and allow for a succinct description of complex reachability properties of pointer structures.

In order to validate a recursive program annotated with assertions in our dialect of propositional dynamic logic, we introduce a method for generating verification conditions based on a new calculus which allows us to compute for an arbitrary precondition (expressed in our dynamic logic with nominals) and (field-)assignment (including dynamic allocation) the corresponding strongest postcondition.

B. Gramlich, D. Miller, and U. Sattler (Eds.): IJCAR 2012, LNAI 7364, pp. 149–163, 2012.

Further, we present a new semantic tableaux method and its prototype implementation for validating the computed verification conditions in terms of the entailment relation of the logic. Given an annotated program we can thus fully automate its validation.

Related work. Since the pioneering work of Morris [11], proving correctness of programs with pointers has been and still is one of the main challenges in the research area of program verification. Many approaches and corresponding logics have been introduced of which separation logic [14] is one of the more recent and popular ones. However, most of these logics are undecidable. For example, in [5] it is proven that even the purely propositional fragment of separation logic is undecidable. In [2] a decidable fragment of separation logic is introduced which however is restricted to certain kinds of tree structures.

Since first-order logic cannot express basic properties of unbounded heaps like reachability, most approaches also use a form of second-order logic (i.e, recursively defined predicates). In recent papers like [17,10,4] restrictions of such extensions of first-order logic are introduced to obtain decidability. In [10], decidability is obtained by restricting to recursively defined models. As mentioned in [10], the logic introduced in [4] (and its predecessor) "have very awkward syntax that involve the domain being partially ordered with respect to sorts, and the logics are heavily curtailed so that the decision procedure can move down the sorted structures hierarchically and hence terminate". In [17], which is a generalization of [1], a decidable logic is introduced which does not allow nested modalities, contrary to our logic which supports this by its very nature. Finally in [13] and [9] reachability predicates are introduced, with the specific purpose of reasoning about linked lists.

In contrast to the above approaches our starting point is an application of dynamic logic, which is one of the most fundamental logics introduced in computer science for program verification. The decidability of this logic does not require any (syntactic or semantic) restrictions. Further dynamic logic is particulary tailored towards the specification of arbitrary complex reachability properties, whereas the other approaches, based on first-order logic, by their nature focus on structural properties.

In [16] it is shown how to express pre- and postconditions of low-level transformations of Kripke structures in variants of the modal μ-calculus. For the expression of (weakest) preconditions an undecidable variant is used which includes inverse programs. However, the (strongest) postcondition is shown in [16] to be expressible in a decidable variant which excludes inverse programs. In contrast, in this paper we show how to express the strongest postcondition of assignments (including field updates and dynamic allocation) of a high-level programming language. We do so by applying the standard definition of the strongest postcondition in our dialect of propositional dynamic logic extended with quantification of propositional variables and field names. However, because of the restricted use of these quantifiers we can in fact eliminate them in the generated verification conditions and thus remain within the realm of decidability.

In [15] a decidability procedure based on automata-theoretic techniques has been introduced for a hybrid μ-calculus including converse programs, nominals and a universal program. In this paper we introduce a tableaux method which fully exploits the particular characteristics of our logic. More specifically, we show that the absence of a converse program (and a universal program) and the deterministic nature of the atomic programs (represented by the field names) allows for a modular approach which first resolves the propositional connectives and the modalities, then resolves the nominals and finally checks eventualities. As a consequence the complexity of our dialect coincides with that of basic propositional dynamic logic [8]. Further, we have implemented a prototype in the rewriting logic of Maude[1] [7].

Structure of the paper. In the next section we introduce our dialect of propositional dynamic logic with nominals for the description of heaps. In Section 3 we discuss the syntax and the partial correctness semantics of annotated recursive programs. In Section 4 we show how to specify strongest postconditions of assignments in our dynamic logic. How to use these strongest postconditions in the generation of verification conditions is described in Section 5. In Section 6 we introduce our tableaux method, and in 7 we discuss the results of executing several examples on our Maude prototype implementation. With Section 8 we conclude.

2 Heaps in Dynamic Logic

We assume an infinite set V of variables including a distinguished variable $nil \in V$ and ranged over by x, y, z, and an infinite set of fields F ranged over by f, g. We use the set $\mathbb{N}$ of natural numbers to represent locations, ranged over by n, m. A *heap* H is a pair $\langle v, h\rangle$ of a variable assignment $v : V \to \mathbb{N}$ and a field assignment $h : F \to (\mathbb{N} \to \mathbb{N})$ which is strict, i.e., $h(f)(v(nil)) = v(nil)$ for all $f \in F$. We write $H(x)$ for $v(x)$, and $H(f)$ for $h(f)$. For a set of variables Var and a set of fields Fld we denote by $\mathcal{R}_H(Var, Fld)$ the set of *reachable locations* in H starting from these variables over fields in Fld. For technical convenience we assume that for every heap H, $\mathcal{R}_H(V, F)$ is finite. We denote variable update by $H[x := n]$, global field update and store update by $H[f := \rho]$ and $\rho[n := m]$, respectively, where $\rho : \mathbb{N} \to \mathbb{N}$, and, finally, a local field update by $H[f := H(f)[n := m]]$. We use the standard notation and definition of simultaneous assignments and updates.

A brief discussion is in order. Heaps, as introduced above, have both an infinite domain and an infinite range. The motivation for having an infinite number of variables and fields is to simplify the introduction of logical variables in the definition of strongest postconditions, described below. The reason we chose for natural numbers to model "locations" is for an easy implementation of dynamic allocation. Variable and field mappings are total functions. The requirement that there is a variable $nil \in V$ then allows to easily distinguish between variables (or fields) which are "undefined", and variables which are not.

[1] The interested reader is referred to `http://www.liacs.nl/~jrot/verify` which contains our implementation and the full version of this paper.

We introduce a logic for heap abstraction, based on propositional dynamic logic. The basic modalities are the fields F, and V is the set of propositional variables. The syntax of dynamic logic formulas is defined in a standard way as follows:

$$\varphi ::= \bot \mid x \mid \varphi \vee \varphi \mid \neg\varphi \mid \langle\alpha\rangle\varphi \qquad\qquad \alpha ::= f \mid \alpha;\alpha \mid \alpha+\alpha \mid \alpha^* \mid x?$$

where x ranges over V, and f ranges over F. We define $[\alpha]\varphi = \neg\langle\alpha\rangle\neg\varphi$, and use in addition to the above the standard connectives from propositional logic ($\wedge, \rightarrow, \ldots$). The expressions α are called *navigation expressions*. We define satisfaction of basic formulas φ in a standard way [3]: given a heap H together with a natural number n we have

$$\begin{aligned}
H, n &\not\models \bot\\
H, n &\models x \text{ iff } H(x) = n\\
H, n &\models \phi_1 \vee \phi_2 \text{ iff } H, n \models \phi_1 \text{ or } H, n \models \phi_2\\
H, n &\models \neg\phi \text{ iff } H, n \not\models \phi\\
H, n &\models \langle\alpha\rangle\phi \text{ iff } \exists m \in \mathbb{N} \text{ such that } (n, m) \in H(\alpha) \text{ and } H, m \models \phi
\end{aligned}$$

where (the relation) $H(\alpha)$ is the extension of $H(f)$ to arbitrary expressions α, defined by structural induction in the standard manner. A variable or field z is *fresh* in a formula φ if it does not occur in φ. We define $\mathit{Var}(\varphi)$ as the set of all variables which occur in φ, and $\mathit{Field}(\varphi)$ as the set of all fields occuring in φ.

As a first attempt for a notion of validity we define in a standard way $H \models \varphi$ iff $H, n \models \varphi$ for all $n \in \mathbb{N}$. However this forces the semantics of formulas also to take into account properties of garbage, i.e., unreachable parts of the heap; this we can solve by redefining validity as $H \models \varphi$ iff $H, n \models \varphi$ for all $n \in \mathcal{R}_H(V, F)$. Unfortunately, this notion of validity gives rise to a highly complicated check of the corresponding entailment relation $\phi \models \psi$: it requires the construction of counterexample models in which ϕ is satisfied in every state, which is formalized by the implication $\models [(f_1 + \ldots + f_k)^*]\phi \rightarrow \psi$, where $f_1, \ldots, f_k$ are the basic modalities occuring in ϕ and ψ [8]. In order to make this process more tractable we relativize this general notion of validity to the local view from the variables, by the introduction of so-called *rooted formulas*, given by the following grammar:

$$\Phi ::= @x.\varphi \mid \Phi_1 \wedge \Phi_2$$

Then for a formula $\Phi = @x_1.\varphi_1 \wedge \ldots \wedge @x_n.\varphi_n$ and a heap H we define

$$H \models \Phi \text{ iff } \forall i \leq n : H, H(x_i) \models \varphi_i$$

Note that $H(x_i)$ represents the location referenced by x_i. Now given the above Φ and another rooted formula $\Psi = @y_1.\psi_1 \wedge \ldots \wedge @y_m.\psi_m$, checking the entailment $\Phi \models \Psi$ reduces to the construction of a heap H such that $H \models \Phi$ and $H \models @y_i.\neg\psi_i$ for some $i \leq m$.

Note that we thus have introduced a very restricted use of the binding operator in hybrid logic [3]: we only allow top-level occurrences of this operator (e.g., it is not allowed to occur in the scope of any modal operator). Let us discuss the main features of the logic with some examples. Determinism of fields in heaps leads to the fact that the following entailment holds: $@x.\langle f\rangle y \wedge \langle f\rangle z \models @y.z$. Variables are nominals, i.e., true in exactly one world: $@x.y \models @y.x$. A formula $@x.\langle(f+g)^*\rangle y$ states that y is reachable from x over fields f and g. For example, the following entailment holds: $@x.\langle f^*\rangle y \wedge @y.\langle g\rangle z \models @x.\langle(f+g)^*\rangle z$ but $@x.\langle f^*\rangle y \wedge @y.\langle g\rangle z \models @x.\langle f^*\rangle z$ does not hold. Finally, as an example of a linked data structure, some variable x being the head of a (non-circular) linked list is succinctly modelled as $@x.\langle next^*\rangle nil$. Compare this to the formula $@x.\langle next^*\rangle\neg nil$; heaps (with a finite number of reachable locations) which satisfy that formula must have a loop somewhere, i.e., have x as the head of a linked list with a loop. To specify that x is the head of a linked list which does not contain y, we can write $@x.[next^*](\neg y) \wedge \langle next^*\rangle nil$. Finally to specify x has a path back to itself (via pointers named *prev*) for every location reachable from x, we may write $@x.[(f_1+\ldots+f_k)^*]\langle prev^*\rangle x$, for some sequence of fields $f_1,\ldots,f_k$.

We conclude this section with the definition of *substitution* used in the definition of strongest postconditions. Substitution of a variable z for x is denoted $\Phi[z/x]$, and substitution of a field f for a navigation expression α as $\Phi[\alpha/f]$. Both are defined by structural induction in the standard manner. Of particular interest is the case of the @ operator:

$$(@y.\varphi)[z/x] = @(y[z/x]).(\varphi[z/x]) \qquad (@y.\varphi)[\alpha/f] = @y.(\varphi[\alpha/f])$$

The relation between substitution in formulas and allocation in heaps is formalized in the following adaptation of the standard substitution lemma:

Lemma 1 (Substitution). *Let H be a heap, Φ a formula, x, z variables, f a field and α a navigation expression such that $H(\alpha)$ is a function $H(\alpha) : \mathbb{N} \to \mathbb{N}$. Then*

$$H \models \Phi[z/x] \text{ iff } H[x := H(z)] \models \Phi, \text{ and } H \models \Phi[\alpha/f] \text{ iff } H[f := H(\alpha)] \models \Phi$$

Since we allow substitution of general navigation expressions for fields, in order for the update $H[f := H(\alpha)]$ to be well-defined we require that $H(\alpha)$ is deterministic.

3 Programs

In this section we introduce a simple Turing-complete imperative programming language which supports recursion (without local variables), assignment of variables and update of fields, and dynamic allocation. The language is specifically tailored to manipulate pointer structures, and does not support any other data. Methods are annotated with pre- and postconditions, and while loops are annotated with invariants, all of these expressed as rooted formulas.

In order to proceed we assume given finite sets $V_P \subset V$ and $F_P \subset F$ of *program variables* and *program fields* respectively, such that $nil \in V_P$. An *annotated recursive program* is a collection

$$P = \{\{\Phi_1\}p_1 :: S_1\{\Psi_1\}, \ldots, \{\Phi_n\}p_n :: S_n\{\Psi_n\}\}$$

of procedure declarations $p_i :: S_i$ with pre- and postconditions Φ_i and Ψ_i respectively, such that all variables and fields occuring in P are contained in V_P and F_P. In a declaration $p_i :: S_i$ the name of the procedure is given by p_i, and S_i is a statement (the body of the procedure), of which the syntax is given by the following grammar:

$$S ::= \text{if } B \text{ then } S_1 \text{ else } S_2 \text{ fi} \mid (inv : \Theta) \text{ while } B \text{ do } S \text{ od} \mid p \mid S_1; S_2 \mid A \mid \epsilon$$
$$A ::= x := y \mid x := y.f \mid x.f := y \mid x := \text{new}$$
$$B ::= x = y \mid x \neq y$$

where x, y range over V_P ($x \neq nil$), f ranges over F_P, Θ is a rooted formula, p is procedure identifier and ϵ is the empty statement, which we include for technical convenience. The expressions A are called *assignments*; note that these include dynamic allocation. More general operations on the heap can be encoded using only these basic statements. For example, a statement $x := y.f_{i_1} \dots f_{i_k}$ is encoded as $x := y.f_{i_1}; x := x.f_{i_2}; x := x.f_{i_3}; \dots ; x := x.f_{i_k}$. In fact, the operation $x := y$ is not strictly necessary in presence of the others, as it can be encoded as $z.f := y; x := z.f$.

The operational semantics of this language is described by means of a transition relation between configurations which are pairs $\langle H, S\rangle$ of a heap H and a statement S. The semantics of assignments is described by the following transitions:

$$\langle H, x := y\rangle \to \langle H[x := H(y)], \epsilon\rangle \qquad \langle H, x := y.f\rangle \to \langle H[x := H(f)(y)], \epsilon\rangle$$

Field update $x.f := y$ affects aliases of x. We require that x is not aliased with nil for a correct execution of such an update:

$$\frac{H(x) \neq H(nil)}{\langle H, x.f := y\rangle \to \langle H[f := H(f)[H(x) := H(y)]], \epsilon\rangle}$$

Finally the transition of dynamic allocation is as follows:

$$\langle H, x := \text{new}\rangle \to \langle H[x := n][\bar{f} := \bar{\rho}], \epsilon\rangle$$

where $n \in \mathbb{N} \setminus \mathcal{R}_H(V_P \setminus \{x\}, F_P)$ denotes a location not reachable from other program variables through navigation expressions over program fields, $\bar{f}$ is the sequence of program fields F_P, and $\bar{\rho}$ is the sequence such that for every i: $\rho_i = H(f_i)[n := H(nil)]$. The extension to sequential composition, if-then-else, while loops and procedure calls is defined in a standard way.

Definition 1 (Partial correctness). *Given a statement S, precondition Φ and postcondition Ψ we define a correctness triple $\models \{\Phi\}S\{\Psi\}$ to be* valid *if for all heaps H:*

$$H \models \Phi \text{ and } \langle H, S\rangle \to^* \langle H', \epsilon\rangle \text{ implies } H' \models \Psi$$

A program P is correct, *denoted $\models P$, if for every $\{\Phi\}p :: S\{\Psi\} \in P$: $\models \{\Phi\}S\{\Psi\}$.*

Note that the standard rules of Hoare logic with respect to the program constructs hold for the above definition of partial correctness.

Example 1. We illustrate the introduced concepts with an annotated implementation of an insertion into a (non-circular) linked list:

$$\{@x.\langle next^*\rangle nil\}y := \text{new}; y.next := x; x := y\{@x.\langle next^*\rangle nil\}$$

Remember from Section 2 that a (non-circular) linked list with head x is represented succinctly by the formula $@x.\langle next^*\rangle nil$. Note that according to this annotation the property of non-circularity is preserved by the insertion of a new element. On the other hand, the annotated program

$$\{@x.\langle next^*\rangle nil\}y.next := x; x := y\{@x.\langle next^*\rangle nil\}$$

is clearly incorrect because if y itself is already part of the list then its insertion will introduce a circularity, as y sets the *next* field to point to x.

4 Strongest Postconditions of Assignments

In general, verification of annotated programs is based on the generation of verification conditions. This can be done systematically by computing weakest preconditions or strongest postconditions. It is worthwhile to observe that it is unclear and highly problematic how to generate statically the weakest precondition in our logic. For instance, suppose we want to compute the weakest precondition of $x := y.f$ with respect to the rooted formula $@z.x$. The standard way to do so is to substitute $y.f$ for x. However $y.f$ is a programming language construct, which does not have a counterpart in the logic. Through a semantic analysis we might come up with $@y.\langle f\rangle z$, but it is far from clear how to arrive at an equivalent formula in a systematic way. Therefore, our approach is based on computing strongest postconditions, which in contrast does allow a relatively simple characterization based on substitution. However, the standard way of expressing the strongest postconditions requires the introduction of existential quantification to denote the old value of the updated variable, giving rise to an undecidable logic. In our case the field update $x.f := y$ would be particularly problematic since it requires (second-order) existential quantification over fields. Fortunately, in general (top-level) existential quantifiers can be eliminated in the entailment relation. To show this we first extend our logic with top-level existential quantification:

$$\Phi_E = \exists z.\Phi_E \mid \exists g.\Phi_E \mid \Phi$$

where Φ is a rooted formula, z is a variable and g is a field. With $FVar(\Phi_E)$ we denote all variables not bound by a quantifier, and similarly with $FField(\Phi_E)$ all such fields. The semantics is defined as follows: $H \models \exists z.\Phi$ iff $H[z := m] \models \Phi$, for some $m \in \mathbb{N}$, and $H \models \exists g.\Phi$ iff $H[g := \rho] \models \Phi$, for some $\rho : \mathbb{N} \to \mathbb{N}$.

Now we have the following basic logical property of the entailment relation:

Lemma 2. *Let Φ and Ψ be formulas. For any $z \in V \setminus FVar(\Psi)$: $\exists z.\Phi \models \Psi$ iff $\Phi \models \Psi$. For any $g \in F \setminus FField(\Psi)$: $\exists g.\Phi \models \Psi$ iff $\Phi \models \Psi$.*

For the purpose of generating verification conditions, described in the following section, the above lemma justifies the introduction of fresh variables in the strongest postcondition *without* existential quantification. The strongest postcondition $SP(x := y, \Phi)$ of a variable assignment of the form $x := y$ and a formula Φ therefore is given by the formula

$$\Phi[z/x] \wedge @y[z/x].x$$

where z is a fresh variable, not occuring in Φ or V_P. In the special case of an assignment $x := x$, we should have that z equals x, which is indeed taken care of by the substitution in $@y[z/x].x$. The strongest postcondition $SP(x := y.f, \Phi)$ of a variable assignment $x := y.f$ and a formula Φ is similar to the above update $x := y$:

$$\Phi[z/x] \wedge @y[z/x].\langle f \rangle x$$

where z is as again a fresh variable. The strongest postcondition $SP(x.f := y, \Phi)$ of an assignment $x.f := y$ and a formula Φ is given by the formula

$$\Phi[((x?; g) + (\neg x?; f))/f] \wedge @x.(\neg nil \wedge \langle f \rangle y)$$

where g is a fresh field name not occuring in Φ or F_P. Here g represents the old value of f. The formula $@x.\neg nil$ is required in order to match the operational semantics, which states that field update to an alias of *nil* is not allowed. For example when $\Phi = @x.nil$, the above strongest postcondition is a contradiction, which is correct, since any execution of $x.f := y$ on a heap satisfying Φ would block according to the operational semantics. Finally the strongest postcondition $SP(x := \text{new}, \Phi)$ of dynamic allocation $x := \text{new}$ and a formula Φ is given by the formula

$$\Phi[z/x] \wedge \bigwedge_{f \in F_P} (@x.\langle f \rangle nil) \wedge \bigwedge_{v \in V_P \setminus \{x\}} (@v.[(f_1 + \ldots + f_k)^*]\neg x)$$

where $\{f_1, \ldots, f_k\} = F_P$. Intuitively the above formula states that x is unreachable from any other program variable after being allocated, and that its fields are initialized to point to *nil*. We illustrate the above in terms of the insertion into a linked list of Example 1.

Example 2. For notational convenience we assume that $F_P = \{next\}$ and $V_P = \{y, x, nil\}$. We first observe that $SP(y := \text{new}, @x.\langle next^* \rangle nil)$ equals

$$@x.\langle next^* \rangle nil \wedge @y.\langle next \rangle nil \wedge @x.[next^*]\neg y \wedge @nil.[next^*]\neg y$$

Next we compute the strongest postcondition of the above formula for the assignment $y.next := x$:

$$@x.\langle \pi^* \rangle nil \wedge @y.\langle \pi \rangle nil \wedge @x.[\pi^*]\neg y \wedge @nil.[\pi^*]\neg y \wedge @y.(\neg nil \wedge \langle next \rangle x)$$

where π stands for $(y?; f + \neg y?; next)$, for some new field name f. Finally, we compute the strongest postcondition of this formula for the assignment $x := y$:

$$@w.\langle\pi^*\rangle nil \wedge @y.\langle\pi\rangle nil \wedge @w.[\pi^*]\neg y \wedge @nil.[\pi^*]\neg y \wedge @y.(\neg nil \wedge \langle next\rangle w) \wedge @y.x$$

In Section 7 we show that the prototype implementation of our semantic tableaux method reports that the resulting strongest postcondition entails $@x.\langle next^*\rangle nil$. For now we give an informal argument as to why this is the case. To this end, suppose the above strongest postcondition is true in some heap H. First note that π agrees with $next$ on every world where y does not hold. Thus since $@w.[\pi^*]\neg y$ holds it follows by induction that $@w.[next^*]\neg y$ does, too; but then from $@w.\langle\pi^*\rangle nil$ we may conclude $@w.\langle next^*\rangle nil$, which again can be shown inductively. So w is the head of a linked list; and since $@y.\langle next\rangle w$ holds y is, too. Finally $@y.x$ states that x and y are equal, and so $@x.\langle next^*\rangle nil$ as desired.

We conclude this section with the following theorem, stating that the above strongest postconditions are sound and complete with respect to the operational semantics:

Theorem 1. *For every formula Φ: $SP(A,\Phi) \models \Psi$ iff $\models \{\Phi\}A\{\Psi\}$, where A is an assignment, $Var(\Psi) \subseteq V_P \cup Var(\Phi)$ and $Field(\Psi) \subseteq F_P \cup Field(\Phi)$.*

Note that because we do *not* existentially quantify the introduced fresh variables or field names we do *not* have the usual property $\models \{\Phi\}A\{SP(A,\Phi)\}$. However, as argued above, these variables are in fact implicitly existentially quantified in the entailment relation, i.e., in the verification conditions described in more detail in the next section.

5 Generation of Verification Conditions

Given the syntactic descriptions of the strongest postconditions for assignments we now turn to a method for generating verification conditions. Our method is implemented in terms of a rewriting system in Maude; we give here a high-level description. Procedure declarations of the form $\{\Phi\}p :: S\{\Psi\}$ are rewritten as follows:

$$\{\Phi\}p :: S\{\Psi\} \Rightarrow \{\Phi\}S;\epsilon\{\Psi\}$$

Here the empty statement ϵ allows for an easy treatment of sequential composition. For any assignment A, we simply compute the strongest postcondition as given above:

$$\{\Phi\}A;S\{\Psi\} \Rightarrow \{SP(A,\Phi)\}S\{\Psi\}$$

On an if-then-else statement with condition B, we branch into two threads of execution, one where B is true and one where it is not. Note that B is either an equality or a disequality and thus not strictly a formula in our logic; with $\widehat{B}$ we denote a logical formula corresponding to the expression B, which is defined simply as $\widehat{B} = @x.y$ in case B is of the form $x = y$, and $@x.\neg y$, in case it is of the form $x \neq y$. This is formalized by the following rule:

$$\{\Phi\}\text{if } B \text{ then } S_1 \text{ else } S_2 \text{ fi}; S\{\Psi\} \Rightarrow \{\Phi \wedge \widehat{B}\}S_1;S\{\Psi\}, \{\Phi \wedge \widehat{\neg B}\}S_2;S\{\Psi\}$$

A while loop is treated by proving that the current state entails the invariant, that the invariant is preserved on execution of the body and by continuing after the loop with the invariant and the negated loop condition:

$$\frac{\Phi \models \Theta}{\{\Phi\}(inv : \Theta) \text{ while } B \text{ do } S \text{ od}; S'\{\Psi\} \Rightarrow \{\Theta \wedge \widehat{B}\}S\{\Theta\}, \{\Theta \wedge \widehat{\neg B}\}S'\{\Psi\}}$$

On procedure call, we show that the current state entails the precondition of the callee, and then continue with the callee's postcondition.

$$\frac{\{\Phi_i\}p_i :: S_i\{\Psi_i\} \in P \quad \Phi \models \Phi_i}{\{\Phi\}p_i; S\{\Psi\} \Rightarrow \{\Psi_i\}S\{\Psi\}}$$

Finally we terminate succesfully when the current state entails the postcondition:

$$\frac{\Phi \models \Psi}{\{\Phi\}\epsilon\{\Psi\} \Rightarrow \top}$$

The above rules are combined into a relation $\Rightarrow$ lifted to sets of triples of the form $\{t_1, \ldots, t_n\}$, where each t_i is either of the form $\{\Phi\}S_i\{\Psi\}$ or $\{\Phi\}p_i :: S_i\{\Psi\}$.

Definition 2. *Let $\Rightarrow^*$ denote the transitive closure of $\Rightarrow$. We define*

$$\vdash \{\Phi\}S\{\Psi\} \textit{ iff } \{\{\Phi\}S\{\Psi\}\} \Rightarrow^* \{\top\}$$

For an annotated program $P = \{\{\Phi_1\}p_1 :: S_1\{\Psi_1\}, \ldots, \{\Phi_n\}p_n :: S_n\{\Psi_n\}\}$ we define $\vdash P$ iff $\{P\} \Rightarrow^* \{\top\}$.

We proceed to discuss the correctness of the above approach. Unfortunately, we can not generalize Theorem 1 to statements S as it is impossible to achieve completeness. The reason for this is that given the combination of a decidable logic for assertions and a Turing-complete programming language, there is no complete Hoare-style axiom system [6]. We continue to show that our method is sound. A technical difficulty is that for basic assignments $\models \{\Phi\}A\{SP(A,\Phi)\}$ does not hold, since $SP(A,\Phi)$ introduces a fresh variable or field. To overcome this problem we first define an alternative version SP_E of the strongest postconditions of assignments as follows. For any A of type $x := y$, $x := y.f$ or $x :=$ new:

$$SP_E(A,\Phi) = \exists z.SP(A,\Phi)$$

where z is the fresh variable introduced in $SP(S,\Phi)$. Similarly we define

$$SP_E(x.f := y, \Phi) = \exists g.SP(S,\Phi)$$

Now by Lemma 2 and Theorem 1 we have the following:

Corollary 1. *For any formula Φ and assignment A: $\models \{\Phi\}A\{SP_E(A,\Phi)\}$.*

We are now ready to prove the correctness of the method:

Theorem 2 (Soundness). *For any program P: if $\vdash P$, then $\models P$.*

Proof. We prove that for any triple $\{\Phi\}S\{\Psi\}$ such that $Var(\Psi) \subseteq V_P$ and $Field(\Psi) \subseteq F_P$: $\{\{\Phi\}S;\epsilon\{\Psi\}\} \Rightarrow^* \{\top\}$ implies $\models \{\Phi\}S\{\Psi\}$ by induction on the size of statements.

The base case, for the empty statement, follows directly from the definition of correctness triples and the termination rule above.

Let $S; S'$ be a statement such that S is not a composition, $\{\Phi\}S; S'\{\Psi\} \Rightarrow^* \top$, and suppose our claim holds for any statement smaller than $S; S'$. We treat the case that S is an assignment and the case that S is an if-then-else statement.

- Suppose S is an assignment. Then $\{\Phi\}S; S'\{\Psi\}$ can only be rewritten by above rule for action prefixing, so $\{\{SP(S,\Phi)\}S'\{\Psi\}\} \Rightarrow^* \{\top\}$ and since S' is smaller than S, by the induction hypothesis $\models \{SP(S,\Phi)\}S'\{\Psi\}$. Since $Var(\Psi) \subseteq V_P$ and $Field(\Psi) \subseteq F_P$ but the introduced existentially quantified variable or field in $SP_E(S,\Phi)$ is not in V_P or F_P respectively, it is easy to show that $\models \{SP_E(S,\Phi)\}S'\{\Psi\}$ holds – this is in fact a generalized version of a standard rule in Hoare logic for the introduction of existential quantifiers. Further by Corollary 1 we have $\models \{\Phi\}S\{SP_E(S,\Phi)\}$. It follows that $\models \{\Phi\}S; S'\{\Psi\}$ holds.
- Suppose $S =$ if B then S_1 else S_2 fi for some S_1, S_2 and B. Then the if-then-else rule is applied, so both $\{\{\Phi \wedge \widehat{B}\}S_1; S\{\Psi\}\} \Rightarrow^* \{\top\}$ and $\{\{\Phi \wedge \widehat{\neg B}\}S_2; S\{\Psi\}\} \Rightarrow^* \{\top\}$. By the induction hypothesis $\models \{\Phi \wedge \widehat{B}\}S_1; S\{\Psi\}$ and $\models \{\Phi \wedge \widehat{\neg B}\}S_2; S\{\Psi\}$ hold, and thus also $\models \{\Phi\}$if B then S_1 else S_2 fi; $S\{\Psi\}$.

□

6 Deciding Entailment with Semantic Tableaux

In this section we sketch the main characteristics of our tableaux method for deciding entailment between rooted formulas and its prototype implementation in the rewriting logic engine of Maude. To check an entailment $\Phi \models \Psi$ we search for a counterexample, that is, a heap which satisfies Φ but does not satisfy Ψ. In our case, for a given $\Phi = @x_1.\varphi_1 \wedge \ldots \wedge @x_n.\varphi_n$ and $\Psi = @y_1.\psi_1 \wedge \ldots \wedge @y_m.\psi_m$ this means the construction of a heap H such that $H \models \Phi$ and $H \models @y_i.\neg\psi_i$ for some $i \leq m$.

To this end we introduce in Maude the sorts *World*, *Model*, *NewWorld* and *Config*. A term of sort *World* is constructed by the operation

$$op \ \_\_ \ : \ Nat \ Form \rightarrow World$$

where *Form* denotes the sort of dynamic logic formulas. The sort of natural numbers *Nat* is used to identify "worlds", as we will call such terms in the sequel. A heap is represented by a term of sort *Model* which is constructed by

$$op \ \_;\_;\_ \ : \ Worlds \ Transitions \ Nat \rightarrow Model$$

where *Worlds* denotes a sequence of terms of sort *World*, a term of sort *Transitions* denotes a set of labelled transitions between worlds represented by their natural numbers. For technical convenience *Nat* is used to represent a bound on the number of worlds in the model. The standard (structural) tableaux rules for propositional dynamic logic operate on the terms of the sort *NewWorlds* constructed by

$$op\ [\_,\_,\_,\_,\_]\ :\ Nat\ Field\ Form\ Form\ Next \rightarrow NewWorld$$

where

1. the first argument represents its originating world,
2. the second argument represents the transition from that world to this new world "under construction",
3. the third argument is a conjunction of formulas which are supposed to hold in this new world but are still to be processed, whereas
4. the fourth argument represents a conjunction of formulas which are already processed and therefore do hold; finally
5. the last argument *Next* is a set of formulas labelled by fields, indicating the worlds supposed to originate from this one.

Finally a term of sort *Config* is constructed by

$$op\ (\_|\_)\ :\ NewWorlds\ Model \rightarrow Config$$

where *NewWorlds* denotes a sequence of terms of sort *NewWorld.* A term of sort *Config* thus represents a set of new worlds which are to be finalized and a term of sort *Model* which contains *finalized* worlds.

For a proper treatment of the identification of nominals our tableaux method operates on a list of configurations. For example, a disjunction $\varphi \vee \psi$ gives rise to a *split* of the current configuration into two new configurations corresponding to the two disjuncts φ and ψ, respectively. The entailment relation $\Phi \models \Psi$ (Φ and Ψ as above) is therefore represented by a sequence of configurations $C_1, \ldots, C_m$ where each C_i consists of an empty model and a set of worlds under construction representing the formulas:

$$x_1 \wedge \varphi_1, \ldots, x_n \wedge \varphi_n, y_i \wedge \neg\psi_i$$

Our method distinguishes the following three (disjoint) sets of rewrite rules to be applied to individual configurations.

1. First the standard (structural) tableaux rules for dynamic logic are applied on the worlds under construction. A literal conjunct for example is simply transferred to the set of processed formulas. As described above, a disjunction splits the current configuration into two. A formula $\langle\alpha^*\rangle\varphi$ at this stage is treated as a disjunction $\varphi \vee \langle\alpha\rangle\langle\alpha^*\rangle\varphi$. A formula $\langle f\rangle\varphi$ is transferred both to the processed formulas and to the set of next worlds. A formula $[f]\varphi$ is treated similarly (note that f is deterministic). All other formulas are dealt with by the usual reduction axioms of dynamic logic.
2. The second phase consists of the *merging* of finalized worlds containing the same nominal, and its propagation to the transitions to maintain the deterministic nature of fields. Here strictness of *nil* is enforced by adding transitions from *nil* to itself for every field occuring in some finalized world.

3. The final phase consists of a *star track*, which checks if the eventualities of the form $\langle\alpha^*\rangle\varphi$ are indeed validated. This phase is implemented by computing α^* using the given bound on the number of worlds.

Every configuration which passes through these three phases and the consistency check on finalized worlds is a valid counterexample. Other configurations are deleted. We denote by $\Rightarrow_T$ the consecutive application of the above sets of rewrite rules. We have the following main theorem.

Theorem 3. *Let Φ and Ψ be rooted formulas as defined above. Further let $C_1, \ldots, C_m$ be a corresponding sequence of configurations, also as defined above. We have*

$$\Phi \models \Psi \text{ iff } C_1, \ldots, C_m \Rightarrow_T^* \epsilon$$

where $\Rightarrow_T^*$ denotes the transitive closure of $\Rightarrow_T$ and ϵ denotes the empty list.

The consecutive application of the above three (disjoint) sets of rewrite rules, i.e., (1) standard structural rules, (2) propagation of nominal identification and (3) checking eventualites, yields a modular method which allows for a transparent proof. Because of space limitations we only observe here that at the heart of the above theorem lies the basic fact that the consecutive application of the above sets of rewrite rules is indeed correct because of the deterministic interpretation of the field names. For example, adding the converse operator on fields, i.e., f^{-1}, merging two worlds (because of a common nominal) would in general require a renewed validation (and corresponding propagation) of a formula $[f^{-1}]\phi$ of the resulting world. On the other hand, a termination proof for our tableaux method *requires* this consecutive application of the above rewrite rules. Intuitively, no rule in the above first set of (structural) rewrite rules is applicable (to a term of sort *Config*) when all terms of sort *NewWorld* have been processed, i.e., transformed into terms of sort *World*. Distinguishing, in a configuration, between new worlds and finalized worlds allows a simple check whether a processed new world already appears as a final world in the model. In case the processed world already exists we only need to possibly add a new transition (from the originating world to the corresponding final world). Nominal identification at the first phase on finalized worlds may lead to divergence. We illustrate this with an informal example.

Example 3. Suppose in a given configuration there exists a final world $\varphi \wedge x$ for some φ. Furthermore suppose there exists a world under construction in which the formula $[f^*]x$ is to be made true. Processing this formula leads to the introduction of a final world in which $x \wedge [f][f^*]x$ is true (assuming that such a world does not yet exist) and a new world orginating from this fresh final world in which again $[f^*]x$ is to be made true. Because of the nominal x occuring both in $\varphi \wedge x$ and $x \wedge [f][f^*]x$ these worlds can then be identified. As a consequence the above scenario will repeat itself.

7 Example Runs of Prototype Implementation

In this section we provide the output resulting from running some of the introduced examples on our Maude implementation of the tableaux method and the automated verification. For example the following output

```
rewrite in TABLO : ((x . < f * > y),y . < g > z) |= x . < (f + g) * > z .
rewrites: 1444 in 8ms cpu (10ms real) (168082 rewrites/second)
result DelimConfig: {empty}
```

tells us no counterexample could be found, and thus that the entailment on the first line holds (in the implementation we write a rooted formula $@x_1.\varphi_1 \wedge \ldots \wedge @x_n.\varphi_n$ as $x_1.\varphi_1, \ldots, x_n.\varphi_n$). In contrast, the following output shows the generation of a counterexample:

```
rewrite in TABLO : ((x . < f * > y),y . < g > z) |= x . < f * > z .
rewrites: 312 in 0ms cpu (1ms real) (355353 rewrites/second)
result DelimConfig: {(true ; (0 . x & y & - z & (< g > z) & (< f * > y)
 & ([f][f *]- z) & [f *]- z),(1 . - z & ([f][f *]- z) & [f *]- z),3 . z
; < f,0,1 >,< f,1,1 >,< g,0,3 > ; 4),(...)}
```

The above output represents a model with worlds $0 : x \wedge y \wedge \ldots$, $1.\neg z \wedge \ldots$ and $3.z$ and transitions $f : 0 \mapsto 1$, $f : 1 \mapsto 1$ and $g : 0 \mapsto 3$. In this model indeed $@x.\langle f^* \rangle y$ holds (in world 0), and $@y.\langle g \rangle z$ holds since z is true in world 3, but $@x.\langle f^* \rangle z$ does not hold in world 0, since it can not reach z through f-transitions.

As an example of the full implementation of the automated verification based on generating verification conditions and proving them using the entailment checker we try to verify first the following incorrect list insertion in Maude:

```
rewrite in VERIFY :
 ver(m1 . x.< next* >nil . ((y..next) := x); x := y . x.< next* >nil) .
rewrites: 53413 in 91ms cpu (92ms real) (582050 rewrites/second)
result PConfig: (...) {(true ; (0 . x & y & - nil & x ' & (...)
 3 . nil & < ((y ; next ') + (- y ; next)) * > nil ;
 < next,0,0 >,< next',0,3 > ; 7),(...)
```

This returns as a counterexample a model in which x points to a circular list; in fact, $x.next = x$. Interesting to see is that $next'$, the fresh field introduced to represent $next$ before the assignment $y.next := x$, does map world 0 to world 3 (in which nil holds); thus, we can see that $y.next := x$ was exactly the statement introducing the circularity. The corrected version is indeed verified, as expected:

```
rewrite in VERIFY :
  ver(m1 . x.< next * > nil . (y := new[next | x y]) ; ((y .. f) := x) ;
  x := y . x . < f * > nil) .
rewrites: 13378832 in 25262ms cpu (25270ms real) (529585 rewrites/second)
result PConfig: success
```

8 Conclusions and Future Work

We presented the first tool-supported formal method for the automated validation of recursive programs with pointers annotated with assertions in dynamic logic. Future work of interest concerns the extension to recursive programs with local variables. Such programs require the introduction of specific adaptation rules [12] suited for automation. Further work concerns optimizations of the prototype implementation and extensive testing on challenging case studies. Of particular interest in this context is the integration of an automated method for the elimination of the implicitly existentially quantified fresh variables (and field names) in the strongest postconditions using the tableaux method itself. Such a method could possibly be applied to the automated generation of loop invariants. Finally it would be interesting to investigate the integration of the separating conjunction [14] in our logic. Unfortunately, reasoning with the frame rule [14] would be very hard to automate.

References

1. Benedikt, M., Reps, T., Sagiv, M.: A Decidable Logic for Describing Linked Data Structures. In: Swierstra, S.D. (ed.) ESOP 1999. LNCS, vol. 1576, pp. 2–19. Springer, Heidelberg (1999)
2. Berdine, J., Calcagno, C., O'Hearn, P.W.: A Decidable Fragment of Separation Logic. In: Lodaya, K., Mahajan, M. (eds.) FSTTCS 2004. LNCS, vol. 3328, pp. 97–109. Springer, Heidelberg (2004)
3. Blackburn, P., de Rijke, M., Venema, Y.: Modal logic. Cambridge University Press (2001)
4. Bouajjani, A., Drăgoi, C., Enea, C., Sighireanu, M.: A Logic-Based Framework for Reasoning about Composite Data Structures. In: Bravetti, M., Zavattaro, G. (eds.) CONCUR 2009. LNCS, vol. 5710, pp. 178–195. Springer, Heidelberg (2009)
5. Brotherston, J., Kanovich, M.I.: Undecidability of propositional separation logic and its neighbours. In: LICS 2010, pp. 130–139. IEEE (2010)
6. Clarke, E.M.: Programming language constructs for which it is impossible to obtain good hoare-like axioms. Journal of the ACM 26, 126–147 (1979)
7. Clavel, M., Eker, S., Lincoln, P., Meseguer, J.: Principles of maude. ENTCS, vol. 4. Elsevier (2000)
8. Harel, D., Kozen, D., Tiuryn, J.: Dynamic Logic. MIT Press (2000)
9. Lahiri, S.K., Qadeer, S.: Verifying properties of well-founded linked lists. In: POPL 2006, pp. 115–126. ACM (2006)
10. Madhusudan, P., Parlato, G., Qiu, X.: Decidable logics combining heap structures and data. In: POPL 2011, pp. 611–622. ACM (2011)
11. Morris, J.M.: Assignment and linked data structures. In: Theoretical Foundations of Programming Methodology (1982)
12. Naumann, D.A.: Calculating sharp adaptation rules. Information Processing Letters 77 (2000)
13. Nelson, G.: Verifying Reachability Invariants of Linked Structures. In: POPL 1983, pp. 38–47. ACM (1983)
14. Reynolds, J.C.: Separation logic: a logic for shared mutable data structures. In: LICS 2002, pp. 55–74. IEEE (2002)
15. Sattler, U., Vardi, M.Y.: The hybrid μ-calculus. In: Goré, R.P., Leitsch, A., Nipkow, T. (eds.) IJCAR 2001. LNCS (LNAI), vol. 2083, pp. 76–91. Springer, Heidelberg (2001)
16. Tanabe, Y., Sekizawa, T., Yuasa, Y., Takahashi, K.: Pre- and post-conditions expressed in variants of the modal μ-calculus. IEICE Transactions (2009)
17. Yorsh, G., Rabinovich, A.M., Sagiv, M., Meyer, A., Bouajjani, A.: A Logic of Reachable Patterns in Linked Data-Structures. In: Aceto, L., Ingólfsdóttir, A. (eds.) FOSSACS 2006. LNCS, vol. 3921, pp. 94–110. Springer, Heidelberg (2006)

Security Protocols, Constraint Systems, and Group Theories*

Stéphanie Delaune[1], Steve Kremer[2], and Daniel Pasaila[1,3]

[1] LSV, CNRS & ENS Cachan & INRIA Saclay Île-de-France, France
[2] LORIA, INRIA Nancy Grand Est, France
[3] Google, Inc.

Abstract. When formally analyzing security protocols it is often important to express properties in terms of an adversary's inability to distinguish two protocols. It has been shown that this problem amounts to deciding the equivalence of two constraint systems, *i.e.*, whether they have the same set of solutions. In this paper we study this equivalence problem when cryptographic primitives are modeled using a group equational theory, a special case of monoidal equational theories. The results strongly rely on the isomorphism between group theories and rings. This allows us to reduce the problem under study to the problem of solving systems of equations over rings. We provide several new decidability and complexity results, notably for equational theories which have applications in security protocols, such as exclusive or and Abelian groups which may additionally admit a unary, homomorphic symbol.

1 Introduction

Automated verification methods used for the analysis of security protocols have been shown extremely successful in the last years. They have for instance been able to discover flaws in the Single Sign On Protocols used in Google Apps [5]. In 2001, J. Millen and V. Shmatikov [19] have shown that confidentiality properties can be encoded as satisfiability of a constraint system. This approach has been widely studied and extended both in terms of the supported cryptographic primitives and security properties (*e.g.* [11,7]).

Recently, many works have concentrated on indistinguishability properties, which state that two slightly different protocols *look the same* to an adversary who interacts with either one of the protocols. The notion of indistinguishability can be modelled using equivalences from cryptographic calculi (*e.g.* [3,2]) and are useful to model a variety of properties such as resistance to guessing attacks in password based protocols [7] as well as anonymity like properties in various applications [16,4]. More generally, indistinguishability allows one to model security by the means of ideal systems, which are correct by construction [3].

* The research leading to these results has received funding from the European Research Council under the European Union's Seventh Framework Programme (FP7/2007-2013) / ERC grant agreement n° 258865, project ProSecure, and the project JCJC VIP ANR-11-JS02-006.

B. Gramlich, D. Miller, and U. Sattler (Eds.): IJCAR 2012, LNAI 7364, pp. 164–178, 2012.

In 2005, M. Baudet has shown that the equivalence of traces can again be encoded using constraint systems: instead of deciding whether a constraint system is satisfiable one needs to decide whether two constraint systems have the same set of solutions. M. Baudet [7], and later Y. Chevalier and M. Rusinowitch [10], have proven the equivalence of two constraint systems decidable when cryptographic primitives are modelled by a subterm convergent equational theory. Subsequently more practical procedures have been implemented in prototype tools [8,22].

Our contributions. We continue the study of the problem of deciding the equivalence of constraint systems used to model security protocols. In particular we consider the case where cryptographic primitives are modelled using a *group theory.* Group theories are a special case of monoidal theories which have been extensively studied by F. Baader and W. Nutt [20,6] who have provided a complete survey of unification in these theories. Group theories include theories for exclusive or and Abelian groups. These theories are useful to model many security protocols (see [13]), as well as for modeling low level properties of encryption schemes and chaining modes.

More precisely we provide several new decidability and complexity results for the equivalence of constraint systems. We consider exclusive or and Abelian Groups which may also contain a unary homomorphic symbol. Our results rely on an encoding of the problem in systems of equations on a ring associated to the equational theory under study.

We may note that these equational theories have been previously studied for deciding the satisfiability of constraint systems [17] and for the static equivalence problem [12]. To the best of our knowledge these are however the first results to decide equivalence of constraint systems for these theories, which in contrast to static equivalence considers the presence of a fully active adversary. We also note that studying group theories may seem very restricted since they do not contain the equational theories for classical operators like encryption or signatures. However, combination results for disjoint equational theories for the problems of satisfiability of constraint systems [9] and static equivalence [12] have already been developed and we are confident that similar results can be obtained for equivalence properties.

Outline of the paper. In Section 2 we recall some basic notation and the central notion of group theory. Then, in Section 3, we introduce the notion of constraint systems and define the two problems we are interested in. The sections 4, 5, and 6 are devoted to the study of the satisfiability and equivalence problems. Our results are summarized in Section 6. Detailed proofs of our results can be found in [15].

2 Preliminaries

2.1 Terms

A *signature* Σ consists of a finite set of *function symbols*, each with an arity. A function symbol with arity 0 is a *constant symbol.* We assume that $\mathcal{N}$ is an infinite set of *names* and $\mathcal{X}$ an infinite set of *variables*. The concept of names is borrowed from the applied pi calculus [2] and is used to model fresh, secret values. Let $\mathcal{A}$ be a set of atoms which may consist of names and variables. We denote by $\mathcal{T}(\Sigma, \mathcal{A})$ the set of *terms* over $\Sigma \cup \mathcal{A}$. We write $n(t)$ (resp. $v(t)$) for the set of names (resp. variables) that occur in the term t. A term is *ground* if it does not contain any variable. A *substitution* σ is a mapping from a finite subset of $\mathcal{X}$ called its domain and written $dom(\sigma)$ to $\mathcal{T}(\Sigma, \mathcal{N} \cup \mathcal{X})$. Substitutions are extended to endomorphisms of $\mathcal{T}(\Sigma, \mathcal{X})$ as usual. We use a postfix notation for their application.

2.2 Group Theories

Equational theories are very useful for modeling the algebraic properties of the cryptographic primitives. Given a signature Σ, an equational theory E is a set of equations (*i.e.*, a set of unordered pairs of terms in $\mathcal{T}(\Sigma, \mathcal{X})$). Given two terms u and v such that $u, v \in \mathcal{T}(\Sigma, \mathcal{N} \cup \mathcal{X})$, we write $u =_{\mathsf{E}} v$ if the equation $u = v$ is a consequence of E. In this paper, we are particularly interested in the class of group theories, a special case of monoidal theories introduced by F. Baader [6] and W. Nutt [20]. It captures many theories with AC properties, which are known to be difficult to deal with.

Definition 1 (group theory). *A theory* E *over* Σ *is called a* group theory *if it satisfies the following properties:*

1. *The signature* Σ *contains a binary function symbol* $+$, *a unary symbol* $-$ *and a constant symbol* 0. *All other function symbols in* Σ *are unary.*
2. *The symbol* $+$ *is associative-commutative with unit* 0 *and inverse* $-$. *This means that the equations* $x + (y + z) = (x + y) + z$, $x + y = y + x$, $x + 0 = x$ *and* $x + (-x) = 0$ *are in* E.
3. *Every unary function symbol* $\mathsf{h} \in \Sigma$ *is an endomorphism for* $+$ *and* 0, i.e. $\mathsf{h}(x + y) = \mathsf{h}(x) + \mathsf{h}(y)$ *and* $\mathsf{h}(0) = 0$.

Note that a group theory on a given signature Σ may contain arbitrary additional equalities over Σ. The only requirement is, that at least the laws given above hold. By abuse of notation we sometimes write $t_1 - t_2$ for $t_1 + (-t_2)$.

Example 1. Suppose $+$ is a binary function symbol and 0 a constant. Moreover assume that the others symbols, *i.e* $-$, h, are unary symbols. The equational theories below are group theories.

- The theory ACUN (*exclusive or*) over $\Sigma = \{+, 0\}$ which consist of the axioms for associativity $(x + y) + z = x + (y + z)$ and commutativity $x + y = y + x$ (AC), unit $x + 0 = x$ (U) and Nilpotency $x + x = 0$ (N).[1]
- The theory AG (*Abelian groups*) over $\Sigma = \{+, -, 0\}$ which is generated by the axioms (AC), (U) and $x + -(x) = 0$ (Inv). Note that the equations $-(x + y) = -(x) + -(y)$ and $-0 = 0$ are consequences of the others.

[1] We here omit to explicit the inverse symbol $-$ as it acts as the identity, *i.e.* $-x = x$.

- The theories ACUNh over $\Sigma = \{+, \mathsf{h}, 0\}$ and AGh over $\Sigma = \{+, -, \mathsf{h}, 0\}$: these theories correspond to the ones described above extended by the homomorphism laws (h) for the symbol h, *i.e.*, $\mathsf{h}(x + y) = \mathsf{h}(x) + \mathsf{h}(y)$ and $\mathsf{h}(0) = 0$.

Other examples of monoidal and group theories can be found in [20].

2.3 Group Theories Define Rings

Group theories have an algebraic structure which are *rings*.

Definition 2 (ring). *A* ring *is a set $\mathcal{R}$ (called the* universe *of the ring) with distinct elements* 0 *and* 1 *that is equipped with two binary operations* + *and* · *such that $(\mathcal{R}, +, 0)$ is an Abelian group, $(\mathcal{R}, \cdot, 1)$ is a monoid, and the following identities hold for all $\alpha, \beta, \gamma \in \mathcal{R}$:*

- $(\alpha + \beta) \cdot \gamma = \alpha \cdot \gamma + \beta \cdot \gamma$ *(right distributivity)*
- $\alpha \cdot (\beta + \gamma) = \alpha \cdot \beta + \alpha \cdot \gamma$ *(left distributivity)*

We call the binary operations + and · respectively the *addition* and the *multiplication* of the ring. The elements 0 and 1 are called respectively *zero* and *unit.* The (additive) inverse of an element $a \in \mathcal{R}$ is denoted $-a$. A ring is *commutative* if its multiplication is commutative.

It has been shown in [20] that for any group theory E there exists a corresponding ring $\mathcal{R}_{\mathsf{E}}$. We can rephrase the definition of $\mathcal{R}_{\mathsf{E}}$ as follows. Let a be a name ($a \in \mathcal{N}$), the universe of $\mathcal{R}_{\mathsf{E}}$ is $\mathcal{T}(\Sigma, \{a\})/\mathsf{E}$, that is the set of equivalence classes of terms built over Σ and a under equivalence by the equational axioms E. The constant 0, the sum + and the additive inverse − of the ring are defined as in the algebra $\mathcal{T}(\Sigma, \{a\})/\mathsf{E}$.

Given an element of the ring $\mathcal{R}_{\mathsf{E}}$, and a term v, multiplication in the ring is defined by $u \cdot v := u[a \mapsto v]$ where $u[a \mapsto v]$ denotes the term u where any occurrence of a has been replaced by v. It can be shown [20] that $\mathcal{R}_{\mathsf{E}}$ is commutative if, and only if, E has commuting homomorphisms, *i.e.*, $\mathsf{h}_1(\mathsf{h}_2(x)) =_{\mathsf{E}} \mathsf{h}_2(\mathsf{h}_1(x))$ for any two homomorphisms h_1 and h_2. For instance, we have that:

- The ring $\mathcal{R}_{\mathsf{ACUN}}$ consists of the two elements 0 and 1 and we have $0 + 1 = 1 + 0 = 1$, $0 + 0 = 1 + 1 = 0$, $0 \cdot 0 = 1 \cdot 0 = 0 \cdot 1 = 0$, and $1 \cdot 1 = 1$. Hence, $\mathcal{R}_{\mathsf{ACUN}}$ is isomorphic to the commutative ring (field) $\mathbb{Z}/2\mathbb{Z}$.
- The ring $\mathcal{R}_{\mathsf{AGh}}$ is isomorphic to $\mathbb{Z}[\mathsf{h}]$ which is a commutative ring. Note that there are two homomorphisms in the theory AGh, namely − and h and these two homomorphisms commute: $\mathsf{h}(-x) = -(\mathsf{h}(x))$.

By abuse of notation, we often omit the · and we mix up the elements of isomorphomic rings. Thus, we will write $2v$ instead of $(a + a) \cdot v$, and $(\mathsf{h} + \mathsf{h}^2)v$ instead of $(\mathsf{h}(a) + \mathsf{h}^2(a)) \cdot v$.

3 Constraint Systems

As mentioned in the introduction, constraint systems are quite common (see *e.g.* [19,11,7]) to model the possible executions of a protocol once an interleaving has been fixed. We recall here their formalism.

3.1 Definitions

Following the notations of [7], we consider a new set $\mathcal{W}$ of variables, called *parameters* $w_1, w_2, \ldots$ and a new set $\mathcal{X}^2$ of variables called *second-order variables* $X, Y, \ldots$, each variable with an arity, denoted $ar(X)$. We call $\mathcal{T}(\Sigma, \mathcal{W} \cup \mathcal{X}^2)$ the set of second-order terms and $\mathcal{T}(\Sigma, \mathcal{N} \cup \mathcal{X})$ the set of first-order terms, or simply terms. Given a term t we denote by $\mathsf{var}^1(t)$ (resp. $\mathsf{var}^2(t)$) the first-order (resp. second-order) variables of t, *i.e.* $\mathsf{var}^1(t) = v(t) \cap \mathcal{X}$ (resp. $\mathsf{var}^2(t) = v(t) \cap \mathcal{X}^2$). We lift these notations to sets and sequences of terms as expected.

Definition 3. *A* constraint system *is a triple* $(\Phi; \mathcal{D}; \mathcal{E})$ *where:*

- Φ *is a sequence of the form* $\{w_1 \rhd t_1, \ldots, w_\ell \rhd t_\ell\}$ *where* t_i *are terms and* w_i *are parameters;*
- $\mathcal{D}$ *is a set of deducibility constraints of the form* $X \rhd^? x$ *with* $ar(X) < \ell$*;*
- $\mathcal{E}$ *is a set of equalities of the form* $s =^?_{\mathsf{E}} s'$ *where* s, s' *are first-order terms.*

In the following we will, by abuse of notation, confuse sequences $\{w_1 \rhd t_1, \ldots, w_\ell \rhd t_\ell\}$ with corresponding substitutions $\{w_1 \to t_1, \ldots, w_\ell \to t_\ell\}$. We will not formally introduce a language for describing protocols and we only informally describe how a constraint system is associated to an interleaving of a protocol. We refer to [19] for a more detailed description. We simply suppose that protocols may perform three kinds of action:

- A protocol may output terms. These terms correspond to the t_i in Φ and represent the adversary's knowledge after having executed part of the protocol. We call the sequence of terms Φ the *frame*.
- A protocol may input terms which can be computed by the adversary. Each input corresponds to a deducibility constraint $X \rhd^? x \in \mathcal{D}$. The second-order variable X of arity k has to be instantiated by a context over the terms $t_1, \ldots, t_k$. This models the computation, used by the adversary to deduce the first-order term that will instantiate x.
- A protocol may perform tests on inputs to check that the terms match some expected values. These tests are modelled by the equality constraints in $\mathcal{E}$ and may as such contain the variables x which correspond to previously received inputs.

Example 2. Consider the group theory AG and an interleaving of a protocol described by the following sequence:

$$\mathsf{out}(a).\mathsf{out}(b).\mathsf{in}(x_1).\mathsf{out}(c + 2x_1).\mathsf{in}(x_2).[x_1 + x_2 = c]$$

where a, b, and c are names in $\mathcal{N}$, $\mathsf{out}(t)$ models the output of term t, $\mathsf{in}(x)$ the input of a term that will be bound to x and $[t_1 = t_2]$ models the conditional which tests that t_1 and t_2 are equal modulo AG, after having instantiated previous input variables. This protocol yields the constraint system $\mathcal{C} = (\Phi; \mathcal{D}; \mathcal{E})$ where:

- $\Phi = \{w_1 \rhd a, w_2 \rhd b, w_3 \rhd c + 2x_1\}$,
- $\mathcal{D} = \{X_1 \rhd^? x_1, X_2 \rhd^? x_2\}$ with $ar(X_1) = 2$ and $ar(X_2) = 3$, and
- $\mathcal{E} = \{x_1 + x_2 =^?_{\mathsf{E}} c\}$.

Indeed the three elements of the sequence Φ correspond to the three outputs of the protocol. The two deduction constraints in $\mathcal{D}$ model that the adversary needs to provide the inputs. Note that the first input occurs after two outputs. Hence the adversary may refer to w_1 and w_2, but not w_3. This is modelled by setting $ar(X_1) = 2$. As the second input occurs after three outputs we have that $ar(X_2) = 3$. Finally, $\mathcal{E}$ simply consists in the test performed by the protocol.

The *size* of a frame $\Phi = \{w_1 \rhd t_1, \ldots, w_\ell \rhd t_\ell\}$, denoted $|\Phi|$, is its length ℓ. We also assume the following conditions are satisfied on a constraint system:

1. for every $x \in \mathsf{var}^1(\mathcal{C})$, there exists a unique X such that $(X \rhd^? x) \in \mathcal{D}$, and each variable X occurs at most once in $\mathcal{D}$;
2. for every $1 \leq k \leq \ell$, for every $x \in \mathsf{var}^1(t_k)$, there exists $(X \rhd^? x) \in \mathcal{D}$ such that $ar(X) < k$.

These constraints are natural whenever the constraint system models an interleaving of a protocol. Condition 1 simply states that each variable defines a unique input. Condition 2 ensures a form of causality: whenever a term t_k is output it may only use variables that have been input before; the condition $ar(X) < k$ ensures that the adversary when computing the input to be used for x only refers to terms in the frame that have been output before. This second condition is often called *origination property*.

Given a frame $\Phi = \{w_1 \rhd t_1, \ldots, w_n \rhd t_n\}$, and a second-order term T with parameters in $\{w_1, \ldots, w_n\}$ and without second-order variable $T\Phi$ denotes the first-order term obtained from T by replacing each w_i by t_i. We define the *structure* of a constraint system $\mathcal{C} = (\Phi; \mathcal{D}; \mathcal{E})$ to be $|\Phi|$ and $\mathsf{var}^2(\mathcal{D})$ with their arity.

Example 3. Note that the two additional conditions are fulfilled by the constraint system $\mathcal{C}$ given in Example 2. In particular, we have that the variable x_1 that occurs in t_3 has been introduced by the deducibility constraint $X_1 \rhd^? x_1$ and $ar(X_1) = 2 < 3$. Let $\mathcal{C}' = (\Phi'; \mathcal{D}; \mathcal{E}')$ where $\Phi' = \{w_1 \rhd a', w_2 \rhd b', w_3 \rhd c' + x_1\}$, and $\mathcal{E}' = \{x_2 + 2x_1 =^?_{\mathsf{E}} c'\}$. We have that $\mathcal{C}'$ is a constraint system that has the same structure as $\mathcal{C}$. Note that $|\Phi| = 3 = |\Phi'|$ and $\mathsf{var}^2(\mathcal{D}) = \{X_1, X_2\}$.

3.2 Satisfiability and Equivalence Problems

First, we have to define the notion of solution of a constraint system.

Definition 4. *A* pre-solution *of a constraint system $\mathcal{C} = (\Phi; \mathcal{D}; \mathcal{E})$ is a substitution θ such that:*

– *dom*$(\theta) = \mathsf{var}^2(\mathcal{C})$, *and*
– $X\theta \in \mathcal{T}(\Sigma, \{w_1, \ldots, w_k\})$ *for any* $X \in dom(\theta)$ *with* $ar(X) = k$.

The substitution λ *with* $dom(\lambda) = \mathsf{var}^1(\mathcal{C})$ *and such that* $x\lambda = (X\theta)(\Phi\lambda)$ *for any* $X \rhd^? x$ *in* $\mathcal{D}$ *is called the* first-order extension of θ for $\mathcal{C}$.

Intuitively, in the preceding definition the substitution θ stores the computation done by the adversary in order to compute the messages he sends (stored in λ) during the execution. Note that, because of the definition of a constraint system, once θ is fixed, its first-order extension is uniquely defined.

To obtain a solution we need to additionally ensure that the first-order extension λ of a pre-solution θ verifies the equality constraints in $\mathcal{E}$.

Definition 5. *Let* $\mathcal{C} = (\Phi; \mathcal{D}; \mathcal{E})$ *be a constraint system. A* solution *of* $\mathcal{C}$ *is a pre-solution* θ *of* $\mathcal{C}$ *whose first-order extension* λ *satisfies the equalities,* i.e. *for every* $(s =_{\mathsf{E}}^? s') \in \mathcal{E}$, *we have that* $s\lambda =_{\mathsf{E}} s'\lambda$. *In such a case, the substitution* λ *is called the* first-order solution *of* $\mathcal{C}$ *associated to* θ. *The set of solutions of a constraint system* $\mathcal{C}$ *is denoted* $\mathsf{Sol}_{\mathsf{E}}(\mathcal{C})$.

We now define the two problems we are interested in.

Definition 6. *A constraint system* $\mathcal{C} = (\Phi; \mathcal{D}; \mathcal{E})$ *is* satisfiable *if* $\mathsf{Sol}_{\mathsf{E}}(\mathcal{C}) \neq \emptyset$.

In the context of security protocols satisfiability of a constraint system corresponds to the adversary's ability to execute an interleaving of the protocol. This generally corresponds to an attack. For instance confidentiality of some secret term s can be encoded by adding an additional deducibility constraint $X_s \rhd^? x_s$ together with an equality constraint $x_s =_{\mathsf{E}}^? s$ (or equivalently adding a final input $\mathsf{in}(x_s)$ to the protocol and testing that the adversary is able to send the term s by adding $[x_s = s]$).

Definition 7. *Let* $\mathcal{C}_1 = (\Phi_1; \mathcal{D}_1; \mathcal{E}_1)$ *and* $\mathcal{C}_2 = (\Phi_2; \mathcal{D}_2; \mathcal{E}_2)$ *be two constraint systems having the same structure. We say that* $\mathcal{C}_1$ *is included in* $\mathcal{C}_2$, *denoted by* $\mathcal{C}_1 \sqsubseteq \mathcal{C}_2$, *if* $\mathsf{Sol}_{\mathsf{E}}(\mathcal{C}_1) \subseteq \mathsf{Sol}_{\mathsf{E}}(\mathcal{C}_2)$. *They are* equivalent *if* $\mathcal{C}_1 \sqsubseteq \mathcal{C}_2$ *and* $\mathcal{C}_2 \sqsubseteq \mathcal{C}_1$, i.e. $\mathsf{Sol}_{\mathsf{E}}(\mathcal{C}_1) = \mathsf{Sol}_{\mathsf{E}}(\mathcal{C}_2)$.

Again, in the context of security protocols this problem corresponds to the adversary's inability to distinguish whether the protocol participants are executing the interleaving modelled by $\mathcal{C}_1$ or $\mathcal{C}_2$. For the exact encoding we refer the reader to [7].

Example 4. Consider the constraint systems $\mathcal{C}$ and $\mathcal{C}'$ described in Example 2 and Example 3. The substitution $\theta = \{X_1 \mapsto w_1 + w_2, X_2 \mapsto -3w_1 - 3w_2 + w_3\}$ is a pre-solution of both $\mathcal{C}$ and $\mathcal{C}'$. The first-order extension of θ for $\mathcal{C}$ is the substitution $\lambda = \{x_1 \mapsto a + b, x_2 \mapsto -a - b + c\}$ whereas the first-order extension of θ for $\mathcal{C}'$ is the substitution $\lambda' = \{x_1 \mapsto a' + b', x_2 \mapsto -2a' - 2b' + c'\}$. It is easy to check that θ is actually a solution of both $\mathcal{C}$ and $\mathcal{C}'$, and thus both constraint systems are satisfiable. Actually, we have that $\mathcal{C}$ and $\mathcal{C}'$ are equivalent, *i.e.* $\mathsf{Sol}_{\mathsf{AG}}(\mathcal{C}) = \mathsf{Sol}_{\mathsf{AG}}(\mathcal{C}')$.

In what follows, we consider decidability and complexity issues for the satisfiability and equivalence problems for group theories. In particular, we proceed in three main steps:

1. we reduce both problems to the case of simple constraint systems (where the terms t_i that occurs in the frame Φ are ground terms);
2. we show how to encode solutions of a (simple) constraint system in a system of (linear) equations;
3. we conclude by showing how to solve such a system of equations.

4 Towards Simple Constraint Systems

The aim of this section is to show how we can transform constraint systems in order to obtain *simple* constraint systems while preserving satisfiability and inclusion. This transformation has been first introduced in [9] to simplify the satisfiability problem for the exclusive or and Abelian group theories. We reuse it in a more general setting. From now on, we consider a group equational theory (see Definition 1).

Let $\mathcal{C} = (\Phi; \mathcal{D}; \mathcal{E})$ where $\Phi = \{w_1 \rhd t_1, \ldots, w_\ell \rhd t_\ell\}$. Let $\tau = \{w_1 \to w_1 - M_1, \cdots, w_\ell \to w_\ell - M_\ell\}$ be a substitution with $dom(\tau) = \{w_1, \cdots, w_\ell\}$. We say that the substitution τ is *compatible* with $\mathcal{C}$ iff $M_1, \cdots, M_\ell$ are second-order terms that do not contain parameters and such that $\mathsf{var}^2(M_i) \subseteq \{X \in \mathsf{var}^2(\mathcal{C}) \mid ar(X) < i\}$. We define the constraint system $\mathcal{C}_\tau$ as $(\Phi_\tau; \mathcal{D}; \mathcal{E})$ where:

- $\Phi_\tau = \{w_1 \rhd t'_1, \ldots, w_\ell \rhd t'_\ell\}$, and
- $t'_i = t_i + M_i\{X \to x \mid X \rhd^? x \in \mathcal{D}\}$ for all $1 \leq i \leq \ell$.

Notice that, if τ is compatible with $\mathcal{C}$, the origination property is satisfied for $\mathcal{C}_\tau$, thus $\mathcal{C}_\tau$ is a constraint system. Let θ be a pre-solution of $\mathcal{C}$ (or equivalently of $\mathcal{C}_\tau$). We denote by θ_τ the substitution $(\theta \circ \tau)^m$ where $m = \#\mathsf{var}^2(\mathcal{C})$.

Example 5. Consider again the constraint systems $\mathcal{C}$ and $\mathcal{C}'$ described in Example 2 and Example 3. Let $\tau = \{w_1 \to w_1, w_2 \to w_2, w_3 \to w_3 - (-2X_1)\}$. We have that τ is a substitution compatible with $\mathcal{C}$ and $\mathcal{C}'$. Then, following the definition, we have that $\mathcal{C}_\tau$ is $(\Phi_\tau; \mathcal{D}; \mathcal{E})$ where $\Phi_\tau = \{w_1 \rhd a, w_2 \rhd b, w_3 \rhd c\}$ whereas $\mathcal{C}'_\tau$ is $(\Phi'_\tau; \mathcal{D}; \mathcal{E}')$ where $\Phi'_\tau = \{w_1 \rhd a', w_2 \rhd b', w_3 \rhd c' - x_1\}$

Consider the substitution $\theta = \{X_1 \to w_1 + w_2, X_2 \to -3w_1 - 3w_2 + w_3\}$ as defined in Example 4. We have that $(\theta \circ \tau) = \{X_1 \to w_1 + w_2, X_2 \to -3w_1 - 3w_2 + w_3 + 2X_1\}$, thus $\theta_\tau = (\theta \circ \tau)^2 = \{X_1 \to w_1 + w_2, X_2 \to -w_1 - w_2 + w_3\}$.

It follows that $\lambda = \{x_1 \to a + b, x_2 \to -a - b + c\}$ (as defined in Example 4) is also the first-order extension of θ_τ for $\mathcal{C}_\tau$, and thus $\theta_\tau \in \mathsf{Sol}_{\mathsf{AG}}(\mathcal{C}_\tau)$. Similarly, we have that $\lambda' = \{x_1 \to a' + b', x_2 \to -2a' - 2b' + c'\}$ (as defined in Example 4) is also the first-order extension of θ_τ for $\mathcal{C}'_\tau$, and thus $\theta_\tau \in \mathsf{Sol}_{\mathsf{AG}}(\mathcal{C}'_\tau)$.

The fact that the messages computed by the attacker in both cases are the same can be formally shown. This is the purpose of the following lemma that shows that the first-order extensions of θ for $\mathcal{C}$ and of θ_τ for $\mathcal{C}_\tau$ coincide. Actually, the changes made in the frame (Φ is transformed into Φ_τ) are compensated by the computations that are performed by the attacker (θ is transformed into θ_τ). This will be used later on for simplifying the two problems we are interested in.

Lemma 1. *Let $\mathcal{C} = (\Phi; \mathcal{D}; \mathcal{E})$ be a constraint system defined as above and $\tau = \{w_1 \to w_1 - M_1, \cdots, w_\ell \to w_\ell - M_\ell\}$ be a substitution compatible with $\mathcal{C}$. Let θ be a pre-solution of $\mathcal{C}$. Then, the first-order extension of θ for $\mathcal{C}$ is equal to the first-order extension of θ_τ for $\mathcal{C}_\tau$.*

Thanks to this lemma, we are able to establish the following proposition.

Proposition 1. *Let $\mathcal{C} = (\Phi; \mathcal{D}; \mathcal{E})$ and $\mathcal{C}' = (\Phi'; \mathcal{D}'; \mathcal{E}')$ be two constraint systems having the same structure and such that $|\Phi| = |\Phi'| = \ell$. Let $\tau = \{w_1 \to w_1 - M_1, \ldots, w_\ell \to w_\ell - M_\ell\}$ be a substitution compatible with $\mathcal{C}$ (and $\mathcal{C}'$). We have that:*

1. *$\mathcal{C}$ satisfiable if, and only if, $\mathcal{C}_\tau$ satisfiable;*
2. *$\mathcal{C} \sqsubseteq \mathcal{C}'$ if, and only if, $\mathcal{C}_\tau \sqsubseteq \mathcal{C}'_\tau$.*

Let $\mathcal{C} = (\Phi; \mathcal{D}; \mathcal{E})$ with $\Phi = \{w_1 \rhd t_1, \ldots, w_\ell \rhd t_\ell\}$. We say that the constraint system $\mathcal{C}$ is *simple* if the terms $t_1, \ldots, t_\ell$ are ground. We observe that for any constraint system there exists a substitution yielding a simple constraint system. Indeed, for any frame $\Phi = \{w_1 \rhd t_1, \ldots, w_\ell \rhd t_\ell\}$ we have that for all $1 \leq i \leq \ell$ there exist t_i^n and t_i^v such that $t_i =_{\mathsf{E}} t_i^n + t_i^v$, $v(t_i^n) = \emptyset$ and $n(t_i^v) = \emptyset$. Now let $\tau_\mathcal{C} = \{w_1 \to w_1 - M_1, \cdots, w_\ell \to w_\ell - M_\ell\}$, where $M_i = -t_i^v\{x \to X \mid X \rhd^? x \in \mathcal{D}\}$ for all $1 \leq i \leq \ell$. By construction the system $\mathcal{C}_{\tau_\mathcal{C}}$ is simple.

Moreover, we say that constraint systems $\mathcal{C}$ and $\mathcal{C}'$ are *simplifiable* if there exists τ such that both $\mathcal{C}_\tau$ and $\mathcal{C}'_\tau$ are simple. This class of constraint systems is motivated by the fact that when checking *real-or-random* secrecy properties as those studied in [7] we obtain systems that have this property. More precisely when encoding real-or-random properties we obtain systems $\mathcal{C} = (\Phi; \mathcal{D}; \mathcal{E})$ and $\mathcal{C}' = (\Phi'; \mathcal{D}'; \mathcal{E}')$ such that $\Phi = \{w_1 \rhd t_1, \ldots, w_\ell \rhd t_\ell\}$, $\Phi' = \{w_1 \rhd t'_1, \ldots, w_\ell \rhd t'_\ell\}$ and for some $1 \leq k \leq \ell$ we have that $t_i = t'_i$ for $i \leq k$ and t_i, t'_i are ground when $i > k$. It immediately follows that $\tau_\mathcal{C} = \tau_{\mathcal{C}'}$ and hence $\tau_\mathcal{C}$ simplifies both systems.

Using Proposition 1, we can reduce:

1. the satisfiability problem of a general constraint systems to the satisfiability problem of a simple constraint system; and
2. the inclusion problem between solutions of general constraint systems to the inclusion problem between solutions of a simple constraint system and a general one, respectively to the inclusion between solutions of simple constraint systems in the case these constraint systems are simplifiable.

Below, we illustrate how Proposition 1 can be applied.

Example 6. Let $\mathcal{C}$ and $\mathcal{C}'$ be the constraint systems defined in Example 2 and in Example 3. We have that $\tau_\mathcal{C} = \tau$ where $\tau = \{w_1 \to w_1, w_2 \to w_2, w_3 \to w_3 -$

$(-2X_1)\}$ is the substitution as defined in Example 5. Relying on Proposition 1, it follows that $\mathcal{C} \sqsubseteq \mathcal{C}'$ if, and only if, $\mathcal{C}_\tau \sqsubseteq \mathcal{C}'_\tau$ where $\mathcal{C}_\tau$ and $\mathcal{C}'_\tau$ are defined in Example 5. Thus, the equivalence problem between constraint systems $\mathcal{C}$ and $\mathcal{C}'$ is reduced to the equivalence problem between a simple constraint system $\mathcal{C}_\tau$ and a general constraint system $\mathcal{C}'_\tau$.

The purpose of the next section is to show how to decide this simplified problem in a systematic way.

5 Encoding Solutions into Systems of Equations

The purpose of this section is to show how to construct systems of equations that encode solutions of constraint systems.

5.1 General Constraint Systems

Consider a constraint system $\mathcal{C} = (\Phi; \mathcal{D}; \mathcal{E})$ where $\Phi = \{w_1 \rhd t_1, \ldots, w_\ell \rhd t_\ell\}$, $\mathcal{D} = \{X_1 \rhd^? x_1, \ldots, X_m \rhd^? x_m\}$, and $\mathcal{E} = \{s_1 =^?_{\mathsf{E}} s'_1, \ldots, s_n =^?_{\mathsf{E}} s'_n\}$.

Step 1. First, we encode second-order variables as sums of terms containing unknown variables over $\mathcal{R}_{\mathsf{E}}$. Actually, for all $1 \leq i \leq m$, each second-order variable X_i can be seen as a sum $y^i_1 t_1 + \cdots + y^i_{ar(X_i)} t_{ar(X_i)}$, where $y^i_1, \cdots, y^i_{ar(X_i)}$ are unknowns over $\mathcal{R}_{\mathsf{E}}$. Therefore every constraint system $\mathcal{C} = (\Phi; \mathcal{D}; \mathcal{E})$ (as described above) can be brought in the following form:

$$\begin{cases} y^1_1 t_1 + \cdots + y^1_{ar(X_1)} t_{ar(X_1)} = x_1 & s_1 = s'_1 \\ \qquad\cdots & \qquad\cdots \\ y^m_1 t_1 + \cdots + y^m_{ar(X_m)} t_{ar(X_m)} = x_m & s_n = s'_n \end{cases}$$

where for all $1 \leq i \leq m$, $1 \leq j \leq ar(X_i)$, y^i_j are unknowns over $\mathcal{R}_{\mathsf{E}}$, the terms $s_1, s'_1, \ldots, s_n, s'_n$ are first-order terms that contain only variables $x_1, \cdots, x_m$ and for all $1 \leq i \leq m$, the terms $t_1, \ldots, t_{ar(X_i)}$ are first-order terms that contain only variables $x_1, \ldots, x_{i-1}$.

Step 2. Our next goal is to remove variables $x_1, \cdots, x_m$ from the first-order terms $t_1, \cdots, t_{ar(X_m)}$. For each variable x_i, we inductively construct a term $E(x_i)$:

$$E(x_1) = y^1_1 t_1 + \cdots + y^1_{ar(X_1)} t_{ar(X_1)}$$
$$E(x_i) = (y^i_1 t_1 + \cdots + y^i_{ar(X_i)} t_{ar(X_i)})[E(x_1)/x_1, \cdots, E(x_{i-1})/x_{i-1}] \text{ where } i > 1.$$

Clearly, we have that, for all $1 \leq i \leq m$, the term $E(x_i)$ is a term that does not contain variables $x_1, \ldots, x_m$.

Step 3. Finally, we will show how a system of equations can be obtained. Given the constraint system $\mathcal{C}$, let $\mathcal{S}(\mathcal{C})$ denote its associated system of equations that we construct. The variables of $\mathcal{S}(\mathcal{C})$ are $\{y_j^i \mid 1 \leq i \leq m, 1 \leq j \leq ar(X_i)\}$ and each solution σ to $\mathcal{S}(\mathcal{C})$ encodes a second-order substitution $\{X_i \mapsto y_1^i w_1 + \cdots + y_{ar(X_i)}^i w_{ar(X_i)} \mid 1 \leq i \leq n\}$ which is a solution of $\mathcal{C}$.

We take each equation $s_i = s_i'$ in $\mathcal{E}$ and we add a set of equations into the system $\mathcal{S}(\mathcal{C})$. We assume that the equation $s_i = s_i'$ has the form $a_1 x_1 + \cdots + a_m x_m = p_i$, where $a_i \in \mathcal{R}_E$ and p_i is a ground first-order term. Notice that any equation can be brought to this form by bringing factors that contain variables to the left-hand side, and the other factors to the right-hand side. Next, we remove the variables from the left-hand side by replacing them with the terms $E(x_i)$, for all $1 \leq i \leq m$. Thus, we now have the equation $a_1 E(x_1) + \cdots + a_m E(x_m) = p_i$. We obtain an equation for each constant, by taking the corresponding coefficients from the left-hand side and equalizing with the coefficients from the right-hand side. Finally, we add this equation to $\mathcal{S}(\mathcal{C})$. We give below an example to illustrate the construction.

Example 7. Consider the constraint system $\mathcal{C}_\tau'$ defined in Example 5. We have that $\mathcal{C}_\tau' = (\Phi_\tau'; \mathcal{D}; \mathcal{E}')$ where: $\Phi_\tau' = \{w_1 \rhd a', w_2 \rhd b', w_3 \rhd c' - x_1\}$, and $\mathcal{E}' = \{x_2 + 2x_1 =_E^? c'\}$.

Step 1. We rewrite the constraint system $\mathcal{C}_\tau'$ as

$$\begin{cases} y_1^1 a' + y_2^1 b' = x_1 \qquad x_2 + 2x_1 = c' \\ y_1^2 a' + y_2^2 b' + y_3^2(c' - x_1) = x_2 \end{cases}$$

Step 2. We construct the terms $E(x_1)$ and $E(x_2)$:

$$E(x_1) = y_1^1 a' + y_2^1 b' \qquad E(x_2) = (y_1^2 a' + y_2^2 b' + y_3^2(c' - x_1))[E(x_1)/x_1]$$
$$= y_1^2 a' + y_2^2 b' + y_3^2 c' - y_3^2 y_1^1 a' - y_3^2 y_2^1 b'$$

Step 3. We take the equation $x_2 + 2x_1 =_E^? c$ and, by replacing x_2 with $E(x_2)$ and x_1 with $E(x_1)$, we obtain: $a(y_1^2 - y_3^2 y_1^1 + 2y_1^1) + b(y_2^2 - y_3^2 y_2^1 + 2y_2^1) + cy_3^2 = c$. Thus, we obtain the following system of equations $\mathcal{S}(\mathcal{C}_\tau')$:

$$\mathcal{S}(\mathcal{C}_\tau') = \{ y_1^2 - y_3^2 y_1^1 + 2y_1^1 = 0; \quad y_2^2 - y_3^2 y_2^1 + 2y_2^1 = 0; \quad y_3^2 = 1 \}$$

Note that any integer solution over $\mathcal{R}_E$ of the system of equations encodes a solution of the constraint system. For instance, take $y_1^1 = 1, y_2^1 = 1, y_1^2 = -1, y_2^2 = -1, y_3^2 = 1$ which is a solution of $\mathcal{S}(\mathcal{C}_\tau')$. This encodes the substitution $\theta = \{X_1 \to w_1 + w_2, X_2 \to -w_1 - w_2 + w_3\}$, which is a solution of $\mathcal{C}_\tau'$.

Proposition 2. *Let $\mathcal{C}$ and $\mathcal{C}'$ be two constraint systems having the same structure. Let $\mathcal{S}(\mathcal{C})$ and $\mathcal{S}(\mathcal{C}')$ be the systems of equations obtained from $\mathcal{C}$ and $\mathcal{C}'$ using the construction described above. We have that:*

1. *$\mathcal{C}$ is satisfiable if, and only if, $\mathcal{S}(\mathcal{C})$ has a solution;*
2. *$\mathcal{C} \sqsubseteq \mathcal{C}'$ if, and only if, the solutions of $\mathcal{S}(\mathcal{C})$ are also solutions of $\mathcal{S}(\mathcal{C}')$.*

5.2 Simple Constraint Systems

When the constraint system $\mathcal{C}$ is simple, then $\mathcal{S}(\mathcal{C})$ is a system of *linear* equations.

Lemma 2. *Let $\mathcal{C}$ be a simple constraint system. The system $\mathcal{S}(\mathcal{C})$ is a system of linear equations.*

Indeed, in Step 2, substitutions are no longer needed since the terms $t_1, \ldots, t_\ell$ do not contain variables, *i.e.* we simply define $E(x_i) = (y_1^i t_1 + \cdots + y_{ar(X_i)}^i t_{ar(X_i)})$ for $1 \leq i \leq m$. The following example illustrates this fact.

Example 8. Consider the constraint system $\mathcal{C}_\mathcal{T}$ defined in Example 5. We have that $\mathcal{C}_\mathcal{T} = (\Phi_\mathcal{T}; \mathcal{D}; \mathcal{E})$ where $\Phi_\mathcal{T} = \{w_1 \rhd a, w_2 \rhd b, w_3 \rhd c\}$, $\mathcal{D} = \{X_1 \rhd^? x_1, X_2 \rhd^? x_2\}$, and $\mathcal{E} = \{x_1 + x_2 =_\mathsf{E}^? c\}$.

Step 1. Then, we bring this constraint system into the following form:

$$\begin{cases} y_1^1 a + y_2^1 b = x_1 \\ y_1^2 a + y_2^2 b + y_3^2 c = x_2 \end{cases} \qquad x_1 + x_2 = c$$

Step 2. It follows that: $E(x_1) = y_1^1 a + y_2^1 b$ and $E(x_2) = y_1^2 a + y_2^2 b + y_3^2 c$.

Step 3. Thus, taking equation $x_1 + x_2 = c$ and replacing x_1 with $E(x_1)$ and x_2 with $E(x_2)$ we obtain $a(y_1^1 + y_1^2) + b(y_2^1 + y_2^2) + cy_3^2 = c$. Therefore the obtained system of linear equations $\mathcal{S}(\mathcal{C}_\mathcal{T})$ is:

$$\mathcal{S}(\mathcal{C}_\mathcal{T}) = \{\, y_1^1 + y_1^2 = 0; \quad y_2^1 + y_2^2 = 0; \quad y_3^2 = 1 \,\}$$

6 Applications and Discussion

In this section we will show how to use the previous results to decide satisfiability and equivalence of constraint systems for several equational theories of interest. Relying on Propositions 1 and 2, as well as Lemma 2, we have that:

Theorem 1. *Let E be a group theory and $\mathcal{R}_\mathsf{E}$ its associated ring.*

- *The satisfiability problem of a constraint system is reducible in polynomial time to the problem of deciding whether a system of linear equation admits a solution;*
- *The equivalence problem between constraint systems is reducible in polynomial time to the problem of deciding whether the solutions of a system of linear equations are included in the set of solutions of a system of equation. Moreover, if the constraint systems are simplifiable, the latter system can also be assumed to be linear.*

Actually, several interesting group theories induce a ring for which those problems are decidable in PTIME. To prove this, we have shown that:

Proposition 3. *Let $\mathcal{S}_{\mathsf{linear}}$ be a system of linear equations over $\mathbb{Z}/2\mathbb{Z}$ (resp. $\mathbb{Z}$, $\mathbb{Z}[\mathsf{h}]$, $\mathbb{Z}/2\mathbb{Z}[\mathsf{h}]$) and $\mathcal{S}$ be a system of equations over $\mathbb{Z}/2\mathbb{Z}$ (resp. $\mathbb{Z}$, $\mathbb{Z}[\mathsf{h}]$, $\mathbb{Z}/2\mathbb{Z}[\mathsf{h}]$) such that both systems are built on the same set of variables. The problem of deciding whether each solution of $\mathcal{S}_{\mathsf{linear}}$ is a solution of $\mathcal{S}$ is decidable in PTIME.*

Proof. (sketch) Roughly, in case $\mathcal{S}_{\mathsf{linear}}$ is satisfiable (note that otherwise, the inlusion problem is trivial), we first put it in solved form $x_1 = t_1/d_1, \ldots, x_n = t_n/d_n$ where t_i are terms that may contain some additional variables y_j, and d_i are elements in the ring under study. Then, we multiply each equation in $\mathcal{S}$ with a factor that is computed from d_i and $\mathcal{S}$ and we replace in the resulting system each x_i with t_i. Lastly, we check whether these equations are valid or not. If the answer is yes, then this means that the solutions of $\mathcal{S}_{\mathsf{linear}}$ are indeed solutions of $\mathcal{S}$. Otherwise, we can show that the inclusion does not hold. □

Example 9. Consider the system of linear equations $\mathcal{S}(\mathcal{C}_\tau)$ given in Example 8

$$\mathcal{S}(\mathcal{C}_\tau) = \{\, y_1^1 + y_1^2 = 0; \quad y_2^1 + y_2^2 = 0; \quad y_3^2 = 1 \,\}$$

This system can be rewritten into a solved form as:

$$\mathcal{S}(\mathcal{C}_\tau) = \{\, y_1^1 = -y_1^2; \quad y_2^1 = -y_2^2; \quad y_3^2 = 1 \,\}$$

Consider also the system of equations $\mathcal{S}(\mathcal{C}'_\tau)$ given in Example 7:

$$\mathcal{S}(\mathcal{C}'_\tau) = \{\, y_1^2 - y_3^2 y_1^1 + 2y_1^1 = 0; \quad y_2^2 - y_3^2 y_2^1 + 2y_2^1 = 0; \quad y_3^2 = 1 \,\}$$

It can be seen that the solutions of $\mathcal{S}(\mathcal{C}_\tau)$ are also solutions of the equations of $\mathcal{S}(\mathcal{C}'_\tau)$. Indeed, all the terms reduce when replacing y_1^1 with $-y_1^2$, y_2^1 with $-y_2^2$ and y_3^2 with 1, as indicated in the solved form of $\mathcal{S}(\mathcal{C}_\tau)$. Thus, we can finally conclude that $\mathcal{C}_\tau \sqsubseteq \mathcal{C}'_\tau$, and thus $\mathcal{C} \sqsubseteq \mathcal{C}'$ where $\mathcal{C}_\tau, \mathcal{C}'_\tau$ are defined in Example 5 and $\mathcal{C}$ (resp. $\mathcal{C}'$) are defined in Example 2 (resp. Example 3).

Decidability and complexity results are summarized in the table. A brief discussion on each equational theory can be found below.

Theory E	$\mathcal{R}_{\mathsf{E}}$	Satisfiability	Equivalence
ACUN	$\mathbb{Z}/2\mathbb{Z}$	PTIME [9]	PTIME (*new*)
AG	$\mathbb{Z}$	PTIME [9]	PTIME (*new*)
ACUNh	$\mathbb{Z}/2\mathbb{Z}[\mathsf{h}]$	PTIME	PTIME (*new*)
AGh	$\mathbb{Z}[\mathsf{h}]$	PTIME	PTIME (*new*)

Theory ACUN (exclusive or). The ring corresponding to this equational theory is the finite field $\mathbb{Z}/2\mathbb{Z}$. The satisfiability problem for the theory ACUN has already been studied and shown to be decidable in PTIME [9].

However, the equivalence problem has only been studied in a very particular case, the so-called static equivalence problem [1]. Static equivalence models indistinguishability of two frames, *i.e.* the adversary cannot interact with the protocol. In our setting the problem of static equivalence of frames Φ and Φ' can be rephrased as the equivalence between two particular constraint systems $(\Phi; \mathcal{D}; \{x_1 \stackrel{?}{=}_{\mathsf{E}} x_2\})$ and $(\Phi'; \mathcal{D}; \{x_1 \stackrel{?}{=}_{\mathsf{E}} x_2\})$ where:

- Φ and Φ' are arbitrary frames of same size that only contain *ground* terms;
- $\mathcal{D} = \{X_1 \rhd^? x_1; X_2 \rhd^? x_2\}$ where $ar(X_1) = ar(X_2) = |\Phi|$.

The static equivalence problem has been shown to be decidable in PTIME [12]. Here, relying on our reduction result (Theorem 1), we show that we can decide the problem of equivalence of general constraint systems in PTIME as well.

Theory AG (Abelian groups). The ring associated to this equational theory is the ring $\mathbb{Z}$ of all integers. There exist several algorithms to compute solutions of linear equations over $\mathbb{Z}$ and to compute a base of the set of solutions (see for instance [21]). Hence, we easily deduce that the satisfiability problem is decidable in PTIME. This was already observed in [9]. Deciding inclusion of solutions of a system of linear equations in solutions of a system of non-linear equations is more tricky but we have shown that it can be done in PTIME (see Proposition 3).

Theories ACUNh and AGh. For the theory ACUNh (resp. AGh) the associated ring is $\mathbb{Z}/2\mathbb{Z}[\mathsf{h}]$ (resp. $\mathbb{Z}[\mathsf{h}]$), *i.e.* the ring of polynomials in one indeterminate over $\mathbb{Z}/2\mathbb{Z}$ (resp. $\mathbb{Z}$). The satisfiability problem for these equational theories has already been studied in [18], but in a slightly different setting. The intruder deduction problem for these theories has been studied in [14] and shown to be decidable in PTIME. Similar to static equivalence the intruder deduction problem considers a passive attacker which simply asks whether a term can be deduced by an adversary from a frame. In our setting we rephrase this problem whether a ground term t can be deduced from Φ as the satisfiability of the particular constraint system $(\Phi; \mathcal{D}; \{x =^?_{\mathsf{E}} t\})$ where:

- Φ is an arbitrary frame that only contains *ground* terms,
- $\mathcal{D} = \{X \rhd^? x\}$ where $ar(X) = |\Phi|$.

In [14] this problem is reduced to the problem of satisfiability of a system of linear equations. Hence, the techniques for the problem of deciding secrecy for a passive adversary are the same as for an active adversary and we immediately obtain the same PTIME complexity as in [14]. However, results obtained on the equivalence problem are new. We are able to use the same technique as for AG to obtain decidability in PTIME. This generalizes and refines the decidability result (without known complexity) for ACUNh and AGh in the particular case of static equivalence [12].

References

1. Abadi, M., Cortier, V.: Deciding knowledge in security protocols under equational theories. Theoretical Computer Science 387(1-2), 2–32 (2006)
2. Abadi, M., Fournet, C.: Mobile values, new names, and secure communication. In: Proc. 28th ACM Symposium on Principles of Programming Languages (POPL 2001), pp. 104–115. ACM Press (2001)
3. Abadi, M., Gordon, A.D.: A calculus for cryptographic protocols: The spi calculus. Inf. Comput. 148(1), 1–70 (1999)

4. Arapinis, M., Chothia, T., Ritter, E., Ryan, M.D.: Analysing unlinkability and anonymity using the applied pi calculus. In: Proc. 23rd Computer Security Foundations Symposium (CSF 2010), pp. 107–121. IEEE Comp. Soc. Press (2010)
5. Armando, A., Carbone, R., Compagna, L., Cuéllar, J., Tobarra, M.L.: Formal analysis of SAML 2.0 web browser single sign-on: breaking the SAML-based single sign-on for google apps. In: Proc. 6th ACM Workshop on Formal Methods in Security Engineering (FMSE 2008), pp. 1–10. ACM Press (2008)
6. Baader, F.: Unification in commutative theories. Journal of Symbolic Computation 8(5), 479–497 (1989)
7. Baudet, M.: Deciding security of protocols against off-line guessing attacks. In: Proc. 12th Conference on Computer and Communications Security (CCS 2005), pp. 16–25. ACM Press (2005)
8. Cheval, V., Comon-Lundh, H., Delaune, S.: Automating Security Analysis: Symbolic Equivalence of Constraint Systems. In: Giesl, J., Hähnle, R. (eds.) IJCAR 2010. LNCS (LNAI), vol. 6173, pp. 412–426. Springer, Heidelberg (2010)
9. Chevalier, Y., Rusinowitch, M.: Symbolic protocol analysis in the union of disjoint intruder theories: Combining decision procedures. Theoretical Computer Science 411(10), 1261–1282 (2010)
10. Chevalier, Y., Rusinowitch, M.: Decidability of equivalence of symbolic derivations. J. Autom. Reasoning 48(2), 263–292 (2012)
11. Comon-Lundh, H., Cortier, V., Zalinescu, E.: Deciding security properties of cryptographic protocols. application to key cycles. Transaction on Computational Logic 11(2) (2010)
12. Cortier, V., Delaune, S.: Decidability and combination results for two notions of knowledge in security protocols. J. of Autom. Reasoning 48(4), 441–487 (2012)
13. Cortier, V., Delaune, S., Lafourcade, P.: A survey of algebraic properties used in cryptographic protocols. Journal of Computer Security 14(1), 1–43 (2006)
14. Delaune, S.: Easy intruder deduction problems with homomorphisms. Information Processing Letters 97(6), 213–218 (2006)
15. Delaune, S., Kremer, S., Pasaila, D.: Security protocols, constraint systems, and group theories. Research Report LSV-12-06, Laboratoire Spécification et Vérification, ENS Cachan, France, 23 pages (2012)
16. Delaune, S., Kremer, S., Ryan, M.D.: Verifying privacy-type properties of electronic voting protocols. Journal of Computer Security 17(4), 435–487 (2009)
17. Delaune, S., Lafourcade, P., Lugiez, D., Treinen, R.: Symbolic protocol analysis for monoidal equational theories. Inf. Comp. 206(2-4), 312–351 (2008)
18. Lafourcade, P., Lugiez, D., Treinen, R.: Intruder Deduction for AC-Like Equational Theories with Homomorphisms. In: Giesl, J. (ed.) RTA 2005. LNCS, vol. 3467, pp. 308–322. Springer, Heidelberg (2005)
19. Millen, J., Shmatikov, V.: Constraint solving for bounded-process cryptographic protocol analysis. In: Proc. 8th ACM Conference on Computer and Communications Security (CCS 2001). ACM Press (2001)
20. Nutt, W.: Unification in Monoidal Theories. In: Stickel, M.E. (ed.) CADE 1990. LNCS, vol. 449, pp. 618–632. Springer, Heidelberg (1990)
21. Schrijver, A.: Theory of Linear and Integer Programming. Wiley (1986)
22. Tiu, A., Dawson, J.: Automating open bisimulation checking for the spi-calculus. In: Proc. 23rd Computer Security Foundations Symposium (CSF 2010), pp. 307–321. IEEE Comp. Soc. Press (2010)

Taming Past LTL and Flat Counter Systems*

Stéphane Demri[1], Amit Kumar Dhar[2], and Arnaud Sangnier[2]

[1] LSV, CNRS, ENS Cachan, INRIA, France
[2] LIAFA, Univ Paris Diderot, Sorbonne Paris Cité, CNRS, France

Abstract. Reachability and LTL model-checking problems for flat counter systems are known to be decidable but whereas the reachability problem can be shown in NP, the best known complexity upper bound for the latter problem is made of a tower of several exponentials. Herein, we show that the problem is only NP-complete even if LTL admits past-time operators and arithmetical constraints on counters. Actually, the NP upper bound is shown by adequately combining a new stuttering theorem for Past LTL and the property of small integer solutions for quantifier-free Presburger formulae. Other complexity results are proved, for instance for restricted classes of flat counter systems.

1 Introduction

Flat counter systems. Counter systems are finite-state automata equipped with program variables (counters) interpreted over non-negative integers. They are used in many places like, broadcast protocols [9] and programs with pointers [12] to quote a few examples. But, alongwith their large scope of usability, many problems on general counter systems are known to be undecidable. Indeed, this computational model can simulate Turing machines. Decidability of reachability problems or model-checking problems based on temporal logics, can be regained by considering subclasses of counter systems, see e.g. [14]. An important and natural class of counter systems, in which various practical cases of infinite-state systems (e.g. broadcast protocols [11]) can be modelled, are those with a *flat* control graph, i.e, those where no control state occurs in more than one simple cycle, see e.g. [1,5,11,20]. Decidability results on verifying safety and reachability properties on flat counter systems have been obtained in [5,11,3]. However, so far, such properties have been rarely considered in the framework of any formal specification language (see an exception in [4]). In [7], a class of Presburger counter systems is identified for which the local model checking problem for Presburger-CTL* is shown decidable. These are Presburger counter systems defined over flat control graphs with arcs labelled by adequate Presburger formulae. Even though flatness is clearly a substantial restriction, it is shown in [20] that many classes of counter systems with computable Presburger-definable reachability sets are *flattable*, i.e. there exists a flat unfolding of the counter system with identical reachability sets. Hence, the possibility of flattening a counter system is strongly

* Supported by ANR project REACHARD ANR-11-BS02-001.

B. Gramlich, D. Miller, and U. Sattler (Eds.): IJCAR 2012, LNAI 7364, pp. 179–193, 2012.

related to semilinearity of its reachability set. Moreover, in [4] model-checking relational counter systems over LTL formulae is shown decidable when restricted to flat formulae (their translation into automata leads to flat structures).

Towards the complexity of temporal model-checking flat counter systems. In [7], it is shown that CTL* model-checking over the class of so-called *admissible* counter systems is decidable by reduction into the satisfiability problem for Presburger arithmetic, the decidable first-order theory of natural numbers with addition. Obviously CTL* properties are more expressive than reachability properties but this has a cost. However, for the class of counter systems considered in this paper, this provides a very rough complexity upper bound in 4EXPTIME. Herein, our goal is to revisit standard decidability results for subclasses of counter systems obtained by translation into Presburger arithmetic in order to obtain optimal complexity upper bounds.

Our contributions. In the paper, we establish several computational complexity characterizations of model-checking problems restricted to flat counter systems in the presence of a rich LTL-like specification language with arithmetical constraints and past-time operators. Not only we provide an optimal complexity but also, we believe that our proof technique could be reused for further extensions. Indeed, we combine three proof techniques: the general stuttering theorem [17], the property of small integer solutions of equation systems [2] (this latter technique is used since [24]) and the elimination of disjunctions in guards (see Section 5.2). Let us be a bit more precise.

We extend the stuttering principle established in [17] for LTL (without past-time operators) to Past LTL. The stuttering theorem from [17] for LTL without past-time operators has been used to show that LTL model-checking over *weak* Kripke structures is in NP [16] (weakness corresponds to flatness). It is worth noting that another way to show a similar result would be to eliminate past-time operators thanks to Gabbay's Separation Theorem [13] (preserving initial equivalence) but the temporal depth of formulae might increase at least exponentially, which is a crucial parameter in our complexity analysis. We show that the model-checking problem restricted to flat counter systems in the presence of LTL with past-time operators is in NP (Theorem 17) by combining the above-mentioned proof techniques. Apart from the use of the general stuttering theorem (Theorem 3), we take advantage of the other properties stated for instance in Lemma 12 (characterization of runs by quantifier-free Presburger formulae) and Theorem 14 (elimination of disjunctions in guards preserving flatness). In the paper, complexity results for fragments/subproblems are also considered. For instance, we get a sharp lower bound since we establish that the model-checking problem on path schemas (a fundamental structure in flat counter systems) with only 2 loops is already NP-hard (see Lemma 11). A summary table can be found in Section 6.

Omitted proofs can be found in [6].

2 Flat Counter Systems and Its LTL Dialect

We write $\mathbb{N}$ [resp. $\mathbb{Z}$] to denote the set of natural numbers [resp. integers] and $[i, j]$ to denote $\{k \in \mathbb{Z} : i \leq k \text{ and } k \leq j\}$. For $\boldsymbol{v} \in \mathbb{Z}^n$, $\boldsymbol{v}[i]$ denotes the ith element of $\boldsymbol{v}$ for every $i \in [1, n]$. For some n-ary tuple t, we write $\pi_j(t)$ to denote the jth element of t ($j \leq n$). In the sequel, integers are encoded with a binary representation. For a finite alphabet Σ, Σ^* represents the set of finite words over Σ, Σ^+ the set of finite non-empty words over Σ and Σ^ω the set of ω-words over Σ. For a finite word $w = a_1 \ldots a_k$ over Σ, we write $\text{len}(w)$ to denote its *length* k. For $0 \leq i < \text{len}(w)$, $w(i)$ represents the $(i + 1)$-th letter of the word, here a_{i+1}.

2.1 Counter Systems

Let $\mathtt{C} = \{\mathsf{x}_1, \mathsf{x}_2, \ldots\}$ be a countably infinite set of *counters* (variables interpreted over non-negative integers) and $\mathrm{AT} = \{p_1, p_2, \ldots\}$ be a countable infinite set of propositional variables (abstract properties about program points). We write $\mathtt{C}_n$ to denote $\{\mathsf{x}_1, \mathsf{x}_2, \ldots, \mathsf{x}_n\}$. The set $\mathtt{G}(\mathtt{C}_n)$ of *guards* (arithmetical constraints on counters in $\mathtt{C}_n$) is defined inductively as follows: $\mathtt{t} ::= a.\mathsf{x} \mid \mathtt{t} + \mathtt{t}$ and $\mathtt{g} ::= \mathtt{t} \sim b \mid \mathtt{g} \wedge \mathtt{g} \mid \mathtt{g} \vee \mathtt{g}$, where $\mathsf{x} \in \mathtt{C}_n$, $a \in \mathbb{Z}$, $b \in \mathbb{Z}$ and $\sim \in \{=, \leq, \geq, <, >\}$. Such guards are closed under negations (but negation is not part of the logical connectives) and the truth constants $\top$ and $\bot$ can be easily defined too. Given $\mathtt{g} \in \mathtt{G}(\mathtt{C}_n)$ and a vector $\mathbf{v} \in \mathbb{N}^n$, we say that $\mathbf{v}$ satisfies $\mathtt{g}$, written $\mathbf{v} \models \mathtt{g}$, if the formula obtained by replacing each x_i by $\boldsymbol{v}[i]$ holds.

Definition 1 (Counter system). *For $n \geq 1$, a counter system S is a tuple $\langle Q, \mathtt{C}_n, \Delta, \mathbf{l} \rangle$ where Q is a finite set of* control states, *$\mathbf{l} : Q \to 2^{\mathrm{AT}}$ is a* labelling function *and $\Delta \subseteq Q \times \mathtt{G}(\mathtt{C}_n) \times \mathbb{Z}^n \times Q$ is a finite set of edges labeled by guards and updates of the counter values (*transitions*).*

For $\delta = (q, \mathtt{g}, \mathbf{u}, q')$ in Δ, we use the following notations $\mathit{source}(\delta) = q$, $\mathit{target}(\delta) = q'$, $\mathit{guard}(\delta) = \mathtt{g}$ and $\mathit{update}(\delta) = \mathbf{u}$. As usual, to a counter system $S = \langle Q, \mathtt{C}_n, \Delta, \mathbf{l} \rangle$, we associate a labeled transition system $TS(S) = \langle C, \to \rangle$ where $C = Q \times \mathbb{N}^n$ is the set of *configurations* and $\to \subseteq C \times \Delta \times C$ is the *transition relation* defined by: $\langle \langle q, \mathbf{v} \rangle, \delta, \langle q', \mathbf{v}' \rangle \rangle \in \to$ (also written $(q, \mathbf{v}) \xrightarrow{\delta} (q', \mathbf{v}')$) iff $q = \mathit{source}(\delta)$, $q' = \mathit{target}(\delta)$, $\mathbf{v} \models \mathit{guard}(\delta)$ and $\mathbf{v}' = \mathbf{v} + \mathit{update}(\delta)$. In such a transition system, the counter values are non-negative since $C = Q \times \mathbb{N}^n$. We extend the transition relation $\to$ to finite words of transitions in Δ^+ as follows. For each $w = \delta_1 \delta_2 \ldots \delta_\alpha \in \Delta^+$, we have $\langle q, \mathbf{v} \rangle \xrightarrow{w} \langle q', \mathbf{v}' \rangle$ if there are $c_0, c_1, \ldots, c_{\alpha+1} \in C$ such that $c_i \xrightarrow{\delta_i} c_{i+1}$ for all $i \in [0, \alpha]$, $c_0 = (q, \mathbf{v})$ and $c_{\alpha+1} = \langle q', \mathbf{v}' \rangle$. We say that an ω-word $w \in \Delta^\omega$ is *fireable* in S from a configuration $c_0 \in Q \times \mathbb{N}^n$ if for all finite prefixes w' of w there exists a configuration $c \in Q \times \mathbb{N}^n$ such that $c_0 \xrightarrow{w'} c$. We write $\mathit{lab}(c_0)$ to denote the set of ω-words (*labels*) which are fireable from c_0 in S.

Given a configuration $c_0 \in Q \times \mathbb{N}^n$, a *run* ρ starting from c_0 in S is an infinite path in the associated transition system $TS(S)$ denoted as: $\rho := c_0 \xrightarrow{\delta_0} \cdots \xrightarrow{\delta_{\alpha-1}} c_\alpha \xrightarrow{\delta_\alpha} \cdots$ where $c_i \in Q \times \mathbb{N}^n$ and $\delta_i \in \Delta$ for all $i \in \mathbb{N}$. Let $\mathit{lab}(\rho)$ be the ω-word

$\delta_0\delta_1\ldots$ associated to the run ρ. Note that by definition we have $lab(\rho) \in lab(c_0)$. When E is an ω-regular expression over the finite alphabet Δ and c_0 is an initial configuration, $lab(E, c_0)$ is defined as the set of labels of infinite runs ρ starting at c_0 such that $lab(\rho)$ belongs to the language defined by E. So $lab(E, c_0) \subseteq lab(c_0)$.

We say that a counter system is *flat* if every node in the underlying graph belongs to at most one simple cycle (a cycle being simple if no edge is repeated twice in it) [5]. In a flat counter system, simple cycles can be organized as a DAG where two simple cycles are in the relation whenever there is path between a node of the first cycle and a node of the second cycle. We denote by $\mathcal{CFS}$ the class of flat counter systems.

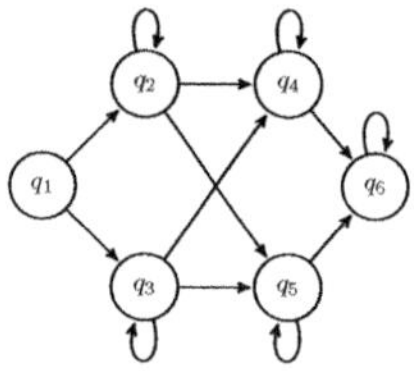

On the left, we present the control graph of a flat counter system (guards and updates are omitted). A *Kripke structure* S is a tuple $\langle Q, \Delta, \mathbf{l}\rangle$ where $\Delta \subseteq Q \times Q$ and $\mathbf{l}$ is labelling. It can be viewed as a degenerate form of counter systems without counters (in the sequel, we take the freedom to see them as counter systems). All standard notions on counter systems naturally apply to Kripke structures too (configuration, run, flatness, etc.). In the sequel, we shall also investigate the complexity of model-checking problems on flat Kripke structures (such a class is denoted by $\mathcal{KFS}$).

2.2 Linear Temporal Logic with Past and Arithmetical Constraints

Model-checking problem for Past LTL over finite state systems is known to be PSPACE-complete. In spite of this nice feature, a propositional variable p only represents an abstract property about the current configuration of the system. A more satisfactory solution is to include in the logical language the possibility to express directly constraints between variables of the program, whence giving up the standard abstraction made with propositional variables. We define below a version of LTL dedicated to counter systems in which the atomic formulae are linear constraints; this is analogous to the use of concrete domains in description logics [21]. Note that capacity constraints from [8] are arithmetical constraints different from those defined below. Formulae of PLTL[$\mathtt{C}$] are defined from $\phi ::= p \mid \mathtt{g} \mid \neg\phi \mid \phi\wedge\phi \mid \phi\vee\phi \mid \mathtt{X}\phi \mid \phi\mathtt{U}\phi \mid \mathtt{X}^{-1}\phi \mid \phi\mathtt{S}\phi$ where $p \in$ AT and $\mathtt{g} \in \mathtt{G}(\mathtt{C}_n)$ for some n. We may use the standard abbreviations $\mathtt{F}$, $\mathtt{G}$, $\mathtt{G}^{-1}$ etc. For instance, the formula $\mathtt{GF}(\mathtt{x}_1 + 2 \geq \mathtt{x}_2)$ states that infinitely often the value of counter 1 plus 2 is greater than the value of counter 2. The past-time operators $\mathtt{S}$ and $\mathtt{X}^{-1}$ do not add expressive power to the logic itself, but it is known that it helps a lot to express properties succinctly, see e.g. [19,18]. The temporal depth of ϕ, written $td(\phi)$, is defined as the maximal number of imbrications of temporal operators in ϕ. Restriction of PLTL[$\mathtt{C}$] to atomic formulae from AT only is written PLTL[$\emptyset$], standard version of LTL with past-time operators. Models of PLTL[$\mathtt{C}$] are essentially abstractions of runs from counter systems, i.e. ω-sequences $\sigma : \mathbb{N} \to 2^{\mathrm{AT}} \times \mathbb{N}^{\mathtt{C}}$. Given a model σ and a position $i \in \mathbb{N}$, the satisfaction relation $\models$ for PLTL[C] is defined as follows (other cases can be defined similarly, see e.g. [18]):

- $\sigma, i \models p \overset{\text{def}}{\Leftrightarrow} p \in \pi_1(\sigma(i))$, $\sigma, i \models \mathtt{g} \overset{\text{def}}{\Leftrightarrow} \mathbf{v}_i \models \mathtt{g}$ where $\mathbf{v}_i[j] \overset{\text{def}}{=} \pi_2(\sigma(i))(\mathsf{x}_j)$,
- $\sigma, i \models \mathtt{X}\phi \overset{\text{def}}{\Leftrightarrow} \sigma, i+1 \models \phi$,
- $\sigma, i \models \phi_1 \mathtt{S} \phi_2 \overset{\text{def}}{\Leftrightarrow} \sigma, j \models \phi_2$ for some $0 \leq j \leq i$ s.t. $\sigma, k \models \phi_1, \forall j < k \leq i$.

Given $\langle Q, \mathtt{C}_n, \Delta, \mathbf{l} \rangle$ and a run $\rho := \langle q_0, \mathbf{v}_0 \rangle \xrightarrow{\delta_0} \cdots \xrightarrow{\delta_{p-1}} \langle q_p, \mathbf{v}_p \rangle \xrightarrow{\delta_p} \cdots$, we consider the model $\sigma_\rho : \mathbb{N} \to 2^{\mathrm{AT}} \times \mathbb{N}^{\mathtt{C}}$ such that $\pi_1(\sigma_\rho(i)) \overset{\text{def}}{=} \mathbf{l}(q_i)$ and $\pi_2(\sigma_\rho(i))(\mathsf{x}_j) \overset{\text{def}}{=} \mathbf{v}_i[j]$ for all $j \in [1, n]$ and all $i \in \mathbb{N}$. Note that $\pi_2(\sigma_\rho(i))(\mathsf{x}_j)$ is arbitrary for $j \notin [1, n]$. As expected, we extend the satisfaction relation to runs so that $\rho, i \models \phi \overset{\text{def}}{\Leftrightarrow} \sigma_\rho, i \models \phi$ whenever ϕ is built from counters in $\mathtt{C}_n$.

Given a fragment L of PLTL[C] and a class $\mathcal{C}$ of counter systems, we write MC(L, $\mathcal{C}$) to denote the existential model checking problem: given $S \in \mathcal{C}$, a configuration c_0 and $\phi \in$ L, does there exist ρ starting from c_0 such that $\rho, 0 \models \phi$? In that case, we write $S, c_0 \models \phi$. It is known that for the full class of counter systems, the model-checking problem is undecidable, see e.g. [22]. Some restrictions, such as flatness, can lead to decidability as shown in [7] but the decision procedure there involves an exponential reduction to Presburger Arithmetic, whence the high complexity.

Theorem 2. *[7,16] MC(PLTL[C], $\mathcal{CFS}$) can be solved in* 4EXPTIME. *MC(PLTL [$\emptyset$], $\mathcal{KFS}$) restricted to formulae with temporal operators* U,X *is* NP*-complete.*

Our main goal is to characterize the complexity of MC(PLTL[C], $\mathcal{CFS}$).

3 Stuttering Theorem for PLTL[$\emptyset$]

Stuttering of finite words or single letters has been instrumental to show results about the expressive power of PLTL[$\emptyset$] fragments, see e.g. [23,17]; for instance, PLTL[$\emptyset$] restricted to the temporal operator U characterizes the class of formulae defining classes of models invariant under stuttering. This is refined in [17] for PLTL[$\emptyset$] restricted to U and X, by taking into account not only the U-depth but also the X-depth of formulae and by introducing a principle of stuttering that involves both letter stuttering and word stuttering. In this section, we establish another substantial generalization that involves PLTL[$\emptyset$] with past-time temporal operators. Roughly speaking, we show that if $\sigma_1 \mathbf{s}^M \sigma_2, 0 \models \phi$ where $\sigma_1 \mathbf{s}^M \sigma_2$ is a PLTL[$\emptyset$] model ($\sigma_1, \mathbf{s}$ being finite words), $\phi \in$ PLTL[$\emptyset$], $td(\phi) \leq N$ and $M \geq 2N+1$, then $\sigma_1 \mathbf{s}^{2N+1} \sigma_2, 0 \models \phi$ (and other related properties). This extends a result without past-time operators [16]. Moreover, this turns out to be a key property (Theorem 3) to establish the NP upper bound even in the presence of counters. Note that Theorem 3 below is interesting for its own sake, independently of our investigation on flat counter systems. By lack of space, we state below the main definitions and result.

Given $M, M', N \in \mathbb{N}$, we write $M \approx_N M'$ iff $\mathtt{Min}(M, N) = \mathtt{Min}(M', N)$. Given $w = w_1 u^M w_2, w' = w_1 u^{M'} w_2 \in \Sigma^\omega$ and $i, i' \in \mathbb{N}$, we define an equivalence relation $\langle w, i\rangle \approx_N \langle w', i'\rangle$ (implicitly parameterized by w_1, w_2 and u) such that $\langle w, i\rangle \approx_N \langle w', i'\rangle$ means that the number of copies of u before position i and the number of copies of u before position i' are related by $\approx_N$ and the same applies for the number of copies after the positions. Moreover, if i and i' occur in the part where u is repeated, then they correspond to identical positions in u. More formally, $\langle w, i\rangle \approx_N \langle w', i'\rangle \overset{\text{def}}{\Leftrightarrow} M \approx_{2N} M'$ and one of the conditions holds true: (1) $i, i' < \text{len}(w_1) + N \cdot \text{len}(u)$ and $i = i'$, (2) $i \geq \text{len}(w_1) + (M - N) \cdot \text{len}(u)$, $i' \geq \text{len}(w_1) + (M' - N) \cdot \text{len}(u)$ and $(i - i') = (M - M') \cdot \text{len}(u)$, (3) $\text{len}(w_1) + N \cdot \text{len}(u) \leq i < \text{len}(w_1) + (M - N) \cdot \text{len}(u)$, $\text{len}(w_1) + N \cdot \text{len}(u) \leq i' < \text{len}(w_1) + (M' - N) \cdot \text{len}(u)$ and $|i - i'| = 0 \mod \text{len}(u)$. We state our stuttering theorem for PLTL[$\emptyset$] that is tailored for our future needs.

Theorem 3 (Stuttering). *Let* $\sigma = \sigma_1 \mathbf{s}^M \sigma_2, \sigma' = \sigma_1 \mathbf{s}^{M'} \sigma_2 \in (2^{\text{AT}})^\omega$ *and* $i, i' \in \mathbb{N}$ *such that* $N \geq 2$, $M, M' \geq 2N + 1$ *and* $\langle \sigma, i\rangle \approx_N \langle \sigma', i'\rangle$. *Then, for every* PLTL[$\emptyset$] *formula* ϕ *with* $td(\phi) \leq N$, *we have* $\sigma, i \models \phi$ *iff* $\sigma', i \models \phi$.

Proof. (sketch) The proof is by structural induction on the formula but first we need to establish properties whose proofs can be found in [6]. Let $w = w_1 u^M w_2, w' = w_1 u^{M'} w_2 \in \Sigma^\omega$, $i, i' \in \mathbb{N}$ and $N \geq 2$ such that $M, M' \geq 2N + 1$ and $\langle w, i\rangle \approx_N \langle w', i'\rangle$. We can show the following properties:

(Claim 1) $\langle w, i\rangle \approx_{N-1} \langle w', i'\rangle$ and $w(i) = w'(i')$.
(Claim 2) $\langle w, i + 1\rangle \approx_{N-1} \langle w', i' + 1\rangle$ and $i, i' > 0$ implies $\langle w, i - 1\rangle \approx_{N-1} \langle w', i' - 1\rangle$.
(Claim 3) For all $j \geq i$, there is $j' \geq i'$ such that $\langle w, j\rangle \approx_{N-1} \langle w', j'\rangle$ and for all $k' \in [i', j' - 1]$, there is $k \in [i, j - 1]$ such that $\langle w, k\rangle \approx_{N-1} \langle w', k'\rangle$.
(Claim 4) For all $j \leq i$, there is $j' \leq i'$ such that $\langle w, j\rangle \approx_{N-1} \langle w', j'\rangle$ and for all $k' \in [j' - 1, i']$, there is $k \in [j - 1, i]$ such that $\langle w, k\rangle \approx_{N-1} \langle w', k'\rangle$.

By way of example, let us present the induction step for subformulae of the form $\psi_1 \mathtt{U} \psi_2$. We show that $\sigma, i \models \psi_1 \mathtt{U} \psi_2$ implies $\sigma', i' \models \psi_1 \mathtt{U} \psi_2$. Suppose there is $j \geq i$ such that $\sigma, j \models \psi_2$ and for every $k \in [i, j - 1]$, we have $\sigma, k \models \psi_1$. There is $j' \geq i'$ satisfying (Claim 3). Since $td(\psi_1), td(\psi_2) \leq N - 1$, by (IH), we have $\sigma', j' \models \psi_2$. Moreover, for every $k' \in [i', j' - 1]$, there is $k \in [i, j - 1]$ such that $\langle w, k\rangle \approx_{N-1} \langle w', k'\rangle$ and by (IH), we have $\sigma', k' \models \psi_1$ for every $k' \in [i', j' - 1]$. Hence, $\sigma', i' \models \psi_1 \mathtt{U} \psi_2$. □

An alternative proof consists in using Ehrenfeucht-Fraïssé games [10].

4 Fundamental Structures: Minimal Path Schemas

In this section, we introduce the notion of a fundamental structure for flat counter systems, namely a path schema. Indeed, every flat counter system can be decomposed into a finite set of minimal path schemas and there are only an exponential number of them. So, all our nondeterministic algorithms on flat counter systems have a preliminary step that first guesses a minimal path schema.

4.1 Minimal Path Schemas

Let $S = \langle Q, \mathtt{C}_n, \Delta, \mathbf{l} \rangle$ be a flat counter system. A *path segment* p of S is a finite sequence of transitions from Δ such that $target(p(i)) = source(p(i+1))$ for all $0 \leq i < \text{len}(p) - 1$. We write $first(p)$ [resp. $last(p)$] to denote the first [resp. last] control state of a path segment, in other words $first(p) = source(p(0))$ and $last(p) = target(p(\text{len}(p) - 1))$. We also write $effect(p)$ to denote the sum vector $\sum_{0 \leq i < \text{len}(p)} update(p(i))$ representing the total effect of the updates along the path segment. A path segment p is said to be *simple* if $\text{len}(p) > 0$ and for all $0 \leq i, j < \text{len}(p)$, $p(i) = p(j)$ implies $i = j$ (no repetition of transitions). A *loop* is a simple path segment p such that $first(p) = last(p)$. A *path schema* P is an ω-regular expression built over Δ such that its language represents an overapproximation of the set of labels obtained from infinite runs following the transitions of P. A path schema P is of the form $p_1 l_1^+ p_2 l_2^+ \ldots p_k l_k^\omega$ where (1) l_1, $\ldots$, l_k are loops and (2) $p_1 l_1 p_2 l_2 \ldots p_k l_k$ is a path segment.

We write $\text{len}(P)$ to denote $\text{len}(p_1 l_1 p_2 l_2 \ldots p_k l_k)$ and $\text{nbloops}(P)$ as its number k of loops. Let $\mathcal{L}(P)$ denote the set of infinite words in Δ^ω which belong to the language defined by P. Note that some elements of $\mathcal{L}(P)$ may not correspond to any run because of constraints on counter values. Given $w \in \mathcal{L}(P)$, we write $iter_P(w)$ to denote the unique tuple in $(\mathbb{N} \setminus \{0\})^{k-1}$ such that $w = p_1 l_1^{iter_P(w)[1]} p_2 l_2^{iter_P(w)[2]} \ldots p_k l_k^\omega$. So, for every $i \in [1, k-1]$, $iter_P(w)[i]$ is the number of times the loop l_i is taken. Then, for a configuration c_0, the set $iter_P(c_0)$ is the set of vectors $\{iter_P(w) \in (\mathbb{N} \setminus \{0\})^{k-1} \mid w \in lab(P, c_0)\}$. Finally, we say that a run ρ starting in a configuration c_0 *respects* a path schema P if $lab(\rho) \in lab(P, c_0)$ and for such a run, we write $iter_P(\rho)$ to denote $iter_P(lab(\rho))$. Note that by definition, if ρ respects P, then each loop l_i is visited at least once, and the last one infinitely.

So far, a flat counter system may have an infinite set of path schemas. However, we can impose minimality conditions on path schemas without sacrifying completeness. A path schema $p_1 l_1^+ p_2 l_2^+ \ldots p_k l_k^\omega$ is *minimal* whenever $p_1 \cdots p_k$ is either the empty word or a simple non-loop segment, and l_1, $\ldots$, l_k are loops with disjoint sets of transitions.

Lemma 4. *Given a flat counter system $S = \langle Q, \mathtt{C}_n, \Delta, \mathbf{l} \rangle$, the total number of minimal path schemas of S is finite and is smaller than* $\text{card}(\Delta)^{(2 \times \text{card}(\Delta))}$.

This is a simple consequence of the fact that in a minimal path schema, each transition occurs at most twice. In Figure 1, we present a flat counter system S with a unique counter and one of its minimal path schemas. Each transition δ_i labelled by $+i$ corresponds to a transition with the guard $\top$ and the update value $+i$. The minimal path schema shown in Figure 1 corresponds to the ω-regular expression $\delta_1(\delta_2\delta_3)^+\delta_4\delta_5(\delta_6\delta_5)^\omega$. Note that in the representation of path schemas, a state may occur several times, as it is the case for q_3 (this cannot occur in the representation of counter systems). Minimal path schemas play a crucial role in the sequel. Indeed, given a path schema P, there is a minimal path schema P' such that every run respecting P respects P' too. This can be easily shown since whenever a maximal number of copies of a simple loop is identified

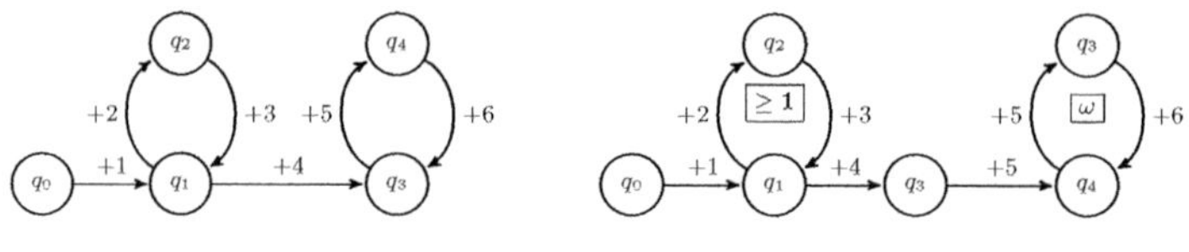

Fig. 1. A flat counter system and one of its minimal path schemas

as a factor of $p_1 l_1 \cdots p_k l_k$, this factor is replaced by the simple loop unless it is already present in the path schema.

Finally, the conditions imposed on the structure of path schemas implies the following corollary which states that the number of minimal path schemas for a given flat counter system is at most exponential in the size of the system (see similar statements in [20]).

Corollary 5. *Given a flat counter system S and a configuration c_0, there is a finite set of minimal path schemas X of cardinality at most* $\text{card}(\Delta)^{(2\times\text{card}(\Delta))}$ *such that $lab(c_0) = lab(\bigcup_{P\in X} P, c_0)$.*

4.2 Complexity Results

We write $\mathcal{CPS}$ [resp. $\mathcal{KPS}$] to denote the class of path schemas from counter systems [resp. the class of path schemas from Kripke structures]. As a preliminary step, we consider the problem MC(PLTL[$\emptyset$], $\mathcal{KPS}$) that takes as inputs a path schema P in $\mathcal{KPS}$, and $\phi \in$ PLTL[$\emptyset$] and asks whether there is a run respecting P that satisfies ϕ. Let ρ and ρ' be runs respecting P. For $\alpha \geq 0$, we write $\rho \equiv_\alpha \rho' \overset{\text{def}}{\Leftrightarrow}$ for every $i \in [1, \text{nbloops}(P) - 1]$, we have $\texttt{Min}(iter_P(\rho)[i], \alpha) = \texttt{Min}(iter_P(\rho')[i], \alpha)$. We state below a result concerning the runs of flat counter systems when respecting the same path schema.

Proposition 6. *Let S be a flat counter system, P be a path schema, and $\phi \in$* PLTL[$\emptyset$]. *For all runs ρ and ρ' respecting P such that $\rho \equiv_{2td(\phi)+5} \rho'$, we have $\rho, 0 \models \phi$ iff $\rho', 0 \models \phi$.*

This property can be proved by applying Theorem 3 repeatedly in order to get rid of the unwanted iterations of the loops.

Our algorithm for MC(PLTL[$\emptyset$], $\mathcal{KPS}$) takes advantage of a result from [18] for model-checking ultimately periodic models with formulae from Past LTL. An *ultimately periodic path* is an infinite word in Δ^ω of the form uv^ω were uv is a path segment and consequently $first(v) = last(v)$. According to [18], given an ultimately periodic path w, and a formula $\phi \in$ PLTL[$\emptyset$], the problem of checking whether there exists a run ρ such that $lab(\rho) = w$ and $\rho, 0 \models \phi$ is in PTime (a tighter bound of NC can be obtained by combining results from [15] and Theorem 3).

Lemma 7. MC(PLTL[$\emptyset$], $\mathcal{KPS}$) *is in* NP.

The proof is a consequence of Proposition 6 and [18]. Indeed, given $\phi \in$ PLTL[$\emptyset$] and $P = p_1 l_1^+ p_2 l_2^+ \ldots p_k l_k^\omega$, first guess $\mathbf{m} \in [1, 2td(\phi) + 5]^{k-1}$ and check whether

$\rho, 0 \models \phi$ where ρ is the obvious ultimately periodic word such that $lab(\rho) = p_1 l_1^{\mathbf{m}[1]} p_2 l_2^{\mathbf{m}[2]} \ldots p_k l_k^{\omega}$. Since $\mathbf{m}$ is of polynomial size and $\rho, 0 \models \phi$ can be checked in polynomial time by [18], we get the NP upper bound.

From [16], we have the lower bound for MC(PLTL[$\emptyset$], $\mathcal{KPS}$).

Lemma 8. *[16]* MC(PLTL[$\emptyset$], $\mathcal{KPS}$) *is* NP*-hard even if restricted to* `X` *and* `F`.

For a fixed $n \in \mathbb{N}$, we write MC(PLTL[$\emptyset$], $\mathcal{KPS}(n)$) to denote the restriction of MC(PLTL[$\emptyset$], $\mathcal{KPS}$) to path schemas with at most n loops. When n is fixed, the number of ultimately periodic paths w in $\mathcal{L}(P)$ such that each loop (except the last one) is visited is at most $2td(\phi)+5$ times is bounded by $(2td(\phi)+5)^n$, which is polynomial in the size of the input (because n is fixed).

Theorem 9. MC(PLTL[$\emptyset$], $\mathcal{KPS}$) *is* NP*-complete.*
Given a fixed $n \in \mathbb{N}$*,* MC(PLTL[$\emptyset$], $\mathcal{KPS}(n)$) *is in* PTIME.

Note that it can be proved that MC(PLTL[$\emptyset$], $\mathcal{KPS}(n)$) is in NC, hence giving a tighter upper bound for the problem. This can be obtained by observing that we can run the NC algorithm for model checking PLTL[$\emptyset$] over ultimately periodic paths parallelly on $(2td(\phi) + 5)^n$ (polynomially many) different paths.

Now, we present how to solve MC(PLTL[$\emptyset$], $\mathcal{KFS}$) using Lemma 7. From Lemma 4, we know that the number of minimal path schemas in a flat Kripke structure $S = \langle Q, \Delta, \mathbf{l} \rangle$ is finite and the length of a minimal path schema is at most $2 \times \text{card}(\Delta)$. Hence, for solving the model-checking problem for a state q and a PLTL[$\emptyset$] formula ϕ, a possible algorithm consists in choosing non-deterministically a minimal path schema P starting at q and then apply the algorithm used to establish Lemma 7. This new algorithm would be in NP. Furthermore, thanks to Corollary 5, we know that if there exists a run ρ of S such that $\rho, 0 \models \phi$ then there exists a minimal path schema P such that ρ respects P. Consequently there is an algorithm in NP to solve MC(PLTL[$\emptyset$], $\mathcal{KFS}$).

Theorem 10. MC(PLTL[$\emptyset$], $\mathcal{KFS}$) *is* NP*-complete.*

NP-hardness can be established as a variant of the proof of Lemma 8.

Similarly, $\mathcal{CPS}(k)$ denotes the class of path schemas obtained from flat counter systems with number of loops bounded by k.

Lemma 11. *For* $k \geq 2$*,* MC(PLTL[`C`], $\mathcal{CPS}(k)$) *is* NP*-hard.*

The proof by reduction from SAT and it is less straightforward than the proof for Lemma 8 or the reduction presented in [16] when path schemas are involved. Indeed, we cannot encode the nondeterminism in the structure itself and the structure has only a constant number of loops. Actually, we cannot use a separate loop for each counter; the reduction is done by encoding the nondeterminism in the (possibly exponential) number of times a single loop is taken, and then using its binary encoding as an assignment for the propositional variables. Hence, the reduction uses in an essential way the counter values and the arithmetical constraints in the formula. By contrast, MC(PLTL[`C`], $\mathcal{CPS}(1)$) can be shown in PTIME.

5 Model-Checking PLTL[C] over Flat Counter Systems

In this section, we provide a nondeterministic polynomial-time algorithm to solve MC(PLTL[C],$\mathcal{CFS}$) (see Algorithm 1). To do so, we combine Theorem 3 with small solutions of constraint systems.

5.1 Characterizing Runs by System of Equations

In this section, we show how to build a system of equations from a path schema P and a configuration c_0 such that the system of equations encodes the set of all runs respecting P from c_0. This can be done for path schemas without disjunctions in guards that satisfy an additional *validity* property. A path schema $P = p_1 l_1^+ p_2 l_2^+ \ldots p_k l_k^\omega$ is *valid* whenever $\mathit{effect}(l_k)[i] \geq 0$ for every $i \in [1, n]$ (see Section 4 for the definition of $\mathit{effect}(l_k)$) and if all the guards in transitions in l_k are conjunctions of atomic guards, then for each guard occurring in the loop l_k of the form $\sum_i a_i \mathsf{x}_i \sim b$ with $\sim \in \{\leq, <\}$ [resp. with $\sim \in \{=\}$, with $\sim \in \{\geq, >\}$] , we have $\sum_i a_i \times \mathit{effect}(l_k)[i] \leq 0$ [resp. $\sum_i a_i \times \mathit{effect}(l_k)[i] = 0$, $\sum_i a_i \times \mathit{effect}(l_k)[i] \geq 0$]. It is easy to check that these conditions are necessary to visit the last loop l_k infinitely. More specifically, if a path schema is not valid, then no infinite run can respect it. Moreover, given a path schema, one can decide in polynomial time whether it is valid.

Now, let us consider a (not necessarily minimal) valid path schema $P = p_1 l_1^+ p_2 l_2^+ \ldots p_k l_k^\omega$ ($k \geq 1$) obtained from a flat counter system S such that all the guards on transitions are conjunctions of atomic guards of the form $\sum_i a_i \mathsf{x}_i \sim b$ where $a_i \in \mathbb{Z}$, $b \in \mathbb{Z}$ and $\sim \in \{=, \leq, \geq, <, >\}$. Hence, disjunctions are *disallowed* in guards. The goal of this section (see Lemma 12 below) is to characterize the set $\mathit{iter}_P(c_0) \subseteq \mathbb{N}^{k-1}$ for some configuration c_0 as the set of solutions of a constraint system. For each loop l_i, we introduce a variable y_i, whence the number of variables of the system/formula is precisely $k - 1$. A *constraint system* $\mathcal{E}$ over the set of variables $\{\mathsf{y}_1, \ldots, \mathsf{y}_n\}$ is a quantifier-free Presburger formula built over $\{\mathsf{y}_1, \ldots, \mathsf{y}_n\}$ as a conjunction of atomic constraints of the form $\sum_i a_i \mathsf{y}_i \sim b$ where $a_i, b \in \mathbb{Z}$ and $\sim \in \{=, \leq, \geq, <, >\}$. Conjunctions of atomic counter constraints and constraint systems are essentially the same objects but the distinction allows to emphasize the different purposes: guard on counters in operational models and symbolic representation of sets of tuples.

Lemma 12. *Let $S = \langle Q, \mathsf{C}_n, \Delta, \mathbf{l} \rangle$ be a flat counter system without disjunctions in guards, P be a valid path schema and c_0 be a configuration. One can compute in polynomial time a constraint system $\mathcal{E}$ such that the set of solutions of $\mathcal{E}$ is equal to $\mathit{iter}_P(c_0)$, $\mathcal{E}$ has* nbloops$(P) - 1$ *variables, $\mathcal{E}$ has at most* len$(P) \times 2 \times$ size$(S)^2$ *conjuncts and the greatest absolute value from constants in $\mathcal{E}$ is bounded by $n \times$* nbloops$(P) \times K^4 \times$ len$(P)^3$ *where K is the greatest absolute value for constants occurring in S.*

5.2 Elimination of Arithmetical Constraints and Disjunctions

As stated in Lemma 12, the procedure for characterizing infinite runs in a counter system by a system of equations works only for a flat counter system with no disjunction in guards (convexity of guards is essential). In this section, we show how to obtain such a system from a general flat counter system. Given a flat counter system $S = \langle Q, \mathtt{C}_n, \Delta, \mathbf{l}\rangle$, a configuration $c_0 = \langle q_0, \boldsymbol{v_0}\rangle$ and a minimal path schema P starting from the configuration c_0, we show that it is possible to build a finite set Y_P of path schemas such that (1) each path schema in Y_P has transitions without disjunctions in guards, (2) existence of a run ρ respecting P is equivalent to the existence of a path schema in Y_P having a run similar to ρ respecting it and (3) each path schema in Y_P is obtained from P by unfolding loops so that the terms in each loop satisfy the same atomic guards. Note that disjunctions could be easily eliminated at the cost of adding new transitions between states but this type of transformation may easily destroy flatness. Hence, the necessity to present a more sophisticated elimination procedure

We first introduce a few definitions. A (syntactic) *resource* R is a triple $\langle X, T, B\rangle$ such that X is a finite set of propositional variables, T is a finite set of terms $\mathtt{t}$ appearing in some guards of the form $\mathtt{t} \sim b$ (with $b \in \mathbb{Z}$) and B is a finite set of integers. We say that a resource $\mathsf{R} = \langle X, T, B\rangle$ is *coherent* with a counter system S [resp. with a path schema P] if B contains all the constants b occurring in guards of S [resp. of P] of the form $\mathtt{t} \sim b$ and T contains all the terms $\mathtt{t}$ occurring in guards of S [resp. of P] of the form $\mathtt{t} \sim b$. The resource R is coherent with a formula $\phi \in$ PLTL[C], whenever the atomic formulae of ϕ are either of the form $p \in X$ or $\mathtt{t} \sim b$ with $\mathtt{t} \in T$ and $b \in B$. In the sequel, we assume that the considered resource is always coherent with S.

Assuming that $B = \{b_1, \ldots, b_m\}$ with $b_1 < \cdots < b_m$, we write I to denote the finite set of intervals $I = \{(-\infty, b_1 - 1], [b_1, b_1], [b_1 + 1, b_2 - 1], [b_2, b_2], \cdots, [b_m, b_m], [b_m + 1, \infty)\}$. Note that I contains exactly $2m+1$ intervals. A *term map* $\boldsymbol{m}$ is a map $\boldsymbol{m} : T \to I$ that abstracts term values. A *footprint* is an abstraction of a model for PLTL[C] restricted to elements from the resource R: it is of the form $\mathsf{ft} : \mathbb{N} \to 2^X \times I^T$ where I is the set of intervals built from B. The satisfaction relation $\models$ involving models or runs can be adapted to footprints as follows (formulae and footprints are from the same resource):

- $\mathsf{ft}, i \models_{\mathsf{symb}} p \overset{\text{def}}{\Leftrightarrow} p \in \pi_1(\mathsf{ft}(i))$; $\mathsf{ft}, i \models_{\mathsf{symb}} \mathtt{t} \geq b \overset{\text{def}}{\Leftrightarrow} \pi_2(\mathsf{ft}(i))(\mathtt{t}) \subseteq [b, +\infty)$,
- $\mathsf{ft}, i \models_{\mathsf{symb}} \mathtt{t} \leq b \overset{\text{def}}{\Leftrightarrow} \pi_2(\mathsf{ft}(i))(\mathtt{t}) \subseteq (-\infty, +b]$,
- $\mathsf{ft}, i \models_{\mathsf{symb}} \mathtt{X}\phi \overset{\text{def}}{\Leftrightarrow} \mathsf{ft}, i+1 \models_{\mathsf{symb}} \phi$,
- $\mathsf{ft}, i \models_{\mathsf{symb}} \phi \mathtt{U} \psi \overset{\text{def}}{\Leftrightarrow} \exists j \geq i$ s.t. $\mathsf{ft}, j \models_{\mathsf{symb}} \psi$ and $\forall j' \in [i, j-1]$, $\mathsf{ft}, j' \models_{\mathsf{symb}} \phi$.

We omit the other obvious clauses. $\models_{\mathsf{symb}}$ is the satisfaction relation for Past LTL when arithmetical constraints are understood as abstract propositions. Let $\mathsf{R} = \langle X, T, B\rangle$ be a resource and $\rho = \langle q_0, \boldsymbol{v_0}\rangle, \langle q_1, \boldsymbol{v_1}\rangle \cdots$ be an infinite run of S. The *footprint* of ρ with respect to R is the footprint $\mathsf{ft}(\rho)$ such that for $i \geq 0$, we have $\mathsf{ft}(\rho)(i) \overset{\text{def}}{=} \langle \mathbf{l}(q_i) \cap X, \boldsymbol{m}_i\rangle$ where for every term $\mathtt{t} = \sum_j a_j \mathtt{x}_j \in T$, we have $\sum_j a_j \boldsymbol{v_i}[j] \in \boldsymbol{m}_i(\mathtt{t})$. Note that $\sum_j a_j \boldsymbol{v_i}[j]$ belongs to a unique element of I since I is a partition of $\mathbb{Z}$. Hence, this definition makes sense. Lemma 13 below roughly states that satisfaction of a formula on a run can be checked symbolically from the footprint (this is useful for the correctness of forthcoming Algorithm 1).

Lemma 13. *Let ϕ be in* PLTL[C], $\mathsf{R} = \langle X, T, B\rangle$ *be coherent with* ϕ, $\rho = \langle q_0, \boldsymbol{v_0}\rangle, \langle q_1, \boldsymbol{v_1}\rangle \cdots$ *be an infinite run and* $i \geq 0$. *(I) Then* $\rho, i \models \phi$ *iff* $\mathsf{ft}(\rho), i \models_{\mathsf{symb}} \phi$. *(II) If* ρ' *is an infinite run s.t.* $\mathsf{ft}(\rho) = \mathsf{ft}(\rho')$, *then* $\rho, i \models \phi$ *iff* $\rho', i \models \phi$.

In [6] we explain in details how to build a set Y_P of path schemas without disjunctions from a minimal path schema P, an initial configuration $\langle q_0, \boldsymbol{v_0}\rangle$ and a resource R. The main idea of this construction consists in adding to the control states of path schemas some information on the intervals to which belongs each term of T. In fact, in the transitions appearing in path schemas of Y_P the states belong to $Q' = Q \times I^T$. Before stating the properties of Y_P, we introduce some notations. Given $\mathtt{t} = \sum_j a_j \mathsf{x}_j \in T$, $\mathbf{u} \in \mathbb{Z}^n$ and a term map $\boldsymbol{m}$, we write $\psi(\mathtt{t}, \mathbf{u}, \boldsymbol{m}(\mathtt{t}))$ to denote the formula below ($b, b' \in B$): $\psi(\mathtt{t}, \mathbf{u}, (-\infty, b]) \stackrel{\text{def}}{=} \sum_j a_j(\mathsf{x}_j + \mathbf{u}(j)) \leq b$; $\psi(\mathtt{t}, \mathbf{u}, [b, +\infty)) \stackrel{\text{def}}{=} \sum_j a_j(\mathsf{x}_j + \mathbf{u}(j)) \geq b$ and $\psi(\mathtt{t}, \mathbf{u}, [b, b']) = ((\sum_j a_j(\mathsf{x}_j + \mathbf{u}(j)) \leq b') \wedge ((\sum_j a_j(\mathsf{x}_j + \mathbf{u}(j)) \geq b)$. We write $\mathtt{G}^\star(T, B, U)$ to denote the set of guards of the form $\psi(\mathtt{t}, \mathbf{u}, \boldsymbol{m}(\mathtt{t}))$ where $\mathtt{t} \in T$, U is the finite set of updates from P and $\boldsymbol{m} : T \to I$. Each guard in $\mathtt{G}^\star(T, B, U)$ is of linear size in the size of P. We denote $\tilde{\Delta}$ the set of transitions $Q' \times \mathtt{G}^\star(T, B, U) \times U \times Q'$. Note that the transitions in $\tilde{\Delta}$ do not contain guards with disjunctions and $\tilde{\Delta}$ is finite. We also define a function proj which associates to $w \in \tilde{\Delta}^\omega$ the ω-sequence $\mathsf{proj}(w) : \mathbb{N} \to 2^X \times I^T$ such that for all $i \in \mathbb{N}$, if $w(i) = \langle\langle q, \boldsymbol{m}\rangle, \mathtt{g}, \mathbf{u}, \langle q', \boldsymbol{m}'\rangle\rangle$ and $\mathbf{l}(q) \cap X = L$ then $\mathsf{proj}(w)(i) \stackrel{\text{def}}{=} \langle L, \boldsymbol{m}\rangle$.

We show that it is possible to build a finite set Y_P of path schemas over $\tilde{\Delta}$ such that if $P' = p'_1(l'_1)^+ p'_2(l'_2)^+ \ldots p'_{k'}(l'_{k'})^\omega$ is a path schema in Y_P and ρ is a run $\langle\langle q_0, \boldsymbol{m}_0\rangle, \boldsymbol{v_0}\rangle \to \langle\langle q_1, \boldsymbol{m}_1\rangle, \boldsymbol{v_1}\rangle \to \langle\langle q_2, \boldsymbol{m}_2\rangle, \boldsymbol{v_2}\rangle \cdots$ respecting P' we have that $\mathsf{proj}(lab(\rho)) = \mathsf{ft}(\rho)$. This point will be useful for Algorithm 1. The following theorem lists the main properties of the set Y_P.

Theorem 14. *Given a flat counter system S, a minimal path schema P, a resource* $\mathsf{R} = \langle X, T, B\rangle$ *coherent with P and a configuration* $\langle q_0, \boldsymbol{v_0}\rangle$, *there is a finite set of path schemas Y_P over $\tilde{\Delta}$ satisfying (1)–(6) below.*

1. *No path schema in Y_P contains guards with disjunctions in it.*
2. *There exists a polynomial $q^\star(\cdot)$ such that for every $P' \in Y_P$,* $\mathrm{len}(P') \leq q^\star(\mathrm{len}(P) + \mathrm{card}(T) + \mathrm{card}(B))$.
3. *Checking whether a path schema P' over $\tilde{\Delta}$ belongs to Y_P can be done in polynomial time in* $\mathrm{size}(P) + \mathrm{card}(T) + \mathrm{card}(B)$.
4. *For every run ρ respecting P and starting at $\langle q_0, \boldsymbol{v_0}\rangle$, we can find a run ρ' respecting some $P' \in Y_P$ such that $\rho \models \phi$ iff $\rho' \models \phi$ for every ϕ built over* R.
5. *For every run ρ' respecting some $P' \in Y_P$ with initial values $\boldsymbol{v_0}$, we can find a run ρ respecting P such that $\rho \models \phi$ iff $\rho' \models \phi$ for every ϕ built over* R.
6. *For every ultimately periodic word $w \cdot u^\omega \in \mathcal{L}(P')$, for every ϕ built over* R *checking whether* $\mathsf{proj}(w \cdot u^\omega), 0 \models_{\mathsf{symb}} \phi$ *can be done in polynomial time in the size of $w \cdot u$ and in the size of ϕ.*

5.3 Main Algorithm

In Algorithm 1 below, a polynomial $p^\star(\cdot)$ is used. We can define polynomial $p^\star(\cdot)$ using the small solutions for constraint systems [2], see details in [6]. Note that

Algorithm 1. The main algorithm in NP with inputs S, $c_0 = \langle q, \boldsymbol{v_0} \rangle$, ϕ

1: guess a minimal path schema P of S
2: build a resource $\mathsf{R} = \langle X, T, B \rangle$ coherent with P and ϕ
3: guess a valid schema $P' = p_1 l_1^+ p_2 l_2^+ \dots p_k l_k^\omega$ such that $\mathrm{len}(P') \leq q^\star(\mathrm{len}(P) + \mathrm{card}(T) + \mathrm{card}(B))$
4: guess $\boldsymbol{y} \in [1, 2td(\phi) + 5]^{k-1}$; guess $\boldsymbol{y'} \in [1, 2^{p^\star(\mathrm{size}(S) + \mathrm{size}(c_0) + \mathrm{size}(\phi))}]^{k-1}$
5: check that P' belongs to Y_P
6: check that $\mathsf{proj}(p_1 l_1^{\boldsymbol{y}[1]} p_2 l_2^{\boldsymbol{y}[2]} \dots l_{k-1}^{\boldsymbol{y}[k-1]} p_k l_k^\omega), 0 \models_{\mathsf{symb}} \phi$
7: build $\mathcal{E}$ over $\mathsf{y}_1, \dots, \mathsf{y}_{k-1}$ for P' with initial values $\boldsymbol{v_0}$ (obtained from Lemma 12)
8: **for** $i = 1 \rightarrow k - 1$ **do**
9:     **if** $\boldsymbol{y}[i] = 2td(\phi) + 5$ **then** $\psi_i \leftarrow$ "$\mathsf{y}_i \geq 2td(\phi) + 5$" **else** $\psi_i \leftarrow$ "$\mathsf{y}_i = \boldsymbol{y}[i]$"
10: **end for**
11: check that $\boldsymbol{y'} \models \mathcal{E} \wedge \psi_1 \wedge \dots \wedge \psi_{k-1}$

$\boldsymbol{y'}$ is a refinement of $\boldsymbol{y}$ (for all i, we have $\boldsymbol{y'}[i] \approx_{2td(\phi)+5} \boldsymbol{y}[i]$) in which counter values are taken into account.

Algorithm 1 starts by guessing a path schema P (line 1) and an unfolded path schema $P' = p_1 l_1^+ p_2 l_2^+ \dots p_k l_k^\omega$ (line 3) and check whether P' belongs to Y_P (line 5). It remains to check whether there is a run ρ respecting P' such that $\rho \models \phi$. Suppose there is such a run ρ; let $\boldsymbol{y}$ be the unique tuple in $[1, 2td(\phi) + 5]^{k-1}$ such that $\boldsymbol{y} \approx_{2td(\phi)+5} iter_{P'}(\rho)$. By Proposition 6, we have $\mathsf{proj}(p_1 l_1^{\boldsymbol{y}[1]} p_2 l_2^{\boldsymbol{y}[2]} \dots l_{k-1}^{\boldsymbol{y}[k-1]} p_k l_k^\omega), 0 \models_{\mathsf{symb}} \phi$. Since the set of tuples of the form $iter_{P'}(\rho)$ is characterized by a system of equations, by the existence of small solutions from [2], we can assume that $iter_{P'}(\rho)$ contains only small values. Hence line 4 guesses $\boldsymbol{y}$ and $\boldsymbol{y'}$ (corresponding to $iter_{P'}(\rho)$ with small values). Line 6 precisely checks $\mathsf{proj}(p_1 l_1^{\boldsymbol{y}[1]} p_2 l_2^{\boldsymbol{y}[2]} \dots l_{k-1}^{\boldsymbol{y}[k-1]} p_k l_k^\omega), 0 \models_{\mathsf{symb}} \phi$ whereas line 11 checks whether $\boldsymbol{y'}$ encodes a run respecting P' with $\boldsymbol{y'} \approx_{2td(\phi)+5} \boldsymbol{y}$.

Lemma 15. *Algorithm 1 runs in nondeterministic polynomial time.*

It remains to check that Algorithm 1 is correct.

Lemma 16. *$S, c_0 \models \phi$ iff Algorithm 1 on inputs S, c_0, ϕ has an accepting run.*

In the proof of Lemma 16, we take advantage of all our preliminary results.

Proof. By way of example, we show that if Algorithm 1 on inputs $S, c_0 = \langle q_0, \boldsymbol{v_0} \rangle$, ϕ has an accepting computation, then $S, c_0 \models \phi$. This means that there are P, P', $\boldsymbol{y}$, $\boldsymbol{y'}$ that satisfy all the checks. Let $w = p_1 l_1^{\boldsymbol{y'}[1]} \cdots p_{k-1} l_{k-1}^{\boldsymbol{y'}[k-1]} p_k l_k^\omega$ and $\rho = \langle \langle q_0, \boldsymbol{m_0} \rangle, \boldsymbol{v_0} \rangle \langle \langle q_1, \boldsymbol{m_1} \rangle, \boldsymbol{x_1} \rangle \langle \langle q_2, \boldsymbol{m_2} \rangle, \boldsymbol{x_2} \rangle \dots \in (Q' \times \mathbb{Z}^n)^\omega$ be defined as follows: for every $i \geq 0$, $q_i \stackrel{\text{def}}{=} \pi_1(source(w(i)))$, and for every $i \geq 1$, we have $\boldsymbol{x_i} \stackrel{\text{def}}{=} \boldsymbol{x_{i-1}} + update(w(i))$. By Lemma 12, since $\boldsymbol{y'} \models \mathcal{E} \wedge \psi_1 \wedge \dots \wedge \psi_{k-1}$, ρ is a run respecting P' starting at the configuration $\langle \langle q_0, \boldsymbol{m_0} \rangle, \boldsymbol{v_0} \rangle$. Since $\boldsymbol{y'} \models \psi_1 \wedge \dots \wedge \psi_{k-1}$ and $\boldsymbol{y} \models \psi_1 \wedge \dots \wedge \psi_{k-1}$, by Proposition 6, (✠) $\mathsf{proj}(p_1 l_1^{\boldsymbol{y}[1]} p_2 l_2^{\boldsymbol{y}[2]} \dots l_{k-1}^{\boldsymbol{y}[k-1]} p_k l_k^\omega), 0 \models_{\mathsf{symb}} \phi$, iff (✠✠) $\mathsf{proj}(p_1 l_1^{\boldsymbol{y'}[1]} p_2 l_2^{\boldsymbol{y'}[2]} \dots l_{k-1}^{\boldsymbol{y'}[k-1]} p_k l_k^\omega), 0 \models_{\mathsf{symb}} \phi$. Algorithm 1 guarantees that $\mathsf{proj}(p_1 l_1^{\boldsymbol{y}[1]} p_2 l_2^{\boldsymbol{y}[2]} \dots l_{k-1}^{\boldsymbol{y}[k-1]} p_k l_k^\omega), 0 \models_{\mathsf{symb}} \phi$, whence we have (✠✠). Since

$\mathsf{proj}(p_1 l_1^{\boldsymbol{y}'[1]} p_2 l_2^{\boldsymbol{y}'[2]} \ldots l_{k-1}^{\boldsymbol{y}'[k-1]} p_k l_k^{\omega}) = \mathsf{ft}(\rho)$, by Lemma 13, we deduce that $\rho, 0 \models \phi$. By Theorem 14(5), there is an infinite run ρ', starting at the configuration $\langle q_0, \boldsymbol{v_0} \rangle$ and respecting P, such that $\rho', 0 \models \phi$.

For the other direction, see [6]. □

As a corollary, we can state the main result of the paper.

Theorem 17. MC(PLTL[C], $\mathcal{CFS}$) *is* NP*-complete.*

6 Conclusion

We have investigated the computational complexity of the model-checking problem for flat counter systems with formulae from an enriched version of LTL. Our main result is the NP-completeness of MC(PLTL[C], $\mathcal{CFS}$), significantly improving the complexity upper bound from [7]. This also improves the results about the effective semilinearity of the reachability relations for such flat counter systems from [5,11] and it extends the recent result on the NP-completeness of model-checking flat Kripke structures with LTL from [16] by adding counters and past-time operators. Our main results are presented above and compared to the reachability problem.

Classes of Systems	PLTL[∅]	PLTL[C]	Reachability
$\mathcal{KPS}$	NP-complete See [16] for X and U	—	PTIME
$\mathcal{CPS}$	NP-complete	NP-complete (Theo. 17)	NP-complete
$\mathcal{KPS}(n)$	PTIME (Theo. 9)	—	PTIME
$\mathcal{CPS}(n)$, $n > 1$	??	NP-complete (Lem. 11)	??
$\mathcal{CPS}(1)$	PTIME	PTIME	PTIME
$\mathcal{KFS}$	NP-complete See [16] for X and U	—	PTIME
$\mathcal{CFS}$	NP-complete	**NP-complete** (Theo. 17)	NP-complete

As far as the proof technique is concerned, the NP upper bound is obtained as a combination of a general stuttering property for LTL with past-time operators (a result extending what is done in [17] with past-time operators) and the use of small integer solutions for quantifier-free Presburger formulae [2]. There are several related problems which are not addressed in the paper. For instance, the extension of the model-checking problem to full CTL* is known to be decidable [7] but the characterization of its exact complexity is open.

References

1. Boigelot, B.: Symbolic methods for exploring infinite state spaces. PhD thesis, Université de Liège (1998)
2. Borosh, I., Treybig, L.: Bounds on positive integral solutions of linear Diophantine equations. American Mathematical Society 55, 299–304 (1976)
3. Bozga, M., Iosif, R., Konečný, F.: Fast Acceleration of Ultimately Periodic Relations. In: Touili, T., Cook, B., Jackson, P. (eds.) CAV 2010. LNCS, vol. 6174, pp. 227–242. Springer, Heidelberg (2010)

4. Comon, H., Cortier, V.: Flatness Is Not a Weakness. In: Clote, P.G., Schwichtenberg, H. (eds.) CSL 2000. LNCS, vol. 1862, pp. 262–276. Springer, Heidelberg (2000)
5. Comon, H., Jurski, Y.: Multiple Counter Automata, Safety Analysis and PA. In: Vardi, M.Y. (ed.) CAV 1998. LNCS, vol. 1427, pp. 268–279. Springer, Heidelberg (1998)
6. Demri, S., Dhar, A., Sangnier, A.: Taming past LTL and flat counter systems. Technical report (2012), ArXiv
7. Demri, S., Finkel, A., Goranko, V., van Drimmelen, G.: Model-checking CTL* over flat Presburger counter systems. JANCL 20(4), 313–344 (2010)
8. Dixon, C., Fisher, M., Konev, B.: Temporal Logic with Capacity Constraints. In: Konev, B., Wolter, F. (eds.) FroCos 2007. LNCS (LNAI), vol. 4720, pp. 163–177. Springer, Heidelberg (2007)
9. Esparza, J., Finkel, A., Mayr, R.: On the verification of broadcast protocols. In: LICS 1999, pp. 352–359 (1999)
10. Etessami, K., Wilke, T.: An until hierarchy and other applications of an Ehrenfeucht-Fraïssé game for temporal logic. I&C 160(1–2), 88–108 (2000)
11. Finkel, A., Leroux, J.: How to Compose Presburger-Accelerations: Applications to Broadcast Protocols. In: Agrawal, M., Seth, A.K. (eds.) FSTTCS 2002. LNCS, vol. 2556, pp. 145–156. Springer, Heidelberg (2002)
12. Finkel, A., Lozes, É., Sangnier, A.: Towards Model-Checking Programs with Lists. In: Archibald, M., Brattka, V., Goranko, V., Löwe, B. (eds.) ILC 2007. LNCS (LNAI), vol. 5489, pp. 56–86. Springer, Heidelberg (2009)
13. Gabbay, D.: The Declarative Past and Imperative Future. In: Banieqbal, B., Pnueli, A., Barringer, H. (eds.) Temporal Logic in Specification. LNCS, vol. 398, pp. 409–448. Springer, Heidelberg (1989)
14. Haase, C., Kreutzer, S., Ouaknine, J., Worrell, J.: Reachability in Succinct and Parametric One-Counter Automata. In: Bravetti, M., Zavattaro, G. (eds.) CONCUR 2009. LNCS, vol. 5710, pp. 369–383. Springer, Heidelberg (2009)
15. Kuhtz, L.: Model Checking Finite Paths and Trees. PhD thesis, Universität des Saarlandes (2010)
16. Kuhtz, L., Finkbeiner, B.: Weak Kripke Structures and LTL. In: Katoen, J.-P., König, B. (eds.) CONCUR 2011. LNCS, vol. 6901, pp. 419–433. Springer, Heidelberg (2011)
17. Kučera, A., Strejček, J.: The stuttering principle revisited. Acta Informatica 41(7–8), 415–434 (2005)
18. Laroussinie, F., Markey, N., Schnoebelen, P.: Temporal logic with forgettable past. In: LICS 2002, pp. 383–392. IEEE (2002)
19. Laroussinie, F., Schnoebelen, P.: Specification in CTL + past for verification in CTL. I&C 156, 236–263 (2000)
20. Leroux, J., Sutre, G.: Flat Counter Automata Almost Everywhere! In: Peled, D.A., Tsay, Y.-K. (eds.) ATVA 2005. LNCS, vol. 3707, pp. 489–503. Springer, Heidelberg (2005)
21. Lutz, C.: NExpTime-Complete Description Logics with Concrete Domains. In: Goré, R.P., Leitsch, A., Nipkow, T. (eds.) IJCAR 2001. LNCS (LNAI), vol. 2083, pp. 45–60. Springer, Heidelberg (2001)
22. Minsky, M.: Computation, Finite and Infinite Machines. Prentice Hall (1967)
23. Peled, D., Wilke, T.: Stutter-invariant temporal properties are expressible without the next-time operator. IPL 63, 243–246 (1997)
24. Rackoff, C.: The covering and boundedness problems for vector addition systems. TCS 6(2), 223–231 (1978)

A Calculus for Generating Ground Explanations*

Mnacho Echenim and Nicolas Peltier

University of Grenoble, (LIG, Grenoble INP/CNRS)
{Mnacho.Echenim,Nicolas.Peltier}@imag.fr

Abstract. We present a modification of the superposition calculus that is meant to generate explanations why a set of clauses is satisfiable. This process is related to abductive reasoning, and the explanations generated are clauses constructed over so-called abductive constants. We prove the correctness and completeness of the calculus in the presence of redundancy elimination rules, and develop a sufficient condition guaranteeing its termination; this sufficient condition is then used to prove that all possible explanations can be generated in finite time for several classes of clause sets, including many of interest to the SMT community. We propose a procedure that generates a set of explanations that should be useful to a human user and conclude by suggesting several extensions to this novel approach.

1 Introduction

The verification of complex systems is generally based on proving the validity, or, dually, the satisfiability of a logical formula. The standard practice consists in translating the behavior of the system to be verified into a logical formula, and proving that the negation of the formula is unsatisfiable. These formulas may be domain-specific, so that it is only necessary to test the satisfiability of the formula modulo some background theory, whence the name *Satisfiability Modulo Theories problems*, or *SMT problems*. If the formula is actually satisfiable, this means the system is not error-free, and any model can be viewed as a trace that generates an error. The models of a satisfiable formula can therefore help the designers of the system guess the origin of the errors and deduce how they can be corrected. Yet, this still requires some work. Indeed, there are generally many interpretations on different domains that satisfy the formula, and it is necessary to further analyze these models to understand where the error(s) may come from.

We present what is, to the best of our knowledge, a novel approach to this debugging problem: we argue that rather than studying one model of a formula, more valuable information can be extracted from the properties that hold in *all*

* This work has been partly funded by the project ASAP of the French *Agence Nationale de la Recherche* (ANR-09-BLAN-0407-01).

B. Gramlich, D. Miller, and U. Sattler (Eds.): IJCAR 2012, LNAI 7364, pp. 194–209, 2012.

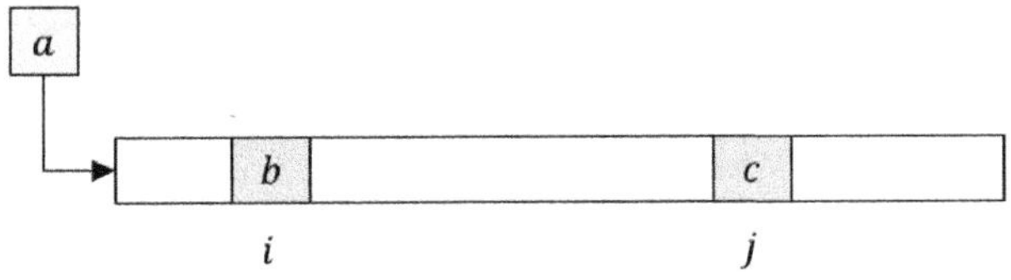

Fig. 1. Insertion into array a of element b at position i and element c at position j

the models of the formula. For instance, consider the theory of arrays, which is axiomatized as follows (as introduced by [13]):

$$\forall x, z, v.\ \text{select}(\text{store}(x, z, v), z) \simeq v, \tag{1}$$
$$\forall x, z, w, v.\ z \simeq w \vee \text{select}(\text{store}(x, z, v), w) \simeq \text{select}(x, w). \tag{2}$$

These axioms state that if element v is inserted into array x at position z, then the resulting array contains v at position z, and the same elements as in x elsewhere. Assume that to verify that the order in which elements are inserted into a given array does not matter, the satisfiability of the following formula is tested (see also Figure 1):

$$\text{select}(\text{store}(\text{store}(a, i, b), j, c), k) \not\simeq \text{select}(\text{store}(\text{store}(a, j, c), i, b), k).$$

This formula asserts that there is a position k that holds different values in the array obtained from a by first inserting element b at position i and then element c at position j, and in the array obtained from a by first inserting element c at position j and then element b at position i. It turns out that this formula is actually satisfiable, which in this case means that some hypotheses are missing. State of the art SMT solvers such as Yices [16] can help find out what hypotheses are missing by outputting a model of the formula. In this particular case, Yices outputs `(= b 1) (= c 3) (= i 2) (= k 2) (= j 2)`, and for this simple example, such a model may be sufficient to quickly understand where the error comes from. However, a simpler and more natural way to determine what hypotheses are missing would be to have a tool that, when fed the formula above, outputs $i \simeq j \wedge b \not\simeq c$, stating that the formula can only be true when elements b and c are distinct, and are inserted at the *same* position in a. This information permits to know immediately what additional hypotheses must be made for the formula to be unsatisfiable. In this example, there are two possible hypotheses that can be added: $i \not\simeq j$ or $b \simeq c$.

In this paper, we investigate what information should be provided to the user and how it can be obtained, by distinguishing a set of constants on which additional hypotheses are allowed to be made. These constants are called *abducible constants* or simply *abducibles*, and the problem boils down to determining what ground clauses containing only abducibles are logically entailed by the formula under consideration, since the negation of any of these clauses can be viewed as a set of additional hypotheses that make the formula unsatisfiable.

Outline. This paper begins by summarizing all necessary background, and then a calculus specially designed for abductive reasoning is defined. This calculus is closely related to the *superposition calculus* $\mathcal{SP}$, and we rely on completeness and termination results for $\mathcal{SP}$ to prove similar results for the new calculus. We also propose a method for generating clauses containing only abducible constants, that can help a user quickly detect where an error comes from, and decide what additional hypotheses should be added to fix the faulty formula.

Due to the space restrictions, several intermediate results and proofs are omitted. A full version of this work containing all proofs is available in [8].

2 Preliminaries

The general framework of this paper is first-order logic with equality. Most of the presentation in this section is standard, and we refer the reader to [14] for details. Given a finite signature Σ and an integer $i \geq 0$, Σ^i stands for the set of function symbols in Σ of arity i. In particular, Σ^0 denotes the set of constants in Σ. We assume the standard definitions of terms, predicates, literals and clauses, all of which are constructed over a set of variables $\mathcal{X}$. Interpretations are defined as usual, $\models$ stands for logical entailment and $\equiv$ stands for logical equivalence. We also consider the standard definitions of positions in terms, predicates, literals or clauses. A term, predicate, literal or clause containing no variable is *ground*. As usual, clauses are assumed to be variable-disjoint. The symbol $\simeq$ stands for unordered equality, $\bowtie$ is either $\simeq$ or $\not\simeq$. If L is a literal, then L^c denotes the complementary literal of L, i.e., $(t \simeq s)^c \stackrel{\text{def}}{=} (t \not\simeq s)$ and $(t \not\simeq s)^c \stackrel{\text{def}}{=} (t \simeq s)$. A literal is *flat* if it only contains constants or variables[1], and a clause is *flat* if it only contains flat literals. The letters l, r, s, u, v and t denote terms, w, x, y, z variables, and all other lower-case letters denote constants or function symbols.

Definition 1. *Given a ground clause C, we denote by $\neg C$ the following set of literals:* $\neg C \stackrel{\text{def}}{=} \{L^c \mid L \in C\}$.

A *substitution* is a function mapping variables to terms. Given a substitution σ, the set of variables x such that $x\sigma \neq x$ is called the *domain* of σ and denoted by $dom(\sigma)$. If σ is a substitution and V is a set of variables, then $\sigma_{|V}$ is the substitution with domain $dom(\sigma) \cap V$, that matches σ on this domain. As usual, a substitution can be extended into a homomorphism on terms, atoms, literals and clauses. The image of an expression $\mathcal{E}$ by a substitution σ will be denoted by $\mathcal{E}\sigma$. If E is a set of expressions, then $E\sigma$ denotes the set $\{\mathcal{E}\sigma \mid \mathcal{E} \in E\}$. The composition of two substitutions σ and θ is denoted by $\sigma\theta$. A substitution σ is *more general* than θ if there exists a substitution η such that $\theta = \sigma\eta$. The substitution σ is a *renaming* if it is injective and $\forall x \in dom(\sigma), x\sigma \in \mathcal{X}$; and it is a *unifier* of two terms t, s if $t\sigma = s\sigma$. Any unifiable pair of terms (t, s) has a most general unifier, unique up to a renaming, and denoted by $\text{mgu}(t, s)$. A substitution σ is *ground* if $x\sigma$ is ground, for every variable x in its domain.

---

[1] Note that we depart from the terminology in [2,1], where flat positive literals can contain a term of depth 1.

Superposition	$\dfrac{C \vee l[u'] \simeq r \quad D \vee u \simeq t}{(C \vee D \vee l[t] \simeq r)\sigma}$	$(i), (ii), (iii), (iv)$
Paramodulation	$\dfrac{C \vee l[u'] \not\simeq r \quad D \vee u \simeq t}{(C \vee D \vee l[t] \not\simeq r)\sigma}$	$(i), (ii), (iii), (iv)$
Reflection	$\dfrac{C \vee u' \not\simeq u}{C\sigma}$	(v)
Equational Factoring	$\dfrac{C \vee u \simeq t \vee u' \simeq t'}{(C \vee t \not\simeq t' \vee u \simeq t')\sigma}$	$(i), (vi)$

where the notation $l[u']$ means that u' appears as a subterm in l, σ is the most general unifier (mgu) of u and u', u' is not a variable in *Superposition* and *Paramodulation*, and the following abbreviations hold:

(i): $u\sigma \not\preceq t\sigma$;
(ii): $\forall L \in D : (u \simeq t)\sigma \not\preceq L\sigma$;
(iii): $l[u']\sigma \not\preceq r\sigma$;
(iv): $\forall L \in C : (l[u'] \bowtie r)\sigma \not\preceq L\sigma$;
(v): $\forall L \in C : (u' \simeq u)\sigma \not\prec L\sigma$;
(vi): $\forall L \in \{u' \simeq t'\} \cup C : (u \simeq t)\sigma \not\prec L\sigma$.

Fig. 2. Inference rules of $\mathcal{SP}$: the clause below the inference line is added to the clause set containing the clauses above the inference line

A *simplification ordering* $\prec$ is an ordering that is stable under substitutions, monotonic, and contains the subterm ordering: if $s \prec t$, then $c[s]\sigma \prec c[t]\sigma$ for any context c and substitution σ, and if s is a strict subterm of t then $s \prec t$. A *complete simplification ordering*, or CSO, is a simplification ordering that is total on ground terms. Similarly to [7], in the sequel, we shall assume that any CSO under consideration is *good*:

Definition 2. *A CSO* $\prec$ *is* good *if for all ground compound terms* t *and constants* c*, we have* $c \prec t$.

The *superposition calculus*, or $\mathcal{SP}$ (see, e.g., [14]), is a refutationally complete rewrite-based inference system for first-order logic with equality. It consists of the inference rules summarized in Fig. 2: each rule contains *premises* which are above the inference line, and generates a *conclusion*, which is below the inference line. If a clause D is generated from premises C, C', then we write $C, C' \vdash D$. The superposition calculus is based on a CSO on terms, which is extended to literals and clauses in a standard way (see, e.g., [3]), and we may write $\mathcal{SP}_\prec$ and $\vdash_\prec$ to specify the ordering. A ground clause C is $\prec$*-redundant in* S, or simply redundant, if there exists a set of ground clauses S' such that $S' \models C$, and for every $D \in S'$, D is an instance of a clause in S and $D \prec C$. A non-ground clause C is $\prec$*-redundant in* S if all its instances are $\prec$-redundant in S. In particular, every strictly subsumed clause and every tautological clause is redundant. A set of clauses S is *saturated* if every clause $C \notin S$ generated from premises in S is redundant in S. A saturated set of clauses that does not contain $\square$ is satisfiable [14]. In practice, it is necessary to use a decidable approximation of this notion of redundancy: for example, a clause is redundant if it can be reduced by some demodulation steps to either a tautology or to a subsumed clause.

In the sequel, it will be necessary to forbid the occurrence of clauses containing maximal literals of the form $x \simeq t$, where $x \not\preceq t$:

Definition 3. *A clause is* variable-eligible w.r.t. $\prec$ *if it contains a maximal literal of the form $x \simeq t$, where $x \not\preceq t$. A set of clauses is* variable-inactive *(see [1]) if no non-redundant clause generated from S is variable-eligible.*

For technical reasons we have chosen to present a slightly relaxed version of the superposition calculus, in which the standard strict maximality conditions have been replaced by non-strict maximality conditions. For instance in Condition (i), $u\sigma \not\preceq t\sigma$ is replaced by $u\sigma \not\prec t\sigma$: it is not forbidden for u and t to be identical in *Paramodulation* and *Superposition* inferences. It is clear that the clauses generated in the case where there is an equality actually turn out to be redundant.

3 A Calculus for Handling Abducible Constants

As explained in the Introduction, the aim of this paper is to start with a formula F and a set of axioms A, and generate a formula H which logically entails F modulo A, i.e., such that $H, A \models F$ (where $H \wedge A$ is satisfiable). As usual in abductive reasoning (see for instance [9]), we actually consider the contrapositive: since $H, A \models F$ is equivalent to $\neg F, A \models \neg H$, the original problem can be solved by generating logical consequences of the formula $\neg F \wedge A$. For the sake of simplicity, the formula $\neg F$ is added to the axioms which are assumed to be in clausal form, and we have the following definition:

Definition 4. *A clause C is an* implicate *of a set of clauses S iff $S \models C$.*

It is clear that after H is generated, one must verify that it is satisfiable modulo A. For instance, if a is some constant, then an explanation such as $a \simeq 0 \wedge a \simeq 1$ or even $0 \simeq 1$ does not provide any information since it contradicts the axioms of Presburger arithmetic. Testing this satisfiability can be done using standard decision procedures. There are many possible candidate sets of implicates, which may be more or less informative. For instance, it is possible to take $C \in S$, but this is obviously of no use. Thus it is necessary to provide additional information in order to restrict the class of formulas that are searched for. In (propositional) abductive reasoning, this is usually done by considering clauses built on a given set of literals: the *abducible literals*. A more natural possibility in the context of this paper is to consider clauses built on a given set of ground terms. We may assume with no loss of generality that each of these terms is replaced by a constant symbol, by applying the usual flattening operation, see, e.g., [2,7]. For example, the term $\mathrm{select}(\mathrm{store}(a, i, b), j)$ may be replaced by a new constant d, along with the axioms: $d \simeq \mathrm{select}(d', j) \wedge d' \simeq \mathrm{store}(a, i, b))$. We thus consider a distinguished set of constants $\mathcal{A} \subseteq \Sigma^0$, called the *set of abducible constants*, and restrict ourselves to explanations that are conjunctions of literals built upon abducible constants. This is formalized with the following definition of an $\mathcal{A}$-implicate:

Definition 5. *Let S be a set of clauses. A clause C is an* $\mathcal{A}$-implicate *of S iff every term occurring in C is also in $\mathcal{A}$ and $S \models C$.*

As in propositional abductive reasoning, the set $\mathcal{A}$ must be provided by the user. Given a set of clauses S containing both the axioms A and the clauses corresponding to the conjunctive normal form of $\neg F$, we investigate how to generate the set of flat ground clauses C built on $\mathcal{A}$, that are logical consequences of S. Since $\mathcal{SP}$ is only *refutationally* complete, this cannot be done directly using this calculus (except in some very particular cases, see for instance [15]). For example, it is clear that $f(a) \not\simeq f(b) \models a \not\simeq b$, but $a \not\simeq b$ cannot be generated from the antecedent clause. In principle, it is possible to enumerate all possible clauses C built on $\mathcal{A}$ and then use the superposition calculus to check whether $S \cup \neg C$ is unsatisfiable, however, this yields a very inefficient procedure. An alternate method consists in replacing the superposition calculus by a less restrictive calculus, such as the Resolution calculus [11] together with the equality axioms. For instance in the previous case, the clause $f(a) \not\simeq f(b)$ and the substitutivity axiom $x \not\simeq y \vee f(x) \simeq f(y)$ permit to generate by the Resolution rule: $a \not\simeq b$. However, again, this calculus is not efficient, and in particular all the termination properties of the superposition calculus on many interesting subclasses of first-order logic [4,2,1] are lost. In this section, we provide a variant of the superposition calculus which is able to *directly* generate, from a set of clauses S, a set of logical consequences of S that are built on a given set of constant symbols $\mathcal{A}$. The calculus is thus parameterized both by the term ordering $\prec$ and by the set of abducible constants $\mathcal{A}$. We shall show that the calculus is complete, in the sense that if $S \models C$ and if C is an $\mathcal{A}$-implicate of S, then C is a logical consequence of other clauses built on $\mathcal{A}$ that are generated from S. We will also prove that the calculus terminates on many classes of interest in the SMT community.

We will thus consider clauses of a particular form and a slight variation of the superposition calculus in order to be able to reason on abducible constants. The principle behind this calculus is similar to that of [5] for the combination of *hierarchic theories*, with the difference that in this framework, abducible constants can potentially interact with other terms, whereas in the framework of [5], such an interaction is prevented by sortedness. In both settings however, a same *abstraction* principle is used to delay the reasoning on the objects of interest (in this case, the abducible constants).

From now on we assume that the set of variables $\mathcal{X}$ is of the form $\mathcal{X} = \mathcal{V} \uplus \mathcal{V}_{\mathcal{A}}$. The elements in $\mathcal{V}$ are ordinary variables and the elements in $\mathcal{V}_{\mathcal{A}}$ are called *abducible variables*, and they will serve as placeholders for abducible constants in terms and clauses. In the sequel, when we mention *standard* terms, literals or clauses, we assume that all the variables they contain are in $\mathcal{V}$.

Definition 6. *An* $\mathcal{A}$-literal *is a literal of the form $t \bowtie s$, where $t, s \in \mathcal{V}_{\mathcal{A}} \cup \mathcal{A}$. An* $\mathcal{A}$-clause *is a disjunction of $\mathcal{A}$-literals. Given a clause C, we denote by $\Delta(C)$*

the disjunction of $\mathcal{A}$-literals in C and by $\overline{\Delta}(C)$ the disjunction of non-$\mathcal{A}$-literals in C. We define $\mathrm{Var}_{\mathcal{A}}(C) \stackrel{\mathrm{def}}{=} \mathrm{Var}(C) \cap \mathcal{V}_{\mathcal{A}}$.

A first step towards reasoning on abducible constants will consist in extracting them from the terms in which they occur, and replacing them by abducible variables. Then, to ensure that such a property is preserved by inferences, every substitution mapping an abducible variable to anything other than an abducible variable will be discarded. More formally:

Definition 7. *A term is* abstracted *if it contains no abducible constant. A literal $t \bowtie s$ is* abstracted *if t and s are both abstracted. A clause is* abstracted *if all non-abstracted literals in C are $\mathcal{A}$-literals.*

If t is an abstracted term, then not every instance of t is also abstracted. We define a condition on substitutions that guarantees such a stability result.

Definition 8. *A substitution σ is* $\mathcal{A}$-compliant *if for all $x \in dom(\sigma)$, $x\sigma$ is abstracted, and for all $x \in dom(\sigma) \cap \mathcal{V}_{\mathcal{A}}$, $x\sigma \in \mathcal{V}_{\mathcal{A}}$. Two abstracted terms are* $\mathcal{A}$-unifiable *if they are unifiable and admit an $\mathcal{A}$-compliant mgu.*

In the sequel, every time abstracted terms are $\mathcal{A}$-unifiable, we will assume the corresponding mgu is $\mathcal{A}$-compliant.

Definition 9. *Let $<_{\mathcal{A}}$ be a total ordering on $\mathcal{A}$ and a_0 denote the smallest abducible in $\mathcal{A}$. Given a term t, we denote by $t_{\downarrow\mathcal{A}}$ the term obtained by replacing every abducible constant occurring in t by a_0. The term t is* $\mathcal{A}$-reduced *if $t_{\downarrow\mathcal{A}} = t$. The previous notation and this definition extend to literals, clauses and sets of clauses.*

Example 1. Let $C = f(b,c) \simeq g(d) \vee x \not\simeq b \vee f(a,b) \not\simeq f(c,d)$, where $\mathcal{A} = \{a,b,c\}$ and $a \prec b \prec c$. Then $C_{\downarrow\mathcal{A}} = f(a,a) \simeq g(d) \vee x \not\simeq a \vee f(a,a) \not\simeq f(a,d)$, and this clause is an $\mathcal{A}$-reduced clause.

It is clear that if all abducible constants are replaced by abducible variables in a standard clause, then the resulting abstracted clause is not equivalent to the former one. However, equivalence can be regained by adding so-called $\mathcal{V}_{\mathcal{A}}$-*constraint literals* to the resulting abstracted clause.

Definition 10. *A* $\mathcal{V}_{\mathcal{A}}$-constraint literal *is a literal of the form $x \not\simeq a$, where $x \in \mathcal{V}_{\mathcal{A}}$ and $a \in \mathcal{A}$. For all clauses C, we denote by $\Gamma(C)$ the disjunction of $\mathcal{V}_{\mathcal{A}}$-constraint literals in C. A* $\mathcal{V}_{\mathcal{A}}$-constraint clause *is a disjunction of $\mathcal{V}_{\mathcal{A}}$-constraint literals. Given a $\mathcal{V}_{\mathcal{A}}$-constraint clause $A = \bigvee_{i=1}^{k} x_i \not\simeq a_i$, the* substitution associated to A *is denoted by ν_A and defined as follows: $dom(\nu_A) = \{x_1, \dots, x_k\}$, and for all $x \in dom(\nu_A)$, $x\nu_A = \min_{<_{\mathcal{A}}} \{a_i \mid i \in [1,k] \wedge x_i = x\}$.*

For readability, if B is a clause then we will write ν_B instead of $\nu_{\Gamma(B)}$. If S is a set of abstracted clauses, then S_ν is the set $S_\nu = \{C\nu_C \mid C \in S\}$.

Example 2. Assume $\mathcal{A} = \{a,b,c\}$, where $a <_{\mathcal{A}} b <_{\mathcal{A}} c$, and let $A = x \not\simeq a \vee x \not\simeq c \vee y \not\simeq b \vee z \not\simeq a \vee y \not\simeq c$. Then $\nu_A = \{x \mapsto a, y \mapsto b, z \mapsto a\}$.

Note that by definition, $C \equiv C\nu_C$ and $S \equiv S_\nu$. As mentioned earlier, abducible variables are meant to be placeholders for abducible constants. In general, it will be necessary to keep some information permitting to know what abducible constants an abducible variable could be replaced by. Such a requirement is satisfied by imposing that every abducible variable occurs in at least one $\mathcal{V}_\mathcal{A}$-constraint literal, which intuitively specifies its value.

Definition 11. *A clause C is* $\mathcal{V}_\mathcal{A}$-stable *if* $\mathrm{Var}_\mathcal{A}(C) \subseteq \mathrm{Var}_\mathcal{A}(\Gamma(C))$. *A set of clauses is* $\mathcal{V}_\mathcal{A}$-stable *if every clause it contains is $\mathcal{V}_\mathcal{A}$-stable.*

Given a set of standard clauses, it is easy to construct an equivalent set of abstracted and $\mathcal{V}_\mathcal{A}$-stable clauses. It suffices to replace every abducible a occurring in a non-$\mathcal{A}$-literal by a fresh variable $x \in \mathcal{V}_\mathcal{A}$, and to add the literal $x \not\simeq a$ to the clause. For instance, if $\mathcal{A} = \{a, b\}$ then the clause $a \simeq b \vee a \simeq c \vee f(b, d, x) \not\simeq g(b, y)$ is replaced by $x_1 \not\simeq a \vee x_2 \not\simeq b \vee x_3 \not\simeq b \vee a \simeq b \vee x_1 \simeq c \vee f(x_2, d, x) \not\simeq g(x_3, y)$.

Definition of the calculus. We introduce a calculus for generating $\mathcal{A}$-implicates. It is a modified version of the superposition calculus, and consists of inference rules that are meant to be applied to abstracted clauses. In particular, it is based on orderings that are suitable for abstracted terms, literals and clauses: the order between two terms t and s should not depend on the abducible constants occurring in t and s, and maximal terms and literals in abstracted clauses should be related to maximal terms and literals in standard clauses, in a sense that will be made precise later. We thus define particular orderings for standard clauses, from which we define suitable orderings for abstracted clauses.

Definition 12. *We consider a good CSO $\prec$ such that*[2]*:*

1. *for all $a, b \in \mathcal{A}$, $a \prec b$ if and only if $a <_\mathcal{A} b$;*
2. *for all $a \in \mathcal{A}$ and for all non-variable terms $t \notin \mathcal{A}$, $a \prec t$;*
3. *for all ground terms t, s not in $\mathcal{A}$, if $t \prec s$ then $t_{\downarrow\mathcal{A}} \preceq s_{\downarrow\mathcal{A}}$, and if $t_{\downarrow\mathcal{A}} \prec s_{\downarrow\mathcal{A}}$ then $t \prec s$.*

We let γ_0 denote the ground substitution of domain $\mathcal{V}_\mathcal{A}$ such that for all $x \in \mathcal{V}_\mathcal{A}$, $x\gamma_0 = a_0$. Given abstracted terms t, s, we define $\prec_\mathcal{A}$ as follows: $t \prec_\mathcal{A} s$ iff $t\gamma_0 \prec s\gamma_0$. This definition extends to literals and clauses in a standard way. A term is $\mathcal{A}$-maximal *if it is maximal for $\prec_\mathcal{A}$; this definition also extends to literals and clauses.*

Definition 13. *We denote by $\mathcal{SP}_\mathcal{A}$ the calculus such that for all clause sets S, we have $S \vdash^\mathcal{A} D$ if $S \vdash_{\prec_\mathcal{A}} D$ and the mgu involved in the $\mathcal{SP}$-inference is $\mathcal{A}$-compliant.*

By construction, $\mathcal{SP}$ and $\mathcal{SP}_\mathcal{A}$ coincide on ground $\mathcal{A}$-clauses. We define a particular notion of redundancy for abstracted clauses, that is related to redundancy

[2] It is not difficult to see that there exist orderings fulfilling these properties (see [8]).

for standard clauses. The main difference with the standard definition is that the redundancy test is performed modulo the substitution ν_C that replaces the abstracted variables in C by the abducible constants they denote.

Definition 14. *Consider a set of abstracted clauses S and an abstracted clause C such that* $\mathrm{Var}(C) \subseteq \mathcal{V}_{\mathcal{A}}$. *The clause C is* $\mathcal{A}$-redundant *in S if:*

- *C is an $\mathcal{A}$-clause, $\nu_C \neq \mathrm{id}$ and $C\nu_C$ either occurs or is $\mathcal{A}$-redundant in S,*
- *or there exists a set of ground clauses S' such that $S' \models C$, every $D \in S'$ is an instance of a clause in S_ν and $D \prec C\nu_C$.*

If C is an abstracted clause such that $\mathrm{Var}(C) \not\subseteq \mathcal{V}_{\mathcal{A}}$, *then C is* $\mathcal{A}$-redundant *in S if for all ground substitutions σ with a domain in $\mathcal{V}$, $C\sigma$ is $\mathcal{A}$-redundant in S. The set S is* $\mathcal{A}$-saturated *if every clause $C \notin S$ generated by an $\mathcal{SP}_{\mathcal{A}}$-inference with premises in S is $\mathcal{A}$-redundant in S.*

This notion of redundancy permits to add the standard contraction rules of the superposition calculus to $\mathcal{SP}_{\mathcal{A}}$ (subsumption, simplification, elimination of tautologies, etc). The following contraction inference rule is also added to $\mathcal{SP}_{\mathcal{A}}$:

$$\mathcal{A}\text{-reduction}: \quad \frac{C}{\overline{C\nu_C}} \quad \text{if } C \text{ is an } \mathcal{A}\text{-clause and } \nu_C \neq \mathrm{id}.$$

After any application of the $\mathcal{A}$-reduction rule, the premise becomes $\mathcal{A}$-redundant and can be deleted.

Theorem 1. *If S is a variable-inactive (w.r.t. $\prec_{\mathcal{A}}$) set of abstracted clauses that are $\mathcal{V}_{\mathcal{A}}$-stable, then every non-redundant clause generated from S by $\mathcal{SP}_{\mathcal{A}}$ is abstracted and $\mathcal{V}_{\mathcal{A}}$-stable. Also, if one of the premises of a binary $\mathcal{SP}_{\mathcal{A}}$-inference is an $\mathcal{A}$-clause, then the other premise is also an $\mathcal{A}$-clause.*

The variable-inactive condition, which ensures that all generated clauses are variable-eligible, prevents non-abstracted clauses from being generated from abstracted ones. For example, if S contains the unit clauses $\{a \simeq b, x \simeq y\}$ with $\{a, b\} \in \mathcal{A}$ and $\{x, y\} \in \mathcal{V}$, then $S \vdash^{\mathcal{A}} y \simeq b$, and the latter is *not* abstracted. In what follows, we will prove completeness and termination results for $\mathcal{SP}_{\mathcal{A}}$. The completeness result guarantees that $\mathcal{SP}_{\mathcal{A}}$ generates the required information about existing abducibles for any abstracted set of clauses, while the termination result relies on termination results for $\mathcal{SP}$, and will be used to verify without any additional effort that our technique can be used as a decision procedure for reasoning about abducibles in SMT problems with several theories of interest.

4 Completeness of the Calculus

This section is devoted to showing that if S is an unsatisfiable set of abstracted clauses that is $\mathcal{A}$-saturated, then $\square \in S$ (due to space restrictions, we only provide a sketch of the proof, see [8] for details). Note that this result does *not* follow from the refutational completeness of the superposition calculus: indeed, the ordering $\prec_{\mathcal{A}}$ is not a simplification ordering (it is not stable by substitution), and all inferences in which non-$\mathcal{A}$-compliant unifiers are involved are ignored. However, the proof is based on the refutational completeness of $\mathcal{SP}$, and requires determining relationships between $\mathcal{SP}$-inferences and $\mathcal{SP}_{\mathcal{A}}$-inferences.

Let S be a $\mathcal{V}_{\mathcal{A}}$-stable and $\mathcal{A}$-saturated set of clauses, with no variable-eligible clause. We will show that S is satisfiable by constructing a set of standard clauses whose satisfiability will entail that of S. The set we construct will be saturated under $\mathcal{SP}_{\prec}$-inferences, and it will not contain the empty clause; we will conclude that it must be satisfiable, and hence that so must S.

Let T be the set of $\mathcal{A}$-clauses in S. Since S is $\mathcal{V}_{\mathcal{A}}$-stable and $\mathcal{A}$-saturated by hypothesis, T can only contain ground $\mathcal{A}$-clauses, because if a non-ground clause occurs in T then $\mathcal{A}$-reduction applies. Since $\mathcal{SP}$ and $\mathcal{SP}_{\mathcal{A}}$ coincide on ground $\mathcal{A}$-clauses, T must also be saturated under $\mathcal{SP}_{\prec}$-inferences and cannot contain $\square$; this set is therefore satisfiable. We consider a fixed interpretation I that is a model of T.

Definition 15. *We define the ground set* $U_I = \{a \simeq b \mid a, b \in \mathcal{A}, a^I = b^I\} \cup \{a \not\simeq b \mid a, b \in \mathcal{A}, a^I \neq b^I\}$. *We inductively define the notion of an* I-reduction*:*

- *For all* $a \in \mathcal{A}$, $a_{\|I} = \min_{\prec} \{b \in \mathcal{A} \mid b^I = a^I\}$.
- $f(t_1, \ldots, t_n)_{\|I} = f(t_{1\|I}, \ldots, t_{n\|I})$.

This definition extends to standard literals and clauses.

The I-reduction procedure is used to define a set whose satisfiability entails that of S, and that turns out to be saturated:

Definition 16. *Let* $S_I = U_I \cup \{\overline{\Delta}(C_{\|I}) \mid C \in S_\nu \wedge U_I \models \neg\Delta(C)\}$.

Proposition 1. *If* S_I *is satisfiable then so is* S_ν*, and therefore so is* S.

Lemma 1. S_I *is saturated for* $\mathcal{SP}_{\prec}$.

Since S_I is saturated for the standard superposition calculus $\mathcal{SP}_{\prec}$ and contains no occurrence of the empty clause, we deduce that it is satisfiable.

Theorem 2. *Let* S *be a set of abstracted clauses that is* $\mathcal{V}_{\mathcal{A}}$*-stable and contains no variable-eligible clause. If* S *is* $\mathcal{A}$*-saturated and does not contain the empty clause, then* S *is satisfiable.*

This theorem proves the refutational completeness of $\mathcal{SP}_{\mathcal{A}}$ together with contraction rules that eliminate $\mathcal{A}$-redundant clauses, for those sets of abstracted clauses S whose saturation is guaranteed to meet the requirements of the theorem. The first two requirements are not restrictive: the abstraction of a set of standard clauses described right before Section 3 produces a set of abstracted and $\mathcal{V}_{\mathcal{A}}$-stable clauses, and the saturation of this set is guaranteed to only contain abstracted and $\mathcal{V}_{\mathcal{A}}$-stable clauses by Theorem 1. The fact that S contains no variable-eligible clause cannot be imposed that easily, but such a condition is guaranteed if S is variable-inactive, which is the case for many classes of clause sets of interest [2,1].

Note that this completeness result is not – by itself – sufficient for our purpose, since our goal is not merely to test the satisfiability of clause sets but rather to generate flat consequences they logically entail. The next section shows how the calculus $\mathcal{SP}_{\mathcal{A}}$ can be employed to reach this goal.

5 A Generation of Explanations

We return to the problem of explaining why a set of clauses is satisfiable, and show how $\mathcal{SP}_{\mathcal{A}}$ can be used to generate explanations relating abducibles to one another. Given a satisfiable set of clauses S', we denote by $\mathrm{I}_{\mathcal{A}}(S')$ the set of all $\mathcal{A}$-implicates of S': $\mathrm{I}_{\mathcal{A}}(S') \stackrel{\text{def}}{=} \{C \text{ an } \mathcal{A}\text{-clause} \mid C \text{ is ground and } S' \models C\}$.

It is clear that all the information about abducible constants that is entailed by S' is contained in $\mathrm{I}_{\mathcal{A}}('S)$. However this set can be very large and it contains a lot of non-pertinent information, for example all logical tautologies, or all instances of the equality axioms. It therefore does not seem reasonable to return this entire set to a user. Another solution could be to return a subset $T \subseteq \mathrm{I}_{\mathcal{A}}(S')$ such that $T \vdash \mathrm{I}_{\mathcal{A}}(S')$, but again, such a set might be large and contain unnecessary information.

The solution we choose is to return a minimal subset $T' \subseteq \mathrm{I}_{\mathcal{A}}(S')$ satisfying the following property: for all $C \in \mathrm{I}_{\mathcal{A}}(S')$ that is not a tautology, there exists a clause $C' \in T'$ such that $C' \models C$. The clauses in T' are the *prime implicates* of S'. The notion of prime implicates plays a central rôle in many applications of computer science and artificial intelligence, and several approaches have been proposed for computing the prime implicates of a given propositional formula (see, e.g., [10]). Some extensions to first-order logic have also been considered, such as, e.g., [12]. In what follows, we define an algorithm that computes prime implicates for sets of flat equational clauses.

It turns out that $\mathcal{SP}_{\mathcal{A}}$ cannot be used to determine the set T'. For instance, if $S' = \{a \simeq b, c \not\simeq d\}$, then the clause $a \not\simeq c \vee b \not\simeq d$ must be in $\mathrm{I}_{\mathcal{A}}(S')$. Since it is subsumed by no clause in $\mathrm{I}_{\mathcal{A}}(S')$ but itself, it must also be in T', but no $\mathcal{SP}_{\mathcal{A}}$-inference rule (or $\mathcal{SP}$-inference rule for that matter) can be applied to S' to generate such a clause. In the sequel, we will show how, starting with a set of $\mathcal{A}$-clauses that logically entails $\mathrm{I}_{\mathcal{A}}(S')$, it is possible to generate a set T' using the *Resolution calculus*, denoted by $\mathcal{R}$ (we refer the reader to [11] for details on the Resolution calculus). From now on, S' denotes a satisfiable set of standard clauses, and S is a set of abstracted clauses such that $S_{\nu} = S'$. Thus, S and S' are equivalent. The first step towards this construction is the definition of a set of $\mathcal{A}$-clauses that logically entails $\mathrm{I}_{\mathcal{A}}(S')$. The (finite) set of all $\mathcal{A}$-clauses in the saturated set generated from S using $\mathcal{SP}_{\mathcal{A}}$ will satisfy this requirement.

Definition 17. *We denote by T_∞ the set of $\mathcal{A}$-clauses in the $\mathcal{A}$-saturated set generated from S by $\mathcal{SP}_{\mathcal{A}}$.*

The key result that makes the generation of $\mathcal{A}$-implicates possible is that all the $\mathcal{A}$-clauses that are entailed by S are actually logical consequences of T_∞:

EXPLAIN($S', \mathcal{A}$) =
  $S := $ ABSTRACT(S')
  $S := \mathcal{SP}_{\mathcal{A}}$-saturation($S$)
  $T_\infty := \{C \in S \mid C \text{ is an } \mathcal{A}\text{-clause}\}$
  return $\mathcal{R}$-saturation($T_\infty \cup Eq_{\mathcal{A}}$)

Fig. 3. Generation of a set of explanations

Proposition 2. $T_\infty \models \mathrm{I}_{\mathcal{A}}(S')$.

Let Eq be the set of axioms stating that $\simeq$ is an equivalence relation[3]: $Eq = \{x \simeq x,\ x \not\simeq y \vee y \simeq x,\ x \not\simeq y \vee y \not\simeq z \vee x \simeq z\}$, and let $Eq_{\mathcal{A}}$ be the set consisting of all instantiations of the axioms in Eq by the elements in $\mathcal{A}$. The result we show is that the $\mathcal{R}$-closure of the set $T_\infty \cup Eq_{\mathcal{A}}$ satisfies the requirements for the set of $\mathcal{A}$-clauses that is searched for.

Theorem 3. *Let $T = T_\infty \cup Eq_{\mathcal{A}}$, and let C be a non-tautological ground clause in $\mathrm{I}_{\mathcal{A}}(S)$. Then there is a derivation from T of a clause C' such that $C' \models C$.*

To summarize, given a set of clauses S' that is satisfiable and a set of abducible constants $\mathcal{A}$, the simple algorithm in pseudo-code described in Figure 3 returns a set of clauses constructed over $\mathcal{A}$ that can be viewed as explanations why S' is satisfiable. Note that $\mathcal{R}$-saturation can be performed on the fly: it is clear that it is not necessary to wait until $\mathcal{SP}_{\mathcal{A}}$-saturation($S$) is computed to start generating the clauses in $\mathcal{R}$-saturation($T_\infty \cup Eq_{\mathcal{A}}$). Thus even in case of non-termination, all the prime implicates can eventually be generated. After the set $\mathcal{R}$-saturation($T_\infty \cup Eq_{\mathcal{A}}$) is computed, it is possible to remove from this set all the clauses that can be inferred from other prime implicates. This solution yields a more compact representation. However, this is possible only in case of termination, since the deleted clauses may be involved in the generation of other prime implicates. A termination result for $\mathcal{SP}_{\mathcal{A}}$ will be presented in the following section. By putting all the previous results together, we obtain the following theorem, stating the soundness and completeness of the procedure EXPLAIN.

Theorem 4. *Let S be a set of clauses. Every clause $C \in$ EXPLAIN($S', \mathcal{A}$) is an $\mathcal{A}$-implicate of S, and for every $\mathcal{A}$-implicate C of S that is not a tautology, there exists a clause $C' \in$ EXPLAIN($S', \mathcal{A}$) such that $C' \models C$.*

Example 3. We return to the problem mentioned in the Introduction. After flattening, we get the following set of clauses:

1	$\mathrm{select}(\mathrm{store}(x,z,v),z) \simeq v$	4	$d_2 \simeq \mathrm{store}(d_1,j,c)$
2	$z \simeq w \vee \mathrm{select}(\mathrm{store}(x,z,v),w) \simeq \mathrm{select}(x,w)$	5	$d_3 \simeq \mathrm{store}(a,j,c)$
3	$d_1 \simeq \mathrm{store}(a,i,b)$	6	$d_4 \simeq \mathrm{store}(d_3,i,b)$
7	$\mathrm{select}(d_2,k) \not\simeq \mathrm{select}(d_4,k)$		

[3] There will be no need to consider the congruence axiom, since all the clauses in T_∞ only contain constants.

Assume that $\mathcal{A} = \{i, j, b, c\}$. Then Clauses 3, 4, 5, 6 are abstracted as follows:

$$\begin{array}{ll}
3' & x' \not\simeq i \vee y' \not\simeq b \vee d_1 \simeq \mathsf{store}(a, x', y') \\
4' & x'' \not\simeq j \vee y'' \not\simeq c \vee d_2 \simeq \mathsf{store}(d_1, x'', y'') \\
5' & x'' \not\simeq j \vee y'' \not\simeq c \vee d_3 \simeq \mathsf{store}(a, x'', y'') \\
6' & x' \not\simeq i \vee y' \not\simeq b \vee d_4 \simeq \mathsf{store}(d_3, x', y')
\end{array}$$

$\mathcal{SP}_{\mathcal{A}}$ generates the following clauses[4]:

8	$x' \not\simeq i \vee w \simeq x' \vee \mathsf{select}(d_1, w) \simeq \mathsf{select}(a, w)$	$(3',2)$
9	$x'' \not\simeq j \vee w \simeq x'' \vee \mathsf{select}(d_2, w) \simeq \mathsf{select}(d_1, w)$	$(4',2)$
10	$x'' \not\simeq j \vee w \simeq x'' \vee \mathsf{select}(d_3, w) \simeq \mathsf{select}(a, w)$	$(5',2)$
11	$x' \not\simeq i \vee w \simeq x' \vee \mathsf{select}(d_4, w) \simeq \mathsf{select}(d_3, w)$	$(6',2)$
12	$x' \not\simeq i \vee y' \not\simeq b \vee \mathsf{select}(d_1, x') \simeq y'$	$(3',1)$
13	$x'' \not\simeq j \vee y'' \not\simeq c \vee \mathsf{select}(d_2, x'') \simeq y''$	$(4',1)$
14	$x'' \not\simeq j \vee y'' \not\simeq c \vee \mathsf{select}(d_3, x'') \simeq y''$	$(5',1)$
16	$x' \not\simeq i \vee y' \not\simeq b \vee \mathsf{select}(d_4, x') \simeq y'$	$(6',1)$
17	$x' \not\simeq i \vee k \simeq x' \vee \mathsf{select}(d_2, k) \not\simeq \mathsf{select}(d_3, k)$	(11, 7)
18	$x' \not\simeq i \vee k \simeq x' \vee x'' \not\simeq j \vee k \simeq x'' \vee \mathsf{select}(d_2, k) \not\simeq \mathsf{select}(a, k)$	(10, 17)
19	$x' \not\simeq i \vee k \simeq x' \vee x'' \not\simeq j \vee k \simeq x'' \vee \mathsf{select}(d_1, k) \not\simeq \mathsf{select}(a, k)$	(9, 18)
20	$x' \not\simeq i \vee x'' \not\simeq j \vee k \simeq x' \vee k \simeq x''$	(8,19)
21	$x' \not\simeq i \vee x'' \not\simeq j \vee k \simeq x' \vee \mathsf{select}(d_2, k) \not\simeq \mathsf{select}(d_4, x'')$	(20,7)
22	$x' \not\simeq i \vee x'' \not\simeq j \vee k \simeq x' \vee x'' \simeq x' \vee \mathsf{select}(d_2, k) \not\simeq \mathsf{select}(d_3, x'')$	(11,21)
23	$x' \not\simeq i \vee x'' \not\simeq j \vee y'' \not\simeq c \vee k \simeq x' \vee x'' \simeq x' \vee \mathsf{select}(d_2, k) \not\simeq y''$	(14,22)
24	$x' \not\simeq i \vee x'' \not\simeq j \vee y'' \not\simeq c \vee k \simeq x' \vee x'' \simeq x' \vee \mathsf{select}(d_2, x'') \not\simeq y''$	(20,23)
25	$x' \not\simeq i \vee x'' \not\simeq j \vee k \simeq x' \vee x'' \simeq x'$	(13,24)
26	$x' \not\simeq i \vee x'' \not\simeq j \vee x'' \simeq x' \vee \mathsf{select}(d_2, k) \not\simeq \mathsf{select}(d_4, x')$	(25,7)
27	$x' \not\simeq i \vee x'' \not\simeq j \vee y' \not\simeq b \vee x'' \simeq x' \vee \mathsf{select}(d_2, k) \not\simeq y'$	(16,26)
28	$x' \not\simeq i \vee x'' \not\simeq j \vee y' \not\simeq b \vee x'' \simeq x' \vee \mathsf{select}(d_2, x') \not\simeq y'$	(25,27)
29	$x' \not\simeq i \vee x'' \not\simeq j \vee y' \not\simeq b \vee x'' \simeq x' \vee \mathsf{select}(d_1, x') \not\simeq y'$	(9,28)
30	$i \simeq j$	(12,29)
31	$x' \not\simeq i \vee x'' \not\simeq j \vee x' \not\simeq x'' \vee k \simeq x'$	(20)
33	$x' \not\simeq i \vee x'' \not\simeq j \vee x' \not\simeq x'' \vee \mathsf{select}(d_2, k) \not\simeq \mathsf{select}(d_4, x')$	(31,7)
34	$x' \not\simeq i \vee x'' \not\simeq j \vee x' \not\simeq x'' \vee y' \not\simeq b \vee \mathsf{select}(d_2, k) \not\simeq y'$	(16,34)
35	$x' \not\simeq i \vee x'' \not\simeq j \vee x' \not\simeq x'' \vee y' \not\simeq b \vee \mathsf{select}(d_2, x') \not\simeq y'$	(31,34)
36	$i \not\simeq j \vee b \not\simeq c$	(13,35)

By Resolution, from 30 and 36, we get $c \not\simeq b$, which subsumes 36. We obtain the A-implicates $\{i \simeq j, b \not\simeq c\}$, yielding the explanation $i \not\simeq j \vee b \simeq c$.

6 A Termination Result for $\mathcal{SP}_{\mathcal{A}}$

We now prove a result that relates the termination of $\mathcal{SP}$ on a set of standard clauses S to the termination of $\mathcal{SP}_{\mathcal{A}}$ on an abstracted version of S. This shows that many existing results about the termination of the superposition calculus for subclasses of first-order logic carry over to $\mathcal{SP}_{\mathcal{A}}$. We relate standard and

[4] For readability we simply drop irrelevant disequations, i.e. $x \not\simeq a \vee C$ is replaced by C if x does not occur in C and $x \not\simeq a \vee x' \not\simeq a \vee C$ is replaced by $x \not\simeq a \vee C\{x' \mapsto x\}$.

abstracted terms by defining a so-called relation of $\mathcal{A}$-relaxation. This relation will be used afterwards to relate the forms of the clauses generated by $\mathcal{SP}$-inferences and those generated by $\mathcal{SP}_{\mathcal{A}}$-inferences in a more precise manner.

Definition 18. *The relation of* $\mathcal{A}$-relaxation *relates an abstracted term* t *to a standard one* t' *and is defined as follows:* $t \trianglelefteq_{\mathcal{A}} t'$ *if and only if* $t\gamma_0 = t'_{\downarrow\mathcal{A}}$.

Given an abstracted clause C *and a standard clause* C'*, we write* $C \trianglelefteq_{\mathcal{A}} C'$ *if and only if* $\overline{\Delta}(C\gamma_0) = \overline{\Delta}(C'_{\downarrow\mathcal{A}})$*. This relation is extended to sets of clauses in a straightforward manner.*

Example 4. Assume $\mathcal{A} = \{a, b\}$, let $C = x \not\simeq a \vee a \simeq b \vee f(x, x, d) \simeq g(y) \vee g(y) \simeq d$ and $C' = a \not\simeq b \vee f(a, b, d) \simeq g(b) \vee g(a) \simeq d$. Then $C \trianglelefteq_{\mathcal{A}} C'$.

We define a notion of redundancy that is meant to hold no matter what abducible constants occur in the clause under consideration.

Definition 19. *An* $\mathcal{A}$*-reduced clause* C' *is* P-redundant *in an* $\mathcal{A}$*-reduced set of clauses* S' *if for all sets of abstracted clauses* S *such that* $(S_\nu)_{\downarrow\mathcal{A}} \equiv S'$ *and for every abstracted clause* D *such that* $(D\nu_D)_{\downarrow\mathcal{A}} \equiv C'$*, clause* D *is* $\mathcal{A}$*-redundant in* S*. An* $\mathcal{A}$*-reduced set of clauses* S' *is* P-saturated *if every clause generated with premises in* S' *either occurs in* S' *or is* P*-redundant in* S'*.*

This notion permits to eliminate clauses that are redundant in the usual sense and do not contain any abducible constant. Notice, however, that P-redundant clauses can possibly contain abducible constants. For example if $\mathcal{A} = \{a, b\}$ and $S' = \{f(c) \not\simeq f(d)\}$, then $C' = g(a, c) \simeq h(a) \vee f(c) \not\simeq f(d)$ is P-redundant in S'.

Theorem 5. *Let* S' *be a set of* $\mathcal{A}$*-reduced clauses, and let* T *be the* P*-saturated set of clauses generated from* S'*. If* T *is finite and* S *is a set of abstracted clauses that is* $\mathcal{V}_{\mathcal{A}}$*-stable, variable-inactive and such that* $S \trianglelefteq_{\mathcal{A}} S'$*, then the set of non-redundant clauses generated from* S *is finite.*

Theorem 5 guarantees that $\mathcal{SP}_{\mathcal{A}}$ (and thus EXPLAIN) terminates on several classes of clause sets, in particular for clause sets related to SMT problems. The authors of [2] and [1] prove that sets of the form $\mathcal{T} \cup U$, where $\mathcal{T}$ is a theory and U a set of ground unit clauses, generate finite saturated sets. This result is extended to clause sets of the form $\mathcal{T} \cup U'$, where U' is an arbitrary set of ground clauses, in [6]. An inspection of the finiteness results of [2,1,6] shows that they hold not only for saturated sets but also for P-saturated sets, since the redundant clauses that are deleted are actually P-redundant: they do not contain any constant at all. Thus, $\mathcal{SP}_{\mathcal{A}}$ terminates for clause sets of the form $\mathcal{T} \cup U'$, where U' is the abstraction of a set of ground clauses, and $\mathcal{T}$ is the axiomatization of any of the following theories: records, integer offsets, possibly empty lists, arrays...

7 Discussion

We have presented a calculus that permits to reason on the relations involving abducible constants, that are logical consequences of a satisfiable set of clauses.

These relations can be viewed as explanations of why the set is satisfiable, since any of their negations, when added to the original clause set, renders the latter unsatisfiable. We proved a completeness result for the calculus, along with a sufficient condition guaranteeing its termination on classes of clause sets, among which SMT problems in several theories of interest. To the best of our knowledge, this approach is novel and there are many interesting directions to explore. One first direction is to investigate what set of clauses can be considered as a *good* set of explanations, and determine what a good trade-off might be between a small set of explanations that may hide too many details, and a large set of explanations that may carry too much unnecessary information. Another line of work that is currently under investigation is the search for a more efficient way to compute explanations from the generated $\mathcal{A}$-clauses. Indeed, the saturation with the Resolution calculus in the presence of the equality axioms is not entirely satisfactory as far as efficiency is concerned, and it would be interesting to see how the calculus $\mathcal{SP}_{\mathcal{A}}$ can be enhanced to directly produce the required set of explanations. As far as other extensions are concerned, we plan to investigate how to extend these results to *abducible terms* and not only abducible constants, by allowing the occurrence of function symbols in $\mathcal{A}$. This would allow the derivation of non-ground explanations. Another possibility is to consider mixed literals, containing both abducible and non-abducible symbols. It would then be possible to generate explanations of the form $a \simeq 0$ without having to declare 0 as an abducible constant. We also plan on devising a calculus capable of efficiently generating explanations with abducibles interpreted in a particular theory, such as, e.g., arithmetic.

References

1. Armando, A., Bonacina, M.P., Ranise, S., Schulz, S.: New results on rewrite-based satisfiability procedures. ACM Transactions on Computational Logic 10(1), 129–179 (2009)
2. Armando, A., Ranise, S., Rusinowitch, M.: A rewriting approach to satisfiability procedures. Information and Computation 183(2), 140–164 (2003)
3. Baader, F., Nipkow, T.: Term Rewriting and All That. Cambridge University Press (1998)
4. Bachmair, L., Ganzinger, H., Waldmann, U.: Superposition with Simplification as a Decision Procedure for the Monadic Class with Equality. In: Mundici, D., Gottlob, G., Leitsch, A. (eds.) KGC 1993. LNCS, vol. 713, pp. 83–96. Springer, Heidelberg (1993)
5. Bachmair, L., Ganzinger, H., Waldmann, U.: Refutational theorem proving for hierachic first-order theories. Appl. Algebra Eng. Commun. Comput. 5, 193–212 (1994)
6. Bonacina, M.P., Echenim, M.: On variable-inactivity and polynomial T-satisfiability procedures. Journal of Logic and Computation 18(1), 77–96 (2008)
7. Bonacina, M.P., Echenim, M.: Theory decision by decomposition. Journal of Symbolic Computation 45(2), 229–260 (2010)
8. Echenim, M., Peltier, N.: A calculus for generating ground explanations (technical report). CoRR, abs/1201.5954 (2012), `http://arxiv.org/1201.5954`

9. Eiter, T., Gottlob, G.: The complexity of logic-based abduction. J. ACM 42(1), 3–42 (1995)
10. Jackson, P.: Computing prime implicates. In: ACM Conference on Computer Science, pp. 65–72 (1992)
11. Leitsch, A.: The resolution calculus. Texts in Theoretical Computer Science. Springer (1997)
12. Marquis, P.: Extending Abduction from Propositional to First-order Logic. In: Jorrand, P., Kelemen, J. (eds.) FAIR 1991. LNCS, vol. 535, pp. 141–155. Springer, Heidelberg (1991)
13. McCarthy, J.: Computer programs for checking mathematical proofs. In: Recursive Function Theory, pp. 219–228. Providence, Rhode Island (1962); Proc. of Symposia in Pure Mathematics, vol. 5. American Mathematical Society
14. Nieuwenhuis, R., Rubio, A.: Paramodulation-based theorem proving. In: Robinson, J.A., Voronkov, A. (eds.) Handbook of Automated Reasoning, pp. 371–443. Elsevier and MIT Press (2001)
15. Tran, D.-K., Ringeissen, C., Ranise, S., Kirchner, H.: Combination of convex theories: Modularity, deduction completeness, and explanation. J. Symb. Comput. 45(2), 261–286 (2010)
16. YICES, `http://yices.csl.sri.com`

EPR-Based Bounded Model Checking at Word Level*

Moshe Emmer[1], Zurab Khasidashvili[1], Konstantin Korovin[2], Christoph Sticksel[2], and Andrei Voronkov[2]

[1] Intel Israel Design Center, Haifa 31015, Israel
{memmer,zurabk}@iil.intel.com
[2] The University of Manchester, School of Computer Science, UK
{korovin,sticksel}@cs.man.ac.uk, andrei@voronkov.com

Abstract. We propose a word level, bounded model checking (BMC) algorithm based on translation into the effectively propositional fragment (EPR) of first-order logic. This approach to BMC allows for succinct representation of unrolled transition systems and facilitates reasoning at a higher level of abstraction. We show that the proposed approach can be scaled to industrial hardware model checking problems involving memories and bit-vectors. Another contribution of this work is in generating challenging benchmarks for first-order theorem provers based on the proposed encoding of real-life hardware verification problems into EPR. We report experimental results for these problems for several provers known to be strong in EPR problem solving. A number of these benchmarks have already been released to the TPTP library.

1 Introduction

SAT-based Bounded Model Checking (BMC) [4] is currently the most widespread formal verification method in the hardware industry used for bug finding. Despite the rapid development of SMT [21] and first-order Theorem Proving (TP) [26] techniques for model-checking at word level, their positive impact has been mainly seen on software verification. So far, hardware verification has benefited far less from word-level verification, and applying SMT and TP techniques to hardware verification remains a difficult challenge to the verification community. This is mainly due to the fact that most of the hardware descriptions are written at very low-level, e.g., without explicit usage of arithmetic operations. Nevertheless, there are natural word-level components in hardware designs, in particular memories and bit-vectors, which are challenging for the bit-level verification due to the size of their bit-level representations. Efficient reasoning, at word level, with bit-vectors and arrays is an active research area [28,8,1,14,24,6,7].

This paper focuses on an encoding of the BMC problem with memories and bit-vectors into first-order logic (FOL) and in particular into the Effectively PRopositional (EPR) fragment. The EPR fragment, also called the Bernays-Schönfinkel-Ramsey fragment, consists of first-order formulas with no occurrences of function symbols other than constants and which when written in prenex normal form have the quantifier prefix $\exists^*\forall^*$. Skolemization applied to EPR formulas can introduce only constant function

* This work is partially supported by EPSRC, the Royal Society and a grant from Intel.

B. Gramlich, D. Miller, and U. Sattler (Eds.): IJCAR 2012, LNAI 7364, pp. 210–224, 2012.

symbols, this can be used to show decidability of the EPR fragment. There are a number of efficient solvers [3,10,19,25] for this fragment as demonstrated at the annual CASC TP competition [27]. Several important verification problems have been encoded into EPR [23,17,13,2], benefiting from succinct representations possible in this fragment. Solvers are becoming increasingly scalable to industrial size problems and therefore it is promising to develop efficient encodings of Model Checking (MC) [9] into EPR.

The first encoding of BMC into EPR was proposed in [23], covering entire linear temporal logic. The transition relation and the initial and final states are specified there via Boolean constraints, and when encoding the unrolled transition relation, the state variables are treated as predicates over states (or over time). This enables a succinct representation of the unrolled system. The main contribution of that paper is theoretical, and no experimental results comparing the method with SAT-based BMC were reported.

Another, completely orthogonal, way to encode the MC problem into EPR was studied in [17,13]. These works explore encodings of hardware MC problems at word level into EPR. In particular in the so-called *relational encoding* approach, bit-vectors are modeled as unary predicates over bits (or bit-indexes), addresses are modeled as terms, and memories are modeled as binary predicates over addresses and bits. Appropriate axiomatization of bits and bit-ranges allows for a sound and complete encoding of hardware MC problems with bit-vector and memory operations into EPR. These papers report initial results of experimental evaluation of EPR-based first-order verification compared with SMT-based verification on equivalence checking problems at Intel.

In this paper we present an encoding scheme which allows us to retain the strengths of both previous approaches. An ad-hoc combination of the previous two approaches would yield an encoding which in most cases will generate problems outside of the EPR fragment. One of the main issues here is that memory addresses on the one hand occur as arguments in memory predicates and therefore should be treated as terms, and on the other hand addresses depend on the state of the transition system and therefore are functions of state. Presence of non-constant functions in the encoding brings the resulting specification of the transition system outside EPR. Even very small non-EPR problems originating from toy model-checking examples are very hard for the strongest theorem provers. Among the main contributions of this paper are techniques allowing one to keep the translated verification problems within the EPR fragment. In particular, i) we introduce *address unrolling* to eliminate address functions and ii) we use *inlining* to eliminate definitions which after Skolemization result in non-EPR formulas. We evaluate our EPR-based encoding on model checking problems obtained from industrial hardware designs used at Intel. We show that in many cases EPR-based BMC can reach higher unrolling bounds than traditional SAT-based BMC.

The rest of the paper is structured as follows. In the next section we present a generic translation scheme that, given a specification of transition relation, initial state constraints and final state constraints in first-order logic, produces a description of the unrolled system up to a bound k that faithfully models transition paths of length k from the initial to the final states. While FOL is closed under the translation, EPR is not. The encoding that we describe below in Section 3 can be seen as a result of this translation applied to the specification of hardware at word level in EPR as described in [13]. This encoding brings the specification of unrolled system outside the EPR fragment.

Therefore, in Section 4 we describe basic transformations allowing the resulting unrolled system to be formalized within EPR. A number of further optimizations that help generating CNFs that are much simpler to solve are described in Section 5. We remark on incremental solving in Section 6. Experimental results are reported in Section 7. Conclusions appear in Section 8.

2 Translation

Let Σ be a signature consisting of constants, function and predicate symbols. We consider constants as function symbols of arity 0. We assume Σ is partitioned into Σ_c, Σ_s and $\Sigma_{s'}$ where Σ_c consists of symbols whose interpretation does not depend on a state, Σ_s consists of *current-state* symbols and $\Sigma_{s'}$ consists of *next-state* symbols. We assume that for every current-state symbol p in Σ_s there is a corresponding next-state symbol p' in $\Sigma_{s'}$ with the same arity as p and vice versa.

A transition system can be symbolically represented by three closed FOL formulas in, trans, and fin, respectively expressing facts about initial states, encoding the transition relation and expressing facts about final states. We assume that in and fin are formulas in $\Sigma_c \cup \Sigma_s$ and trans is a formula in Σ. In order to adapt such a representation for bounded model checking in the EPR fragment we define the following transformation on formulas. First, we replace each current-state and next-state function or predicate symbol p of arity n in Σ_s and $\Sigma_{s'}$, respectively, with a *transient symbol* of arity $n+1$, which has an extra argument for representing transitions over states. Let Σ_t be the signature consisting of all transient symbols corresponding to symbols in Σ_s and by $\overline{\Sigma}$ the signature $\Sigma_c \cup \Sigma_t$.

Let $\mathtt{S}, \mathtt{S'}$ be two fresh variables. Let us define a translation $\mathcal{T}$ of Σ terms to $\overline{\Sigma}$ terms and Σ formulas to $\overline{\Sigma}$ formulas by induction as follows. Let $r_1, \ldots, r_n$ denote terms, let $t_i = \mathcal{T}(r_i)$ for all $i = 1, \ldots, n$, and let $\mathbf{r}, \mathbf{t}$ denote the sequences $r_1, \ldots, r_n$ and $t_1, \ldots, t_n$, respectively. Then:

- For any n-ary function or predicate symbol p define:

$$\mathcal{T}(p(\mathbf{r})) \stackrel{\text{def}}{=} \begin{cases} p(\mathbf{t}), & \text{if } p \in \Sigma_c; \\ p_t(\mathtt{S}, \mathbf{t}), & \text{if } p \in \Sigma_s; \\ p_t(\mathtt{S'}, \mathbf{t}), & \text{if } p \in \Sigma_{s'}. \end{cases}$$

- $\mathcal{T}(F_1 \wedge F_2) \stackrel{\text{def}}{=} \mathcal{T}(F_1) \wedge \mathcal{T}(F_2)$, and similarly for other connectives in place of $\wedge$.
- $\mathcal{T}(\forall x\, F) \stackrel{\text{def}}{=} \forall x\, \mathcal{T}(F)$ and similarly for $\exists$ in place of $\forall$. Recall, we assume that the variables $\mathtt{S}, \mathtt{S'}$ are fresh, this implies that $\mathtt{S}, \mathtt{S'}$ are distinct from x.

For every closed formula F, the only free variables of $\mathcal{T}(F)$ are $\mathtt{S}$ and $\mathtt{S'}$. Moreover, if F uses no next-state symbols, as it is the case for the formulas in and fin, then $\mathcal{T}(F)$ does not contain $\mathtt{S'}$. Let us denote the formulas $\mathcal{T}(\mathit{in})$, $\mathcal{T}(\mathit{trans})$ and $\mathcal{T}(\mathit{fin})$ respectively as $\mathit{In}(\mathtt{S})$, $\mathit{Trans}(\mathtt{S}, \mathtt{S'})$ and $\mathit{Fin}(\mathtt{S})$, parametrized by their free variables.

Let n be a non-negative integer. We define the *n-step unrolling of the transition system* as follows. Take new constants $s_0, \ldots, s_n$ and a new binary predicate next. The n-step unrolling of the transition system is defined as the set of formulas

$$In(s_0); Fin(s_n); \forall \mathtt{S}, \mathtt{S'} (next(\mathtt{S}, \mathtt{S'}) \rightarrow Trans(\mathtt{S}, \mathtt{S'}));$$
$$next(s_0, s_1); next(s_1, s_2); \ldots next(s_{n-1}, s_n).$$

Theorem 1. *There exists an n-step computation of the transition system leading from a state satisfying in to a state satisfying fin if and only if the n-step unrolling of the transition system is satisfiable.*

Note that the n-step unrolling of the system contains only one copy of the transition relation $Trans$. This explains the name we have chosen for our encoding of BMC into FOL: *BMC1*. It stands for *BMC with one copy of the transition relation*. Unlike in SAT-based BMC [4], there is no need to create a new copy of the transition relation for each unrolling bound.

Unfortunately, this translation is not EPR preserving if the original system contains constants, which become transient functions of states after the translation. Such transient functions are essential in memory specifications representing, e.g., addresses which change during transitions. In later sections we show how to restore the EPR representation.

3 Encoding Hardware Specifications into FOL

Let us show how to encode a hardware verification problem into EPR, using a simple yet realistic word-level hardware design shown in Fig. 1. This example contains typical word-level components: a memory, bit-vectors and addresses.

The memory $\mathtt{mem}$ has 32 rows and 64 columns, each cell containing one bit. When both the write enable signal wren and the clock signal clock are true, bits 0 to 63 (written as $[63:0]$ in hardware notation) of the bit-vector wrdata$[63:0]$ are written into $\mathtt{mem}$ at the address given in the bit-vector wraddr$[5:0]$. In order to prevent read and write from happening simultaneously, only if clock is false and the read enable signal rden is true, the value of the memory at address rdaddr$[5:0]$ is read into the bit-vector rddata$[63:0]$.

The circuit also contains a cache line in the 64 bit bit-vector cacheline$[63:0]$, the component sel that compares two bit-vectors bit-wise and a multiplexing device mux which selects one of its inputs depending on the output value of sel. The final output of the circuit is either the bit-wise negated bit-vector rddata$[63:0]$ if wraddr$[63:0]$ and rdaddr$[63:0]$ are equal, or the bit-vector cacheline$[63:0]$ otherwise.

3.1 Encoding of Bit-Vectors and Memories

With any bit-vector we associate a binary predicate. For example, atom wrdata(S, B) denotes the Boolean value of bit B in the write data vector wrdata in state S. Similarly, with a memory $\mathtt{mem}$ we associate a ternary predicate $\mathtt{mem}$, where an atom $\mathtt{mem}(\mathtt{S}, \mathtt{A}, \mathtt{B})$ denotes the Boolean value of $\mathtt{mem}$ in row A and column B, in state S.

In our encoding, there are bit-vectors that in addition to this predicate representation also require a functional representation, we call them *functional bit-vectors*. There are

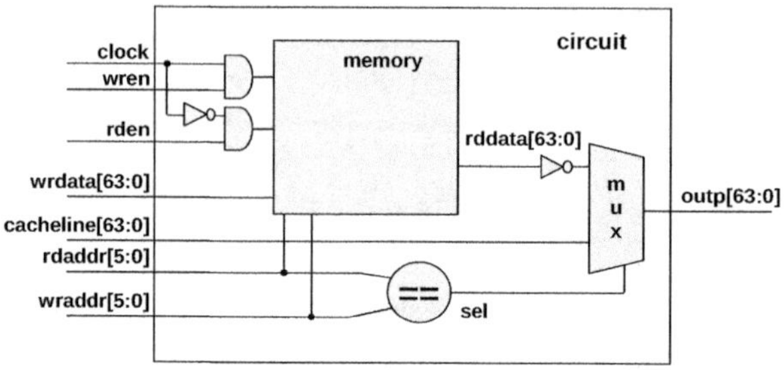

Fig. 1. Running example: a word-level hardware design

two main sources of such bit-vectors. The first consists of bit-vectors representing addresses which are used as arguments in memory predicates. The second consists of bit-vectors that are used in comparisons such as in the `sel` component in our running example in Fig. 1. With a functional bit-vector, in addition to associating a binary predicate over states and bit-indexes, we associate a function over states. We represent the value of an address `addr` in state `S` by the term `addrFunc(S)`. Thus the value of `mem` at address `addr` and bit (column) `B` is represented by the atom `mem(S, addrFunc(S), B)`. We use similar notation for functional bit-vectors which are not addresses.

We often need to refer to particular bits of a bit-vector. We use the constant $\mathtt{bitInd}_i$ to denote the i-th bit. Similarly, we use the constant $\mathtt{s}_j$ to denote the j-th state. Thus, the atom $\mathtt{mem}(\mathtt{s}_0, \mathtt{addrFunc}(\mathtt{s}_0), \mathtt{bitInd}_5)$ represents the value of bit 5 at row $\mathtt{addrFunc}(\mathtt{s}_0)$ in memory `mem` in state $\mathtt{s}_0$.

The reader may have noticed that bit-vector width and memory dimension information is not directly encoded. For example, a predicate representing a bit-vector of width 64 does not carry the width information. This loss of information is recovered, if necessary, when specifying bit-vector operations, which will be explained below. Similarly, the functions associated with addresses and other functional bit-vectors do not carry the width information of the corresponding bit-vector.

For addresses, the only information we need is whether they are equal. For two addresses `wraddr` and `rdaddr`, we can axiomatize this using:

$$\begin{aligned}&\mathtt{wraddrFunc(S)} = \mathtt{rdaddrFunc(S)} \leftrightarrow \\ &\quad(\mathtt{wraddr(S,bitInd_5)} \leftrightarrow \mathtt{rdaddr(S,bitInd_5)} \wedge \ldots \wedge \\ &\quad\mathtt{wraddr(S,bitInd_0)} \leftrightarrow \mathtt{rdaddr(S,bitInd_0)}).\end{aligned} \tag{1}$$

We assume that free variables are implicitly universally quantified.

By using the predicate $\mathtt{less}_k$, which defines bit-indexes in the range of $[0, k-1]$, the above formulas can be written more concisely, without referring explicitly to all of the bits 5 to 0, as follows:

$$\begin{aligned}&\mathtt{wraddrFunc(S)} = \mathtt{rdaddrFunc(S)} \leftrightarrow \\ &\quad\forall\mathtt{B}(\mathtt{less_6(B)} \rightarrow (\mathtt{wraddr(S,B)} \leftrightarrow \mathtt{rdaddr(S,B)})).\end{aligned} \tag{2}$$

We axiomatize the $\mathtt{less}_k$ predicates explicitly as in [13]:

$$\mathtt{less}_k(x) \leftrightarrow (x = \mathtt{bitInd}_0 \vee \ldots \vee x = \mathtt{bitInd}_{k-1}). \tag{3}$$

In addition we need axioms stating that all bit-indexes are different:

$$\texttt{bitInd}_i \neq \texttt{bitInd}_j \text{ for } 1 \leq i < j \leq n, \text{ where } n \text{ is the size of the bit-index domain.} \quad (4)$$

In practical hardware examples the domain size of bit-indexes is in order of thousands and therefore adding such axioms can be a bottleneck. Fortunately, the number of different $\texttt{less}_k$ predicates is usually not very big compared to the number of bit-indexes. Since equality over bit-indexes occurs only in formulas such as (3) we can replace axioms (4) by axioms:

$$\begin{array}{l} \texttt{less}_k(\texttt{bitInd}_j) \;//\; \textit{if } j < k \\ \neg\texttt{less}_k(\texttt{bitInd}_j) \;//\; \textit{otherwise}, \end{array} \quad (5)$$

where $\texttt{less}_k$ occurs in the problem instance.

In our encoding we also frequently use predicates of the form $\texttt{range}_{[m,k]}$ which defines bit-indexes in the range $[m, k]$. The range predicates can be defined either using less predicates $(\texttt{less}_k(\texttt{B}) \wedge \neg\texttt{less}_m(\texttt{B})) \leftrightarrow \texttt{range}_{[m,k-1]}(\texttt{B})$ or explicitly:

$$\texttt{range}_{[m,k]}(x) \leftrightarrow (x = \texttt{bitInd}_m \vee \ldots \vee x = \texttt{bitInd}_k). \quad (6)$$

Explicit representation of range predicates has several advantages: i) we can replace $\texttt{less}_k$ predicates using $\texttt{range}_{[0,k-1]}$ predicates, ii) in our encoding, after such replacement all non-ground occurrences of the range predicates will be negative, iii) based on ii), instead of having both positive and negative axioms for $\texttt{range}_{[0,k-1]}$ as we have for $\texttt{less}_k$ in (5), it is sufficient to introduce positive axioms: $\texttt{range}_{[m,k]}(\texttt{bitInd}_j)$ for $m \leq j \leq k$. Wlog we can assume that range predicates do not overlap: we can factor out intersections by introducing corresponding range predicates. This can enable further higher level reasoning at the interval level. Another way of representing ranges is using integer arithmetic. Experiments with these representations are presented in Section 7.1.

In general, we may need to refer to constant addresses as well, say row 0 specified as bit-vector $b000000$ (using 6 bits). This constant address is represented in our encoding using a term, denoted by $t000000$. Note that since a constant address does not depend on the state of the hardware, we do not need to treat it as a function on states – it is a constant. We can define when $\texttt{wraddrFunc}(\texttt{S})$ and $t000000$ refer to the same row by

$$\begin{array}{l} \texttt{wraddrFunc}(\texttt{S}) = t000000 \leftrightarrow \\ \quad (\texttt{wraddr}(\texttt{S}, \texttt{bitInd}_5) \leftrightarrow \textit{false} \wedge \ldots \wedge \texttt{wraddr}(\texttt{S}, \texttt{bitInd}_0) \leftrightarrow \textit{false}). \end{array} \quad (7)$$

Thus, by $\mathrm{mem}(\mathrm{s}_0, t000000, \texttt{bitInd}_5)$, we can refer to the value of bit 5 in row 0 of mem, in state s_0. In Section 4.1 we consider different approaches for defining equality over functional address.

3.2 Encoding of Bit-Vector Operations

In our running example bit-vectors `wraddr` and `rdaddr` are compared by `sel`, and therefore we treat them as functional bit-vectors. We define `sel` as follows:

$$\texttt{sel}(\texttt{S}) \leftrightarrow \texttt{wraddrFunc}(\texttt{S}) = \texttt{rdaddrFunc}(\texttt{S}), \quad (8)$$

which in predicate representation can be rewritten as:

$$\texttt{sel}(\texttt{S}) \leftrightarrow \forall \texttt{B}(\texttt{less}_6(\texttt{B}) \rightarrow (\texttt{wraddr}(\texttt{S}, \texttt{B}) \leftrightarrow \texttt{rdaddr}(\texttt{S}, \texttt{B}))). \quad (9)$$

Similarly, the logic of `outp` can then be defined as follows:

$$\begin{aligned}&\texttt{sel(S)} \rightarrow \forall \texttt{B}(\texttt{less}_{64}(\texttt{B}) \rightarrow (\texttt{outp(S,B)} \leftrightarrow \neg\texttt{rddata(S,B)}))\\ &\neg\texttt{sel(S)} \rightarrow \forall \texttt{B}(\texttt{less}_{64}(\texttt{B}) \rightarrow (\texttt{outp(S,B)} \leftrightarrow \texttt{cacheline(S,B)})).\end{aligned} \tag{10}$$

3.3 Encoding of the Transition Relation

To express the next state functions, we use $\texttt{S}$ to denote the current state and $\texttt{S'}$ to denote the next state. The predicate $next(\texttt{S},\texttt{S'})$ denotes the transition relation, and its axiomatization is described next on our running example (Fig. 1). Recall that in our design, write is enabled when $\texttt{wren} \wedge \texttt{clock}$ holds, and read is enabled when $\texttt{rden} \wedge \neg\texttt{clock}$ holds. The transition relation for the write and read operations are therefore written as follows:

$$\begin{aligned}
&\forall \texttt{S},\texttt{S'}(next(\texttt{S},\texttt{S'}) \rightarrow \quad \text{// write is enabled}\\
&\quad \forall \texttt{A}((\texttt{clock(S')} \wedge \texttt{wren(S')} \wedge \texttt{A} = \texttt{wraddrFunc(S')}) \rightarrow\\
&\quad\quad \forall \texttt{B}(\texttt{range}_{[0,63]}(\texttt{B}) \rightarrow (\texttt{mem(S',A,B)} \leftrightarrow \texttt{wrdata(S',B)}))));\\
&\forall \texttt{S},\texttt{S'}(next(\texttt{S},\texttt{S'}) \rightarrow \quad \text{// write is disabled}\\
&\quad \forall \texttt{A}(\neg(\texttt{clock(S')} \wedge \texttt{wren(S')} \wedge \texttt{A} = \texttt{wraddrFunc(S')}) \rightarrow\\
&\quad\quad \forall \texttt{B}(\texttt{range}_{[0,63]}(\texttt{B}) \rightarrow (\texttt{mem(S',A,B)} \leftrightarrow \texttt{mem(S,A,B)}))));\\
&\forall \texttt{S},\texttt{S'}(next(\texttt{S},\texttt{S'}) \rightarrow \quad \text{// read is enabled}\\
&\quad \forall \texttt{A}((\neg\texttt{clock(S')} \wedge \texttt{rden(S')} \wedge \texttt{A} = \texttt{rdaddrFunc(S')}) \rightarrow\\
&\quad\quad \forall \texttt{B}(\texttt{range}_{[0,63]}(\texttt{B}) \rightarrow (\texttt{rddata(S',B)} \leftrightarrow \texttt{mem(S',A,B)}))));\\
&\forall \texttt{S},\texttt{S'}(next(\texttt{S},\texttt{S'}) \rightarrow \quad \text{// read is disabled}\\
&\quad (\neg(\neg\texttt{clock(S')} \wedge \texttt{rden(S')}) \rightarrow\\
&\quad\quad \forall \texttt{B}(\texttt{range}_{[0,63]}(\texttt{B}) \rightarrow (\texttt{rddata(S',B)} \leftrightarrow \texttt{rddata(S,B)})))).
\end{aligned} \tag{11}$$

Note that in the formulas above, we have assumed that the cells in the memory $\texttt{mem}$ and the read data $\texttt{rddata}$ are modeled as latches rather than flip-flops. We assume that the next-state value of a latch is updated by the next-state value of its input if the next-state value of its enable logic is true. For flip-flops, the next-state value is updated by the current-state value of the input data, when the current-state value of the enable logic is true.

3.4 Encoding of Initial and Final State Constraints

If in the initial state $\texttt{s}_0$ the memory $\texttt{mem}$ is reset (with 0 in each cell), we write this condition as follows:

$$\forall \texttt{A},\texttt{B}(\texttt{less}_{64}(\texttt{B}) \rightarrow \neg\texttt{mem}(\texttt{s}_0,\texttt{A},\texttt{B})). \tag{12}$$

If memory cells are initialized with different values, we write the initial state constraints for it bit-wise. For example, the next formula states that the value of $\texttt{mem}$ in row 0 and column 5 is 0.

$$\neg\texttt{mem}(\texttt{s}_0, t000000, \texttt{bitInd}_5). \tag{13}$$

Initial state values for bit-vectors are specified similarly.

Suppose an assertion $\texttt{prop}$ that we want to verify states that the values of $\texttt{outp}$ and $\texttt{cacheline}$ coincide. We write $\texttt{prop}$ as:

$$\texttt{prop(S)} \leftrightarrow \forall \texttt{B}(\texttt{less}_{64}(\texttt{B}) \rightarrow (\texttt{outp(S,B)} \leftrightarrow \texttt{cacheline(S,B)})). \tag{14}$$

and in order to show correctness of the design we try to refute the negated conjecture

$$\exists \texttt{S}\ \neg\texttt{prop(S)}. \tag{15}$$

3.5 Encoding the BMC Problem

Finally, having the encoding for transition relation, initial and final state constraints, we encode the BMC problem as it is presented in Section 2. The issue we are left with is restoring the EPR encoding which we consider in the next section.

To summarize our word-level encoding, note that, unlike the word-level encoding scheme to EPR in [13], and similar to the *bit-level* encoding in [23], the unrolling bound is explicitly represented. There is no need for unrolling in the sense of BMC [4] in order to refer to a bit-vector or a memory at a desired bound. For this reason, our encoding has a higher potential for abstraction (i.e., removing irrelevant parts of the assertion formula before passing it to a solver engine), and we can use both the initial and final state constraints to simplify the assertion formula with constant propagation and other advanced pre- or in-processing techniques. Sequential ATPG [16] also avoids unrolling via backward time-frame expansion, however, it cannot efficiently use the initial state constraints to simplify the assertion formula. Thus we can combine the strengths of forward and backward reachability analysis in one algorithm. Another advantage of our approach is that, thanks to explicit treatment of time, we can infer invariant properties of the system by pure first-order reasoning without using any form of induction, in the spirit of [20]. A new approach that avoids explicit unrolling for incremental SAT-based MC is proposed in [5]; it is unclear at present how this approach relates to ours.

4 Back to EPR

There are two important problems to be solved in order to obtain an EPR encoding of BMC1: one is with *functional bit-vectors* which in BMC1 are functions of states, and the other is with *naming* of subformulas that result in non-EPR after clausification. The two subsections describe a way out – back to EPR: We solve the first problem during the encoding, by proposing a smart way to deal with addresses; functional bit-vectors occurring in bit-vector comparison can be treated similarly. We solve the second problem as part of the pre-processing of the *entire problem instance* and by improving the clausification (this requires the global view of the entire problem instance).

4.1 Unrolling Addresses

Since non-constant addresses become unary functions in our encoding, the BMC1 problem instances for hardware designs with bit-vectors and memories are outside of the EPR fragment. In order to recover an EPR encoding we apply the following transformation. Consider a non-constant address addr, which is transformed into a unary function addrFunc(S), denoting the value of addr in state S. We introduce new constants $\mathtt{addr}_0, \ldots, \mathtt{addr}_n$ where n is the unrolling bound, and a binary predicate $\mathtt{Assoc}_{\mathtt{addr}}(x, y)$. We add axioms

$$\mathtt{Assoc}_{\mathtt{addr}}(s_0, \mathtt{addr}_0) \wedge \ldots \wedge \mathtt{Assoc}_{\mathtt{addr}}(s_n, \mathtt{addr}_n),$$

which associate each constant $\mathtt{addr}_i$ with the state s_i for $0 \leq i \leq n$. We also transform each formula $\phi[\mathtt{addr}(x)]$ into a formula

$$\forall y(\mathtt{Assoc}_{\mathtt{addr}}(x, y) \rightarrow \phi[y]).$$

Consider bit-vectors of a fixed length, say $k > 0$. We introduce a binary predicate $\mathrm{Val}(x, y)$ which defines values of functional bit-vectors, e.g., the value of $\mathrm{Val}(b, i)$ represents the value of the bit-vector b at the index i. We use a unary predicate $\mathcal{A}_k(x)$ to represent the set of all functional bit-vectors of length k used in the hardware model. We define equality between two functional bit-vectors of length k as follows.

$$\begin{aligned}&\forall x, y\, [\mathcal{A}_k(x) \wedge \mathcal{A}_k(y) \rightarrow \\ &(x = y \leftrightarrow \forall \mathtt{B}(\mathtt{range}_{[0,k-1]}(\mathtt{B}) \rightarrow (\mathrm{Val}(x, \mathtt{B}) \leftrightarrow \mathrm{Val}(y, \mathtt{B}))))].\end{aligned} \tag{16}$$

We thus replace address equality axioms like (1) and (7) discussed earlier by (16). After Skolemizing (16) and some simplifications we obtain:

$$\begin{aligned}&\forall x, y\, [\mathcal{A}_k(x) \wedge \mathcal{A}_k(y) \rightarrow \\ &(x = y \vee (\mathtt{range}_{[0,k-1]}(df_k(x, y)) \wedge (\mathrm{Val}(x, df_k(x, y)) \leftrightarrow \neg\mathrm{Val}(y, df_k(x, y)))))].\end{aligned} \tag{17}$$

Informally, this formula states that if two bit-vectors are different then the Skolem function $df_k(x, y)$ gives an index, within the range $[0, k-1]$, witnessing the difference. Unfortunately, (17) is outside of the EPR class. In order to get back to an EPR encoding, we represent the function df_k using a new predicate Df_k as follows.

$$\forall x, y\, [Df_k(x, y, 0) \vee \ldots \vee Df_k(x, y, k-1)]. \tag{18}$$

$$\begin{aligned}\forall x, y, \mathtt{B}\, [&\mathcal{A}_k(x) \wedge \mathcal{A}_k(y) \wedge Df_k(x, y, \mathtt{B}) \rightarrow \\ &(x = y \vee (\mathrm{Val}(x, \mathtt{B}) \leftrightarrow \neg\mathrm{Val}(y, \mathtt{B}))))].\end{aligned} \tag{19}$$

In addition, for each constant c of length k representing an address, such as addresses $\mathtt{addr}_i$ discussed above, or Skolem constants representing addresses, we need an axiom $\mathcal{A}_k(c)$. In the many-sorted setting the formulas can be simplified assuming we have a sort for all bit-vectors of length k.

Another approach to eliminating the function df_k is to introduce for each pair of address constants a_i, a_j, an index constant $d_{i,j}$ which witnesses the difference of a_i and a_j if they are different in the interpretation. We axiomatize this as follows. Let $a_0, \ldots, a_m$ be the list of all address constants of length k occurring in our problem, including the Skolem constants. Then define the following set of axioms, where $0 \leq i < j \leq m$:

$$a_i = a_j \vee (\mathtt{range}_{[0,k-1]}(d_{ij}) \wedge (\mathrm{Val}(a_i, d_{ij}) \leftrightarrow \neg\mathrm{Val}(a_j, d_{ij}))). \tag{20}$$

This alternative encoding can be used when we have a relatively small number of bit-vector constants but of a large bit-vector size.

As an example, the next-state axiom for the write operation (11) will be:

$$\begin{aligned}&\forall \mathtt{S}, \mathtt{S'}(\mathit{next}(\mathtt{S}, \mathtt{S'}) \rightarrow \qquad /\!/ \text{ write is enabled} \\ &\;(\forall y(\mathtt{Assoc}_{\mathtt{wraddr}}(\mathtt{S'}, y) \rightarrow \\ &\;\;(\forall \mathtt{A}((\mathtt{clock}(\mathtt{S'}) \wedge \mathtt{wren}(\mathtt{S'}) \wedge \mathtt{A} = y) \rightarrow \\ &\;\;\;(\forall \mathtt{B}(\mathtt{range}_{[0,63]}(\mathtt{B}) \rightarrow (\mathtt{mem}(\mathtt{S'}, \mathtt{A}, \mathtt{B}) \leftrightarrow \mathtt{wrdata}(\mathtt{S'}, \mathtt{B})))))))).\end{aligned} \tag{21}$$

Constant addresses which are non-transient, are represented as constants rather than functions of states, thus associated address constants for them are not introduced.

4.2 Pre-processing and Clausification

Consider the defining axiom (14) for the property prop. If we apply the standard clausification algorithm to (14) then we obtain a non-EPR formula due to the negative occurrence of the universal quantifier in the '$\leftarrow$' direction of the outer equivalence. Our first observation comes from a well-known idea used in the optimized CNF transformation: if the defined predicate occurs only positively in the rest of the formula then we can replace the outer equivalence in the definition with '$\rightarrow$' implication. The new simplified axiom can now be safely transformed into an EPR formula. Unfortunately, this is not always the case in the verification examples we have tried. On the other hand such negative occurrences are usually limited. For example, assume that prop occurs negatively only in a negated conjecture (15). Our next idea is to *inline* the definition of prop into such negative occurrences of prop, i.e., replace prop by its definition. After inlining (14) into (15) we obtain:

$$\exists \mathtt{S}(\exists \mathtt{B}(\mathtt{less}_{64}(\mathtt{B}) \land \neg(\mathtt{outp}(\mathtt{S},\mathtt{B}) \leftrightarrow \mathtt{cacheline}(\mathtt{S},\mathtt{B})))). \tag{22}$$

Let us note that after inlining we have i) obtained an equivalent formula and ii) eliminated one negative occurrence of the defined predicate. In this way we can remove all negative occurrences of the defined predicate. We can see that Skolemization applied to the new formula (22) produces an EPR formula. Likewise, since we removed all negative occurrences of prop we can now simplify the definition of prop as above, which after Skolemization also becomes an EPR formula. It is still possible (albeit infrequent in practice) that inlining fails to restore EPR clausification or results in a large increase in the formula size. In these cases we can apply techniques as in Section 4.1 to the Skolem functions, restoring EPR. For further *EPR-restoring* pre-processing and clausification techniques we refer to [15]. Inlining and EPR-restoring pre-processing is implemented in Vampire's clausifier. [1]

5 Other Optimizations in the Encoding

Writing Next-state Formulas for All States. Hardware is driven into its initial state (or states) after applying a reset sequence. Therefore the initial state is such that, for each latch, if its enable is true, its input and output have the same value. We use this assumption to simplify the next-state functions for latches in this case (i.e., when the latch is updated). For example, instead of the next-state axioms (21) for the write operation (which is a latch vector) we write

$$\begin{array}{l}\forall \mathtt{S}(\forall y(\mathtt{Assoc}_{\mathtt{wraddr}}(\mathtt{S},y) \rightarrow \qquad /\!/ \text{ write is enabled}\\ \quad (\forall \mathtt{A}((\mathtt{clock}(\mathtt{S}) \land \mathtt{wren}(\mathtt{S}) \land \mathtt{A} = y) \rightarrow\\ \qquad (\forall \mathtt{B}(\mathtt{range}_{[0,63]}(\mathtt{B}) \rightarrow (\mathtt{mem}(\mathtt{S},\mathtt{A},\mathtt{B}) \leftrightarrow \mathtt{wrdata}(\mathtt{S},\mathtt{B}))))))).\end{array}$$

The latter formula is much easier for theorem provers: it can be applied to any state constant, no need to (constructively) derive that it is a next-state for some other state.

When a latch retains its previous value, we still need to refer to both current and next states in the next-state function. A similar optimization is made for memories whose cells are implemented as latches. However, this optimization does not apply to flip-flops and memories whose cells are implemented as flip-flops.

[1] Available at http://www.vprover.org/

Abstracting Bit-Vector Widths. Consider a subformula quantified over a bit-index variable within a range, say

$$\forall \mathtt{B}(\mathtt{range}_{[0,63]}(\mathtt{B}) \rightarrow (bv_1(\mathtt{S},\mathtt{B}) \leftrightarrow \neg bv_2(\mathtt{S},\mathtt{B}))). \tag{23}$$

Let us call such a subformula a *range* subformula, of range $[63 : 0]$. Such subformulas might also contain free occurrences of address variables.

If a range subformula occurs positively and the full ranges of all involved bit-vectors coincide with the range of the formula, then we transform the subformula into a simpler one, by omitting the relevant range of B. For example, positive occurrences of subformula (23) will be transformed into:

$$\forall \mathtt{B}(bv_1(\mathtt{S},\mathtt{B}) \leftrightarrow \neg bv_2(\mathtt{S},\mathtt{B})). \tag{24}$$

The latter formula is easier for theorem provers, since in order to use it in the inference one does not need to know the range of B.

Adding Sorts. Sorts (types) are now supported in the TPTP standard for FOL problems. We work with three sorts: addresses, states, and bit-indexes. This simplifies the encoding, makes solving faster, and improves the representation of models (counter-examples) since constants of different sorts are not mixed any more in the models.

6 Incremental Bounds

Given the BMC1 encoding of a transition system it is desirable to search for falsifying paths incrementally, bound after bound, avoiding repeated computations. We have implemented such an incremental algorithm in our instantiation-based automated reasoning system iProver [19]. In a nutshell, iProver generates instances of the first-order input clauses in a smart way in an attempt to approximate a ground model. The ground reasoning is delegated to a solver for propositional satisfiability, currently MiniSAT [11].

iProver supports incrementality based on propositional unit assumptions as follows. We can add and retract propositional unit assumptions without repeating calculations which were not based on these assumptions. For each bound k we introduce a propositional variable p_k and use unit assumptions to activate and deactivate bound dependent axioms. For example, consider a bound k and bound dependent axioms for reachable states:

$$\mathrm{RState}(s_0) \wedge \cdots \wedge \mathrm{RState}(s_k). \tag{25}$$

$$p_k \rightarrow \forall x(\mathrm{RState}(x) \rightarrow x = s_0 \vee \ldots \vee x = s_k). \tag{26}$$

Then we can add the unit assumption p_k which activates the state axiom (26) above. Optionally, we can also add axioms $\neg p_0, \ldots, \neg p_{k-1}$ which would be used by the SAT solver to ignore all state axioms for the previous bounds $1, \ldots, k-1$. The only other bound dependent axioms are those defining the $next$ predicate (see Section 2) and unrolling of addresses (see Section 4.1). Let us note that specifications of the transition relation and initial/final state constraints are independent from the unrolling bounds and remain unchanged.

7 Benchmarks and Experimental Results

To evaluate our encoding we have generated two sets of benchmarks in TPTP format, where the first is already available as part of the TPTP library and the second is about to be released. We use iProver and Z3 in our experiments since these solvers performed best on the first set of examples (the results of other solvers are available as part of the TPTP library). Z3 has a dedicated EPR algorithm [25] and is also among the best on quantified SMT problems with arithmetic, while iProver has won the EPR division in several recent CASC competitions [27].

The first set of benchmarks was generated from a simple finite-state machine model called "Robot" and has been released as part of the TPTP library $v5.3.0$. The unrolling bounds were chosen so that the problem instances fit the required level of complexity for the competition, that is, each problem can be solved by at least one prover within the timeout. In the TPTP library the problems have been named HWV039 to HWV047 and are available in up to four variants in four forms each: as first-order formulas (FOF), in clausal normal form (CNF), as typed first-order formulas (TFF) and as typed first-order formulas with interpreted arithmetic symbols (TFA). In the last form, we treated bit-indexes as integers and modeled `less` and `range` predicates with the $<$ predicate.

The second set of benchmarks originate from real-life hardware verification problems on Intel designs containing memories. We are in the process of releasing scrambled versions of these benchmarks into the TPTP library. For the evaluation in this paper we focus on the real-life benchmark problems which are challenging due to a large number of word-level components.

7.1 Comparison of Encodings of Bit-Ranges

We evaluated three different encodings of bit-ranges on the second set of industrial BMC problems. In the first two encodings ranges are modeled with the $\mathtt{range}_{[m,k]}$ and $\mathtt{less}_k$ predicates as described in Section 3.1. In the third encoding ranges are straightforwardly modeled using integer arithmetic.

We ran Z3 and iProver on problems unrolled to several bounds, on Intel Xeon Quad Core machines with 12 GB of memory with 20000s timeout. iProver accepts only CNF format with sorts and therefore for iProver the problems were clausified by Vampire; for Z3 we used problems in the original non-clausified sorted TFF and TFA formats.

Let us discuss experimental results shown in Table 1. The first and fourth problems can be solved by iProver and not by Z3; on the third and fifth problems, while iProver succeeded on some bounds, Z3 can reach higher bounds. We can observe that on higher unrolling bounds i) the performance of Z3 on the $\mathtt{range}_{[m,k]}$ encoding is considerably better than on the $\mathtt{less}_k$ encoding, ii) the range encoding is on a par with the arithmetic encoding. The arithmetic encoding is better on smaller bounds. iProver performs similarly on $\mathtt{range}_{[m,k]}$ and $\mathtt{less}_k$ encodings, with $\mathtt{range}_{[m,k]}$ encoding reaching higher bounds on one problem. To conclude, the results suggest that the range encoding is a reasonable alternative to the arithmetic encoding and for future work we investigate ways of combining iProver and SMT solvers for better reasoning with ranges.

Table 1. Different encodings of bit-indexes and bit-ranges

Problem (bound)	Z3			iProver	
	`less`$_k$	`range`$_{[m,k]}$	arithmetic	`less`$_k$	`range`$_{[m,k]}$
BPB (bound 2)	—	—	—	42s	**41s**
BPB (bound 4)	—	—	—	**634s**	669s
DCC (bound 2)	78s	56s	**29s**	55s	79s
DCC (bound 4)	1204s	636s	**157s**	266s	238s
DCC (bound 6)	8540s	3396s	3512s	—	**1407s**
PMS (bound 2)	44s	1266s	**9s**	161s	163s
PMS (bound 4)	638s	**149s**	188s	1295s	1298s
PMS (bound 6)	2898s	5730s	**4564s**	—	—
PMS (bound 8)	12303s	**3062s**	—	—	—
ROB (bound 2)	—	—	—	**250s**	282s
SCD (bound 2)	167s	119s	178s	**15s**	**15s**
SCD (bound 4)	434s	316s	346s	**276s**	277s
SCD (bound 6)	886s	548s	699s	**635s**	**635s**
SCD (bound 8)	2037s	**1017s**	1497s	—	—

7.2 Comparison with SAT-Based BMC

In Table 2, we compare incremental SAT-based BMC [12] (column incBMC) with EPR-based incremental BMC1 (column incBMC1), on the second set of Intel benchmarks. The column #memories reports the number of memories in the cone of the property and their collective size in terms of number of memory cells (bits). Similarly, columns #BVs give the number of transient and constant bit-vectors, respectively (including bit-vectors of size 1), and their collective size in terms of bits. These two columns show how "word-level" the cone of the property really is, and the cone size. Columns incBMC and incBMC1 report the maximal bound reached by the respective algorithms within 10000 seconds time limit and unrolling bound limit 50.

We used Intel's SAT-based model checker to perform experiments with incremental BMC [18]. It has a state-of-the-art implementation of incremental BMC, and its SAT solver is especially tuned on problem instances originating from formal verification problems on Intel designs. In [18], it is shown that Intel SAT-based BMC tool is on a par with a leading academic model checking tool ABC [22].

Table 2. Comparing SAT-based incremental BMC with EPR-based BMC1

Problem	# Memories	# Transient BVs	# Const. BVs	incBMC	incBMC1
PMS1	8 (46080 bits)	1486 (6109 bits)	3 (47 bits)	2	**10**
SCD1	2 (16384 bits)	556 (1923 bits)	5 (45 bits)	4	**12**
SCD2	2 (16384 bits)	80 (756 bits)	3 (10 bits)	4	**14**
BPB2	4 (10240 bits)	550 (4955 bits)	6 (42 bits)	**50**	11
DCI1	32 (9216 bits)	3625 (6496 bits)	3 (9 bits)	**6**	4
DCC2	4 (8960 bits)	426 (1844 bits)	2 (2 bits)	8	**11**
DCC1	4 (8960 bits)	1827 (5294 bits)	5 (106 bits)	7	**8**
ROB2	2 (4704 bits)	255 (3479 bits)	26 (129 bits)	**50**	8

Experimental results show that although on smaller memories SAT-based BMC is faster, when memory sizes increase, the advantage of EPR-based BMC1 becomes evident (rows in Table 2 are ordered in decreasing memory size). These results show that EPR-based model checking is a promising alternative to SAT-based model checking at word-level, scalable to industrial hardware designs. We refer to [13] for a comparison with SMT solvers supporting bit-vectors and arrays on unrolled BMC instances.

8 Conclusions and Future Work

In this paper we presented an encoding of bounded model checking at word level into the EPR fragment of first-order logic. The EPR-based encoding allows us to i) represent memories and bit-vectors at word-level and ii) succinctly specify the transition relation and state constraints, independently from the unrolling bound. Due to the presence of memories and bit-vectors, a naive encoding of the BMC problem into first-order logic would result in problems outside of the EPR fragment. We show how to restore the EPR encoding by introducing two methods: i) address unrolling and ii) definition inlining.

Another contribution of this work is in generating challenging benchmarks for first-order theorem provers based on real-life hardware designs used at Intel. We hope this will encourage further research into EPR-based model checking and EPR decision procedures. We have evaluated our encoding on these benchmarks using general purpose theorem provers iProver and Z3 which are not optimized for such problems. Our experimental results show that already at this stage our approach is scalable to industrial verification problems and on large memories can reach higher unrolling bounds compared to optimized SAT-based BMC.

There are many directions for future work, and we mention only few of them here. First, we intend to develop an abstraction-refinement approach to EPR-based model checking by providing the EPR solver with some bit-vector related information (e,g., the bit-vector width) via attributes. Second, we intend to investigate further incremental solving in the EPR-based BMC1 and how derived information on lower bounds can be exploited for reasoning on higher bounds. Finally, we believe that EPR-based BMC1 can be extended to efficiently work with arithmetic operations at word level, by building-in efficient arithmetic reasoning in the EPR decision procedures.

References

1. Abu-Haimed, H., Dill, D.L., Berezin, S.: A refinement method for validity checking of quantified first-order formulas in hardware verification. In: FMCAD 2006 (2006)
2. Alberti, F., Armando, A., Ranise, S.: ASASP: Automated Symbolic Analysis of Security Policies. In: Bjørner, N., Sofronie-Stokkermans, V. (eds.) CADE 2011. LNCS, vol. 6803, pp. 26–33. Springer, Heidelberg (2011)
3. Baumgartner, P., Fuchs, A., Tinelli, C.: Implementing the Model Evolution Calculus. Inter. J. on Artificial Intelligence Tools 15(1), 21–52 (2006)
4. Biere, A., Cimatti, A., Clarke, E., Zhu, Y.: Symbolic Model Checking without BDDs. In: Cleaveland, W.R. (ed.) TACAS 1999. LNCS, vol. 1579, pp. 193–207. Springer, Heidelberg (1999)
5. Bradley, A.R.: SAT-Based Model Checking without Unrolling. In: Jhala, R., Schmidt, D. (eds.) VMCAI 2011. LNCS, vol. 6538, pp. 70–87. Springer, Heidelberg (2011)

6. Bradley, A.R., Manna, Z., Sipma, H.B.: What's Decidable About Arrays? In: Emerson, E.A., Namjoshi, K.S. (eds.) VMCAI 2006. LNCS, vol. 3855, pp. 427–442. Springer, Heidelberg (2005)
7. Brummayer, R., Biere, A.: Boolector: An Efficient SMT Solver for Bit-Vectors and Arrays. In: Kowalewski, S., Philippou, A. (eds.) TACAS 2009. LNCS, vol. 5505, pp. 174–177. Springer, Heidelberg (2009)
8. Bryant, R.E., Lahiri, S.K., Seshia, S.A.: Modeling and Verifying Systems Using a Logic of Counter Arithmetic with Lambda Expressions and Uninterpreted Functions. In: Brinksma, E., Larsen, K.G. (eds.) CAV 2002. LNCS, vol. 2404, pp. 78–92. Springer, Heidelberg (2002)
9. Clarke, E.M., Grumberg, O., Peled, D.A.: Model Checking. MIT Press (1999)
10. Claessen, K., Sörensson, N.: New techniques that improve MACE-style model finding. In: Workshop on Model Computation, MODEL (2003)
11. Eén, N., Sörensson, N.: An Extensible SAT-solver. In: Giunchiglia, E., Tacchella, A. (eds.) SAT 2003. LNCS, vol. 2919, pp. 502–518. Springer, Heidelberg (2004)
12. Eén, N., Sörensson, N.: Temporal induction by incremental SAT solving. ENTCS 89(4) (2003)
13. Emmer, M., Khasidashvili, Z., Korovin, K., Voronkov, A.: Encoding Industrial Hardware Verification Problems into Effectively Propositional Logic. In: FMCAD 2010 (2010)
14. Ghilardi, S., Nicolini, E., Ranise, S., Zucchelli, D.: Decision procedures for extensions of the theory of arrays. Annals of Mathematics and Artificial Intelligence, AMAI (2006)
15. Hoder, K., Khasidashvili, Z., Korovin, K., Voronkov, A.: Preprocessing techniques for first-order clausification (in preparation)
16. Huang, S.-Y., Cheng, K.-T.: Formal Equivalence Checking and Design Debugging. Kluwer (1998)
17. Khasidashvili, Z., Kinanah, M., Voronkov, A.: Verifying Equivalence of Memories Using a First Order Logic Theorem Prover. In: FMCAD 2009 (2009)
18. Khasidashvili, Z., Nadel, A.: Implicative simultaneous satisfiability and applications. In: HVC 2011 (2011)
19. Korovin, K.: iProver – An Instantiation-Based Theorem Prover for First-Order Logic (System Description). In: Armando, A., Baumgartner, P., Dowek, G. (eds.) IJCAR 2008. LNCS (LNAI), vol. 5195, pp. 292–298. Springer, Heidelberg (2008)
20. Kovács, L., Voronkov, A.: Finding Loop Invariants for Programs over Arrays Using a Theorem Prover. In: Chechik, M., Wirsing, M. (eds.) FASE 2009. LNCS, vol. 5503, pp. 470–485. Springer, Heidelberg (2009)
21. Kroening, D., Strichman, O.: Decision Procedures. EATCS. Springer (2008)
22. Mishchenko, A., Chatterjee, S., Brayton, R., Een, N.: Improvements to combinational equivalence checking. In: ICCAD (2006)
23. Navarro-Pérez, J.A., Voronkov, A.: Encodings of Bounded LTL Model Checking in Effectively Propositional Logic. In: Pfenning, F. (ed.) CADE 2007. LNCS (LNAI), vol. 4603, pp. 346–361. Springer, Heidelberg (2007)
24. Manolios, P., Srinivasan, S.K., Vroon, D.: Automatic memory reductions for RTL model verification. In: ICCAD (2006)
25. Piskac, R., de Moura, L., Bjørner, N.: Deciding Effectively Propositional Logic Using DPLL and Substitution Sets. J. Autom. Reasoning (2010)
26. Robinson, A., Voronkov, A. (eds.): Handbook of Automated Reasoning. Elsevier and MIT Press (2001)
27. Sutcliffe, G.: The 5th IJCAR automated theorem proving system competition CASC-J5. AI Communications 24(1), 75–89 (2011)
28. Velev, M.N., Bryant, R.E.: Verification of Pipelined Microprocessors by Comparing Memory Execution Sequences in Symbolic Simulation. In: Shyamasundar, R.K. (ed.) ASIAN 1997. LNCS, vol. 1345, pp. 18–31. Springer, Heidelberg (1997)

Proving Non-looping Non-termination Automatically*

Fabian Emmes, Tim Enger, and Jürgen Giesl

LuFG Informatik 2, RWTH Aachen University, Germany

Abstract. We introduce a technique to prove non-termination of term rewrite systems automatically. Our technique improves over previous approaches substantially, as it can also detect non-looping non-termination.

1 Introduction

Approaches to prove termination of term rewrite systems (TRSs) have been studied for decades and there exist several techniques to prove termination of programs via a translation to TRSs. In contrast, techniques to *disprove* termination of TRSs have received much less attention, although this is highly relevant to detect bugs during program development. To prove non-termination of a TRS, one has to provide a finite description of an infinite rewrite sequence.

The most common way for this is to find a *loop*, i.e., a finite rewrite sequence $s \to_{\mathcal{R}}^{+} C[s\mu]$ for some term s, context C, and substitution μ. Indeed, any loop gives rise to an infinite rewrite sequence $s \to_{\mathcal{R}}^{n} C[s\mu] \to_{\mathcal{R}}^{n} C[C\mu[s\mu^2]] \to_{\mathcal{R}}^{n} \dots$ for some $n > 0$. While this is a very intuitive way to prove non-termination, it cannot capture non-periodic infinite rewrite sequences.

For instance, consider the imperative program fragment on the side which does not terminate if `x > y` and `x > 0`. However, if `gt` (greater than) and `dbl` (double) are user-defined, then the number of evaluation steps needed for `gt` and `dbl` increases in each loop iteration. Hence, this is a non-periodic form of non-termination.

```
while (gt(x,y)){
  x = dbl(x);
  y = y + 1;    }
```

The following TRS $\mathcal{R}$ corresponds to the imperative program fragment above.

$$\begin{aligned} \mathsf{f}(\mathsf{tt},x,y) &\to \mathsf{f}(\mathsf{gt}(x,y),\ \mathsf{dbl}(x),\ \mathsf{s}(y)) & \mathsf{dbl}(x) &\to \mathsf{times}(\mathsf{s}(\mathsf{s}(0)),x)\\ \mathsf{gt}(\mathsf{s}(x),0) &\to \mathsf{tt} & \mathsf{times}(x,0) &\to 0\\ \mathsf{gt}(0,y) &\to \mathsf{ff} & \mathsf{times}(x,\mathsf{s}(y)) &\to \mathsf{plus}(\mathsf{times}(x,y),x)\\ \mathsf{gt}(\mathsf{s}(x),\mathsf{s}(y)) &\to \mathsf{gt}(x,y) & \mathsf{plus}(x,0) &\to x\\ & & \mathsf{plus}(x,\mathsf{s}(y)) &\to \mathsf{plus}(\mathsf{s}(x),y) \end{aligned}$$

This TRS is non-terminating, but not looping. For $n > m$, we have

$$\begin{array}{lll} \mathsf{f}(\mathsf{tt},\mathsf{s}^n(0),\mathsf{s}^m(0)) & \to_{\mathcal{R}} \mathsf{f}(\mathsf{gt}(\mathsf{s}^n(0),\mathsf{s}^m(0)),\mathsf{dbl}(\mathsf{s}^n(0)),\mathsf{s}^{m+1}(0)) & \to_{\mathcal{R}}^{m+1}\\ \mathsf{f}(\mathsf{tt},\mathsf{dbl}(\mathsf{s}^n(0)),\mathsf{s}^{m+1}(0)) & \to_{\mathcal{R}} \mathsf{f}(\mathsf{tt},\mathsf{times}(\mathsf{s}(\mathsf{s}(0)),\mathsf{s}^n(0)),\mathsf{s}^{m+1}(0)) & \to_{\mathcal{R}}^{4\cdot n}\\ \mathsf{f}(\mathsf{tt},\mathsf{s}^{2\cdot n}(\mathsf{times}(\mathsf{s}(\mathsf{s}(0)),0)),\mathsf{s}^{m+1}(0)) & \to_{\mathcal{R}} \mathsf{f}(\mathsf{tt},\mathsf{s}^{2\cdot n}(0),\mathsf{s}^{m+1}(0)) & \to_{\mathcal{R}} \dots \end{array}$$

* Supported by the DFG grant GI 274/5-3.

B. Gramlich, D. Miller, and U. Sattler (Eds.): IJCAR 2012, LNAI 7364, pp. 225–240, 2012.

Since the number of steps required to evaluate gt and dbl increases in every iteration, this is a non-periodic sequence that cannot be represented as a loop.

While interesting classes of non-looping TRSs were identified in earlier papers (e.g., [3,14]), up to now virtually all methods to prove non-termination of TRSs automatically were restricted to loops (e.g., [4,5,11,13,15,16]).[1] A notable exception is a technique and tool for non-termination of non-looping *string* rewrite systems (SRSs) in [10]. To represent rewrite sequences, this approach uses rules between string patterns of the form $u\,v^n\,w$. Here, u, v, w are strings and n can be instantiated by any natural number. We will extend this idea in order to prove non-termination of (possibly non-looping) *term* rewrite systems automatically.

To detect loops, one can start with a rule and repeatedly *narrow* it using other rules, until it has the form of a loop. To handle non-looping TRSs as well, we generate *pattern rules* which represent a whole set of rewrite sequences and also allow narrowing with pattern rules. In this way, one can create more and more pattern rules until one obtains a pattern rule that is obviously non-terminating. In Sect. 2, we define pattern rules formally and introduce a set of inference rules to derive pattern rules from a TRS automatically. Sect. 2 also contains a criterion to detect pattern rules that are obviously non-terminating. In Sect. 3 we present a strategy for the application of our inference rules. We implemented our contributions in the automated termination tool AProVE [6] and in Sect. 4, we present an experimental evaluation of our technique.

2 Pattern Rules

To represent rewrite sequences, we extend the idea of [10] from SRSs to TRSs and define *pattern terms* and *pattern rules* which are parameterized over $\mathbb{N}$.

A *pattern term* describes a set of terms.[2] Formally, a pattern term is a mapping from natural numbers to terms which are constructed from a *base term*, a *pumping substitution* that is applied multiple times to the base term, and a *closing substitution* that is applied once to "close" the term. For example, to represent $\mathsf{gt}(\mathsf{s}^2(x), \mathsf{s}(0))$, $\mathsf{gt}(\mathsf{s}^3(x), \mathsf{s}^2(0))$, $\mathsf{gt}(\mathsf{s}^4(x), \mathsf{s}^3(0))$, ..., we use the pattern term $n \mapsto \mathsf{gt}(\mathsf{s}(x), \mathsf{s}(y))\ [x/\mathsf{s}(x), y/\mathsf{s}(y)]^n\ [x/\mathsf{s}(x), y/0]$, where $\mathsf{gt}(\mathsf{s}(x), \mathsf{s}(y))$ is the base term, $[x/\mathsf{s}(x), y/\mathsf{s}(y)]$ is the pumping substitution, and $[x/\mathsf{s}(x), y/0]$ is the closing substitution. For $n = 0$ this pattern term evaluates to $\mathsf{gt}(\mathsf{s}^2(x), \mathsf{s}(0))$, for $n = 1$ to $\mathsf{gt}(\mathsf{s}^3(x), \mathsf{s}^2(0))$, etc. In the following, $\mathcal{T}(\Sigma, \mathcal{V})$ denotes the set of terms over the underlying signature Σ and the infinite set of variables $\mathcal{V}$.

---

[1] Similarly, most existing automated approaches for non-termination of *programs* also just detect loops. For Java Bytecode, we recently presented an approach that can also prove non-periodic non-termination, provided that there are no sub-loops and that non-termination is due to operations on integers [2]. However, this approach is not suitable for TRSs where one treats terms instead of integers and where sub-loops (i.e., recursively defined auxiliary functions like gt and times) are common.

[2] In contrast to *tree automata*, pattern terms can also describe non-regular sets.

Definition 1 (Pattern Terms and Rules). *A function* $\mathbb{N} \to \mathcal{T}(\Sigma, \mathcal{V})$ *is a* pattern term *if it is a mapping* $n \mapsto t\sigma^n\mu$ *where* $t \in \mathcal{T}(\Sigma, \mathcal{V})$ *and* σ, μ *are substitutions. For readability, we omit "*$n \mapsto$*" if it is clear that we refer to a pattern term. For a pattern term* $p = t\sigma^n\mu$*, its* base term *is* $\text{base}(p) = t$*, its* pumping substitution *is* σ*, and its* closing substitution *is* μ*. We also say that* σ, μ *are its* pattern substitutions*. Its* domain variables *are* $\text{dv}(p) = \text{dom}(\sigma) \cup \text{dom}(\mu)$.

If p, q *are pattern terms, then* $p \hookrightarrow q$ *is a* pattern rule*. A pattern rule* $p \hookrightarrow q$ *is* correct w.r.t. a TRS $\mathcal{R}$ *if* $p(n) \to^+_{\mathcal{R}} q(n)$ *holds for all* $n \in \mathbb{N}$.

As an example, consider the pattern rule

$$\mathsf{gt}(\mathsf{s}(x), \mathsf{s}(y))\ [x/\mathsf{s}(x), y/\mathsf{s}(y)]^n\ [x/\mathsf{s}(x), y/\mathsf{0}] \quad \hookrightarrow \quad \mathsf{tt}\ \varnothing^n\ \varnothing, \tag{1}$$

where $\varnothing$ denotes the empty (identical) substitution. This pattern rule is correct w.r.t. the TRS $\mathcal{R}$ in Sect. 1, since $\mathsf{gt}(\mathsf{s}^{n+2}(x), \mathsf{s}^{n+1}(\mathsf{0})) \to^+_{\mathcal{R}} \mathsf{tt}$ holds for all $n \in \mathbb{N}$. Thus, a pattern rule describes a set of rewrite sequences of arbitrary length.

In the following, we present 9 inference rules to derive correct pattern rules from a TRS automatically. As soon as one finds a correct pattern rule that is obviously non-terminating, one has proved non-termination of the original TRS.

The inference rules have the form $\frac{p_1 \hookrightarrow q_1 \quad \ldots \quad p_k \hookrightarrow q_k}{p \hookrightarrow q}$. In Thm. 7 we will prove their soundness, i.e., if all the pattern rules $p_1 \hookrightarrow q_1, \ldots, p_k \hookrightarrow q_k$ are correct w.r.t. a TRS $\mathcal{R}$, then the pattern rule $p \hookrightarrow q$ is also correct w.r.t. $\mathcal{R}$.

The inference rules in Sect. 2.1 create initial pattern rules from a TRS. Sect. 2.2 shows how to modify the pattern terms in a pattern rule without changing the represented set of terms. Sect. 2.3 introduces inference rules in order to instantiate pattern rules and to combine them by narrowing. Finally, Sect. 2.4 shows how to detect whether a pattern rule directly leads to non-termination.

2.1 Creating Pattern Rules

The first inference rule converts rules from the TRS to equivalent pattern rules by simply using the identity $\varnothing$ as pattern substitution. Since a pattern term $\ell\,\varnothing^n\,\varnothing$ just represents the (ordinary) term ℓ, this inference rule is clearly sound. So by applying **(I)** to the recursive gt-rule from Sect. 1, we obtain the pattern rule

> **(I) Pattern Rule from TRS**
>
> $$\frac{}{\ell\,\varnothing^n\,\varnothing \;\hookrightarrow\; r\,\varnothing^n\,\varnothing} \quad \text{if } \ell \to r \in \mathcal{R}$$

$$\mathsf{gt}(\mathsf{s}(x), \mathsf{s}(y))\ \varnothing^n\ \varnothing \quad \hookrightarrow \quad \mathsf{gt}(x, y)\ \varnothing^n\ \varnothing. \tag{2}$$

The next inference rule generates pattern rules that represent the repeated application of a rewrite sequence at the same position. Here, we say that two substitutions θ and σ *commute* iff $x\theta\sigma = x\sigma\theta$ holds for all variables $x \in \mathcal{V}$. When applying **(II)** to Rule (2), we have $s = \mathsf{gt}(\mathsf{s}(x), \mathsf{s}(y))$ and $t = \mathsf{gt}(x, y)$. By choosing $\theta = \varnothing$ and

> **(II) Pattern Creation 1**
>
> $$\frac{s\,\varnothing^n\,\varnothing \;\hookrightarrow\; t\,\varnothing^n\,\varnothing}{s\,\sigma^n\,\varnothing \;\hookrightarrow\; t\,\theta^n\,\varnothing} \quad \text{if } s\theta = t\sigma, \text{ and } \theta \text{ commutes with } \sigma$$

$\sigma = [x/\mathsf{s}(x), y/\mathsf{s}(y)]$, we obtain $s\theta = t\sigma$. Moreover since θ is the identical substitution, θ and σ obviously commute. Hence, by **(II)** we obtain the following new pattern rule which describes how (2) can be applied repeatedly on terms of the form $\mathsf{gt}(\mathsf{s}^n(x), \mathsf{s}^n(y))$.

$$\mathsf{gt}(\mathsf{s}(x), \mathsf{s}(y))\ [x/\mathsf{s}(x), y/\mathsf{s}(y)]^n\ \varnothing \quad \hookrightarrow \quad \mathsf{gt}(x, y)\ \varnothing^n\ \varnothing \tag{3}$$

To see why commutation of θ and σ is needed for the soundness of Rule **(II)**, consider $s = \mathsf{f}(x, \mathsf{a})$ and $t = \mathsf{f}(\mathsf{b}, x)$ for a TRS $\mathcal{R}' = \{s \to t\}$. Then for $\theta = [x/\mathsf{b}]$ and $\sigma = [x/\mathsf{a}]$ we have $s\theta = t\sigma$. But θ and σ do not commute and $s\sigma = \mathsf{f}(\mathsf{a}, \mathsf{a}) \not\to^+_{\mathcal{R}'} \mathsf{f}(\mathsf{b}, \mathsf{b}) = t\theta$. Thus, $s\,\sigma^n\,\varnothing \hookrightarrow t\,\theta^n\,\varnothing$ is not correct w.r.t. $\mathcal{R}'$.

To automate the application of inference rule **(II)**, one has to find substitutions θ and σ that satisfy the conditions for its applicability. In our implementation, we use a sufficient criterion which proved useful in our experiments: We first apply unification to find the most general substitutions θ and σ such that $s\theta = t\sigma$. Then we check whether θ and σ commute. More precisely, to find θ and σ with $s\theta = t\sigma$, we use a variable renaming ρ which renames all variables in $\mathcal{V}(s)$ to fresh ones. If there exists $\tau = \mathrm{mgu}(s\rho, t)$, then we set $\theta = (\rho\,\tau\,\rho^{-1})|_{\mathcal{V}(s)}$ and $\sigma = (\tau\rho^{-1})|_{\mathcal{V}(t)}$. Now we have $s\theta = s\rho\,\tau\,\rho^{-1} = t\,\tau\,\rho^{-1} = t\sigma$ and thus, it remains to check whether θ commutes with σ. So in our example, we use a renaming ρ with $x\rho = x'$ and $y\rho = y'$. The mgu of $s\rho = \mathsf{gt}(\mathsf{s}(x'), \mathsf{s}(y'))$ and $t = \mathsf{gt}(x, y)$ is $\tau = [x/\mathsf{s}(x'), y/\mathsf{s}(y')]$. Hence, we obtain $x\theta = x\rho\,\tau\,\rho^{-1} = x$, $y\theta = y$, $x\sigma = x\,\tau\,\rho^{-1} = \mathsf{s}(x)$, and $y\sigma = y\,\tau\,\rho^{-1} = \mathsf{s}(y)$. Here, θ and σ obviously commute.

The next inference rule generates pattern rules to represent rewrite sequences where the context around the redex increases in each iteration. For instance, the times-rule of Sect. 1 can be applied repeatedly to rewrite terms of the form $\mathsf{times}(x, \mathsf{s}^n(y))$ to $\mathsf{plus}(\mathsf{plus}(\ldots \mathsf{plus}(\mathsf{times}(x, y), x), \ldots, x), x)$. But since these rewrite steps (except for the first) occur below the root, instead of **(II)** we need Rule **(III)**. As usual, $t[z]_\pi$ results from replacing the subterm at position π by z. Moreover, $\sigma \cup [z/t[z]_\pi]$ is the extension of the substitution σ which maps the fresh variable z to $t[z]_\pi$.

(III) Pattern Creation 2

$$\frac{s\,\varnothing^n\,\varnothing \;\hookrightarrow\; t\,\varnothing^n\,\varnothing}{s\,\sigma^n\,\varnothing \;\hookrightarrow\; t[z]_\pi\ (\sigma \cup [z/t[z]_\pi])^n\ [z/t|_\pi]} \quad \text{if } \pi \in \mathcal{P}os(t),\ s = t|_\pi\,\sigma, \text{ and } z \in \mathcal{V} \text{ is fresh}$$

Rule **(III)** can easily be automated, since one only has to check whether some subterm[3] of t matches s. For example, regard the pattern rule $\mathsf{times}(x, \mathsf{s}(y))\,\varnothing^n\,\varnothing \hookrightarrow \mathsf{plus}(\mathsf{times}(x, y), x)\,\varnothing^n\,\varnothing$ resulting from the times-rule. Here, $s = \mathsf{times}(x, \mathsf{s}(y))$ and $t = \mathsf{plus}(\mathsf{times}(x, y), x)$. For the subterm $t|_\pi = \mathsf{times}(x, y)$ at position $\pi = 1$ we have $s = t|_\pi\,\sigma$ with $\sigma = [y/\mathsf{s}(y)]$. Hence, by **(III)** we obtain the pattern rule

$$\mathsf{times}(x, \mathsf{s}(y))\ [y/\mathsf{s}(y)]^n\ \varnothing \;\hookrightarrow\; \mathsf{plus}(z, x)\ [y/\mathsf{s}(y), z/\mathsf{plus}(z, x)]^n\ [z/\mathsf{times}(x, y)]. \tag{4}$$

[3] In the automation, we restrict Rule **(III)** to non-variable subterms $t|_\pi$ in order to obtain pattern rules with "small" terms in the ranges of the pumping substitutions.

Note that if π is the root position, then inference rule **(III)** is the special case of inference rule **(II)** where θ is the identity. In this case, both inference rules create a pattern rule equivalent to $s\,\sigma^n\,\varnothing \hookrightarrow t\,\varnothing^n\,\varnothing$.

2.2 Using Equivalence of Pattern Terms

As mentioned in the introduction, a common technique to prove that a TRS is looping is to construct loops via repeated narrowing operations. Narrowing is similar to rewriting, but uses unification instead of matching.

For instance, to narrow the right-hand side of the recursive rule $\mathsf{gt}(\mathsf{s}(x), \mathsf{s}(y)) \to \mathsf{gt}(x, y)$ with the rule $\mathsf{gt}(\mathsf{s}(x), \mathsf{0}) \to \mathsf{tt}$, one could first instantiate the recursive rule using the substitution $[x/\mathsf{s}(x), y/\mathsf{0}]$, which yields $\mathsf{gt}(\mathsf{s}(\mathsf{s}(x)), \mathsf{s}(\mathsf{0})) \to \mathsf{gt}(\mathsf{s}(x), \mathsf{0})$. Now its right-hand side can be rewritten by the non-recursive gt-rule, which results in the new rule $\mathsf{gt}(\mathsf{s}(\mathsf{s}(x)), \mathsf{s}(\mathsf{0})) \to \mathsf{tt}$.

Our goal is to extend this concept to pattern rules. However, the problem is that the pattern terms in the rules may have different pattern substitutions. Thus, to narrow the right-hand side of a pattern rule $p \hookrightarrow q$ with another pattern rule $p' \hookrightarrow q'$, we first transform the rules such that the pattern substitutions in all four terms p, q, p', q' are the same. Then $p \hookrightarrow q$ and $p' \hookrightarrow q'$ have the form $s\,\sigma^n\,\mu \hookrightarrow t\,\sigma^n\,\mu$ and $u\,\sigma^n\,\mu \hookrightarrow v\,\sigma^n\,\mu$, respectively (i.e., the *same* pattern substitutions σ and μ are used on both sides of both pattern rules). To achieve that, it is often useful to modify the pattern terms in the rules appropriately without changing the set of terms represented by the pattern terms.

Definition 2 (Equivalent Pattern Terms). *We say that two pattern terms p and p' are* equivalent *iff $p(n) = p'(n)$ holds for all $n \in \mathbb{N}$.*

Based on Def. 2, we immediately obtain inference rule **(IV)** that allows us to replace pattern terms by equivalent other pattern terms. To apply rule **(IV)** automatically, in Lemmas 4, 6, and 9 we will present three criteria for equivalence of pattern terms.

(IV) Equivalence	
$\dfrac{p \hookrightarrow q}{p' \hookrightarrow q'}$	if p is equivalent to p' and q is equivalent to q'

The first criterion allows us to rename the *domain variables* in the pattern substitutions. For example, in the pattern term $\mathsf{gt}(\mathsf{s}(x), \mathsf{s}(y))\ [x/\mathsf{s}(x), y/\mathsf{s}(y)]^n\ \varnothing$ one can rename its domain variables x and y to x' and y'. This results in the pattern term $\mathsf{gt}(\mathsf{s}(x'), \mathsf{s}(y'))\ [x'/\mathsf{s}(x'), y'/\mathsf{s}(y')]^n\ [x'/x, y'/y]$ which is equivalent, since for every n, both pattern terms represent $\mathsf{gt}(\mathsf{s}^n(x), \mathsf{s}^n(y))$.

Definition 3 (Domain Renamings). *For any substitution σ, let* $\mathrm{range}(\sigma) = \{x\sigma \mid x \in \mathrm{dom}(\sigma)\}$ *and $\mathcal{V}(\sigma) = \mathrm{dom}(\sigma) \cup \mathcal{V}(\mathrm{range}(\sigma))$. Let ρ be a variable renaming on $\mathrm{dom}(\rho)$, i.e., $\mathrm{range}(\rho) \subseteq \mathcal{V}$ and ρ is injective on $\mathrm{dom}(\rho)$. This allows us to define ρ^{-1} as $\rho^{-1}(y) = x$ if there is some $x \in \mathrm{dom}(\rho)$ with $x\rho = y$ and as $\rho^{-1}(y) = y$, otherwise. Note that $x\rho\rho^{-1} = x$ holds for all $x \in \mathrm{dom}(\rho)$ and also for all $x \notin \mathrm{range}(\rho)$. For any pattern term $p = t\sigma^n\mu$, we define its variables as $\mathcal{V}(p) = \mathcal{V}(t) \cup \mathcal{V}(\sigma) \cup \mathcal{V}(\mu)$. We say that a variable renaming ρ is a* domain renaming *for a pattern term p if $\mathrm{dom}(\rho) \subseteq \mathrm{dv}(p)$ and $\mathrm{range}(\rho) \cap \mathcal{V}(p) = \varnothing$. For* *a pattern term* $p = t\sigma^n \mu$, *we define the result of renaming* p *by* ρ *as* $p^\rho = t' \, \sigma'^n \, \mu'$ *where* $t' = t\rho$, $\sigma' = [\, x\rho/s\rho \mid x/s \in \sigma \,]$, *and* $\mu' = [\, x\rho/s \mid x/s \in \mu \,] \, \rho^{-1}$.

To illustrate Def. 3, consider $\rho = [x/x', y/y']$. This is indeed a variable renaming on $\mathrm{dom}(\rho) = \{x, y\}$ and we have $\rho^{-1} = [x'/x, y'/y]$. Moreover, we regard the pattern term $p = \mathsf{gt}(\mathsf{s}(x), \mathsf{s}(y)) \; [x/\mathsf{s}(x), y/\mathsf{s}(y)]^n \; \varnothing$. Thus, its base term is $t = \mathsf{gt}(\mathsf{s}(x), \mathsf{s}(y))$, and it has the pattern substitutions $\sigma = [x/\mathsf{s}(x), y/\mathsf{s}(y)]$ and $\mu = \varnothing$. Hence, ρ is a domain renaming for p since $\mathrm{dom}(\rho) \subseteq \mathrm{dv}(p) = \{x, y\}$ and since $\mathrm{range}(\rho) = \{x', y'\}$ is disjoint from $\mathcal{V}(p) = \mathcal{V}(t) \cup \mathcal{V}(\sigma) \cup \mathcal{V}(\mu) = \{x, y\}$. Thus, the result of renaming p by ρ is $p^\rho = \mathsf{gt}(\mathsf{s}(x'), \mathsf{s}(y')) \; [x'/\mathsf{s}(x'), y'/\mathsf{s}(y')]^n \; [x'/x, y'/y]$. Lemma 4 gives the first criterion for obtaining equivalent pattern terms (in order to apply inference rule **(IV)** automatically).

Lemma 4 (Equivalence by Domain Renaming). *Let* p *be a pattern term and let* ρ *be a domain renaming for* p. *Then* p *is equivalent to* p^ρ.

Proof. Let $p = t \, \sigma^n \, \mu$, $\sigma' = [\, x\rho/s\rho \mid x/s \in \sigma \,]$, and $\mu' = [\, x\rho/s \mid x/s \in \mu \,] \, \rho^{-1}$. We first show the following conjecture:

$$x \, \sigma \rho = x \, \rho \, \sigma' \text{ for all } x \in \mathcal{V}(p) \tag{5}$$

For (5), let $x \in \mathcal{V}(p)$. If $x \in \mathrm{dom}(\sigma)$, then $x \, \rho \, \sigma' = x \, \sigma \, \rho$ by the definition of σ'. If $x \notin \mathrm{dom}(\sigma)$, then $x\rho \notin \mathrm{dom}(\sigma')$. Thus, $x \, \rho \, \sigma' = x\rho = x \, \sigma \, \rho$, which proves (5).

Moreover, we show the following conjecture:

$$x \, \mu = x \, \rho \, \mu' \text{ for all } x \in \mathcal{V}(p) \tag{6}$$

For (6), let $x \in \mathcal{V}(p)$. If $x \in \mathrm{dom}(\mu)$, then $x \, \rho \, \mu' = x \, \mu \, \rho^{-1}$ by the definition of μ'. Since $\mathcal{V}(x\mu) \subseteq \mathcal{V}(p)$, we have $\mathrm{range}(\rho) \cap \mathcal{V}(x\mu) = \varnothing$. Thus, $x \, \rho \, \mu' = x \, \mu \, \rho^{-1} = x\mu$.

Otherwise, if $x \notin \mathrm{dom}(\mu)$, then $x\mu = x$ and $x \, \rho \, \mu' = x \, \rho \, \rho^{-1} = x$. This concludes the proof of Conjecture (6).

Now we show the lemma. We have $p(n) = t \, \sigma^n \, \mu$. By (6), this is equal to $t \, \sigma^n \, \rho \, \mu'$. Using Conjecture (5) n times, we get $t \, \sigma^n \, \rho \, \mu' = t \, \rho \, \sigma'^n \, \mu' = p^\rho(n)$. □

Thus, we can apply inference rule **(IV)** (using Lemma 4 with the domain renaming $\rho = [x/x', y/y']$) to obtain the following pattern rule from Rule (3).

$$\mathsf{gt}(\mathsf{s}(x'), \mathsf{s}(y')) \; [x'/\mathsf{s}(x'), y'/\mathsf{s}(y')]^n \; [x'/x, y'/y] \quad \hookrightarrow \quad \mathsf{gt}(x, y) \; \varnothing^n \; \varnothing \tag{7}$$

Recall that to perform narrowing of pattern rules, we would like to have the same pattern substitutions on both sides of the rule. So the above domain renaming has the advantage that the variables x', y' used for "pumping" are now different from the variables x, y occurring in the final term. This allows us to add the pattern substitutions also on the right-hand side of the rule, since they only concern variables x', y' that are not *relevant* in the right-hand side up to now.

Definition 5 (Relevant Variables). *For a pattern term* $p = t\sigma^n \mu$, *we define its* relevant variables *as* $\mathrm{rv}(p) = \mathcal{V}(\{t, t\sigma, t\sigma^2, \ldots\})$, *i.e.,* $\mathrm{rv}(p)$ *is the smallest set such that* $\mathcal{V}(t) \subseteq \mathrm{rv}(p)$ *and such that* $\mathcal{V}(x\sigma) \subseteq \mathrm{rv}(p)$ *holds for all* $x \in \mathrm{rv}(p)$.

So the relevant variables of the pattern term $\mathsf{gt}(x, y)\ \varnothing^n\ \varnothing$ are x and y. In contrast, a pattern term $\mathsf{gt}(x, y)\ [x/\mathsf{s}(x'), y'/\mathsf{s}(y')]^n\ \varnothing$ would have the relevant variables x, x', and y. Lemma 6 states that one can modify pattern substitutions as long as this only concerns variables that are not relevant in the pattern term.

Lemma 6 (Equivalence by Irrelevant Pattern Substitutions). *Let $p = t\,\sigma^n\,\mu$ be a pattern term and let σ' and μ' be substitutions such that $x\sigma = x\sigma'$ and $x\mu = x\mu'$ holds for all $x \in \mathrm{rv}(p)$. Then p is equivalent to $t\,\sigma'^n\,\mu'$.*

Proof. We prove $t\sigma^n = t\sigma'^n$ by induction on n. For $n = 0$ this is trivial. For $n > 0$, the induction hypothesis implies $t\sigma^{n-1} = t\sigma'^{n-1}$, and since $\mathcal{V}(t\sigma^{n-1}) \subseteq \mathrm{rv}(p)$, we also obtain $t\sigma^n = t\sigma'^n$. Finally, $\mathcal{V}(t\sigma^n) \subseteq \mathrm{rv}(p)$ implies $t\sigma^n\mu = t\sigma'^n\mu'$. □

Hence, since x', y' are not relevant in the pattern term $\mathsf{gt}(x, y)\ \varnothing^n\ \varnothing$, we can add the pattern substitutions from the left-hand side of Rule (7) also on its right-hand side. Thus, by applying **(IV)** (using Lemma 6) to (7), we obtain

$$\begin{array}{lll} & \mathsf{gt}(\mathsf{s}(x'), \mathsf{s}(y'))\ [x'/\mathsf{s}(x'), y'/\mathsf{s}(y')]^n\ [x'/x, y'/y] \\ \hookrightarrow & \mathsf{gt}(x, y) \qquad\qquad\quad [x'/\mathsf{s}(x'), y'/\mathsf{s}(y')]^n\ [x'/x, y'/y]. \end{array} \tag{8}$$

Recall that our goal was to narrow the recursive gt-rule (resp. (8)) with the non-recursive gt-rule $\mathsf{gt}(\mathsf{s}(x), \mathsf{0}) \to \mathsf{tt}$. As a first step towards this goal, we now made the pattern substitutions on both sides of (8) equal.

2.3 Modifying Pattern Rules by Instantiation and Narrowing

For the desired narrowing, we have to instantiate the recursive pattern rule (8) such that the base term of its right-hand side contains the left-hand side of the rule $\mathsf{gt}(\mathsf{s}(x), \mathsf{0}) \to \mathsf{tt}$. To this end, we use inference rule **(V)**.

(V) Instantiation

$$\frac{s\,\sigma_s^n\,\mu_s \ \hookrightarrow\ t\,\sigma_t^n\,\mu_t}{(s\rho)\ (\sigma_s)_\rho^n\ (\mu_s)_\rho \ \hookrightarrow\ (t\rho)\ (\sigma_t)_\rho^n\ (\mu_t)_\rho} \quad \text{if } \mathcal{V}(\rho) \cap (\mathrm{dom}(\sigma_s) \cup \mathrm{dom}(\mu_s) \cup \mathrm{dom}(\sigma_t) \cup \mathrm{dom}(\mu_t)) = \varnothing$$

For any two substitutions σ and ρ, let σ_ρ result from the composition of σ and ρ, but restricted to the domain of σ. Thus, $\sigma_\rho = [x/s\rho \mid x/s \in \sigma]$.

Hence, we now apply inference rule **(V)** on the pattern rule (8) using $\rho = [x/\mathsf{s}(x), y/\mathsf{0}]$. The domain variables of (8) are x' and y'. Thus, due to the domain renaming in Sect. 2.2 they are disjoint from $\mathcal{V}(\rho) = \{x, y\}$. In the resulting pattern rule, the base terms are instantiated with ρ and the new pattern substitutions result from composing the previous pattern substitutions with ρ (restricted to the domains of the previous substitutions). So for $\sigma = [x'/\mathsf{s}(x'), y'/\mathsf{s}(y')]$ we have $\sigma_\rho = \sigma$ and for $\mu = [x'/x, y'/y]$, we obtain $\mu_\rho = [x'/\mathsf{s}(x), y'/\mathsf{0}]$. This yields

$$\begin{array}{lll} & \mathsf{gt}(\mathsf{s}(x'), \mathsf{s}(y'))\ [x'/\mathsf{s}(x'), y'/\mathsf{s}(y')]^n\ [x'/\mathsf{s}(x), y'/\mathsf{0}] \\ \hookrightarrow & \mathsf{gt}(\mathsf{s}(x), \mathsf{0}) \qquad\quad [x'/\mathsf{s}(x'), y'/\mathsf{s}(y')]^n\ [x'/\mathsf{s}(x), y'/\mathsf{0}] \end{array} \tag{9}$$

For the narrowing, the original rule $\mathsf{gt}(\mathsf{s}(x), \mathsf{0}) \to \mathsf{tt}$ of the TRS can be transformed to a pattern rule $\mathsf{gt}(\mathsf{s}(x), \mathsf{0})\ \varnothing^n\ \varnothing \ \hookrightarrow\ \mathsf{tt}\ \varnothing^n\ \varnothing$ by **(I)**. Afterwards, one can add the pattern substitutions of (9) by Rule **(IV)** using Lemma 6, since x', y' are not relevant in the pattern rule:

$$\begin{array}{ll} & \mathsf{gt}(\mathsf{s}(x),\mathsf{0})\ [x'/\mathsf{s}(x'),y'/\mathsf{s}(y')]^n\ [x'/\mathsf{s}(x),y'/\mathsf{0}] \\ \hookrightarrow & \mathsf{tt}\qquad\qquad\ \ [x'/\mathsf{s}(x'),y'/\mathsf{s}(y')]^n\ [x'/\mathsf{s}(x),y'/\mathsf{0}] \end{array} \tag{10}$$

Now all pattern terms in (9) and (10) have the same pattern substitutions.

Hence, we can apply the narrowing rule **(VI)** which rewrites the right-hand side of one pattern rule with another pattern rule, if the pattern substitutions of all pattern terms coincide.

(VI) Narrowing

$$\frac{s\,\sigma^n\,\mu \hookrightarrow t\,\sigma^n\,\mu \qquad u\,\sigma^n\,\mu \hookrightarrow v\,\sigma^n\,\mu}{s\,\sigma^n\,\mu \hookrightarrow t[v]_\pi\,\sigma^n\,\mu} \quad \text{if } t|_\pi = u$$

In our example, $s\,\sigma^n\,\mu \hookrightarrow t\,\sigma^n\,\mu$ is the pattern rule (9) and $u\,\sigma^n\,\mu \hookrightarrow v\,\sigma^n\,\mu$ is the pattern rule (10). Thus, we have $t = \mathsf{gt}(\mathsf{s}(x),\mathsf{0}) = u$ and we obtain the following new pattern rule (which corresponds to Rule (1) in the introduction).

$$\begin{array}{ll} & \mathsf{gt}(\mathsf{s}(x'),\mathsf{s}(y'))\ [x'/\mathsf{s}(x'),y'/\mathsf{s}(y')]^n\ [x'/\mathsf{s}(x),y'/\mathsf{0}] \\ \hookrightarrow & \mathsf{tt}\qquad\qquad\qquad\ [x'/\mathsf{s}(x'),y'/\mathsf{s}(y')]^n\ [x'/\mathsf{s}(x),y'/\mathsf{0}] \end{array} \tag{11}$$

In general, to make the narrowing rule **(VI)** applicable for two rules $s\,\sigma_s^n\,\mu_s \hookrightarrow t\,\sigma_t^n\,\mu_t$ and $u\,\sigma_u^n\,\mu_u \hookrightarrow v\,\sigma_v^n\,\mu_v$, one should first instantiate the base terms t, u such that t contains u. Then one should try to make the substitutions $\sigma_s, \sigma_t, \sigma_u, \sigma_v$ equal and finally, one should try to make $\mu_s, \mu_t, \mu_u, \mu_v$ identical.

To illustrate that, let us try to narrow the pattern rule $\mathsf{f}(\mathsf{tt},x,y)\,\varnothing^n\,\varnothing \hookrightarrow \mathsf{f}(\mathsf{gt}(x,y),\mathsf{dbl}(x),\mathsf{s}(y))\,\varnothing^n\,\varnothing$ resulting from the f-rule with the above pattern rule (11) for gt. To let the base term $\mathsf{gt}(\mathsf{s}(x'),\mathsf{s}(y'))$ of (11)'s left-hand side occur in the right-hand side of f's pattern rule, we instantiate the latter with the substitution $[x/\mathsf{s}(x'), y/\mathsf{s}(y')]$. Thus, inference rule **(V)** yields

$$\mathsf{f}(\mathsf{tt},\mathsf{s}(x'),\mathsf{s}(y'))\,\varnothing^n\,\varnothing \hookrightarrow \mathsf{f}(\mathsf{gt}(\mathsf{s}(x'),\mathsf{s}(y')),\mathsf{dbl}(\mathsf{s}(x')),\mathsf{s}^2(y'))\,\varnothing^n\,\varnothing. \tag{12}$$

Now we try to replace the current pumping substitution σ of Rule (12) by the one of (11). To this end, we use inference rule **(VII)** which allows us to instantiate pumping substitutions.

(VII) Instantiating σ

$$\frac{s\,\sigma_s^n\,\mu_s \hookrightarrow t\,\sigma_t^n\,\mu_t}{s\,(\sigma_s\rho)^n\,\mu_s \hookrightarrow t\,(\sigma_t\rho)^n\,\mu_t} \quad \text{if } \rho \text{ commutes with } \sigma_s, \mu_s, \sigma_t, \text{ and } \mu_t$$

So in our example, we apply inference rule **(VII)** to the pattern rule (12) using the substitution $\rho = [x'/\mathsf{s}(x'), y'/\mathsf{s}(y')]$. Since the pattern substitutions of (12) are just $\varnothing$, ρ trivially commutes with them. Hence, we obtain

$$\begin{array}{ll} & \mathsf{f}(\mathsf{tt},\mathsf{s}(x'),\mathsf{s}(y'))\qquad\qquad\qquad\qquad\qquad\quad [x'/\mathsf{s}(x'),y'/\mathsf{s}(y')]^n\,\varnothing \\ \hookrightarrow & \mathsf{f}(\mathsf{gt}(\mathsf{s}(x'),\mathsf{s}(y')),\mathsf{dbl}(\mathsf{s}(x')),\mathsf{s}^2(y'))\ [x'/\mathsf{s}(x'),y'/\mathsf{s}(y')]^n\,\varnothing. \end{array} \tag{13}$$

Note that **(VII)** differs from the previous instantiation rule **(V)** which does not add new variables to the domains of the pattern substitutions (i.e., with **(V)** we would not have been able to modify the pattern substitutions of (12)).

To make also the closing substitutions of the f-rule (13) and the gt-rule (11) identical, we use inference rule **(VIII)** which allows arbitrary instantiations of pattern rules (i.e., in contrast to **(V)** and **(VII)**, here we impose no conditions on ρ).

(VIII) Instantiating μ

$$\frac{s\,\sigma_s^n\,\mu_s \;\hookrightarrow\; t\,\sigma_t^n\,\mu_t}{s\,\sigma_s^n\,(\mu_s\rho) \;\hookrightarrow\; t\,\sigma_t^n\,(\mu_t\rho)}$$

Applying inference rule **(VIII)** to Rule (13) with $\rho = [x'/\mathsf{s}(x), y'/\mathsf{0}]$ yields

$$\begin{aligned} &\mathsf{f}(\mathsf{tt}, \mathsf{s}(x'), \mathsf{s}(y')) \; [x'/\mathsf{s}(x'), y'/\mathsf{s}(y')]^n \; [x'/\mathsf{s}(x), y'/\mathsf{0}] \\ \hookrightarrow\; &\mathsf{f}(\mathsf{gt}(\mathsf{s}(x'), \mathsf{s}(y')), \mathsf{dbl}(\mathsf{s}(x')), \mathsf{s}^2(y')) \; [x'/\mathsf{s}(x'), y'/\mathsf{s}(y')]^n \; [x'/\mathsf{s}(x), y'/\mathsf{0}]. \end{aligned} \tag{14}$$

By **(VI)**, now one can narrow (14) with the gt-rule (11) which yields

$$\begin{aligned} &\mathsf{f}(\mathsf{tt}, \mathsf{s}(x'), \mathsf{s}(y')) \; [x'/\mathsf{s}(x'), y'/\mathsf{s}(y')]^n \; [x'/\mathsf{s}(x), y'/\mathsf{0}] \\ \hookrightarrow\; &\mathsf{f}(\mathsf{tt}, \mathsf{dbl}(\mathsf{s}(x')), \mathsf{s}^2(y')) \; [x'/\mathsf{s}(x'), y'/\mathsf{s}(y')]^n \; [x'/\mathsf{s}(x), y'/\mathsf{0}]. \end{aligned} \tag{15}$$

So to narrow a pattern rule with another one, we require identical pattern substitutions. Moreover, we only allow narrowing of the *base term* (i.e., the narrowing rule **(VI)** does not modify terms in the ranges of the pattern substitutions). In contrast, rewriting with ordinary rules is also allowed in the pattern substitutions and moreover, here the two pattern terms in the pattern rule may also have different pattern substitutions.

(IX) Rewriting

$$\frac{p \;\hookrightarrow\; t\,\sigma^n\,\mu}{p \;\hookrightarrow\; t'\,\sigma'^n\,\mu'} \quad \begin{array}{l}\text{if } t \to_{\mathcal{R}}^* t',\ \forall x \in \mathcal{V} : x\sigma \to_{\mathcal{R}}^* x\sigma', \\ \text{and } \forall x \in \mathcal{V} : x\mu \to_{\mathcal{R}}^* x\mu'\end{array}$$

While no rewriting is possible for the terms in the ranges of the pattern substitutions of (15), one can rewrite the base term using the dbl-rule:

$$\begin{aligned} &\mathsf{f}(\mathsf{tt}, \mathsf{s}(x'), \mathsf{s}(y')) \; [x'/\mathsf{s}(x'), y'/\mathsf{s}(y')]^n \; [x'/\mathsf{s}(x), y'/\mathsf{0}] \\ \hookrightarrow\; &\mathsf{f}(\mathsf{tt}, \mathsf{times}(\mathsf{s}^2(\mathsf{0}), \mathsf{s}(x')), \mathsf{s}^2(y')) \; [x'/\mathsf{s}(x'), y'/\mathsf{s}(y')]^n \; [x'/\mathsf{s}(x), y'/\mathsf{0}] \end{aligned} \tag{16}$$

To continue our example further, we now want to narrow the above f-rule (16) with the pattern rule (4) for times. To make the narrowing rule **(VI)** applicable, the base term of (4)'s left-hand side must occur in (16) and all four pattern terms in the rules must have the same pattern substitutions. Thus, one first has to transform the pattern rules by the equivalence rule **(IV)** (using Lemmas 4 and 6) and instantiations (using **(V)**, **(VII)**, and **(VIII)**). After the narrowing, one can simplify the resulting pattern rule by rewriting (Rule **(IX)**) and by removing irrelevant parts of substitutions (Rule **(IV)** using Lemma 6), which yields

$$\begin{aligned} &\mathsf{f}(\mathsf{tt}, \mathsf{s}(x'), \mathsf{s}(y')) \; [x'/\mathsf{s}(x'), y'/\mathsf{s}(y')]^n \; [x'/\mathsf{s}(x), y'/\mathsf{0}] \\ \hookrightarrow\; &\mathsf{f}(\mathsf{tt}, \mathsf{s}^2(z), \mathsf{s}^2(y')) \; [y'/\mathsf{s}(y'), z/\mathsf{s}^2(z)]^n \; [y'/\mathsf{0}, z/\mathsf{times}(\mathsf{s}^2(\mathsf{0}), \mathsf{s}(x))]. \end{aligned} \tag{17}$$

The following theorem shows that all our inference rules are sound.

Theorem 7 (Soundness of Inference Rules). *For all inference rules* **(I)** - **(IX)** *of the form* $\frac{p_1 \hookrightarrow q_1 \quad \dots \quad p_k \hookrightarrow q_k}{p \hookrightarrow q}$, *if all pattern rules* $p_1 \hookrightarrow q_1, \dots, p_k \hookrightarrow q_k$ *are correct w.r.t. a TRS* $\mathcal{R}$, *then the pattern rule* $p \hookrightarrow q$ *is also correct w.r.t.* $\mathcal{R}$.

Proof. Soundness of Rule **(I)** is trivial. Soundness of Rule **(II)** is proved by induction on n. For $n = 0$, we have $s\,\sigma^0 = s \to_{\mathcal{R}}^+ t = t\,\theta^0$, since $s\,\varnothing^n\,\varnothing \hookrightarrow t\,\varnothing^n\,\varnothing$ is correct w.r.t. $\mathcal{R}$. For $n > 0$, we obtain $s\,\sigma^n \to_{\mathcal{R}}^+ t\,\theta^{n-1}\,\sigma$ by the induction hypothesis. Since θ and σ commute, we have $t\,\theta^{n-1}\,\sigma = t\sigma\,\theta^{n-1} = s\,\theta^n \to_{\mathcal{R}}^+ t\,\theta^n$.

Soundness of Rule **(III)** is also proved by induction on n. For $n = 0$, we have $s\,\sigma^0 = s \to_{\mathcal{R}}^+ t = t[z]_\pi\,[z/t|_\pi] = t[z]_\pi\,(\sigma \cup [z/t[z]_\pi])^0\,[z/t|_\pi]$. For $n > 0$, we obtain

$$
\begin{array}{rll}
s\,\sigma^n & = s\,\sigma^{n-1}\,\sigma & \\
& \to_{\mathcal{R}}^+ t[z]_\pi\,(\sigma \cup [z/t[z]_\pi])^{n-1}\,[z/t|_\pi]\,\sigma & \text{by induction hypothesis} \\
& = t[z]_\pi\,(\sigma \cup [z/t[z]_\pi])^{n-1}\,(\sigma \cup [z/t|_\pi\sigma]) & \text{since } z \notin \mathrm{dom}(\sigma) \\
& = t[z]_\pi\,(\sigma \cup [z/t[z]_\pi])^{n-1}\,(\sigma \cup [z/s]) & \\
& \to_{\mathcal{R}}^+ t[z]_\pi\,(\sigma \cup [z/t[z]_\pi])^{n-1}\,(\sigma \cup [z/t]) & \\
& = t[z]_\pi\,(\sigma \cup [z/t[z]_\pi])^{n-1}\,(\sigma \cup [z/t[z]_\pi])\,[z/t|_\pi] & \text{since } z \notin \mathrm{range}(\sigma) \\
& = t[z]_\pi\,(\sigma \cup [z/t[z]_\pi])^{n}\,[z/t|_\pi] &
\end{array}
$$

Rule **(IV)** is trivially sound. For Rule **(V)**, note that correctness of $s\,\sigma_s^n\,\mu_s \hookrightarrow t\,\sigma_t^n\,\mu_t$ also implies correctness of $s\,\sigma_s^n\,(\mu_s\rho) \hookrightarrow t\,\sigma_t^n\,(\mu_t\rho)$. But we have

$$
\begin{array}{rll}
s\,\sigma_s^n\,(\mu_s\,\rho) & = s\,\sigma_s^n\,\rho\,\mu_{s_\rho} & \text{since } \mathcal{V}(\rho) \cap \mathrm{dom}(\mu_s) = \varnothing \\
& = (s\rho)\,(\sigma_s)_\rho^n\,(\mu_s)_\rho & \text{since } \mathcal{V}(\rho) \cap \mathrm{dom}(\sigma_s) = \varnothing.
\end{array}
$$

Similarly, $t\,\sigma_t^n\,(\mu_t\,\rho) = (t\rho)\,(\sigma_t)_\rho^n\,(\mu_t)_\rho$, which implies soundness of Rule **(V)**.

Soundness of Rule **(VI)** is trivial. For soundness of Rule **(VII)**, correctness of $s\,\sigma_s^n\,\mu_s \hookrightarrow t\,\sigma_t^n\,\mu_t$ also implies correctness of $s\,\sigma_s^n\,(\mu_s\,\rho^n) \hookrightarrow t\,\sigma_t^n\,(\mu_t\,\rho^n)$. As ρ commutes with σ_s, μ_s, σ_t, μ_t, this is equivalent to $s\,(\sigma_s\,\rho)^n\,\mu_s \hookrightarrow t\,(\sigma_t\,\rho)^n\,\mu_t$. Soundness of Rules **(VIII)** and **(IX)** is again straightforward. □

2.4 Detecting Non-termination

Thm. 8 introduces a criterion to detect pattern rules that directly lead to non-termination. Hence, whenever we have inferred a new pattern rule that satisfies this criterion, we can conclude non-termination of our TRS.

For a pattern rule $s\,\sigma^n\,\mu \hookrightarrow t\,\sigma_t^n\,\mu_t$, we check whether the pattern substitutions of the right-hand side are specializations of the pattern substitutions of the left-hand side. More precisely, there must be an $m \in \mathbb{N}$ such that $\sigma_t = \sigma^m\,\sigma'$ and $\mu_t = \mu\,\mu'$ for some σ' and μ', where σ' commutes with σ and μ. Then one only has to check whether there is a $b \in \mathbb{N}$ such that $s\,\sigma^b$ is equal to some subterm of t.

Theorem 8 (Detecting Non-termination). *Let $s\,\sigma^n\,\mu \hookrightarrow t\,\sigma_t^n\,\mu_t$ be correct w.r.t. a TRS $\mathcal{R}$ and let there be an $m \in \mathbb{N}$ such that $\sigma_t = \sigma^m\,\sigma'$ and $\mu_t = \mu\,\mu'$ for some substitutions σ' and μ', where σ' commutes with both σ and μ. If there is a $\pi \in \mathcal{P}os(t)$ and some $b \in \mathbb{N}$ such that $s\,\sigma^b = t|_\pi$, then $\mathcal{R}$ is non-terminating.*

Proof. We show that for all $n \in \mathbb{N}$, the term $s\,\sigma^n\,\mu$ rewrites to a term containing an instance of $s\,\sigma^{m \cdot n + b}\mu$. By repeating these rewrite steps on this subterm, we obtain an infinite rewrite sequence. Here, $\trianglerighteq$ denotes the superterm relation.

$$\begin{aligned} s\,\sigma^n\,\mu &\to_{\mathcal{R}}^{+} t\,\sigma_t^n\,\mu_t && \text{since } s\,\sigma^n\,\mu \hookrightarrow t\,\sigma_t^n\,\mu_t \text{ is correct} \\ &\trianglerighteq\ t|_\pi\,\sigma_t^n\,\mu_t \\ &=\ s\,\sigma^b\,\sigma_t^n\,\mu_t \\ &=\ s\,\sigma^b\,(\sigma^m\,\sigma')^n\,(\mu\,\mu') \\ &=\ s\,\sigma^{m\cdot n+b}\,\mu\,\sigma'^n\,\mu' && \text{since } \sigma' \text{ commutes with both } \sigma \text{ and } \mu \qquad \square \end{aligned}$$

To apply Thm. 8 to the pattern rule (17) obtained in our example, we have to transform the rule such that the pattern substitutions on the right-hand side become specializations of the pattern substitutions on the left-hand side. Thus, we use a domain renaming for the right-hand side to rename the variable z to x' (using Rule **(IV)** with Lemma 4). Moreover, we would like to get rid of the closing substitution $[x'/\mathsf{s}(x)]$ on the left-hand side. To this end, we first apply $[x/x']$ to the whole pattern rule (using inference rule **(VIII)**) and remove irrelevant parts of the pattern substitutions (Rule **(IV)** with Lemma 6), which yields

$$\begin{aligned} &\mathsf{f}(\mathsf{tt},\mathsf{s}(x'),\mathsf{s}(y'))\quad [x'/\mathsf{s}(x'),y'/\mathsf{s}(y')]^n\quad [x'/\mathsf{s}(x'),y'/\mathsf{0}] \\ \hookrightarrow\ &\mathsf{f}(\mathsf{tt},\mathsf{s}^2(x'),\mathsf{s}^2(y'))\ [x'/\mathsf{s}^2(x'),y'/\mathsf{s}(y')]^n\ [x'/\mathsf{times}(\mathsf{s}^2(\mathsf{0}),\mathsf{s}(x')),y'/\mathsf{0}]. \end{aligned} \tag{18}$$

Now the closing substitution $[x'/\mathsf{s}(x')]$ on the left-hand side of the rule can be moved from the closing substitution to the base term. This is stated by the following lemma, which can be used in addition to Lemmas 4 and 6 in order to transform pattern terms to equivalent other pattern terms in inference rule **(IV)**.

Lemma 9 (Equivalence by Simplifying μ). *Let $p = t\,\sigma^n\,\mu$ be a pattern term and let $\mu = \mu_1\,\mu_2$ where μ_1 commutes with σ. Then p is equivalent to $(t\,\mu_1)\,\sigma^n\,\mu_2$.*

Proof. For any n, $t\,\sigma^n\,\mu = t\,\sigma^n\,\mu_1\mu_2 = t\mu_1\,\sigma^n\,\mu_2$, as μ_1 commutes with σ. □

The closing substitution μ of (18)'s left-hand side has the form $\mu = \mu_1\,\mu_2$ for $\mu_1 = [x'/\mathsf{s}(x')]$ and $\mu_2 = [y'/\mathsf{0}]$. Since μ_1 commutes with $\sigma = [x'/\mathsf{s}(x'),y'/\mathsf{s}(y')]$, by inference rule **(IV)** and Lemma 9, we can replace the left-hand side of (18) by the equivalent pattern term $\mathsf{f}(\mathsf{tt},\mathsf{s}^2(x'),\mathsf{s}(y'))\ [x'/\mathsf{s}(x'),y'/\mathsf{s}(y')]^n\ [y'/\mathsf{0}]$.

Moreover, by rewriting $\mathsf{times}(\mathsf{s}^2(\mathsf{0}),\mathsf{s}(x'))$ on the right-hand side using Rule **(IX)**, the right-hand side is transformed to $\mathsf{f}(\mathsf{tt},\mathsf{s}^2(x'),\mathsf{s}^2(y'))\ [x'/\mathsf{s}^2(x'),y'/\mathsf{s}(y')]^n$ $[x'/\mathsf{s}^2(\mathsf{times}(\mathsf{s}^2(\mathsf{0}),x')),y'/\mathsf{0}]$. So now its closing substitution μ' has the form $\mu' = \mu'_1\,\mu'_2$ for $\mu'_1 = [x'/\mathsf{s}(x')]$ and $\mu'_2 = [x'/\mathsf{s}(\mathsf{times}(\mathsf{s}^2(\mathsf{0}),x')),y'/\mathsf{0}]$. Since μ'_1 commutes with the pumping substitution $\sigma' = [x'/\mathsf{s}^2(x'),y'/\mathsf{s}(y')]$, by applying inference rule **(IV)** and Lemma 9 also on the right-hand side, we get

$$\begin{aligned} &\mathsf{f}(\mathsf{tt},\mathsf{s}^2(x'),\mathsf{s}(y'))\quad [x'/\mathsf{s}(x'),y'/\mathsf{s}(y')]^n\quad [y'/\mathsf{0}] \\ \hookrightarrow\ &\mathsf{f}(\mathsf{tt},\mathsf{s}^3(x'),\mathsf{s}^2(y'))\ [x'/\mathsf{s}^2(x'),y'/\mathsf{s}(y')]^n\ [x'/\mathsf{s}(\mathsf{times}(\mathsf{s}^2(\mathsf{0}),x')),y'/\mathsf{0}]. \end{aligned} \tag{19}$$

The resulting rule (19) satisfies the conditions of Thm. 8, i.e., one can directly detect its non-termination. It has the form $s\,\sigma^n\,\mu \hookrightarrow t\,\sigma_t^n\,\mu_t$ with

$\sigma = [x'/\mathsf{s}(x'), y'/\mathsf{s}(y')]$ and $\mu = [y'/\mathsf{0}]$, where $\sigma_t = \sigma\,\sigma'$ for $\sigma' = [x'/\mathsf{s}(x')]$ and $\mu_t = \mu\,\mu'$ for $\mu' = [x'/\mathsf{s}(\mathsf{times}(\mathsf{s}^2(\mathsf{0}), x'))]$. Clearly σ' commutes with σ and μ and moreover, $s\sigma = t$. Thus, non-termination of the TRS in Sect. 1 is proved.

Note that with our inference rules and the criterion of Thm. 8, one can also prove non-termination of any looping TRS $\mathcal{R}$. The reason is that then there is also a loop $s \to_{\mathcal{R}}^{+} C[s\mu]$ where the first rewrite step is on the root position. By translating the rules of the TRS to pattern rules (Rule **(I)**) and by performing instantiation (Rule **(V)**) followed by narrowing or rewriting (Rule **(VI)** or **(IX)**) repeatedly, we can also obtain a corresponding pattern rule $s\,\varnothing^n\,\varnothing \hookrightarrow C[s\mu]\,\varnothing^n\,\varnothing$. To detect its non-termination by Thm. 8, we replace the closing substitution $\varnothing$ by μ (using Rule **(VIII)**) which yields $s\,\varnothing^n\,\mu \hookrightarrow C[s\mu]\,\varnothing^n\,\mu$. Simplifying the closing substitution on the left-hand side (Rule **(IV)** with Lemma 9) yields $(s\mu)\,\varnothing^n\,\varnothing \hookrightarrow C[s\mu]\,\varnothing^n\,\mu$. Since the closing substitution μ on the right-hand side is a specialization of the closing substitution $\varnothing$ on the left-hand side and since $s\mu$ is equal to a subterm of $C[s\mu]$, Thm. 8 now detects non-termination.

3 A Strategy to Prove Non-termination Automatically

The inference rules in Sect. 2 constitute a powerful calculus to prove non-termination. We now present a strategy for their automated application which turned out to be successful in our implementation in the tool AProVE, cf. Sect. 4.

The strategy first transforms all rules of the TRS[4] into pattern rules using Rule **(I)** and if possible, one uses Rules **(II)** and **(III)** afterwards to obtain pattern rules with non-empty pattern substitutions. Then for every pattern rule $p \hookrightarrow q$, one repeatedly tries to rewrite its right-hand side (Rule **(IX)**) or to narrow it with every pattern rule $p' \hookrightarrow q'$ (see below). Whenever a new pattern rule is obtained, one checks whether it satisfies the non-termination criterion of Thm. 8.[5] In this case, the procedure stops and non-termination has been proved.

Before trying to narrow $p \hookrightarrow q$ with $p' \hookrightarrow q'$ at some $\pi \in \mathcal{P}os(\mathrm{base}(q))$, to avoid conflicting instantiations of variables, one uses domain renamings to ensure that $\mathrm{dv}(p)$, $\mathrm{dv}(q)$, $\mathrm{dv}(p')$, and $\mathrm{dv}(q')$ are pairwise disjoint (Rule **(IV)** with Lemma 4). Moreover, pattern rules are made variable-disjoint (using Rule **(V)**). Then the strategy proceeds by the following steps to make the narrowing rule **(VI)** applicable. After presenting the strategy, we illustrate it by an example.

1. Make $\mathrm{base}(q)|_\pi$ equal to $\mathrm{base}(p')$: If $\mathrm{base}(q)|_\pi$ and $\mathrm{base}(p')$ do not unify, then abort with failure. If $\mathrm{base}(q)|_\pi = \mathrm{base}(p')$, then go to Step 2. Otherwise, let $\theta = \mathrm{mgu}(\mathrm{base}(q)|_\pi, \mathrm{base}(p'))$, let $x \in \mathrm{dom}(\theta)$, and let $s = \theta(x)$. W.l.o.g. we assume $x \in \mathcal{V}(p')$ (the case where $x \in \mathcal{V}(q)$ works analogously).

[4] It is preferable to check non-termination within the *dependency pair framework* [5,7,8]. In this way, one can automatically decompose the TRS into parts where termination can easily be proved and into parts which can potentially cause non-termination.

[5] To this end, one tries to transform the pattern rule using Rules **(IV)** and **(VIII)** such that the pattern substitutions on the right-hand sides become specializations of the corresponding pattern substitutions on the left-hand sides.

(a) If $x \notin \text{dv}(p')$ and $s \notin \text{dv}(p')$, then let s' result from s by renaming all variables from $\text{dv}(p')$ occurring in s by pairwise different fresh variables. Instantiate $p' \hookrightarrow q'$ with $\rho = [x/s']$ (Rule **(V)**) and go back to Step 1.

(b) If $x \notin \text{dv}(p')$ and $s \in \text{dv}(p')$, then use Rule **(VII)** to add x to the domain of p''s pumping substitution, such that it operates on x as it operates on s. To make Rule **(VII)** applicable, some pre-processing with Rules **(VIII)** and **(IV)** may be required. Then go back to Step 1 (resp. to case (c)). The case where $x \in \text{dv}(p')$ and $s \in \mathcal{V}(p') \setminus \text{dv}(p')$ is analogous.

(c) If both $x, s \in \text{dv}(p')$ and $[x/s]$ commutes with p''s pumping substitution, then apply **(VIII)** on $p' \hookrightarrow q'$ such that p''s closing substitution gets the form $[x/s]\,\mu$ for some μ. Then, move $[x/s]$ from p''s closing substitution to p''s base term with Rule **(IV)** (using Lemma 9) and go to Step 1.

(d) If $x \in \text{dv}(p')$ and $s \in \mathcal{V} \setminus \mathcal{V}(p')$, then apply Rule **(IV)** (using Lemma 4) with the domain renaming $[x/s]$ on $p' \hookrightarrow q'$ and go back to Step 1.

(e) Otherwise, abort with failure.

2. Make the pumping substitutions of p, q, p', and q' equal (without changing $\text{base}(q), \text{base}(p')$): resolve all conflicts using Rules **(VII)** and **(IV)**.
3. Make the closing substitutions of p, q, p', q' equal (without changing pumping substitutions or $\text{base}(q), \text{base}(p')$): resolve conflicts by **(VIII)** and **(IV)**.
4. Apply narrowing according to Rule **(VI)**.

To illustrate the strategy, consider the TRS with the plus-rules of Sect. 1 and

$$\mathsf{f}(\mathsf{tt}, x) \to \mathsf{f}(\mathsf{isNat}(x), \mathsf{plus}(x, x)), \qquad \mathsf{isNat}(\mathsf{0}) \to \mathsf{tt}, \qquad \mathsf{isNat}(\mathsf{s}(y)) \to \mathsf{isNat}(y).$$

After creating pattern rules for f, isNat, and plus, we narrow the recursive isNat- and plus-rules with the non-recursive ones. For plus, this results in

$$\mathsf{plus}(x, \mathsf{s}(y'))\,[y'/\mathsf{s}(y')]^n\,[y'/\mathsf{0}] \hookrightarrow \mathsf{s}(x')\,[x'/\mathsf{s}(x')]^n\,[x'/x]. \tag{20}$$

Moreover, we use the resulting isNat-rule to narrow the f-rule, which yields

$$\mathsf{f}(\mathsf{tt}, \mathsf{s}(y))\,[y/\mathsf{s}(y)]^n\,[y/\mathsf{0}] \hookrightarrow \mathsf{f}(\mathsf{tt}, \mathsf{plus}(\mathsf{s}(y), \mathsf{s}(y)))\,[y/\mathsf{s}(y)]^n\,[y/\mathsf{0}]. \tag{21}$$

Now our goal is to narrow the f-rule (21) with the plus-rule (20). We begin with Step 1 in the strategy. The mgu of $\mathsf{plus}(\mathsf{s}(y), \mathsf{s}(y))$ (in (21)'s right-hand side q) and $\mathsf{plus}(x, \mathsf{s}(y'))$ (in (20)'s left-hand side p') is $\theta = [y'/y, x/\mathsf{s}(y)]$. Let us first regard the variable y'. Since $y' \in \text{dv}(p')$ and $y \in \mathcal{V} \setminus \mathcal{V}(p')$, we are in Case (d). Thus, we apply the domain renaming $[y'/y]$ to (20) (with Rule **(IV)**) and obtain

$$\mathsf{plus}(x, \mathsf{s}(y))\,[y/\mathsf{s}(y)]^n\,[y/\mathsf{0}] \hookrightarrow \mathsf{s}(x')\,[x'/\mathsf{s}(x')]^n\,[x'/x]. \tag{22}$$

Now $\theta = \text{mgu}(\mathsf{plus}(\mathsf{s}(y), \mathsf{s}(y)), \mathsf{plus}(x, \mathsf{s}(y))) = [x/\mathsf{s}(y)]$. Since x is no domain variable of (22)'s left-hand side and $\mathsf{s}(y) \notin \mathcal{V}$, we are in Case (a). Thus, we apply **(V)** with $\rho = [x/\mathsf{s}(z)]$ for a fresh $z \in \mathcal{V}$. After simplification with **(IV)**, we get

$$\mathsf{plus}(\mathsf{s}(z), \mathsf{s}(y))\,[y/\mathsf{s}(y)]^n\,[y/\mathsf{0}] \hookrightarrow \mathsf{s}(x')\,[x'/\mathsf{s}(x')]^n\,[x'/\mathsf{s}(z)]. \tag{23}$$

Now $\theta = \text{mgu}(\mathsf{plus}(\mathsf{s}(y), \mathsf{s}(y)), \mathsf{plus}(\mathsf{s}(z), \mathsf{s}(y))) = [z/y]$. Since z is no domain variable of (23)'s left-hand side, but y is, we are in Case (b). Hence, our goal is to extend the pumping substitution $[y/\mathsf{s}(y)]$ to operate on z as on y (i.e., we want to add $[z/\mathsf{s}(z)]$). To make Rule **(VII)** applicable, we have to remove the closing substitution $[x'/\mathsf{s}(z)]$ on (23)'s right-hand side which does not commute with $[z/\mathsf{s}(z)]$. To this end, we instantiate (23)'s closing substitutions with $[z/x']$ (Rule **(VIII)**) and simplify both sides of (23) using Rule **(IV)** with Lemmas 9 and 6.

$$\mathsf{plus}(\mathsf{s}(x'), \mathsf{s}(y))\,[y/\mathsf{s}(y)]^n\,[y/\mathsf{0}] \hookrightarrow \mathsf{s}^2(x')\,[x'/\mathsf{s}(x')]^n\,\varnothing \tag{24}$$

Now $\theta = \text{mgu}(\mathsf{plus}(\mathsf{s}(y), \mathsf{s}(y)), \mathsf{plus}(\mathsf{s}(x'), \mathsf{s}(y))) = [x'/y]$ for the non-domain variable x' and the domain variable y. Thus, we can proceed according to Case (b) and add $[x'/\mathsf{s}(x')]$ to the pumping substitutions of (24) using Rule **(VII)**.

$$\mathsf{plus}(\mathsf{s}(x'), \mathsf{s}(y))\,[x'/\mathsf{s}(x'), y/\mathsf{s}(y)]^n\,[y/\mathsf{0}] \hookrightarrow \mathsf{s}^2(x')\,[x'/\mathsf{s}^2(x')]^n\,\varnothing \tag{25}$$

We still have $\theta = \text{mgu}(\mathsf{plus}(\mathsf{s}(y), \mathsf{s}(y)), \mathsf{plus}(\mathsf{s}(x'), \mathsf{s}(y))) = [x'/y]$. But now both x', y are domain variables of (25)'s left-hand side, i.e., we are in Case (c). Indeed, now $[x'/y]$ commutes with the pumping substitution $[x'/\mathsf{s}(x'), y/\mathsf{s}(y)]$. So we instantiate the closing substitutions of (25) with $\rho = [x'/\mathsf{0}]$ (Rule **(VIII)**). Then the closing substitution $[y/\mathsf{0}, x'/\mathsf{0}]$ of (25)'s left-hand side has the form $[x'/y][y/\mathsf{0}]$ and hence, Rule **(IV)** with Lemma 9 yields

$$\mathsf{plus}(\mathsf{s}(y), \mathsf{s}(y))\,[x'/\mathsf{s}(x'), y/\mathsf{s}(y)]^n\,[y/\mathsf{0}] \hookrightarrow \mathsf{s}^2(x')\,[x'/\mathsf{s}^2(x')]^n\,[x'/\mathsf{0}]. \tag{26}$$

Thus, now the term $\mathsf{plus}(\mathsf{s}(y), \mathsf{s}(y))$ from the right-hand side of (21) also occurs on the left-hand side of (26), i.e., Step 1 is finished. In Step 2 of the strategy, we have to make the pumping substitutions of (21) and (26) equal. By Rule **(IV)** with Lemma 6 we first remove the irrelevant substitution $[x'/\mathsf{s}(x')]$ from the left-hand side of (26) and then extend the pumping substitutions by new irrelevant parts such that they all become $[x'/\mathsf{s}^2(x'), y/\mathsf{s}(y)]$. Similarly, in Step 3 of the strategy, all closing substitutions are extended to $[x'/\mathsf{0}, y/\mathsf{0}]$ by Rule **(IV)** with Lemma 6. Now narrowing the f- with the plus-rule (by Rule **(VI)**) and subsequent removal of irrelevant substitutions (by Rule **(IV)** with Lemma 6) yields

$$\mathsf{f}(\mathsf{tt}, \mathsf{s}(y))\,[y/\mathsf{s}(y)]^n\,[y/\mathsf{0}] \hookrightarrow \mathsf{f}(\mathsf{tt}, \mathsf{s}^2(x'))\,[x'/\mathsf{s}^2(x')]^n\,[x'/\mathsf{0}]. \tag{27}$$

Hence, we now have to check whether (27) leads to non-termination due to Thm. 8. As in Footnote 5, to this end we apply a domain renaming $[x'/y]$ to (27)'s right-hand side in order to turn the pattern substitutions on the right-hand side into a specialization of the pattern substitutions on the left-hand side.

$$\mathsf{f}(\mathsf{tt}, \mathsf{s}(y))\,[y/\mathsf{s}(y)]^n\,[y/\mathsf{0}] \hookrightarrow \mathsf{f}(\mathsf{tt}, \mathsf{s}^2(y))\,[y/\mathsf{s}^2(y)]^n\,[y/\mathsf{0}]. \tag{28}$$

Rule (28) satisfies the criterion of Thm. 8. If σ is the pumping substitution $[y/\mathsf{s}(y)]$ of (28)'s left-hand side, then (28)'s right-hand side has the pumping substitution $\sigma\,\sigma$. Moreover, if s resp. t are the base terms of the two sides, then $s\sigma = t$. Thus, non-termination of the original (non-looping) TRS is proved.

4 Evaluation and Conclusion

We introduced a new technique to prove non-termination of possibly non-looping TRSs automatically. To this end, we adapted an idea of [10] from string to term rewriting and introduced *pattern rules* which represent a whole class of rewrite sequences. Afterwards, we presented 9 inference rules to deduce new pattern rules, a strategy for the application of these rules, and a criterion to detect non-terminating pattern rules. In this way, one can now repeatedly generate pattern rules until one obtains a rule which is detected to be non-terminating.

We implemented our contributions in the tool AProVE [6] and compared the new version AProVE-NL (for non-loop) with the previous version AProVE '11 and 3 other powerful tools for non-termination of TRSs (NTI [11], $\mathsf{T_TT_2}$ [9], VMTL [12]). We ran the tools on the 1438 TRSs of the *Termination Problem Data Base* (*TPDB*) used in the annual *International Termination Competition*.[6] In the table, we consider those 241 TRSs of the TPDB where at least one tool proved non-termination. Moreover, we also tested the tools on 58 typical non-looping non-terminating TRSs obtained from actual programs and other sources (*"nl"*). We used a time-out of 1 minute for each example. "N" indicates how often **N**on-termination was proved and "R" gives the average **R**untime in seconds for each example. Thus, AProVE-NL could solve 75.9 % of the non-looping examples without compromising its power on looping examples, whereas the other tools cannot handle non-looping non-termination. To access our implementation via a web interface and for further details on our experiments, we refer to [1].

	TPDB		*nl*	
	N	**R**	**N**	**R**
AProVE-NL	232	6.6	44	5.2
AProVE '11	228	6.6	0	60.0
NTI	214	7.3	0	60.0
$\mathsf{T_TT_2}$	194	2.5	0	10.4
VMTL	95	16.5	0	42.8

Future work will be concerned with (i) improving our strategy for applying inference rules and with (ii) extending the notion of pattern rules. To motivate (i), we compared AProVE-NL with the tools Knocked for Loops (KFL) [15], Matchbox [13], and nonloop [10] for non-termination of *string* rewriting on the 1316 SRSs of the TPDB. The table regards those 156 SRSs where at least one tool proved non-termination. Only AProVE-NL and nonloop handle non-looping non-terminating SRSs, and AProVE-NL succeeds whenever nonloop succeeds. However, some looping SRSs are found by other tools, but not by our current strategy which mainly focuses on *term* rewriting.

	N	**R**
KFL	147	6.2
AProVE-NL	120	19.0
Matchbox	111	22.0
AProVE '11	97	31.1
nonloop	95	26.3
NTI	67	37.1
$\mathsf{T_TT_2}$	24	51.4
VMTL	0	56.8

For (ii), while our approach is "complete" for looping TRSs, there are TRSs whose non-termination cannot be proved with our inference rules. An example is the TRS with rules for isNat, double, and $\mathsf{f}(\mathsf{tt},\mathsf{tt},x,\mathsf{s}(y)) \to \mathsf{f}(\mathsf{isNat}(x),\mathsf{isNat}(y),\mathsf{s}(x),\mathsf{double}(\mathsf{s}(y)))$. Here, one needs the rule $\mathsf{f}(\mathsf{tt},\mathsf{tt},x,\mathsf{s}(y))\,[x/\mathsf{s}(x)]^n\,[y/\mathsf{s}(y)]^m\,[x/\mathsf{0},y/\mathsf{0}] \hookrightarrow \mathsf{f}(\mathsf{tt},\mathsf{tt},\mathsf{s}(x),\mathsf{s}(\mathsf{s}(y)))\,[x/\mathsf{s}(x)]^n\,[y/\mathsf{s}(\mathsf{s}(y))]^m\,[x/\mathsf{0},y/\mathsf{0}]$ with *two parameters* n and m, which goes beyond our current notion of pattern rules.

[6] See `http://termination-portal.org/wiki/Termination_Competition`

References

1. http://aprove.informatik.rwth-aachen.de/eval/NonLooping/
2. Brockschmidt, M., Ströder, T., Otto, C., Giesl, J.: Automated detection of non-termination and NullPointerExceptions for Java Bytecode. In: Damiani, F., Gurov, D. (eds.) FoVeOOS 2011. LNCS. Springer, Heidelberg (to appear, 2012) Available from [1]
3. Geser, A., Zantema, H.: Non-looping string rewriting. Informatique Théorique et Applications 33(3), 279–302 (1999)
4. Geser, A., Hofbauer, D., Waldmann, J.: Termination proofs for string rewriting systems via inverse match-bounds. J. Automated Reasoning 34(4), 365–385 (2005)
5. Giesl, J., Thiemann, R., Schneider-Kamp, P.: Proving and Disproving Termination of Higher-Order Functions. In: Gramlich, B. (ed.) FroCos 2005. LNCS (LNAI), vol. 3717, pp. 216–231. Springer, Heidelberg (2005)
6. Giesl, J., Schneider-Kamp, P., Thiemann, R.: AProVE 1.2: Automatic Termination Proofs in the Dependency Pair Framework. In: Furbach, U., Shankar, N. (eds.) IJCAR 2006. LNCS (LNAI), vol. 4130, pp. 281–286. Springer, Heidelberg (2006)
7. Giesl, J., Thiemann, R., Schneider-Kamp, P., Falke, S.: Mechanizing and improving dependency pairs. Journal of Automated Reasoning 37(3), 155–203 (2006)
8. Hirokawa, N., Middeldorp, A.: Automating the dependency pair method. Information and Computation 199(1-2), 172–199 (2005)
9. Korp, M., Sternagel, C., Zankl, H., Middeldorp, A.: Tyrolean Termination Tool 2. In: Treinen, R. (ed.) RTA 2009. LNCS, vol. 5595, pp. 295–304. Springer, Heidelberg (2009)
10. Oppelt, M.: Automatische Erkennung von Ableitungsmustern in nichtterminierenden Wortersetzungssystemen, Diploma Thesis, HTWK Leipzig, Germany (2008)
11. Payet, É.: Loop detection in term rewriting using the eliminating unfoldings. Theoretical Computer Science 403, 307–327 (2008)
12. Schernhammer, F., Gramlich, B.: VMTL–A Modular Termination Laboratory. In: Treinen, R. (ed.) RTA 2009. LNCS, vol. 5595, pp. 285–294. Springer, Heidelberg (2009)
13. Waldmann, J.: Matchbox: A Tool for Match-Bounded String Rewriting. In: van Oostrom, V. (ed.) RTA 2004. LNCS, vol. 3091, pp. 85–94. Springer, Heidelberg (2004)
14. Wang, Y., Sakai, M.: On non-looping term rewriting. In: WST 2006, pp. 17-21 (2006)
15. Zankl, H., Sternagel, C., Hofbauer, D., Middeldorp, A.: Finding and Certifying Loops. In: van Leeuwen, J., Muscholl, A., Peleg, D., Pokorný, J., Rumpe, B. (eds.) SOFSEM 2010. LNCS, vol. 5901, pp. 755–766. Springer, Heidelberg (2010)
16. Zantema, H.: Termination of string rewriting proved automatically. Journal of Automated Reasoning 34, 105–139 (2005)

Rewriting Induction + Linear Arithmetic = Decision Procedure*

Stephan Falke[1] and Deepak Kapur[2]

[1] Inst. for Theoretical Computer Science, KIT, Germany
stephan.falke@kit.edu
[2] Dept. of Computer Science, University of New Mexico, USA
kapur@cs.unm.edu

Abstract. This paper presents new results on the decidability of inductive validity of conjectures. For these results, a class of term rewrite systems (TRSs) with built-in linear integer arithmetic is introduced and it is shown how these TRSs can be used in the context of inductive theorem proving. The proof method developed for inductive theorem proving couples (implicit) inductive reasoning with a decision procedure for the theory of linear integer arithmetic with (free) constructors. The effectiveness of the new decidability results on a large class of conjectures is demonstrated by an evaluation of the prototype implementation Sail2.

1 Introduction

Reasoning about the partial correctness of programs often requires proofs by induction, in particular for reasoning about recursive functions. There are two commonly used paradigms for inductive theorem proving: *explicit induction* and *implicit induction*. In explicit induction (see, e.g., [8,21,10]), an induction scheme is computed for each conjecture, and the subsequent reasoning is based on this induction scheme. Here, an induction scheme explicitly gives the base cases and the step cases. In implicit induction (see, e.g., [17,18,23,6,22]), the induction scheme is not constructed a priori but implicitly during the proof attempt.

Implicit induction is typically based on term rewriting. While ordinary term rewrite systems (TRSs) are a well-understood formalism for modelling algorithms, they lack in expressivity since they don't support built-in data structures such as integers. In the first part of this paper, an expressive class of TRSs (called $\mathbb{Z}$*-TRSs*) with built-in linear integer arithmetic is introduced. The semantics of integers can be utilized in the form of linear integer arithmetic constraints (LIA-*constraints*). Next, an inductive proof method for $\mathbb{Z}$-TRSs is developed. This method couples implicit induction with a decision procedure for the theory of linear integer arithmetic with the (free) constructors of the $\mathbb{Z}$-TRS.

While inductive proof methods can be automated, they do not provide a decision procedure since proof attempts may diverge, fail, or need intermediate

* This work was supported in part by the "Concept for the Future" of Karlsruhe Institute of Technology within the framework of the German Excellence Initiative and by NSF awards CCF-0729097 and CNS-0905222.

B. Gramlich, D. Miller, and U. Sattler (Eds.): IJCAR 2012, LNAI 7364, pp. 241–255, 2012.

lemmas. In program verification, a decision procedure that can be used as a "black box" is preferable since an interactive use of inductive reasoning methods is typically only possible by trained experts. The goal of the second part of this paper is to derive conditions on $\mathbb{Z}$-TRSs and conjectures under which the proof method can be used as a decision procedure, i.e., will always produce a proof or disproof. These conditions are based on properties of the rewrite rules in a $\mathbb{Z}$-TRS that can be pre-computed during parsing. As we show experimentally in this paper, checking whether a conjecture satisfies the conditions is easily possible and requires much less time than attempting a proof or disproof.

Work on identifying conditions under which (explicit) inductive theorem proving provides a decision procedure was initiated in [20,15,16]. These previous papers impose strong restrictions on both the TRSs and the conjectures. The functions defined by the TRS have to be given in such a way that any function f may only make recursive calls to the function f itself. Often, it is necessary to allow calls to other auxiliary functions or even mutually recursive definitions. Both of these possibilities are supported in this paper. All of [20,15,16] impose the restriction that the conjectures contain a subterm of the form $f(x_1, \ldots, x_n)$ for a defined function f and pairwise distinct variables $x_1, \ldots, x_n$. This term is then chosen for the construction of the induction scheme. In this paper, much like in [13], this restriction is relaxed by making it possible to base the induction proof on a subterm where the arguments are not necessarily pairwise distinct variables. The result in this paper are however much more general than [13] since that paper did not yet consider $\mathbb{Z}$-TRSs. As a result, many conjectures which could not be handled previously can now be decided.

The integration of decision procedures for the theory of linear arithmetic on natural numbers or integers into inductive reasoning has been previously considered in [9,19,3], with the main focus on contextual rewriting which integrates rewriting with decision procedures. The proof method developed in this paper is in general incomparable to these methods, though, since instead of the more complex contextual rewriting without constraints, regular constrained rewriting is employed. This gives rise to an elegant and intuitive proof method. Inductive theorem proving for TRSs with constraints has been investigated in [7]. That method, however, does not support LIA-constraints and is thus incomparable to the method presented below. Another method for inductive theorem proving with constrained TRSs has been presented in [24], but that paper is only available in Japanese, making it impossible for us to compare the methods.

All proofs missing from this paper can be found in the full version [14].

2 $\mathbb{Z}$-TRSs

This section introduces $\mathbb{Z}$-TRSs, a class of constrained TRSs that contains linear integer arithmetic as the built-in constraint theory. This class of TRSs can be seen as a simplified special case of the constrained equational rewrite systems (CERSs) from [12]. We assume familiarity with the notions and concepts from (many-sorted) term rewriting and refer to [4] for details.

For $\mathbb{Z}$-TRSs, built-in integers are modeled using the function symbols $\mathcal{F}_{\mathsf{LIA}} = \{0 : \to \mathtt{int},\ 1 : \to \mathtt{int},\ - : \mathtt{int} \to \mathtt{int},\ + : \mathtt{int} \times \mathtt{int} \to \mathtt{int}\}$ where we use a simplified notation for terms built using $+$ and $-$. Recall that $\mathbb{Z}$ is an Abelian group with unit 0 that is generated using the element 1. Integers thus satisfy the following properties $\mathcal{E}_{\mathsf{LIA}}$:

$$x + y \approx y + x \qquad x + (y + z) \approx (x + y) + z$$
$$x + 0 \approx x \qquad x + (-x) \approx 0$$

An *atomic* LIA-*constraint* has the form $t_1\ P\ t_2$ for a predicate symbol $P \in \{>, \geq, \simeq\}$ and terms $t_1, t_2 \in \mathcal{T}(\mathcal{F}_{\mathsf{LIA}}, \mathcal{V})$ (where $\mathcal{V}$ is a set of variables). The set of LIA-*constraints* is the closure of the set of atomic LIA-constraints under $\top$ (truth), $\neg$, and $\wedge$. The Boolean connectives $\vee$, $\Rightarrow$, and $\Leftrightarrow$ can be defined as usual. Also, LIA-constraints have the expected semantics. The main interest is in LIA-*satisfiability* and LIA-*validity*. Both of these properties are decidable.

For $\mathbb{Z}$-TRSs, $\mathcal{F}_{\mathsf{LIA}}$ is extended by a signature $\mathcal{F}$ of function symbols. In the following, *terms* denote members of $\mathcal{T}(\mathcal{F} \cup \mathcal{F}_{\mathsf{LIA}}, \mathcal{V})$ unless otherwise noted. A sequence $t_1, \ldots, t_n$ of terms is also denoted by t^*.

Definition 1 ($\mathbb{Z}$-Free Terms). *A term t is* $\mathbb{Z}$-free *iff it does not contain any occurrences of function symbols from $\mathcal{F}_{\mathsf{LIA}}$.*

Definition 2 ($\mathbb{Z}$-TRSs, Defined Symbols, Constructors). *A* $\mathbb{Z}$-rule *is a rewrite rule of the form $l \to r[\![\varphi]\!]$ where l and r are terms with* $\mathrm{sort}(l) = \mathrm{sort}(r)$ *such that l is $\mathbb{Z}$-free and φ is a* LIA*-constraint. If $\varphi = \top$, then $l \to r$ can be written instead of $l \to r[\![\varphi]\!]$. A $\mathbb{Z}$-TRS is a finite set of $\mathbb{Z}$-rules. The set of* defined symbols *is $\mathcal{D}(\mathcal{R}) = \{f \mid f = \mathrm{root}(l)$ for some $l \to r[\![\varphi]\!] \in \mathcal{R}\}$. The set $\mathcal{R}(f) = \{l \to r[\![\varphi]\!] \in \mathcal{R} \mid \mathrm{root}(l) = f\}$ are the rules* defining f. *The set $\mathcal{C}(\mathcal{R}) = \mathcal{F} - \mathcal{D}(\mathcal{R})$ denotes the* constructors *of $\mathcal{R}$.*

It is assumed in the following that $\mathcal{C}(\mathcal{R})$ does not contain any function symbol with resulting sort `int`, i.e., no new constructors for $\mathbb{Z}$ are added. This is a natural assumption and not a restriction. The assumption that l is $\mathbb{Z}$-free is not severe in practice since an occurrence of a term $t \in \mathcal{T}(\mathcal{F}_{\mathsf{LIA}}, \mathcal{V})$ in the left-hand side can be replaced by a fresh variable x_t if $x_t \simeq t$ is added to the constraint.

Example 3. The following rules determine whether x is a divisor of y:

$$\begin{array}{lll}
\mathsf{divides}(x,y) \to \mathsf{divides}(-x,y) & [\![x < 0 \wedge y \geq 0]\!] & (1)\\
\mathsf{divides}(x,y) \to \mathsf{divides}(x,-y) & [\![x \geq 0 \wedge y < 0]\!] & (2)\\
\mathsf{divides}(x,y) \to \mathsf{divides}(-x,-y) & [\![x < 0 \wedge y < 0]\!] & (3)\\
\mathsf{divides}(x,y) \to \mathsf{true} & [\![x \geq 0 \wedge y \simeq 0]\!] & (4)\\
\mathsf{divides}(x,y) \to \mathsf{false} & [\![x \simeq 0 \wedge y > 0]\!] & (5)\\
\mathsf{divides}(x,y) \to \mathsf{false} & [\![x > 0 \wedge y > 0 \wedge x > y]\!] & (6)\\
\mathsf{divides}(x,y) \to \mathsf{divides}(x,y-x) & [\![x > 0 \wedge y > 0 \wedge y \geq x]\!] & (7)
\end{array}$$

Then $\mathcal{D}(\mathcal{R}) = \{\mathsf{divides}\}$ and $\mathcal{C}(\mathcal{R}) = \{\mathsf{true}, \mathsf{false}\}$. ◊

The restriction that left-hand sides of rules are $\mathbb{Z}$-free allows for a simple definition of the rewrite relation of a $\mathbb{Z}$-TRS since LIA can be disregarded for matching. Notice that the matching substitution needs to be restricted in order to make sure that validity of the instantiated LIA-constraint can be decided.

Definition 4 ($\mathbb{Z}$-Based Substitutions, Rewrite Relation of a $\mathbb{Z}$-TRS). *Let $\mathcal{R}$ be a $\mathbb{Z}$-TRS and let s be a term. Then $s \to_{\mathcal{R},\mathbb{Z}} t$ iff there exist a constrained rewrite rule $l \to r[\![\varphi]\!] \in \mathcal{R}$, a position $p \in \mathcal{P}os(s)$, and a $\mathbb{Z}$-based substitution σ such that (1) $s|_p = l\sigma$, (2) $\varphi\sigma$ is* LIA*-valid, and (3) $t = s[r\sigma]_p$. Here, a substitution σ is* $\mathbb{Z}$-based *iff $\sigma(x) \in \mathcal{T}(\mathcal{F}_{\mathsf{LIA}}, \mathcal{V})$ for all variables x of sort* `int`.

Example 5. Using the $\mathbb{Z}$-TRS from Exa. 3, $\mathsf{divides}(2, -6) \to_{\mathcal{R},\mathbb{Z}} \mathsf{divides}(2, 6) \to_{\mathcal{R},\mathbb{Z}} \mathsf{divides}(2, 4) \to_{\mathcal{R},\mathbb{Z}} \mathsf{divides}(2, 2) \to_{\mathcal{R},\mathbb{Z}} \mathsf{divides}(2, 0) \to_{\mathcal{R},\mathbb{Z}} \mathsf{true}$ using the rewrite rules (2), (7) (three times), and (4). ◊

$\mathbb{Z}$-TRSs that are to be used for inductive theorem proving need to satisfy certain properties. The first property is termination, i.e., well-foundedness of $\to_{\mathcal{R},\mathbb{Z}}$. While this is in general undecidable, the methods for proving termination of CERSs developed in [12] are applicable since $\mathbb{Z}$-TRSs are a restricted kind of CERSs. These methods are based on the dependency pair approach and have been implemented in the termination tool AProVE.

The defined functions of a $\mathbb{Z}$-TRS need to be total, i.e., result in a constructor ground term when applied to constructor ground terms as their arguments.

Definition 6 (Constructor Ground Terms, Quasi-Reductivity). *A $\mathbb{Z}$-TRS $\mathcal{R}$ is* quasi-reductive *iff every ground term of the form $f(t^*)$ with $f \in \mathcal{D}(\mathcal{R})$ and constructor ground terms t^* is reducible by $\to_{\mathcal{R},\mathbb{Z}}$. Here, a* constructor ground term *is a term from $\mathcal{T}(\mathcal{C}(\mathcal{R}) \cup \mathcal{F}_{\mathsf{LIA}})$. Furthermore, a* constructor ground substitution *is a substitution that maps all variables to constructor ground terms.*

The final property required of a $\mathbb{Z}$-TRS is *confluence.*

Definition 7 (Confluence). *A $\mathbb{Z}$-TRS $\mathcal{R}$ is* confluent *iff $\leftarrow^*_{\mathcal{R},\mathbb{Z}} \circ \to^*_{\mathcal{R},\mathbb{Z}} \subseteq \to^*_{\mathcal{R},\mathbb{Z}} \circ \leftarrow^*_{\mathcal{R},\mathbb{Z}}$.*

Checking whether a $\mathbb{Z}$-TRS is quasi-reductive and confluent is a hard problem in general. Thus, a restricted class of $\mathbb{Z}$-TRSs is considered in the following. For this class, checking for quasi-reductivity is easily possible. Furthermore, $\mathbb{Z}$-TRSs from this class will always be confluent. It is required that the rules are left-linear, constructor-based, and "disjoint" in the sense that at most one rule is applicable to each position in any term. Notice that two rules might have identical left-hand sides as long as the conjunction of their LIA-constraints is unsatisfiable.

Definition 8 (Orthogonal $\mathbb{Z}$-TRSs). *A $\mathbb{Z}$-TRS $\mathcal{R}$ is* orthogonal *iff*

1. *For all $l \to r[\![\varphi]\!] \in \mathcal{R}$, l is linear and of the form $f(l^*)$ with $l^* \in \mathcal{T}(\mathcal{C}(\mathcal{R}), \mathcal{V})$.*
2. *For any two rules $l_1 \to r_1[\![\varphi_1]\!], l_2 \to r_2[\![\varphi_2]\!]$, either $l_1 = l_2$ or l_1, l_2 are not unifiable after their variables have been renamed apart.*
3. *For any two non-identical rules $l_1 \to r_1[\![\varphi_1]\!], l_2 \to r_2[\![\varphi_2]\!]$ with $l_1 = l_2$, the constraint $\varphi_1 \wedge \varphi_2$ is* LIA*-unsatisfiable.*
4. *Whenever $l_1 \to r_1[\![\varphi_1]\!], \ldots, l_n \to r_n[\![\varphi_n]\!]$ are all rules with identical left-hand sides, then the constraint $\varphi_1 \vee \ldots \vee \varphi_n$ is* LIA*-valid.*

Example 9. The $\mathbb{Z}$-TRS from Exa. 3 is orthogonal. ◊

Using this definition, quasi-reductivity of orthogonal $\mathbb{Z}$-TRSs can be reduced to quasi-reductivity of ordinary TRSs. Furthermore, orthogonal $\mathbb{Z}$-TRSs are always confluent, regardless of whether they are terminating or not.

Theorem 10. *Let $\mathcal{R}$ be an orthogonal $\mathbb{Z}$-TRS. Then it is decidable whether $\mathcal{R}$ is quasi-reductive. Furthermore, $\mathcal{R}$ is confluent.*

3 Inductive Theorem Proving with $\mathbb{Z}$-TRSs

In the following, it is assumed that $\mathcal{R}$ is a terminating quasi-reductive orthogonal $\mathbb{Z}$-TRS, which implies that $\mathcal{R}$ is confluent. The atomic conjectures in inductive theorem proving are equalities between terms. In this paper, a generalized form of these atomic conjectures is used that also incorporates a LIA-constraint.

Definition 11 (Atomic Conjectures). *An* atomic conjecture *has the form $s \equiv t[\![\varphi]\!]$ where s and t are terms with* $\mathrm{sort}(s) = \mathrm{sort}(t)$ *and φ is a* LIA*-constraint. If $\varphi = \top$, then $s \equiv t$ can be written instead of $s \equiv t[\![\varphi]\!]$. An atomic conjecture $s \equiv t[\![\varphi]\!]$ is an* inductive theorem *iff $s\sigma \leftrightarrow^*_{\mathcal{R}\cup\mathcal{E}_{\mathsf{LIA}},\mathbb{Z}} t\sigma$ for all constructor ground substitutions σ such that $\varphi\sigma$ is* LIA*-valid. A set of atomic conjectures is an* inductive theorem *iff all of its elements are inductive theorems.*

The inductive theorem proving method for $\mathbb{Z}$-TRSs is based on Reddy's *term rewriting induction* [23]. The presentation follows [2]. The main idea of this method is to *expand* subterms of atomic conjectures using narrowing.

Definition 12 (Expd)**.** *A $\mathbb{Z}$-free term t is* basic *iff $t = f(t^*)$ where $f \in \mathcal{D}(\mathcal{R})$ and $t^* \in \mathcal{T}(\mathcal{C}(\mathcal{R}), \mathcal{V})$. For an atomic conjecture $s \equiv t[\![\varphi]\!]$ and a basic term u such that $s = C[u]$, the set* $\mathrm{Expd}_u(s,t,\varphi)$ *is defined as*

$$\mathrm{Expd}_u(s,t,\varphi) = \{C[r]\sigma \equiv t\sigma[\![\varphi\sigma \wedge \psi\sigma]\!] \mid l \to r[\![\psi]\!] \in \mathcal{R},\ \sigma = \mathrm{mgu}(u,l) \textit{ exists, and } \varphi\sigma \wedge \psi\sigma \textit{ is } \mathsf{LIA}\textit{-satisfiable}\}$$

Here, it has been assumed that the variables of $l \to r[\![\psi]\!]$ have been renamed to be disjoint from the variables of $s \equiv t[\![\varphi]\!]$.

Example 13. Consider the atomic conjecture $\mathsf{divides}(x,x) \equiv \mathsf{true}$ in the context of the $\mathbb{Z}$-TRS from Exa. 3. For $s = \mathsf{divides}(x,x)$, $t = \mathsf{true}$, $u = s$, and $\varphi = \top$, $\mathrm{Expd}_u(s,t,\varphi) = \{\mathsf{divides}(-x,-x) \equiv \mathsf{true}\ [\![x < \mathsf{0} \wedge x < \mathsf{0}]\!],\ \mathsf{true} \equiv \mathsf{true}\ [\![x \geq \mathsf{0} \wedge x \simeq \mathsf{0}]\!],\ \mathsf{divides}(x, x-x) \equiv \mathsf{true}\ [\![x > \mathsf{0} \wedge x > \mathsf{0} \wedge x \geq x]\!]\}$. ◊

The inductive proof method for $\mathbb{Z}$-TRSs is formulated using the inference system $\mathcal{I}$ given in Fig. 1. Here, $s \doteq t[\![\varphi]\!]$ denotes one of $s \equiv t[\![\varphi]\!]$ and $t \equiv s[\![\varphi]\!]$ and "$\uplus$" denotes a disjoint union. The inference rules operate on tuples $\langle E, H\rangle$, where E consists of atomic conjectures that are to be proven and H consists of atomic conjectures that have been oriented as rewrite rules. Instances of these rules constitute the *hypotheses* in a proof by induction. The goal of an inductive proof is to obtain a tuple of the form $\langle \emptyset, H\rangle$ starting from the tuple $\langle E, \emptyset\rangle$. If none of the inference rules is applicable to $\langle E', H'\rangle$ where $E' \neq \emptyset$, then the inductive proof fails. Finally, an inductive proof may also diverge or end in $\bot$.

Expand	$$\frac{\langle E \uplus \{s \doteq t[\![\varphi]\!]\},\ H\rangle}{\langle E \cup \mathrm{Expd}_u(s,t,\varphi),\ H \cup \{s \to t[\![\varphi]\!]\}\rangle}$$	if $\mathcal{R} \cup H \cup \{s \to t[\![\varphi]\!]\}$ terminates, u is basic, and s is $\mathbb{Z}$-free
Simplify	$$\frac{\langle E \uplus \{s \doteq t[\![\varphi]\!]\},\ H\rangle}{\langle E \cup \{s' \doteq t[\![\varphi]\!]\},\ H\rangle}$$	if $s[\![\varphi]\!] \to_{\mathcal{R}\cup H,\mathbb{Z}} s'[\![\varphi]\!]$
Case-Simplify	$$\frac{\langle E \uplus \{s \doteq t[\![\varphi]\!]\},\ H\rangle}{\langle E \cup \{s' \doteq t[\![\varphi']\!] \mid s'[\![\varphi']\!] \in \mathrm{Case}_p(s,\varphi)\},\ H\rangle}$$	
Delete	$$\frac{\langle E \uplus \{s \doteq t[\![\varphi]\!]\},\ H\rangle}{\langle E,\ H\rangle}$$	if $s \leftrightarrow^*_{\mathcal{E}_{\mathsf{LIA}},\mathbb{Z}} t$ or φ is LIA-unsatisfiable
Theory$_\top$	$$\frac{\langle E \uplus \{s \doteq t[\![\varphi]\!]\},\ H\rangle}{\langle E,\ H\rangle}$$	if s, t do not contain symbols from $\mathcal{D}(\mathcal{R})$ and $\varphi \Rightarrow s \simeq t$ is LIAC-valid
Theory$_\bot$	$$\frac{\langle E \uplus \{s \doteq t[\![\varphi]\!]\},\ H\rangle}{\bot}$$	if s, t do not contain symbols from $\mathcal{D}(\mathcal{R})$ and $\varphi \Rightarrow s \simeq t$ is not LIAC-valid

Fig. 1. The inference system $\mathcal{I}$

The inference rule Expand uses Def. 12 to expand a basic subterm of an atomic conjecture. Then, this atomic conjecture is oriented as a rewrite rule and added to the set H of hypotheses. Notice that this addition is only allowed if the $\mathbb{Z}$-TRS consisting of $\mathcal{R} \cup H$ and this newly obtained rule is terminating. This restriction is needed in order to obtain a sound inductive proof method.

The rule Simplify uses simplification with $\mathcal{R}$ and the hypotheses in H. For this, the constraint of the atomic conjecture is taken into account by considering the following rewrite relation. It only differs from Def. 4 in the second condition.

Definition 14 (Rewrite Relation of a $\mathbb{Z}$-TRS on Constrained Terms). *Let $\mathcal{R}$ be a $\mathbb{Z}$-TRS, let s be a term, and let ψ be a* LIA*-constraint. Then $s[\![\psi]\!] \to_{\mathcal{R},\mathbb{Z}} t[\![\psi]\!]$ iff there exist a constrained rewrite rule $l \to r[\![\varphi]\!] \in \mathcal{R}$, a position $p \in \mathcal{P}os(s)$, and a $\mathbb{Z}$-based substitution σ such that (1) $s|_p = l\sigma$, (2) $\psi \Rightarrow \varphi\sigma$ is* LIA*-valid, and (3) $t = s[r\sigma]_p$.*

The inference rule Case-Simplify combines a case split with simplification using $\mathcal{R}$ (but not using H). It makes use of the following definition which refines the constraint in a case distinction in order to make a rewrite rule applicable.

Definition 15 (Case). *For a term s, a* LIA*-constraint φ, and a position $p \in \mathcal{P}os(s)$, the set $\mathrm{Case}_p(s,\varphi)$ is defined as*

$$\mathrm{Case}_p(s,\varphi) = \{s[r_i\sigma_i]_p[\![\varphi \wedge \psi_i\sigma_i]\!] \mid l_i \to r_i[\![\psi_i]\!] \in \mathcal{R},\ s|_p = l_i\sigma_i, \text{ and } \varphi \wedge \psi_i\sigma_i \text{ is LIA-satisfiable } \}$$

The construction is only performed if all σ_i are $\mathbb{Z}$-based.

The rule Delete removes trivial atomic conjectures, and the rules Theory$_\top$ and Theory$_\bot$ apply to atomic conjectures that do not contain any defined symbols and make use of a decision procedure for the theory LIAC that combines the linear theory of integers with the (free) constructors from $\mathcal{C}(\mathcal{R})$. LIAC-validity and LIAC-satisfiability are decidable and decision procedures have been implemented, for instance in the SMT-solver CVC3 [5].

The following lemma and theorem state properties of $\mathcal{I}$, where $\langle E, H\rangle \vdash_{\mathcal{I}} \langle E', H'\rangle$ denotes application of one of the inference rules.

Lemma 16. *1. If $\langle E_n, H_n\rangle \vdash_{\mathcal{I}} \langle E_{n+1}, H_{n+1}\rangle$ using a rule other than* $\mathsf{Theory}_\bot$, *then $\leftrightarrow^*_{\mathcal{R}\cup E_n\cup H_n\cup\mathcal{E}_{\mathsf{LIA}},\mathbb{Z}} = \leftrightarrow^*_{\mathcal{R}\cup E_{n+1}\cup H_{n+1}\cup\mathcal{E}_{\mathsf{LIA}},\mathbb{Z}}$ on ground terms.*
2. If $\langle E, \emptyset\rangle \vdash^*_{\mathcal{I}} \langle \emptyset, H\rangle$ using inference rules other than $\mathsf{Theory}_\bot$, *then $\rightarrow_{H,\mathbb{Z}} \subseteq \rightarrow_{\mathcal{R},\mathbb{Z}} \circ \rightarrow^*_{\mathcal{R}\cup H,\mathbb{Z}} \circ \leftrightarrow^*_{\mathcal{E}_{\mathsf{LIA}},\mathbb{Z}} \circ \leftarrow^*_{\mathcal{R}\cup H,\mathbb{Z}}$ on ground terms.*

Theorem 17. *If $\langle E, \emptyset\rangle \vdash^*_{\mathcal{I}} \langle \emptyset, H\rangle$, then E is an inductive theorem.*

Proof. By Lem. 16.1, $\leftrightarrow^*_{\mathcal{R}\cup E\cup\mathcal{E}_{\mathsf{LIA}},\mathbb{Z}} = \leftrightarrow^*_{\mathcal{R}\cup H\cup\mathcal{E}_{\mathsf{LIA}},\mathbb{Z}}$ on ground terms. Thus, it suffices to show that $\leftrightarrow^*_{\mathcal{R}\cup H\cup\mathcal{E}_{\mathsf{LIA}},\mathbb{Z}} = \leftrightarrow^*_{\mathcal{R}\cup\mathcal{E}_{\mathsf{LIA}},\mathbb{Z}}$ on ground terms. For this, the following principle is used:

Assume that the following conditions are satisfied:
1. $\rightarrow_{\mathcal{R}\cup H,Z}$ is terminating on ground terms.
2. $\rightarrow_{H,\mathbb{Z}} \subseteq \rightarrow_{\mathcal{R},\mathbb{Z}} \circ \rightarrow^*_{\mathcal{R}\cup H,Z} \circ \leftrightarrow^*_{\mathcal{E}_{\mathsf{LIA}},\mathbb{Z}} \circ \leftarrow^*_{\mathcal{R}\cup H,Z}$ on ground terms.

Then $\leftrightarrow^*_{\mathcal{R}\cup\mathcal{E}_{\mathsf{LIA}},\mathbb{Z}} = \leftrightarrow^*_{\mathcal{R}\cup H\cup\mathcal{E}_{\mathsf{LIA}},\mathbb{Z}}$ on ground terms.

This principle is quite similar to an abstract principle due to Koike and Toyama as reported in [2] but differs from that principle by incorporating $\mathcal{E}_{\mathsf{LIA}}$.

With this principle, the statement of the theorem can be shown. The first condition follows from the assumption on $\mathcal{R}$ and from the condition of the inference rule Expand. The second condition is the property from Lem. 16.2. □

Example 18. This example considers the definition of divides from Exa. 3 and adds the following rules defining gcd:

$$\begin{array}{ll}
\mathsf{gcd}(x,y) \rightarrow \mathsf{gcd}(-x,y) & [\![x < 0 \wedge y \geq 0]\!]\\
\mathsf{gcd}(x,y) \rightarrow \mathsf{gcd}(x,-y) & [\![x \geq 0 \wedge y < 0]\!]\\
\mathsf{gcd}(x,y) \rightarrow \mathsf{gcd}(-x,-y) & [\![x < 0 \wedge y < 0]\!]\\
\mathsf{gcd}(x,y) \rightarrow y & [\![x \simeq 0 \wedge y \geq 0]\!]\\
\mathsf{gcd}(x,y) \rightarrow x & [\![x > 0 \wedge y \simeq 0]\!]\\
\mathsf{gcd}(x,y) \rightarrow \mathsf{gcd}(x-y,y) & [\![x > 0 \wedge y > 0 \wedge x \geq y]\!]\\
\mathsf{gcd}(x,y) \rightarrow \mathsf{gcd}(x,y-x) & [\![x > 0 \wedge y > 0 \wedge y > x]\!]
\end{array}$$

Then, the conjectures

$$\begin{array}{rr}
\mathsf{divides}(x,x) \equiv \mathsf{true} & \mathsf{divides}(x,y) \equiv \mathsf{true}\ [\![x \simeq -y]\!]\\
\mathsf{divides}(x,y) \equiv \mathsf{true}\ [\![x \simeq 1]\!] & \mathsf{gcd}(x,x) \equiv x\ [\![x \geq 0]\!]\\
\mathsf{gcd}(x,x) \equiv -x\ [\![x \leq 0]\!] & \mathsf{gcd}(x,y) \equiv 1\ [\![y \simeq 1]\!]
\end{array}$$

can be proved fully automatically using the inference system $\mathcal{I}$. ◊

The inference system $\mathcal{I}$ cannot only be used in order to prove inductive theorems but also in order to disprove conjectures.

Theorem 19. *If $\langle E, \emptyset\rangle \vdash^*_{\mathcal{I}} \bot$, then E is not an inductive theorem.*

Example 20. For the function gcd from Exa. 18, the conjecture $\mathsf{gcd}(x,x) \equiv x$ can be disproved fully automatically using the inference system $\mathcal{I}$ (since $\mathsf{gcd}(x,x) \equiv -x$ if x is negative). ◊

4 Inductive Theorem Proving as a Decision Procedure

In this section, we derive conditions on $\mathbb{Z}$-TRSs and conjectures under which the inference system $\mathcal{I}$ can be used as a decision procedure, i.e., will always produce a proof or disproof of a conjecture if a suitable strategy on the use of the inference rules is employed. These conditions are based on properties of the rewrite rules in a $\mathbb{Z}$-TRS that can be pre-computed during parsing. Thus, checking whether a conjecture satisfies the conditions under which $\mathcal{I}$ provides a decision procedure is easily possible and requires much less time than attempting a proof or disproof. Much of the material presented in this section has appeared in preliminary form in [13]. The use of $\mathbb{Z}$-TRSs and constrained rewriting is a significant generalization, however, since [13] was based on ordinary rewriting and did not support the combination of integers with (free) constructors (but was restricted to *either* natural numbers *or* (free) constructors).

In order to simplify notation and presentation, this section is restricted to the case where all function symbols are at most binary (the general case is discussed in [14]). We say that a binary function symbol f is *equal-sorted* iff both arguments of f have the same sort.

4.1 Simple Decidable Conjectures

For the purpose of this section, a simple class of function definitions is considered. In its simplest form, functions may only make recursive calls to themselves. Furthermore, nesting of recursive calls is not permitted. This is captured by the following definition (related to the definition of $\mathcal{T}$-based functions in [20,16]).

Definition 21 (LIAC-Based Functions). *A function $g \in \mathcal{D}(\mathcal{R})$ is* LIAC-based *iff all right-hand sides of rules in $\mathcal{R}(g)$ have the form $C[g(r_1^*), \ldots, g(r_m^*)]$ for a context C over $\mathcal{C}(\mathcal{R}) \cup \mathcal{F}_{\mathsf{LIA}}$ such that $r_k^* \in \mathcal{T}(\mathcal{C}(\mathcal{R}) \cup \mathcal{F}_{\mathsf{LIA}}, \mathcal{V})$ for all $1 \leq k \leq m$.*

In order to ensure that a non-linear hypothesis is applicable to all recursive calls of a LIAC-based functions after application of the Expand rule, it needs to be ensured that the corresponding arguments of the recursive calls are "equal". This needs to be required only under the assumption that these arguments are equal in the left-hand side of the rule since Expand does otherwise not create any new atomic conjectures to which the hypothesis needs to be applied. This property depends only on the rules in $\mathcal{R}(g)$ and is independent of the conjecture.

Definition 22 ($\mathcal{I}mp\mathcal{E}q$). *Let g be equal-sorted* LIAC*-based. Then $g \in \mathcal{I}mp\mathcal{E}q$ iff for all $g(l^*) \to C[g(r_1^*), \ldots, g(r_m^*)][\![\varphi]\!] \in \mathcal{R}(g)$ for which $\varphi \wedge l_1 \simeq l_2$ is* LIAC*-satisfiable, the terms $r_{k,1}, r_{k,2}$ are $\mathbb{Z}$-free for all $1 \leq k \leq m$ and $\varphi \wedge l_1 \simeq l_2 \Rightarrow \bigwedge_{k=1}^{m} r_{k,1} \simeq r_{k,2}$ is* LIAC*-valid.*

Hence, if a term of the form $g(l^*)\sigma$ is simplified using the rule $g(l^*) \to C[g(r_1^*), \ldots, g(r_m^*)][\![\varphi]\!]$ and $g \in \mathcal{I}mp\mathcal{E}q$, then $r_{k,1}\sigma = r_{k,2}\sigma$ for all $1 \leq k \leq m$ whenever $l_1\sigma = l_2\sigma$. The set $\mathcal{I}mp\mathcal{E}q$ can easily be computed from the rules defining g with the help of a decision procedure for LIAC.

Example 23. The following orthogonal $\mathbb{Z}$-TRS determines whether a list is pointwise bigger than another list of the same length:

$$\begin{aligned}
\mathsf{ptwise}(\mathsf{nil},\mathsf{nil}) &\to \mathsf{true} \\
\mathsf{ptwise}(\mathsf{nil},\mathsf{cons}(y,ys)) &\to \mathsf{false} \\
\mathsf{ptwise}(\mathsf{cons}(x,xs),\mathsf{nil}) &\to \mathsf{false} \\
\mathsf{ptwise}(\mathsf{cons}(x,xs),\mathsf{cons}(y,ys)) &\to \mathsf{ptwise}(xs,ys)\ [\![x \geq y]\!] \\
\mathsf{ptwise}(\mathsf{cons}(x,xs),\mathsf{cons}(y,ys)) &\to \mathsf{false} \qquad\qquad\ \ [\![y > x]\!]
\end{aligned}$$

Then $\mathsf{ptwise} \in \mathcal{I}mp\mathcal{E}q$. To see this, notice that the implications from Def. 22 are trivially true for the first, second, third, and fifth rules since these rules do not contain any recursive calls. For the fourth rule, the LIAC-validity of the LIAC-constraint $x \geq y \wedge \mathsf{cons}(x,xs) \simeq \mathsf{cons}(y,ys) \Rightarrow xs \simeq ys$ is easily shown. ◊

The definition of $\mathcal{I}mp\mathcal{E}q$ requires that the argument are equal in all recursive calls in all rules. Using rewriting with constraints, this requirement can be relaxed so that more function definitions satisfy it. For recursive calls that can already be simplified to a term not containing the defined symbol, the $\mathcal{I}mp\mathcal{E}q$ requirement does not need to be satisfied since the inductive hypothesis does not need to be applied to this recursive call.

Example 24. Consider the function divides from Exa. 3 again. Then, the $\mathcal{I}mp\mathcal{E}q$ requirement is satisfied for the first through sixth rules. For the final rule, the $\mathcal{I}mp\mathcal{E}q$ requirement is not satisfied since $x > 0 \wedge y > 0 \wedge y \geq x \wedge x \simeq y \Rightarrow x \simeq y - x$ is *not* LIAC-valid. However, the atomic conjecture $\mathsf{divides}(x, x - x) \equiv \mathsf{true}\ [\![x > 0 \wedge x > 0 \wedge x \geq x]\!]$ generated in a proof attempt of $\mathsf{divides}(x,x) \equiv \mathsf{true}$ simplifies using $\to_{\mathcal{R},\mathbb{Z}}$ to $\mathsf{true} \equiv \mathsf{true}\ [\![x > 0 \wedge x > 0 \wedge x \geq x]\!]$ using the rewrite rule $\mathsf{divides}(x,y) \to \mathsf{true}\ [\![x \geq 0 \wedge y \simeq 0]\!]$. Thus, the inductive hypothesis does not need to be applied for this obligation. ◊

In order to be as general as possible, we consider simplification using the inference rule Case-Simplify (since Case-Simplify subsumes Simplify for rewriting using $\mathcal{R}$). The effect of repeated applications of Case-Simplify is captured by the following definition.

Definition 25 (Simplification Trees). *Let s be a term and let φ be a* LIA*-constraint. A* simplification tree *for $s[\![\varphi]\!]$ is a non-empty tree whose nodes are labelled with terms and* LIA*-constraints and whose root is labelled with $s[\![\varphi]\!]$ such that for every internal node labelled with $t[\![\psi]\!]$, the node has one child for every $t'[\![\psi']\!] \in \mathrm{Case}_p(t,\psi)$ and this child is labelled with $t'[\![\psi']\!]$.*

Using simplification trees, the set $\mathcal{I}mp\mathcal{E}q$ can be relaxed as follows.

Definition 26 ($\mathcal{I}mp\mathcal{E}q'$)**.** *Let g be equal-sorted* LIAC*-based. Then $g \in \mathcal{I}mp\mathcal{E}q'$ iff for all rules $g(l^*) \to C[g(r_1^*),\ldots,g(r_m^*)][\![\varphi]\!] \in \mathcal{R}(g)$ and all $1 \leq k \leq m$, either (1) $\varphi \wedge l_1 \simeq l_2$ is* LIAC*-unsatisfiable, or (2) $r_{k,1}, r_{k,2}$ are $\mathbb{Z}$-free and $\varphi \wedge l_1 \simeq l_2 \Rightarrow r_{k,1} \simeq r_{k,2}$ is* LIAC*-valid, or (3) there exists a simplification tree for $g(r_k^*)[\![\varphi \wedge \theta]\!]$ such that all leaves in this tree have labels of the form $t[\![\psi]\!]$ for a term $t \in \mathcal{T}(\mathcal{C}(\mathcal{R}) \cup \mathcal{F}_{\mathsf{LIA}}, \mathcal{V})$ and a* LIA*-constraint ψ, where $\theta = l_1 \simeq l_2$ if the arguments of g have sort* `int` *and $\theta = \top$ otherwise.*

Example 27. Continuing Exa. 24, $\mathsf{divides}(x, y-x)\ [\![x > 0 \land y > 0 \land y \geq x \land x \simeq y]\!]$ — $\mathsf{true}\ [\![x > 0 \land y > 0 \land y \geq x \land x \simeq y \land x \geq 0 \land y - x \simeq 0]\!]$ is a simplification tree for $\mathsf{divides}(x, y-x)[\![x > 0 \land y > 0 \land y \geq x \land x \simeq y]\!]$ that satisfies the conditions from case *2* in Def. 26. Thus, $\mathsf{divides} \in \mathcal{I}mp\mathcal{E}q'$. Similarly, $\mathsf{gcd} \in \mathcal{I}mp\mathcal{E}q'$. ◊

Notice that $\mathcal{I}mp\mathcal{E}q'$ strictly subsumes $\mathcal{I}mp\mathcal{E}q$ and remains easily computable. The first version of decidable conjectures is now given as follows. Only a simple form of basic terms is allowed, but non-linearity is possible.

Definition 28 (Simple Conjectures). *A* simple conjecture *is an atomic conjecture of the form $g(x^*) \equiv t$ such that the following conditions are satisfied: (1) $\mathcal{R} \cup \{g(x^*) \to t\}$ is terminating, (2) g is* LIAC*-based, (3) x^* consists of variables and $t \in \mathcal{T}(\mathcal{C}(\mathcal{R}) \cup \mathcal{F}_{\mathsf{LIA}}, \mathcal{V})$ and (4) if g is equal-sorted and $x_1 = x_2$, then $g \in \mathcal{I}mp\mathcal{E}q'$.*

Example 29. For the $\mathbb{Z}$-TRS from Exa. 23, the conjecture $\mathsf{ptwise}(xs, xs) \equiv \mathsf{true}$ is simple. For the $\mathbb{Z}$-TRS from Exa. 18, the conjectures $\mathsf{divides}(x, x) \equiv \mathsf{true}$ and $\mathsf{gcd}(x, x) \equiv x$ from Exa. 20 are simple. ◊

Theorem 30. *Using the strategy* $\mathsf{Expand} \cdot \mathsf{Case\text{-}Simplify}^* \cdot \mathsf{Simplify}^* \cdot (\mathsf{Theory}_\top \cup \mathsf{Theory}_\bot)^*$*, where* Simplify *uses only hypotheses from H, it is decidable whether a simple conjecture is an inductive theorem.*

Proof. Let $g(x^*) \equiv t$ be a simple conjecture and consider a rewrite rule $g(l^*) \to C[g(r_1^*), \ldots, g(r_m^*)][\![\varphi]\!] \in \mathcal{R}(g)$. Provided $g(x^*)$ and $g(l^*)$ are unifiable and $\varphi\sigma$ is LIA-satisfiable for $\sigma = \mathrm{mgu}(g(x^*), g(l^*))$, application of Expand to $g(x^*) \equiv t$ produces (amongst others) $C\sigma[g(r_1^*)\sigma, \ldots, g(r_m^*)\sigma] \equiv t\sigma[\![\varphi\sigma]\!]$. After the application of Expand, the set H of hypotheses consists of the oriented conjecture $g(x^*) \to t$.

Now, if g is equal-sorted and $x_1 = x_2$, then $g \in \mathcal{I}mp\mathcal{E}q'$. Thus, since $x_1 = x_2$ implies $l_1\sigma = l_2\sigma$, the definition of $\mathcal{I}mp\mathcal{E}q'$ yields, for all $1 \leq k \leq m$, that either $r_{k,1}\sigma = r_{k,2}\sigma$ or there exists a simplification tree for $g(r_k^*)[\![\varphi \land l_1 \simeq l_2]\!]$ such that all leaves have labels of the form $t[\![\psi]\!]$ for a $t \in \mathcal{T}(\mathcal{C}(\mathcal{R}) \cup \mathcal{F}_{\mathsf{LIA}}, \mathcal{V})$ and a LIA-constraint ψ. Hence, Case-Simplify and/or Simplify using $g(x^*) \to t \in H$ can be applied to $C\sigma[g(r_1^*)\sigma, \ldots, g(r_m^*)\sigma] \equiv t\sigma[\![\varphi\sigma]\!]$ to obtain $C\sigma[q_1, \ldots, q_m] \equiv t\sigma[\![\psi]\!]$, where q_i is either from $\mathcal{T}(\mathcal{C}(\mathcal{R}) \cup \mathcal{F}_{\mathsf{LIA}}, \mathcal{V})$ or $q_i = t\tau_i$ with $\tau_i = \{x^* \mapsto r_i^*\sigma\}$. Now, either $\mathsf{Theory}_\top$ or $\mathsf{Theory}_\bot$ can be applied. □

The concept of LIAC-based functions is quite restrictive since a LIAC-based function may only make recursive calls to itself and not to any other function. The next definition considers a *set* of function symbols that may make recursive calls to each other (these are essentially the jointly $\mathcal{T}$-based functions from [13]).

Definition 31 (LIAC-Based Functions–Version 2). *A set of functions $\mathcal{G} = \{g_1, \ldots, g_n\} \subseteq \mathcal{D}(\mathcal{R})$ is* LIAC-based *iff all right-hand sides of rules in $\mathcal{R}(\mathcal{G})$ have the form $C[g_{k_1}(r_1^*), \ldots, g_{k_m}(r_m^*)]$ for some context C over $\mathcal{C}(\mathcal{R}) \cup \mathcal{F}_{\mathsf{LIA}}$ such that $r_i^* \in \mathcal{T}(\mathcal{C}(\mathcal{R}) \cup \mathcal{F}_{\mathsf{LIA}}, \mathcal{V})$ and $g_{k_i} \in \mathcal{G}$ for all $1 \leq i \leq m$.*

Example 32. This example computes the pointwise average of two lists (stopping as soon as either list is empty) and uses the auxiliary function avg:

$$\begin{aligned}
\mathsf{avg}(x,y) &\to \mathsf{avg}(y,x) && [\![x > y]\!]\\
\mathsf{avg}(x,y) &\to x && [\![y \geq x \wedge y - x \leq 1]\!]\\
\mathsf{avg}(x,y) &\to \mathsf{avg}(x+1,y-1) && [\![y \geq x \wedge y - x > 1]\!]\\
\mathsf{avglist}(xs,\mathsf{nil}) &\to \mathsf{nil}\\
\mathsf{avglist}(\mathsf{nil},\mathsf{cons}(y,ys)) &\to \mathsf{nil}\\
\mathsf{avglist}(\mathsf{cons}(x,xs),\mathsf{cons}(y,ys)) &\to \mathsf{cons}(\mathsf{avg}(x,y),\mathsf{avglist}(xs,ys))
\end{aligned}$$

Then avglist is not LIAC-based, but the set $\{\mathsf{avg}, \mathsf{avglist}\}$ is LIAC-based. $\Diamond$

In order to ensure that non-linear hypotheses are still applicable, the definition of $\mathcal{I}mp\mathcal{E}q$ and $\mathcal{I}mp\mathcal{E}q'$ needs to be adapted as well. For this, the idea is to collect a subset of a LIAC-based set of functions such that recursive calls to one of these functions have equal arguments or can be rewritten to terms from $\mathcal{T}(\mathcal{C}(\mathcal{R}) \cup \mathcal{F}_{\mathsf{LIA}}, \mathcal{V})$ if all members of the subset have equal arguments.

Definition 33 ($\mathcal{I}mp\mathcal{E}q'$–Version 2). *Let $\mathcal{G} = \{g_1, \ldots, g_n\}$ be a* LIAC*-based set of functions. Then $\langle g, \Gamma\rangle \in \mathcal{I}mp\mathcal{E}q'$ for an equal-sorted $g \in \mathcal{G}$ iff $\Gamma \subseteq \mathcal{G}$ such that all members of Γ are equal-sorted and for all rewrite rules $g_k(l^*) \to C[g_{k_1}(r_1^*), \ldots, g_{k_m}(r_m^*)][\![\varphi]\!] \in \mathcal{R}(\mathcal{G})$ and all $1 \leq i \leq m$ with $g_{k_i} = g$, either (1) $\varphi \wedge \theta$ is* LIAC*-unsatisfiable where $\theta = l_1 \simeq l_2$ if $g_k \in \Gamma$ and $\theta = \top$ if $g_k \notin \Gamma$, or (2) $r_{i,1}, r_{i,2}$ are $\mathbb{Z}$-free and $\varphi \wedge \theta \Rightarrow r_{i,1} \simeq r_{i,2}$ is* LIAC*-valid where $\theta = l_1 \simeq l_2$ if $g_k \in \Gamma$ and $\theta = \top$ if $g_k \notin \Gamma$, or (3) there exists a simplification tree for $g(r_i^*)[\![\varphi \wedge \theta]\!]$ such that all leaves in this tree have labels of the form $t[\![\psi]\!]$ for a term $t \in \mathcal{T}(\mathcal{C}(\mathcal{R}) \cup \mathcal{F}_{\mathsf{LIA}}, \mathcal{V})$ and a* LIA*-constraint ψ, where $\theta = l_1 \simeq l_2$ if the arguments of g_k have sort* `int` *and $g_k \in \Gamma$, and $\theta = \top$ otherwise.*

Simple conjectures immediately generalize to LIAC-based sets $\mathcal{G}$ of functions. Now, an atomic conjecture for each member of the set $\mathcal{G}$ needs to the proved.

Definition 34 (Simple Conjectures–Version 2). *A* simple conjecture *is a set of the form $\{g_1(x_1^*) \equiv t_1, \ldots, g_n(x_n^*) \equiv t_n\}$ such that the following conditions are satisfied: (1) $\mathcal{R} \cup \{g_1(x_1^*) \to t_1, \ldots, g_n(x_n^*) \to t_n\}$ is terminating, (2) $\mathcal{G} = \{g_1, \ldots, g_n\}$ is* LIAC*-based, (3) x_i^* consists of variables and $t_i \in \mathcal{T}(\mathcal{C}(\mathcal{R}) \cup \mathcal{F}_{\mathsf{LIA}}, \mathcal{V})$ for all $1 \leq i \leq n$, and (4) if g_k is equal-sorted and $x_{k,1} = x_{k,2}$, then there exists an $\langle g_k, \Gamma\rangle \in \mathcal{I}mp\mathcal{E}q'$ such that $x_{k',1} = x_{k',2}$ for all $g_{k'} \in \Gamma$.*

Example 35. For the $\mathbb{Z}$-TRS from Exa. 32, $\{\mathsf{avg}(x,x) \equiv x,\ \mathsf{avglist}(xs,xs) \equiv xs\}$ is a simple conjecture. To see this, notice that both $\langle \mathsf{avg}, \{\mathsf{avg}, \mathsf{avglist}\}\rangle \in \mathcal{I}mp\mathcal{E}q'$ and $\langle \mathsf{avglist}, \{\mathsf{avglist}\}\rangle \in \mathcal{I}mp\mathcal{E}q'$ due to the LIAC-validity of $\mathsf{cons}(x,xs) \simeq \mathsf{cons}(y,ys) \Rightarrow xs \simeq ys$ and $\mathsf{cons}(x,xs) \simeq \mathsf{cons}(y,ys) \Rightarrow x \simeq y$. $\Diamond$

Theorem 36. *Using the strategy* $\mathsf{Expand}^* \cdot \mathsf{Case\text{-}Simplify}^* \cdot \mathsf{Simplify}^* \cdot (\mathsf{Theory}_\top \cup \mathsf{Theory}_\bot)^*$*, where* Expand *is applied once to each member of the set and* Simplify *uses only H, it is decidable whether a simple conjecture is an inductive theorem.*

4.2 Decidable Conjectures with Nesting

One restriction of the simple decidable conjectures from Sect. 4.1 is that nesting of defined function symbols is not permitted. This restriction was imposed in order to ensure that the inductive hypotheses are always applicable, resulting in an atomic conjecture whose validity can be decided using $\mathsf{Theory}_\top$ or $\mathsf{Theory}_\bot$.

For atomic conjectures with nested defined function symbols, this is not always the case since Expand might introduce contexts from the right-hand sides of rules around the recursive calls. This context must be removed before the inductive hypotheses can be applied. This observation leads to the concept of *compatibility*, meaning that the $\mathbb{Z}$-TRS can handle the contexts introduced in right-hand sides of rules. The presentation in this section is influenced by the presentation in [16], which presents similar results for ordinary TRSs. In contrast to [16], the notion of compatibility is more complex and powerful in this paper since it is based on the inference rule Case-Simplify, i.e., it uses simplification trees. This makes the notion of compatibility more general, see Exa. 38 below.

Definition 37 (Compatibility). *Let g be* LIAC*-based, let $1 \leq j \leq \text{arity}(g)$, and let f be* LIAC*-based. Then g* is compatible with f on argument j *iff the jth argument of g has the same sort as f and for all rules $f(l^*) \to C[f(r_1^*), \ldots, f(r_n^*)][\![\varphi]\!]$ there is a simplification tree for $g(x_1, \ldots, x_{j-1}, C[z_1, \ldots, z_n], x_{j+1}, \ldots, x_m)[\![\varphi]\!]$ such that all leaves in this tree have labels of the form*

$$D[g(x_1, \ldots, x_{j-1}, z_{i_1}, x_{j+1}, \ldots, x_m), \ldots, g(x_1, \ldots, x_{j-1}, z_{i_k}, x_{j+1}, \ldots, x_m)][\![\varphi \wedge \psi]\!]$$

for a LIA*-constraint ψ, a context D over $\mathcal{C}(\mathcal{R}) \cup \mathcal{F}_{\mathsf{LIA}}$, and $i_1, \ldots, i_k \in \{1, \ldots, n\}$ such that $z_i \notin \mathcal{V}(D)$ for all $1 \leq i \leq n$.*

Example 38. Consider the $\mathbb{Z}$-TRS consisting of the following rules:

$$\begin{aligned}
\mathsf{zip}(xs, \mathsf{nil}) &\to \mathsf{pnil} \\
\mathsf{zip}(\mathsf{nil}, \mathsf{cons}(y, ys)) &\to \mathsf{pnil} \\
\mathsf{zip}(\mathsf{cons}(x, xs), \mathsf{cons}(y, ys)) &\to \mathsf{pcons}(\mathsf{pair}(x, y), \mathsf{zip}(xs, ys)) \\
\mathsf{maxpair}(\mathsf{pnil}) &\to \mathsf{nil} \\
\mathsf{maxpair}(\mathsf{pcons}(\mathsf{pair}(x, y), zs)) &\to \mathsf{cons}(x, \mathsf{maxpair}(zs)) & [\![x \geq y]\!] \\
\mathsf{maxpair}(\mathsf{pcons}(\mathsf{pair}(x, y), zs)) &\to \mathsf{cons}(y, \mathsf{maxpair}(zs)) & [\![y > x]\!]
\end{aligned}$$

Then maxpair is compatible with zip on argument 1. For the first two zip-rules, C is pnil (a context without holes), and maxpair(pnil) — nil is a simplification tree for maxpair(pnil). The leaf has the required form by letting $D = \mathsf{nil}$. For the third zip-rule, C is $\mathsf{pcons}(\mathsf{pair}(x, y), \square)$ and

$$\dfrac{\mathsf{maxpair}(\mathsf{pcons}(\mathsf{pair}(x, y), z_1))}{\mathsf{cons}(x, \mathsf{maxpair}(z_1))\ [\![x \geq y]\!] \quad \mathsf{cons}(y, \mathsf{maxpair}(z_1))\ [\![y > x]\!]}$$

is a simplification tree for $\mathsf{maxpair}(\mathsf{pcons}(\mathsf{pair}(x, y), z_1))$. Both leaves have the required form by letting $D = \mathsf{cons}(x, \square)$ or $D = \mathsf{cons}(y, \square)$, respectively.

Notice that the use of simplification trees is essential, i.e., maxpair is *not* compatible with zip on argument 1 using compatibility as in [16] since the term $\mathsf{maxpair}(\mathsf{pcons}(\mathsf{pair}(x, y), z_1))$ cannot be simplified without a case split. ◊

The concept of compatibility can be extended to arbitrarily deep nestings of functions, resulting in *compatibility sequences*.

Definition 39 (Compatibility Sequences). *The sequence $\langle f_1, \ldots, f_d \rangle$ where $f_1, \ldots, f_{d-1}, f_d$ are* LIAC*-based is a* compatibility sequence on argument positions $\langle j_1, \ldots, j_{d-1} \rangle$ *iff* f_i *is compatible with* f_{i+1} *on argument* j_i *for all* $1 \leq i \leq d-1$. *A term* s has *this compatibility sequence iff*

$$s = f_1(p_1^*, f_2(p_2^*, \ldots f_{d-1}(p_{d-1}^*, f_d(x^*), q_{d-1}^*) \ldots, q_2^*), q_1^*)$$

such that the variables in x^* *do not occur elsewhere in* s, *the* p_i^* *and* q_i^* *are from* $\mathcal{T}(\mathcal{C}(\mathcal{R}) \cup \mathcal{F}_{\mathsf{LIA}}, \mathcal{V})$, *and* $f_i(p_i^*, f_{i+1}(\ldots), q_i^*)|_{j_i} = f_{i+1}(\ldots)$ *for all* $1 \leq i \leq d-1$.

Nested conjectures generalize simple conjectures as introduced in Sect. 4.1 by allowing nested defined functions on the left-hand side.

Definition 40 (Nested Conjectures). *A* nested conjecture *is an atomic conjecture of the form* $D[g(x^*)] \equiv t$ *such that the following conditions are satisfied: (1)* $\mathcal{R} \cup \{D[g(x^*)] \to t\}$ *is terminating, (2)* $D[g(x^*)]$ *has a compatibility sequence, (3)* x^* *consists of variables and* $t \in \mathcal{T}(\mathcal{C}(\mathcal{R}) \cup \mathcal{F}_{\mathsf{LIA}}, \mathcal{V})$, *and (4) if* g *is equal-sorted and* $x_1 = x_2$, *then* $g \in \mathcal{I}mp\mathcal{E}q$.

Example 41. Continuing Example 38, the term $\mathsf{maxpair}(\mathsf{zip}(xs, xs))$ has the compatibility sequence $\langle \mathsf{maxpair}, \mathsf{zip} \rangle$ on arguments $\langle 1 \rangle$. Also, $\mathsf{zip} \in \mathcal{I}mp\mathcal{E}q$. Thus, $\mathsf{maxpair}(\mathsf{zip}(xs, xs)) \equiv xs$ is a nested conjecture. ◊

Theorem 42. *Using the strategy* $\mathsf{Expand} \cdot \mathsf{Case\text{-}Simplify}^* \cdot \mathsf{Simplify}^* \cdot (\mathsf{Theory}_\top \cup \mathsf{Theory}_\bot)^*$, *where* Simplify *uses only hypotheses from* H, *it is decidable whether a nested conjecture is an inductive theorem.*

Of course, the concept of nested conjectures can be extended from LIAC-based functions to LIAC-based sets of functions, similarly to how this was done for simple conjectures in Sect. 4.1 (see [14] for details).

5 Implementation and Evaluation

The inductive proof method based on the inference system $\mathcal{I}$ has been implemented in the prototype Sail2, the successor of Sail [13]. Functions for checking whether a conjecture is simple or nested have been implemented in Sail2 as well. In order to perform these checks as efficiently as possible, the following are pre-computed while parsing the $\mathbb{Z}$-TRS:

1. $\mathcal{I}mp\mathcal{E}q$ and $\mathcal{I}mp\mathcal{E}q'$ (using a decision procedure for LIAC-validity).
2. Information on the compatibility between function symbols.

In order to check for termination of $\mathcal{R} \cup H$ as needed in the side condition of Expand, the implementation of the methods for proving termination of CERSs developed in [12] in the termination tool AProVE is used. For validity and satisfiability checking of LIA- and LIAC-constraints, the external tools Yices [11] (for LIA-constraints) and CVC3 [5] (for LIAC-constraints) are used.

The implementation has been tested on 57 examples. This collection contains conjectures which can be shown to be decidable using the results from Sect. 4 and conjectures where this is not the case. The time spent for checking whether a conjecture is decidable as well as the time needed for (dis-)proving it have been recorded. Recall that a proof attempt requires calls to AProVE in order to prove termination and to SMT-solvers in order to determine validity and satisfiability of constraints. The following table contains average times, the detailed results can be found at `http://baldur.iti.kit.edu/~falke/sail2/`.

Checking Time	SMT Time	Termination Time within AProVE	Other Time	Total Time
0.019 msec	16.451 msec	102.934 msec	0.554 msec	119.957 msec

As is immediate by inspection, the most time-consuming part (over 90% of the total time) is the termination check using AProVE. In contrast, checking whether a conjecture is a member of the class of decidable conjectures is orders of magnitude faster than proving or disproving it (i.e., SMT time + other time).

6 Conclusions

We have presented new results on the decidability of validity for a class of conjectures that requires inductive reasoning. An implementation in the prototype Sail2 has been successfully evaluated on a large collection of examples. This evaluation confirms that checking whether the inductive validity of a conjecture is decidable is indeed much faster than attempting to prove or disprove it.

The new decidability results reported in this paper were obtained using $\mathbb{Z}$-TRSs for which an inductive proof method based on implicit induction coupled with a decision procedure for the theory LIAC is given. The development of this proof method is a contribution in itself, independent of the decidability results about inductive validity. The inductive proof method not only makes it possible to prove inductive conjectures but also allows disproving false conjectures.

There are several independent directions for future work. We are interested in developing an inductive proof method for more general classes of CERSs as defined in [12]. In contrast to $\mathbb{Z}$-TRSs, these CERSs also support (non-free) collection data structures such as sets or multisets and thus provide an even more expressive kind of term rewrite systems. In addition, we are planning to identify classes of conjectures with constraints and classes of conjectures containing nested function symbols on both sides whose inductive validity can be decided. This may require techniques similar to [15,16] which automatically generate suitable generalization lemmas that are needed for deciding validity. Finally, Reddy's term rewriting induction [23] has recently been extended in order to support non-orientable atomic conjectures that cannot be turned into terminating rewrite rules [1], and we are interested in adding these capabilities to our proof method for $\mathbb{Z}$-TRSs.

References

1. Aoto, T.: Dealing with Non-orientable Equations in Rewriting Induction. In: Pfenning, F. (ed.) RTA 2006. LNCS, vol. 4098, pp. 242–256. Springer, Heidelberg (2006)
2. Aoto, T.: Soundness of rewriting induction based on an abstract principle. IPSJ Courier 4, 58–68 (2008)

3. Armando, A., Rusinowitch, M., Stratulat, S.: Incorporating decision procedures in implicit induction. JSC 34(4), 241–258 (2002)
4. Baader, F., Nipkow, T.: Term Rewriting and All That. Cambridge University Press (1998)
5. Barrett, C.W., Tinelli, C.: CVC3. In: Damm, W., Hermanns, H. (eds.) CAV 2007. LNCS, vol. 4590, pp. 298–302. Springer, Heidelberg (2007)
6. Bouhoula, A.: Automated theorem proving by test set induction. JSC 23(1), 47–77 (1997)
7. Bouhoula, A., Jacquemard, F.: Automated Induction with Constrained Tree Automata. In: Armando, A., Baumgartner, P., Dowek, G. (eds.) IJCAR 2008. LNCS (LNAI), vol. 5195, pp. 539–554. Springer, Heidelberg (2008)
8. Boyer, R.S., Moore, J.S.: A Computational Logic. Academic Press (1979)
9. Boyer, R.S., Moore, J.S.: Integrating decision procedures into heuristic theorem provers. In: Machine Intelligence 11. pp. 83–124. Oxford University Press (1988)
10. Bundy, A.: The automation of proof by mathematical induction. In: Handbook of Automated Reasoning, pp. 845–911. Elsevier (2001)
11. Dutertre, B., de Moura, L.: A Fast Linear-Arithmetic Solver for DPLL(T). In: Ball, T., Jones, R.B. (eds.) CAV 2006. LNCS, vol. 4144, pp. 81–94. Springer, Heidelberg (2006)
12. Falke, S.: Term Rewriting with Built-In Numbers and Collection Data Structures. Ph.D. thesis, University of New Mexico, Albuquerque, NM, USA (2009)
13. Falke, S., Kapur, D.: Inductive Decidability Using Implicit Induction. In: Hermann, M., Voronkov, A. (eds.) LPAR 2006. LNCS (LNAI), vol. 4246, pp. 45–59. Springer, Heidelberg (2006)
14. Falke, S., Kapur, D.: Rewriting Induction + Linear Arithmetic = Decision Procedure. Karlsuhe Report in Informatics 2012-2, KIT (2012)
15. Giesl, J., Kapur, D.: Decidable Classes of Inductive Theorems. In: Goré, R.P., Leitsch, A., Nipkow, T. (eds.) IJCAR 2001. LNCS (LNAI), vol. 2083, pp. 469–484. Springer, Heidelberg (2001)
16. Giesl, J., Kapur, D.: Deciding Inductive Validity of Equations. In: Baader, F. (ed.) CADE 2003. LNCS (LNAI), vol. 2741, pp. 17–31. Springer, Heidelberg (2003)
17. Huet, G.P., Hullot, J.M.: Proofs by induction in equational theories with constructors. JCSS 25(2), 239–266 (1982)
18. Kapur, D., Narendran, P., Zhang, H.: Automating inductionless induction using test sets. JSC 11(1–2), 81–111 (1991)
19. Kapur, D., Subramaniam, M.: New uses of linear arithmetic in automated theorem proving by induction. JAR 16(1–2), 39–78 (1996)
20. Kapur, D., Subramaniam, M.: Extending Decision Procedures with Induction Schemes. In: McAllester, D. (ed.) CADE 2000. LNCS (LNAI), vol. 1831, pp. 324–345. Springer, Heidelberg (2000)
21. Kapur, D., Zhang, H.: An overview of Rewrite Rule Laboratory (RRL). Computers & Mathematics with Applications 29(2), 91–114 (1995)
22. Nahon, F., Kirchner, C., Kirchner, H., Brauner, P.: Inductive proof search modulo. AMAI 55(1–2), 123–154 (2009)
23. Reddy, U.S.: Term Rewriting Induction. In: Stickel, M.E. (ed.) CADE 1990. LNCS, vol. 449, pp. 162–177. Springer, Heidelberg (1990)
24. Sakata, T., Nishida, N., Sakabe, T., Sakai, M., Kusakari, K.: Rewriting induction for constrained term rewriting systems. IPSJ Programming 2(2), 80–96 (2009) (in Japanese)

Combination of Disjoint Theories: Beyond Decidability

Pascal Fontaine[1], Stephan Merz[2], and Christoph Weidenbach[3]

[1] Université de Lorraine & LORIA, Nancy, France
[2] INRIA Nancy & LORIA, Nancy, France
[3] Max-Planck-Institut für Informatik, Saarbrücken, Germany

Abstract. Combination of theories underlies the design of satisfiability modulo theories (SMT) solvers. The Nelson-Oppen framework can be used to build a decision procedure for the combination of two disjoint decidable stably infinite theories.

We here study combinations involving an arbitrary first-order theory. Decidability is lost, but refutational completeness is preserved. We consider two cases and provide complete (semi-)algorithms for them. First, we show that it is possible under minor technical conditions to combine a decidable (not necessarily stably infinite) theory and a disjoint finitely axiomatized theory, obtaining a refutationally complete procedure. Second, we provide a refutationally complete procedure for the union of two disjoint finitely axiomatized theories, that uses the assumed procedures for the underlying theories without modifying them.

1 Introduction

The problem of combining *decidable* first-order theories has been widely studied (e.g., [9,10,13]). The fundamental result due to Nelson and Oppen yields a decision procedure for the satisfiability (or dually, validity) problem concerning *quantifier-free* formulas in the union of the languages of two decidable theories, provided these theories are *disjoint* (i.e., they only share the equality symbol) and *stably infinite* (i.e., every satisfiable set of literals has an infinite model). This result, and its extensions, underly the design of automated reasoners known as SMT (Satisfiability Modulo Theories [2]) solvers.

The problem of combining theories for which there exist refutationally complete semi-decision procedures for the validity (unsatisfiability) problem has received less attention. In this paper, we will show that the fundamental results about combinations of disjoint decidable theories extend naturally to semi-decidable theories, and that refutationally complete procedures for such theories can be combined to yield a refutationally complete procedure for their union.

From a theoretical point of view, our observation may appear trivial. In particular, consider two theories presented by finitely many first-order axioms: complete first-order theorem provers provide semi-decision procedures for them. A refutationally complete procedure for the union of these theories is simply obtained by running the same prover on the union of the axioms. We believe

B. Gramlich, D. Miller, and U. Sattler (Eds.): IJCAR 2012, LNAI 7364, pp. 256–270, 2012.

however that combining theories à la Nelson-Oppen is still valuable in the two scenarios in Sections 5 and 6. Specialized efficient decision procedures exist for some theories of high practical relevance such as arithmetic fragments, uninterpreted symbols or arrays. In Section 5 we consider the combination of a decidable theory with a disjoint finitely axiomatized theory (without further restriction on cardinalities or on the form of this theory). Using these results, the usual language of SMT solvers can be extended with symbols defined by finitely axiomatized theories, preserving refutational completeness. In Section 6 we consider combining two disjoint finitely axiomatized theories. Here the interest lies in the fact that the refutationally complete procedures for the theories in the combination share very little information; the procedure is essentially parallel.

There has already been work on extending automated first-order theorem provers in order to accommodate interpreted symbols from decidable theories, such as fragments of arithmetic, for which an encoding by first-order axioms does not yield a decision procedure [1,3]. Bonacina et al. [5] give a calculus for a refutationally complete combination of superposition and SMT solvers. Instantiation-based frameworks (see [2] for more information) also have interesting completeness results. In [8], a complete instantiation procedure is given, even for some cases where theories are not disjoint. Compared to this approach, ours handles a less expressive fragment, but allows working in standard models. Also, our approach imposes no restrictions on the first-order theory and is independent of the actual presentation of the underlying theories or the nature of the semi-decision procedures. It uses the underlying semi-decision procedures as "black boxes" for the combination, in the spirit of the Nelson-Oppen approach.

Our results rely on two main restrictions inherited from Nelson-Oppen. First, we consider unsatisfiability of *quantifier-free* formulas, and second, those formulas are studied in the union of *disjoint* theories. Both restrictions appear crucial, specifically for theories that are not finitely axiomatized. Consider combining Presburger arithmetic (which is decidable) with a first-order and finitely axiomatizable but *non-disjoint* theory defining multiplication in terms of addition. One would expect the result to be non-linear arithmetic on naturals. Because of the unsolvability of Hilbert's tenth problem, there exists no refutationally complete decision procedure for this fragment. Consider also the disjoint union of Presburger arithmetic and the empty theory for uninterpreted symbols (both decidable); considering *quantified* formulas on the union of the languages, it is easy to define multiplication and hence to encode the Hilbert's tenth problem: there cannot be a refutationally complete decision procedure (on the standard model) for arbitrary quantified formulas in the union of these theories. We additionally assume as a reasonable simplification hypothesis that the theories are either decidable or are finitely axiomatized.

Outline. Section 2 fixes basic notations and introduces a pseudo-code language. Sections 3 and 4 present elementary results on combining theories, including the simple case of combining refutationally complete procedures for disjoint stably infinite theories. Lifting the restriction on cardinalities, Section 5 considers combinations of a decidable theory with a disjoint, finitely axiomatized theory.

Finally, we propose in section 6 an algorithm to combine two disjoint finitely axiomatized theories. The results in this paper can of course be used modularly to build complex combinations, involving several decidable theories and several finitely axiomatized theories.

2 Notations

A first-order language $\mathcal{L} = \langle \mathcal{V}, \mathcal{F}, \mathcal{P} \rangle$ consists of an enumerable set $\mathcal{V}$ of variables and enumerable sets $\mathcal{F}$ and $\mathcal{P}$ of function and predicate symbols, associated with their arities. Nullary function symbols are called constant symbols.

Terms and formulas over the language $\mathcal{L}$ are defined in the usual way. An atomic formula is either an equality statement $(t = t')$ where t and t' are terms, or a predicate symbol applied to the corresponding number of terms. Formulas are built from atomic formulas using Boolean connectives ($\neg$, $\wedge$, $\vee$, $\Rightarrow$, $\equiv$) and quantifiers ($\forall$, $\exists$). A literal is an atomic formula or the negation of an atomic formula. A formula with no free variables is closed.

An interpretation $\mathcal{I}$ for a first-order language $\mathcal{L}$ provides a non-empty domain D, a total function $\mathcal{I}[f] : D^r \to D$ of appropriate arity for every function symbol f, a predicate $\mathcal{I}[p] : D^r \to \{\top, \bot\}$ of appropriate arity for every predicate symbol p, and an element $\mathcal{I}[x] \in D$ for every variable x. By extension, an interpretation defines a value in D for every term, and a truth value for every formula. The cardinality of an interpretation is the cardinality of its domain. The notation $\mathcal{I}_{x_1/d_1,\ldots,x_n/d_n}$ where $x_1, \ldots, x_n$ are different variables (or constants) denotes the interpretation that agrees with $\mathcal{I}$, except that it associates $d_i \in D$ to the variable (resp. constant) x_i, for $1 \leq i \leq n$. A model of a formula (resp., a set of formulas) is an interpretation in which the formula (resp., every formula in the set) evaluates to true. A formula is satisfiable if it has a model, and it is unsatisfiable otherwise.

Given an interpretation $\mathcal{I}$ for a first-order language $\mathcal{L} = \langle \mathcal{V}, \mathcal{F}, \mathcal{P} \rangle$, the restriction $\mathcal{I}'$ of $\mathcal{I}$ to $\mathcal{L}' = \langle \mathcal{V}', \mathcal{F}', \mathcal{P}' \rangle$ with $\mathcal{V}' \subseteq \mathcal{V}$, $\mathcal{F}' \subseteq \mathcal{F}$, $\mathcal{P}' \subseteq \mathcal{P}$, is the unique interpretation for $\mathcal{L}'$ similar to $\mathcal{I}$, i.e. $\mathcal{I}'$ and $\mathcal{I}$ have the same domain and assign the same value to symbols in $\mathcal{V}'$, $\mathcal{F}'$ and $\mathcal{P}'$.

A theory $\mathcal{T}$ in a first-order language is a set of interpretations such that, for every interpretation $\mathcal{I} \in \mathcal{T}$, every variable x of the language, and every element d of the domain, $\mathcal{I}_{x/d} \in \mathcal{T}$. A theory may also be defined by a set of closed formulas, in which case it is the set of all the models of the set of formulas. A finite theory or a finitely axiomatized theory is the set of models of a finite set of closed formulas. A constant a is uninterpreted in a theory if for every interpretation $\mathcal{I} \in \mathcal{T}$, and every element d of the domain, $\mathcal{I}_{a/d} \in \mathcal{T}$. The *spectrum* of a theory $\mathcal{T}$, denoted spectrum($\mathcal{T}$), is the set of all (finite or infinite) cardinalities of the interpretations in $\mathcal{T}$. A theory is satisfiable if it is a non-empty set of interpretations; it is unsatisfiable otherwise.

Two theories $\mathcal{T}_1$ and $\mathcal{T}_2$ in languages $\mathcal{L}_1 = \langle \mathcal{V}_1, \mathcal{F}_1, \mathcal{P}_1 \rangle$ and $\mathcal{L}_2 = \langle \mathcal{V}_2, \mathcal{F}_2, \mathcal{P}_2 \rangle$ respectively are disjoint if $\mathcal{P}_1 \cap \mathcal{P}_2 = \emptyset$ and if $\mathcal{F}_1 \cap \mathcal{F}_2$ only contains constants that are uninterpreted in both $\mathcal{T}_1$ and $\mathcal{T}_2$. The union $\mathcal{T}_1 \cup \mathcal{T}_2$ of two theories $\mathcal{T}_1$

and $\mathcal{T}_2$ (respectively in languages $\mathcal{L}_1 = \langle \mathcal{V}_1, \mathcal{F}_1, \mathcal{P}_1 \rangle$ and $\mathcal{L}_2 = \langle \mathcal{V}_2, \mathcal{F}_2, \mathcal{P}_2 \rangle$) is the largest set of interpretations for language $\mathcal{L} = \langle \mathcal{V}_1 \cup \mathcal{V}_2, \mathcal{F}_1 \cup \mathcal{F}_2, \mathcal{P}_1 \cup \mathcal{P}_2 \rangle$ such that for every $\mathcal{I} \in \mathcal{T}_1 \cup \mathcal{T}_2$, $\mathcal{I}$ restricted to $\mathcal{L}_1$ ($\mathcal{L}_2$) belongs to $\mathcal{T}_1$ (resp., $\mathcal{T}_2$). Notice that the union of two theories defined by sets C_1 and C_2 of closed formulas is exactly the theory defined by the union $C_1 \cup C_2$.

A $\mathcal{T}$-model of a formula G is an interpretation in $\mathcal{T}$ which is a model of G. A formula G is $\mathcal{T}$-satisfiable if it has a $\mathcal{T}$-model, and it is $\mathcal{T}$-unsatisfiable otherwise. A decidable theory $\mathcal{T}$ is a theory such that the $\mathcal{T}$-satisfiability problem for finite sets of ground literals in the language of $\mathcal{T}$ is decidable.

A refutationally complete procedure for $\mathcal{T}$ is a (semi-)algorithm for the $\mathcal{T}$-unsatisfiability problem that will always terminate on an unsatisfiable formula by stating that it is unsatisfiable. Given a satisfiable formula, it may either terminate by stating that it is satisfiable or continue running forever. Thus, a refutationally complete procedure is a decision procedure if and only if it always terminates. A refutationally complete theory $\mathcal{T}$ is a theory such that there exists a refutationally complete procedure for the $\mathcal{T}$-unsatisfiability problem for sets of ground literals in the language of $\mathcal{T}$.

A theory is stably infinite if every $\mathcal{T}$-satisfiable set of literals has an infinite model of cardinality $\aleph_0$.[1]

For convenience, we define the notation $\mathrm{card}_{\geq}(n)$ that denotes a set of literals satisfiable only on models of cardinality at least n (where n is a natural number). Such a cardinality constraint can be enforced by augmenting the set of literals by the set of disequalities $\{a_i \neq a_j \mid 1 \leq i < j \leq n\}$ for fresh constants a_i.

Pseudocode

We will describe our algorithms using pseudocode. Beyond familiar constructs whose semantics is well known, we use the construct **execute in parallel**, which spawns several child processes, explicitly identified using the **process** keyword. These processes can execute truly in parallel, or be subject to any fair interleaving. The **execute in parallel** construct and all its child processes terminate as soon as some child process terminates. Similarly, if child processes are spawned inside a function, any **return** e instruction executed by a child processes will terminate all child processes and make the function return the result e. In our applications, we never have to consider race conditions such as two sibling processes potentially returning different values.

For synchronization, processes can use the instruction **wait** C, where C is a Boolean expression. This instruction blocks the process while C is false, and allows the process to resume as soon as C becomes true.[2] In particular, we use

[1] Traditionally, a theory is said to be stably infinite if every $\mathcal{T}$-satisfiable set of literals has an infinite model. In fact, *a set of first-order formulas* in a countable language (i.e. with a enumerable set of variables, functions, and predicates) has a model with cardinality $\aleph_0$ if it has an infinite model, thanks to the Löwenheim-Skolem theorem.

[2] In our algorithms, we only use synchronization expressions C that never become false after being true, hence the underlying implementation of the **wait** mechanism is unimportant.

the **wait** instruction as syntactic sugar for **wait false** in order to definitely block processes. However, we assume that the **execute in parallel** construct and all its child process terminate if all child processes are waiting. Controlling concurrent accesses to shared variables will be explicitly specified when required.

3 Combining Models

In the following, we will restrict our attention to the (un)satisfiability of finite sets of literals. Recall that the $\mathcal{T}$-satisfiability of quantifier-free formulas can be reduced to a series of $\mathcal{T}$-satisfiability checks for finite sets of literals [2]. For example, and disregarding efficient techniques used in SMT solvers, a quantifier-free formula G is satisfiable if and only if the set of literals corresponding to one of the cubes (i.e. one of the conjunctions of literals) in the disjunctive normal form (DNF) of G is satisfiable.

Assume that $\mathcal{T}$ is the union of two disjoint theories $\mathcal{T}_1$ and $\mathcal{T}_2$, respectively in languages $\mathcal{L}_1$ and $\mathcal{L}_2$. By introducing new uninterpreted constants, it is possible to *purify* any finite set of literals L into a $\mathcal{T}$-equisatisfiable set of literals $L_1 \cup L_2$ where each L_i is a set of literals in language $\mathcal{L}_i$ (see e.g. [10]).

Definition 1. *An arrangement $\mathcal{A}$ for a set of constant symbols S is a maximal satisfiable set of equalities and inequalities $a = b$ or $a \neq b$, with $a, b \in S$.*

That is, an arrangement $\mathcal{A}$ for S cannot be consistently extended with any equality or disequality over S which is not already a consequence of $\mathcal{A}$. Obviously, there exist only finitely many arrangements for a finite set of constants.

The following theorem (see also [11,12,7]) underlies the completeness proof of combinations of decision procedures. It is also the cornerstone of the results for combining refutationally complete decision procedures.

Theorem 2. *Consider disjoint theories $\mathcal{T}_1$ and $\mathcal{T}_2$, and finite sets of literals L_1 and L_2, respectively in languages $\mathcal{L}_1$ and $\mathcal{L}_2$. $L_1 \cup L_2$ is $\mathcal{T}_1 \cup \mathcal{T}_2$-satisfiable if and only if there exist an arrangement $\mathcal{A}$ of constants shared in L_1 and L_2, a (finite or infinite) cardinality κ, and models $\mathcal{M}_1$ and $\mathcal{M}_2$ of cardinality κ, such that $\mathcal{M}_1$ is a $\mathcal{T}_1$-model of $\mathcal{A} \cup L_1$ and $\mathcal{M}_2$ is a $\mathcal{T}_2$-model of $\mathcal{A} \cup L_2$.*

Intuitively, if a set of literals is satisfiable in the combination of theories, a model of this set defines in a straightforward way an arrangement and two models with the same cardinality for the two sets of literals. The converse is also true: from models of the set of literals augmented with the arrangement, it is possible to build a model for the union, since both models agree on the cardinality and on the interpretation of the shared constants (thanks to the arrangement). The cardinality condition is essential to be able to map elements in the domains of the individual models into a unique domain.

Corollary 3. *Consider disjoint theories $\mathcal{T}_1$ and $\mathcal{T}_2$, and finite sets of literals L_1 and L_2, respectively in languages $\mathcal{L}_1$ and $\mathcal{L}_2$. $L_1 \cup L_2$ is $\mathcal{T}_1 \cup \mathcal{T}_2$-satisfiable if and only if there exists an arrangement $\mathcal{A}$ of constants shared in L_1 and L_2 such that the spectra of $\mathcal{T}_1 \cup L_1 \cup \mathcal{A}$ and $\mathcal{T}_2 \cup L_2 \cup \mathcal{A}$ have a non-empty intersection.*

Function check_sat(L_1, L_2)

1 **foreach** arrangement $\mathcal{A}$ of shared constants of L_1 and L_2 **do**
2   **if** check_sat_arrangement($\mathcal{A}$, L_1, L_2) = SAT **then**
3   **return** SAT;
4 **return** UNSAT;

Function check_sat_arrangement($\mathcal{A}$, L_1, L_2)

1 **if** $\mathcal{A} \cup L_1$ is $\mathcal{T}_1$-unsatisfiable **then**
2   **return** UNSAT;
3 **if** $\mathcal{A} \cup L_2$ is $\mathcal{T}_2$-unsatisfiable **then**
4   **return** UNSAT;
5 **if** spectrum($\mathcal{T}_1 \cup \mathcal{A} \cup L_1$) $\cap$ spectrum($\mathcal{T}_2 \cup \mathcal{A} \cup L_2$) $= \emptyset$ **then**
6   **return** UNSAT;
7 **return** SAT;

Algorithm 1. Combination of decidable theories: a generic algorithm

For combining decision procedures, the cardinality or spectrum requirements of the above theorems are usually fulfilled by assuming properties of the theories in the combination. In the classical combination scheme, the theories are supposed to be stably infinite: if $\mathcal{A} \cup L_i$ has a $\mathcal{T}_i$-model, it also has a $\mathcal{T}_i$-model of infinite cardinality (more precisely, of cardinality $\aleph_0$), and thus the cardinality requirement is trivially fulfilled.

Theorem 2 and Corollary 3 do not require decision procedures to exist, and also apply to refutationally complete theories.

4 Combinations: Decidable Theories and Beyond

Consider two decidable disjoint theories $\mathcal{T}_1$ and $\mathcal{T}_2$ in languages $\mathcal{L}_1$ and $\mathcal{L}_2$, such that, given sets of literals L_1 and L_2 (in $\mathcal{L}_1$ and $\mathcal{L}_2$ respectively), it is computable whether spectrum($\mathcal{T}_1 \cup L_1$)$\cap$spectrum($\mathcal{T}_2 \cup L_2$) is empty or not. Algorithm 1 is the generic combination algorithm for $\mathcal{T}_1 \cup \mathcal{T}_2$, based on a straightforward application of the theorem in the previous section. Notice that the code at lines 1–4 in function check_sat_arrangement is not required because of the spectrum test at lines 5–6: the combined theory is unsatisfiable if the intersection of the spectra is empty. In case of stably infinite theories however, the spectrum condition at line 5 is guaranteed to be false, and thus lines 5 and 6 are not necessary; without these lines, the algorithm corresponds to the Nelson-Oppen combination framework [9,10]. If the theories are not stably infinite, there exist many specific results (e.g. [12,7]) for which it is possible to compute the condition at line 5.

Assume now that the theories in the combination are refutationally complete but not decidable. For the combination procedure to be refutationally complete, it is necessary and sufficient that check_sat_arrangement($\mathcal{A}$, L_1, L_2) terminates

```
Function check_sat_arrangement(A, L1, L2)
1  execute in parallel
2      process
3          if A ∪ L1 is T1-unsatisfiable then
4              return UNSAT;
5          wait;
6      process
7          if A ∪ L2 is T2-unsatisfiable then
8              return UNSAT;
9          wait;
10 return SAT;
```

Algorithm 2. Nelson-Oppen for refutationally complete procedures.

if $\mathcal{A} \cup L_1 \cup L_2$ is $\mathcal{T}_1 \cup \mathcal{T}_2$-unsatisfiable. Otherwise one could not guarantee that a call to check_sat($\mathcal{A} \cup L_1$, $\mathcal{A} \cup L_2$) would return. The following algorithms for refutationally complete procedures are all based on the above check_sat function, but differ in the check_sat_arrangement function. In the following, we say that a function check_sat_arrangement is a refutationally complete procedure for a theory, if function check_sat together with the considered function yields a refutationally complete procedure for a theory.

The Nelson-Oppen schema traditionally gets rid of the spectrum condition in Algorithm 1 by considering only stably infinite theories. As a first step, we also restrict attention to stably infinite theories. Thus the intersection of the spectra is empty only if one of $\mathcal{T}_1 \cup \mathcal{A} \cup L_1$ or $\mathcal{T}_2 \cup \mathcal{A} \cup L_2$ is unsatisfiable.

In the case of refutationally complete theories, the sequentiality of Algorithm 1 may cause a completeness problem. Indeed it may happen that $\mathcal{A} \cup L_1$ is $\mathcal{T}_1$-satisfiable and the test at line 1 in function check_sat_arrangement never terminates; $\mathcal{A} \cup L_2$ would never be checked for unsatisfiability. This behavior breaks completeness. A natural way to circumvent this problem is to run the unsatisfiability tests in parallel, as in Algorithm 2.

Theorem 4. *Assume $\mathcal{T}_1$ and $\mathcal{T}_2$ are stably infinite and disjoint theories with refutationally complete procedures for finite sets of literals. Algorithm 2 yields a refutationally complete procedure for $\mathcal{T}_1 \cup \mathcal{T}_2$ for finite sets $L_1 \cup L_2$ of literals, where L_1 and L_2 are respectively literals in the language of $\mathcal{T}_1$ and $\mathcal{T}_2$. If $\mathcal{T}_1$ and $\mathcal{T}_2$ are furthermore decidable, Algorithm 2 is a decision procedure.*

Proof. The proof is similar to that for the Nelson-Oppen combination framework for disjoint stably infinite decidable theories. First notice that soundness is a direct consequence of Theorem 2: whenever the algorithm terminates, it provides the right answer. We only need to ensure termination in the unsatisfiable case.

The algorithm terminates for $\mathcal{T}_1 \cup \mathcal{T}_2$-unsatisfiable sets of literals. The finite sets of L_1 and L_2 only share a finite set of constants S. There exist only a finite number of arrangements $\mathcal{A}_1, \ldots \mathcal{A}_n$ of S, and these can be checked in sequence for unsatisfiability. If every call to function check_sat_arrangement($\mathcal{A}$, L_1, L_2) terminates when $\mathcal{A} \cup L_1 \cup L_2$ is $\mathcal{T}_1 \cup \mathcal{T}_2$-unsatisfiable, then the algorithm is a refutationally complete procedure for $\mathcal{T}_1 \cup \mathcal{T}_2$.

A call to check_sat_arrangement($\mathcal{A}$, L_1, L_2) does not terminate for some $\mathcal{A} \cup L_1 \cup L_2$ only if $\mathcal{A} \cup L_1$ is $\mathcal{T}_1$-satisfiable and $\mathcal{A} \cup L_2$ is $\mathcal{T}_2$-satisfiable. In that case, there exist a $\mathcal{T}_1$-model $\mathcal{M}_1$ of $\mathcal{A} \cup L_1$ and a $\mathcal{T}_2$-model $\mathcal{M}_2$ of $\mathcal{A} \cup L_2$. The cardinality of the models can be assumed to be $\aleph_0$. Thus, according to Theorem 2, $\mathcal{A} \cup L_1 \cup L_2$ is $\mathcal{T}_1 \cup \mathcal{T}_2$-satisfiable. □

As a concrete example, consider Presburger arithmetic – which is decidable and has only models of infinite cardinality – and a finite set of first-order formulas with only infinite models (e.g. including axioms for dense order), for which refutationally complete procedures exist (any complete first-order logic prover is a suitable procedure). The above theorem yields a refutationally complete procedure for quantifier-free formulas in the union of the languages.

The above theorem imposes two major constraints, one on the cardinalities, the other on the disjointness. For decision procedures, relaxing the cardinality constraints is possible using asymmetric combinations, where one theory in the combination has strong properties that allow to relax the cardinality property on the other one. We investigate this solution in the next section.

5 Combining a Decidable and an Arbitrary Theory

The restriction to stably infinite theories has proved to be useful in the decidable case: the Nelson-Oppen framework is simple and efficient, and several important decidable theories are indeed stably infinite. Still, being stably infinite is a strong constraint (e.g., the theory $\forall x \,.\, x = a \vee x = b$ is not stably infinite), and there is no general procedure to check whether a theory is stably infinite. In practice, most theories (and all theories actually implemented in SMT solvers) furthermore have other spectral properties that allow for less restrictive combinations.

When theories are not stably infinite, the spectrum condition of lines 5-6 in Algorithm 1 becomes important. Indeed, even if both $\mathcal{T}_1 \cup \mathcal{A} \cup L_1$ and $\mathcal{T}_2 \cup \mathcal{A} \cup L_2$ are satisfiable, $L_1 \cup L_2$ may still be $\mathcal{T}_1 \cup \mathcal{T}_2$-unsatisfiable because the spectra of the models are disjoint. However, the condition is not directly implementable in general. In this section, we consider the case of a decidable theory for which the spectrum is computable. It can then be translated into suitable constraints for the procedure for the (refutationally complete) theory $\mathcal{T}_2$.

Lemma 5. *Assume $\mathcal{T}$ is a finitely axiomatized theory. There exists a refutationally complete procedure for $\mathcal{T}$ restricted to models of cardinalities belonging to a given set of the following nature:*

1. *one or all infinite cardinalities;*
2. *a finite set of finite cardinalities;*
3. *all cardinalities larger than a fixed finite cardinality;*
4. *the complement of a finite set of finite cardinalities.*

Function check_sat_arrangement($\mathcal{A}$, L_1, L_2)

```
if A ∪ L1 is T1-unsatisfiable then
    return UNSAT;
if spectrum(T1 ∪ A ∪ L1) ∩ spectrum(T2 ∪ A ∪ L2) = ∅ then
    return UNSAT;
return SAT;
```

Algorithm 3. Combination of a decidable theory with a refutationally complete theory.

It is a decision procedure for the second case.

Proof. It suffices to show that the $\mathcal{T}$-unsatisfiability of a finite set of literals with the given cardinality constraints can be reduced to checking the unsatisfiability for another finitely axiomatized theory $\mathcal{T}'$. The theory $\mathcal{T}'$ would simply be the union of $\mathcal{T}$, the set of literals, and a formula encoding the restriction on the cardinality.

It is easy to restrict the unsatisfiability check to infinite cardinalities, e.g. by adding a formula of the form

$$\forall x\, \neg R(x,x) \wedge \forall x \exists y \left(R(x,y) \wedge \forall z \left(R(y,z) \Rightarrow R(x,z) \right) \right)$$

to the set, where R is a fresh relation symbol. To restrict the unsatisfiability check to a given finite cardinality n, it suffices to add a constraint of the form

$$\bigwedge_{1 \leq i < j \leq n} a_i \neq a_j \wedge \forall x \bigvee_{1 \leq i \leq n} x = a_i$$

where the a_i are fresh constants. In fact, checking the satisfiability of a finitely axiomatized theory with a given finite cardinality is trivially a decidable problem since there are essentially finitely many interpretations for a finite language with a given finite domain. For a finite set of finite cardinalities, it suffices to take the disjunction of constraints for each finite cardinality in the set. Again, this constitutes a finite set of decidable problems, which is therefore itself decidable.

The third case is simply handled by adding a constraint $\mathrm{card}_{\geq}(n)$. The complement of a finite set of finite cardinalities is the union of a finite set of finite cardinalities and all cardinalities larger than a finite cardinality. A suitable constraint for the final case is thus the disjunction of a constraint for the third case and a constraint for the second case. □

Theorem 6. *Assume $\mathcal{T}_2$ is a finitely axiomatized theory, and $\mathcal{T}_1$ is a decidable theory for which the spectrum is computable and falls in the cases referred in Lemma 5. Then $\mathcal{T}_1 \cup \mathcal{T}_2$ is refutationally complete, and Algorithm 3 yields a refutationally complete procedure for it. The procedure is decidable in the second case. It is also decidable in the two last cases if $\mathcal{T}_2$ is decidable.*

Proof. By the previous lemma, the test at line 3 is implementable by an unsatisfiability procedure that terminates whenever $\mathcal{T}_1 \cup \mathcal{T}_2 \cup \mathcal{A} \cup L_1 \cup L_2$ is unsatisfiable. Furthermore, the test at line 3 is guaranteed to terminate if the spectrum of $\mathcal{T}_1 \cup \mathcal{A} \cup L_1$ is a finite set of finite cardinalities. Finally, in the two last cases, checking that spectrum($\mathcal{T}_2 \cup \mathcal{A} \cup L_2$) has a non-empty intersection with a given co-finite set can be reduced to checking the satisfiability of a finite collection of sets of literals $\mathcal{T}_2 \cup \mathcal{A} \cup L_2 \cup C$ where C is a set of literals encoding a constraint on cardinality; this is decidable if the satisfiability problem for sets of literals in $\mathcal{T}_2$ is decidable. □

The above conditions on the spectrum of the decidable theories appear reasonable: the decidable theories considered in combinations of theories usually fall in one of the categories of the theorem. Gentle theories [6] have a spectrum which is computable and either a finite set of finite cardinalities or a co-finite set of cardinalities. Shiny theories [12] have a computable spectrum that falls into case (3). Linear arithmetic on integers or reals obviously belongs to the first category.

6 Parallel Refutation of a Union of Disjoint Theories

In the previous section, we concentrated on combining a decidable theory with another for which a refutationally complete decision procedure exists. In this section, we study the combination of two refutationally complete theories, dropping the cardinality requirement imposed in Section 4. In the context of SMT, it seems fairly natural to restrict our study to theories that can be represented by a finite number of first-order axioms. These theories not only have well-known refutationally complete procedures in the form of complete theorem provers for first-order logic, but it is also decidable if they have a model of a given finite cardinality. Another property of finitely axiomatized theories that we will use is the Skolem-Löwenheim Theorem: such theories have either a finite number of finite models, or they have models for all infinite cardinalities.

Algorithm 4 presents a refutationally complete procedure for the combination of two disjoint first-order theories. It basically interleaves or parallelizes the run of both a refutationally complete procedure and a finite model finder for a set of literals, for $\mathcal{T}_1$ and $\mathcal{T}_2$. The task of the finite model finder is to check if the set of literals is satisfiable on a model in the theory with a given finite cardinality. Very schematically, the finite model finders and the refutationally complete procedures may not terminate in case $\mathcal{A} \cup L_1$ and $\mathcal{A} \cup L_2$ are respectively $\mathcal{T}_1$- and $\mathcal{T}_2$-satisfiable and have no finite model. In such a case they must have infinite models; thanks to the Löwenheim-Skolem theorems one can ensure that the spectra have a non-empty intersection and so $\mathcal{A} \cup \mathcal{T}_1 \cup \mathcal{T}_2$ must be satisfiable.

The difficulties come from the facts that:

- it is necessary to stop and restart the unsatisfiability checker whenever a model is found for some cardinality. The unsatisfiability checker may run forever in that case, and it may be required to check if the set is unsatisfiable for *larger* cardinalities.

```
Function check_sat_arrangement(𝒜, L₁, L₂)
 1  k₁ := 1; 𝒮₁ := ∅;
 2  k₂ := 1; 𝒮₂ := ∅;
 3  execute in parallel
 4      process
 5          for ever do
 6              k₁′ := k₁;
 7              execute in parallel
 8                  process
 9                      while ¬find_model(k₁, 𝒯₁, 𝒜 ∪ L₁) do
10                          ⌊ k₁ := k₁ + 1;
11                  process
12                      if 𝒜 ∪ L₁ ∪ card≥(k₁′) is 𝒯₁-unsatisfiable then
13                          wait k₂ ≥ k₁′;
14                          return UNSAT;
15                  wait;
16              𝒮₁ := 𝒮₁ ∪ {k₁};
17              if k₁ ∈ 𝒮₂ then
18                  ⌊ return SAT;
19              k₁ := k₁ + 1;
20      process
21          for ever do
22              k₂′ := k₂;
23              execute in parallel
24                  process
25                      while ¬find_model(k₂, 𝒯₂, 𝒜 ∪ L₂) do
26                          ⌊ k₂ := k₂ + 1;
27                  process
28                      if 𝒜 ∪ L₂ ∪ card≥(k₂′) is 𝒯₁-unsatisfiable then
29                          wait k₁ ≥ k₂′;
30                          return UNSAT;
31                  wait;
32              𝒮₂ := 𝒮₂ ∪ {k₂};
33              if k₂ ∈ 𝒮₁ then
34                  ⌊ return SAT;
35              k₂ := k₂ + 1;
```

Algorithm 4. Combination of two finitely axiomatized theories.

- for completeness, it is however necessary to eventually leave the unsatisfiability checker run undisturbed for ever longer periods;
- if one theory is found not to have models with cardinality greater than k, it is necessary, before returning the answer "unsatisfiable", to wait for the other procedure to check all interpretations of cardinality up to k.

The finite model finder is called at lines 9 and 25; find_model($k_1, \mathcal{T}_1, \mathcal{A} \cup L_1$) returns true if and only if $\mathcal{A} \cup L_1$ has a $\mathcal{T}_1$-model of cardinality k_1. A call to the finite model finder should eventually terminate. It is not required for the unsatisfiability checks at lines 12 and 28 to return, in case the considered formulas are satisfiable.

The shared variables are k_1, k_2, $\mathcal{S}_1$ and $\mathcal{S}_2$. The value of k_2 read by the process for $\mathcal{T}_1$ does not have to be up to date. It is however mandatory for the correctness of the algorithm that the age of the value (i.e. the difference between the current time and the last time for which k_2 had this value for the process for $\mathcal{T}_2$) stays bounded. The symmetric requirement exists for k_1. We assume that there is a critical section wrapping the reading and writing of $\mathcal{S}_1$ and $\mathcal{S}_2$, and that both processes have the same view of those variables. The order of lines 16 and 17–18 and of lines 32 and 33–34 matters. A completeness bug would result from switching those lines.

Theorem 7. *Assume $\mathcal{T}_1$ and $\mathcal{T}_2$ are two disjoint finitely axiomatized theories, then Algorithm 4 yields a refutationally complete procedure for $\mathcal{T}_1 \cup \mathcal{T}_2$.*

Proof. In the following, the process for $\mathcal{T}_1$ denotes lines 5 to 19, and the process for $\mathcal{T}_2$ denotes lines 21 to 35.

It is useful to show that the following invariant properties hold throughout the execution of the algorithm after the initialization at lines 1-2:

1. $I_1 : \mathcal{S}_1 \cap \{k \mid k \geq k_1\} = \emptyset$, at all times except when process for $\mathcal{T}_1$ is executing instructions at lines 16-19
2. $I_2 : \mathcal{S}_2 \cap \{k \mid k \geq k_2\} = \emptyset$, at all times except when process for $\mathcal{T}_2$ is executing instructions at lines 32-35
3. $I_3 : \mathcal{S}_1 \cap \{1, \ldots, k_1 - 1\} = \text{spectrum}(\mathcal{T}_1 \cup \mathcal{A} \cup L_1) \cap \{1, \ldots, k_1 - 1\}$
4. $I_4 : \mathcal{S}_2 \cap \{1, \ldots, k_2 - 1\} = \text{spectrum}(\mathcal{T}_2 \cup \mathcal{A} \cup L_2) \cap \{1, \ldots, k_2 - 1\}$
5. $I_5 : \mathcal{S}_1 \cap \mathcal{S}_2 = \emptyset$, at all times except when process for $\mathcal{T}_1$ (or $\mathcal{T}_2$) is executing instructions at lines 16-18 (resp. instructions at lines 32-34)

Note that all the above invariants are established by the initialization.

The first two invariants I_1 and I_2 are fairly easy to prove. To prove I_1 notice that every addition (of k_1) to $\mathcal{S}_1$ is done at line 16, and immediately followed (otherwise the function terminates) at line 19 by incrementing k_1. Symmetrically for I_2.

For invariant I_3, it is sufficient to show that, if the property is true before executing line 10, it is true after, and if it is true before executing line 19, it is true after. Adding k_1 to $\mathcal{S}_1$ (at line 16) cannot modify the truth status of I_1 since $\mathcal{S}_1 \cap \{1, \ldots, k_1 - 1\}$ does not change. At line 10, k_1 is incremented only if $\mathcal{A} \cup L_1$ does not have any $\mathcal{T}_1$-models with k_1 elements, i.e. only if $k_1 \notin$

spectrum($\mathcal{T}_1 \cup \mathcal{A} \cup L_1$); thanks to I_1, $k_1 \notin \mathcal{S}_1$. And incrementing k_1 at line 19 is only done if the loop at lines 9-10 terminates, that is, if a $\mathcal{T}_1$-model with cardinality k_1 is found for $\mathcal{A} \cup L_1$. Symmetrically for I_4.

To show that I_5 is true, it is sufficient to consider the group of lines 16–18 and 32–34, where $\mathcal{S}_1$ and $\mathcal{S}_2$ are modified. Assume I_5 holds before executing lines at 16–18, and assume the process for $\mathcal{T}_2$ is not running the section between lines 32 and 34. While adding k_1 to $\mathcal{S}_1$, if k_1 also belongs to $\mathcal{S}_2$, the function will return at line 18. The argument is similar for modifications of $\mathcal{S}_2$. Notice however that it is necessary to guarantee that the values of $\mathcal{S}_1$ and $\mathcal{S}_2$ read by the processes for $\mathcal{T}_2$ and $\mathcal{T}_1$ respectively are up to date. Otherwise, the algorithm may "miss" the check of some cardinality.

As a consequence of invariants I_3 and I_4, and thanks to Corollary 3, the algorithm terminates by returning SAT only if $\mathcal{A} \cup L_1 \cup L_2$ is indeed $\mathcal{T}_1 \cup \mathcal{T}_2$-satisfiable (with a model of finite cardinality).

As a consequence of invariants I_3, I_4 and I_5, it can be deduced at line 14 that the spectra are disjoint. If the algorithm returns UNSAT at line 14, then $\mathcal{A} \cup L_1 \cup L_2$ is indeed $\mathcal{T}_1 \cup \mathcal{T}_2$-unsatisfiable. The discussion is symmetric for line 30. Thus the algorithm is sound: a returned answer is always correct. It remains to show that the algorithm is complete, that is, it eventually terminates if $\mathcal{T}_1 \cup \mathcal{T}_2 \cup \mathcal{A} \cup L_1 \cup L_2$ is unsatisfiable.

We prove termination by contradiction, and assume the function never terminates, although the spectra for $\mathcal{T}_1 \cup \mathcal{A} \cup L_1$ and $\mathcal{T}_2 \cup \mathcal{A} \cup L_2$ are disjoint. Remember the Löwenheim-Skolem Theorem that states that a finitely axiomatized theory with (a) an infinite spectrum has models of all infinite cardinalities; (b) an infinite model has models for every infinite cardinality. Thus, if spectrum($\mathcal{T}_1 \cup \mathcal{A} \cup L_1$) and spectrum($\mathcal{T}_2 \cup \mathcal{A} \cup L_2$) are disjoint, at least one of those sets must be a finite set of finite cardinalities. Assume without loss of generality that spectrum($\mathcal{T}_1 \cup \mathcal{A} \cup L_1$) is finite. Then, there exists some k (assume k is the smallest integer) such that spectrum($\mathcal{T}_1 \cup \mathcal{A} \cup L_1$) $\cap \{k' \mid k' \geq k\} = \emptyset$, that is, such that $\mathcal{A} \cup L_1 \cup \mathrm{card}_{\geq}(k)$ is $\mathcal{T}_1$-unsatisfiable.

Notice that if $\mathcal{T}_1 \cup \mathcal{A} \cup L_1$ has a model of finite cardinality $k - 1$ then k_1 will eventually reach k (provided the finite model finder at line 9 is terminating). After that, the process for $\mathcal{T}_1$ will let the finite model finder run forever searching for non-existent models of cardinality greater than or equal to k, while in the meantime, the refutationally complete procedure at line 12 for the $\mathcal{T}_1$-unsatisfiability of $\mathcal{A} \cup L_1 \cup \mathrm{card}_{\geq}(k'_1)$ will run undisturbed (that is, will never be killed) for the amount of time necessary for it to terminate. The process for $\mathcal{T}_1$ will eventually reach line 13. If spectrum($\mathcal{T}_2 \cup \mathcal{A} \cup L_2$) is infinite, k_2 will grow to infinity and will eventually be greater than k'_1. The waiting process for $\mathcal{T}_1$ will eventually be leaving the waiting state and return UNSAT.

Notice (symmetrically as before) that if $\mathcal{T}_2 \cup \mathcal{A} \cup L_2$ has a model of finite cardinality then k_2 will eventually reach and overstep this cardinality (provided the finite model finder at line 26 is terminating). If spectrum($\mathcal{T}_2 \cup \mathcal{A} \cup L_2$) is finite, and if its maximal element is greater or equal than the value k'_1 reached by process $\mathcal{T}_1$ on the waiting state, the waiting process for $\mathcal{T}_1$ will eventually be leaving the waiting state and return UNSAT. The case where its maximal element is strictly smaller than value $k'_1 - 1$ does not need to be considered, by symmetry.

Let us consider finally the degenerated case for which both spectra have the same maximal value, and thus $k'_1 = k'_2$, and suppose that both processes are waiting, at instruction 13 and 29 respectively. k'_1 and k'_2 must then both be strictly greater than 1. This cannot happen: spectrum($\mathcal{T}_1 \cup \mathcal{A} \cup L_1$) should contain $k'_1 - 1$ and spectrum($\mathcal{T}_2 \cup \mathcal{A} \cup L_2$) should contain $k'_2 - 1$, i.e. $k'_1 - 1$, which contradicts invariant I_5. □

Algorithm 4 eventually terminates if $\mathcal{A} \cup L_1 \cup L_2$ is $\mathcal{T}_1 \cup \mathcal{T}_2$-unsatisfiable, or if $\mathcal{A} \cup L_1 \cup L_2$ is $\mathcal{T}_1 \cup \mathcal{T}_2$-satisfiable in a finite model. It is both a complete refutation procedure and a finite model finder.

7 Conclusion

We studied two cases of refutationally complete combination of disjoint theories. In the first case, we considered combining a decidable theory with an arbitrary finitely axiomatized theory. In the second, we provided an algorithm to combine two finitely axiomatized theories. Both these algorithms are not yet efficiently implemented and would require further techniques and heuristics to be turned into useful solvers. Just as in the case of decidable theories, it is not realistic to consider every arrangement separately: techniques developed for the combination of decision procedures — e.g. cooperation by equality exchange, or delayed theory combination (see e.g. [2]) — will have to be transposed, but will require further research since it can not be expected that the components in the combinations will eventually terminate. Similarly, the negotiation of suitable cardinalities will require specific methods and heuristics.

Bonacina et al. [4] show that, in the case of decidable and universal finitely axiomatized theories, it is possible (and sufficient for some combinations) to extract cardinality constraints from the saturation of a superposition solver. We here considered refutationally complete procedures as black boxes, but, among the potential heuristics to make the approach work in practice, one could consider extracting cardinality hints from the saturation provers, to improve the cardinality negotiation in Algorithm 4.

In [14], tableaux are used to combine in a refutationally complete way two theories sharing a dense order. In this case, the difficulty of the combination does not lie in the agreement on cardinalities (since both theories can only have infinite models), but in the non-disjoint signature containing this order predicate. As an extension of the present work, it would be interesting to find practical ways to combine loosely connected rather than disjoint theories, such as theories sharing a few unary predicates.

Acknowledgment. We would like to thank Christophe Ringeissen and Michaël Rusinowitch for pointing out some related works. We also thank the anonymous reviewers for their comments.

References

1. Althaus, E., Kruglov, E., Weidenbach, C.: Superposition Modulo Linear Arithmetic SUP(LA). In: Ghilardi, S., Sebastiani, R. (eds.) FroCoS 2009. LNCS, vol. 5749, pp. 84–99. Springer, Heidelberg (2009)
2. Barrett, C., Sebastiani, R., Seshia, S.A., Tinelli, C.: Satisfiability modulo theories. In: Biere, A., Heule, M.J.H., van Maaren, H., Walsh, T. (eds.) Handbook of Satisfiability. Frontiers in Artificial Intelligence and Applications, vol. 185, ch. 26, pp. 825–885. IOS Press (February 2009)
3. Baumgartner, P., Tinelli, C.: Model Evolution with Equality Modulo Built-in Theories. In: Bjørner, N., Sofronie-Stokkermans, V. (eds.) CADE 2011. LNCS, vol. 6803, pp. 85–100. Springer, Heidelberg (2011)
4. Bonacina, M.P., Ghilardi, S., Nicolini, E., Ranise, S., Zucchelli, D.: Decidability and Undecidability Results for Nelson-Oppen and Rewrite-Based Decision Procedures. In: Furbach, U., Shankar, N. (eds.) IJCAR 2006. LNCS (LNAI), vol. 4130, pp. 513–527. Springer, Heidelberg (2006)
5. Bonacina, M.P., Lynch, C., de Moura, L.: On Deciding Satisfiability by DPLL($\Gamma + \mathcal{T}$) and Unsound Theorem Proving. In: Schmidt, R.A. (ed.) CADE 2009. LNCS, vol. 5663, pp. 35–50. Springer, Heidelberg (2009)
6. Fontaine, P.: Combinations of Theories for Decidable Fragments of First-Order Logic. In: Ghilardi, S., Sebastiani, R. (eds.) FroCoS 2009. LNCS, vol. 5749, pp. 263–278. Springer, Heidelberg (2009)
7. Fontaine, P., Gribomont, E.P.: Combining non-stably infinite, non-first order theories. In: Ahrendt, W., Baumgartner, P., de Nivelle, H., Ranise, S., Tinelli, C. (eds.) Selected Papers from the Workshops on Disproving and the Second International Workshop on Pragmatics of Decision Procedures (PDPAR 2004). ENTCS, vol. 125, pp. 37–51 (July 2005)
8. Ge, Y., de Moura, L.: Complete Instantiation for Quantified Formulas in Satisfiabiliby Modulo Theories. In: Bouajjani, A., Maler, O. (eds.) CAV 2009. LNCS, vol. 5643, pp. 306–320. Springer, Heidelberg (2009)
9. Nelson, G., Oppen, D.C.: Simplifications by cooperating decision procedures. ACM Trans. on Programming Languages and Systems 1(2), 245–257 (1979)
10. Tinelli, C., Harandi, M.T.: A new correctness proof of the Nelson–Oppen combination procedure. In: Baader, F., Schulz, K.U. (eds.) FroCoS, Applied Logic, pp. 103–120. Kluwer Academic Publishers (March 1996)
11. Tinelli, C., Ringeissen, C.: Unions of non-disjoint theories and combinations of satisfiability procedures. Theoretical Computer Science 290(1), 291–353 (2003)
12. Tinelli, C., Zarba, C.G.: Combining non-stably infinite theories. Journal of Automated Reasoning 34(3), 209–238 (2005)
13. Wies, T., Piskac, R., Kuncak, V.: Combining Theories with Shared Set Operations. In: Ghilardi, S., Sebastiani, R. (eds.) FroCoS 2009. LNCS, vol. 5749, pp. 366–382. Springer, Heidelberg (2009)
14. Zarba, C.G., Manna, Z., Sipma, H.B.: Combining theories sharing dense orders. In: TABLEAUX, Position Papers and Tutorials, pp. 83–98 (2003), Technical Report RT-DIA-80-2003

Automated Analysis of Regular Algebra

Simon Foster and Georg Struth

Department of Computer Science, The University of Sheffield
{s.foster,g.struth}@dcs.shef.ac.uk

Abstract. Regular algebras axiomatise the equational theory of regular expressions. We use Isabelle/HOL's automated theorem provers and counterexample generators to study the regular algebras of Boffa, Conway, Kozen and Salomaa, formalise their soundness and completeness (relative to a deep result by Krob) and engineer their hierarchy. Proofs range from fully automatic axiomatic and inductive calculations to integrated higher-order reasoning with numbers, sets and monoid submorphisms. In combination with Isabelle's simplifiers and structuring mechanisms, automated deduction provides powerful support to the working mathematician beyond first-order reasoning.

1 Introduction

Regular languages, regular expressions and finite automata belong to the foundations of computing. Regular algebras are the mathematical structures that underly these formalisms. Originally proposed for axiomatising the equational theory of regular expressions, they have since found wide applications in various fields of computing.

Work on regular algebras has spanned decades. Salomaa gave two axiom systems, proved completeness of the first and conjectured it of the second [12]. Conway, in his influential monograph, conjectured completeness of several alternative axiomatisations [6]. Krob gave a long and intricate completeness proof of Conway's so-called classical axioms extended by a system of monoid identities [10]. Boffa proved completeness of two particularly simple algebras relative to Krob's result [3,4]. Relative to Boffa's algebras, Krob, in turn, verified some of Conway's remaining conjectures. Kozen proved completeness of a simplified algebra of Conway [8], which under the name *Kleene algebra* has been widely studied and applied since. Boffa, in turn, showed completeness of a simplified version of Kleene algebra.

Within the programme of enhancing mathematics by theorem provers, regular algebras yield an interesting test case: they include pure first-order as well as higher-order structures axiomatised by inductive families of identities and with elements generated by finite monoids via submorphisms. Proofs include equational calculations and integrated higher-order reasoning about algebra, numbers, sets (of lists), infinite suprema and functions. Our main motivation is the following question: *How far can first-order automated theorem provers support the working mathematician in such a heterogeneous environment?*

B. Gramlich, D. Miller, and U. Sattler (Eds.): IJCAR 2012, LNAI 7364, pp. 271–285, 2012.

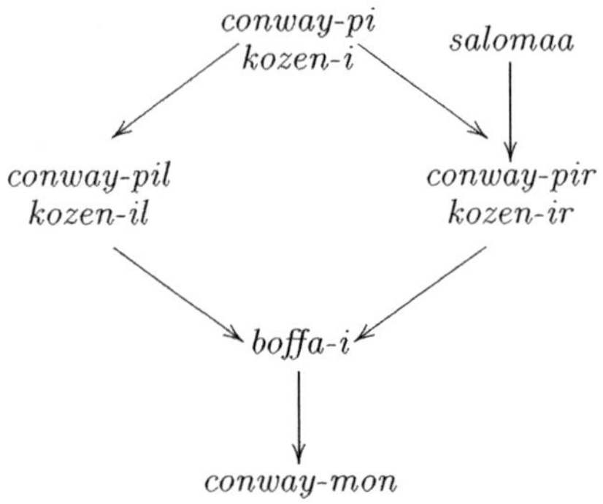

Fig. 1. Fine structure of regular algebra

Automated theorem proving alone, is of course too limited for our study. It is only possible due to Isabelle/HOL's [11] recent integration of first-order proof and counterexample search technology into a higher-order interactive theorem proving environment (cf. [2] for an overview). In a nutshell, Isabelle's Sledgehammer tool delegates proof goals to external automated theorem provers (ATPs) and satisfiability modulo theories (SMT) solvers. A relevance filter gathers hypotheses for the external tools. Their proof outputs are reconstructed within Isabelle to increase trustworthiness. ATP in Isabelle is complemented by the Nitpick and Quickcheck counterexample search tools. This integration supports a very natural new style of computer enhanced mathematics. Traditionally, work with Isabelle was driven by its simplifiers and direct applications of theorems from its libraries. Now, paper and pencil proofs can be typed directly into Isabelle's proof scripting language and verified step by step by an ATP system using the hypotheses it gathers. With this approach, an Isabelle repository for Kleene and relation algebras with more than 2000 facts has already been implemented[1]. But the fine structure of regular algebras with their higher-order features has not yet been considered. Our main contributions are as follows:

We implement the algebras of Boffa, Conway (without monoid identities), Kozen and Salomaa as abstract type classes in Isabelle and develop a library of regular identities and auxiliary concepts for Boffa's algebras.

We use Isabelle's locale mechanism in combination with Nitpick to capture meta-theorems that relate these algebras. We reconstruct known completeness results for regular algebras and add some new ones: equipollence (mutual deducibility) of Boffa's algebras, and of some of Conway's algebras and Kleene algebras, a simple completeness proof for Salomaa's first algebra, a gap in Boffa's completeness proof for his second one, and proofs that various subclasses are proper. The main relationships are shown in Figure 1. Nodes represent equipollent algebras; arrows the implication preorder. All completeness proofs are relative to Krob's result; they are based on implications between axiom systems.

[1] `http://www.dcs.shef.ac.uk/~georg/isa`

We establish soundness of regular algebras relative to regular languages. Soundness of Salomaa's algebra and Kleene algebra is automatically propagated down the hierarchy by Isabelle's sublocale mechanism.

We reconstruct Boffa's completeness result relative to Conway's classical axioms with monoid identities. This requires an alternative implementation of Boffa's algebras with explicit carrier sets and additional theory infrastructure. In this case, the sublocale mechanism propagates completeness up the hierarchy.

Most subclass and equipollence proofs are fully automatic. This demonstrates the impressive power of ATP in algebraic reasoning. Automating more complex results requires specific elimination rules for higher-order structure and an interplay with Isabelle's simplifier. As soon as supporting libraries were developed, all proofs could be implemented at least at textbook-level granularity in a natural mathematical style. Some formalisation tasks, in particular the construction of infinite counterexamples, are deliberately left open to demonstrate not only the potential, but also the limitations of our lightweight ATP-based approach.

This paper can only highlight some main features of our work. The complete Isabelle implementation can be accessed through our repository. We must also assume familiarity with the basics of Isabelle. We refer to the excellent online documentation, in particular the locale tutorial [1] and the references given therein, for further information. The paper itself has been processed by Isabelle's document preparation system, including the verification of its technical results. The following numbers underpin the success of ATP in analysing regular algebra: our implementation contains 303 proof goals. 242 were fully automatic (apart perhaps from calling an induction or case analysis tactic); 35 were fully automatic after invoking a simplifier; 26 required moderate user interaction.

2 Dioids, Powers and Finite Sums

All regular algebras can be based on dioids or idempotent semirings. Implementations of these structures and a library of facts can be found in the repository.

Formally, a *semiring* is a structure $(S, +, \cdot, 0, 1)$ where $(S, +, 0)$ is a commutative monoid, $(S, \cdot, 1)$ is a monoid, and the distributivity laws $x \cdot (y + z) = x \cdot y + x \cdot z$ and $(x + y) \cdot z = x \cdot z + y \cdot z$, and annihilation laws $x \cdot 0 = 0$ and $0 \cdot x = 0$ hold. A semiring is *idempotent*—a *dioid*—if $x + x = x$. In this case the reduct $(S, +)$ forms a semilattice, and can be endowed with the usual semilattice order $x \leq y \leftrightarrow x + y = y$. The least element of this order is 0 and the operations of addition and multiplication are isotone. An important concept in semiring theory is duality with respect to opposition. It is based on the opposite multiplication $x \circ y = y \cdot x$. We have implemented this duality in Isabelle and shown that $(S, +, \circ, 0, 1)$ is a dioid whenever $(S, +, \cdot, 0, 1)$ is. Duals of theorems in dioids are available for free in Isabelle. This also yields automatic completeness proofs for the duals of all structures in this paper (e.g. the righthanded algebras in Figure 1).

For most of the development in this paper, implementing algebras by axiomatic type classes is sufficient. Consequently, their carrier sets are left implicit. This is common mathematical practice, beneficial to automation, but insufficient for more advanced mathematics (cf. Section 9). Some axiomatisations of regular algebras require powers and finite sums. Powers can be defined recursively.

primrec *power* :: *'a* ⇒ *nat* ⇒ *'a* (-ˆ [101,50] 100)
  where $x^0 = 1$
  | $x^{Suc\ n} = x{\cdot}x^n$

We have developed a basic library for powers. Typical facts are $x^n \cdot x = x \cdot x^n$ or $y \cdot x \leq y \rightarrow y \cdot x^n \leq y$. Apart from induction, proofs are mostly automatic. The following example illustrates the style of reasoning.

Lemma *power-add*: $x^m{\cdot}x^n = x^{m+n}$
Proof (*induct m*)
  case *0* **show** *?case* **by** (*metis add-0-left mult-onel power.simps*(*1*))
  case (*Suc m*) **show** *?case* **by** (*smt Suc add-Suc mult-assoc power.simps*(*2*))
qed

Isabelle's induction tactic is called to generate proof obligations for the base case and the induction step. Sledgehammer is then called on both cases. The first case is discharged by Metis, an internally verified ATP system. The second one uses SMT proof reconstruction. The proof uses the clauses in the definition of *power*, induction hypothesis *Suc* and facts about dioids and numbers. All have been gathered by the relevance filter.

Next we define a function that sums up powers: $x_m^n = \sum_{i=m}^{n+m} x^i$. Avoiding Isabelle's library function *setsum* yields better control over proof automation, but ultimately, an integration with existing Isabelle libraries is desirable (cf. [7]).

primrec *powsum* :: *'a* ⇒ *nat* ⇒ *nat* ⇒ *'a* (-_ˆ [101,50,50] 100)
  where ${x_n}^0 = x^n$
  | ${x_n}^{Suc\ m} = {x_n}^m{+}x^{n+Suc\ m}$

Again we have proved a number of basic facts by ATP, often by induction, and sometimes calling Isabelle's simplifier before Sledgehammer.

3 Conway's Classical Axioms

Regular algebras are dioids expanded by the regular operation *. We implement Conway's classical axioms (p.25 in his monograph) using Isabelle's axiomatic type classes; hence again without explicit carrier sets.

Class *regalg-base* = *dioid-one-zero* + *star-op* + *plus-ord* +
  assumes *C11*: $(x{+}y)^* = (x^*{\cdot}y)^*{\cdot}x^*$
  and *C12*: $(x{\cdot}y)^* = 1{+}x{\cdot}(y{\cdot}x)^*{\cdot}y$

Class *conway* = *regalg-base* +
  assumes *C13*: $(x^*)^* = x^*$

Class *conway-classical* = *conway* +
  assumes *C14*: $x^* = (x^{n+1})^*{\cdot}{x_0}^n$

The class *regalg-base* is reused for Boffa's first axiomatisation. In class *conway*, axiom schema $C14$—also called *powerstar* axiom—has been removed from the classical axioms, since it is not needed for most of our results.

Conway himself uses semirings instead of dioids. He shows that $x + x = x$ can be derived from that basis; hence both variants are equipollent. We use dioids for the sake of uniformity across the paper. Conway has shown that the classical axioms are incomplete with respect to (the equational theory of) regular languages (p. 118). He has also analysed the role of axiom $C13$ (p. 104). We could easily automate his analysis with Nitpick: a 3-element counterexample shows irredundancy of $C13$ in the semiring setting; in its absence, $x+x = x$ (3-element counterexample) and $x^* \cdot x^* = x^*$ (5-element counterexample) could be refuted. In the dioid setting, however, we could neither prove nor refute $x^* \cdot x^* = x^*$ automatically in the absence of $C13$ within Isabelle's default time limits. In the presence of powerstar, Nitpick uniformly failed. In fact, Conway constructs an infinite model of a semiring in which the classical axioms except $C13$ hold and in which $C13$ fails (p.104). We have not attemtped to formalise his model.

4 Boffa's Axioms

Boffa [3,4] presented two axiom systems for regular algebra. His first axiomatisation adds a very simple quasi-identity to Conway's classical axioms. In his second paper he shows that some of Conway's axioms—including powerstar—are redundant. He also shows that his second axiomatisation implies the first. We can base the first axiomatisation on *regalg-base*.

Class *boffa-1* = *regalg-base* +
  assumes *R*: $x{\cdot}x = x \to x^* = 1{+}x$

Class *boffa-2* = *dioid-one-zero* + *star-op* +
  assumes *B1*: $1{+}x \leq x^*$
  and *B2*: $x^*{\cdot}x^* = x^*$
  and *B3*: $1{+}x \leq y \wedge y{\cdot}y = y \to x^* \leq y$

Boffa algebras are closed under duality since all axioms are self-dual.

We first show that *boffa-1* and *boffa-2* are equipollent (*boffa-1* = *boffa-2*). Boffa has already shown that *boffa-2* ⊆ *boffa-1*—the first is a subclass of the second—whereas the converse inclusion is new. Following Boffa, we then relate Boffa's algebras with Conway's classical axioms. In Isabelle, subclass relationships can be captured by subclass or sublocale proofs. We use sublocales simply because the associated syntax leads to more readable statements. In general, an understanding of Isabelle's subclass and locale mechanisms is not needed to grasp the mathematical statements in this paper.

Sublocale *boffa-1* ⊆ *boffa-2*

Isabelle dictates the proof obligations: all *boffa-2* axioms must be derived from *boffa-1*. In fact, only $B1$-$B3$ need to be verified. Isabelle recognises that both algebras extend the class *dioid-one-zero*. All proof obligations were discharged by ATP. All theorems for *boffa-2* are now automatically available for *boffa-1*.

Proving the converse sublocale relationship is more involved. A direct automated proof was impossible within Isabelle's time limits. First, we therefore verified all regular identities that have been proved for Kleene algebras in the repository in the weaker context of *boffa-2*. These 46 facts include well known identities such as $1 \leq x^*$, $x \leq x^*$, $x^* \cdot x^* = x^*$, $x^{**} = x^*$, $1^* = 1$, $0^* = 1$, $1+x \cdot x^* = x^*$, $(x \cdot y)^* \cdot x = x \cdot (x \cdot y)^*$, and $(x+y)^* = x^* \cdot (y \cdot x^*)^*$. 41 proofs were automatic; for the remaining ones, paper and pencil proofs could be translated. Consider the following proof of $C12$ as an example.

Proof –
  have $\forall x\ y.\ 1{+}x{\cdot}(y{\cdot}x)^*{\cdot}y = (1{+}x{\cdot}(y{\cdot}x)^*{\cdot}y){\cdot}(1{+}x{\cdot}(y{\cdot}x)^*{\cdot}y)$ — by smt
  hence $\forall x\ y.\ (x{\cdot}y)^* \leq 1{+}x{\cdot}(y{\cdot}x)^*{\cdot}y$ — by metis
  hence $1{+}x{\cdot}(y{\cdot}x)^*{\cdot}y \leq 1{+}x{\cdot}y{+}x{\cdot}y{\cdot}(x{\cdot}y)^*{\cdot}(x{\cdot}y)$ — by smt
  hence $1{+}x{\cdot}(y{\cdot}x)^*{\cdot}y \leq (x{\cdot}y)^*$ — by smt ...
  thus *?thesis* — by metis ...
qed

The remaining half of *boffa-1* = *boffa-2* is then fully automatic.

Sublocale *boffa-2* $\subseteq$ *boffa-1*

All regular identities are now available also in *boffa-1*.

Deriving Conway's classical axioms from Boffa's algebras again requires some preparation. We need a few general lemmas about the interaction of the star with (sums of) powers, for instance, that $x^n \leq x^*$, $x^k \cdot (x^n)^* = (x^n)^* \cdot x^k$ and $x_0^k \cdot (x^n)^* = (x^n)^* \cdot x_0^k$ for $k \leq n$, and $x_m^n \leq x^*$. Most of them are automatic up to induction. Finally, to derive *powerstar* from $B3$, it suffices to prove the following two facts.

Lemma *conway-powerstar1*: $(x^{n+1})^*{\cdot}x_0{}^n{\cdot}(x^{n+1})^*{\cdot}x_0{}^n = (x^{n+1})^*{\cdot}x_0{}^n$

Lemma *conway-powerstar2*: $1{+}x \leq (x^{n+1})^*{\cdot}x_0{}^n$

Their proofs require a case analysis on n. While the $n = 0$ cases are automatic, those for $n \neq 0$ translate paper and pencil proofs. *powerstar* can then be derived automatically in two steps ($\leq$ and $\geq$), and the desired sublocale statement is automatic as well.

Theorem *powerstar*: $x^* = (x^{n+1})^*{\cdot}x_0{}^n$

Sublocale *boffa-2* $\subseteq$ *conway-classical*

All theorems of Boffa's algebras are now available for Conway's classical axioms.

The subclass relationship is strict. Boffa has shown that his algebras are complete (relative to Krob's result, cf. Section 9); Conway has shown that his classical axioms are not (p. 118). This implies that R cannot be derived in *conway-classical*. We have tried unsuccessfully to test this fact with Sledgehammer and Nitpick. This is not surprising because Conway's counterexample is constructed inductively. Again we have not further attempted to formalise Conway's proof.

5 Conway's Conjectures

Conway presents several extensions of his classical axioms and conjectures their completeness (p. 103). Boffa has verified one of them, Krob the remaining ones relative to *boffa-1* (p. 329f). All completeness results are relative to Krob's completeness proofs of Conway's classical axioms with monoid indentities. Following Boffa, these axioms are derived from Boffa's algebras in Section 9, which shows that Boffa's algebras are complete as well. The (relative) completeness results in this section are obtained by deriving Boffa's axioms. We automatically reconstruct Boffa and Krob's results in the weaker setting of *conway* without *powerstar* by deriving the axioms of *boffa-1* from them. We also establish new equipollence results for Conway's variants. Conway considers dual lefthanded and righthanded variants as well as their combinations. Here we only present the lefthanded ones. Their duals and all dual statements can be found in the repository.

Class *conway-p0* = *conway* +
assumes *P0*: $x \cdot y = y \cdot z \rightarrow x^* \cdot y = y \cdot z^*$

Class *conway-p1l* = *conway* +
assumes *P1l*: $x \cdot y \leq y \cdot z \rightarrow x^* \cdot y \leq y \cdot z^*$

Class *conway-p2l* = *conway* +
assumes *P2l*: $x = y \cdot x \rightarrow x = y^* \cdot x$

Class *conway-p3l* = *conway* +
assumes *P3l*: $x \cdot y \leq y \rightarrow x^* \cdot y \leq y$

The rule $P3l$ and its dual will reappear in Kozen's axiomatisation.

We establish two results. First, we show that *conway-p2l* is complete. Second, we prove that all lefthanded variants are equipollent, hence complete as well. The following result is automatic:

Sublocale *conway-p2l* $\subseteq$ *boffa-1*

The question whether *conway-p2l* = *boffa-1* remains open. We could neither prove nor refute the remaining inclusion within Isabelle's default time limits, despite the fact that all our regular identities are available in Boffa's algebras.

The regular identities can now be used in *conway-p2l* to prove equipollence of Conway's variants in a completely automatic fashion. As usual, the sublocale mechanism takes care of metalogical aspects such as theorem propagation.

Sublocale *conway-p2l* $\subseteq$ *conway-p3l*

Sublocale *conway-p3l* $\subseteq$ *conway-p1l*

Sublocale *conway-p1l* $\subseteq$ *conway-p2l*

Finally we show for $i = 1, 2, 3$ that the combination of *conway-pil* and *conway-pir* is equipollent to *conway-p0*. Here we only present the result for $i = 2$.

Class *conway-p2* = *conway-p2l* + *conway-p2r*

Sublocale *conway-p0* $\subseteq$ *conway-p2*

Sublocale *conway-p2* $\subseteq$ *conway-p0*

6 Kozen's Kleene Algebras

Kozen's Kleene algebras are essentially *conway-p3* with $C11$-$C13$ replaced by a simpler axiom. Kozen gave an elementary completeness proof for his variant based on Conway's trick of encoding finite automata in terms of a matrix regular algebra over a regular algebra. This proof has recently been formalised in the proof assistant Coq [5]. Boffa proved completeness for left Kleene algebras, where axiom $P3r$ is absent, relative to *boffa-2* (it seems that Kozen's proof does not go through in this weaker context).

We reconstruct Boffa's completeness result and prove new results that establish equipollence of Kleene algebras and Conway's variants. Finally, we reproduce well known equipollence results between two variants of Kleene algebra introduced by Kozen. As usual, we stick to the left. Dual classes and statements can be found in the repository.

Class *kozen-base-l* = *dioid-one-zero* + *star-op* +
assumes *star-unfoldl'*: $1{+}x{\cdot}x^* \leq x^*$

Class *kozen-1l* = *kozen-base-l* +
assumes *star-inductl*: $x{\cdot}y \leq y \rightarrow x^*{\cdot}y \leq y$

Class *kozen-2l* = *kozen-base-l* +
assumes *star-inductl-var*: $z{+}x{\cdot}y \leq y \rightarrow x^*{\cdot}z \leq y$

Class *kozen* = *kozen-1l* + *kozen-1r*

Conceptually, completeness of *kozen-1l* (and its dual) follows from the equipollence results below. Technically, however, the corresponding sublocale proof is particularly simple and automatic; it also brings the regular identities into the scope of *kozen-1l* for equipollence proofs.

Sublocale *kozen-1l* $\subseteq$ *boffa-2*

Sublocale *kozen-1l* $\subseteq$ *conway-p2l*

Sublocale *conway-p2l* $\subseteq$ *kozen-1l*

Sublocale *kozen* $\subseteq$ *conway-p0*

Sublocale *conway-p0* $\subseteq$ *kozen*

All proofs are fully automatic. They show that *conway-pil* = *kozen-1l* and *conway-pi* = *kozen*. Finally, we establish equipollence of Kozen's variants.

Sublocale *kozen-1l* $\subseteq$ *kozen-2l*

Sublocale *kozen-2l* $\subseteq$ *kozen-1l*

Once more we were unsuccessful in testing whether Kozen's algebras are equipollent to Boffa's within Isabelle's default time limits.

7 Salomaa's Axioms

Salomaa's axioms are based on dioids without 1, since in the presence of the Kleene star, 1 can be defined as 0^*. Boffa has observed that idempotency is redundant in this setting. As before we base Salomaa's axiomatisation on dioids to keep the development simple and uniform.

Salomaa presents two axiom systems, proves completeness for the first and conjectures that property for the second one. His completeness proof uses an algebraic abstraction of Arden's well known rule for solving linear equations over regular languages (axiom *salomaa*). Since a precondition of Arden's rule is the absence of the empty word property—some language must not contain the empty word—Salomaa inductively defines the negation of property *ewp* for regular algebra terms (or regular expressions). Due to this, one of his axioms is not defined for first-order variables, but for substitution instances of terms.

To circumvent this complication we define *ewp* abstractly with respect to a property that holds in the case of regular languages, as we show in the next section. This property suffices for our completeness proof. It can safely be replaced by stronger (inductive) properties that imply it.

Class *salomaa-ewp* = *dioid-one-zero* + *star-op* +
  fixes *ewp* :: $'a \Rightarrow$ *bool*
  assumes *S11*: $(1{+}x)^* = x^*$
  and *S12*: $x^* = 1{+}x^*{\cdot}x$
  and *ewp-form* : $ewp\ x \leftrightarrow (\exists\, y.\ x = 1{+}y \wedge \neg\ ewp\ y)$

Class *salomaa* = *salomaa-ewp* +
  assumes *salomaa* : $(\neg\ ewp\ y) \wedge x = x{\cdot}y{+}z \rightarrow x = z{\cdot}y^*$

Class *salomaa-conj* = *salomaa-ewp* +
  assumes *salomaa-small* : $(\neg\ ewp\ y) \wedge x = x{\cdot}y{+}1 \rightarrow x = y^*$

Property *ewp-form* states that the empty word can be isolated from every language that contains it. We can easily reconstruct the following relationship [3].

Sublocale *salomaa* $\subseteq$ *salomaa-conj*

salomaa = *salomaa-conj* could be refuted by a 3-element counterexample. We have not tested whether this would still hold for stronger variants of *ewp*.

Boffa has presented a completeness proof of *salomaa-conj* relative to *boffa-1*. We provide a new direct completeness proof of *salomaa* relative to *kozen-1r* and briefly argue why Boffa's proof contains a gap.

Proving *star-inductr-var* from *salomaa* yields completeness automatically.

Lemma *kozen-induct*: $y{\cdot}x{+}z \leq y \rightarrow z{\cdot}x^{*} \leq y$
Proof (*cases ewp x*)
  case *False* **thus** *?thesis* — one step by metis
next
  case *True* **thus** *?thesis* — several steps by metis and smt, using *ewp-form*
qed

Sublocale *salomaa* $\subseteq$ *kozen-2r*

The proof of *kozen-induct* illustrates the fact that reasoning with Salomaa's axioms typically requires case analyses on *ewp* and the trick of using *ewp-form* to reduce the negative case to one where *salomaa* can again be applied.

Such a case analysis is needed in the completeness proof of *salomaa-conj*, but omitted by Boffa [3]. We have attempted a complete case analysis for $C11$ but failed with manual proofs based on Boffa's paper as well as with automated and interactive attempts. Also, Nitpick could not find a counterexample. As far as we can tell, completeness of *salomaa-conj* therefore remains open.

8 Soundness

We now prove soundness of Salomaa's axioms and Kleene algebras, which in this context means that the regular languages form models of these axioms. By our sublocale relationships this implies soundness of all the other regular algebras investigated (cf. Figure 1). The main step is proving Arden's lemma (i.e. axiom *salomaa*) at the language level, for which we could have reused a previous formalisation in Isabelle [9]. Access to the algebraic level, however, significantly simplifies this previous development. Only a few non-automatic non-algebraic proofs are needed.

As usual in Isabelle, words are represented as lists; @ denotes word concatenation. To enhance automation we introduce elimination rules for higher-order concepts. They can be used for simplification before calling Sledgehammer.

type-synonym *'a lan = 'a list set*

Definition *l-prod* :: *'a lan ⇒ 'a lan ⇒ 'a lan* (**infixr** · *75*)
  where $X{\cdot}Y = \{v@w \mid v\ w.\ v{\in}X \wedge w{\in}Y\}$

Lemma *l-prod-elim*: $w{\in}X{\cdot}Y \leftrightarrow (\exists\, u\ v.\ w = u@v \wedge u{\in}X \wedge v{\in}Y)$

We can directly show by an interpretation statement that regular languages form dioids (though that might not be immediately evident from Isabelle's syntax).

Interpretation *dioid-one-zero* (*op* ∪) *l-prod* (*op* ⊆) (*op* ⊂) {[]} {}

We can now use the function *power* from *dioid-one-zero* to define the Kleene star of a language as usual (*powsum* would only yield finite sums). We also define the empty word property in the obvious way.

Definition *star* :: *'a lan ⇒ 'a lan* (-* *[101] 100*)

where $X^* = (\bigcup n.\ X^n)$

Definition *l-ewp* $X \leftrightarrow \{[]\} \subseteq X$

Lemma *star-elim*: $x{\in}X^* \leftrightarrow (\exists k.\ x{\in}X^k)$

To show that regular languages form Kleene algebras, only two continuity properties are needed. Both are automatic after calling Isabelle's simplifier.

Lemma *star-contl*: $X{\cdot}Y^* = (\bigcup n.\ X{\cdot}Y^n)$

Lemma *star-contr*: $X^*{\cdot}Y = (\bigcup n.\ X^n{\cdot}Y)$

Interpretation *kozen* $(op \cup)$ *l-prod* $(op \subseteq)$ $(op \subset)$ $\{[]\}$ $\{\}$ *star*

Only the verification of the unfold rules required a few interactions. All regular identities are now available for regular languages and can be used in the remaining step; the derivation of Arden's rule, which verifies axiom *salomaa*.

In fact, only an inequality remains to be shown since one half of the proof is already covered by axiom *star-inductr-var* of Kleene algebra. Part of this inequality can be captured at the abstract algebraic level as well.

Lemma (in *boffa-1*) *arden-aux*: $y \leq y{\cdot}x{+}z \rightarrow y \leq y{\cdot}x^{Suc\ n}{+}z{\cdot}x^*$

Its proof translates an inductive paper and pencil argument. It now suffices to show that—under the conditions of axiom *salomaa* interpreted in regular languages— the term $Y \cdot X^{Suc\ n}$ vanishes. Following the textbook proofs of Arden's lemma, this is the case since the length of minimal words in $Y \cdot X^n$ grows proportionally to n, hence all words in Y die out in $Y \cdot X^n$ for n sufficiently large. We formalise this using two elementary facts about lower bounds of word lengths in languages.

Lemma *prod-lb*: $(\forall w{\in}X.\ m{\leq}|w|\ ) \rightarrow (\forall w{\in}Y.\ n{\leq}|w|\ ) \rightarrow (\forall w{\in}X{\cdot}Y.\ m{+}n{\leq}|w|\ )$

Lemma *power-lb*: $(\forall w{\in}X.\ k{\leq}|w|\ ) \rightarrow (\forall w.\ w{\in}X^{Suc\ n} \rightarrow n{*}k{\leq}|w|\ )$

Lemma *word-suicide*: $\neg$ *l-ewp* $X \rightarrow Y \neq \{\} \rightarrow (\forall w{\in}Y.\ \exists n.\ w{\notin}Y{\cdot}X^{Suc\ n})$

Only *power-lb* requires induction and some user interaction in the induction step. The proof of *word-suicide* is calculational with 3 intermediate steps. Together with *arden-aux* it is used in the following soundness result, which now is completely automatic.

Interpretation *salomaa op* $\cup$ *l-prod op* $\subseteq$ *op* $\subset$ $\{[]\}$ $\{\}$ *star l-ewp*

9 Relative Completeness

Krob has proved completeness of Conway's classical axioms extended by the following rule: If $x_i \cdot x_j \leq x_{i\circ j}$ and $(x_{i,i})^* = x_{i,i}$ hold for all $i, j \in I$, then $(\sum x_i)^* = \sum x_i$ (p. 116 of Conway's monograph). In this definition, I is a finite monoid, $\sum$ indicates summation over I and $x_{i,j} = \sum_{ik=j} x_k$. The discussion of this schematic rule—which has been called *monoid identities* by Krob—and of Krob's proof requires group theory beyond the scope of this paper; a short sketch can be found in Conway's monograph.

Perhaps surprisingly, Boffa has shown that the monoid identities are derivable from *boffa-1* by purely elementary reasoning. Relative to Krob's result, this establishes completeness of *boffa-1* (hence of all algebras in Figure 1).

We now reconstruct Boffa's proof in Isabelle. While his original proof covers just a few lines, a certain amount of theory infrastructure must be developed in Isabelle beforehand. First, abstract axiomatic reasoning—as in the previous sections—is no longer sufficient; an axiomatisation of Boffa's algebras based on carrier sets is needed. Second, finite sums need to be implemented for algebras with carrier sets since they are not available in Isabelle's standard library. Third, elements of Boffa's algebras must be modelled as functions from a finite monoid into a Boffa algebra in order to capture indexing.

Algebras with explicit carrier sets can be found in the Isabelle library, however, the associated syntax is not well documented and some constructions used in this section may therefore remain somewhat obscure. We have implemented dioids along these lines and proved some essential properties.

locale *dioid* = *weak-partial-order D* **for** *D* (**structure**) +
  assumes *add-closed*: ⟦*x*∈*carrier D*; *y*∈*carrier D*⟧ ⇒ *x*+*y*∈*carrier D*
  — and further closure conditions
  and *mult-assoc*: ⟦*x*∈*carrier D*; *y*∈*carrier D*; *z*∈*carrier D*⟧ ⇒ *x*·(*y*·*z*) = (*x*·*y*)·*z*
  — and the remaining dioid axioms

Algebraic structures are now parametrised with respect to their carrier set, and closure conditions for all operations must be added. ATP systems must check these additional conditions, which involve some simple set expressions. At the level of dioids, however, this has little impact on their performance.

The most natural way of defining finite sums over dioids with carrier sets would be using a fold function, as does Isabelle's *setsum* operator without carriers. For automated theorem proving, however, it turns out to be much simpler to define this (partial) recursive function by locale extension.

locale *dioid-finsup* = *dioid D* **for** *D* (**structure**) +
  assumes *finsup-closed* : ⟦*finite A*; *A* ⊆ *carrier D*⟧ ⇒ *ΣA*∈*carrier D*
  and *finsup-empty*: *Σ*{} = **0**
  and *finsup-insert*: ⟦*A* ⊆ *carrier D*; *finite A*; *x*∈*carrier D*⟧ ⇒ *Σ*(*insert x A*) = *x*+*ΣA*

We have developed a basic library for sums, in particular for their interaction with the dioid operations. Typical examples are $\sum A \leq y \leftrightarrow \forall x \in A. x \leq y$, $\sum(A \cup B) = (\sum A) + (\sum B)$, and $(\sum A) \cdot (\sum B) = \sum\{a \cdot b \mid a \in A, b \in B\}$, whenever A and B are finite sets. Their proofs are the least automatic ones in the paper, since side conditions on the elements and sets involved need to be processed. All individual proof steps, however, could still be discharged automatically, sometimes after simplifying. We expect that the degree of automation can significantly be increased in a more thoroughly designed library.

Next we axiomatise *boffa1* with carrier sets.

locale *boffa1* = *dioid-finsup B* **for** *B* (**structure**) +

assumes *star-closed*: $x \in carrier\ B \Rightarrow x^* \in carrier\ B$
and *C11*: $[\![x \in carrier\ B;\ y \in carrier\ B]\!] \Rightarrow (x+y)^* = (x^* \cdot y)^* \cdot x^*$
and *C12*: $[\![x \in carrier\ B;\ y \in carrier\ B]\!] \Rightarrow (x \cdot y)^* = \mathbf{1} + x \cdot (y \cdot x)^* \cdot y$
and *R*: $x \in carrier\ B \Rightarrow x \cdot x = x \rightarrow x^* = \mathbf{1} + x$

We now link this algebra with the index monoid I. First, we define I—parametrised by the carrier of the algebra—as an arbitrary set that is mapped by a function x—again parametrised by the carrier—into the regular algebra. The record *'a boffa* provides the signature for the locale *boffa1*.

record *('a, 'b) boffa-gen* = *'a boffa* +
gen-set :: *'b set* (I_1)
gen :: *'b* ⇒ *'a* ($x_{1\_}$)

locale *boffa-gen* = *boffa1 G* **for** *G* (**structure**) +
assumes *gen-closed*: $i \in I \Rightarrow x_i \in carrier\ G$

We can then impose the monoid structure and finiteness constraint on I.

record *('a, 'b) boffa-monoid* = *('a, 'b) boffa-gen* +
comp :: *['b, 'b]* ⇒ *'b* (**infix** $\circ_1$ *80*)
unit :: *'b* (e_1)

locale *boffa-monoid* = *boffa-gen G* **for** *G* (**structure**) +
assumes *gen-finite*: *finite I*
and *comp-closed*: $[\![i \in I;\ j \in I]\!] \Rightarrow i \circ j \in I$
and *unit-closed*: $e \in I$
and *comp-assoc*: $[\![i \in I;\ j \in I;\ k \in I]\!] \Rightarrow i \circ (j \circ k) = (i \circ j) \circ k$
and *unit-left*: $i \in I \Rightarrow e \circ i = i$
and *unit-right*: $i \in I \Rightarrow i \circ e = i$

This infrastructure allows us to write down Conway's monoid identities in Isabelle. Deriving them requires about 10 additional lemmas on the interaction of the monoid and the regular algebra. To shorten expressions we write $\{x_i\}$ instead of $\{x_i \mid i \in I\}$ and similarly $\{x_i \cdot x_j\}$ or $\{x_{i \circ j}\}$ when indices range over I. We have shown, for instance, that the set $\{x_i \mid i \in A\}$ is a finite subset of the carrier G of our algebra and that $\sum\{x_i \mid i \in A\} \in G$, for every $A \subseteq I$. Another example is that $\{x_i \cdot x_j\}$ is a finite subset of G and the sum over this set an element of G. Finally, we have shown that the image of the monoidal unit e under x can be isolated from sums: $\sum\{x_i\} = x_e + \sum\{x_i \mid i \in (I - \{e\})\}$. Most corresponding proofs are fully automatic.

The final missing step is the implementation of the pair notation $x_{i,j}$.

Definition *mon-pair* :: *('a, 'b, 'c) boffa-monoid-scheme* ⇒ *'b* ⇒ *'b* ⇒ *'a* ($x_{1\_,\_}$)
where $x_{G i,j} = \Sigma_G\{x_{G k} \mid k.\ k \in I_G \wedge i \circ_G k = j\}$

For syntactic reasons, the index G refers to the underlying carrier set. The following lemma corresponds to the first step in Boffa's proof [3].

Lemma *mon-pair-split*: $(\forall i \in I.\ \forall j \in I.\ {x_{i,j}}^* = x_{i,j}) \Rightarrow \Sigma\{x_i\} = \mathbf{1} + \Sigma\{x_i\}$

Its proof translates Boffa's reasoning more or less directly. The remaining two lemmas formalise properties that have been left implicit in Boffa's next steps.

Lemma *aux1*: $\{x_{i \circ j}\} = \{x_i\}$

Lemma *aux2*: $(\forall i{\in}I.\ \forall j{\in}I.\ x_i{\cdot}x_j \leq x_{i \circ j}) \Rightarrow \Sigma\{x_i{\cdot}x_j\} \leq \Sigma\{x_{i \circ j}\}$

By Lemma *aux1*, summing over all elements $i \circ j$ of I means summing over all elements i. Lemma *aux2* helps to lift the assumption in Conway's rule that x is a submorphism to the level of suprema. Therefore, the map x from the monoid I into the Boffa algebra B is "almost" an embedding.

Finally, these three lemmas allow us to feed Boffa's remaining proof of Conway's rule directly into Isabelle, verifying all his proof steps automatically.

Theorem *mon-id*: $(\forall i{\in}I.\ \forall j{\in}I.\ x_i{\cdot}x_j \leq x_{i \circ j} \wedge {x_{i,j}}^* = x_{i,j}) \Rightarrow (\Sigma\{x_i\})^* = \Sigma\{x_i\}$
Proof –
  assume $\forall i{\in}I.\ \forall j{\in}I.\ x_i{\cdot}x_j \leq x_{i \circ j} \wedge {x_{i,j}}^* = x_{i,j}$
  — preparatory steps on the assumption
  have $(\Sigma\{x_i\}){\cdot}(\Sigma\{x_i\}) = (\mathbf{1}{+}\Sigma\{x_i\}){\cdot}(\mathbf{1}{+}\Sigma\{x_i\})$ — by smt
  also have ... $= \mathbf{1}{+}(\Sigma\{x_i\}){+}(\Sigma\{x_i\}){\cdot}(\Sigma\{x_i\})$ — by smt
  also have ... $= \mathbf{1}{+}(\Sigma\{x_i\}{+}\Sigma\{x_i{\cdot}x_j\})$ — by simplification
  ultimately have $(\Sigma\{x_i\}){\cdot}(\Sigma\{x_i\}) = \Sigma\{x_i\}$ — by smt
  thus $(\Sigma\{x_i\})^* = \Sigma\{x_i\}$ — by smt, essentially R and *mon-pair-split*
qed

This last theorem establishes completeness of all regular algebras in our hierarchy relative to Krob's proof (cf. Figure 1). Formalising this result fully in Isabelle would require linking our abstract implementations of algebras with the carrier based ones. Unfortunately, to our knowledge, this is impossible. Alternatively, we could have based the entire development on carrier sets. But that seems mathematically rather unnatural and it hampers proof automation.

10 Conclusion

We have reconstructed the fine structure of regular algebras within Isabelle based on the Sledgehammer tool for automated theorem proving and on automated counterexample search. The main emphasis was on known completeness results, yet some new findings clarify the overall picture in Figure 1.

As an exercise in computer enhanced mathematics, our study underlines the impressive potential of integrated automated and interactive proof technology for the working mathematician. Automation of axiomatic algebraic reasoning left little to desire; that of moderately difficult higher-order and integrated reasoning (e.g. by induction, with algebra, numbers or sets) was still reasonably high. The most complex proofs could be translated directly and rather quickly from paper and pencil proofs and automated step by step. The hardest work was certainly in library design. Overall, formalising regular algebras in this new kind of integrated environment seems reasonably lightweight and natural from a mathematician's point of view. Results that eminent scientists found worth publishing could be reconstructed with relative ease and a high degree of automation.

We end with some remarks on proof technology. Isabelle proof reconstruction often requires proof search. This remains a bottleneck. Standardised detailed

ATP output would support fast microstep proof reconstruction even when proof search takes time. Standardised type support for ATP seems desirable for heterogeneous mathematical reasoning. Sledgehammer calls five ATP systems and the SMT solver Z3 (cf. [2]). Having them all is certainly a gain, but Z3 showed definitely the most consistent performance. In Isabelle, the gap between abstract and carrier-based structures inhibits smooth mathematical reasoning. A less rigid proof scripting language could yield simpler and less verbose ATP-based proofs: assumption contexts are managed by the relevance filter; hence detailed control at command level—which determines the scripting syntax— seems unnecessary.

Acknowledgements. We are grateful to Geoff Sutcliffe and the München Isabelle group for making ATP/SMT systems freely available over the Internet.

References

1. Ballarin, C.: Tutorial to locales and locale interpretation. In: Lambán, L., Romero, A., Rubio, J. (eds.) Contribuciones Científicas en honor de Mirian Andrés. Servicio de Publicationes de la Universidad de La Rioja (2010)
2. Blanchette, J.C., Bulwahn, L., Nipkow, T.: Automatic Proof and Disproof in Isabelle/HOL. In: Tinelli, C., Sofronie-Stokkermans, V. (eds.) FroCos 2011. LNCS (LNAI), vol. 6989, pp. 12–27. Springer, Heidelberg (2011)
3. Boffa, M.: Une remarque sur les systèmes complets d'identités rationnelles. Informatique théorique et Applications 24(4), 419–423 (1990)
4. Boffa, M.: Une condition impliquant toutes les identités rationnelles. Informatique théorique et Applications 29(6), 515–518 (1995)
5. Braibant, T., Pous, D.: An Efficient Coq Tactic for Deciding Kleene Algebras. In: Kaufmann, M., Paulson, L.C. (eds.) ITP 2010. LNCS, vol. 6172, pp. 163–178. Springer, Heidelberg (2010)
6. Conway, J.H.: Regular Algebra and Finite Machines. Chapman and Hall (1971)
7. Guttmann, W., Struth, G., Weber, T.: Automating Algebraic Methods in Isabelle. In: Qin, S., Qiu, Z. (eds.) ICFEM 2011. LNCS, vol. 6991, pp. 617–632. Springer, Heidelberg (2011)
8. Kozen, D.: A completeness theorem for Kleene algebras and the algebra of regular events. Information and Computation 110(2), 366–390 (1994)
9. Krauss, A., Nipkow, T.: Proof pearl: Regular expression equivalence and relation algebra. J. Autom. Reasoning (2011)
10. Krob, D.: Complete systems of $\mathcal{B}$-rational identities. Theoretical Computer Science 89, 207–343 (1991)
11. Nipkow, T., Paulson, L.C., Wenzel, M.: Isabelle/HOL. LNCS, vol. 2283. Springer, Heidelberg (2002)
12. Salomaa, A.: Two complete axiom systems for the algebra of regular events. J. ACM 13(1), 158–169 (1966)

δ-Complete Decision Procedures for Satisfiability over the Reals*

Sicun Gao, Jeremy Avigad, and Edmund M. Clarke

Carnegie Mellon University, Pittsburgh, PA 15213

Abstract. We introduce the notion of "δ-complete decision procedures" for solving SMT problems over the real numbers, with the aim of handling a wide range of nonlinear functions including transcendental functions and solutions of Lipschitz-continuous ODEs. Given an SMT problem φ and a positive rational number δ, a δ-complete decision procedure determines either that φ is unsatisfiable, or that the "δ-weakening" of φ is satisfiable. Here, the δ-weakening of φ is a variant of φ that allows δ-bounded numerical perturbations on φ. We establish the existence and complexity of δ-complete decision procedures for bounded SMT over reals with functions mentioned above. We propose to use δ-completeness as an ideal requirement for numerically-driven decision procedures. As a concrete example, we formally analyze the DPLL$\langle$ICP$\rangle$ framework, which integrates Interval Constraint Propagation in DPLL(T), and establish necessary and sufficient conditions for its δ-completeness. We discuss practical applications of δ-complete decision procedures for correctness-critical applications including formal verification and theorem proving.

1 Introduction

Given a first-order signature $\mathcal{L}$ and a structure $\mathcal{M}$, the *Satisfiability Modulo Theories* (SMT) problem asks whether a quantifier-free $\mathcal{L}$-formula is satisfiable over $\mathcal{M}$, or equivalently, whether an existential $\mathcal{L}$-sentence is true in $\mathcal{M}$. Solvers for SMT problems have become the key enabling technology in formal verification and related areas. SMT problems over the real numbers are of particular interest, because of their importance in verification and design of hybrid systems, as well as in theorem proving. While efficient algorithms [10] exist for deciding SMT problems with only linear real arithmetic, practical problems normally contain nonlinear polynomials, transcendental functions, and differential equations. Solving formulas with these functions is inherently intractable. Decision algorithms [9] for formulas with nonlinear polynomials have very high complexity [6]. When the sine function is involved, the SMT problem is undecidable, and only partial algorithms can be developed [2,1].

* This research was sponsored by the National Science Foundation grants no. DMS1068829, no. CNS0926181, and no. CNS0931985, the GSRC under contract no. 1041377 (Princeton University), the Semiconductor Research Corporation under contract no. 2005TJ1366, General Motors under contract no. GMCMUCRLNV301, and the Office of Naval Research under award no. N000141010188.

B. Gramlich, D. Miller, and U. Sattler (Eds.): IJCAR 2012, LNAI 7364, pp. 286–300, 2012.

Recently much attention has been given to developing practical solvers that incorporate scalable numerical computations. Examples of numerical algorithms that have been exploited include optimization algorithms [4,28], interval-based algorithms [13,11,12,17], Bernstein polynomials [26], and linearization algorithms [14]. These solvers have shown promising results on various nonlinear benchmarks in terms of scalability.

However, for correctness-critical problems, there is always the concern that numerical errors can result in incorrect answers from numerically-driven solvers. For example, safety problems for hybrid systems can not be decided by numerical methods [29]. The problem is compounded by, for instance, the difficulty in understanding the effect of floating-point arithmetic in place of exact computation. There are two common ways of addressing these concerns. One is to use exact versions of the numerical algorithms, replacing floating-point operations by exact symbolic arithmetic [26]; the other is to use post-processing (validation) procedures to ensure that only correct results are returned. Both options reduce the full power of numerical algorithms and are usually hard to implement as well. For instance, in the Flyspeck project [19] for the formal proof of the Kepler conjecture, validating the numerical procedures used in the original proof turns out to be the hardest computational part (and unfinished yet). In general, there has been no framework for understanding the actual performance guarantees of numerical algorithms in the context of decision problems.

In this paper we aim to fill this gap by formally establishing the applicability of numerical algorithms in decision procedures, and the correctness guarantees they can actually provide. We do this as follows.

First, we introduce "the δ-SMT problem" over the real numbers, to capture what can in fact be *correctly* solved by numerically-driven procedures. Given an SMT formula φ, and any positive rational number δ, the δ-SMT problem asks for one of the following decisions:

- unsat: φ is unsatisfiable.
- δ-sat: The δ-*weakening* of φ is satisfiable.

Here, the δ-weakening of φ is defined as a numerical relaxation of the original formula. For instance, the δ-weakening of $x = 0$ is $|x| \leq \delta$. Note that if a formula is satisfiable, its δ-weakening is always satisfiable. Thus, when a formula is δ-sat, either it is indeed satisfiable, or it is unsatisfiable but a δ-perturbation on its numerical terms would make it satisfiable. The effect of this slight relaxation is significant. In sharp contrast to the undecidability of SMT for any signature extending real arithmetic by sine, we show that the bounded δ-SMT problem for a wide range of nonlinear functions is decidable. In fact, we show that the bounded δ-SMT problem for the theory with exponentiation and trigonometric functions is NP-complete, and PSPACE-complete for theories with Lipschitz-continuous ODEs. We use techniques from computable analysis [31,5]. These results provide the theoretical basis for our analysis of numerically-driven procedures.

Next, if a decision algorithm can solve the δ-SMT problem correctly, we say it is "δ-complete". We propose to use δ-completeness as the ideal correctness requirement on numerically-driven procedures, replacing the conventional notion of complete solvers (which can never be met in this context). This new notion makes it worthwhile to formally analyze numerical methods for decision problems and compare their strength, instead of viewing them as partial heuristics. As an example, we study DPLL⟨ICP⟩, the integration of Interval Constraint Propagation (ICP) [20] in DPLL(T) [25]. It is a general solving framework for nonlinear formulas and has shown promising results [13,17,12]. We obtain conditions that are sufficient and necessary for the δ-completeness of DPLL⟨ICP⟩.

Further, we show the applicability of δ-complete procedures in correctness-critical practical problems. In bounded model checking [7,8], using a δ-complete solver we return one of the following answers: either a system is absolutely safe up to some depth (unsat answers), or it would *become unsafe* under some δ-bounded numerical perturbations (δ-sat answers). Since δ can be made very small, in the latter case the algorithm is essentially detecting robustness problems in the system: If a system would be unsafe under some small perturbations, it can hardly be regarded as safe in practice. Similar guarantees can be given for invariant validation and theorem proving. The conclusion is that, under suitable interpretations, the answers of numerically-driven decision procedures can indeed be relied on in correctness-critical applications, as long as they are δ-complete.

Related Work. Our goal is to provide a formal basis for the promising trend of numerically-driven decision procedures [4,28,13,11,12,17,26,14]. Related attempts can be seen in Ratschan's work [30], in which he investigated the stability of first-order constraints under numerical perturbations. Our approach is, instead, to take numerical perturbations as a given and study its implications in practical applications. Results in this paper are related to our more theoretical results [16] for arbitrarily-quantified sentences, where we do not analyze practical procedures. A preliminary notion of δ-completeness was proposed by us earlier in [17], in which only polynomials are considered.

The paper is organized as follows. In Section 2 and 3 we define the bounded δ-SMT problem and establish its decidability and complexity. In Section 4 we formally analyze DPLL⟨ICP⟩ and discuss applications in Section 5.

2 SMT with Type 2 Computable Functions

2.1 Basics of Computable Analysis

Real numbers can be encoded as infinite strings, and a computability theory of real functions can be developed with oracle machines that perform operations using oracles encoding real numbers. This is the approach developed in computable analysis (Type 2 Computability) [31,23,5]. We briefly review results of importance to us.

Throughout the paper $||\cdot||$ denotes $||\cdot||_\infty$ over $\mathbb{R}^n$ for various n.

Definition 2.1 (Names). *A name of $a \in \mathbb{R}$ is any function $\gamma_a : \mathbb{N} \to \mathbb{Q}$ satisfying that for every $i \in \mathbb{N}$, $|\gamma_a(i) - a| < 2^{-i}$. For $\boldsymbol{a} \in \mathbb{R}^n$, $\gamma_{\boldsymbol{a}}(i) = \langle \gamma_{a_1}(i), ..., \gamma_{a_n}(i)\rangle$.*

Thus the name of a real number is a sequence of rational numbers converging to it. For $\boldsymbol{a} \in \mathbb{R}^n$, we write $\Gamma(\boldsymbol{a}) = \{\gamma : \gamma \text{ is a name of } \boldsymbol{a}\}$.

A real function f is computable if there is an oracle Turing machine that can take any argument x of f as an oracle, and output the value of $f(x)$ up to an arbitrary precision.

Definition 2.2 (Computable Functions). *We say $f :\subseteq \mathbb{R}^n \to \mathbb{R}$ is computable if there exists an oracle Turing machine $\mathcal{M}_f$, outputting rational numbers, such that*

$$\forall \boldsymbol{x} \in \mathrm{dom}(f)\ \forall \gamma_{\boldsymbol{x}} \in \Gamma(\boldsymbol{x})\ \forall i \in \mathbb{N}\ |M_f^{\gamma_{\boldsymbol{x}}}(i) - f(\boldsymbol{x})| < 2^{-i}.$$

In the definition, i specifies the desired error bound on the output of M_f with respect to $f(\boldsymbol{x})$. For any $\boldsymbol{x} \in \mathrm{dom}(f)$, M_f has access to an oracle encoding the name $\gamma_{\boldsymbol{x}}$ of $\boldsymbol{x}$, and output a 2^{-i}-approximation of $f(\boldsymbol{x})$. In other words, the sequence $M_f^{\gamma_{\boldsymbol{x}}}(1), M_f^{\gamma_{\boldsymbol{x}}}(2), ...$ is a name of $f(\boldsymbol{x})$. A key property of this notion of computability is that computable functions over the reals are continuous [31]. Moreover, over any compact set $D \subseteq \mathbb{R}^n$, computable functions are uniformly continuous with a *computable modulus of continuity* defined as follows.

Definition 2.3 (Uniform Modulus of Continuity). *Let $f :\subseteq \mathbb{R}^n \to \mathbb{R}$ be a function and $D \subseteq \mathrm{dom}(f)$ a compact set. The function $m_f : \mathbb{N} \to \mathbb{N}$ is called a uniform modulus of continuity of f on D, if*

$$\forall \boldsymbol{x}, \boldsymbol{y} \in D\ \forall i \in \mathbb{N}\ ||\boldsymbol{x} - \boldsymbol{y}|| < 2^{-m_f(i)} \to |f(\boldsymbol{x}) - f(\boldsymbol{y})| < 2^{-i}.$$

Proposition 2.1 ([31]). *Let $f :\subseteq \mathbb{R}^n \to \mathbb{R}$ be computable and $D \subseteq \mathrm{dom}(f)$ a compact set. Then f has a computable uniform modulus of continuity over D.*

Intuitively, if a function has a computable uniform modulus of continuity, then fixing any desired error bound 2^{-i} on the outputs, we can compute a *global* precision $2^{-m_f(i)}$ on the inputs from D such that using any $2^{-m_f(i)}$-approximation of any $\boldsymbol{x} \in D$, $f(\boldsymbol{x})$ can be computed within the error bound.

Most common continuous real functions are computable [31]: Addition, multiplication, absolute value, min, max, exp, sin and solutions of Lipschitz-continuous ordinary differential equations are all computable functions. Compositions of computable functions are computable.

Moreover, complexity of real functions can be defined over compact domains.

Definition 2.4 ([24]). *Let $D \subseteq \mathbb{R}^n$ be compact. A real function $f : D \to \mathbb{R}$ is* P*-computable (*PSPACE*-computable), if it is computable by an oracle Turing machine $M_f^{\gamma(\boldsymbol{x})}(i)$ that halts in polynomial-time (polynomial-space) for every $i \in \mathbb{N}$ and every $\boldsymbol{x} \in \mathrm{dom}(f)$.*

We say f is in Type 2 complexity class C if it is C-computable. f is C-complete if it is C-computable and C-hard [23]. If $f : D \to \mathbb{R}$ is C-computable, then it has a C-computable modulus of continuity over D. Polynomials, exp, and sin are all P-computable functions. A recent result [22] established that the complexity of computing solutions of Lipschitz-continuous ODEs over compact domains is a PSPACE-complete problem.

2.2 Bounded SMT over $\mathbb{R}_{\mathcal{F}}$

We now let $\mathcal{F}$ denote any finite collection of Type 2 computable functions. $\mathcal{L}_{\mathcal{F}}$ denotes the first-order signature and $\mathbb{R}_{\mathcal{F}}$ is the standard structure $\langle \mathbb{R}, \mathcal{F}\rangle$. We can then consider the SMT problem over $\mathbb{R}_{\mathcal{F}}$, namely, satisfiability of quantifier-free $\mathcal{L}_{\mathcal{F}}$-formulas over $\mathbb{R}_{\mathcal{F}}$. We consider formulas whose variables take values from bounded intervals. Because of this, it is more convenient to directly write the bounds on existential quantifiers and express bounded SMT problems as Σ_1-sentences with bounded quantifiers.

Definition 2.5 (Bounded Σ_1-Sentences). *A bounded Σ_1-sentence in $\mathcal{L}_{\mathcal{F}}$ is*

$$\varphi: \exists^{I_1} x_1 \cdots \exists^{I_n} x_n . \psi(x_1, ..., x_n).$$

- *For all i, $I_i \subseteq \mathbb{R}$ is a bounded (open or closed) interval with rational end-points.*
- *Each bounded quantifier $\exists^{I_i} x_i . \phi$ denotes $\exists x_i . (x_i \in I_i \wedge \phi)$.*
- *$\psi(x_1, ..., x_n)$ is a quantifier-free $\mathcal{L}_{\mathcal{F}}$-formula, i.e., a Boolean combination of atomic formulas of the form $f(x_1, ..., x_n) \circ 0$, where f is a composition of functions in $\mathcal{F}$ and $\circ \in \{<, \leq, >, \geq, =, \neq\}$.*
- *We write* $\mathrm{dom}(\varphi) = I_1 \times \cdots \times I_n$, *and require that all the functions occurring in $\psi(\boldsymbol{x})$ are defined everywhere over its closure $\overline{\mathrm{dom}(\varphi)}$.*

We can write a bounded Σ_1-sentence as $\exists^{\boldsymbol{I}} \boldsymbol{x} . \psi(\boldsymbol{x})$ for short.

Lemma 2.1 (Standard Form). *Any bounded Σ_1-sentence φ in $\mathcal{L}_{\mathcal{F}}$ is equivalent over $\mathbb{R}_{\mathcal{F}}$ to a sentence of the following form:*

$$\exists^{I_1} x_1 \cdots \exists^{I_n} x_n \bigwedge_{i=1}^{m} (\bigvee_{j=1}^{k_i} f_{ij}(\boldsymbol{x}) = 0).$$

Proof. Assume that φ is originally $\exists^{\boldsymbol{I}} \boldsymbol{x} \bigwedge_{i=1}^{m} (\bigvee_{j=1}^{k_i} g_{ij}(\boldsymbol{x}) \circ 0)$, where $\circ \in \{<, \leq, >, \geq, =, \neq\}$. We apply the following transformations:

1. **(Eliminate $\neq$)** Substitute each atomic formula of the form $g_{ij} \neq 0$ by $g_{ij} < 0 \vee g_{ij} > 0$.

2. **(Eliminate $\leq, <$)** Substitute $g_{ij} \leq 0$ by $-g_{ij} \geq 0$, and $g_{ij} < 0$ by $-g_{ij} > 0$. Now the formula is rewritten to $\exists^{\boldsymbol{I}} \boldsymbol{x} . \bigwedge_{i=1}^{m} (\bigvee_{j=1}^{k_i} g'_{ij}(\boldsymbol{x}) \circ 0)$, where $\circ \in \{>, \geq, =\}$. ($g'_{ij} = -g_{ij}$ if the inequality is reversed; otherwise $g'_{ij} = g_{ij}$.)

3. **(Eliminate $\geq, >$)** Substitute $g'_{ij} \geq 0$ (or $g'_{ij} > 0$) by $g'_{ij} - v_{ij} = 0$, where v_{ij} is a newly introduced variable, and add an innermost bounded existential quantifier $\exists v_{ij} \in I_{v_{ij}}$, where $I_{v_{ij}} = [0, m_{v_{ij}}]$ ($I_v = (0, m_{v_{ij}}]$). Here, $m_{v_{ij}} \in \mathbb{Q}$ is any value greater than the maximum of g'_{ij} over $\overline{\mathrm{dom}(\varphi)}$. Note that such maximum of g'_{ij} always exists over $\overline{\mathrm{dom}(\varphi)}$, since g'_{ij} is continuous on $\overline{\mathrm{dom}(\varphi)}$, which is a compact, and is computable [23].

The formula is now in the form $\exists^{\boldsymbol{I}} \boldsymbol{x} \exists^{\boldsymbol{I}_v} \boldsymbol{v} . \bigwedge_{i=1}^{m} (\bigvee_{j=1}^{k_i} f_{ij}(\boldsymbol{x}, \boldsymbol{v}) = 0)$, where $f_{ij} = g'_{ij} - v_{ij}$ if v_{ij} has been introduced in the previous step; otherwise, $f_{ij} = g'_{ij}$. The new formula is in the standard form and equivalent to the original one. □

Example 2.1. A standard form of $\exists^{[-1,1]} x \exists^{[-1,1]} y \exists^{[-1,1]} z \, (e^z < x \rightarrow y < \sin(x))$ is $\exists^{[-1,1]} x \exists^{[-1,1]} y \exists^{[-1,1]} z \exists^{[0,10]} u \exists^{(0,10]} v \, (e^z - x - u = 0) \vee (\sin(x) - y - v = 0)$.

3 The Bounded δ-SMT Problem

The key for bridging numerical procedures and SMT problems is to introduce syntactic perturbations on Σ_1-sentences in $\mathcal{L}_{\mathcal{F}}$.

Definition 3.1 (δ-Weakening and Perturbations). *Let $\delta \in \mathbb{Q}^+ \cup \{0\}$ be a constant and φ be a Σ_1-sentence in the standard form:*

$$\varphi := \exists^{\boldsymbol{I}} \boldsymbol{x}. \bigwedge_{i=1}^{m} (\bigvee_{j=1}^{k_i} f_{ij}(\boldsymbol{x}) = 0).$$

The δ-weakening of φ defined as:

$$\varphi^{\delta} := \exists^{\boldsymbol{I}} \boldsymbol{x}. \bigwedge_{i=1}^{m} (\bigvee_{j=1}^{k_i} |f_{ij}(\boldsymbol{x})| \leq \delta).$$

Also, a δ-perturbation is a constant vector $\mathbf{c} = (c_{11}, ..., c_{mk_m})$, $c_{ij} \in \mathbb{Q}$, satisfying $||\mathbf{c}|| \leq \delta$, such that the $\mathbf{c}$-perturbed form of φ is given by:

$$\varphi^{\mathbf{c}} := \exists^{\boldsymbol{I}} \boldsymbol{x}. \bigwedge_{i=1}^{m} (\bigvee_{j=1}^{k_i} f_{ij}(\boldsymbol{x}) = c_{ij}).$$

Proposition 3.1. *φ^{δ} is true iff there exists a δ-perturbation $\mathbf{c}$ such that $\varphi^{\mathbf{c}}$ is true. In particular, $\mathbf{c}$ can be the zero vector, and thus $\varphi \to \varphi^{\delta}$.*

We now define the bounded δ-SMT problem. We follow the convention that SMT solvers return sat/unsat, which is equivalent to the corresponding Σ_1-sentence being true/false.

Definition 3.2 (Bounded δ-SMT). *Let $\mathcal{F}$ be a finite collection of Type 2 computable functions. Let φ be a bounded Σ_1-sentence in $\mathcal{L}_{\mathcal{F}}$ in standard form, and $\delta \in \mathbb{Q}^+$. The* bounded δ-SMT problem *asks for one of the following decisions:*

- unsat : *φ is false.*
- δ-sat : *φ^{δ} is true.*

When the two cases overlap, either decision can be returned.

Our main theoretical claim is that the bounded δ-SMT problem is decidable for $\delta \in \mathbb{Q}^+$. This is essentially a special case of our more general results for arbitrarily-quantified $\mathcal{L}_{\mathcal{F}}$-sentences [16]. However, different from [16], here we defined the standard forms of SMT problems to contain only equalities in the matrix, on which the original proof does not work directly. Also, in [16] we relied on results from computable analysis that are not needed here. We now give a direct proof for the decidability of δ-SMT and analyze its complexity.

Theorem 3.1 (Decidability). *Let $\mathcal{F}$ be a finite collection of Type 2 computable functions, and $\delta \in \mathbb{Q}^+$ be given. The bounded δ-SMT problem in $\mathcal{L}_{\mathcal{F}}$ is decidable.*

Proof. We describe a decision procedure which, given any bounded Σ_1-sentence φ in $\mathcal{L}_{\mathcal{F}}$ and $\delta \in \mathbb{Q}^+$, decides either φ is false or φ^δ is true. Assume that φ is in the form of Definition 3.1.

First, we need a uniform bound on all the variables so that a modulus of continuity for each function can be computed. Suppose each x_i is bounded by I_i, whose closure is $\overline{I_i} = [l_i, u_i]$. We write

$$\overline{\varphi} := \exists^{[0,1]}x_1 \cdots \exists^{[0,1]}x_n \bigwedge_{i=1}^{m}(\bigvee_{j=1}^{k_i} f_{ij}(l_1 + (u_1 - l_1)x_1, ..., l_n + (u_n - l_n)x_n) = 0).$$

From now on, $g_{ij} = f_{ij}(l_1 + (u_1 - l_1)x_1, ..., l_n + (u_n - l_n)x_n)$. After the transformation, we have $\overline{\mathrm{dom}(\varphi)} = [0,1] \times \cdots \times [0,1]$, on which each g_{ij} is computable and has a computable modulus of continuity $m_{g_{ij}}$. We write $\psi(\boldsymbol{x})$ to denote the matrix of φ after the transformation.

Choose $r \in \mathbb{N}$ such that $2^{-r} < \delta/4$. Then for each g_{ij}, we use $m_{g_{ij}}$ to obtain $e_{ij} = m_{g_{ij}}(r)$. Choose $e \in \mathbb{N}$ such that $e \geq \max(e_{11}, ..., e_{mk_m})$ and write $\varepsilon = 2^{-e}$. We then have

$$\forall \boldsymbol{x}, \boldsymbol{y} \in \overline{\mathrm{dom}(\varphi)} \ (||\boldsymbol{x} - \boldsymbol{y}|| < \varepsilon \rightarrow |g_{ij}(\boldsymbol{x}) - g_{ij}(\boldsymbol{y})| < \delta/4). \tag{1}$$

We now consider a finite ε-net of $\overline{\mathrm{dom}(\varphi)}$, i.e., a finite $S_\varepsilon \subseteq \overline{\mathrm{dom}(\varphi)}$, satisfying

$$\forall \boldsymbol{x} \in \overline{\mathrm{dom}(\varphi)} \ \exists \boldsymbol{a} \in S_\varepsilon \ ||\boldsymbol{x} - \boldsymbol{a}|| < \varepsilon. \tag{2}$$

In fact, S_ε can be explicitly defined as

$$S_\varepsilon = \{(a_1, ..., a_n) : a_i = k \cdot \varepsilon, \text{ where } k \in \mathbb{N}, 0 \leq k \leq 2^e\}.$$

Next, we evaluate the matrix $\psi(\boldsymbol{x})$ on each point in S_ε, as follows. Let $\boldsymbol{a} \in S_\varepsilon$ be arbitrary. For each g_{ij} in ψ, we compute $g_{ij}(\boldsymbol{a})$ up to an error bound of $\delta/8$, and write the result of the evaluation as $\overline{g_{ij}(\boldsymbol{a})}^{\delta/8}$. Then $|g_{ij}(\boldsymbol{a}) - \overline{g_{ij}(\boldsymbol{a})}^{\delta/8}| < \delta/8$. Note $\overline{g_{ij}(\boldsymbol{a})}^{\delta/8}$ is a rational number. We then define

$$\widehat{\psi}(\boldsymbol{x}) := \bigwedge_{i=1}^{m} \bigvee_{j=1}^{k_i} |\overline{g_{ij}(\boldsymbol{x})}^{\delta/8}| < \delta/2.$$

Then for each $\boldsymbol{a}$, evaluating $\widehat{\psi}(\boldsymbol{a})$ only involves comparison of rational numbers and Boolean evaluation, and $\widehat{\psi}(\boldsymbol{a})$ is either true or false. Now, by collecting the value of $\widehat{\psi}$ on every point in S_ε, we have the following two cases.

- Case 1: For some $\boldsymbol{a} \in S_\varepsilon$, $\widehat{\psi}(\boldsymbol{a})$ is true. We show that φ^δ is true. Note that

$$\widehat{\psi}(\boldsymbol{a}) \Rightarrow \bigwedge_{i=1}^{m} \bigvee_{j=1}^{k_i} |\overline{g_{ij}(\boldsymbol{a})}^{\delta/8}| < \delta/2 \Rightarrow \bigwedge_{i=1}^{m} \bigvee_{j=1}^{k_i} |g_{ij}(\boldsymbol{a})| < \delta \cdot 5/8.$$

We need to be careful about $\boldsymbol{a}$, since it is an element in $\overline{\mathrm{dom}(\varphi)}$, not $\mathrm{dom}(\varphi)$. If $\boldsymbol{a} \in \mathrm{dom}(\varphi)$, then φ^δ is true, witnessed by $\boldsymbol{a}$. Otherwise, $\boldsymbol{a} \in \partial(\mathrm{dom}(\varphi))$. Then by continuity of g_{ij}, there exists $\boldsymbol{a}' \in \mathrm{dom}(\varphi)$ such that $\bigwedge_{i=1}^{m} \bigvee_{j=1}^{k_i} |g_{ij}(\boldsymbol{a}')| < \delta$. (Just let a small enough ball around $\boldsymbol{a}$ intersect $\mathrm{dom}(\varphi)$ at $\boldsymbol{a}'$.) That means φ^δ is also true in this case, witnessed by $\boldsymbol{a}'$.

- Case 2: For every $\boldsymbol{a} \in S_\varepsilon$, $\widehat{\psi}(\boldsymbol{a})$ is false. We show that φ is false. Note that

$$\neg\widehat{\psi}(\boldsymbol{a}) \Rightarrow \bigvee_{i=1}^{m} \bigwedge_{j=1}^{k_i} |\overline{g_{ij}(\boldsymbol{a})}^{\delta/8}| \geq \delta/2 \Rightarrow \bigvee_{i=1}^{m} \bigwedge_{j=1}^{k_i} |g_{ij}(\boldsymbol{a})| \geq \delta \cdot 3/8.$$

Now recall conditions (1) and (2). For an arbitrary $\boldsymbol{x} \in \mathrm{dom}(\varphi)$, there exists $\boldsymbol{a} \in S_\varepsilon$ such that $|g_{ij}(\boldsymbol{x}) - g_{ij}(\boldsymbol{a})| < \delta/4$ for every g_{ij}. Consequently, we have $|g_{ij}(\boldsymbol{x})| \geq \delta \cdot 3/8 - \delta/4 = \delta/8$. Thus, $\forall \boldsymbol{x} \in \mathrm{dom}(\varphi), \bigvee_{i=1}^{m} \bigwedge_{j=1}^{k_i} |g_{ij}(\boldsymbol{x})| > 0$. This means $\neg\varphi$ is true, and φ is false.

In all, the procedure decides either that φ^δ is true, or that φ is false. □

We now analyze the complexity of the δ-SMT problem. The decision procedure given above essentially evaluates the formula on each sample point. Thus, using an oracle for evaluating the functions, we can construct a nondeterministic Turing machine that randomly picks the sample points and decides the formula. Most of the functions we are interested in (exp, sin, ODEs) are in Type 2 complexity class P or PSPACE. In this case, the oracle only uses polynomial space on the query tape (Proposition 3.2 below), and all the computations can be done in polynomial-time. Thus, it should be clear that the δ-SMT problem is in NP^{C}, where C is the complexity of the computable functions in the formula.

Formally, to prove interesting complexity results, a technical restriction is that we need to bound the number of function compositions in a formula, because otherwise evaluating nested polynomial-time functions can be exponential in the number of nesting. Formally we define:

Definition 3.3 (Uniformly Bounded Σ_1-class). *Let $\mathcal{F}$ be a finite set of Type 2 computable functions, and S a class of bounded Σ_1-sentences in $\mathcal{L}_{\mathcal{F}}$. Let $l, u \in \mathbb{Q}$ satisfy $l \leq u$. We say S is uniformly $(l, u, \mathcal{F})$-bounded, if for all $\varphi \in S$ of the form $\exists^{I_1} x_1 \cdots \exists^{I_n} x_n \bigwedge_{i=1}^{m} \bigvee_{j=1}^{k_i} f_{ij}(\boldsymbol{x}) = 0$ we have:*

- *$\forall 1 \leq i \leq n$, $I_i \subseteq [l, u]$.*
- *Each $f_{ij}(\boldsymbol{x})$ is contained in $\mathcal{F}$.*

Proposition 3.2 ([23]). *Let* C *be a Type 2 complexity class contained in* PSPACE. *Then given any compact domain D, a* C*-computable function has a uniform modulus of continuity over D given by a polynomial function.*

The main complexity claim is as follows. We have sketched the intuition above and a detailed proof is given in [15].

Theorem 3.2 (Complexity). *Let $\mathcal{F}$ be a finite set of functions in Type 2 complexity class* C, P ⊆ C ⊆ PSPACE. *The δ-SMT problem for uniformly bounded Σ_1-classes in $\mathcal{L}_{\mathcal{F}}$ is in* NP^{C}.

Corollary 3.1. *Let $\mathcal{F}$ be a finite set of* P*-time computable real functions, such as* $\{+, \times, \exp, \sin\}$*. The uniformly-bounded δ-SMT problem for $\mathcal{L}_{\mathcal{F}}$ is* NP*-complete.*

Corollary 3.2. *Let $\mathcal{F}$ be a finite set of Lipschitz-continuous ODEs over compact domains. Then the uniformly-bounded δ-SMT problem in $\mathcal{L}_{\mathcal{F}}$ is in* PSPACE*, and there exists $\mathcal{L}_{\mathcal{F}}$ such that it is* PSPACE*-complete.*

4 δ-Completeness of the DPLL⟨ICP⟩ Framework

We now give a formal analysis of the integration of ICP and DPLL(T) for solving bounded δ-SMT with nonlinear functions. Our goal is to establish sufficient and necessary conditions under which such an integration is δ-complete.

4.1 Interval Constraint Propagation

The method of Interval Constraint Propagation (ICP) [3] finds solutions of real constraints using a "branch-and-prune" method, combining interval arithmetic and constraint propagation. The idea is to use interval extensions of functions to "prune" out sets of points that are not in the solution set, and "branch" on intervals when such pruning can not be done, until a small enough box that may contain a solution is found. A high-level description of the decision version of ICP is given in Algorithm 1, and we give formal definitions below.

Definition 4.1 (Floating-Point Intervals and Hulls). *Let $\mathbb{F}$ denote the finite set of all floating point numbers with symbols $-\infty$ and $+\infty$ under the conventional order $<$. Let $\mathbb{IF} = \{[a,b] \subseteq \mathbb{R} : a, b \in \mathbb{F}, a \leq b\}$ denote the set of closed real intervals with floating-point endpoints, and $\mathbb{BF} = \bigcup_{n=1}^{\infty} \mathbb{IF}^n$ the set of* boxes *with these intervals. Let $S \subseteq \mathbb{R}$ be any set of real numbers, the hull of S is written as* $\text{Hull}(S) = \bigcap\{I \in \mathbb{IF} : S \subseteq I\}$.

For $I = [a,b] \in \mathbb{IF}$, we write $|I| = |b - a|$ to denote its size.

Definition 4.2 (Interval Extension (cf. [3])). *Let $f :\subseteq \mathbb{R}^n \to \mathbb{R}$ be a real function. An interval extension operator $\sharp(\cdot)$ maps f to a function $\sharp f :\subseteq \mathbb{BF} \to \mathbb{IF}$, such that $\forall B \in \mathbb{BF} \cap \text{dom}(\sharp f), \{f(\boldsymbol{x}) : \boldsymbol{x} \in B\} \subseteq \sharp f(B)$.*

Example 4.1. The natural extension of $f = 2{\cdot}(x{+}y){\cdot}z$ is given by $\sharp f = [2,2]{\cdot}(I_x{+}I_y){\cdot}I_z$, where the interval operations are defined as $[a_1,b_1]{+}[a_2,b_2] = [a_1{+}a_2, b_1{+}b_2]$ and $[a_1,b_1]\cdot[a_2,b_2] = [\min(a_1a_2, a_1b_2, b_1a_2, b_1b_2), \max(a_1a_2, a_1b_2, b_1a_2, b_1b_2)]$.

In Algorithm 1, Branch(B, i) is an operator that returns two smaller boxes $B' = I_1 \times \cdots \times I_i' \times \cdots \times I_n$ and $B'' = I_1 \times \cdots \times I_i'' \times \cdots \times I_n$, where $I_i \subseteq I_i' \cup I_i''$. To ensure termination it is assumed that there exists some unifrom constant $0 < c < 1$ such that in every branching operation, $c \cdot |I_i| \leq |I_i'|$ and $c \cdot |I_i| \leq |I_i''|$.

The key component of the algorithm is the Prune(B, f) operation. A simple example of a pruning operation is as follows.

Algorithm 1. High-Level ICP$_\varepsilon$ (decision version of Branch-and-Prune)

input : Constraints $f_1(x_1, ..., x_n) = 0, ..., f_m(x_1, ..., x_n) = 0$, initial box $B^0 = I_1^0 \times \cdots \times I_n^0$, box stack $S = \emptyset$, and precision $\varepsilon \in \mathbb{Q}^+$.
output: sat or unsat.

1 S.push(B_0);
2 **while** $S \neq \emptyset$ **do**
3     $B \leftarrow S$.pop() ;
4   **while** $\exists 1 \leq i \leq m, B \neq$ Prune(B, f_i) **do**
5         $B \leftarrow$ Prune(B, f_i) ;
6   **end**
7   **if** $B \neq \emptyset$ **then**
8   **if** $\exists 1 \leq i \leq n, |I_i| \geq \varepsilon$ **then**
9             $\{B', B''\} \leftarrow$ Branch(B, i);
10   S.push($\{B', B''\}$);
11   **end**
12   return sat;
13   **end**
14 **end**
15 return unsat;

Example 4.2. Consider $x - y^2 = 0$ with initial intervals $x \in [1, 2]$ and $y \in [0, 4]$. Let $\sharp f(I_x, I_y) = I_x - I_y^2$ be the natural interval extension of the left hand side. Since we know $0 \notin \sharp f([1, 2], [2, 4])$, we can contract the interval on y from $[0, 4]$ to $[0, 2]$ in one pruning step.

In principle, any operation that contracts the intervals on variables can be seen as pruning. However, for correctness we need several formal requirements on the pruning operator in ICP$_\varepsilon$.

Notation 4.1 *For any* $f : \mathbb{R}^n \rightarrow \mathbb{R}$, *we write* $Z_f = \{\boldsymbol{a} \in \mathbb{R}^n : f(\boldsymbol{a}) = 0\}$.

Definition 4.3 (Well-defined Pruning Operators). *Let* $\mathcal{F}$ *be a collection of real functions, and* $\sharp$ *be an interval extension operator on* $\mathcal{F}$. *A* well-defined (equality) pruning operator *with respect to* $\sharp$ *is a partial function* Prune$_\sharp$ $:\subseteq \mathbb{BF} \times \mathcal{F} \rightarrow \mathbb{BF}$, *such that for all* $f \in \mathcal{F}$ *and* $B \in \mathbb{BF}$,

- *(W1)* Prune$_\sharp(B, f) \subseteq B$;
- *(W2) If* (Prune$_\sharp(B, f)) \neq \emptyset$, *then* $0 \in \sharp f($Prune$_\sharp(B, f))$.
- *(W3)* $B \cap Z_f \subseteq$ Prune$_\sharp(B, f)$;

When $\sharp$ is clear, we simply write Prune. It specifies the following conditions. (W1) requires contraction, so that the algorithm always makes progress: branching always decreases the size of boxes, and pruning never increases them. (W2) requires that the result of a pruning is always a reasonable box that may contain a zero. Otherwise B should have been pruned out. (W3) ensures that the real solutions are never discarded in pruning (called "completeness" in [3]). We use Prune($B, f_1, ..., f_m$) to denote the iterative application of Prune($\cdot, f_i$) on B for all $1 \leq i \leq m$, until a fixed-point is reached. (Line 4-6 in Algorithm 1.)

Proposition 4.1. *For all i, $Prune(B, f_1, ..., f_m) \subseteq Prune(B, f_i)$.*

Lemma 4.1. *Algorithm 1 always terminates. If it returns* sat *then there exists nonempty boxes $B, B' \subseteq B_0$, such that $||B|| < \varepsilon$ and $B = \text{Prune}(B', f_1, ..., f_m)$. If it returns* unsat *then for every $\boldsymbol{a} \in B_0$, there exists $B \subseteq B_0$ such that $\boldsymbol{a} \in B$ and $\text{Prune}(B, f_1, ..., f_m) = \emptyset$.*

Remark 4.1. It is important to see that in sat answers, B is a result of pruning on some B' instead of the output of a simple branching.

Theorem 4.2 (δ-Completeness of ICP$_\varepsilon$). *Let $\delta \in \mathbb{Q}^+$ be arbitrary. We can find an $\varepsilon \in \mathbb{Q}^+$ such that the ICP$_\varepsilon$ algorithm is δ-complete for conjunctive Σ_1-sentences in $\mathcal{L}_\mathcal{F}$ (where* sat *is interpreted as δ-*sat*) if and only if the pruning operator in ICP$_\varepsilon$ is well-defined.*

Proof. We consider an arbitrary bounded existential $\mathcal{L}_\mathcal{F}$-sentence containing only conjunctions, written as $\varphi : \exists^{\boldsymbol{I}}\boldsymbol{x}. \bigwedge_{i=1}^{m} f_i(\boldsymbol{x}) = 0$. Let $B_0 = \boldsymbol{I}$ be the initial bounding box.

Since all the functions in φ are computable over B_0, each f_i has a uniform modulus of continuity over B_0, which we write as m_{f_i}. Choose any $k \in \mathbb{N}$ such that $2^{-k} < \delta$. Then for any $\varepsilon_i < m_{f_i}(k)$, we have

$$\forall \boldsymbol{x}, \boldsymbol{y} \in B_0, ||\boldsymbol{x} - \boldsymbol{y}|| < \varepsilon_i \to |f_i(\boldsymbol{x}) - f_i(\boldsymbol{y})| < \delta. \tag{3}$$

We now fix ε to be any positive rational number smaller than $\min(\varepsilon_1, ..., \varepsilon_m)$.

By the previous lemma, we know ICP$_\varepsilon$ terminates and returns either sat or unsat. We now prove the two directions of the biconditional.

$\Leftarrow$: Suppose the pruning operator in ICP$_\varepsilon$ is well-defined.

Suppose ICP$_\varepsilon$ returns "δ-sat", then by Lemma 4.1, there exist $B, B' \subseteq B_0$ such that $B = \text{Prune}(B', f_1, ..., f_m)$ and $||B|| < \varepsilon$. Then by the (W2), we know that $0 \in \sharp f_i(B)$ for every f_i. Now, by the definition of ε, we know from (3) that for every i, $\forall \boldsymbol{a} \in B, |f_i(\boldsymbol{a}) - 0| < \delta$. Namely, any $\boldsymbol{a} \in B$ is a witness for $\varphi^\delta : \exists^{\boldsymbol{I}}\boldsymbol{x}\ |f(\boldsymbol{x})| < \delta$. Thus the δ-weakening of φ is true.

Suppose ICP$_\varepsilon$ returns "unsat". Suppose φ is in fact satisfiable. Then there is a point $\boldsymbol{a} \in B_0$ such that $\psi(\boldsymbol{a})$ is true. However, following Lemma 4.1, $\boldsymbol{a} \in B$ for some $B \subseteq B_0$ and $\text{Prune}(B_0, f_1, ..., f_m) = \emptyset$. However, this contradicts condition (W3) of the pruning operator.

$\Rightarrow$: We only need to show that without any one of the three conditions in Definition 4.3, we can define a pruning operator that fails δ-completeness.

Without (W1), we define a pruning operator that always outputs intervals bigger than ε (such as the initial intervals). Then the procedure never terminates. Note that the other two conditions are trivially satisfied in this case (for any f and B_0 satisfying $0 \in \sharp f(B_0)$). Without (W2), consider the function $f(x) = x^2+1$ with $x \in [-1, 1]$. We can define a pruning operator such that $\text{Prune}([-1, 1], f) = [1, 1]$. This operator satisfies the other two conditions. However, the returned result $[1, 1]$ fails δ-completeness for any δ smaller than 2, since $f(1) = 2$. Without (W3), we simply prune any set to $\emptyset$ and always return unsat. This violates δ-completeness, which requires that if unsat is returned the formula must be indeed unsatisfiable. The other two conditions are also satisfied in this case. □

In practice, pruning operators are defined based on *consistency conditions* from constraint propagation techniques. Many pruning operators are used in practice [3]. Following Theorem 4.2, we only need to prove their well-definedness to ensure δ-completeness. For instance:

Definition 4.4 (Box-consistent Pruning [20]). *We say* $\pi_B : \mathbb{BF} \times \mathcal{F} \to \mathbb{BF}$ *is box-consistent, if for all* $f \in \mathcal{F}$ *and* $B = I_1 \times \cdots \times I_n \subseteq \mathrm{dom}(f)$*, the i-th interval of* $\pi_B(B, f)$ *is* $I_i \cap \mathrm{Hull}\big(\{a_i \in \mathbb{R} : 0 \in \sharp f(I_1, ..., \mathrm{Hull}(\{a_i\}), ..., I_n\}\big)$.

Proposition 4.2. *The Box-consistent Pruning operator is well-defined.*

4.2 Handling ODEs

In this section we expand our language to consider solutions of the initial value problems (IVP) of Lipschitz-continuous ODEs. Let $t_0, T \in \mathbb{R}$ and $g : \mathbb{R}^n \to \mathbb{R}$ be a Lipschitz-continuous function, i.e., for all $\boldsymbol{x}_1, \boldsymbol{x}_2 \in \mathbb{R}^n$, $|g(\boldsymbol{x}_1) - g(\boldsymbol{x}_2)| \leq c||\boldsymbol{x}_1 - \boldsymbol{x}_2||$ for some constant c. Let $t_0, T \in \mathbb{R}$ satisfy $t_0 \leq T$ and $\boldsymbol{y}_0 \in \mathbb{R}^n$. An (autonomous) IVP problem is given by

$$\frac{d\boldsymbol{y}}{dt} = g(\boldsymbol{y}(t)) \text{ and } \boldsymbol{y}(t_0) = \boldsymbol{y}_0, \text{ where } t \in [t_0, T].$$

where $\boldsymbol{y} : [t_0, T] \to \mathbb{R}^n$ is called the *solution* of the IVP. Consider $\boldsymbol{y}(t)$ as $(y_1(t), ..., y_n(t))$, then each component $y_i : [t, T] \to \mathbb{R}$ is a Type 2 computable function, and can appear in some signature $\mathcal{F}$. In fact, we can also regard $\boldsymbol{y}_0$ as an argument of y_i and write $y_i(t_0, \boldsymbol{y}_0)$. This does not change computability properties of y_i, since following the Picard-Lindelöf representation $\boldsymbol{y}(t) = \int_{t_0}^{t} g(\boldsymbol{y}(s))ds + \boldsymbol{y}_0$, $y_i(t)$ is only linearly dependent on $\boldsymbol{y}_0$.

In practice, with an ICP framework, we can exploit interval solvers for IVP problems [27], for pruning intervals on variables that appear in constraints involving ODEs. This direction has received much recent attention [12,11,18,21].

Consider the IVP problem defined above, with $\boldsymbol{y}_0$ contained in a box $B_{t_0} \subseteq \mathbb{R}^n$. Let $t_0 \leq t_1 \leq ... \leq t_m = T$ be a set of points in $[t_0, T]$. An interval-based ODE solver returns a set of boxes $B_{t_1}, ..., B_{t_m}$ such that

$$\forall i \in \{1, ..., m\},\ \{\boldsymbol{y}(t) : t_{i-1} \leq t \leq t_i, \boldsymbol{y}_0 \in B_{\boldsymbol{y}_0}\} \subseteq B_{t_i}.$$

Now let $y_i : [t_0, T] \times B_0 \to \mathbb{R}$ be the i-th component of the solution $\boldsymbol{y}$ of an IVP problem. Then interval-based ODE solvers compute interval extensions of y_i. Thus, pruning operators that respect the interval extension computed by interval ODE solvers can be defined. It can be concluded from Theorem 4.2 that ICP_ε is δ-complete for equalities involving ODEs, as long as the pruning operator is well-defined. A simplest strategy is just to prune out any set of points outside the interval extension:

Proposition 4.3 (Simple ODE-Pruning). *Let* $y_i = f(t, \boldsymbol{y}_0)$ *be the i-th component function of an IVP problem. Suppose* $\sharp f$ *is computed by an interval ODE solver. Then the pruning operator* $\mathrm{Prune}(I_{y_i}, f) = I_{y_i} \cap \sharp f(I_t, B_{\boldsymbol{y}_0})$ *is well-defined, where* I_{y_i} *is an interval on* y_i *and* I_t *is an interval on* t.

4.3 DPLL⟨ICP⟩

Now consider the integration of ICP into the framework of DPLL(T), so that the full δ-SMT problem can be solved. Given a formula φ, a DPLL⟨ICP⟩ solver uses SAT solvers to enumerate solutions to the Boolean abstraction φ^B of the formula, and uses ICP_ε to decide the satisfiability of conjunctions of atomic formulas. DPLL⟨ICP⟩ returns sat when ICP_ε returns sat to some conjunction of theory atoms witnessing the satisfiability of φ^B, and returns unsat when ICP_ε returns unsat on all the solutions to φ^B. Thus, it follows naturally that using a δ-complete theory solver ICP_ε, DPLL⟨ICP⟩ is also δ-complete.

Corollary 4.1 (δ-Completeness of DPLL⟨ICP⟩). *Let $\mathcal{F}$ be a set of real functions. Then the pruning operators in ICP_ε are well-defined for $\mathcal{F}$, if and only if, DPLL⟨ICP⟩ using ICP_ε is δ-complete for bounded Σ_1-sentences in $\mathcal{L}_\mathcal{F}$.*

In practice, correctness of numerical solvers is always a major concern. For complete trustworthiness, it is important for numerically-driven decision procedures to return certificates for their decisions δ-sat and sat. We outline methods for producing certificates in DPLL⟨ICP⟩ in [15].

5 Applications

δ-Complete solvers return answers that allow one-sided, δ-bounded errors. The framework allows us to easily understand the implications of such errors in practical problems. Indeed, δ-complete solvers can be *directly* used in the following correctness-critical problems.

Bounded Model Checking and Invariant Validation. Let $S = \langle X, \mathsf{Init}, \mathsf{Trans}\rangle$ be a transition system over X, which can by continuous or hybrid. Then given a subset $U \subseteq X$, the bounded model checking problem asks whether $\varphi_n := \exists \boldsymbol{x}_0, ..., \boldsymbol{x}_n(\boldsymbol{x}_0 \wedge \bigwedge_{i=0}^{n-1} \mathsf{Trans}(\boldsymbol{x}_i, \boldsymbol{x}_{i+1}) \wedge \boldsymbol{x}_n \in U)$ is true. Here U denotes the "unsafe" values of the system, and we say S is safe up to n if φ_n is false. Thus, using a δ-complete solver for φ_n, we can determine the following: If φ_n is unsat, then S is indeed safe up to n; on the other hand, if φ_n is δ-sat, then either the system is unsafe, or it would be unsafe under a δ-perturbation, and a counterexample is provided by the certificate for δ-sat. This δ can be set by the user based on the intended tolerance of errors of the system. Thus, a δ-complete solver can be directly used.

For invariant validation, a proposed invariant Inv can prove safety if the sentence $\varphi := \forall \boldsymbol{x}, \boldsymbol{x}'((\mathsf{Init}(\boldsymbol{x}) \rightarrow \mathsf{Inv}(\boldsymbol{x})) \wedge (\mathsf{Inv}(\boldsymbol{x}) \wedge \mathsf{Trans}(\boldsymbol{x}, \boldsymbol{x}') \rightarrow \mathsf{Inv}(\boldsymbol{x}')) \wedge \mathsf{Inv}(\boldsymbol{x}) \rightarrow \neg(U(\boldsymbol{x})))$ is true. We then use a δ-complete solver on $\neg\varphi$, which is existential. When unsat is returned, Inv is indeed an inductive invariant proving safety. When δ-sat is returned, either Inv is not an inductive invariant, or under a small numerical perturbation, Inv would violate the inductive conditions.

Theorem Proving. For theorem proving, one-sided errors are not directly useful since no robustness problem is involved. We can still approach a statement φ by making δ-decisions on $\neg\varphi$, and refine δ when needed. Starting from any δ, whenever unsat is returned, φ is proved; when δ-sat, we can try a smaller δ. This reflects the common practice in proving these statements.

6 Conclusion

We introduced the notion of "δ-complete decision procedures" for solving SMT problems the over real numbers. Our aim is to provide a general framework for solving a wide range of nonlinear functions including transcendental functions and solutions of Lipschitz-continuous ODEs. δ-Completeness serves as a replacement of the conventional completeness requirement on exact solvers, which is impossible to satisfy in this domain. We proved the existence of δ-complete decision procedures for bounded SMT with Type 2 computable functions and showed the complexity of the problem. We use δ-completeness as the standard correctness requirement on numerically-driven decision procedures, and formally analyzed the solving framework DPLL⟨ICP⟩. We proved sufficient and necessary conditions for its δ-completeness. We believe our results serve as a foundation for the development of scalable numerically-driven decision procedures and their application in formal verification and theorem proving.

Acknowledgement. We are grateful for many important suggestions from Lenore Blum and the anonymous reviewers.

References

1. Akbarpour, B., Paulson, L.C.: Metitarski: An automatic theorem prover for real-valued special functions. J. Autom. Reasoning 44(3), 175–205 (2010)
2. Avigad, J., Friedman, H.: Combining decision procedures for the reals. Logical Methods in Computer Science, 2(4) (2006)
3. Benhamou, F., Granvilliers, L.: Continuous and Interval Constraints. In: Rossi, F., van Beek, P., Walsh, T. (eds.) Handbook of Constraint Programming, ch. 16. Elsevier (2006)
4. Borralleras, C., Lucas, S., Navarro-Marset, R., Rodríguez-Carbonell, E., Rubio, A.: Solving Non-linear Polynomial Arithmetic via SAT Modulo Linear Arithmetic. In: Schmidt, R.A. (ed.) CADE 2009. LNCS, vol. 5663, pp. 294–305. Springer, Heidelberg (2009)
5. Brattka, V., Hertling, P., Weihrauch, K.: A tutorial on computable analysis. In: Cooper, S.B., Löwe, B., Sorbi, A. (eds.) New Computational Paradigms, pp. 425–491. Springer, New York (2008)
6. Brown, C.W., Davenport, J.H.: The complexity of quantifier elimination and cylindrical algebraic decomposition. In: ISSAC 2007 (2007)
7. Clarke, E.M., Biere, A., Raimi, R., Zhu, Y.: Bounded model checking using satisfiability solving. Formal Methods in System Design 19(1), 7–34 (2001)
8. Clarke, E.M., Grumberg, O., Peled, D.: Model checking. MIT Press (2001)
9. Collins, G.E.: Hauptvortrag: Quantifier Elimination for Real Closed Fields by Cylindrical Algebraic Decomposition. In: Brakhage, H. (ed.) GI-Fachtagung 1975. LNCS, vol. 33, pp. 134–183. Springer, Heidelberg (1975)
10. Dutertre, B., de Moura, L.: A Fast Linear-Arithmetic Solver for DPLL(T). In: Ball, T., Jones, R.B. (eds.) CAV 2006. LNCS, vol. 4144, pp. 81–94. Springer, Heidelberg (2006)
11. Eggers, A., Fränzle, M., Herde, C.: SAT Modulo ODE: A Direct SAT Approach to Hybrid Systems. In: Cha, S(S.), Choi, J.-Y., Kim, M., Lee, I., Viswanathan, M. (eds.) ATVA 2008. LNCS, vol. 5311, pp. 171–185. Springer, Heidelberg (2008)

12. Eggers, A., Ramdani, N., Nedialkov, N., Fränzle, M.: Improving SAT Modulo ODE for Hybrid Systems Analysis by Combining Different Enclosure Methods. In: Barthe, G., Pardo, A., Schneider, G. (eds.) SEFM 2011. LNCS, vol. 7041, pp. 172–187. Springer, Heidelberg (2011)
13. Fränzle, M., Herde, C., Teige, T., Ratschan, S., Schubert, T.: Efficient solving of large non-linear arithmetic constraint systems with complex boolean structure. JSAT 1(3-4), 209–236 (2007)
14. Ganai, M.K., Ivančić, F.: Efficient decision procedure for non-linear arithmetic constraints using cordic. In: Formal Methods in Computer Aided Design, FMCAD (2009)
15. Gao, S., Avigad, J., Clarke, E.: δ-Decision procedures for satisfiability over the reals. Extended version, `http://arxiv.org/abs/1204.3513`
16. Gao, S., Avigad, J., Clarke, E.: δ-Decidability over the reals. In: Logic in Computer Science, LICS (2012)
17. Gao, S., Ganai, M., Ivancic, F., Gupta, A., Sankaranarayanan, S., Clarke, E.: Integrating ICP and LRA solvers for deciding nonlinear real arithmetic. In: FMCAD (2010)
18. Goldsztejn, A., Mullier, O., Eveillard, D., Hosobe, H.: Including Ordinary Differential Equations Based Constraints in the Standard CP Framework. In: Cohen, D. (ed.) CP 2010. LNCS, vol. 6308, pp. 221–235. Springer, Heidelberg (2010)
19. Hales, T.C.: Introduction to the flyspeck project. In: Mathematics, Algorithms, Proofs (2005)
20. Hentenryck, P.V., McAllester, D., Kapur, D.: Solving polynomial systems using a branch and prune approach. SIAM Journal on Numerical Analysis 34(2), 797–827 (1997)
21. Ishii, D., Ueda, K., Hosobe, H.: An interval-based sat modulo ode solver for model checking nonlinear hybrid systems. STTT 13(5), 449–461 (2011)
22. Kawamura, A.: Lipschitz continuous ordinary differential equations are polynomial-space complete. In: IEEE Conference on Computational Complexity, pp. 149–160. IEEE Computer Society (2009)
23. Ko, K.-I.: Complexity Theory of Real Functions. Birkhäuser (1991)
24. Ko, K.-I.: On the computational complexity of integral equations. Ann. Pure Appl. Logic 58(3), 201–228 (1992)
25. Kroening, D., Strichman, O.: Decision Procedures: An Algorithmic Point of View. Springer (2008)
26. Munoz, C., Narkawicz, A.: Formalization of an efficient representation of Bernstein polynomials and applications to global optimization, `http://shemesh.larc.nasa.gov/people/cam/Bernstein/`
27. Nedialkov, N.S., Jackson, K.R., Corliss, G.F.: Validated solutions of initial value problems for ordinary differential equations. Applied Mathematics and Computation 105(1), 21–68 (1999)
28. Nuzzo, P., Puggelli, A., Seshia, S.A., Sangiovanni-Vincentelli, A.L.: Calcs: Smt solving for non-linear convex constraints. In: Bloem, R., Sharygina, N. (eds.) FMCAD, pp. 71–79. IEEE (2010)
29. Platzer, A., Clarke, E.M.: The Image Computation Problem in Hybrid Systems Model Checking. In: Bemporad, A., Bicchi, A., Buttazzo, G. (eds.) HSCC 2007. LNCS, vol. 4416, pp. 473–486. Springer, Heidelberg (2007)
30. Ratschan, S.: Quantified constraints under perturbation. J. Symb. Comput. 33(4), 493–505 (2002)
31. Weihrauch, K.: Computable Analysis: An Introduction (2000)

BDD-Based Automated Reasoning for Propositional Bi-Intuitionistic Tense Logics

Rajeev Goré and Jimmy Thomson

Logic and Computation Group,
Research School of Computer Science, The Australian National University

Abstract. We give Binary Decision Diagram (BDD) based methods for deciding validity and satisfiability of propositional Intuitionistic Logic **Int** and Bi-intuitionistic Tense Logic **BiKt**. We handle intuitionistic implication and bi-intuitionistic exclusion by treating them as modalities, but the move to an intuitionistic basis requires careful analysis for handling the reflexivity, transitivity and antisymmetry of the underlying Kripke relation. **BiKt** requires a further extension to handle the interactions between the intuitionistic and modal binary relations, and their converses. We explain our methodology for using the Kripke semantics of these logics to constrain the underlying least and greatest fixpoint approaches of the finite model construction. With some optimisations this technique is competitive with the state of the art theorem provers for Intuitionistic Logic using the ILTP benchmark and randomly generated formulae.

1   Introduction

For many logics, we can decide the validity of a given formula φ_0 by constructing the set of all subsets of some closure $cl(\varphi_0)$, and checking whether these subsets can support a (counter) model that makes φ_0 false. If no such model exists, then we can safely declare φ_0 to be valid using this finite model property (fmp).

At first sight, this "fmp method" seems impractical since the first step requires us to "construct" the set of all (exponentially many) subsets of $cl(\varphi_0)$, thus giving a procedure whose worst case and best case complexity is always of order $O(2^{|cl(\varphi_0)|})$. However, Pan et al. [12] and Marrero [9] have shown that Binary Decision Diagrams (BDDs) can be used to represent the required subsets efficiently, without actually "constructing" them explicitly for **K** and **CTL**.

We investigate the potential of this BDD-based method for Intuitionistic Propositional Logic (**Int**) and its extensions Bi-Intuitionistic Logic (**BiInt**) and Bi-Intuitionistic Tense Logic (**BiKt**). These logics introduce various complications over **K** and **CTL**: the logic **Int** has an intuitionistic rather than a classical basis; the logic **BiInt** has an operator whose semantics uses the converse of the Kripke binary relation; the logic **BiKt** has two binary relations $R_\square$ and $R_\lozenge$ so that $\square$ and $\lozenge$ are not De Morgan duals, has their converses to handle $\blacklozenge$ and $\blacksquare$ and has two further interaction conditions. A priori, it is not obvious how to

B. Gramlich, D. Miller, and U. Sattler (Eds.): IJCAR 2012, LNAI 7364, pp. 301–315, 2012.
© Springer-Verlag Berlin Heidelberg 2012

$w \not\Vdash \bot$		$w \Vdash p$	iff $\rho(w,p) = t$
$w \Vdash \varphi \wedge \psi$	iff $w \Vdash \varphi$ and $w \Vdash \psi$	$w \Vdash \varphi \vee \psi$	iff $w \Vdash \varphi$ or $w \Vdash \psi$
$w \Vdash \varphi \rightarrow \psi$	iff $\forall v \sqsupseteq w. v \not\Vdash \varphi$ or $v \Vdash \psi$	$w \Vdash \varphi \prec \psi$	iff $\exists v \sqsubseteq w. v \Vdash \varphi$ and $v \not\Vdash \psi$
$w \Vdash \Diamond\varphi$	iff $\exists v. w R_\Diamond v$ and $v \Vdash \varphi$	$w \Vdash \blacklozenge\varphi$	iff $\exists v. w R_\Box^{-1} v$ and $v \Vdash \varphi$
$w \Vdash \Box\varphi$	iff $\forall z \forall v. w \sqsubseteq z R_\Box v \Rightarrow v \Vdash \varphi$	$w \Vdash \blacksquare\varphi$	iff $\forall z \forall v. w \sqsubseteq z R_\Diamond^{-1} v \Rightarrow v \Vdash \varphi$

Fig. 1. Kripke Semantics for **BiKt** in model $\mathcal{M} = (W, \sqsubseteq, R_\Box, R_\Diamond, \rho)$ and $w \in W$

handle all of these complications using the BDD method, and indeed, we find that the least fixpoint approach for BDDs does not work for all of our logics.

We show how to adapt the BDD-method to **Int**, extend it to **BiInt** and **BiKt**, and describe some useful optimisations. We also compare our implementation with the state of the art theorem provers for **Int** (`PITP` [1] and `Imogen` [11]), and `DBiKt` [15], the only theorem prover for **BiKt** that we are aware of.

Our results show that with the help of some optimisations, this method is competitive with state-of-the-art theorem provers for **Int**, and still works well for some of its tense extensions. Its biggest advantage is its versatility.

1.1 Related Work

Current state of the art theorem provers for **Int** are based on an optimised tableau method [1] or a heuristically guided, focused, polarised, inverse method [11]. Pointers to other theorem provers for **Int** can be found on the ILTP Benchmark website [16]: most of them are based upon tableaux or sequent calculi.

Various sequent calculi for **BiInt** exist [4, 6, 13, 14, 15]. Some of them allow backward proof-search, and some have been extended to handle **BiKt** [7]. However, we know of only one implementation for both of these logics [15].

Pan et al. [12] give a BDD-based algorithm for deciding **K**, the simplest propositional classical normal modal logic. They show how to handle a single binary relation using BDDs, but do not need to consider multiple interacting "converse" relations, nor further frame conditions like reflexivity, transitivity and anti-symmetry, as we do. They also experiment with some potential optimisations, some of which are not limited to **K**. We make use of some of these optimisations, as well as describing some new optimisations appropriate for **Int**.

Marrero [9] gives a BDD-based algorithm for deciding computation tree logic **CTL**, a propositional modal temporal logic with fixpoints. He provides a way of handling the transitive closure of a discrete and serial relation by explicitly calculating a least fixpoint, which he uses to deal with eventualities. For our logics, the relation itself is required to be transitive, so we use a different method.

2 Syntax and Semantics of Bi-Intuitionistic Tense Logics

Formulae of **BiKt** [7] are defined from a set *Prp* of primitive propositions as:

$$\varphi ::= p \in Prp \mid \bot \mid \varphi \wedge \varphi \mid \varphi \vee \varphi \mid \varphi \rightarrow \varphi \mid \varphi \prec \varphi \mid \Diamond\varphi \mid \Box\varphi \mid \blacklozenge\varphi \mid \blacksquare\varphi$$

Models of **BiKt** are structures $\mathcal{M} = (W, \sqsubseteq, R_\Box, R_\Diamond, \rho)$ where W is a non-empty set of worlds; $\sqsubseteq$ is a reflexive, transitive and antisymmetric binary relation on W; both $R_\Box$ and $R_\Diamond$ are binary relations on W satisfying the "zig-zag" frame conditions $(F1)$ and $(F2)$ below; and $\rho : W \times Prp \mapsto \{t, f\}$ is a valuation which obeys the persistence property:

$(F1)$:	If $x \sqsubseteq y$ and $xR_\Diamond z$ then $\exists w.yR_\Diamond w$ and $z \sqsubseteq w$
$(F2)$:	If $xR_\Box y$ and $y \sqsubseteq z$ then $\exists w.x \sqsubseteq w$ and $wR_\Box z$
Persistence:	If $\rho(w,p) = t$ and $w \sqsubseteq v$ then $\rho(v,p) = t$.

Given $\mathcal{M} = (W, \sqsubseteq, R_\Box, R_\Diamond, \rho)$ and $w \in W$, the semantics of **BiKt** are given in Figure 1. We use $\rightarrow$ for intuitionistic implication while we use $\Rightarrow$ for classical implication in the meta-logic. We define intuitionistic negation $\neg\varphi$ as $\varphi \rightarrow \bot$. Note that $\Box$ and $\Diamond$, and $\blacksquare$ and $\blacklozenge$, are not de Morgan duals via negation.

BiInt [18] is the $\{\wedge, \vee, \rightarrow, -\!\!<, \bot\}$-fragment of **BiKt** and models of **BiInt** thus do not need the $R_\Box$ and $R_\Diamond$ relations. **Int** is **BiInt** without $-\!\!<$ -formulae.

A formula φ is **L**-valid if for all **L**-models $\mathcal{M}$, and for all worlds $w \in \mathcal{M}$ we have $\mathcal{M}, w \Vdash \varphi$. Dually, φ is **L**-satisfiable if there is some **L**-model $\mathcal{M}$ with some world $w \in \mathcal{M}$ such that $\mathcal{M}, w \Vdash \varphi$. A formula is **L**-falsifiable iff it is not **L**-valid. We define global logical consequence for **BiKt** and fragments as follows where $\mathcal{M} = (W, \sqsubseteq, R_\Box, R_\Diamond, \rho)$ and Γ is a finite set of "global assumptions":

$$\Gamma \models \varphi \text{ iff } \forall \mathcal{M}.\ (\forall w \in W. \mathcal{M}, w \Vdash \Gamma) \Rightarrow \forall w \in W.\ \mathcal{M}, w \Vdash \varphi.$$

3 A BDD Perspective of the Finite Model Method

For each of our logics, our goal is to construct a finite model $\mathcal{M} = (W_f, \preceq_f, R_\Box, R_\Diamond, \rho)$, as appropriate, similar to Pan et al., by constructing a sequence of frames $(W_0, \preceq_0, R^0_\Box, R^0_\Diamond)$, $(W_1, \preceq_1, R^1_\Box, R^1_\Diamond)$, …, $(W_f, \preceq_f, R^f_\Box, R^f_\Diamond)$ such that the final frame gives a model which is "canonical" in two senses: if φ_0 is satisfiable (falsifiable) then some world of W_f satisfies (falsifies) φ_0. Given such a finite "canonical" model, we can decide whether a given φ_0 is satisfiable or valid by checking whether such worlds exist.

For a given formula φ_0, and a closure $cl(\varphi_0)$ we first define a set of atoms $Atm \subseteq cl(\varphi_0)$, as appropriate for the logic. Each subset of Atm is a classical (bi-valent) valuation on these atoms, where membership means truth-hood. The set of potentially good worlds $\mathcal{W} = 2^{Atm}$ is thus an upper bound on each W_i above, and the binary relation $\mathcal{W} \times \mathcal{W}$ is an upper bound on each $\preceq_i$.

We next use the Kripke semantics to extract necessary constraints to construct a relation $\preceq_{\max} \subseteq \mathcal{W} \times \mathcal{W}$ that is maximal in that it throws out only the edges which break these constraints. We then monotonically refine an initial approximation W_0 towards W_f, using the constructed $\preceq_{\max}$ relation to enforce the correct modal interpretation of the elements of $cl(\varphi_0)$ in all the worlds. Once W_f has been computed, the final step is to determine which, if any, worlds in W_f satisfy and falsify φ_0, giving the satisfiability and validity of φ_0.

3.1 A Better Basis for W_f

For our logics, $cl(\varphi_0) = sub(\varphi_0)$, the set of all subformulae of φ_0 including φ_0. The naive way to construct W_f is simply to use the set of all subsets of $cl(\varphi_0)$. We instead use only the "sensible subsets" following Pan et al.'s "lean" representation and Marrero's choice of BDD variables. We represent each primitive proposition and implication from $cl(\varphi_0)$ as an explicit BDD-variable, and compute the "denotation" of an arbitrary formula from $cl(\varphi_0)$ as follows:

$$Atm = (Prp \cap sub(\varphi_0)) \cup \{\varphi \to \psi \mid \varphi \to \psi \in sub(\varphi_0)\}$$

$$\mathcal{W} = 2^{Atm} \quad [\![\bot]\!] = \emptyset \qquad [\![a]\!] = \{w \in \mathcal{W} \mid a \in w\}$$

$$[\![\phi \wedge \psi]\!] = [\![\phi]\!] \cap [\![\psi]\!] \qquad [\![\phi \vee \psi]\!] = [\![\phi]\!] \cup [\![\psi]\!]$$

Thus $\mathcal{W}$ is finite, $w \in \mathcal{W}$ corresponds to a classical binary valuation on our BDD-variables, and for every $\psi \in cl(\varphi_0)$, the world w claims to satisfy ψ if $w \in [\![\psi]\!]$, and claims to falsify ψ if $w \in \overline{[\![\psi]\!]}$, where $\overline{[\![\psi]\!]} = \mathcal{W} \setminus [\![\psi]\!]$.

The set $\mathcal{W}$ is smaller than $2^{cl(\varphi_0)}$, and does not contains worlds which behave inappropriately with respect to conjunction and disjunction. We are thus left with worlds containing primitive propositions and implications. The semantics of an intuitionistic implication refers to $\sqsubseteq$. We therefore use an explicit representation $\preceq$ of the $\sqsubseteq$ relation as a finite set of ordered pairs from $\mathcal{W} \times \mathcal{W}$.

3.2 Constructing the Maximal $\preceq$ Relation

Our eventual goal is to construct a $W_f \subseteq \mathcal{W}$ and a binary relation $\preceq_f$ over W_f which obeys all of the semantic restrictions of intuitionistic models. We now show how to construct an over-approximation $\preceq_{\max}$ over $\mathcal{W}$ which is persistent, transitive and anti-symmetric, and which also obeys one half of the semantics of implication. These restrictions on the binary relation are not required for **K** or **CTL**, and so are not considered by Pan et al. [12] and Marrero [9].

Persistence. For any particular primitive proposition $p \in Atm$, the persistence condition can be expressed in terms of denotations as below:

$$\forall w, v \in \mathcal{W}.\ w \in [\![p]\!] \ \&\ w \preceq v \Rightarrow v \in [\![p]\!] \tag{1}$$

Alternatively, dropping universal quantifiers, we can write it as either of:

$$w \preceq v \Rightarrow w \in \overline{[\![p]\!]} \text{ or } v \in [\![p]\!] \qquad w \in [\![p]\!] \ \&\ v \in \overline{[\![p]\!]} \Rightarrow w \not\preceq v \tag{2}$$

The constraint obtained from (2) is expressed in terms of set notation as:

$$\preceq \subseteq (\overline{[\![p]\!]} \times \mathcal{W}) \cup (\mathcal{W} \times [\![p]\!]) \qquad ([\![p]\!] \times \mathcal{W}) \cap (\mathcal{W} \times \overline{[\![p]\!]}) \subseteq \not\preceq \tag{3}$$

That is, an upper bound on $\preceq$ is the set of ordered pairs from $\mathcal{W} \times \mathcal{W}$ where the first world is not in the denotation of p or the second is in the denotation of p. Alternately, a pair of worlds from $\mathcal{W} \times \mathcal{W}$ is forbidden from being in $\preceq$ if the first is in the denotation of p and the second is not. Taking the conjunction over all $p \in cl(\varphi_0)$ gives our final over-approximation from persistence:

$$\preceq \;\subseteq \bigcap_{p \in Prp \cap Atm} (\overline{[\![p]\!]} \times \mathcal{W}) \cup (\mathcal{W} \times [\![p]\!]) \tag{4}$$

Semantics of implication. Since $\sqsubseteq$ is transitive, if $w \sqsubseteq v$ then all successors of v are successors of w as well: thus if $\mathcal{M}, w \Vdash \phi \to \psi$, then $\mathcal{M}, v \Vdash \phi \to \psi$ and implications persist across $\sqsubseteq$. We mimic this by extending (4) from just the primitive propositions to all atoms ψ in Atm:

$$\preceq \;\subseteq \bigcap_{\psi \in Atm} (\overline{[\![\psi]\!]} \times \mathcal{W}) \cup (\mathcal{W} \times [\![\psi]\!]) \tag{5}$$

For any particular implication $\phi \to \psi$, the "only if" part of the semantics of implication can be expressed using denotations by dropping quantifiers as either:

$$w \in [\![\phi \to \psi]\!] \;\&\; w \preceq v \;\Rightarrow\; v \in \overline{[\![\phi]\!]} \cup [\![\psi]\!] \tag{6}$$

$$w \preceq v \;\Rightarrow\; w \in \overline{[\![\phi \to \psi]\!]} \text{ or } v \in \overline{[\![\phi]\!]} \cup [\![\psi]\!] \tag{7}$$

Just as (1) became (4), constraint (7) becomes the following in terms of sets:

$$\preceq \;\subseteq \bigcap_{\phi \to \psi \in Atm} (\overline{[\![\phi \to \psi]\!]} \times \mathcal{W}) \cup (\mathcal{W} \times (\overline{[\![\phi]\!]} \cup [\![\psi]\!])) \tag{8}$$

The conjunction of (5) and (8) gives $\preceq_{\max}$, an upper bound on $\preceq$, as:

$$\preceq_{\max} = RHS(5) \cap RHS(8) \tag{9}$$

Transitivity and Antisymmetry

Lemma 1 *The relation $\preceq_{\max}$ is transitive: $(\preceq_{\max} \circ \preceq_{\max}) \subseteq \preceq_{\max}$.*

Proof. For a contradiction, pick any $(x, y) \in \preceq_{\max} \circ \preceq_{\max}$ and suppose (x, y) fails the persistence condition (5): thus for some $\psi \in Atm$, we have $x \in [\![\psi]\!]$ and $y \in \overline{[\![\psi]\!]}$. By the definition of $\circ$ there must be some "midpoint" z such that $(x, z) \in \preceq_{\max}$ and $(z, y) \in \preceq_{\max}$. Since $x \preceq_{\max} z$ and $x \in [\![\psi]\!]$ we must have $z \in [\![\psi]\!]$ by persistence of $\preceq_{\max}$. Then $z \preceq_{\max} y$ gives $y \in [\![\psi]\!]$: contradiction.

Suppose then that (x, y) fails condition (8): thus $x \in [\![\phi \to \psi]\!]$ and $y \in [\![\phi]\!]$ and $y \in \overline{[\![\psi]\!]}$. As before, the midpoint $z \in [\![\phi \to \psi]\!]$. By (8), if $(z, y) \in \preceq_{\max}$ then $y \in \overline{[\![\phi]\!]} \cup [\![\psi]\!]$, but this again contradicts our earlier assumption. Thus any pair in $\preceq_{\max} \circ \preceq_{\max}$ must obey (9). So $\preceq_{\max}$ is transitively closed.

Lemma 2 *If $w \preceq_{\max} v$ and $v \preceq_{\max} w$ then $w = v$: thus $\preceq_{\max}$ is antisymmetric.*

Proof. Let $x \preceq_{\max} y$ and $y \preceq_{\max} x$. Suppose they differ on some atom a. If $a \in x$, then $a \in y$ by persistence, and vice-versa. Thus x and y cannot be distinct.

The relation $\preceq_{\max}$ may not be reflexive since $\mathcal{W}$ may contain a $w \in [\![\phi \to \psi]\!] \cap [\![\phi]\!] \cap \overline{[\![\psi]\!]}$, meaning that $(w, w) \notin \preceq_{\max}$.

3.3 Using $\preceq_{\max}$ to Construct W_f

We now have a set of "sensible" worlds $\mathcal{W}$ and an over-approximation $\preceq_{\max}$ of $\sqsubseteq$ that is persistent, transitive and anti-symmetric (but not necessarily reflexive). The structure $(\mathcal{W}, \preceq_{\max})$ may still contain "bad" worlds that do not obey the semantics: for example, a world $w \in \mathcal{W}$ with $w \in \overline{[\![\phi \rightarrow \psi]\!]}$ which lacks a $v \in [\![\phi]\!] \cap \overline{[\![\psi]\!]}$ with $w \preceq_{\max} v$. We can refine this structure into a model in two ways: by starting with $W_0 = \mathcal{W}$ as the set of all "potentially good" worlds and removing only "bad" worlds, or by starting with $W_0 = \emptyset$ as the set of all "known good" worlds and adding only "good" worlds. The greatest fixpoint of the first way, and the least fixpoint of the second way gives the W_f we seek.

At each stage, $\preceq_i$ is just the restriction of $\preceq_{\max}$ to W_i. These restrictions maintain persistence and antisymmetry of $\preceq_i$ because no new edges are added. We maintain transitivity ($x \preceq_i y$ & $y \preceq_i z \Rightarrow x \preceq_i z$) because each restriction removes worlds rather than edges, thus the only way we can lose an existing edge $x \preceq_i z$, is by removing x or z, whence we also lose $x \preceq_i y$ or $y \preceq_i z$.

Regaining Reflexivity. The restriction of $\preceq_{\max}$ to W is reflexive if the formula below left holds, so the maximal subset of $\mathcal{W}$ on which $\preceq_{\max}$ is reflexive is *Refl*:

$$\forall w \in W.\ w \preceq_{\max} w \qquad\qquad \mathit{Refl} = \{w \in \mathcal{W} \mid (w, w) \in \preceq_{\max}\}$$

Any $w \in \mathcal{W}$ such that $w \notin \mathit{Refl}$ is not permitted to be reflexive by our constraints on $\preceq_{\max}$, and thus must not appear in any model and so $W_f \subseteq \mathit{Refl}$.

Enforcing the Semantics of Implications. The remaining aspect of the semantics to consider is the (contra-position of the) "if" component of implication:

$$w \in \overline{[\![\phi \rightarrow \psi]\!]} \Rightarrow \exists v \in W_i.\ w \preceq v \ \&\ v \in [\![\phi]\!] \cap \overline{[\![\psi]\!]} \tag{10}$$

Given the current "potentially good" or "known good" worlds W_i, the potential witnesses $V_i^{\phi\rightarrow\psi}$ that falsify a particular $\phi \rightarrow \psi \in \mathit{Atm}$ are found by:

$$V_i^{\phi\rightarrow\psi} = W_i \cap [\![\phi]\!] \cap \overline{[\![\psi]\!]}.$$

From this, we can identify the worlds which can reach such a witness using the $\preceq_{\max}$ pre-image of $V_i^{\phi\rightarrow\psi}$, and complete the representation:

$$W_i^{\phi\rightarrow\psi} = [\![\phi \rightarrow \psi]\!] \cup \{x \mid \exists y \in V_i^{\phi\rightarrow\psi}\ .\ (x, y) \in \preceq_{\max}\} \tag{11}$$

The pre-image is found using existential quantification of BDDs, as described by Pan et al. and Marrero. We now show how to find the set W_f using fixpoints.

The GFP Method. We start with $W_0 = \mathcal{W}$, as the set of all "potentially good" worlds and prune bad worlds by computing the greatest fixpoint of:

$$W_{i+1} = W_i \cap \mathit{Refl} \cap \bigcap_{\phi\rightarrow\psi \in \mathit{Atm}} W_i^{\phi\rightarrow\psi} \tag{12}$$

Since *Refl* does not depend on W_i, we can instead just start with $W_0 = Refl$. Formula (12) is monotonic decreasing, and since $\mathcal{W}$ is finite, finding a fixpoint by repeated iteration is guaranteed to terminate in exponential time.

The LFP Method. We start with $W_0 = \emptyset$ and add only "good" worlds. Unlike **K**, where the least- and greatest-fixpoint iterated formulae are essentially the same, we must account for reflexivity at each iteration to handle the case where $x = y$ in (11). A solution to this is to explicitly allow for reflexivity:

$$W^+ = \overline{[\![\phi]\!]} \cup [\![\psi]\!] \qquad W_i^- = ([\![\phi]\!] \cap \overline{[\![\psi]\!]}) \cup \{x \mid \exists y \in V_i^{\phi\to\psi} \,.\, (x,y) \in \preceq_{\max}\}$$

$$W_{i+1} = W_i \cup \bigcap_{\phi\to\psi\in Atm} (\overline{[\![\phi \to \psi]\!]} \cap W_i^-]) \cup ([\![\phi \to \psi]\!] \cap W^+) \tag{13}$$

Here the formula for W^+ captures the worlds that satisfy $\phi \to \psi$ locally. The formula for W_i^- captures the worlds that falsify $\phi \to \psi$ locally, or by having some other successor which falsifies $\phi \to \psi$.

In the first iteration, $W_0 = \emptyset$, so $V_0^{\phi\to\psi} = \emptyset$ and $W_i^- = [\![\phi]\!] \cap \overline{[\![\psi]\!]}$. Thus

$$W_1 = \bigcap_{\phi\to\psi\in Atm} \left(\overline{[\![\phi \to \psi]\!]} \cap [\![\phi]\!] \cap \overline{[\![\psi]\!]}\right) \cup \left([\![\phi \to \psi]\!] \cap (\overline{[\![\phi]\!]} \cup [\![\psi]\!])\right)$$

That is, W_1 contains all worlds that satisfy/falsify all their implications locally. Then, W_2 will be the worlds which satisfy/falsify all their implications either locally or in the worlds of W_1, and so on. Since equation (13) is monotonically increasing, and $\mathcal{W}$ is finite, this least fixpoint computation terminates.

Deciding Satisfiability, Falsifiability and Validity. Once W_f is constructed, the model is $\mathcal{M}_f = (W_f, \preceq_f, \rho)$ where, for all $w \in W_f$, we put $\rho(w,p) = t$ iff $p \in w$. We can lift this valuation to $cl(\varphi_0)$ by showing that $\mathcal{M}_f, w \Vdash \psi$ iff $w \in [\![\psi]\!]$ for all $\psi \in cl(\varphi_0)$, giving us soundness.

For completeness, we have to show that the witness w_0 which satisfies or falsifies φ_0 in some model $\mathcal{M}$ is also represented in W_f. Since the fmp guarantees that only members of $cl(\varphi_0)$ are relevant, w_0 is represented by $w_0' \in \mathcal{W}$ as the subset $w_0' = \{\psi \in Atm \mid \mathcal{M}, w_0 \Vdash \psi\}$. For the greatest fixpoint method, $w_0' \in W_0 = \mathcal{W}$ and we prove that after all refinements, it is in W_f. For the least fixpoint method, $W_0 = \emptyset$ so we prove that w_0' is added to W_i, for some $i > 0$.

Theorem 3 *φ_0 is satisfiable iff $[\![\varphi_0]\!] \cap W_f \neq \emptyset$ and φ_0 is valid iff $\overline{[\![\varphi_0]\!]} \cap W_f = \emptyset$.*

Since we construct a representation of a model, we can relatively easily create a concrete example model illustrating satisfiability or falsifiability. But we deduce unsatisfiability and validity by the absence of certain worlds, so a convincing proof or "reason" for unsatisfiability or validity is more difficult to produce.

Global Assumptions. The greatest fixpoint method can easily handle global assumptions to decide whether $\Gamma \models \varphi_0$ by using $W_0 = [\![\Gamma]\!]$ instead of $W_0 = \mathcal{W}$, thus immediately considering only those worlds that satisfy Γ. For the least fixpoint method, we must assert the global assumptions Γ at each iteration.

3.4 Extension to Bi-Intuitionistic Logic BiInt

We now show that the greatest fixpoint BDD-method extends easily to handle converse, but the least fixpoint one does not. Our outline follows the methodology we set out for **Int**, but we no longer explicitly distinguish between $\preceq$, $\preceq_i$ and $\preceq_{\max}$. Strictly speaking, the same distinctions as for **Int** apply.

The denotation of $-\!\!<$ -formulae uses the (converse of the) semantic binary relation $\sqsubseteq$, so as for $\to$-formulae, we add all $-\!\!<$ -formulae from $cl(\varphi_0)$ to Atm.

The semantics of $-\!\!<$ are handled similarly to $\to$. Transitivity of the underlying relation $\sqsubseteq$ means that if $\mathcal{M}, w \Vdash \phi -\!\!< \psi$ and $w \sqsubseteq v$ then $\mathcal{M}, v \Vdash \phi -\!\!< \psi$, so exclusions persist. Thus we demand that all atoms still persist across $\preceq$.

The (contra-position of the) "if" component of the semantics of $\phi -\!\!< \psi$ is:

$$w \in \overline{[\![\phi -\!\!< \psi]\!]} \ \ \& \ \ w \preceq^{-1} v \quad \Rightarrow \quad v \in \overline{[\![\phi]\!]} \cup [\![\psi]\!] \tag{14}$$

Equation (14) transforms to a constraint on $\preceq^{-1}$ and hence $\preceq$:

$$\preceq^{-1} \ \subseteq \ ([\![\phi -\!\!< \psi]\!] \times \mathcal{W}) \cup (\mathcal{W} \times (\overline{[\![\phi]\!]} \cup [\![\psi]\!])) \tag{15}$$

$$\preceq \ \subseteq \ (\mathcal{W} \times [\![\phi -\!\!< \psi]\!]) \cup ((\overline{[\![\phi]\!]} \cup [\![\psi]\!]) \times \mathcal{W}) \tag{16}$$

The "only if" component of the semantics of exclusion is:

$$w \in [\![\phi -\!\!< \psi]\!] \Rightarrow \exists v.\ w \preceq^{-1} v \ \& \ v \in [\![\phi]\!] \cap \overline{[\![\psi]\!]} \tag{17}$$

We now have to modify the fixpoint formula. For greatest fixpoints, we first calculate the witnesses $V_i^{\phi -\!\!< \psi}$ to the existential of (17), as for implication earlier, and then determine the worlds $W_i^{\phi -\!\!< \psi}$ which reach the witness via $\preceq^{-1}$:

$$V_i^{\phi -\!\!< \psi} = W_i \cap [\![\phi]\!] \cap \overline{[\![\psi]\!]} \quad W_i^{\phi -\!\!< \psi} = \overline{[\![\phi -\!\!< \psi]\!]} \cup \{y \mid \exists x \in V_i^{\phi -\!\!< \psi}.\ (x, y) \in \preceq\}$$

The greatest fixpoint simply extends the one for **Int** with this new constraint:

$$W_{i+1} = W_i \ \cap \ \mathit{Refl} \ \cap \bigcap_{\phi \to \psi \in Atm} W_i^{\phi \to \psi} \ \cap \bigcap_{\phi -\!\!< \psi \in Atm} W_i^{\phi -\!\!< \psi}$$

On the other hand, it is not clear that there can be a least fixpoint approach for **BiInt**. For example, the formula $p \wedge (((p \to \bot) \to \bot) -\!\!< p)$ is satisfiable, but only in models containing a group of worlds which simultaneously require the existence of each other. Such worlds lead to a non-well-founded ordering on the inclusion of worlds in the least fixpoint, meaning that $W_0 = \emptyset$ does not suffice.

3.5 Extension to Bi-Intuitionistic Tense Logic BiKt

Moving from **BiInt** to **BiKt** presents more of a challenge. In addition to the 4 new connectives $\Box, \Diamond, \blacksquare$ and $\blacklozenge$, we must handle the two frame conditions $(F1)$ and $(F2)$. These conditions are difficult to capture directly as they refer to both the intuitionistic and modal relations and are existential in nature. However, their purpose is to ensure that truth persists over $\sqsubseteq$, and this is easier to use.

Theorem 4 (Persistence of BiKt) *For all* **BiKt** *models* $\mathcal{M}$, *if* $\mathcal{M}, w \Vdash \varphi$ *and* $w \sqsubseteq v$ *then* $\mathcal{M}, v \Vdash \varphi$.

The proof proceeds by induction on the size of φ and relies on $(F1)$ and $(F2)$ for the persistence of $\Diamond$- and $\blacklozenge$-formulae. Thus $(F1)$ and $(F2)$ cause persistence.

Suppose now that we have a structure which fails $(F1)$ or $(F2)$, but in which all formulae persist across $\sqsubseteq$. We can soundly add the missing $R_\Box$ or $R_\Diamond$ edges, without changing the satisfaction relation, to obtain a **BiKt**-model (which obeys $(F1)$ and $(F2)$) as encapsulated in the next theorem.

Theorem 5 *By adding* $R_\Diamond$ *and* $R_\Box$ *edges, a structure* $\mathcal{M}_1 = (W, \sqsubseteq, R^1_\Diamond, R^1_\Box, \rho)$ *which is persistent can be converted to a structure* $\mathcal{M}_n = (W, \sqsubseteq, R^n_\Diamond, R^n_\Box, \rho)$ *which satisfies* $(F1)$ *and* $(F2)$, *and such that* $\mathcal{M}_1, w \Vdash \varphi$ *iff* $\mathcal{M}_n, w \Vdash \varphi$, *for all* $w \in W$.

Thus, considering all persistent structures is sufficient. We must first extend *Atm* by adding all formulae with a main connective from $\{\Box, \Diamond, \blacklozenge, \blacksquare\}$ from $cl(\varphi_0)$. We can then enforce persistence as for **Int** and **BiInt** via (5).

Having handled the frame conditions, we handle the semantics for $\Diamond$ and $\blacklozenge$ using Pan et al.'s methods for **K**. The $\Diamond$-formulae impose a restriction on $R_\Diamond$, while the $\blacklozenge$-formulae impose a similar restriction on $R_\Box^{-1}$, and hence upon $R_\Box$:

$$R_\Diamond \subseteq (\llbracket \Diamond\psi \rrbracket \times \mathcal{W}) \cup (\mathcal{W} \times \overline{\llbracket \psi \rrbracket}) \tag{18}$$

$$R_\Box^{-1} \subseteq (\llbracket \blacklozenge\psi \rrbracket \times \mathcal{W}) \cup (\mathcal{W} \times \overline{\llbracket \psi \rrbracket}) \tag{19}$$

For the greatest fixpoint method, we also need:

$$W_i^{\Diamond\psi} = \overline{\llbracket \Diamond\psi \rrbracket} \cup \{x \mid \exists y \in (W_i \cap \llbracket \psi \rrbracket) \,.\, (x, y) \in R_\Diamond\}$$

The $\Box$- and $\blacksquare$-formulae are more complicated to represent since $R_\Box$ and $R_\Diamond$ interact with $\preceq$. The contra-positive of the "if" part of the semantics for $\Box$ is:

$$w \in \overline{\llbracket \Box\psi \rrbracket} \quad \Rightarrow \quad \exists z.\, w \preceq z \;\&\; \exists v.\, zR_\Box v \;\&\; v \in \overline{\llbracket \psi \rrbracket}$$

This has two existentials, which can be handled by computing two pre-images as follows. Let $Z_i^{\Box\psi} = \{z \mid \exists y \in (W_i \cap \overline{\llbracket \psi \rrbracket}) \,.\, (z, y) \in R_\Box\}$ and let

$$W_i^{\Box\psi} = \llbracket \Box\psi \rrbracket \cup \{x \mid \exists z \in (W_i \cap Z_i^{\Box\psi}) \,.\, (x, z) \in \preceq_{\max}\}$$

For the "only if" component, the interactions of the relations are more troublesome. But since $\preceq$ is required to be reflexive, the following are essential:

$$wR_\Box v \Rightarrow w \in \overline{\llbracket \Box\psi \rrbracket} \vee v \in \llbracket \psi \rrbracket \tag{20}$$

$$vR_\Diamond w \Rightarrow w \in \overline{\llbracket \blacksquare\psi \rrbracket} \vee v \in \llbracket \psi \rrbracket \tag{21}$$

Additionally, because (5) enforces persistence, if $u \in \llbracket \Box\psi \rrbracket$, then any w such that $u \preceq w$ must also satisfy $w \in \llbracket \Box\psi \rrbracket$. By induction this will force all $\preceq$-successors w of u to satisfy (20), and thus to satisfy the original semantics.

The constraints on $R_\Box$ are thus (19) and (20):

$$R_\Box \subseteq (\overline{[\![\Box\psi]\!]} \times \mathcal{W}) \cup (\mathcal{W} \times [\![\psi]\!]) \qquad (\mathcal{M}, w \Vdash \Box\psi \Rightarrow \mathcal{M}, v \Vdash \psi)$$
$$R_\Box \subseteq (\mathcal{W} \times [\![\blacklozenge\psi]\!]) \cup (\overline{[\![\psi]\!]} \times \mathcal{W}) \qquad (\mathcal{M}, w \Vdash \psi \Rightarrow \mathcal{M}, v \Vdash \blacklozenge\psi)$$

Similarly, the constraints on $R_\Diamond$ are (18) and (21):

$$R_\Diamond \subseteq ([\![\Diamond\psi]\!] \times \mathcal{W}) \cup (\mathcal{W} \times \overline{[\![\psi]\!]}) \qquad (\mathcal{M}, v \Vdash \psi \Rightarrow \mathcal{M}, w \Vdash \Diamond\psi)$$
$$R_\Diamond \subseteq (\mathcal{W} \times \overline{[\![\blacksquare\psi]\!]}) \cup ([\![\psi]\!] \times \mathcal{W}) \qquad (\mathcal{M}, v \Vdash \blacksquare\psi \Rightarrow \mathcal{M}, w \Vdash \psi)$$

To complete the decision procedure, the greatest fixpoint calculation is:

$$W_{i+1} = W_i \cap \mathit{Refl} \cap W_i^{\phi\to\psi} \cap W_i^{\phi-\!\!<\,\psi} \cap W_i^{\Box\psi} \cap W_i^{\Diamond\psi} \cap W_i^{\blacklozenge\psi} \cap W_i^{\blacksquare\psi}$$

3.6 Optimisations

Variable ordering. The choice of variable ordering is critical when using ordered BDDs. We chose the following ordering after minimal experimentation since its preliminary results are encouraging. For each member of *Atm* whose main connective is non-classical (i.e. implication, exclusion, diamond or box), we do a pre-order traversal of the formation tree stopping at other members of *Atm*. The first time any member of *Atm* is encountered, it is appended to the current ordering. Pre-image computations require copying *Atm*, so the copy a' of a appears immediately after a in the ordering. Using an ordering which puts all copies at the end of the ordering is particularly bad.

Since determining the best ordering is a difficult problem in itself, BDD packages allow us to dynamically reorder the BDD variables. There is a trade-off between the quality of a reordering and the time taken to perform the reordering, so the main question is when to use this feature. For the greatest-fixpoint method, we provide an option which uses this feature once only to find a good ordering after computing W_0 using *Refl* and any global assumptions.

Normalisation. Another component of complexity is the number of BDD variables. We use techniques such as constant propagation ($\top \wedge \varphi = \varphi$, $\bot \to \varphi = \top$, etc.) to reduce formula size, and possibly reduce the number of atoms. We use syntactic equality to check whether two formulae are equivalent when determining the set of atoms. Normalising wrt an arbitrary fixed ordering $<$ on formulae improves the efficiency of this equality check. For example, putting $p < q$ collapses $\{(p \wedge q) \to \bot, (q \wedge p) \to \bot\}$ to $\{(p \wedge q) \to \bot\}$, requiring fewer atoms.

Early Termination. If we only want to check satisfiability or validity, then early termination is possible. In the greatest fixpoint approach, the sets W_i are strictly decreasing. If any $W_i \subseteq [\![\varphi_0]\!]$ then $W_f \subseteq [\![\varphi_0]\!]$, so φ_0 is valid, and if any $W_i \cap [\![\varphi_0]\!] = \emptyset$ then $W_f \cap [\![\varphi_0]\!] = \emptyset$, so φ_0 is unsatisfiable. In the least fixpoint approach, W_i is strictly increasing. If any $W_i \cap [\![\varphi_0]\!] \neq \emptyset$ then $W_f \cap [\![\varphi_0]\!] \neq \emptyset$, so φ_0 is satisfiable, and if any $W_i \cap \overline{[\![\varphi_0]\!]} \neq \emptyset$ then $W_f \cap \overline{[\![\varphi_0]\!]} \neq \emptyset$, so φ_0 is not valid.

Explicit representation of $\leftrightarrow$. Expanding bi-implications $\leftrightarrow$ into two implications can lead to an exponential blowup in the size of the formula. We therefore gave a direct semantics for $\leftrightarrow$ and added it to the set of atoms.

Explicit global assumptions. When determining **Int**-validity of a formula $\varphi_0 = (\gamma \to \varphi)$, any counterexample must make γ true in all states reachable from the root. The formula is valid iff all models where γ is true globally must make φ true. Thus, in the greatest fixpoint approach, we can start with $W_0 = [\![\gamma]\!]$, rather than $W_0 = \mathcal{W}$. By translating top-level implications to global assumptions in this manner, there are fewer atoms to consider, and the global assumptions may restrict the search space resulting in fewer iterations before reaching the fixpoint. This optimisation cannot be used for **BiInt** because $-\!\!<$ allows us to look "backwards" along $\sqsubseteq$, so γ cannot be turned into a global assumption.

4 Experimental Results

We used the ILTP propositional benchmarks [17, 16] and randomly generated formulae. All tests were performed on 32bit Ubuntu with a Core 2 Duo 3.0GHz processor, 3 GB RAM and a timeout of 600 seconds for each problem instance.

Benchmarks. The ILTP benchmarks consists of several categories of structured intuitionistic formulae. Some are "uninteresting" since they are easy for all provers. The remaining "interesting" benchmarks are split into 12 problem sets with 20 instances each, parametrised by a size n, consisting of zero or more axioms $\{\gamma_0, \cdots, \gamma_k\}$ and a single conjecture C giving $(\gamma_0 \wedge \cdots \wedge \gamma_k) \to C$.

The random benchmarks are generated to have a fixed number of symbols (treating $\neg\varphi = \varphi \to \bot$ as only one additional symbol) and a maximum ratio of distinct propositions to formula size. Formulae for **Int** use connectives $\wedge, \vee, \neg, \to$ and $\leftrightarrow$ while formulae for **BiKt** add in connectives $-\!\!<$, $\Box$, $\blacklozenge$, $\Diamond$ and $\blacksquare$. We used 1000 instances of each size from 10 through to 90 in steps of 5, which are available here: `http://users.cecs.anu.edu.au/~rpg/BDDBiKtProver/`

Theorem provers. According to the ILTP benchmark [16], the two best provers for propositional intuitionistic logic are `PITP`/`PITPINV`, and `Imogen`.

`PITP` and `PITPINV` [1] implement a signed tableau calculus to determine **Int** validity. The tableau rules are divided into 6 categories based on the branching factor and whether or not they are invertible. `PITPINV` attempts a non-invertible branch of one category before the invertible branch, while `PITP` attempts the invertible branch first. `PITP` and `PITPINV` are written in C++, and make use of optimisations such as dynamic formula simplification.

`Imogen` [11] uses a focused polarised inverse method to determine **Int** validity. Given a formula, `Imogen` performs a pre-processing step to assign polarities to each subformula. It then makes use of focusing based on the polarities to generate inference rules, and these rules are used (and extended) by the inverse method in a saturation phase to attempt to construct a sequent proving the original formula. Imogen is written in ML, and uses heuristics when assigning polarities to try to minimise the search space. When given 600 seconds, it tries one heuristic for 2 seconds and then, if needed, tries an alternate heuristic for 598 seconds.

DBiKt [15] is the only theorem prover we are aware of for **BiKt**. It uses a deep-inference nested sequent calculus for **BiKt** and is implemented in Java. It has not been heavily optimised, but is intended as a proof of concept.

BDDBiKt is our Ocaml theorem prover using the Buddy [3] BDD library. It is available here: `http://users.cecs.anu.edu.au/~rpg/BDDBiKtProver/`

Results. The numbers reported here for PITP differ from those on the ILTP website for two reasons. The first is that we use different hardware. The second is that formula SYN007+1.0014 expands to 4GB when converted to the input format of PITP, which did not allow it to be converted in-memory in the initial comparison. We instead write the formula to disk during the conversion, which allows the conversion to finish and hence allows PITP to solve it.

Our numbers for PITPINV differ from the ILTP website because we discovered a bug during the comparison on randomly generated formulae. The authors of PITPINV corrected the bug, and this has impacted its performance.

We analysed the impact of the implemented optimisations by testing the following versions of our own implementation:

GFP:	Greatest-fixpoint with early termination and explicit handling of $\leftrightarrow$
Ga:	GFP, with explicit global assumptions
Gn:	GFP, with normalisation.
Gna:	GFP with both explicit global assumptions and normalisation
Gnar:	Gna with dynamic variable reordering.
LFP:	Least-fixpoint with the same optimisations as GFP
Ln:	LFP with normalisation.

In Figure 2, "sum" is the sum out of the 240 "interesting" problems shown individually, while "total" is out of the whole 274 instance benchmark.

With all optimisations enabled, our BDD-method (Gnar) solved the second highest number of instances on the ILTP benchmark. Unlike the other theorem provers, when the BDD-method fails, it usually runs out of memory rather than time. Experiments on the same hardware with a 64bit OS and 8GB RAM showed that no instance caused Gnar to run out of memory and BDD times were improved, but only one additional problem was solved by Gnar, while Imogen performed notably slower. We now discuss the effects of each optimisation.

Explicit Assumptions. Converting top-level implications to explicit global assumptions has the largest impact. All of the benchmark formulae with axioms were helped by this optimisation, and some were trivialised by it.

Normalisation. Normalising the input formula was not as beneficial as explicit assumptions. In some cases it helps significantly: for example it rewrites SYJ206 to $\top$. In others it is detrimental because changing the formula structure changes our heuristically chosen BDD order into a worse one.

	GFP	Gn	Ga	Gna	Gnar	LFP	Ln	Imogen	PITP	PITPINV	Out of
SYJ201	6	6	20	20	20	6	6	20	20	20	20
SYJ202	12	10	12	10	8	12	10	8	10	10	20
SYJ203	17	18	20	20	20	18	19	20	20	20	20
SYJ204	17	19	20	20	20	18	20	20	20	20	20
SYJ205	14	14	14	19	19	12	12	20	20	11	20
SYJ206	15	20	15	20	20	19	20	20	20	20	20
SYJ207	6	6	20	20	20	7	7	20	7	8	20
SYJ208	7	7	9	10	8	9	9	19	20	20	20
SYJ209	17	18	20	20	20	18	19	20	10	10	20
SYJ210	17	18	20	20	20	19	20	20	20	20	20
SYJ211	6	6	9	8	15	9	9	20	20	9	20
SYJ212	13	20	13	20	20	18	20	20	20	20	20
sum of above	147	162	192	207	210	165	171	227	206	187	240
total over all	181	196	226	241	244	199	205	261	240	221	274

Fig. 2. Number of instances solved in the ILTP benchmark

Assumptions + Normalisation. Combining both optimisations works well on the whole. For class SYJ205, the formula is a conjunction of two semantically equivalent implications which are syntactically reordered. Normalisation combines the two implications into one, which is then converted into a global assumption. No assumptions can be made explicit without normalisation, and normalisation alone only removes one implication from the closure.

Dynamic Variable Reordering. Adding dynamic variable ordering is a mixed bag. In most cases its overhead is significant, while the benefits are small. For SYJ211 this is reversed, with reordering taking little time but giving significant improvement. We speculate that our relatively naive ordering performs reasonably well on most of the benchmarks, possibly because of the prevalence of lexicographically ordered sequences of propositional variables, so in general the dynamic ordering does not give a big benefit. However when the initial ordering is bad, the dynamic ordering can assist.

GFP vs LFP. LFP performs similarly to GFP in many cases, although it is not compatible with some of the helpful optimisations. The small differences between LFP and GFP arise from their different fixpoint formulae. LFP generally has fewer iterations, but each iteration is a more complex formula than the one used by GFP and thus takes longer to compute.

The results of random **Int** tell a different story. Now `Imogen` performs considerably worse than all other provers, failing to solve many cases. GFP scales reasonably, but is still significantly worse than `PITP`. Gnar is consistently slower than Gna, however at size 90 the non-reordering version runs out of memory on 7 formulae, while the reordering version times out on only 5. It seems that reordering may not help very often, but it makes the method more robust.

LFP does quite well, though not as well as `PITP`. The majority of the randomly generated formulae are invalid, so LFP can terminate early. Since each iteration is quite expensive, performing fewer iterations on the invalid formulae here gives a large benefit. On the valid formulae, it performs worse than GFP.

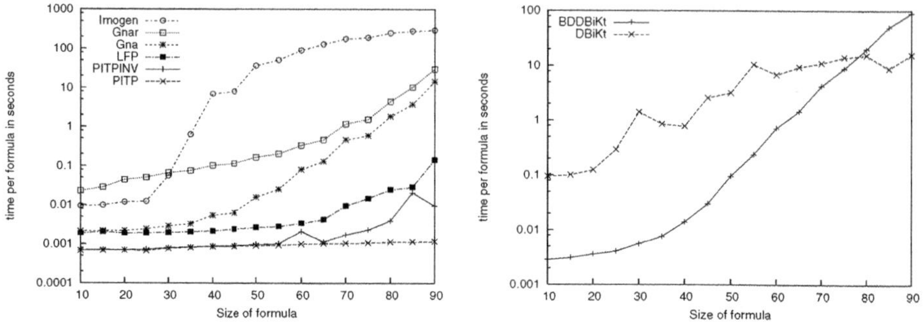

Fig. 3. Average time taken per random **Int** and **BiKt** instance with a timeout of 600s

The comparison with `DBiKt` for **BiKt** shows that each theorem prover can handle some formulae that the other could not. For sizes up to 75, GFP solved all instances, and did it faster than `DBiKt`. Past that point, `DBiKt` solved an increasing number of problems that GFP could not solve due to time or memory limits, and the time taken by GFP increased significantly above that of `DBiKt`. In general, all but 5 or so of the randomly generated formulae were invalid, but some of the few valid instances proved difficult for `DBiKt` and not GFP.

5 Conclusion and Further Work

Our optimised BDD-method `Gnar` for **Int** is competitive with the state-of-the-art provers `Imogen` and `PITP` in the following sense: on the ILTP benchmarks, it solves more problems than `PITP` but less problems than `Imogen`, and on randomly generated formulae, it performs better than `Imogen` but worse than `PITP`.

Unlike the other methods, BDD-methods are "memory hungry" so adding memory is likely to improve their relative performance. Indeed, moving from a 32bit OS to a 64bit OS gave a small improvement, but not as much as we hoped since the bottleneck just moved from memory to time.

To some extent, our implementation is naive, and further optimisations from the model checking community need to be investigated. In particular, we need to ascertain whether the BDD method is relatively brittle to variable ordering heuristics, or robust over many potential choices.

We are currently extending this method to handle all 15 basic modal logics obtained by combinations of reflexivity, transitivity, seriality, euclideaness, and symmetry, as well as to the modal mu-calculus. We are also extending the implementation to generate explicit (counter) models. A characterisation of when and how the method works would also be nice. Finally, can the BDD method be extended to predicate logics, possibly using instantiation-based methods?

The biggest advantage of the BDD-method is the ease with which it extends from **Int** to **BiInt** to **BiKt** compared to tableaux and inverse methods. For example, handling a "converse" operator to give **BiInt** using tableaux requires significant methodological extensions [2, 8]. Similarly, the inverse method has not been extended to handle "converse" as far as we know. McLaughlin and Pfenning [10] have implemented an inverse method for intuitionistic modal logics which do not require the complications of converse. We can handle these intuitionistic modal logics using our BDD-method for **BiKt** by just dropping $-\!\!<$, $\blacksquare$ and $\blacklozenge$, and replacing $R_\Box$ and $R_\Diamond$ with a single modal relation R.

References

[1] Avellone, A., Fiorino, G., Moscato, U.: Optimization techniques for propositional intuitionistic logic and their implementation. Theoretical Computer Science 409(1), 41–58 (2008)

[2] Baader, F., Calvanese, D., McGuinness, D.L., Nardi, D., Patel-Schneider, P.F. (eds.): The Description Logic Handbook: Theory, Implementation, and Applications. Cambridge University Press (2003)

[3] Buddy (2011), `http://sourceforge.net/projects/buddy/`

[4] Buisman, L., Goré, R.: A Cut-Free Sequent Calculus for Bi-intuitionistic Logic. In: Olivetti, N. (ed.) TABLEAUX 2007. LNCS (LNAI), vol. 4548, pp. 90–106. Springer, Heidelberg (2007)

[5] Goré, R., Thomson, J., Widmann, F.: An experimental comparison of theorem provers for CTL. In: TIME 2011: Eighteenth International Symposium on Temporal Representation and Reasoning, pp. 49–56 (September 2011)

[6] Goré, R., Postniece, L., Tiu, A.: Cut-elimination and proof-search for bi-intuitionistic logic using nested sequents. In: AiML 2008, pp. 43–66 (2008)

[7] Goré, R., Postniece, L., Tiu, A.: Cut-elimination and proof search for bi-intuitionistic tense logic. In: Advances in Modal Logic, pp. 156–177 (2010)

[8] Goré, R., Widmann, F.: Sound Global State Caching for *ALC* with Inverse Roles. In: Giese, M., Waaler, A. (eds.) TABLEAUX 2009. LNCS, vol. 5607, pp. 205–219. Springer, Heidelberg (2009)

[9] Marrero, W.: Using BDDs to Decide CTL. In: Halbwachs, N., Zuck, L.D. (eds.) TACAS 2005. LNCS, vol. 3440, pp. 222–236. Springer, Heidelberg (2005)

[10] McLaughlin, S., Pfenning, F.: The focused constraint inverse method for intuitionistic modal logics (2010), (unpublished manuscript) (accessed January 31, 2012)

[11] McLaughlin, S., Pfenning, F.: Imogen: Focusing the Polarized Inverse Method for Intuitionistic Propositional Logic. In: Cervesato, I., Veith, H., Voronkov, A. (eds.) LPAR 2008. LNCS (LNAI), vol. 5330, pp. 174–181. Springer, Heidelberg (2008)

[12] Pan, G., Sattler, U., Vardi, M.Y.: BDD-based decision procedures for the modal logic K. Journal of Applied Non-classical Logics 49 (2005)

[13] Pinto, L., Uustalu, T.: Proof Search and Counter-Model Construction for Bi-intuitionistic Propositional Logic with Labelled Sequents. In: Giese, M., Waaler, A. (eds.) TABLEAUX 2009. LNCS, vol. 5607, pp. 295–309. Springer, Heidelberg (2009)

[14] Postniece, L.: Deep Inference in Bi-intuitionistic Logic. In: Ono, H., Kanazawa, M., de Queiroz, R. (eds.) WoLLIC 2009. LNCS, vol. 5514, pp. 320–334. Springer, Heidelberg (2009)

[15] Postniece, L.: Proof Theory and Proof Search of Bi-Intuitionistic and Tense Logic. Ph.D. thesis, Australian National University (2011)

[16] Raths, T., Otten, J.: The ILTP library (2007), `http://www.cs.uni-potsdam.de/ti/iltp/` (accessed January 2012)

[17] Raths, T., Otten, J., Kreitz, C.: The ILTP problem library for intuitionistic logic, release v1.1. Journal of Automated Reasoning (2006)

[18] Rauszer, C.: Applications of Kripke models to Heyting-Brouwer logic. Studia Logica 36, 61–71 (1977)

From Linear Temporal Logic Properties to Rewrite Propositions

Pierre-Cyrille Héam*, Vincent Hugot**, and Olga Kouchnarenko

FEMTO-ST CNRS 6174, University of Franche-Comté & INRIA/CASSIS, France
{pierre-cyrille.heam,vincent.hugot,olga.kouchnarenko}@inria.fr

Abstract. In the regular model-checking framework, reachability analysis can be guided by temporal logic properties, for instance to achieve the counter example guided abstraction refinement (CEGAR) objectives. A way to perform this analysis is to translate a temporal logic formula expressed on maximal rewriting words into a "rewrite proposition" – a propositional formula whose atoms are language comparisons, and then to generate semi-decision procedures based on (approximations of) the rewrite proposition. This approach has recently been studied using a non-automatic translation method. The extent to which such a translation can be systematised needs to be investigated, as well as the applicability of approximated methods wherever no exact translation can be effected. This paper presents contributions to that effect: **(1)** we investigate suitable semantics for LTL on maximal rewriting words and their influence on the feasibility of a translation, and **(2)** we propose a general scheme providing exact results on a fragment of LTL corresponding mainly to safety formulæ, and approximations on a larger fragment.

1 Introduction and Context

Term rewriting and rewriting logic have been intensively and successfully used for solving equational problems in automated deduction, for programming language definitions, for model transformations and generation of efficient interpreters as well as for specification and verification in software engineering. In this last context, system states are modelled by languages, while rewrite rules stand for *actions* of the system; for instance procedure or method calls. This technique has been successfully used to prove the security of cryptographic protocols [11] and Java Bytecode programs [3]. When proving security, reachability analysis over sets of terms can be guided by temporal logic properties, like e.g., in [7,6].

In [7], three specific Linear Temporal Logic (LTL) formulæ – chosen for their relevance to model-checking [13], in particular with respect to Java MIDLets, in the framework of the French ANR RAVAJ project – have been translated into what we will call *rewrite propositions*, with respect to straightforward semantics for LTL on finite words. For instance, given a rewrite system $\mathcal{R}$, of which

* This author is supported by the project ANR 2010 BLAN 0202 02 FREC.
** This author is supported by the French DGA (Direction Générale de l'Armement).

B. Gramlich, D. Miller, and U. Sattler (Eds.): IJCAR 2012, LNAI 7364, pp. 316–331, 2012.

$X, Y \subseteq \mathcal{R}$ are subsets, and an initial language Π, the LTL property $\square(X \Rightarrow \bullet Y)$ signifies that whenever an accessible term is rewritten by some rewrite rule in X, then the resulting term can be rewritten by some rule in Y, and not by any other rule. As shown in [7], that property is satisfied if and only if the following rewrite proposition holds: $[\mathcal{R} \setminus Y]\left(X\left(\mathcal{R}^*(\Pi)\right)\right) = \varnothing \wedge X\left(\mathcal{R}^*(\Pi)\right) \subseteq Y^{-1}(\mathcal{T}(\mathbb{A}))$, where $\mathcal{R}^*(\Pi)$ is the transitive-reflexive forward closure of Π by $\mathcal{R}$, and $\mathcal{T}(\mathbb{A})$ is the set of all trees. The point of translating satisfaction in terms of rewrite propositions is that they present a more tractable intermediary form which can itself be translated into automata-based (semi-)decision procedures. Indeed, if the initial language Π is regular, then the literature is rife with constructive results concerning questions such as preservation of regularity under a rewriting step, or under forward closure; that is to say, "under which conditions on the rewrite system $\mathcal{R}$ is $\mathcal{R}(\Pi)$ (resp. $\mathcal{R}^*(\Pi)$) still regular?". And when preserving regularity is not an option, one may fall back on more expressive classes of tree automata (TA) such as TAGED [10]. As an example of both aspects, [7, Prop. 5] states that a language given by $\mathcal{R}^{-1}(\mathcal{T}(\mathbb{A}))$ can in all generality be represented by a positive TAGED; furthermore, if $\mathcal{R}$ is left-linear, then regularity is preserved. Such results can be combined with regular approximation techniques; for instance, if $\mathcal{A}$ is a tree automaton, a procedure $\mathtt{Approx}(\mathcal{A}, \mathcal{R})$ in [4] yields another TA $\mathcal{B}$ such that $\mathfrak{Lang}(\mathcal{B}) \supseteq \mathcal{R}^*(\mathfrak{Lang}(\mathcal{A}))$, where $\mathfrak{Lang}(\mathcal{A})$ is the language accepted by $\mathcal{A}$. Put together, those tools provide a framework for building decision and semi-decision procedures from rewrite propositions. For instance, the proposition given above is semi-decided by the conjunction of the procedures $\mathtt{IsEmpty}(\mathtt{OneStep}(\mathcal{R} \setminus Y, \mathtt{Approx}(\mathcal{A}, \mathcal{R})), X)$ and $\mathtt{Subset}(\mathtt{OneStep}(X, \mathtt{Approx}(\mathcal{A}, \mathcal{R})), \mathtt{Backward}(Y))$, where $\mathfrak{Lang}(\mathcal{A}) = \Pi$ and under the additional constraint that Y must be left-linear. Note that this is almost a straightforward reformulation of the original rewrite proposition.

To summarise the above, our approach to model-checking temporal properties of sequences of rewrite rules consists of two phases outlined in [7]: **(1)** translation of a temporal logic formula expressed on maximal rewriting words into a "rewrite proposition" – a propositional formula whose atoms are language comparisons, and **(2)** translation of the rewrite proposition into a semi-decision procedure. To make this approach useful for program verification, both steps must be automated; neither is at present. The *general question* investigated in the present paper is whether – and more specifically how and to what extent – such a translation can be automated for arbitrary temporal properties. More specifically, we focus solely on the *first* step, i.e. translation from temporal logic to rewrite propositions. The second step is an entirely different problem, and is out of the scope of this paper.

Related Work. In recent years, new results in rewriting logic have deeply extended the spectrum of its applications [9,17,5,16], especially in relation with temporal logic for rewriting [14,2]. Unlike [2], where LTL model checking is performed over finite structures, our approach handles temporal formulæ over infinite state systems. In this sense, it is close to [9]. However, in spite of its simplicity for practical applications, it does not permit – in its current state, at least – to consider equational theories. Our viewpoint differs slightly from other regular model-checking approaches such as Regular LTL [6] in that the temporal property relates to sequences of *actions* as opposed to sequences of *states*. It is however very similar to the method presented in [15], when reducing the equational theory to the identity.

Organisation of the Paper. Section 2 presents the notions and notations in use throughout this paper, including the choice of temporal semantics and a precise statement of the problem at hand. Section 3[p320] deals with the main contributions of the paper: the translation rules and the technical tools (signatures, weak/strong intertwined semantics, etc) on which they depend.

2 Preliminaries and Problem Statement

The *extended naturals* are denoted by $\overline{\mathbb{N}} \triangleq \mathbb{N} \cup \{+\infty\}$ and $[\![n, m]\!]$ denotes the integer interval $[n, m] \cap \mathbb{Z}$, with the convention that $[\![0, +\infty]\!] = \mathbb{N}$. For any $k \in \mathbb{N}$, $\mathbb{N}_k \triangleq [\![k, +\infty]\!]$ and $\overline{\mathbb{N}}_k \triangleq \mathbb{N}_k \cup \{+\infty\}$. The powerset of S is written $\wp(S)$. Substitution is written $f[v/X]$, meaning "v replaces X in the expression f".

2.1 Rewrite Words and Maximal Rewrite Words

A comprehensive survey on term rewriting can be found in [8]. Let $\mathcal{T}(\mathbb{A})$ be the set of all terms on a ranked alphabet $\mathbb{A}$, let $\mathcal{R}$ be a finite rewrite system, and $\Pi \subseteq \mathcal{T}(\mathbb{A})$ any set of terms. A *finite or infinite word on* $\mathcal{R}$ is an element of

$$\mathcal{W} \triangleq \bigcup_{n \in \overline{\mathbb{N}}} ([\![1, n]\!] \to \mathcal{R}) \ .$$

The length $\#w \in \overline{\mathbb{N}}$ of a word w is defined as $\mathrm{Card}(\mathrm{dom}\, w)$. Note that the empty function – of graph $\varnothing \times \mathcal{R} = \varnothing$ – is a word, which we call the *empty word*, denoted by λ. Let $w \in \mathcal{W}$ be a word of domain $[\![1, n]\!]$, for $n \in \overline{\mathbb{N}}$, and let $m \in \mathbb{N}_1$; then the *m-suffix of* w is the word denoted by w^m, such that

$$w^m \triangleq \left| \begin{array}{ccc} [\![1, n - m + 1]\!] & \longrightarrow & \mathcal{R} \\ k & \longmapsto & w(k + m - 1) \end{array} \right. .$$

Note that $w^1 = w$, for any word w. The intuitive meaning that we attach to a word w is a sequence of rewrite rules of $\mathcal{R}$, called in succession – in other words, it represents a "run" of the TRS $\mathcal{R}$. Of course, there is nothing in the above definition of words that guarantees that such a sequence is in any way feasible, and such a notion only makes sense with respect to initial terms to be rewritten. Thus we now define the *maximal rewrite words of* $\mathcal{R}$*, originating in* Π:

$${}^{\mathcal{R}}(\!|\Pi|\!) \triangleq \left\{ w \in \mathcal{W} \;\middle|\; \begin{array}{c} \exists u_0 \in \Pi : \forall k \in \mathrm{dom}\, w,\ \exists u_k \in \mathcal{T}(\mathbb{A}) : u_{k-1} \xrightarrow{w(k)} u_k \\ \wedge\ \#w \in \mathbb{N} \Rightarrow \nexists \rho \in \mathcal{R} : \exists v \in \mathcal{T}(\mathbb{A}) : u_{\#w} \xrightarrow{\rho} v \end{array} \right\} .$$

Note the potential presence of the empty word in that set. Informally, a word w is in ${}^{\mathcal{R}}(\!|\Pi|\!)$ if and only if the rewrite rules $w(1), \ldots, w(n), \ldots$ can be activated in succession, starting from a term $u_0 \in \Pi$, and the word w is "maximal" in the sense that it cannot be extended. That is to say, w ends only when no further rewrite rule can be activated. Thus ${}^{\mathcal{R}}(\!|\Pi|\!)$ captures the behaviours (or runs) of $\mathcal{R}$, starting from Π; this notion corresponds to the full paths of the rewrite graph described in [7].

2.2 Defining Temporal Semantics on Rewrite Words

Choice of LTL and Syntax. Before starting to think about translating temporal logic formulæ on rewrite words, we need to define precisely the kind of temporal formulæ under consideration, and their semantics. Given that prior work in [7] was done on LTL, and that our aim is to *generalise* this work, LTL – with subsets of $\mathcal{R}$ as atomic proposition – seems a reasonable choice. In practice we shall use a slight variant with generalised weak and strong next operators; the reasons for this choice will be discussed when the semantics are examined. A formula $\varphi \in \text{LTL}$ is generated by the following grammar:

$$\begin{aligned} \varphi := X \mid \neg\varphi \mid \varphi \wedge \varphi \mid \bullet^m \varphi \mid \circ^m \varphi \mid \varphi \,\mathbf{U}\, \varphi \qquad & X \in \wp(\mathcal{R}) \\ \top \mid \bot \mid \varphi \vee \varphi \mid \varphi \Rightarrow \varphi \mid \Diamond \varphi \mid \Box \varphi \qquad & m \in \mathbb{N}\,. \end{aligned}$$

Note that the operators which appear on the first line are functionally complete; the remaining operators are defined syntactically as: $\top \triangleq \mathcal{R} \vee \neg\mathcal{R}$, $\bot \triangleq \neg\top$, $\varphi \vee \psi \triangleq \neg(\neg\varphi \wedge \neg\psi)$, $\varphi \Rightarrow \psi \triangleq \neg\varphi \vee \psi$, $\Diamond \varphi \triangleq \top \,\mathbf{U}\, \varphi$ and $\Box \varphi \triangleq \neg \Diamond \neg \varphi$.

Choice of Semantics. In the literature, the semantics of LTL are defined and well-understood for ω-words; however the words of ${}^{\mathcal{R}}(\!|\Pi|\!)$ may be infinite *or* finite, or even empty, which corresponds to the fact that, depending on its input, a rewrite system may either not terminate, terminate after some rewrite operations, or terminate immediately. Therefore we need semantics capable of accommodating both ω-words and finite words, as well as the edge-case of the empty word. In contrast to the classical case of ω-words, there are several ways to define (two-valued) semantics for LTL on finite, maximal words. One such way found in the literature is Finite-LTL (F-LTL) [13], which complements the long-standing use of a "strong" *next* operator introduced in [12] by coining a "weak" *next* variant. Figure 1[p320] presents our choice of semantics for this paper, which is essentially F-LTL with generalised next operators and the added twist that words *may* be infinite or empty. Note that $\bullet^1$ and $\circ^1$ correspond exactly to the classical strong and weak next operators, and that for $m \geqslant 1$, $\bullet^m$ (resp. $\circ^m$) can trivially be obtained by repeating $\bullet^1$ (resp. $\circ^1$) m times. So the only non-trivial difference here is the existence of $\bullet^0$ and $\circ^0$; this will prove quite convenient when we deal with the translation of $\Box$, using the following lemma.

***Lemma 1** (Weak-Next & Always). Let $\varphi \in \text{LTL}$, $w \in \mathcal{W}$, $k \in \mathbb{N}$ and $i \in \mathbb{N}_1$; it holds that* **(1)** $(w, i) \models \Box\varphi$ *iff* $(w, i) \models \bigwedge_{m=0}^{\infty} \circ^m \varphi$ *and* **(2)** $(w, i) \models \Box\varphi$ *iff* $(w, i) \models \bigwedge_{m=0}^{k-1} (\circ^m \varphi) \wedge \circ^k \Box \varphi$.

Before moving on, let us stress that the choice of semantics, or even the choice of LTL for that matter, should by no means be considered as etched in stone;

$$
\begin{array}{lll}
(w,i) \models X & \text{iff} & i \in \operatorname{dom} w \text{ and } w(i) \in X \\
(w,i) \models \neg\varphi & \text{iff} & (w,i) \not\models \varphi \\
(w,i) \models (\varphi \wedge \psi) & \text{iff} & (w,i) \models \varphi \text{ and } (w,i) \models \psi \\
(w,i) \models \bullet^m \varphi & \text{iff} & i+m \in \operatorname{dom} w \text{ and } (w,i+m) \models \varphi \\
(w,i) \models \circ^m \varphi & \text{iff} & i+m \notin \operatorname{dom} w \text{ or } (w,i+m) \models \varphi \\
(w,i) \models \varphi \,\mathbf{U}\, \psi & \text{iff} & \exists j \in \operatorname{dom} w : j \geqslant i \wedge \begin{cases} (w,j) \models \psi \quad \wedge \\ \forall k \in [\![i, j-1]\!],\ (w,k) \models \varphi \end{cases}
\end{array}
$$

For any $w \in \mathcal{W}$, $i \in \mathbb{N}_1$, $m \in \mathbb{N}$ and $X \in \wp(\mathcal{R})$.

Fig. 1. LTL Semantics on Maximal Rewrite Words

it is very much a variable of the general problem. However it will henceforth be considered as data for the purposes of this paper.

TRS and LTL. Let φ be an LTL formula. We say that a word w satisfies/is a model of φ (denoted by $w \models \varphi$) iff $(w,1) \models \varphi$. Alternatively, we have $(w,i) \models \varphi$ iff $w^i \models \varphi$. We say that the rewrite system $\mathcal{R}$, with initial language Π, satisfies/is a model of φ (denoted by $\mathcal{R}, \Pi \models \varphi$) iff $\forall w \in {}^{\mathcal{R}}(\!|\Pi|\!)$, $w \models \varphi$.

2.3 Rewrite Propositions and Problem Statement

A *rewrite proposition on* $\mathcal{R}$*, from* Π is a formula of propositional logic whose atoms are language or rewrite systems comparisons. More specifically, a rewrite proposition π is generated by the following grammar:

$$
\pi := \gamma \mid \gamma \wedge \gamma \mid \gamma \vee \gamma \quad \gamma := \ell = \varnothing \mid X \subseteq X \mid \ell \subseteq \ell \qquad X \in \wp(\mathcal{R}) .
$$
$$
\ell := \Pi \mid \mathcal{T}(\mathbb{A}) \mid X(\ell) \mid X^{-1}(\ell) \mid X^*(\ell)
$$

Since the comparisons γ have obvious truth values, the interpretation of rewrite propositions is trivial; thus we will not introduce any notation for it, and automatically confuse π with its truth value in the remainder of this paper. Note that while other operators for propositional logic could be added, conjunction and disjunction will be enough for our purposes.

Problem Statement. We have now done enough groundwork to state our problem more formally. Given a rewrite system $\mathcal{R}$, a temporal formula φ in LTL (or some fragment of LTL), and an initial language $\Pi \subseteq \mathcal{T}(\mathbb{A})$, we search for an algorithmic method of building a rewrite proposition π such that $\mathcal{R}, \Pi \models \varphi$ if and only if π holds. We call such a method, as well as its result, an *exact translation* of φ, and say that π translates φ. If π is only a sufficient (resp. necessary) condition, then it is an *under-approximated* (resp. *over-approximated*) translation.

3 Building Translation Rules

3.1 Overview and Intuitions of the Translation

The Base Cases. Counterintuitively, $\varphi = \neg X$ is actually a simpler case than $\varphi = X$ as far as the translation is concerned, so we will consider it first. CASE 1: NEGA-

TIVE LITERAL. Suppose $\mathcal{R}, \Pi \models \neg X$. Recalling the semantics in Fig. 1[p320], this means that no term of Π can be rewritten by a rule in X. They *may* or *may not* be rewritable by rules *not* in X, though. Consider now $\pi_1 \equiv X(\Pi) = \varnothing$; it is easy to become convinced that this is an exact translation. CASE 2: POSITIVE LITERAL. Let $\varphi = X$. A first intuition would be that this is *roughly* the same case as before, but with the complement of X wrt. $\mathcal{R}$. So we write $\pi_2 \equiv [\mathcal{R} \setminus X](\Pi) = \varnothing$. This, however, is not strong enough. It translates the fact that *only* rules of X can rewrite Π. But again, while X *may* in fact rewrite Π, there is nothing in π_2 to enforce that. Looking at the semantics, *all* possible words of ${}^{\mathcal{R}}(\!|\Pi|\!)$ *must* have at least one move (i.e. $1 \in \operatorname{dom} w$); this condition must be translated. It is equivalent to saying that all terms of Π are rewritable, which is expressed by $\Pi \subseteq \mathcal{R}^{-1}(\mathcal{T}(\mathbb{A}))$. More specifically, since we already impose that they are not rewritable by $\mathcal{R} \setminus X$, we can even write directly that they are rewritable by X, i.e. $\Pi \subseteq X^{-1}(\mathcal{T}(\mathbb{A}))$. Putting those two conditions together, we obtain $\pi_2' \equiv [\mathcal{R} \setminus X](\Pi) = \varnothing \wedge \Pi \subseteq X^{-1}(\mathcal{T}(\mathbb{A}))$, and this is an exact translation.

Of Strength and Weakness. Let us reflect on the previous cases for a minute; the immediate intuition is that X is *stronger* than $\neg X$, in the sense that whenever we see X, we must write an additional clause – enforcing rewritability – compared to $\neg X$. This actually depends on the context, as the next example will show. CASE 3: ALWAYS NEGATIVE. Let $\varphi = \Box\neg X$. This means that neither the terms of Π nor their successors can be rewritten by X; in other words $\pi_3 \equiv X\left(\mathcal{R}^*(\Pi)\right) = \varnothing$. The translation is almost the same as for $\neg X$, the only difference being the use of $\mathcal{R}^*(\Pi)$ (Π and successors) instead of just Π as in π_1. More formally, $\pi_3 \equiv \pi_1[\mathcal{R}^*(\Pi)/\Pi]$. CASE 4: ALWAYS POSITIVE. Seeing this, one is tempted to infer that the same relationship that exists between the translations of $\neg X$ and $\Box\neg X$ exists as well between those of X and $\Box X$. In the case $\varphi = \Box X$, this would yield $\pi_4 \equiv \pi_2'[\mathcal{R}^*(\Pi)/\Pi] \equiv [\mathcal{R}\setminus X]\left(\mathcal{R}^*(\Pi)\right) = \varnothing \wedge \mathcal{R}^*(\Pi) \subseteq X^{-1}(\mathcal{T}(\mathbb{A}))$. But clearly this translation is much too strong as its second part implies that every term of Π can be rewritten by X, and so can all of the successors; consequently, ${}^{\mathcal{R}}(\!|\Pi|\!)$ must form an ω-language. Yet we have for instance $\lambda \models \Box X$ —note incidentally that $\lambda \models \Box\psi$ holds vacuously for any ψ. In general, under the semantics for $\Box$, words of any length, infinite, finite or nought, may satisfy $\Box X$. Thus the correct translation was simply $\pi_4' \equiv [\mathcal{R}\setminus X]\left(\mathcal{R}^*(\Pi)\right) = \varnothing$. So, unlike Cases 1 and 2, X is *not* in any sense stronger than $\neg X$ when behind a $\Box$. This is an important point which we shall need to keep track of during the translation; that necessary bookkeeping is the reason for the introduction of the weak and strong intertwined semantics described in Section 3.2[p322].

Conjunction, Disjunction and Negation. CASE 5: AND & OR. It is pretty clear that if π_5 translates φ and π_5' translates ψ, then $\pi_5 \wedge \pi_5'$ translates $\varphi \wedge \psi$. This holds thanks to the implicit universal quantifier, as we have $(\mathcal{R}, \Pi \models \varphi \wedge \psi) \iff (\mathcal{R}, \Pi \models \varphi) \wedge (\mathcal{R}, \Pi \models \psi)$. Contrariwise, the same does not hold for the disjunction, and we have no general solution[1] to handle it. Given that

[1] There are however special cases where disjunction can be translated exactly; see rules $(\vee^{\Rightarrow}_{\wedge})$[p328] and $(\vee^{\neg}_{\Rightarrow})$.

one of the implications still holds, namely $(\mathcal{R}, \Pi \models \varphi \vee \psi) \Longleftarrow (\mathcal{R}, \Pi \models \varphi) \vee (\mathcal{R}, \Pi \models \psi)$, a crude under-approximation can still be given if all else fails: $\pi_5 \vee \pi_5' \implies \mathcal{R}, \Pi \models \varphi \vee \psi$. CASE 6: NEGATION. Although we have seen in Case 1 that a negative literal can easily be translated, negation cannot be handled in all generality by our method. Note that, because of the universal quantification, $\mathcal{R}, \Pi \not\models \varphi \neq \mathcal{R}, \Pi \models \neg\varphi$; thus the fact that π_6 translates φ does *not* a priori imply that $\neg\pi_6$ translates $\neg\varphi$. This is why we will assume in practice that input formulæ are provided in Negative Normal Form, which is licit as the presence of *both* weak and strong next operators enables any formula to be put in NNF.

Handling Material Implication. CASE 7. We have just seen in Cases 5 and 6 that we can provide exact translations for neither negation nor disjunction. Inasmuch as $\varphi \Rightarrow \psi$ is defined as $\neg\varphi \vee \psi$, must material implication be forgone as well? An example involving an implication has been given in the introduction (page 316), so it would seem that a translation can be provided in at least *some* cases. Let us take the simple example $X \Rightarrow \bullet Y$. Assuming that any term $u \in \Pi$ is rewritten into some u' by a rule in X, then u' must be rewritable by Y, and only by Y. The set of X-successors of Π being $X(\Pi)$, those conditions yield the translation $\pi_7 \equiv X(\Pi) \subseteq Y^{-1}\left(\mathcal{T}(\mathbb{A})\right) \wedge [\mathcal{R} \setminus Y]\left(X(\Pi)\right) = \varnothing$. Note that the way in which implication has been handled here is very different from the approach taken for the other binary operators, which essentially consists in splitting the formula around the operator and translating the two subparts separately. In contrast, the antecedent of the implication was "assumed", whilst the consequent was translated as usual. In fact, recalling that π_2' translates X, and thus $\pi_2'' \equiv \pi_2'[Y/X]$ translates Y, we have $\pi_7 \equiv \pi_2''[X(\Pi)/\Pi]$. So, "assuming" the antecedent consisted simply in changing our set of reachable terms —which we will from now on call the *past*, hence the notation Π. This is not an isolated observation; if π_0 denotes the translation of $\square(X \Rightarrow \bullet Y)$ given in the introduction, then $\pi_0 \equiv \pi_7[\mathcal{R}^*(X(\Pi))/X(\Pi)]$. Thus "updating" the past is enough of a tool to deal with some simple uses of $\square$ and implication... but consider the following formula: $\bullet Y \Rightarrow X$. In that case the antecedent lies in the future, relatively to the consequent. Therefore, in order to deal with all cases, we need some means of making assumptions about both past and future. This is the goal of the *signatures* presented in Section 3.3[p323].

3.2   Weak and Strong Semantics for LTL

Restricting the Fragment. As mentioned in Cases 3 and 4 of the previous section, we will in practice be restricted to working with formulæ provided in Negative Normal Form. Furthermore, there are operators, such as $\lozenge$, for which we think that no translation *can* be provided, because rewrite propositions are not expressive enough —in particular, $\mathcal{R}^*(\Pi)$ hides all information regarding finite or infinite traces. If this is the case, then none of the operators of the "Until" family $\{\lozenge, \mathbf{U}, \mathbf{W}, \mathbf{R}, \dots\}$ can be dealt with. Consequently, we are restricted to the following fragment of LTL, which will be denoted by $\mathcal{R}$-LTL:

$$\varphi := X \mid \neg X \mid \varphi \wedge \varphi \mid \varphi \vee \varphi \mid \varphi \Rightarrow \varphi \mid \qquad X \in \wp(\mathcal{R})$$
$$\bullet^m \varphi \mid \circ^m \varphi \mid \Box \varphi \qquad m \in \mathbb{N}\,.$$

Bookkeeping. (cf. Sec. 3.1$_{[\mathrm{p320}]}$, case 4) In order to address the question of whether the translation of an atom X should be "strong" – enforce rewritability – or "weak", information is needed from the context. Namely, does the atom appear in the direct scope of a $\Box$? We solve this by introducing intertwined *weak semantics* – written $\models^{\mathbf{w}}$ – and *strong semantics* – written $\models^{\mathbf{s}}$, given in Fig. 2. For $\mu \in \{\,\mathbf{w}, \mathbf{s}\,\}$ the notations $w \models^{\mu} \varphi$ and $\mathcal{R}, \Pi \models^{\mu} \varphi$ are defined in the same way as for $\models$. How those semantics are used will become clearer in section 3.4$_{[\mathrm{p328}]}$, where the translation rules are given. The important point for now is that the strong semantics are equivalent to the normal semantics of LTL on the fragment $\mathcal{R}$-LTL, which is shown by Lemma 3.

$$
\begin{array}{lll}
(w,i) \models^{\mu} \top & (w,i) \not\models^{\mu} \bot & \\
(w,i) \models^{\mathbf{s}} X & \text{iff} & i \in \operatorname{dom} w \text{ and } w(i) \in X \\
(w,i) \models^{\mathbf{w}} X & \text{iff} & i \notin \operatorname{dom} w \text{ or } w(i) \in X \\
(w,i) \models^{\mu} \neg X & \text{iff} & i \notin \operatorname{dom} w \text{ or } w(i) \notin X \\
(w,i) \models^{\mu} (\varphi \vee \psi) & \text{iff} & (w,i) \models^{\mu} \varphi \text{ or } (w,i) \models^{\mu} \psi \\
(w,i) \models^{\mu} (\varphi \wedge \psi) & \text{iff} & (w,i) \models^{\mu} \varphi \text{ and } (w,i) \models^{\mu} \psi \\
(w,i) \models^{\mu} (\varphi \Rightarrow \psi) & \text{iff} & (w,i) \models^{\mathbf{s}} \varphi \implies (w,i) \models^{\mathbf{s}} \psi \\
(w,i) \models^{\mu} \bullet^m \varphi & \text{iff} & i+m \in \operatorname{dom} w \text{ and } (w,i+m) \models^{\mathbf{s}} \varphi \\
(w,i) \models^{\mu} \circ^m \varphi & \text{iff} & i+m \notin \operatorname{dom} w \text{ or } (w,i+m) \models^{\mathbf{w}} \varphi \\
(w,i) \models^{\mu} \Box \varphi & \text{iff} & \forall j \in \operatorname{dom} w,\ j \geqslant i \Rightarrow (w,j) \models^{\mathbf{w}} \varphi \\
\hline
\multicolumn{3}{c}{\text{For any } m \in \mathbb{N}, \mu \in \{\,\mathbf{w}, \mathbf{s}\,\}}
\end{array}
$$

Fig. 2. $\mathcal{R}$-LTL Weak & Strong Semantics

Lemma 2 *(Strong-Weak Domain-Equivalence). For all w, φ, i, it holds that $i \in \operatorname{dom} w \implies (w,i) \models^{\mathbf{s}} \varphi \Leftrightarrow (w,i) \models^{\mathbf{w}} \varphi$.*

Lemma 3 *(Strong Semantics). For all words $w \in \mathcal{W}$ and all formulæ $\varphi \in \mathcal{R}$-LTL, we have $\forall i \in \mathbb{N}_1,\ (w,i) \models^{\mathbf{s}} \varphi \iff (w,i) \models \varphi$.*

3.3 Girdling the Future: Signatures

As discussed in Sec. 3.1$_{[\mathrm{p320}]}$, Case 7, implication is handled by converting the antecedent φ of a formula $\varphi \Rightarrow \psi$ into "assumptions". Concretely, this consists in building a model of φ – called a *signature* of φ, written $\xi(\varphi)$ – which can be manipulated during the translation. The variety of signatures defined hereafter handles formulæ φ within the fragment $\mathcal{A}$-LTL ($\mathcal{A}$ for **a**ntecedent), which is $\mathcal{R}$-LTL without $\vee$ or $\Rightarrow$. This section covers the technical tools needed for building signatures (Fig. 3$_{[\mathrm{p326}]}$) and understanding the translation rules (Sec. 3.4$_{[\mathrm{p328}]}$).

Definitions. SIGNATURES. A *signature* σ is an element of the space

$$\Sigma = \bigcup_{n \in \mathbb{N}} \Big[(\llbracket 1, n \rrbracket \cup \{\omega\}) \to \wp(\mathcal{R}) \Big] \times \wp(\overline{\mathbb{N}}) \ .$$

CORE, SUPPORT, DOMAIN, CARDINAL. Let $\sigma = (f, S)$; then the function f is called the *core of* σ, denoted by $\partial\sigma$, and S is called its *support*, written $\nabla\sigma$. The *domain of* σ is defined as $\operatorname{dom}\sigma \triangleq \operatorname{dom} f \setminus \{\omega\}$, and its *cardinal* is $\#\sigma \triangleq \operatorname{Card}(\operatorname{dom}\sigma)$. SPECIAL NOTATIONS, EMPTY SIGNATURE. A signature $\sigma = (f, S)$ will be written either compactly as $\sigma = \wr f \mid S \wr$, or *in extenso* as $\wr f(1), f(2), \ldots, f(\#\sigma) \,\S\, f(\omega) \mid S \wr$. We denote by $\varepsilon \triangleq \wr \S \mathcal{R} \mid \overline{\mathbb{N}} \wr$ the *empty signature*. Let $k \in \mathbb{N}_1 \cup \{\omega\}$, then we write

$$\sigma[k] \triangleq \begin{cases} f(k) & \text{if } k \in \operatorname{dom}\sigma \\ f(\omega) & \text{if } k \notin \operatorname{dom}\sigma \end{cases} .$$

SIGNATURE PRODUCT. Let σ and σ' two signatures; then their *product* is another signature defined as $\sigma \otimes \sigma' \triangleq \wr g \mid \nabla\sigma \cap \nabla\sigma' \wr$, where

$$g \triangleq \left| \begin{array}{ccc} \operatorname{dom}\partial\sigma \cup \operatorname{dom}\partial\sigma' & \longrightarrow & \wp(\mathcal{R}) \\ k & \longmapsto & \sigma[k] \cap \sigma'[k] \end{array} \right. .$$

Note that as a consequence, $\forall k \in \mathbb{N}_1$, $(\sigma \otimes \sigma')[k] = \sigma[k] \cap \sigma'[k]$. **(e.g.** Let $\sigma = \wr X, Y \,\S\, Z \mid \mathbb{N}_2 \wr$ and $\rho = \wr X' \,\S\, Z' \mid \mathbb{N}_3 \wr$; then $\sigma \otimes \rho = \wr X \cap X', Y \cap Z' \,\S\, Z \cap Z' \mid \mathbb{N}_3 \wr$.)

Remark 4 (Summation Notation). The set of signatures Σ, equipped with the signature-product $\otimes$, forms a commutative monoid whose neutral element is ε.

CONVERGENCE. Let $\rho = (\sigma_n)_{n \in \mathbb{N}}$ be an infinite sequence of signatures. It is *convergent* if **(1)** the sequence $(\nabla\sigma_n)_{n \in \mathbb{N}}$ converges towards a limit $\nabla\sigma_\infty$, and **(2)** for all $k \in \mathbb{N}_1$, the sequence $(\sigma_n[k])_{n \in \mathbb{N}}$ converges towards a limit $\sigma_\infty[k]$, and **(3)** the sequence of limits $(\sigma_\infty[k])_{k \in \mathbb{N}_1}$ itself converges towards a limit $\sigma_\infty[\infty]$. We call this sequence the *limit core*. It is not directly in the form of a bona fide signature core. However, its co-domain being $\wp(\mathcal{R})$, which is finite, there exists a rank $N \geqslant 0$ such that for all $k > N$, $\sigma_\infty[k] = \sigma_\infty[\infty]$, and thus, taking the smallest such N, we define $\wr \sigma_\infty[1], \ldots, \sigma_\infty[N] \,\S\, \sigma_\infty[\infty] \mid \nabla\sigma_\infty \wr$ to be the *limit* of ρ, which we denote by $\lim \rho$ or $\lim_{n \to \infty} \sigma_n$, or more simply by σ_∞. Note that the core of the limit is equivalent to the limit core, in the intuitive sense that they define the same constrained words. Otherwise ρ is *divergent*, and its limit is left undefined. **(e.g.** The sequence $(\wr \mathcal{R}_1, \ldots \mathcal{R}_n, X \,\S\, \mathcal{R} \mid \llbracket 1, n \rrbracket \wr)_{n \in \mathbb{N}}$, with $\mathcal{R}_i = \mathcal{R}\ \forall i$, converges towards $\wr \S X \mid \mathbb{N} \wr$.) INFINITE PRODUCTS. Remark 4 legitimates the use of a Sigma-notation $\bigotimes_{k=l}^{m} \sigma_k$ for $\sigma_l \otimes \sigma_{l+1} \otimes \cdots \otimes \sigma_m$, with the usual properties. We define a notion of *infinite product* of signatures as well, in the classical way: the infinite product $\bigotimes_{k=l}^{\infty} \sigma_k$ converges if and only if the associated sequence of partial products $(\bigotimes_{k=l}^{n} \sigma_k)_{n \in \mathbb{N}_l}$ converges, and in that case

$$\bigotimes_{k=l}^{\infty} \sigma_k \triangleq \lim_{n \to \infty} \bigotimes_{k=l}^{n} \sigma_k \ .$$

Constrained Words. The *words of $\mathcal{R}$, originating in Π and constrained by* σ are defined by ${}^{\mathcal{R}}\llparenthesis \Pi \fatsemi \sigma \rrparenthesis \triangleq \{ w \in {}^{\mathcal{R}}\llparenthesis \Pi \rrparenthesis \mid \#w \in \nabla\sigma \ \wedge\ \forall k \in \operatorname{dom} w,\ w(k) \in \sigma[k] \}$. (**e.g.** Let $\sigma = \wr X, Y \fatsemi Z \mid \mathbb{N}_2 \wr$; then its core is the function $\partial\sigma = \{ 1 \mapsto X, 2 \mapsto Y, \omega \mapsto Z \}$, its domain is $\operatorname{dom}\sigma = \llbracket 1, 2 \rrbracket$, its support is $\nabla\sigma = \mathbb{N}_2$, its cardinal is $\#\sigma = 2$, and we have $\sigma[1] = X$, $\sigma[2] = Y$, $\sigma[3] = \sigma[4] = \cdots = \sigma[\omega] = Z$. Its constrained words are the maximal words of length at least 2, whose first two letters are in X and Y, respectively, and whose other letters are all in Z.) Lemma 5 serves in the base cases of signature-building, and Lem. 6 in the construction of $\xi(\varphi \wedge \psi)$ and rule $(\Rightarrow_\Sigma)_{[\mathrm{p328}]}$;

Lemma 5 (*No Constraints*). *We have* ${}^{\mathcal{R}}\llparenthesis \Pi \fatsemi \varepsilon \rrparenthesis = {}^{\mathcal{R}}\llparenthesis \Pi \rrparenthesis$.

Lemma 6 (*Breaking Products*). *For any signatures $\sigma, \sigma' \in \Sigma$, and any language Π, we have* (1) ${}^{\mathcal{R}}\llparenthesis \Pi \fatsemi \sigma \otimes \sigma' \rrparenthesis = {}^{\mathcal{R}}\llparenthesis \Pi \fatsemi \sigma \rrparenthesis \cap {}^{\mathcal{R}}\llparenthesis \Pi \fatsemi \sigma' \rrparenthesis$. *Furthermore, this generalises to infinitary cases:* (2) *given a sequence $(\sigma_n)_{n\in\mathbb{N}}$ such that the infinite product $\bigotimes_{n=0}^{\infty} \sigma_n$ converges, it holds that* ${}^{\mathcal{R}}\llparenthesis \Pi \fatsemi \bigotimes_{n=0}^{\infty} \sigma_n \rrparenthesis = \bigcap_{n=0}^{\infty} {}^{\mathcal{R}}\llparenthesis \Pi \fatsemi \sigma_n \rrparenthesis$.

Arithmetic Overloading. We overload the operators $+$ and $-$ on the profile $\wp(\overline{\mathbb{N}}) \times \mathbb{N} \to \wp(\overline{\mathbb{N}})$ such that, for any $S \in \wp(\overline{\mathbb{N}})$ and $n \in \mathbb{N}$, we have $S + n \triangleq \{ k + n \mid k \in S \}$ and $S - n \triangleq \{ k - n \mid k \in S \} \cap \overline{\mathbb{N}}$. Shifts Left & Right. Let $m \in \mathbb{N}$; then we define the *strong m-left shift of σ* as $\sigma \blacktriangleleft m \triangleq \wr \partial\sigma(m+1), \ldots, \partial\sigma(\#\sigma) \fatsemi \partial\sigma(\omega) \mid (\nabla\sigma - m) \setminus \{0\} \wr$ and the *weak m-left shift of σ* as $\sigma \lhd m \triangleq \wr \partial\sigma(m+1), \ldots, \partial\sigma(\#\sigma) \fatsemi \partial\sigma(\omega) \mid \nabla\sigma - m \wr$. Conversely, the *strong m-right shift of σ* is $\sigma \blacktriangleright m \triangleq \wr \mathcal{R}_1, \ldots, \mathcal{R}_m, \partial\sigma(1), \ldots, \partial\sigma(\#\sigma) \fatsemi \partial\sigma(\omega) \mid (\nabla\sigma \setminus \{0\}) + m \wr$, while the *weak m-right shift of σ* is $\sigma \rhd m \triangleq \wr \mathcal{R}_1, \ldots, \mathcal{R}_m, \partial\sigma(1), \ldots, \partial\sigma(\#\sigma) \fatsemi \partial\sigma(\omega) \mid \llbracket 0, m \rrbracket \cup (\nabla\sigma + m) \wr$, with $\mathcal{R}_1 = \mathcal{R}, \ldots, \mathcal{R}_m = \mathcal{R}$. Note that for all $m \in \mathbb{N}$ and all $k \in \mathbb{N}_1$, $(\sigma \lhd m)[k] = \sigma[k+m]$, for all $k \leqslant m$, $(\sigma \blacktriangleright m)[k] = (\sigma \rhd m)[k] = \mathcal{R}$ and for all $k > m$, $(\sigma \blacktriangleright m)[k] = (\sigma \rhd m)[k] = \sigma[k-m]$. (**e.g.** Let $\sigma = \wr X, Y \fatsemi Z \mid \mathbb{N}_2 \wr$; then $\sigma \blacktriangleleft 1 = \sigma \lhd 1 = \wr Y \fatsemi Z \mid \mathbb{N}_1 \wr$, $\sigma \blacktriangleright 1 = \wr \mathcal{R}, X, Y \fatsemi Z \mid \mathbb{N}_3 \wr$, and $\sigma \rhd 1 = \wr \mathcal{R}, X, Y \fatsemi Z \mid \mathbb{N} \setminus \{2\} \wr$.)

Lemma 7 justifies the fact that the computation of $\xi(\Box\varphi)$ always yields a useable signature; a closed form of the limit is given in the proof.

Lemma 7 (*Automatic Convergences*). *Let $(\sigma_n)_{n\in\mathbb{N}}$ be any sequence of signatures, and $(\rho_n)_{n\in\mathbb{N}}$ its associated sequence of partial products $(\bigotimes_{i=0}^{n} \sigma_i)_{n\in\mathbb{N}}$. Then $(\rho_n)_{n\in\mathbb{N}}$ satisfies convergence criteria* (1) *and* (2). *Furthermore, if σ is a given signature and $\sigma_i = \sigma \blacktriangleright i$ or $\sigma_i = \sigma \rhd i$, for any $i \in \mathbb{N}$, then criterion* (3) *is satisfied as well, and the infinite product $\bigotimes_{n=0}^{\infty} \sigma_n$ converges.*

Proof. (1) For all $n \in \mathbb{N}$, $\nabla\rho_n = \bigcap_{i=0}^{n} \nabla\sigma_i$, thus it is clear that $\nabla\rho_n = \bigcap_{i=0}^{n} \nabla\sigma_i \supseteq \bigcap_{i=0}^{n+1} \nabla\sigma_i = \nabla\rho_{n+1}$ or, in other words, $(\nabla\rho_n)_{n\in\mathbb{N}}$ is a (trivial) contracting sequence of finite sets. Therefore it converges towards $\bigcap_{i=0}^{\infty} \nabla\sigma_i$. (2) Let $k \in \mathbb{N}_1$; we have

$$\rho_n[k] = \left(\bigotimes_{i=0}^{n} \sigma_i \right)[k] = \bigcap_{i=0}^{n} \sigma_i[k] \ ,$$

and thus $\rho_n[k] = \bigcap_{i=0}^{n} \sigma_i[k] \supseteq \bigcap_{i=0}^{n+1} \sigma_i[k] = \rho_{n+1}[k]$ and again, $(\rho_n[k])_{n\in\mathbb{N}}$ is a trivial contracting sequence of finite sets; therefore it converges towards a limit which we denote by $\rho_\infty[k] = \bigcap_{i=0}^{\infty} \sigma_i[k]$. **(3)** Suppose now that $\sigma_i = \sigma \rhd i$ (resp. $\sigma_i = \sigma \blacktriangleright i$, the computation will be unchanged), we have

$$\rho_\infty[k] = \bigcap_{i=0}^{\infty} \sigma_i[k] = \bigcap_{i=0}^{\infty} (\sigma \rhd i)[k] = \left(\bigcap_{i=0}^{k-1} (\sigma \rhd i)[k] \right) \cap \left(\bigcap_{i=k}^{\infty} (\sigma \rhd i)[k] \right)$$
$$= \left(\bigcap_{i=0}^{k-1} \sigma[k-i] \right) \cap \left(\bigcap_{i=k}^{\infty} \mathcal{R} \right) = \bigcap_{i=0}^{k-1} \sigma[k-i] = \bigcap_{i=1}^{k} \sigma[i] \, .$$

Given that for all $i > \#\sigma$, $\sigma[i] = \sigma[\omega]$, it follows that for all $k > \#\sigma$, $\rho_\infty[k] = \bigcap_{i=1}^{\#\sigma+1} \sigma[i]$. Thus $(\rho_\infty[k])_{k \in \mathbb{N}_1}$ converges. This shows that the infinite product $\bigotimes_{n=0}^{\infty} \sigma_n$ is convergent. □

Building Signatures. Figure $3_{[\text{p326}]}$ defines the function $\xi(\cdot) : \mathcal{A}\text{-LTL} \to \Sigma$. As Theorem 8 shows, the signature $\xi(\varphi)$ essentially captures a model of φ.

$$\xi(\top) \triangleq \wr \, ⨾ \mathcal{R} \mid \overline{\mathbb{N}} \, \wr = \varepsilon \qquad \xi(\bot) \triangleq \wr \, ⨾ \varnothing \mid \varnothing \, \wr$$
$$\xi(X) \triangleq \wr X ⨾ \mathcal{R} \mid \overline{\mathbb{N}}_1 \wr \qquad \xi(\neg X) \triangleq \wr \mathcal{R} \setminus X ⨾ \mathcal{R} \mid \overline{\mathbb{N}} \wr$$
$$\xi(\bullet^m \varphi) \triangleq \xi(\varphi) \blacktriangleright m \qquad \xi(\circ^m \varphi) \triangleq \xi(\varphi) \rhd m$$
$$\xi(\varphi \wedge \psi) \triangleq \xi(\varphi) \otimes \xi(\psi) \qquad \xi(\Box \varphi) \triangleq \bigotimes_{m=0}^{\infty} \left[\xi(\varphi) \rhd m \right]$$

Fig. 3. Building Signatures on $\mathcal{A}$-LTL

Theorem 8 (*Signatures*). *For any $\Pi \subseteq \mathcal{T}(\mathbb{A})$ and any $\varphi \in \mathcal{A}$-LTL,*

$$^{\mathcal{R}}\llparenthesis \Pi ⨾ \xi(\varphi) \rrparenthesis = \left\{ w \in {}^{\mathcal{R}}\llparenthesis \Pi \rrparenthesis \mid w \models \varphi \right\} .$$

3.4 The Translation Rules

Now that the main technical tools are in place, there remains to define what is meant by "translation rule", and to state the rules themselves. For any $\mu \in \{\mathbf{w}, \mathbf{s}\}$, $\varphi \in$ LTL, $\Pi \subseteq \mathcal{T}(\mathbb{A})$, $\sigma \in \Sigma$, we define $\langle \Pi ⨾ \sigma \Vdash^{\mu} \varphi \rangle$ as shorthand for $\forall w \in {}^{\mathcal{R}}\llparenthesis \Pi ⨾ \sigma \rrparenthesis$, $w \models^{\mu} \varphi$. We call such a notation $\langle \Pi ⨾ \sigma \Vdash^{\mu} \varphi \rangle$ a *translation block*. A *translation rule* is of the form

$$\updownarrow \frac{A \qquad P(\sigma, \varphi)}{E} \quad \text{or} \quad \uparrow \frac{A \qquad P(\sigma, \varphi)}{E} \quad \text{or} \quad ?\frac{A \qquad \uparrow P(\sigma, \varphi) \ \updownarrow Q(\sigma, \varphi)}{E} \, ,$$

where A stands for some translation block $\langle \Pi ⨾ \sigma \Vdash^{\mu} \varphi \rangle$, $P, Q \in \Sigma \times \mathcal{R}\text{-LTL} \to \mathcal{B}$ are predicates on signatures and formulæ, and E is a mixed translation/reachability proposition. More precisely, E is generated by the grammar given in

Sec. $2.3_{[p320]}$, with the added production $\gamma := \Upsilon$, where Υ is a translation block. The $\updownarrow$-rules (exact translations) are defined to hold iff $P(\sigma,\varphi) \implies (A \Leftrightarrow E)$, the $\uparrow$-rules (under-approximations) hold iff $P(\sigma,\varphi) \implies (E \Rightarrow A)$, and the ?-rules hold iff $P(\sigma,\varphi) \implies (E \Rightarrow A)$ and $(P(\sigma,\varphi) \wedge Q(\sigma,\varphi)) \implies (A \Leftrightarrow E)$. When omitted, P is assumed to be $\top$.

Theorem 9 entails that any derivation (i.e. tree of rule applications with no translation blocks left in the leaves) starting with $\langle \Pi \mathbin{;} \varepsilon \Vdash^{\mathsf{s}} \varphi\rangle$ yields an exact translation of φ (if only exact rules are involved), or an under-approximation (if some $\uparrow$-rules are used).

Theorem 9 *(Translation Satisfaction).* $\langle \Pi \mathbin{;} \varepsilon \Vdash^{\mathsf{s}} \varphi\rangle \iff \mathcal{R}, \Pi \models \varphi$.

A few additional definitions and results about signatures are needed in order to justify some translation rules: Remark 10 is needed by rule $(\Box\, \hbar)_{[p328]}$; Lem. 13 intervenes in rules $(\bullet^m)_{[p328]}$ and $(\circ^m)$; Lem. 11 and Cor. 12 in rule $(\bullet^m)$; Rmk. 14 and Lem. 15 in rule $(\Box\, *)$; Prp. $16_{[p328]}$ justifies that rule $(\Box\, \hbar)$ eventually terminates. Signature Iteration. Let $\Pi \subseteq \mathcal{T}(\mathbb{A})$ a language, and $\sigma \in \Sigma$ a signature; then for $n \in \mathbb{N}$ we let $\Pi^n_\sigma \triangleq \sigma[n]\left(\sigma[n-1]\left(\cdots\sigma[1]\left(\Pi\right)\cdots\right)\right)$ be the *n-iteration* of the signature σ. More formally, it is defined recursively such that $\Pi^0_\sigma \triangleq \Pi$ and $\Pi^{n+1}_\sigma \triangleq \sigma[n+1]\left(\Pi^n_\sigma\right)$. Length Rejector. For $n \in \mathbb{N}$, the rewrite proposition $\Psi^\sigma_\Pi(n)$ is called the *n-length rejector*, and defined as $\Psi^\sigma_\Pi(n) \triangleq \Pi^n_\sigma \subseteq \sigma[n+1]^{-1}\left(\mathcal{T}(\mathbb{A})\right)$. Strengthening. If σ is a signature, then $\star\sigma \triangleq \wr \partial\sigma \mid \nabla\sigma \setminus \{0\} \int$ is its *strengthening*. Note that $(\sigma \blacktriangleleft m) = \star(\sigma \lhd m)$, for all m.

Remark 10 (Strengthening of Always). Let $\Pi \subseteq \mathcal{T}(\mathbb{A})$, $\sigma \in \Sigma$, and $\varphi \in \mathcal{R}$-LTL. Then $\langle \Pi \mathbin{;} \sigma \Vdash^\mu \Box\varphi\rangle \iff \langle \Pi \mathbin{;} \star\sigma \Vdash^\mu \Box\varphi\rangle$.

Lemma 11. *Let σ be a signature and $\Pi \subseteq \mathcal{T}(\mathbb{A})$ a language; then for any $n \in \mathbb{N}$, the proposition $\Psi^\sigma_\Pi(n)$ holds iff for all $w \in {}^{\mathcal{R}}(\!|\Pi \mathbin{;} \sigma|\!)$, $\#w \neq n$.*

Corollary 12 *(Length Rejection). Let $S \in \wp(\mathbb{N})$, σ a signature and Π a language; the rewrite proposition $\bigwedge_{n\in S\cap\nabla\sigma} \Psi^\sigma_\Pi(n)$ holds iff for all $w \in {}^{\mathcal{R}}(\!|\Pi \mathbin{;} \sigma|\!)$, $\#w \notin S$.*

Lemma 13 *(Shifting Words). Let σ be a signature and $\Pi \subseteq \mathcal{T}(\mathbb{A})$ a language; then ${}^{\mathcal{R}}(\!|\Pi^m_\sigma \mathbin{;} \sigma \lhd m|\!) = \left\{ w^{m+1} \;\middle|\; w \in {}^{\mathcal{R}}(\!|\Pi \mathbin{;} \sigma|\!) \wedge \#w \geqslant m \right\}$.*

Remark 14 (Constrained Union). Let $\sigma \in \Sigma$, $I \subseteq \mathbb{N}$, and for each $i \in I$, $\Pi_i \subseteq \mathcal{T}(\mathbb{A})$. Then $\bigcup_{i\in I} {}^{\mathcal{R}}(\!|\Pi_i \mathbin{;} \sigma|\!) = {}^{\mathcal{R}}(\!|\bigcup_{i\in I} \Pi_i \mathbin{;} \sigma|\!)$.

Stability. A signature $\sigma \in \Sigma$ is called *stable* if $\sigma \lhd 1 = \sigma$; this is equivalent to the condition $\#\sigma = 0$ and $\nabla\sigma \in \{\varnothing, \{+\infty\}, \mathbb{N}, \overline{\mathbb{N}}\}$, and also to the condition $\forall n \in \mathbb{N},\ \sigma \lhd n = \sigma$. High Point. The *high point* $\hbar\sigma$ of a signature σ is the smallest $h \in \mathbb{N}$ such that $\sigma \lhd h$ is stable. Note that σ is stable if and only if $\hbar\sigma = 0$. Given the characterisation of stability given above, an alternative definition of $\hbar\sigma$ would be the smallest $h \geqslant \#\sigma$ such that either $\mathbb{N}_h \subseteq \nabla\sigma$ or $\nabla\sigma \cap \mathbb{N}_h = \varnothing$. If no such h exists[2], we take by convention $\hbar\sigma = +\infty$. Low

[2] Consider a signature σ such that $\nabla\sigma$ is the set of odd numbers, or the set of prime numbers, for instance. Such a signature cannot be stabilised. Fortunately, Proposition $16_{[p328]}$ shows that such exotic cases are irrelevant to this paper.

POINT. The *low point* $\ell\sigma$ of a signature σ is the smallest length authorised by σ; more precisely, it is defined as $\ell\sigma \triangleq \min \nabla\sigma$.

Lemma 15 (*All Suffixes*). *Let σ be a stable signature, and $\Pi \subseteq \mathcal{T}(\mathbb{A})$ a language. Then we have* $\left\{ w^{1+n} \mid n \in \mathbb{N},\ w \in {}^{\mathcal{R}}(\!|\Pi \mathbin{⨾} \sigma|\!),\ \#w \geqslant n \right\} = {}^{\mathcal{R}}(\!|\sigma[\omega]^* (\Pi) \mathbin{⨾} \sigma|\!)$.

Proposition 16 (*Stability of $\xi(\cdot)$*). *The signature of any formula $\varphi \in \mathcal{A}$-LTL is stabilisable; in other words,* $\hbar\xi(\varphi) \in \mathbb{N},\ \forall\varphi \in \mathcal{A}\text{-LTL}$.

Theorem 17 (*Translation*). *All the following translation rules hold.*

$$\updownarrow \frac{\langle \Pi \mathbin{⨾} \sigma \Vdash^{\mu} \top \rangle}{\top} \quad (\top) \qquad\qquad \updownarrow \frac{\langle \Pi \mathbin{⨾} \sigma \Vdash^{\mu} \bot \rangle}{\bot} \quad (\bot)$$

$$\updownarrow \frac{\langle \Pi \mathbin{⨾} \sigma \Vdash^{\mu} X \wedge Y \rangle}{\langle \Pi \mathbin{⨾} \sigma \Vdash^{\mu} X \cap Y \rangle} \quad (\wedge_X) \qquad\qquad \updownarrow \frac{\langle \Pi \mathbin{⨾} \sigma \Vdash^{\mu} X \vee Y \rangle}{\langle \Pi \mathbin{⨾} \sigma \Vdash^{\mu} X \cup Y \rangle} \quad (\vee_X)$$

$$\updownarrow \frac{\langle \Pi \mathbin{⨾} \sigma \Vdash^{\mu} \varphi \wedge \psi \rangle}{\langle \Pi \mathbin{⨾} \sigma \Vdash^{\mu} \varphi \rangle \wedge \langle \Pi \mathbin{⨾} \sigma \Vdash^{\mu} \psi \rangle} \quad (\wedge)$$

$$\updownarrow \frac{\langle \Pi \mathbin{⨾} \sigma \Vdash^{\mu} [\varphi \vee \varphi'] \Rightarrow \psi \rangle}{\langle \Pi \mathbin{⨾} \sigma \Vdash^{\mu} \varphi \Rightarrow \psi \rangle \wedge \langle \Pi \mathbin{⨾} \sigma \Vdash^{\mu} \varphi' \Rightarrow \psi \rangle} \quad (\vee^{\Rightarrow}_{\wedge})$$

$$\updownarrow \frac{\langle \Pi \mathbin{⨾} \sigma \Vdash^{\mu} \varphi \vee \psi \rangle \qquad \neg\varphi \in \mathcal{A}\text{-LTL}}{\langle \Pi \mathbin{⨾} \sigma \Vdash^{\mu} \neg\varphi \Rightarrow \psi \rangle} \quad (\vee^{\neg}_{\Rightarrow})$$

$$\uparrow \frac{\langle \Pi \mathbin{⨾} \sigma \Vdash^{\mu} \varphi \vee \psi \rangle}{\langle \Pi \mathbin{⨾} \sigma \Vdash^{\mu} \varphi \rangle \vee \langle \Pi \mathbin{⨾} \sigma \Vdash^{\mu} \psi \rangle} \quad (\vee_{\uparrow})$$

$$\updownarrow \frac{\langle \Pi \mathbin{⨾} \sigma \Vdash^{\mu} \varphi \Rightarrow \psi \rangle}{\langle \Pi \mathbin{⨾} \sigma \otimes \xi(\varphi) \Vdash^{\mathbf{s}} \psi \rangle} \quad (\Rightarrow_{\Sigma})$$

$$\updownarrow \frac{\langle \Pi \mathbin{⨾} \sigma \Vdash^{\mu} \circ^{m}\varphi \rangle}{\langle \Pi^{m}_{\sigma} \mathbin{⨾} \sigma \blacktriangleleft m \Vdash^{\mathbf{w}} \varphi \rangle} \quad (\circ^{m})$$

$$\updownarrow \frac{\langle \Pi \mathbin{⨾} \sigma \Vdash^{\mu} \bullet^{m}\varphi \rangle}{\langle \Pi \mathbin{⨾} \sigma \Vdash^{\mu} \circ^{m}\varphi \rangle \wedge \bigwedge_{n \in [\![0,m]\!] \cap \nabla\sigma} \Psi^{\sigma}_{\Pi}(n)} \quad (\bullet^{m})$$

$$\updownarrow \frac{\langle \Pi \mathbin{⨾} \sigma \Vdash^{\mu} \Box\varphi \rangle \qquad \sigma \text{ is stable}}{\left\langle \sigma[\omega]^{*}(\Pi) \mathbin{⨾} \star\sigma \Vdash^{\mathbf{w}} \varphi \right\rangle} \quad (\Box_{*})$$

$$\updownarrow \frac{\langle \Pi \mathbin{⨾} \sigma \Vdash^{\mu} \Box\varphi \rangle \qquad \hbar\sigma \in \mathbb{N}_1}{\left\langle \Pi \mathbin{⨾} \sigma \Vdash^{\mu} \bigwedge_{k=0}^{\hbar\sigma - 1} \circ^{k}\varphi \right\rangle \wedge \left\langle \Pi^{\hbar\sigma}_{\sigma} \mathbin{⨾} \sigma \lhd \hbar\sigma \Vdash^{\mu} \Box\varphi \right\rangle} \quad (\Box_{\hbar})$$

$$\updownarrow \frac{\langle \Pi \mathbin{⨾} \sigma \Vdash^{\mu} \neg X \rangle}{\langle \Pi \mathbin{⨾} \sigma \Vdash^{\mathbf{w}} \mathcal{R} \setminus X \rangle} \quad (\neg X)$$

Additionally, the following four rules are being explored as a possible coverage of the difficult case of the atom X. While the main bodies of those rules encompass all the necessary translations, adjusting their exact respective application predicates is still ongoing work, which sets them apart from the proven formulæ of Thm. 17.

$$?\frac{\langle \Pi \fatsemi \sigma \Vdash^{\mathbf{w}} X\rangle \qquad \uparrow \ell\sigma \leqslant 1 \ \updownarrow \sigma \lhd 1 = \varepsilon}{\left[\mathcal{R} \setminus (X \cap \sigma[1])\right](\Pi) = \varnothing} \qquad (X^{\mathbf{w}}_{\ell \leqslant 1})$$

$$?\frac{\langle \Pi \fatsemi \sigma \Vdash^{\mathbf{s}} X\rangle \qquad \uparrow \ell\sigma = 0 \ \updownarrow \sigma \lhd 1 = \varepsilon}{\langle \Pi \fatsemi \sigma \Vdash^{\mathbf{w}} X\rangle \wedge \Pi \subseteq \left(X \cap \sigma[1]\right)^{-1}(\mathcal{T}(\mathbb{A}))} \qquad (X^{\mathbf{s}}_{\ell 0})$$

$$?\frac{\langle \Pi \fatsemi \sigma \Vdash^{\mathbf{s}} X\rangle \qquad \uparrow \ell\sigma = 1 \ \updownarrow \sigma \lhd 1 = \varepsilon}{\langle \Pi \fatsemi \sigma \Vdash^{\mathbf{w}} X\rangle} \qquad (X^{\mathbf{s}}_{\ell 1})$$

$$?\frac{\langle \Pi \fatsemi \sigma \Vdash^{\mu} X\rangle \qquad \uparrow \ell\sigma \geqslant 2 \ \updownarrow \sigma \lhd \ell\sigma = \varepsilon}{\sigma[\ell\sigma]\Big(\cdots \sigma[2]\Big(\left[\mathcal{R} \setminus (X \cap \sigma[1])\right](\Pi)\Big)\cdots\Big) = \varnothing} \qquad (X^{\mu}_{\ell 2})$$

The general derivation algorithm consists in systematically applying the first rule that matches, starting with the block $\langle \Pi \fatsemi \varepsilon \Vdash^{\mathbf{s}} \varphi\rangle$. Let it be noted that not all of the given rules are strictly necessary. For instance $(\vee^{\Rightarrow}_{\wedge})$ corresponds to a basic tautology of propositional logic, which rewrites the formula in a form more amenable to translation. Similarly, rule $(\vee^{\neg}_{\Rightarrow})$ relies on a transformation of the antecedent into $\mathcal{A}$-LTL (which is not always possible, in which case the rule does not apply). While their presence is not fundamental to the system, they extend the number of translatable cases. There are doubtless many other such simplifications not listed here – an obvious one being the commutation of $(\vee^{\neg}_{\Rightarrow})$. This sensitivity of the translation to transformations of the input formula makes it difficult to give an exact characterisation of the supported fragment – it is not simply $\mathcal{R}$-LTL, restricted to $\mathcal{A}$-LTL for antecedents. For instance, even though $\Diamond$ cannot be translated in general, its presence in the NNF of the input φ is not enough in itself to assert that φ cannot be translated: if it appears in, say, $\Diamond X \vee \psi$, it can be handled using rule $(\vee^{\neg}_{\Rightarrow})$. We intend to expand the translatable fragment in future works; this will hopefully make it easier to characterise.

Example. Let us derive the translation of $\varphi = \Box(X \Rightarrow \bullet^1 Y)$.

$$\updownarrow \frac{\langle \Pi \fatsemi \varepsilon \Vdash^{\mathbf{s}} \Box(X \Rightarrow \bullet^1 Y)\rangle \qquad (\Box_*)}{\updownarrow \dfrac{\langle \mathcal{R}^*(\Pi) \fatsemi \star\varepsilon \Vdash^{\mathbf{w}} X \Rightarrow \bullet^1 Y\rangle \qquad (\Rightarrow_{\Sigma})}{\updownarrow \dfrac{\langle \mathcal{R}^*(\Pi) \fatsemi \lbag X \fatsemi \mathcal{R} \mid \overline{\mathbb{N}}_1 \rbag \Vdash^{\mathbf{s}} \bullet^1 Y\rangle \qquad (\bullet^m)}{\Psi^{\lbag X \fatsemi \mathcal{R} \mid \overline{\mathbb{N}}_1 \rbag}_{\mathcal{R}^*(\Pi)}(1) \wedge \updownarrow \dfrac{\langle \mathcal{R}^*(\Pi) \fatsemi \lbag X \fatsemi \mathcal{R} \mid \overline{\mathbb{N}}_1 \rbag \Vdash^{\mathbf{s}} \circ^1 Y\rangle \qquad (\circ^m)}{\updownarrow \dfrac{\langle X(\mathcal{R}^*(\Pi)) \fatsemi \lbag \fatsemi \mathcal{R} \mid \overline{\mathbb{N}}_1 \rbag \Vdash^{\mathbf{w}} Y\rangle \, (X^{\mathbf{w}}_{\ell \leqslant 1})}{\left[\mathcal{R} \setminus Y\right](X(\mathcal{R}^*(\Pi))) = \varnothing}}}}} .$$

This yields the exact translation $\left[\mathcal{R} \setminus Y\right](X(\mathcal{R}^*(\Pi))) = \varnothing \wedge \Psi^{\lbag X \fatsemi \mathcal{R} \mid \overline{\mathbb{N}}_1 \rbag}_{\mathcal{R}^*(\Pi)}(1)$ which, once expanded, yields $\left[\mathcal{R} \setminus Y\right](X(\mathcal{R}^*(\Pi))) = \varnothing \wedge X(\mathcal{R}^*(\Pi)) \subseteq \mathcal{R}^{-1}(\mathcal{T}(\mathbb{A}))$. This is equivalent to $\left[\mathcal{R} \setminus Y\right](X(\mathcal{R}^*(\Pi))) = \varnothing \wedge X(\mathcal{R}^*(\Pi)) \subseteq Y^{-1}(\mathcal{T}(\mathbb{A}))$, which is the expected exact translation.

4 Conclusions and Perspectives

In the term rewriting framework, to perform reachability analysis guided by properties of interest, the present paper addresses the question of a systematic translation of linear temporal logic properties into rewrite propositions. More precisely, we have investigated suitable semantics for LTL on maximal rewriting words and their influence on the feasibility of a translation, and proposed a framework providing exact translations on a fragment of LTL corresponding mainly to safety formulæ, and approximations on a larger fragment.

As a future work, we intend to expand the fragment for which translations and approximations can be provided, and study the feasibility of handling equational theories in the same framework. The present work being a part of a rewrite approximation based analysis, the end goal is the integration of the paper's proposals into the verification chain dedicated to the automatic analysis of security-/safety-critical applications.

References

1. Baader, F. (ed.): RTA 2007. LNCS, vol. 4533. Springer, Heidelberg (2007)
2. Bae, K., Meseguer, J.: The linear temporal logic of rewriting Maude model checker. In: Ölveczky [16], pp. 208–225
3. Boichut, Y., Genet, T., Jensen, T.P., Roux, L.L.: Rewriting approximations for fast prototyping of static analyzers. In: Baader [1], pp. 48–62
4. Boichut, Y., Héam, P.C., Kouchnarenko, O.: Approximation-based tree regular model-checking. Nord. J. Comput. 14(3), 216–241 (2008)
5. Boronat, A., Heckel, R., Meseguer, J.: Rewriting Logic Semantics and Verification of Model Transformations. In: Chechik, M., Wirsing, M. (eds.) FASE 2009. LNCS, vol. 5503, pp. 18–33. Springer, Heidelberg (2009)
6. Boyer, B., Genet, T.: Verifying Temporal Regular Properties of Abstractions of Term Rewriting Systems. In: RULE. EPTCS, vol. 21, pp. 99–108 (2009)
7. Courbis, R., Héam, P.-C., Kouchnarenko, O.: TAGED Approximations for Temporal Properties Model-Checking. In: Maneth, S. (ed.) CIAA 2009. LNCS, vol. 5642, pp. 135–144. Springer, Heidelberg (2009)
8. Dershowitz, N., Jouannaud, J.P.: Rewrite Systems. In: Handbook of Theoretical Computer Science, Volume B: Formal Models and Sematics (B), pp. 243–320 (1990)
9. Escobar, S., Meseguer, J.: Symbolic model checking of infinite-state systems using narrowing. In: Baader [1], pp. 153–168
10. Filiot, E., Talbot, J.-M., Tison, S.: Tree Automata with Global Constraints. In: Ito, M., Toyama, M. (eds.) DLT 2008. LNCS, vol. 5257, pp. 314–326. Springer, Heidelberg (2008)
11. Genet, T., Klay, F.: Rewriting for Cryptographic Protocol Verification. In: McAllester, D. (ed.) CADE 2000. LNCS, vol. 1831, pp. 271–290. Springer, Heidelberg (2000)
12. Kamp, H.W.: Tense Logic and the Theory of Linear Order (1968)

13. Manna, Z., Pnueli, A.: Temporal Verification of Reactive Systems - Safety. Springer (1995)
14. Meseguer, J.: The Temporal Logic of Rewriting: A Gentle Introduction. In: Degano, P., De Nicola, R., Meseguer, J. (eds.) Concurrency, Graphs and Models. LNCS, vol. 5065, pp. 354–382. Springer, Heidelberg (2008)
15. Meseguer, J.: Conditioned Rewriting Logic as a United Model of Concurrency. TCS 96(1), 73–155 (1992)
16. Ölveczky, P.C. (ed.): WRLA 2010. LNCS, vol. 6381. Springer, Heidelberg (2010)
17. Serbanuta, T.F., Rosu, G., Meseguer, J.: A rewriting logic approach to operational semantics. Inf. Comput. 207(2), 305–340 (2009)

Tableaux Modulo Theories Using Superdeduction

An Application to the Verification of B Proof Rules with the Zenon Automated Theorem Prover

Mélanie Jacquel[1], Karim Berkani[1], David Delahaye[2], and Catherine Dubois[3]

[1] Siemens IC-MOL, Châtillon, France
{Melanie.Jacquel,Karim.Berkani}@siemens.com
[2] CEDRIC/CNAM, Paris, France
David.Delahaye@cnam.fr
[3] INRIA/CEDRIC/ENSIIE, Évry, France
dubois@ensiie.fr

Abstract. We propose a method which allows us to develop tableaux modulo theories using the principles of superdeduction, among which the theory is used to enrich the deduction system with new deduction rules. This method is presented in the framework of the Zenon automated theorem prover, and is applied to the set theory of the B method. This allows us to provide another prover to Atelier B, which can be used to verify B proof rules in particular. We also propose some benchmarks, in which this prover is able to automatically verify a part of the rules coming from the database maintained by Siemens IC-MOL.

Keywords: Tableaux, Superdeduction, Zenon, Set Theory, B Method, Proof Rules, Verification.

1 Introduction

In this paper, we propose to integrate superdeduction [3] (a variant of deduction modulo) into the tableau method in order to reason modulo theories (see also [4] for a similar approach). This integration is motivated by an experiment which is managed by Siemens IC-MOL regarding the verification of B proof rules [5]. The B method [1], or B for short, allows engineers to develop correct by design software with high guarantees of confidence. A significant use of B by Siemens IC-MOL has concerned the control system of the driverless metro line 14 in Paris. B is a formal method based on set theory and theorem proving, and which relies on a refinement-based development process. The Atelier B environment is a platform that supports B and offers both automated and interactive provers. To ensure the global correctness of formalized applications, the user must discharge proof obligations. These proof obligations may be proved automatically, but otherwise, they have to be handled manually either by using the interactive prover, or by adding new proof rules that the automated prover can exploit. These new proof rules can be seen as axioms and must be verified by other means.

B. Gramlich, D. Miller, and U. Sattler (Eds.): IJCAR 2012, LNAI 7364, pp. 332–338, 2012.

In [5], we develop an approach based on the use of the Zenon automated theorem prover [2], which relies on classical first order logic with equality and applies the tableau method. In this context, the choice of Zenon is strongly influenced by its ability of producing checkable proof traces under the form of Coq proofs in particular. The method used in this approach consists in first normalizing the formulas to be proved, in order to obtain first order logic formulas containing only the membership set operator, and then calling Zenon on these new formulas. This experiment gives satisfactory results in the sense that it can prove a significant part of the rules coming from the database maintained by Siemens IC-MOL. However, this approach is not complete and suffers from efficiency issues due to the preliminary normalization. To deal with these problems, the idea developed in this paper is to integrate the B set theory into the Zenon proof search method by means of superdeduction rules. This integration can be concretely achieved thanks to the extension mechanism offered by Zenon, which allows us to extend its core of deductive rules to match specific requirements.

The paper is organized as follows: in Section 2, we present the computation of superdeduction rules from axioms in the framework of the tableau method used by Zenon; we then introduce, in Section 3, the superdeduction rules corresponding to the B set theory; finally, in Section 4, we describe the corresponding implementation and provide some benchmarks concerning the verification of B proof rules coming from the database maintained by Siemens IC-MOL.

2 From Axioms to Superdeduction Rules

Reasoning modulo a theory in a tableau method using superdeduction requires to generate new deduction rules from some axioms of the theory. The axioms which can be considered for superdeduction are of the form $\forall \bar{x}\ (P \Leftrightarrow \varphi)$, where P is atomic. This specific form of axiom allows us to introduce an orientation of the axiom from P to φ, and we introduce the notion of proposition rewrite rule (this notion appears in [3], from which we borrow the following notation and definition). The notation $R : P \rightarrow \varphi$ is a proposition rewrite rule and denotes the axiom $\forall \bar{x}\ (P \Leftrightarrow \varphi)$, where R is the name of the rule, P an atomic proposition, φ a proposition, and $\bar{x}$ the free variables of P and φ.

As said in the introduction, one of our main objectives is to develop a proof search procedure for the set theory of the B method using the Zenon automated theorem prover [2]. In the following, we will thus consider the tableau method used by Zenon as the framework in which superdeduction rules will be generated from proposition rewrite rules.

The proof search rules of Zenon are described in detail in [2] and summarized in Figure 1 (for the sake of simplification, we have omitted the relational, unfolding, and extension rules), where ϵ is Hilbert's operator, capital letters are used for metavariables, and R_r and R_s are respectively reflexive and symmetric relations. As hinted by the use of Hilbert's operator, the δ-rules are handled by means of ϵ-terms rather than using Skolemization. What we call here metavariables are often named free variables in the tableau-related literature; they are

Closure and Cut Rules

$$\frac{\bot}{\odot}\ \odot_\bot \qquad \frac{\neg\top}{\odot}\ \odot_{\neg\top} \qquad \frac{}{P \mid \neg P}\ \text{cut}$$

$$\frac{\neg R_r(t,t)}{\odot}\ \odot_r \qquad \frac{P \qquad \neg P}{\odot}\ \odot \qquad \frac{R_s(a,b) \qquad \neg R_s(b,a)}{\odot}\ \odot_s$$

Analytic Rules

$$\frac{\neg\neg P}{P}\ \alpha_{\neg\neg} \qquad \frac{P \Leftrightarrow Q}{\neg P, \neg Q \mid P, Q}\ \beta_\Leftrightarrow \qquad \frac{\neg(P \Leftrightarrow Q)}{\neg P, Q \mid P, \neg Q}\ \beta_{\neg\Leftrightarrow}$$

$$\frac{P \wedge Q}{P, Q}\ \alpha_\wedge \qquad \frac{\neg(P \vee Q)}{\neg P, \neg Q}\ \alpha_{\neg\vee} \qquad \frac{\neg(P \Rightarrow Q)}{P, \neg Q}\ \alpha_{\neg\Rightarrow}$$

$$\frac{P \vee Q}{P \mid Q}\ \beta_\vee \qquad \frac{\neg(P \wedge Q)}{\neg P \mid \neg Q}\ \beta_{\neg\wedge} \qquad \frac{P \Rightarrow Q}{\neg P \mid Q}\ \beta_\Rightarrow$$

$$\frac{\exists x\ P(x)}{P(\epsilon(x).P(x))}\ \delta_\exists \qquad \frac{\neg\forall x\ P(x)}{\neg P(\epsilon(x).\neg P(x))}\ \delta_{\neg\forall}$$

γ-Rules

$$\frac{\forall x\ P(x)}{P(X)}\ \gamma_{\forall M} \qquad \frac{\forall x\ P(x)}{P(t)}\ \gamma_{\forall \text{inst}} \qquad \frac{\neg\exists x\ P(x)}{\neg P(X)}\ \gamma_{\neg\exists M} \qquad \frac{\neg\exists x\ P(x)}{\neg P(t)}\ \gamma_{\neg\exists \text{inst}}$$

Fig. 1. Proof Search Rules of Zenon

not used as variables as they are never substituted. The proof search rules are applied with the normal tableau method: starting from the negation of the goal, apply the rules in a top-down fashion to build a tree. When all branches are closed, the tree is closed, and this closed tree is a proof of the goal.

Let us now describe how the computation of superdeduction rules for Zenon is performed from a given proposition rewrite rule.

Definition 1 (Computation of Superdeduction Rules). *Let $\mathcal{S}$ be a set of rules composed by the subset of the proof search rules of Zenon formed of the closure rules, the analytic rules, as well as the $\gamma_{\forall M}$ and $\gamma_{\neg\exists M}$ rules. Given a proposition rewrite rule $R : P \rightarrow \varphi$, two superdeduction rules (a positive one R and a negative one $\neg R$) are generated.*

To get the positive rule R (resp. the negative rule $\neg R$), initialize the procedure with the formula φ (resp. $\neg\varphi$). Next, apply the rules of $\mathcal{S}$ until there is no open leaf anymore on which they can be applied. Then, collect the premises and the conclusion, and replace φ by P (resp. $\neg\varphi$ by $\neg P$) to obtain the positive rule R (resp. the negative rule $\neg R$).

If the rule R (resp. $\neg R$) involves metavariables, an instantiation rule R_{inst} (resp. $\neg R_{\text{inst}}$) is added, where one or several metavariables can be instantiated.

Axioms

$$(x, y) \in a \times b \Leftrightarrow x \in a \wedge y \in b \qquad a \in \mathbb{P}(b) \Leftrightarrow \forall x\, (x \in a \Rightarrow x \in b)$$
$$x \in \{\, y \mid P(y) \,\} \Leftrightarrow P(x) \qquad a = b \Leftrightarrow \forall x\, (x \in a \Leftrightarrow x \in b)$$

Derived Constructs

$$a \cup b \triangleq \{\, x \mid x \in a \vee x \in b \,\} \qquad a \cap b \triangleq \{\, x \mid x \in a \wedge x \in b \,\}$$
$$a - b \triangleq \{\, x \mid x \in a \wedge x \notin b \,\} \qquad \emptyset \triangleq \mathrm{BIG} - \mathrm{BIG}$$
$$\{\, e_1, \ldots, e_n \,\} \triangleq \{\, x \mid x = e_1 \,\} \cup \ldots \cup \{\, x \mid x = e_n \,\}$$

Binary Relation Constructs: First Series

$$a^{-1} \triangleq \{\, (y, x) \mid (x, y) \in a \,\}$$
$$\mathrm{dom}(a) \triangleq \{\, x \mid \exists y\, (x, y) \in a \,\} \qquad \mathrm{ran}(a) \triangleq \mathrm{dom}(a^{-1})$$
$$a; b \triangleq \{\, (x, z) \mid \exists y\, ((x, y) \in a \wedge (y, z) \in b \,\}$$
$$\mathrm{id}(a) \triangleq \{\, (x, y) \mid (x, y) \in a \times a \wedge x = y \,\}$$
$$a \lhd b \triangleq \mathrm{id}(a); b \qquad a \rhd b \triangleq a; \mathrm{id}(b)$$

Fig. 2. Axioms and Constructs of the B Set Theory

3 Superdeduction Rules for the B Set Theory

The B method [1] is based on a typed set theory, which consists of six axiom schemes defining the basic operators and the extensional equality. The other operators ($\cup$, $\cap$, etc.) are defined using the previous basic ones. Figure 2 gathers a part of the axioms and constructs of the B set theory, where BIG is an infinite set. In this figure, we only consider the four first axioms of the B set theory, as we do not need the two remaining axioms in the rules that we want to verify (see Section 4). Due to space restrictions, we only present the main constructs, even though we can deal with other constructs (like functions) in our superdeduction system. Compared to [1], all type information has been removed from the axioms and constructs thanks to the modularity between the type and proof systems.

To generate the superdeduction rules corresponding to the axioms and constructs defined in Figure 2, we use the algorithm described in Definition 1 of Section 2, and we must therefore identify the proposition rewrite rules. On the one hand, the axioms are of the form $P_i \Leftrightarrow Q_i$, and the associated proposition rewrite rules are $R_i : P_i \rightarrow Q_i$. On the other hand, the constructs are expressed by the definitions $E_i \triangleq F_i$, where E_i and F_i are expressions, and the corresponding proposition rewrite rules are $R_i : x \in E_i \rightarrow x \in F_i$. The superdeduction rules are then generated as described in Figure 3 (except the instantiation rules associated with rules involving metavariables, due to space restrictions). The computation of these superdeduction rules goes further than the one proposed in Section 2, since given a proposition rewrite rule $R : P \rightarrow Q$, we apply to Q not only all the rules considered by Definition 1, but also the new generated superdeduction rules (except the rules for the extensional equality, in order to benefit from the dedicated rules of Zenon for equality) whenever applicable.

Rules for Axioms

$$\frac{(x,y) \in a \times b}{x \in a, y \in b}\ \times \qquad \frac{a \in \mathbb{P}(b)}{X \notin a \mid X \in b}\ \mathbb{P} \qquad \frac{x \in \{\ y \mid P(y)\ \}}{P(x)}\ \{|\}$$

$$\frac{(x,y) \notin a \times b}{x \notin a \mid y \notin b}\ \neg\times \qquad \frac{a \notin \mathbb{P}(b)}{\epsilon_x \in a, \epsilon_x \notin b}\ \neg\mathbb{P} \qquad \frac{x \notin \{\ y \mid P(y)\ \}}{\neg P(x)}\ \neg\{|\}$$

with $\epsilon_x = \epsilon(x).\neg(x \in a \Rightarrow x \in b)$

$$\frac{a = b}{X \notin a, X \notin b \mid X \in a, X \in b}\ = \qquad \frac{a \neq b}{\epsilon_x \notin a, \epsilon_x \in b \mid \epsilon_x \in a, \epsilon_x \notin b}\ \neq$$

with $\epsilon_x = \epsilon(x).\neg(x \in a \Leftrightarrow x \in b)$

Rules for Derived Constructs

$$\frac{x \in a \cup b}{x \in a \mid x \in b}\ \cup \qquad \frac{x \in a \cap b}{x \in a, x \in b}\ \cap \qquad \frac{x \in a - b}{x \in a, x \notin b}\ -$$

$$\frac{x \notin a \cup b}{x \notin a, x \notin b}\ \neg\cup \qquad \frac{x \notin a \cap b}{x \notin a \mid x \notin b}\ \neg\cap \qquad \frac{x \notin a - b}{x \notin a \mid x \in b}\ \neg-$$

$$\frac{x \in \{\ e_1, \ldots, e_n\ \}}{x = e_1 \mid \ \ldots \ \mid x = e_1}\ \{\} \qquad \frac{x \notin \{\ e_1, \ldots, e_n\ \}}{x \neq e_1, \ldots, x \neq e_n}\ \neg\{\} \qquad \frac{x \in \emptyset}{\odot}\ \emptyset$$

Rules for Binary Relation Constructs: First Series

$$\frac{(x,y) \in a^{-1}}{(y,x) \in a}\ a^{-1} \qquad \frac{x \in \mathrm{dom}(a)}{(x, \epsilon_y) \in a}\ \mathrm{dom} \qquad \frac{y \in \mathrm{ran}(a)}{(\epsilon_x, y) \in a}\ \mathrm{ran}$$

with $\epsilon_y = \epsilon(y).((x,y) \in a)$ (dom); with $\epsilon_x = \epsilon(x).((x,y) \in a)$ (ran)

$$\frac{(x,y) \notin a^{-1}}{(y,x) \notin a}\ \neg a^{-1} \qquad \frac{x \notin \mathrm{dom}(a)}{(x, Y) \notin a}\ \neg\mathrm{dom} \qquad \frac{y \notin \mathrm{ran}(a)}{(X, y) \notin a}\ \neg\mathrm{ran}$$

$$\frac{(x,z) \in a;b}{(x, \epsilon_y) \in a, (\epsilon_y, z) \in b}\ ; \qquad \frac{(x,z) \notin a;b}{(x, Y) \notin a \mid (Y, z) \notin b}\ \neg;$$

with $\epsilon_y = \epsilon(y).((x,y) \in a \wedge (y,z) \in b)$

$$\frac{(x,y) \in \mathrm{id}(a)}{x = y, x \in a, y \in a}\ \mathrm{id} \qquad \frac{(x,y) \in a \lhd b}{(x,y) \in b, x \in a}\ \lhd \qquad \frac{(x,y) \in a \rhd b}{(x,y) \in a, y \in b}\ \rhd$$

$$\frac{(x,y) \notin \mathrm{id}(a)}{x \neq y \mid x \notin a \mid y \notin a}\ \neg\mathrm{id} \qquad \frac{(x,y) \notin a \lhd b}{(x,y) \notin b \mid x \notin a}\ \neg\lhd \qquad \frac{(x,y) \notin a \rhd b}{(x,y) \notin a \mid y \notin b}\ \neg\rhd$$

Fig. 3. Superdeduction Rules for the B Set Theory

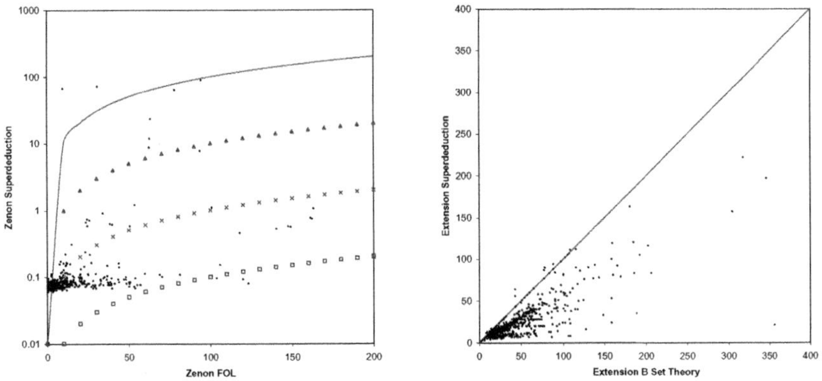

Fig. 4. Proof Time and Proof Size Comparative Benchmarks

4 Implementation and Benchmarks

The extension of Zenon for the B set theory described in Section 3 has been implemented thanks to the ability of Zenon to extend its core of deductive rules. The motivation of this extension is to verify B proof rules of Atelier B, and in particular rules coming from the database maintained by Siemens IC-MOL. Regarding benchmarks, we consider a selection of rules of this database consisting of well-typed and well-defined rules, which involve the B set constructs handled by our extension, i.e. all the constructs of the B-Book [1] until the override construct. This represents 1,397 rules (over a total of 5,281 rules), and we propose two benchmarks whose results are gathered in Figure 4.

The first benchmark aims to compare our extension of Zenon with the approach described in [5], where the set formulas must be preliminarily normalized (in order to obtain first order logic formulas containing only the membership set operator) before calling Zenon. Over the 1,397 selected rules, our extension proves 1,340 rules (96%), while our initial approach proves 1,145 rules (82%). The left-hand side graph of Figure 4 presents a comparison of both approaches in terms of proof time (run on an Intel Core i5-2500K 3.30GHz/12GB computer) for a subset of the 1,397 selected rules, where both approaches succeed in finding a proof (the time measures include the compilation of Coq proofs generated by Zenon), i.e. for 1,145 rules. In this figure, a point represents the result for a rule, and the x/y-axes respectively correspond to the approach with pre-normalization of the formulas and to our extension using superdeduction. On average, the superdeduction proofs are obtained 67 times faster (the best ratio is 1,540).

We propose a second benchmark whose purpose is to compare our extension of Zenon using superdeduction with another extension of Zenon for the B set theory, where the proposition rewrite rules are not computed into superdeduction rules, but just unfolded/folded (like in Prawitz's approach). The comparison consists in computing the number of proof nodes of each proof generated by Zenon. We consider a subset of 1,340 rules, for which both extensions succeed in finding a proof. The results are summarized by the right-hand side graph of Figure 4, where a point represents the result for a rule, and where the x/y-axes respectively correspond to the extension without and with superdeduction. As can be seen, the major part of proofs in the latter are on average 1.6 times shorter than the former proofs (the best ratio is 6.25).

5 Conclusion

We have proposed a method which allows us to develop tableaux modulo theories using superdeduction. This method has been presented in the framework of the Zenon automated theorem prover, and applied to the set theory of the B method. This has allowed us to provide another prover to Atelier B, which can be used to verify B proof rules automatically. We have also proposed some benchmarks using rules coming from the database maintained by Siemens IC-MOL. These benchmarks have emphasized significant speed-ups both in terms of proof time and proof size compared to previous and alternative approaches.

As future work, we first aim to generalize our approach of superdeduction for Zenon and provide a generator of superdeduction rules from proposition rewrite rules. This will allow us to generate automatically a superdeduction prover from a theory, provided that a part of the axioms of this theory can be turned into proposition rewrite rules. We also plan to extend our implementation realized for verifying B proof rules in order to deal with a larger set of rules of the database maintained by Siemens IC-MOL. Finally, we intend to study some properties of this system for the B set theory, such as consistency and completeness.

Acknowledgement. Many thanks to G. Burel and O. Hermant for their detailed comments on this paper, to G. Dowek for seminal discussions of this work, and to D. Doligez for his help in the integration of superdeduction into Zenon.

References

1. Abrial, J.-R.: The B-Book, Assigning Programs to Meanings. Cambridge University Press, Cambridge (1996) ISBN 0521496195
2. Bonichon, R., Delahaye, D., Doligez, D.: Zenon: An Extensible Automated Theorem Prover Producing Checkable Proofs. In: Dershowitz, N., Voronkov, A. (eds.) LPAR 2007. LNCS (LNAI), vol. 4790, pp. 151–165. Springer, Heidelberg (2007)
3. Brauner, P., Houtmann, C., Kirchner, C.: Principles of Superdeduction. In: Ong, L. (ed.) Logic in Computer Science (LICS), Wrocław (Poland), pp. 41–50. IEEE Computer Society Press (July 2007)
4. Houtmann, C.: Axiom Directed Focusing. In: Berardi, S., Damiani, F., de'Liguoro, U. (eds.) TYPES 2008. LNCS, vol. 5497, pp. 169–185. Springer, Heidelberg (2009)
5. Jacquel, M., Berkani, K., Delahaye, D., Dubois, C.: Verifying B Proof Rules Using Deep Embedding and Automated Theorem Proving. In: Barthe, G., Pardo, A., Schneider, G. (eds.) SEFM 2011. LNCS, vol. 7041, pp. 253–268. Springer, Heidelberg (2011)

Solving Non-linear Arithmetic

Dejan Jovanović[1] and Leonardo de Moura[2]

[1] New York University
[2] Microsoft Research

Abstract. We present a new algorithm for deciding satisfiability of non-linear arithmetic constraints. The algorithm performs a Conflict-Driven Clause Learning (CDCL)-style search for a feasible assignment, while using projection operators adapted from cylindrical algebraic decomposition to guide the search away from the conflicting states.

1 Introduction

From the early beginnings in Persian and Chinese mathematics until the present day, polynomial constraints and the algorithmic ways of solving them have been one of the driving forces in the development of mathematics. Though studied for centuries due to the natural elegance they provide in modeling the real world, from resolving simple taxation arguments to modeling planes and hybrid systems, we are still lacking a practical algorithm for solving a system of polynomial constraints. Throughout the history of mathematics, many brilliant minds have studied and algorithmically solved many of the related problems, such as root finding and factorization of polynomials. But, it was not until Alfred Tarski [26] showed that the theory of real closed fields admits elimination of quantifiers that it became clear that a general decision procedure for solving polynomial constraints was possible. Granted a wonderful theoretical result of landmark importance, with its non-elementary complexity, Tarski's procedure was unfortunately totally impractical.

As one would expect, Tarski's procedure consequently has been much improved. Most notably, Collins [10] gave the first relatively effective method of quantifier elimination by cylindrical algebraic decomposition (CAD). The CAD procedure itself has gone through many revisions [8]. However, even with the improvements and various heuristics, its doubly-exponential worst-case behavior has remained as a serious impediment. The CAD algorithm works by decomposing $\mathbb{R}^k$ into connected components such that, in each cell, all of the polynomials from the problem are sign-invariant. To be able to perform such a particular decomposition, CAD first performs a *projection* of the polynomials from the initial problem. This projection includes many new polynomials, derived from the initial ones, and these polynomials carry enough information to ensure that the decomposition is indeed possible. Unfortunately, the size of these projection sets grows exponentially in the number of variables, causing the projection phase, and its consequent impact on the search space, to be a key hurdle to CAD scalability.

B. Gramlich, D. Miller, and U. Sattler (Eds.): IJCAR 2012, LNAI 7364, pp. 339–354, 2012.

We propose a new decision procedure for the existential theory of the reals that tries to alleviate the above problem. As in [16,20,17], the new procedure performs a backtracking search for a model in $\mathbb{R}$, where the backtracking is powered by a novel conflict resolution procedure. Our approach takes advantage of the fact that each conflict encountered during the search is based on the current assignment and generally involves only a few constraints, a *conflicting core*. When in conflict, we project only the polynomials from the conflicting core and explain the conflict in terms of the current model. This means that we use projection conservatively, only for the subsets of polynomials that are involved in the conflict, and even then we reduce it further. As another advantage, the conflict resolution provides the usual benefits of a Conflict-Driven Clause Learning (CDCL)-style [24] search engine, such as non-chronological backtracking and the ability to ignore irrelevant parts of the search space. The projection operators we use as part of the conflict resolution need not be CAD based and, in fact, one can easily adapt projections based on other algorithms (e.g [19,3]).

Due to the lack of space and the volume of algorithms and concepts involved, we concentrate on the details of the decision procedure in this paper and refer the reader to the existing literature for further information [7,8,9].[1]

2 Preliminaries

As usual, we denote the ring of integers with $\mathbb{Z}$, the field of rational numbers with $\mathbb{Q}$, and the field of real numbers as $\mathbb{R}$. Unless stated otherwise, we assume all polynomials take integer coefficients, i.e. a polynomial $f \in \mathbb{Z}[\boldsymbol{y}, x]$ is of the form

$$f(\boldsymbol{y}, x) = a_m \cdot x^{d_m} + a_{m-1} \cdot x^{d_{m-1}} + \cdots + a_1 \cdot x^{d_1} + a_0 \ ,$$

where $0 < d_1 < \cdots < d_m$, and the coefficients a_i are in $\mathbb{Z}[\boldsymbol{y}]$ with $a_m \neq 0$. We call x the *top variable*. The highest degree d_m is the *degree* of the polynomial f in variable x, and we denote it with $\mathsf{deg}(f, x)$. The set of coefficients of f is denoted as $\mathsf{coeff}(f, x)$. We call a_m the *leading coefficient* in variable x, and denote it with $\mathsf{lc}(f, x)$. If we exclude the first k terms of the polynomial f, we obtain the polynomial $\mathsf{R}_k(f, x) = a_{m-k}x^{d_{m-k}} + \cdots + a_0$, called the k-th *reductum* of f. We write $\mathsf{R}^*(f, x)$ for the set $\{\mathsf{R}_0(f, x), \ldots, \mathsf{R}_m(f, x)\}$ containing all reductums. We denote the set of variables appearing in a polynomial f as $\mathsf{vars}(f)$ and call the polynomial *univariate* if $\mathsf{vars}(f) = \{x\}$ for some variable x. Otherwise the polynomial is *multivariate*, or a constant polynomial (if it contains no variables). Given a set of polynomials $A \subset \mathbb{Z}[x_1, \ldots x_n]$, we denote with A_k the subset of polynomials in A that belong to $\mathbb{Z}[x_1, \ldots, x_k]$, i.e. $A_k = A \cap \mathbb{Z}[x_1, \ldots, x_k]$.

[1] The website `http://cs.nyu.edu/~dejan/nonlinear/` contains a technical report, our prototype nlsat, and experimental results. The technical report contains additional examples, proofs of all main theorems, additional references, and implementation details.

A number $\alpha \in \mathbb{R}$ is a *root of the polynomial* $p \in \mathbb{Z}[x]$ iff $f(\alpha) = 0$. We call a real number $\alpha \in \mathbb{R}$ *algebraic* iff it is a root of a univariate polynomial $f \in \mathbb{Z}[x]$, and we denote the field of all real algebraic numbers by $\mathbb{R}_{\mathsf{alg}}$. We can represent any algebraic number α as $(l, u)_f$, with $l, u \in \mathbb{Q}$, where α is a root of a polynomial f, and the only root in the interval (l, u).

Example 1. Consider the univariate polynomial $f_1 = 16x^3 - 8x^2 + x + 16$. This polynomial has only one root, the irrational number $\alpha_1 \approx -0.840661$ and we can represent it as $(-0.9, -0.8)_{f_1}$.

Given a set of variables $X = \{x_1, \ldots, x_n\}$, we call υ a *variable assignment* if it maps each variable x_k to a real algebraic number $\upsilon(x_k)$, the value of x_k under υ. We overload υ, as usual, to obtain the value of a polynomial $f \in \mathbb{Z}[x_1, \ldots, x_n]$ under υ and write it as $\upsilon(f)$. We say that a polynomial f *vanishes* under υ if $\upsilon(f) = 0$. We can update the assignment υ to map a variable x_k to the value α, and we denote this as $\upsilon[x_k \mapsto \alpha]$. Under a variable assignment υ that interprets the variables $\boldsymbol{y}$, some coefficients of a polynomial $f(\boldsymbol{y}, x)$ may vanish. If a_k is the first non-vanishing coefficient of f, i.e., $\upsilon(a_k) \neq 0$, we write $\mathsf{R}(f, x, \upsilon) = a_k x^{d_k} + \cdots + a_0$ for the *reductum of f with respect to υ* (the non-vanishing part). Given any sequence of polynomials $\boldsymbol{f} = (f_1, \ldots, f_s)$ and a variable assignment υ we define the *vanishing signature* of $\boldsymbol{f}$ as the sequence $\mathsf{v\text{-}sig}(\boldsymbol{f}, \upsilon) = (f_1, \ldots, f_k)$, where $k \leq s$ is the minimal number such that $\upsilon(f_k) \neq 0$, or s if they all vanish. For the polynomial $f(\boldsymbol{y}, x)$ as above, we define the *vanishing coefficients signature* as $\mathsf{v\text{-}coeff}(f, x, \upsilon) = \mathsf{v\text{-}sig}(a_m, \ldots, a_0, \upsilon)$.

A *basic polynomial constraint* F is a constraint of the form $f \triangledown 0$ where f is a polynomial and $\triangledown \in \{<, \leq, =, \neq, \geq, >\}$. We denote the polynomial constraint that represents the *negation* of a constraint F with $\neg F$.[2] In order to identify the polynomial f of the constraint F, and the variables of F, we write $\mathsf{poly}(F)$ and $\mathsf{vars}(F)$, respectively. We normalize all constraints over constant polynomials to the dedicated constants true and false with the usual semantics. We write $\upsilon(F)$ to denote the evaluation of F under υ, which is the constraint $\upsilon(f) \triangledown 0$. If f does not evaluate to a constant under υ, then $\upsilon(F)$ evaluates to a new polynomial constraint F', where $\mathsf{poly}(F')$ can contain algebraic coefficients.

Borrowing from the extended Tarski language [4, Chapter 7], in addition to the basic constraints, we will also be working with *extended polynomial constraints*. An extended polynomial constraint F is of the form $x \triangledown_r \mathsf{root}(f, k)$, where $\triangledown_r \in \{<_r, \leq_r, =_r, \neq_r, \geq_r, >_r\}$, f is a polynomial in $\mathbb{Z}[\boldsymbol{y}, \tilde{z}]$, with $x \notin \mathsf{vars}(f)$, and the natural number $k \leq \mathsf{deg}(f, \tilde{z})$ is the *root index*. Variable $\tilde{z}$ is a distinguished free variable that cannot be used outside the root object. To be able to extract the polynomial of the constraint, we define $\mathsf{poly}(F) = f(\boldsymbol{y}, x)$. Note that $\mathsf{poly}(F)$ replaces $\tilde{z}$ with x. The semantics of the predicate $\triangledown_r$ under a variable assignment υ is the following. If the polynomial $\upsilon(f)$ is univariate, and υ assigns x to α, the (Boolean) value of the constraint can be determined as follows. If the univariate polynomial $\upsilon(f) \in \mathbb{R}_{\mathsf{alg}}[\tilde{z}]$ has the roots $\beta_1 < \cdots < \beta_n$, with $k \leq n$, and $\alpha \triangledown \beta_k$

[2] For example $\neg(x^2 + 1 > 0) \equiv x^2 + 1 \leq 0$.

holds, then the predicate evaluates to true. Otherwise it evaluates to false. We denote the number of real roots of a univariate polynomial f as $\mathsf{rootcount}(f)$. Naturally, if F is an extended polynomial constraint, so is the negation $\neg F$.[3]

A *polynomial constraint* is either a basic or an extended one. Given a set of polynomial constraints $\mathcal{F}$, we say that the variable assignment υ satisfies $\mathcal{F}$ if it satisfies each constraint in $\mathcal{F}$. If there is such a variable assignment, we say that $\mathcal{F}$ is *satisfiable*, otherwise it is *unsatisfiable*. A *clause* of polynomial constraints is a disjunction $C = F_1 \vee \ldots \vee F_n$ of polynomial constraints. We use $\mathsf{literals}(C)$ to denote the set $\{F_1, \neg F_1, \ldots, F_n, \neg F_n\}$. We say that the clause C is satisfied under the assignment υ if some polynomial constraint $F_j \in C$ evaluates to true under υ. Finally, a *polynomial constraint problem* is a set of clauses $\mathcal{C}$, and it is satisfiable if there is a variable assignment υ that satisfies all the clauses in $\mathcal{C}$. If the clauses of $\mathcal{C}$ contain the variables $x_1, \ldots, x_n$ then, for $k \leq n$, we denote with $\mathcal{C}_k$ the subset of the clauses that only contains variables $x_1, \ldots, x_k$.

3   An Abstract Decision Procedure

We describe our procedure as an abstract transition system in the spirit of Abstract DPLL [21]. The crucial difference between the system we present is that we depart from viewing the Boolean search engine and the theory reasoning as two separate entities that communicate only through existing literals. Instead, we allow the model that the theory is trying to construct to be involved in the search and in explaining the conflicts, while allowing new literals to be introduced so as to support more complex conflict analyses. The transition system presented here applies to non-linear arithmetic, but it can in general be applied to other theories.

The states in the transition system are indexed pairs of the form $\langle M, \mathcal{C} \rangle_n$, where M is a sequence (usually called a *trail*) of *trail elements*, and $\mathcal{C}$ is a set of clauses. The index n denotes the current *stage* of the state. Trail elements can be decided literals, propagated literals, or a variable assignment. A *decided literal* is a polynomial constraint F that we assume to be true. On the other hand, a *propagated literal*, denoted as $E \rightarrow F$, marks a polynomial constraint $F \in E$ that is implied to be true in the current state by the clause E (the *explanation*). In both cases, we say that the constraint F appears in M, and write this as $F \in M$. We denote the set of polynomial constraints appearing in M with $\mathsf{constraints}(M)$. We say M is *non-redundant* if no polynomial constraint appears in M more than once. A *trail variable assignment*, written as $x \mapsto \alpha$, is an assignment of a single variable to a value $\alpha \in \mathbb{R}_{\mathsf{alg}}$. Given a trail M, containing variable assignments $x_{i_1} \mapsto \alpha_1, \ldots, x_{i_k} \mapsto \alpha_k$, in order, we can construct an assignment $\upsilon[M] = \upsilon_0[x_{i_1} \mapsto \alpha_1] \ldots [x_{i_k} \mapsto \alpha_k]$, where υ_0 is an empty assignment that does not assign any variables.

---

[3] Note that, for example, $\neg(x <_r \mathsf{root}(f, k))$ is not necessarily equivalent to $x \geq_r \mathsf{root}(f, k)$.

We say that the sequence M is *stage increasing* when the sequence is of the form

$$M = [\![N_1, x_1 \mapsto \alpha_1, \ldots, x_{k-1} \mapsto \alpha_{k-1}, N_k, x_k \mapsto \alpha_k, \ldots, x_{n-1} \mapsto \alpha_{n-1}, N_n]\!] \ ,$$

where, for each $k \leq n$, the sequence N_k does not contain any variable assignments, each constraint $F \in \mathsf{constraints}(N_k)$ contains the variable x_k, and (optionally) the variables $x_1, \ldots, x_{k-1}$ (and z̃). In such a sequence M, we denote with $\mathsf{stage}(M) = n$ the *stage* of the sequence. If $\mathcal{F} = \mathsf{constraints}(M)$, we say that M is *feasible*, when the set of univariate polynomial constraints $\upsilon[M](\mathcal{F})$ has a solution. We write $\mathsf{feasible}(M)$ to denote the feasible set of $\upsilon[M](\mathcal{F})$. Given an additional polynomial constraint $F \in \mathbb{Z}[x_1, \ldots, x_n]$, we say that F is *compatible* with the sequence M, when $\mathsf{feasible}([\![M, F]\!]) \neq \emptyset$ and denote this with a predicate $\mathsf{compatible}(F, M)$. The technical report contains additional details on how these procedures are implemented.

Our transition system will work over states that are *well-formed*. Intuitively, in such a state, we commit to the variable assignment, but make sure that the current stage is consistent on the Boolean level. With this in mind, given a polynomial constraint F with $\mathsf{vars}(F) \subseteq \{x_1, \ldots, x_n\}$, and a state M with $\mathsf{stage}(M) = n$, we define the *state value* of F in M as

$$\mathsf{value}(F, M) = \begin{cases} \upsilon[M](F) & x_n \notin \mathsf{vars}(F) \ , \\ \mathsf{true} & F \in \mathsf{constraints}(M) \ , \\ \mathsf{false} & \neg F \in \mathsf{constraints}(M) \ , \\ \mathsf{undef} & \text{otherwise.} \end{cases}$$

Naturally, we overload value to also evaluate clauses of polynomial constraints, and sets of clauses, i.e. for a clause C we define $\mathsf{value}(C, M)$ to be true, if any of the literals evaluates to true, false if all literals evaluate to false, and undef otherwise.

Definition 1 (Well-Formed State). *We say a state* $\langle M, \mathcal{C} \rangle_n$ *is well-formed when* M *is non-redundant, stage increasing with* $\mathsf{stage}(M) = n$, *and all of the following hold.*

1. *Clauses up to stage n are satisfied, i.e. we have that* $\mathsf{value}(\mathcal{C}_{n-1}, M) = \mathsf{true}$.
2. *The state is consistent, i.e.* $\mathsf{feasible}(M) \neq \emptyset$ *and for each* $F \in \mathsf{constraints}(M)$ *we have that that* $\mathsf{value}(F, M) = \mathsf{true}$.
3. *Propagated literals* $E \to F$ *are implied, i.e. for all literals* $F' \neq F$ *in* E, $\mathsf{value}(F', M) = \mathsf{false}$.

We are now ready to define the transition system. We separate the transition rules into three groups: the search rules, the clause processing rules, and the conflict analysis rules. The *search rules* are the main driver of the procedure, with the responsibility for selecting clauses to process, creating the variable assignment while lifting the stages, and detecting Boolean conflicts. The search rules operate on well-formed states $\langle M, \mathcal{C} \rangle_n$. If the search rules select a clause

C to process, we switch to a state $\langle M, \mathcal{C}\rangle_n \vDash C$, where we can apply the set of *clause processing rules*. The notation $\vDash C$ designates that we are performing semantic reasoning in order to assign a value to a literal of C. If the search rules detect that in the current state some clause $C \in \mathcal{C}$ is falsified, we switch to a state $\langle M, \mathcal{C}\rangle_n \vdash C$, where we can apply the *conflict analysis rules*. The notation $\vdash C$ denotes that we are trying to produce a proof of why C is inconsistent in the current state.

Finally, given a polynomial constraint problem $\mathcal{C}$, with $\mathsf{vars}(\mathcal{C}) = \{x_1, \ldots, x_n\}$, the overall goal of the procedure is, starting from an initial state $\langle [\![]\!], \mathcal{C}\rangle_1$, and applying the rules, to end up either in a state $\langle v, \mathsf{sat}\rangle$, indicating that the initial set of clauses $\mathcal{C}$ is satisfiable where the assignment v is the witness, or derive unsat, which indicates that the set $\mathcal{C}$ unsatisfiable.

Search Rules. Fig 1 presents the set of search rules. The SELECT-CLAUSE rule selects one of the clauses of the current stage, whose state value is still undetermined, and transitions into the clause processing mode that will hopefully satisfy the clause. The CONFLICT rule detects if there is a clause of the current stage that is inconsistent in the current state, and transitions into the conflict resolution mode that will explain the conflict and backtrack appropriately. On the other hand, if all the clauses of the current stage are satisfied, we can either transition to the next stage, using the LIFT-STAGE rule, or conclude that our problem is satisfiable, using the SAT rule. Since at this point the current stage is consistent, in addition to formally introducing the new stage, the LIFT-STAGE rule selects a particular value for the current variable from the feasible set of the current stage. Note that once we move to the next stage, all the clauses of previous stages have values in the state, and can never be selected by the SELECT-CLAUSE or the CONFLICT rules. We conclude this set of rules with the FORGET rule that can be used to eliminate any learnt clause (a clause added while analyzing conflicts) from the current set of clauses.

Clause Processing Rules. In this set of rules, presented in Fig 2, we are trying to assign a currently unassigned literal of the given clause C, hoping to satisfy the clause. When one of the clause processing rules is applied, we immediately switch back to the search rules. As usual in a CDCL-style procedure, the simplest way to satisfy the clause C is to perform the Boolean unit propagation, if applicable, by using the $\mathbb{B}$-PROPAGATE rule. We restrict the application of this rule so that adding the constraint to the state keeps it consistent, i.e., it is compatible with the current set of constraints. If this is the case, we add the constraint to the state together with the explanation (clause C itself). To allow more complex propagations, the ones that are valid in $\mathbb{R}$ modulo the current state, we provide the $\mathbb{R}$-PROPAGATE rule. This rule can propagate a constraint from the clause, if assuming the negation would be incompatible with the current state. The $\mathbb{R}$-PROPAGATE rule is equipped with an explanation function explain. The explain function, given a polynomial constraint F, and the trail M, returns the explanation clause $E = \mathsf{explain}(F, M)$ that is valid in $\mathbb{R}$, and implies the constraint F under the current assignment i.e., $F \in E$, and all literals in E but

Select-Clause

$$\langle M, \mathcal{C}\rangle_k \longrightarrow \langle M, \mathcal{C}\rangle_k \vDash C \quad \textbf{if} \quad \begin{array}{l} C \in \mathcal{C}_k \\ \mathsf{value}(C, M) = \mathsf{undef} \end{array}$$

Conflict

$$\langle M, \mathcal{C}\rangle_k \longrightarrow \langle M, \mathcal{C}\rangle_k \vdash C \quad \textbf{if} \quad \begin{array}{l} C \in \mathcal{C}_k \\ \mathsf{value}(C, M) = \mathsf{false} \end{array}$$

Sat

$$\langle M, \mathcal{C}\rangle_k \longrightarrow \langle v[M], \mathsf{sat}\rangle \quad \textbf{if} \quad x_k \notin \mathsf{vars}(\mathcal{C})$$

Lift-Stage

$$\langle M, \mathcal{C}\rangle_k \longrightarrow \langle [\![M, x_k \mapsto \alpha]\!], \mathcal{C}\rangle_{k+1} \quad \textbf{if} \quad \begin{array}{l} x_k \in \mathsf{vars}(\mathcal{C}) \\ \alpha \in \mathsf{feasible}(M) \\ \mathsf{value}(\mathcal{C}_k, M) = \mathsf{true} \end{array}$$

Forget

$$\langle M, \mathcal{C}\rangle_k \longrightarrow \langle M, \mathcal{C} \setminus \{C\}\rangle_k \quad \textbf{if} \quad \begin{array}{l} C \in \mathcal{C} \\ C \text{ is a learnt clause} \end{array}$$

Fig. 1. The search rules

F are false in the state. The clause E may contain new literals that do not occur in $\mathcal{C}$, as long as they evaluate to false in the state. To simplify the presentation, in the $\mathbb{R}$-Propagate rule, the explanation clause E is eagerly generated, but in our actual implementation, we compute them only if they are needed during conflict resolution. Finally, if we cannot deduce the value of an unassigned literal, we can assume a value for such a literal using the Decide-Literal rule.

Conflict Analysis Rules. The conflict analysis rules start from an initial proper state $\langle M, \mathcal{C}\rangle_n \vdash C$, where $C \in \mathcal{C}$ is the conflicting clause. The conflict analysis is a standard Boolean conflict analysis [24] with a model-based twist. As the rules move the state backwards, the goal is to construct a new resolvent clause R, that will explain the conflict and ensure progress in the search. This means that, when we backtrack the sequence M just enough, the addition of R will ensure progress in the search by eliminating the inconsistent part from the state, and thus forcing the search rules to change some of the choices made. On the other hand, if the conflict analysis backtracks the state all the way into an empty state, this will be a signal that the original problem is unsatisfiable. Once the conflict analysis backtracks enough and deduces the resolvent R, then we pass it to the clause processing immediately.[4]

Termination. Our decision procedure consists of all three sets of rules described above. Any derivation will proceed by switching amongst the three distinct modes. Proving termination in the basic CDCL(T) framework is usually a fairly straightforward task, as the new explanation and conflict clauses always contain

[4] This is crucial in order to ensure termination.

Decide-Literal				
$\langle M, \mathcal{C} \rangle_k \models C$	$\longrightarrow$	$\langle [\![M, F_1]\!], \mathcal{C} \rangle_k$	**if**	$F_1, F_2 \in C$ $\forall i : \mathsf{value}(F_i, M) = \mathsf{undef}$ $\mathsf{compatible}(F_1, M)$
$\mathbb{B}$-Propagate				
$\langle M, \mathcal{C} \rangle_k \models C$	$\longrightarrow$	$\langle [\![M, C{\rightarrow}F]\!], \mathcal{C} \rangle_k$	**if**	$C = F_1 \vee \ldots \vee F_m \vee F$ $\mathsf{value}(F, M) = \mathsf{undef}$ $\forall i : \mathsf{value}(F_i, M) = \mathsf{false}$ $\mathsf{compatible}(F, M)$
$\mathbb{R}$-Propagate				
$\langle M, \mathcal{C} \rangle_k \models C$	$\longrightarrow$	$\langle [\![M, E{\rightarrow}F]\!], \mathcal{C} \rangle_k$	**if**	$F \in \mathsf{literals}(C)$ $\mathsf{value}(F, M) = \mathsf{undef}$ $\neg\mathsf{compatible}(\neg F, M)$ $E = \mathsf{explain}(F, M)$

Fig. 2. The clause satisfaction rules

only literals from the finite set of literals in the initial set of constraints. In our case, the main conundrum in proving termination is that we allow the explanations to contain fresh constraints, which, if we are not careful, could lead to non-termination. We therefore require the set of new constraints to be finite. We call an explanation function explain a *finite basis explanation function* with respect to a set of constraints $\mathcal{C}$, when there is a finite set of polynomial constraints $\mathcal{B}$ such that for any derivation of the proof rules, the clauses returned by applications of explain always contain only constraints from the basis $\mathcal{B}$. Having such an explanation function will therefore provide us with a termination argument, and we will provide one such explanation function for the theory of reals in the next section.

Theorem 1. *Given a set of polynomial constraints $\mathcal{C}$, and assuming a finite basis explanation function* explain, *any derivation starting from the initial state $\langle [\![]\!], \mathcal{C} \rangle_1$ will terminate either in a state $\langle v, \mathsf{sat} \rangle$, where the assignment v satisfies the constraints $\mathcal{C}$, or in the* unsat *state. In the later case, the set of constraints $\mathcal{C}$ is unsatisfiable in $\mathbb{R}$.*

4 Producing Explanations

Given a polynomial constraint F with $\mathsf{poly}(F) \in \mathbb{Z}[\boldsymbol{y}, x]$, and a trail M such that $\neg F$ is not compatible with M, the procedure $\mathsf{explain}(F, M)$ returns an explanation clause E that implies F in the current state. In principle, for any theory that admits elimination of quantifiers, it is possible to construct an explanation function explain. In this section, we describe how to produce an explain procedure for theory of the reals based on cylindrical algebraic decomposition (CAD). Before that, we first make a short interlude into the world of CAD.

RESOLVE-PROPAGATION			
$\langle [\![M, E{\to}F]\!], \mathcal{C}\rangle_k \vdash C$	$\longrightarrow$	$\langle M, \mathcal{C}\rangle_k \vdash R$	if $\neg F \in C$, $R = \mathsf{resolve}(C, E, F)$
▷ resolve returns the standard Boolean resolvent			
RESOLVE-DECISION			
$\langle [\![M, F]\!], \mathcal{C}\rangle_k \vdash C$	$\longrightarrow$	$\langle M, \mathcal{C} \cup \{C\}\rangle_k \models C$	if $\neg F \in C$
CONSUME			
$\langle [\![M, F]\!], \mathcal{C}\rangle_k \vdash C$	$\longrightarrow$	$\langle M, \mathcal{C}\rangle_k \vdash C$	if $\neg F \notin C$
$\langle [\![M, E{\to}F]\!], \mathcal{C}\rangle_k \vdash C$	$\longrightarrow$	$\langle M, \mathcal{C}\rangle_k \vdash C$	if $\neg F \notin C$
DROP-STAGE			
$\langle [\![M, x_{k+1}{\mapsto}\alpha]\!], \mathcal{C}\rangle_{k+1} \vdash C$	$\longrightarrow$	$\langle M, \mathcal{C}\rangle_k \vdash C$	if $\mathsf{value}(C, M) = \mathsf{false}$
$\langle [\![M, x_{k+1}{\mapsto}\alpha]\!], \mathcal{C}\rangle_{k+1} \vdash C$	$\longrightarrow$	$\langle M, \mathcal{C} \cup \{C\}\rangle_k \models C$	if $\mathsf{value}(C, M) = \mathsf{undef}$
UNSAT			
$\langle [\![]\!], \mathcal{C}\rangle_1 \vdash C$	$\longrightarrow$	unsat	

Fig. 3. The conflict analysis rules

4.1 Cylindrical Algebraic Decomposition

A crucial role in the theory of CADs and in the construction of our explain procedure is the property of delineability. Following the terminology used in CAD, we say that a connected subset of $\mathbb{R}^k$ is a *region*. A set of polynomials $\{f_1, \ldots f_s\} \subset \mathbb{Z}[\boldsymbol{y}, x]$, with $\boldsymbol{y} = (y_1, \ldots, y_n)$, is said to be *delineable* in a region $S \subseteq \mathbb{R}^n$ if for every f_i (and f_j) from the set, the following properties are invariant for any $\alpha \in S$:

1. the *total number of complex roots* of $f_i(\alpha, x)$;
2. the *number of distinct complex roots* of $f_i(\alpha, x)$;
3. the *number of common complex roots* of $f_i(\alpha, x)$ and $f_j(\alpha, x)$.

Example 2. Consider the polynomial $f = x^2+y^2+z^2-1$, with zeros of f depicted in Fig 4(a) together with two squiggly regions of $\mathbb{R}^2$. In the region S_1 that does not intersect the sphere, polynomial f is delineable, as the number of complex (and real) roots of $f(\alpha, x)$ is 2 for any α in S_1. In the region S_2 that intersects the sphere, f is not delineable, as the number of real roots of f varies from 0 (α's outside the unit circle), 1 (on the circle), and 2 (inside the unit circle).

We will call a *projection operator* any map P that, given a variable x and set of polynomials $A \subset \mathbb{Z}[\boldsymbol{y}, x]$, transforms A into a set of polynomials $\mathsf{P}(A, x) \subset \mathbb{Z}[\boldsymbol{y}]$. We call $\mathsf{P}(A, x)$ the *projection* of A under P with respect to variable x. In his seminal paper [10], Collins introduced a projection operator which we denote with P_c. In order to define the operator P_c, we first need to define some "advanced" operations on polynomials, and we refer the reader to [18,3,6] for a more detailed exposition.

Let $f, g \in \mathbb{Z}[\boldsymbol{y}, x]$ be two polynomials with $n = \min(\deg(f, x), \deg(g, x))$. For $k = 0, \ldots, n-1$, we denote with $\mathsf{S}_k(f, g, x)$ the k-th *subresultant* of f and g. The k-th subresultant is defined as the determinant of the k-th Sylvester-Habicht matrix of f and g, and is a polynomial of degree $\leq k$ in x with coefficients in $\mathbb{Z}[\boldsymbol{y}]$. The matrix in question is a particular matrix containing as elements the coefficients of f and g. Additionally, we denote with $\mathsf{psc}_k(f, g, x)$ the k-th *principal subresultant coefficient* of f and g, which is the coefficient of x^k in the polynomial $\mathsf{S}_k(f, g, x)$, and define $\mathsf{psc}_n(f, g, x) = 1$. We denote the sequence of principal subresultant coefficients as $\mathsf{psc}(f, g, x) = (\mathsf{psc}_0(f, g, x), \ldots, \mathsf{psc}_n(f, g, x))$.

Definition 2. *Given a set of polynomials $A = \{f_1, \ldots, f_m\} \subset \mathbb{Z}[\boldsymbol{y}, x]$ the Collins projector operator $\mathsf{P}_c(A, x)$ is defined as*

$$\bigcup_{f \in A} \mathsf{coeff}(f, x) \cup \bigcup_{\substack{f \in A \\ g \in \mathsf{R}^*(f, x)}} \mathsf{psc}(g, g'_x, x) \cup \bigcup_{\substack{i < j \\ g_i \in \mathsf{R}^*(f_i, x) \\ g_j \in \mathsf{R}^*(f_j, x)}} \mathsf{psc}(g_i, g_j, x) \ .$$

Let $A = \{f_1, \ldots, f_m\} \subset \mathbb{Z}[\boldsymbol{y}]$ be a set of polynomials, where $\boldsymbol{y} = (y_1, \ldots, y_n)$, and S be a region of $\mathbb{R}^n$. If for any assignment υ such that $\upsilon(\boldsymbol{y}) = \boldsymbol{\alpha} \in S$, the polynomials in A have the same sign under υ, we say that A is *sign-invariant* on S.

Theorem 2 (Theorem 4 in [10]). *Given a finite set of polynomials $A \subset \mathbb{Z}[\boldsymbol{y}, x]$, where $\boldsymbol{y} = (y_1, \ldots, y_n)$, and let S be a region of $\mathbb{R}^n$. If $\mathsf{P}_c(A)$ is sign invariant on S, then A is delineable over S.*

A *sign assignment* for a set of polynomials A is a mapping σ, from polynomials in A to $\{-1, 0, 1\}$. Given a set of polynomials $A \subset \mathbb{Z}[\boldsymbol{y}, x]$, we say a sign assignment σ is *realizable* with respect to some $\boldsymbol{\alpha}$ in $\mathbb{R}^n$, if there exists a $\beta \in \mathbb{R}$ such that every $f \in A$ takes the sign corresponding to its sign assignment, i.e., $\mathsf{sgn}(f(\boldsymbol{\alpha}, \beta)) = \sigma(f)$. The function sgn maps a real number to its sign $\{-1, 0, 1\}$. We use $\mathsf{signs}(A, \alpha)$ to denote the set of realizable sign assignments of A with respect to α.

Lemma 1. *If a set of polynomials $A \subset \mathbb{Z}[\boldsymbol{y}, x]$ is delineable over a region S, then $\mathsf{signs}(A, \alpha)$ is invariant over S.*

4.2   Projection-Based Explanations

Suppose that we need to produce an explanation for propagating a polynomial constraint F, i.e. we are in a state such that $\neg\,\mathsf{compatible}(\neg F, M)$, with $\mathsf{poly}(F) \in \mathbb{Z}[\boldsymbol{y}, x]$, where $\boldsymbol{y} = (y_1, \ldots, y_n)$. To simplify the presentation, in the following, we write υ for $\upsilon[M]$. The explanation procedure $\mathsf{explain}(F, M)$ consists of the following steps.

IsolateCore: Find a minimal set $\mathcal{F}$ of literals in M such that $\upsilon(\mathcal{F} \cup \{\neg F\})$ does not allow a solution for x. We call the set $\mathcal{F} \cup \{\neg F\}$ set a *conflicting core*.

Project: Construct a region S of $\mathbb{R}^n$ where $A = \mathsf{poly}(\mathcal{F} \cup \{F\})$ is delineable, and $\upsilon(\boldsymbol{y})$ is in S. Note that, from Lemma 1, $\neg F$ is incompatible with $\mathcal{F}$ for any other $\boldsymbol{\alpha}'$ in S.

Explain: Define the region S using extended polynomial constraints, obtaining a set of constraints $\mathcal{E}$. Then, we define $\mathsf{explain}(F, M) \equiv (\mathcal{E} \wedge \mathcal{F}) \implies F$.

We focus here on the second step of the procedure. To obtain the region S we will use a projection operator which, with insights of Theorem 2, will ensure delineability. Since our procedure requires a region S that contains the current assignment $\upsilon(\boldsymbol{y}) = \boldsymbol{\alpha}$, we add the assignment υ as an additional argument to the projection operator, and call such a projection operator *model-based.* Given a variable assignment υ, we denote the vanishing signature of a principal subresultant sequence as $\mathsf{v\text{-}psc}(f, g, x, \upsilon) = \mathsf{v\text{-}sig}(\mathsf{psc}_0(f, g, x), \ldots, \mathsf{psc}_n(f, g, x), \upsilon)$, and define our model-based projection operator $\mathsf{P}_m(A, x, \upsilon)$ as follows.

Definition 3. *Given a set of polynomials* $A = \{f_1, \ldots, f_m\} \subset \mathbb{Z}[\boldsymbol{y}, x]$ *and a variable assignment* υ, *the model-based Collins projector operator* $\mathsf{P}_m(A, x, \upsilon)$ *is defined as*

$$\bigcup_{f \in A} \mathsf{v\text{-}coeff}(f, x, \upsilon) \cup \bigcup_{\substack{f \in A \\ g = \mathsf{R}(f, x, \upsilon)}} \mathsf{v\text{-}psc}(g, g'_x, x, \upsilon) \cup \bigcup_{\substack{i < j \\ g_i = \mathsf{R}(f_i, x, \upsilon) \\ g_j = \mathsf{R}(f_j, x, \upsilon)}} \mathsf{v\text{-}psc}(g_i, g_j, x, \upsilon) \ .$$

Example 3. Consider the variable assignment υ, with $\upsilon(x) = 0$, and the set A containing two polynomials $f_2 = x^2 + y^2 - 1$ and $f_3 = -4xy - 4x + y - 1$. The projection operator P_m maps the set A into $\mathsf{P}_m(A, y, \upsilon)$

$$\{ \ \underbrace{(16x^3 - 8x^2 + x + 16)}_{f_1}x, \ -4x + 1, \ 4(x + 1)(x - 1), \ 2, \ 1 \ \} \ , \tag{1}$$

where f_1 is the polynomial from Ex. 1. The zeros of f_2 and f_3 are depicted in Fig. 4(b), together with a set of important points $\{-1, \alpha_1, 0, \frac{1}{4}, 1\}$, where α_1 is the algebraic number from Ex. 1. These points are exactly the roots of the projection polynomials (1). It is easy to see that f_2 and f_3 are delineable in the intervals defined by these points. But, considering a polynomial $f_4 = x^3 + 2x^2 + 3y^2 - 5$, we can see that it is not delineable on the interval $(1, +\infty)$.

We will use the projection operator P_m to compute the required region S, and show that A is delineable in S. First, we close the set of polynomials $A \subset \mathbb{Z}[y_1, \ldots, y_n, x]$ under the application of a projection operator P_m. We compute this closure by computing sets of polynomials $\mathcal{P}^n, \ldots, \mathcal{P}^1$ iteratively, starting from $\mathcal{P}^n = \mathsf{P}_m(A, \upsilon, x)$, and then for $k = n, \ldots, 2$, compute the subsequent ones as $\mathcal{P}^{k-1} = \mathsf{P}_m(\mathcal{P}^k, y_k, \upsilon) \cup (\mathcal{P}^k \cap \mathbb{Z}[y_1, \ldots, y_{k-1}])$. Each set of polynomials $\mathcal{P}^k \subseteq \mathbb{Z}[y_1, \ldots, y_k]$ is obtained by projecting the previous set $\mathcal{P}^{k+1}$ and adding all the polynomials from $\mathcal{P}^{k+1}$ that do not involve the variable y_{k+1}.

Now, we can build the region S inductively, in a bottom-up fashion, by constructing a sequence of regions $S^k \subset \mathbb{R}^k$ such that each $\mathcal{P}^k$ is sign invariant in

S^k, and $\mathcal{P}^{k+1}$ is delineable in S^k. Assume that S^{k-1}, and its defining constraints $\mathcal{E}^{k-1}$, have already been constructed. Now, consider the set of root objects

$$R^k = \{ \ \mathsf{root}(f, i) \ | \ f \in \mathcal{P}^k, \ 1 \leq i \leq \mathsf{rootcount}(\upsilon(f)) \ \} \ .$$

Under the assignment υ each of the root objects $\mathsf{root}(f, i)$ is defined and evaluates to some value $\omega_f^i \in \mathbb{R}_{\mathsf{alg}}$. The values ω_f^i partition the real line into maximal intervals where the polynomials $f \in \mathcal{P}^k$ are sign invariant. We pick the one interval that contains $\upsilon(y_k) = \alpha_k$ and construct the defining constraints $\mathcal{E}^k$ of the region S^k by selecting one of the appropriate cases

$$\begin{aligned}
\alpha_k &\in (\omega_f^i, \omega_g^j) && \Longrightarrow \mathcal{E}^k = \mathcal{E}^{k-1} \cup \{ \ y_k >_r \mathsf{root}(f, i), y_k <_r \mathsf{root}(g, j) \ \} \ , \\
\alpha_k &\in (-\infty, \omega_f^i) && \Longrightarrow \mathcal{E}^k = \mathcal{E}^{k-1} \cup \{ \ y_k <_r \mathsf{root}(f, i) \ \} \ , \\
\alpha_k &\in (\omega_f^i, +\infty) && \Longrightarrow \mathcal{E}^k = \mathcal{E}^{k-1} \cup \{ \ y_k >_r \mathsf{root}(f, i) \ \} \ , \\
\alpha_k &= \omega_f^i && \Longrightarrow \mathcal{E}^k = \mathcal{E}^{k-1} \cup \{ \ y_k =_r \mathsf{root}(f, i) \ \} \ .
\end{aligned}$$

Finally, we guarantee that $\mathcal{P}^{k+1}$ is delineable in S^k because polynomials in $\mathcal{P}^* = \mathcal{P}^1 \cup \ldots \cup \mathcal{P}^k$ are by construction sign invariant in S^k. Once we have computed the regions $S^1, \ldots, S^n$, we can use the region $S = S^n$ and the corresponding constraints $\mathcal{E} = \mathcal{E}^n$ to explain why $\neg F$ is incompatible with $\mathcal{F}$. Thus, we set $\mathsf{explain}(F, M) \equiv (\mathcal{E} \wedge \mathcal{F}) \implies F$.

Theorem 3. *The explanation function* $\mathsf{explain}(F, M)$ *is a finite-basis explanation function for the existential theory of real closed fields.*

Example 4. Consider the polynomial $f = x^2 + y^2 + z^2 - 1$, from Ex. 2, and the constraint $f < 0$ corresponding to the interior of the sphere in Fig. 4(a). Under an assignment υ with $\upsilon(x) = \frac{3}{4}$ and $\upsilon(y) = -\frac{3}{4}$ this constraint does not allow a solution for z (it evaluates to $z^2 < -\frac{1}{8}$). In order to explain it, we can compute the projection closure of $A = \{f\}$, using P_m, obtaining $\mathcal{P}_3 = A$ and

$$\mathcal{P}_2 = \{ \ 4x^2 + 4y^2 - 4, \ 2, \ 1 \ \} \ , \qquad \mathcal{P}_1 = \{ \ 256x^2 - 256, \ 8, \ 4, \ 2, \ 1 \ \} \ .$$

The sets of root objects under υ are then

$$\begin{aligned}
R^2 &= \{ \ \mathsf{root}(\tilde{z}^2 + x^2 - 1, 1), \ \mathsf{root}(\tilde{z}^2 + x^2 - 1, 2) \ \} \ , \\
R^1 &= \{ \ \mathsf{root}(\tilde{z}^2 - 1, 1), \ \mathsf{root}(\tilde{z}^2 - 1, 2) \ \} \ .
\end{aligned}$$

Since $\upsilon(x) = \frac{3}{4} = 0.75$ and the root objects of R_1 evaluate to -1 and 1, respectively, the constraints corresponding to the region S^1 are $(x > -1)$ and $(x < 1)$. The root objects of R_2 evaluate to $-\frac{\sqrt{7}}{4} \approx -0.6614$ and $\frac{\sqrt{7}}{4} \approx 0.6614$. Since $\upsilon(y) = -\frac{3}{4} = -0.75$, and we describe the region S^2 with the additional constraint $(y < \mathsf{root}(\tilde{z}^2 - x^2 - 1, 1))$. Using the constraints defining the region S^2 we construct the explanation $\mathsf{explain}(f < 0, \upsilon)$ as

$$(x \leq -1) \vee (x \geq 1) \vee \neg(y < \mathsf{root}(\tilde{z}^2 - x^2 - 1, 1)) \vee (f \geq 0) \ .$$

The explanation clause states that, in order to fix the conflict under υ, we must change υ so as to exit the region $-1 < x < 1$ below (in y) the unit circle.

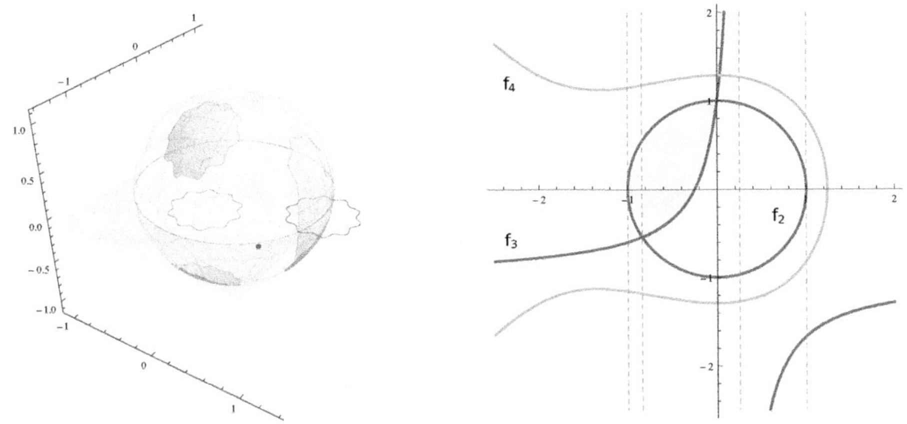

(a) (b)

Fig. 4. (a) The sphere corresponding to the roots of $x^2 + y^2 + z^2 - 1$, and regions of Ex 2 and Ex 4. (b) Solutions of $f_2 = x^2 + y^2 - 1 = 0$, $f_3 = -4xy - 4x + y - 1 = 0$, and $f_4 = x^3 + 2x^2 + 3y^2 - 5 = 0$, with the solution set of $\{f_2 < 0, f_3 > 0, f_4 < 0\}$ emphasized. The dashed lines represent the zeros of the projection set (1).

Isolating the conflicting core. Given a constraint F incompatible with a trail M, we can compute a minimal set of constraints $\mathcal{F}$ from M that is not compatible with F by taking the constraints that that caused the inconsistency and then refine it by trying to eliminate the constraints one by one.

Example 5. Consider the set of polynomial constraints $\mathcal{C} = \{f_2 < 0, f_3 > 0, f_4 < 0\}$, where the polynomials f_2 and f_3 are from Ex. 3. The roots of these polynomials and the feasible region of $\mathcal{C}$ are depicted in Fig. 4(b). Assume the transition is in the state $\langle [\![x \mapsto 0, (f_2 < 0), (f_4 < 0), E \rightarrow (f_3 \leq 0)]\!], \mathcal{C} \rangle_2$, and we need to compute the explanation E of the last propagation. Although the propagation was based on the inconsistency of $\mathcal{C}$ under M, we can pick the subset $\{f_2 < 0, f_3 > 0\}$ to produce the explanation. It is a smaller set, but sufficient, as it is also inconsistent with M. Doing so we reduce the number of polynomials we need to project, which, in CAD settings, is always an improvement.

5 Related Work and Experimental Results

In addition to CAD, a number of other procedures have been developed and implemented in working tools since the 1980s, including Weispfenning's method of virtual term substitution (VTS) [28] (as implemented in Reduce/Redlog), and the Harrison-McLaughlin proof producing version of the Cohen-Hörmander method [19]. Abstract Partial Cylindrical Algebraic Decomposition [22] combines fast, sound but incomplete procedures with CAD. Tiwari [27] presents an approach using Gröbner bases and sign conditions to produce unsatisfiability witnesses for nonlinear constraints. Platzer, Quesel and Rümmer combine Gröbner bases with semidefinite programming [23] for the real Nullstellensatz.

In order to evaluate the new decision procedure we have implemented a new solver nlsat, the implementation being a clean translation of the decision procedure described in this paper. We compare the new solver to the following solvers that have been reported to perform reasonably well on fragments of non-linear arithmetic: the z3 3.2 [11], cvc3 2.4.1 [2], and MiniSmt 0.3 [29] SMT solvers; the quantifier elimination based solvers Mathematica 8.0 [25], QEPCAD 1.65 [5], Redlog-CAD and Redlog-VTS [12]; and the interval based iSAT [13] solver.[5]

We ran all the solvers on several sets of benchmarks, where each benchmark set has particular characteristics that can be problematic for a non-linear solver. The meti-tarski benchmarks are proof obligations extracted from the MetiTarski project [1], where the constraints are of high degree and the polynomials represent approximations of the elementary real functions being analyzed. The keymaera benchmark set contains verification conditions from the Keymaera verification platform [23]. The zankl set of problems are the benchmarks from the QF_NRA category of the SMT-LIB library, with most problems originating from attempts to prove termination of term-rewrite systems [14]. We also have two crafted sets of benchmarks, the hong benchmarks, which are a parametrized generalization of the problem from [15], and the kissing problems that describe some classic kissing number problems, both sets containing instances of increasing dimensions.

Table 1. Experimental results

	meti-tarski (1006)		keymaera (421)		zankl (166)		hong (20)		kissing (45)		all (1658)	
solver	solved	time (s)	solved	time (s)	solved	time (s)	solved	time (s)	solved	time (s)	solved	time (s)
nlsat	1002	343	**420**	**5**	**89**	**234**	10	170	13	95	**1534**	**849**
Mathematica	**1006**	**796**	420	171	50	366	9	208	6	29	1491	1572
QEPCAD	991	2616	368	1331	21	38	6	43	4	5	1390	4036
Redlog-VTS	847	28640	419	78	42	490	6	3	10	275	1324	29488
Redlog-CAD	848	21706	363	730	21	173	6	2	4	0	1242	22613
z3	266	83	379	1216	21	0	1	0	0	0	667	1299
iSAT	203	122	291	16	21	24	**20**	**822**	0	0	535	986
cvc3	150	13	361	5	12	3	0	0	0	0	523	22
MiniSmt	40	697	35	0	46	1370	0	0	**18**	**44**	139	2112

All tests were conducted on an Intel Pentium E2220 2.4 GHz processor, with individual runs limited to 2GB of memory and 900 seconds. The results of our experimental evaluation are presented in Table 1. The rows are associated with the individual solvers, and columns separate the problem sets. For each problem set we write the number of problems that the solver managed to solve within the time limit, and the cumulative time (rounded) for the solved problems.

The results are both revealing and encouraging. On this set of benchmarks, except for nlsat and the quantifier elimination based solvers, all other solvers that we've tried have a niche problem set where they perform well (or reasonably well),

[5] We ran the solvers with default settings, using the `Resolve` command of Mathematica, the `rlcad` command for Redlog-CAD, and the `rlqe` for Redlog-VTS.

whereas on others they perform poorly. The new nlsat solver, on the other hand, is consistently one of the best solvers for each problem set, with impressive running times, and, overall manages to solve the most problems, in much faster time.

6 Conclusion

We proposed a new procedure for solving systems of non-linear polynomial constraints. The new procedure performs a backtracking search for a model, where the backtracking is powered by a novel conflict resolution procedure. In our experiments, our first prototype was consistently one of the best solvers for each problem set we tried, and, overall manages to solve the most problems, in much faster time. We expect even better results after several missing optimizations in the core algorithms are implemented. We see many possible improvements and extensions to our procedure. We plan to design and experiment with different explain procedures. One possible idea is to try explain procedures that are more efficient, but do not guarantee termination. Heuristics for reordering variables and selecting a value from the feasible set should also be tried. Integrating our solver with a Simplex-based procedure is another promising possibility.

Acknowledgements. We would like to thank Grant Passmore for providing valuable feedback, the Meti-Tarski benchmark set, and so many interesting technical discussions. We also would like to thank Clark Barrett for all his support.

References

1. Akbarpour, B., Paulson, L.C.: MetiTarski: An automatic theorem prover for real-valued special functions. Journal of Automated Reasoning 44(3), 175–205 (2010)
2. Barrett, C., Tinelli, C.: CVC3. In: Damm, W., Hermanns, H. (eds.) CAV 2007. LNCS, vol. 4590, pp. 298–302. Springer, Heidelberg (2007)
3. Basu, S., Pollack, R., Roy, M.-F.: Algorithms in real algebraic geometry. Springer (2006)
4. Brown, C.W.: Solution formula construction for truth invariant CAD's. PhD thesis, University of Delaware (1999)
5. Brown, C.W.: QEPCAD B: a program for computing with semi-algebraic sets using CADs. ACM SIGSAM Bulletin 37(4), 97–108 (2003)
6. Brown, W.S., Traub, J.F.: On Euclid's algorithm and the theory of subresultants. Journal of the ACM 18(4), 505–514 (1971)
7. Buchberger, B., Collins, G.E., Loos, R., Albrecht, R. (eds.): Computer algebra. Symbolic and algebraic computation. Springer (1982)
8. Caviness, B.F., Johnson, J.R. (eds.): Quantifier Elimination and Cylindrical Algebraic Decomposition. Texts and Monographs in Symbolic Computation. Springer (2004)
9. Cohen, H.: A Course in Computational Algebraic Number Theory. Springer (1993)
10. Collins, G.E.: Quantifier Elimination for Real Closed Fields by Cylindrical Algebraic Decomposition. In: Brakhage, H. (ed.) GI-Fachtagung 1975. LNCS, vol. 33, pp. 134–183. Springer, Heidelberg (1975)

11. de Moura, L., Bjørner, N.S.: Z3: An Efficient SMT Solver. In: Ramakrishnan, C.R., Rehof, J. (eds.) TACAS 2008. LNCS, vol. 4963, pp. 337–340. Springer, Heidelberg (2008)
12. Dolzmann, A., Sturm, T.: Redlog: Computer algebra meets computer logic. ACM SIGSAM Bulletin 31(2), 2–9 (1997)
13. Fränzle, M., Herde, C., Teige, T., Ratschan, S., Schubert, T.: Efficient solving of large non-linear arithmetic constraint systems with complex Boolean structure. Journal on Satisfiability, Boolean Modeling and Computation 1(3-4), 209–236 (2007)
14. Fuhs, C., Giesl, J., Middeldorp, A., Schneider-Kamp, P., Thiemann, R., Zankl, H.: SAT Solving for Termination Analysis with Polynomial Interpretations. In: Marques-Silva, J., Sakallah, K.A. (eds.) SAT 2007. LNCS, vol. 4501, pp. 340–354. Springer, Heidelberg (2007)
15. Hong, H.: Comparison of several decision algorithms for the existential theory of the reals (1991)
16. Jovanović, D., de Moura, L.: Cutting to the Chase Solving Linear Integer Arithmetic. In: Bjørner, N., Sofronie-Stokkermans, V. (eds.) CADE 2011. LNCS, vol. 6803, pp. 338–353. Springer, Heidelberg (2011)
17. Korovin, K., Tsiskaridze, N., Voronkov, A.: Conflict Resolution. In: Gent, I.P. (ed.) CP 2009. LNCS, vol. 5732, pp. 509–523. Springer, Heidelberg (2009)
18. Loos, R.: Generalized polynomial remainder sequences. Computer Algebra: Symbolic and Algebraic Computation, 115–137 (1982)
19. McLaughlin, S., Harrison, J.V.: A Proof-Producing Decision Procedure for Real Arithmetic. In: Nieuwenhuis, R. (ed.) CADE 2005. LNCS (LNAI), vol. 3632, pp. 295–314. Springer, Heidelberg (2005)
20. McMillan, K.L., Kuehlmann, A., Sagiv, M.: Generalizing DPLL to Richer Logics. In: Bouajjani, A., Maler, O. (eds.) CAV 2009. LNCS, vol. 5643, pp. 462–476. Springer, Heidelberg (2009)
21. Nieuwenhuis, R., Oliveras, A., Tinelli, C.: Solving SAT and SAT modulo theories: From an abstract Davis–Putnam–Logemann–Loveland procedure to DPLL(T). Journal of the ACM 53(6), 937–977 (2006)
22. Passmore, G.O.: Combined Decision Procedures for Nonlinear Arithmetics, Real and Complex. PhD thesis, University of Edinburgh (2011)
23. Platzer, A., Quesel, J.-D., Rümmer, P.: Real World Verification. In: Schmidt, R.A. (ed.) CADE-22. LNCS, vol. 5663, pp. 485–501. Springer, Heidelberg (2009)
24. Silva, J.P.M., Sakallah, K.A.: GRASP: A search algorithm for propositional satisfiability. IEEE Transactions on Computers 48(5), 506–521 (1999)
25. Strzeboński, A.W.: Cylindrical algebraic decomposition using validated numerics. Journal of Symbolic Computation 41(9), 1021–1038 (2006)
26. Tarski, A.: A decision method for elementary algebra and geometry. Technical Report R-109, Rand Corporation (1951)
27. Tiwari, A.: An Algebraic Approach for the Unsatisfiability of Nonlinear Constraints. In: Ong, L. (ed.) CSL 2005. LNCS, vol. 3634, pp. 248–262. Springer, Heidelberg (2005)
28. Weispfenning, V.: Quantifier elimination for real algebra - the quadratic case and beyond. AAECC 8, 85–101 (1993)
29. Zankl, H., Middeldorp, A.: Satisfiability of Non-linear (Ir)rational Arithmetic. In: Clarke, E.M., Voronkov, A. (eds.) LPAR-16 2010. LNCS, vol. 6355, pp. 481–500. Springer, Heidelberg (2010)

Inprocessing Rules*

Matti Järvisalo[1], Marijn J.H. Heule[2,3], and Armin Biere[3]

[1] Department of Computer Science & HIIT, University of Helsinki, Finland
[2] Department of Software Technology, Delft University of Technology, The Netherlands
[3] Institute for Formal Models and Verification, Johannes Kepler University, Linz, Austria

Abstract. Decision procedures for Boolean satisfiability (SAT), especially modern conflict-driven clause learning (CDCL) solvers, act routinely as core solving engines in various real-world applications. Preprocessing, i.e., applying formula rewriting/simplification rules to the input formula before the actual search for satisfiability, has become an essential part of the SAT solving tool chain. Further, some of the strongest SAT solvers today add more reasoning to search by *interleaving* formula simplification and CDCL search. Such *inprocessing SAT solvers* witness the fact that implementing additional deduction rules in CDCL solvers leverages the efficiency of state-of-the-art SAT solving further. In this paper we establish formal underpinnings of inprocessing SAT solving via an abstract inprocessing framework that covers a wide range of modern SAT solving techniques.

1 Introduction

Decision procedures for Boolean satisfiability (SAT), especially modern conflict-driven clause learning (CDCL) [1,2] SAT solvers, act routinely as core solving engines in many industrial and other real-world applications today. Formula simplification techniques such as [3,4,5,6,7,8,9,10,11,12,13,14] applied before the actual satisfiability search, i.e., in preprocessing, have proven integral in enabling efficient conjunctive normal form (CNF) level Boolean satisfiability solving for real-world application domains, and have become an essential part of the SAT solving tool chain. Taking things further, some of the strongest SAT solvers today add more reasoning to search by *interleaving* formula simplification and CDCL search. Such *inprocessing SAT solvers*, including the successful state-of-the-art CDCL SAT solvers PRECOSAT [15], CRYPTOMINISAT [16], and LINGELING [17], witness the fact that implementing additional deduction rules within CDCL solvers leverages the efficiency of state-of-the-art SAT solving further.

To illustrate the usefulness of preprocessing and inprocessing in improving the performance of current state-of-the-art SAT solvers, we modified the 2011 SAT Competition version of the state-of-the-art SAT solver LINGELING that is based on the inprocessing CDCL solver paradigm. The resulting patch[1] allows to either disable all preprocessing or to just disable inprocessing during search. We have run the original version and these two versions on the benchmarks from the application track—the most important competition category from the industrial perspective—of the last two SAT competitions organized in 2009 and 2011. The results are shown in Table 1.

* The 1st author is supported by Academy of Finland (grants 132812 and 251170), 2nd and 3rd authors by Austrian Science Foundation (FWF) NFN Grant S11408-N23 (RiSE).

[1] `http://fmv.jku.at/lingeling/lingeling-587f-disable-pre-and-inprocessing.patch`

B. Gramlich, D. Miller, and U. Sattler (Eds.): IJCAR 2012, LNAI 7364, pp. 355–370, 2012.

Table 1. Results of running the original 2011 competition version 587f of LINGELING on the application instances from 2009 and from 2011, then without inprocessing and in the last row without any pre- nor inprocessing. The experiments were obtained on a cluster with Intel Core 2 Duo Quad Q9550 2.8-GHz processors, 8-GB main memory, running Ubuntu Linux. Memory consumption was limited to 7 GB and run-time to 900 seconds. The single-threaded sequential version of LINGELING was used with one solver instance per processor.

	2009				2011			
LINGELING	solved	SAT	UNSAT	time	solved	SAT	UNSAT	time
original version 587f	196	79	117	114256	164	78	86	144932
only preprocessing	184	72	112	119161	159	77	82	145218
no pre- nor inprocessing	170	68	102	138940	156	78	78	153434

The CNF preprocessor SATELITE introduced in [7] applied *variable elimination*, one of the most effective simplification techniques in state-of-the-art SAT solvers. As already shown in [7] preprocessing can also be extremely useful within *incremental* SAT solving. This form of preprocessing, which is performed at each incremental call to the SAT solver, can be considered as an early form of inprocessing. Fig. 1 confirms this observation in the context of incremental SAT solving for bounded model checking.

However, developing and implementing sound inprocessing solvers in the presence of a wide range of different simplification techniques (including variable elimination, blocked clause elimination, distillation, equivalence reasoning) is highly non-trivial. It requires in-depth understanding on how different techniques can be combined together and interleaved with the CDCL algorithm in a satisfiability-preserving way. Moreover, the fact that many simplification techniques only preserve satisfiability but not logical

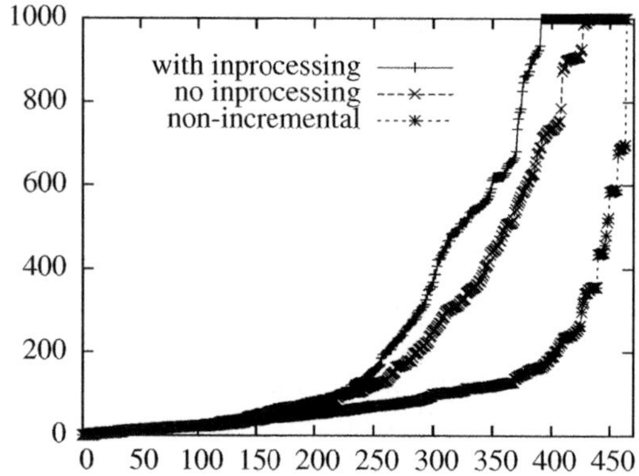

	*	bounds	time
with inpr.	158	153975	304158
no inpr.	115	125436	335393
non-incr.	67	49915	369104

* = #solved + #(bound 1000 reached)

bounds = $\sum$ reached bounds

Time is summed up over all benchmarks, while unfinished runs are counted with the time limit of 900 seconds (same experimental setup as in Tab. 1).

Fig. 1. Running the bounded model checker BLIMC, which is part of the LINGELING distribution with and without inprocessing on the single property benchmarks of the Hardware Model Checking Competition 2011 up to bound 1000. With inprocessing 153975 bounds were reached, while without inprocessing only 125436. The figure shows the maximum bound reached (successfully checked) on the y-axis for each of the 465 benchmark (x-axis). Benchmarks are sorted by the maximum bound. For completeness we also include a run in non-incremental mode, which reaches only 49915 bounds. In this mode a new CNF is generated and checked for each bound with a fresh SAT solver instance separately, but with both pre- and inprocessing enabled.

equivalence poses additional challenges, since in many practical applications of SAT solvers a solution is required for satisfiable formulas, not only the knowledge of the satisfiability of the input formula. Hence, when designing inprocessing SAT solvers for practical purposes, one also has to address the intricate task of solution reconstruction.

In this paper we propose an abstract framework that captures generally the deduction mechanisms applied within inprocessing SAT solvers. The framework consists of four generic and clean deduction rules. Importantly, the rules specify general conditions for sound inprocessing SAT solving, against which specific inprocessing techniques can be checked for correctness. The rules also capture solution reconstruction for a wide range of simplification techniques that do not preserve logical equivalence: while solution reconstruction algorithms have been proposed previously for specific inprocessing techniques [18,11], we show how a simple linear-time algorithm covers solution reconstruction for a wide range of techniques.

Our abstract framework has similarities to the abstract DPLL(T) framework [19] and its extensions [20,21], and the proof strategies approach of [22], in describing deduction via transition systems. However, in addition to inprocessing as built-in feature, our framework captures SAT solving on a more generic level than [19], not being restricted to DPLL-style search procedures, and at the same time it gives a fine-grained view of inprocessing SAT solving. We show how the rules of our framework can be instantiated to obtain both known and novel inprocessing techniques. We give examples of how the correctness of such specific techniques can be checked based on the generic rules in our framework. Furthermore, we show that our rules in the general setting are extremely powerful, even capturing Extended Resolution [23].

Arguing about correctness of combinations of different solving techniques in concrete SAT solver development is tremendously simplified by our framework. One example is the interaction of learned clauses with *variable elimination* [7]. After variable elimination is performed on the irredundant (original) clauses during inprocessing, the question is what to do with learned clauses that still contain eliminated variables. While current implementations simply forget (remove) such learned clauses, it follows easily from our framework that it is sound to keep such learned clauses and use them subsequently for propagation. It is also easy to observe e.g. that one can (selectively) turn eliminated or blocked clauses into learned clauses to preserve propagation power.

Another more intricate example from concrete SAT solver development occurs in the context of *blocked clauses* [12]. An intermediate version of LINGELING contained a simple algorithm for adding new redundant (learned) binary clauses, which are blocked, but only w.r.t. irredundant (original) clauses, thus disregarding already learned clauses. This would be convenient since focusing on irredundant clauses avoids having full occurrence lists for learned clauses. Further, marking the added clauses as redundant implies that they would not have to be considered in consecutive variable eliminations, and thus might enable to eliminate more variables without increasing the number of clauses. However, we found examples that proved this approach to be incorrect. An attempt to fix this problem was to include those added clauses in further blocked clause *removal* and *addition* attempts, and only ignore them during variable elimination. This version was kept in the code for some months without triggering any inconsistencies. However, this is incorrect, and can be easily identified via our formal framework.

After preliminaries (Sect. 2), we review redundancy properties (Sect. 3) and their extensions (Sect. 4) based on different clause elimination and addition procedures. The abstract inprocessing rules are discussed in Sect. 5, followed by an instantiation of the rules using a specific redundancy property and a related generic solution reconstruction approach (Sect. 6). Based on this instantiation of the rules, we show how the rules capture a wide range of modern SAT solving techniques and, via examples, how the rules catch incorrect variations of these techniques (Sect. 7).

2 Preliminaries

For a Boolean variable x, there are two *literals*, the positive literal x and the negative literal $\neg x$. A *clause* is a disjunction of literals and a CNF formula a conjunction of clauses. A clause can be seen as a finite set of literals and a CNF formula as a finite set of clauses. A truth assignment is a function τ that maps literals to $\{0, 1\}$ under the assumption $\tau(x) = v$ if and only if $\tau(\neg x) = 1 - v$. A clause C is satisfied by τ if $\tau(l) = 1$ for some literal $l \in C$. An assignment τ satisfies F if it satisfies every clause in F; such a τ is a *model* of F.

Two formulas are *logically equivalent* if they are satisfied by exactly the same set of assignments, and *satisfiability-equivalent* if both formulas are satisfiable or both unsatisfiable. The length of a clause C is the number of literals in C. A *unit clause* has length one, and a *binary clause* length two. The set of binary clauses in a CNF formula F is denoted by F_2. The resolution rule states that, given two clauses $C_1 = \{l, a_1, \ldots, a_n\}$ and $C_2 = \{\neg l, b_2, \ldots, b_m\}$, the clause $C_1 \otimes C_2 = \{a_1, \ldots, a_n, b_1, \ldots, b_m\}$, called the *resolvent* $C_1 \otimes_l C_2$ (or simply $C_1 \otimes C_2$ when clear from context) of C_1 and C_2, can be inferred by *resolving* on the literal l. For a CNF formula F, let F_l denote the set of clauses in F that contain the literal l. The resolution operator $\otimes_l$ can be lifted to sets of clauses by defining $F_l \otimes_l F_{\neg l} = \{C \otimes_l C' \mid C \in F_l,\, C' \in F_{\neg l}\}$.

3 Clause Elimination and Addition

Clause Elimination Procedures. [11] are an important family of CNF simplification techniques which are to an extent orthogonal with resolution-based techniques [12]. Intuitively, clause elimination refers to removing from CNF formulas clauses that are *redundant* (with respect to some specific redundancy property) in the sense that satisfiability is preserved under removal.

Definition 1. *Given a CNF formula F, a specific clause elimination procedure $\mathcal{P}$E removes clauses that have a specific property $\mathcal{P}$ from F until fixpoint. In other words, $\mathcal{P}$E on input F modifies F by repeating the following until fixpoint: if there is a clause $C \in F$ that has $\mathcal{P}$, let $F := F \setminus \{C\}$.*

Clause Addition Procedures, the dual of clause elimination procedures, add to (instead of removing from) CNF formulas clauses that are *redundant* (with respect to some specific redundancy property) in the sense that satisfiability is preserved under adding.

Definition 2. *Given a CNF formula F, a specific clause addition procedure $\mathcal{P}\mathrm{A}$ adds clauses that have a specific property $\mathcal{P}$ to F until fixpoint. In other words, $\mathcal{P}\mathrm{A}$ on input F modifies F by repeating the following until fixpoint: if there is a clause C that has $\mathcal{P}$, let $F := F \cup \{C\}$.*

While clause elimination procedures have been studied and exploited to a much broader extent than clause addition, the latter has already proven important both from the theoretical and the practical perspectives, as we will discuss further in Sect. 7.

For establishing concrete instantiations of clause elimination and addition procedures, redundancy properties on which such procedures are based on need to be defined. We will now review various such properties, following [11].

3.1 Notions of Redundancy

A clause is a *tautology* if it contains both x and $\neg x$ for some variable x. Given a CNF formula F, a clause $C_1 \in F$ *subsumes* (another) clause $C_2 \in F$ in F if and only if $C_1 \subset C_2$, and then C_2 is *subsumed by* C_1.

Given a CNF formula and a clause $C \in F$, (*hidden literal addition*) $\mathrm{HLA}(F, C)$ is the *unique* clause resulting from repeating the following clause extension steps until fixpoint: if there is a literal $l_0 \in C$ such that there is a clause $(l_0 \vee l) \in F_2 \setminus \{C\}$ for some literal l, let $C := C \cup \{\neg l\}$.

For a clause C, (*asymmetric literal addition*) $\mathrm{ALA}(F, C)$ is the *unique* clause resulting from repeating the following until fixpoint: if $l_1, \ldots, l_k \in C$ and there is a clause $(l_1 \vee \cdots \vee l_k \vee l) \in F \setminus \{C\}$ for some literal l, let $C := C \cup \{\neg l\}$.

Given a CNF formula F and a clause C, a literal $l \in C$ *blocks* C w.r.t. F if (i) for each clause $C' \in F$ with $\neg l \in C'$, the resolvent $C \otimes_l C'$ is a tautology, or (ii) $\neg l \in C$, i.e., C is itself a tautology. A clause C is *blocked* w.r.t. F if there is a literal l that blocks C w.r.t. F. For such an l, we say that C is blocked *on* $l \in C$ w.r.t. F.

What follows is a list of properties based on which various clause elimination procedures [11,12] can be defined.

S	(*subsumption*)	C is subsumed in F.
HS	(*hidden subsumption*)	$\mathrm{HLA}(F, C)$ is subsumed in F.
AS	(*asymmetric subsumption*)	$\mathrm{ALA}(F, C)$ is subsumed in F.
T	(*tautology*)	C is a tautology.
HT	(*hidden tautology*)	$\mathrm{HLA}(F, C)$ is a tautology.
AT	(*asymmetric tautology*)	$\mathrm{ALA}(F, C)$ is a tautology.
BC	(*blocked*)	C is blocked w.r.t. F.
HBC	(*hidden blocked*)	$\mathrm{HLA}(F, C)$ is blocked w.r.t. F.
ABC	(*asymmetric blocked*)	$\mathrm{ALA}(F, C)$ is blocked w.r.t. F.

As concrete examples, BC gives the clause elimination procedure *blocked clause elimination* (BCE) [12], and HT *hidden tautology elimination* (HTE) [11].

A relevant question is how the above-listed properties are related to each other. Especially, if any C having property $\mathcal{P}$ also has property $\mathcal{P}'$, then we know that a clause elimination procedure based on $\mathcal{P}'$ can remove at least the same clauses as a clause elimination procedure based on $\mathcal{P}$ (similarly for clause addition procedures). The relationships between these properties (first analyzed as the *relative effectiveness* in the special case of *clause elimination procedures* in [11]) are illustrated in Fig. 2. The properties prefixed with R are new and will be defined next.

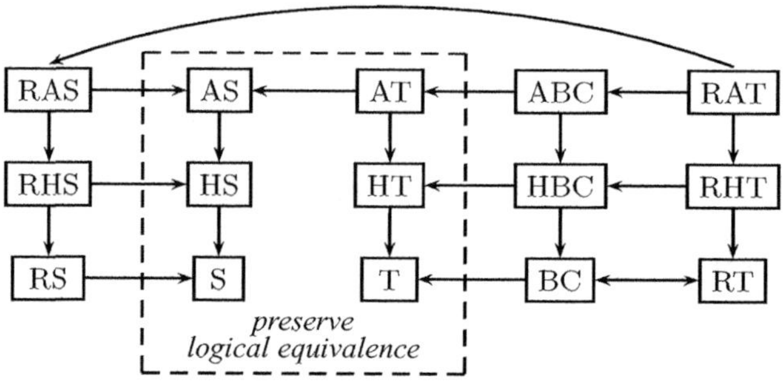

Fig. 2. Relationships between clause redundancy properties. An edge from $\mathcal{P}$ to $\mathcal{P}'$ means that any clause that has property $\mathcal{P}'$ also has property $\mathcal{P}$. A missing edge from $\mathcal{P}$ to $\mathcal{P}'$ means that there are clauses with property $\mathcal{P}'$ that do not have property $\mathcal{P}$. Clause elimination and addition procedures based on the properties inside the *preserve logical equivalence* box preserve logical equivalence under elimination and addition [11].

4 Extended Notions of Redundancy

Clause elimination procedures can be extended by using the resolution rule as a specific kind of "look-ahead step" within the procedures. This turns a specific clause elimination procedure $\mathcal{P}$E based on property $\mathcal{P}$ into the clause elimination procedure R$\mathcal{P}$E based on a property R$\mathcal{P}$. Analogously, a specific clause addition procedure $\mathcal{P}$A based on property $\mathcal{P}$ turns into the clause addition procedure R$\mathcal{P}$A based on a property R$\mathcal{P}$.

Definition 3. *Given a CNF formula F and a clause $C \in F$, C* has property R$\mathcal{P}$ *iff either (i) C has the property $\mathcal{P}$, or (ii) there is a literal $l \in C$ such that for each clause $C' \in F$ with $\neg l \in C'$, each resolvent in $C \otimes_l C'$ has $\mathcal{P}$ (in this case C has* R$\mathcal{P}$ on l*).*

Example 1. Consider the formula $F = (a \vee b \vee x) \wedge (\neg x \vee c \vee d) \wedge (a \vee b \vee c)$. The only resolvent of $(a \vee b \vee x)$ on x is $(a \vee b \vee c \vee d)$ which is subsumed by $(a \vee b \vee c)$. Therefore $(a \vee b \vee x)$ has property RS (resolution subsumption).

The intuition is that the "resolution look-ahead" step can reveal additional redundant clauses, resulting in the hierarchy shown in Fig. 2. Notice that the property RT (resolution tautology) is the same as the property BC (blocked).

Proposition 1. *For any CNF formula F and clause C that has* RAT *on $l \in C$ w.r.t. F, F is satisfiability-equivalent to $F \cup \{C\}$.*

Proof. By definition, since C has RAT on $l \in C$ w.r.t. F, all resolvents $C \otimes_l F_{\neg l}$ are asymmetric tautologies w.r.t. F (and w.r.t. the larger $F \cup (C \otimes_l F_{\neg l})$ as well). Hence F is logically equivalent to $F \cup (C \otimes_l F_{\neg l})$. Now consider a truth assignment τ that satisfies F, but falsifies C. Since C is falsified by τ, and all $C' \in C \otimes_l F_{\neg l}$ are satisfied by τ due to logical equivalence of F and $F \cup (C \otimes_l F_{\neg l})$, τ satisfies at least two literals in each clause in $F_{\neg l}$ (at least one more beside $\neg l$). Hence the truth assignment τ' that is a copy of τ except for $\tau'(l) = 1$ satisfies F and C. □

Proposition 2. *The set of clauses that have* RAS *is a proper subset of the set of clauses that have* RAT.

Proof. Assume a clause C has RAS on $l \in C$ w.r.t. F. If C has AS, then C has AT [11] and hence also RAT. Otherwise, take any resolvent $C' \in C \otimes_l F_{\neg l}$. By definition, C' has AS. Since clauses with AS are a proper subset of the clauses with AT, C has RAT on l w.r.t. F. Moreover, let $F := (a \vee \neg b) \wedge (\neg a \vee b)$. Now $(a \vee \neg b)$ has RAT on a w.r.t. F. However, $(a \vee \neg b)$ does not have RAS w.r.t. F. □

Proposition 3. *The set of clauses that have* ABC *is a proper subset of the set of clauses that have* RAT.

Proof. Let C be clause that has ABC on $l \in C$ w.r.t. F. W.l.o.g. assume C to be non-tautological. By [11, Lemma 19], $l \in C$. Take the resolvent $C' = C \otimes_l C''$ for any $C'' \in F_{\neg l}$. First, we show that $\mathrm{ALA}(F, C) \subseteq \mathrm{ALA}(F, C')$. C' overlaps with C'' in all literals except $\neg l$. W.l.o.g. assume $C' \notin F$ (otherwise C' is subsumed by F and thus also has AT w.r.t. F). Therefore, by the definition of ALA, $l \in \mathrm{ALA}(F, C')$. Hence $C \subseteq \mathrm{ALA}(F, C')$. Due to monotonicity of ALA under the assumption $C' \notin F$, we have $\mathrm{ALA}(F, C) \subseteq \mathrm{ALA}(F, C')$. By definition of ABC, the clause $\mathrm{ALA}(F, C) \otimes_l C''$ is a tautology, and hence there is an $l' \in \mathrm{ALA}(F, C) \setminus \{l\}$ with $\neg l' \in C''$. Now, $l' \in \mathrm{ALA}(F, C')$ since $\mathrm{ALA}(F, C) \subseteq \mathrm{ALA}(F, C')$, and $\neg l' \in \mathrm{ALA}(F, C')$ since $C' = C \otimes_l C''$. Thus C' has AT on l w.r.t. F, which implies that C has RAT w.r.t. F.

For proper containment, consider the formula $F = (a \vee b \vee c \vee d) \wedge (a \vee b \vee x) \wedge (\neg x \vee c \vee d) \wedge (\neg a \vee y \vee z) \wedge (\neg b \vee y \vee \neg z) \wedge (\neg c \vee \neg y \vee z) \wedge (\neg d \vee \neg y \vee \neg z)$. No clause in F has ABC. Yet $(a \vee b \vee x)$ has RAT on x w.r.t. F. □

Proposition 4. *The set of clauses which have* RHT *is a proper subset of the set of clauses that have* RAT.

Proof. Assume C has RHT on $l \in C$ w.r.t. F. If C has HT, then C has AT [11] and hence also RAT. Otherwise, take any $C' \in C \otimes_l F_{\neg l}$. By definition, C' has HT. Since clauses with HT are a proper subset of the clauses with AT, C has RAT on l w.r.t. F. Moreover, let $F := (a \vee b \vee x) \wedge (\neg x \vee c) \wedge (a \vee b \vee c) \wedge (\neg a) \wedge (\neg b) \wedge (\neg c)$. Now $(a \vee b \vee x)$ has RAT on x w.r.t. F, but $(a \vee b \vee x)$ does not have RHT. □

5 Inprocessing as Deduction

We will now introduce generic rules for inprocessing CNF formulas. The rules describe inprocessing as a transition system. States in the transition system are described by tuples of the form $\varphi\,[\,\rho\,]\,\sigma$, where φ and ρ are CNF formulas, and σ is a sequence of literal-clause pairs. For inprocessing a given CNF formula F, the initial state is $F\,[\,\emptyset\,]\,\langle\rangle$, where $\emptyset$ denotes the empty CNF formula, and $\langle\rangle$ the empty sequence.

Generally, a state $\varphi\,[\,\rho\,]\,\sigma$ has the following interpretation.

- φ is a CNF formula that consists of the set of *irredundant* clauses. Irredundant means here that all clauses in φ are considered to be "hard" in the sense that, in order to satisfy the input CNF formula F, all clauses in φ are to be satisfied.

- ρ is a CNF formula that consists of *redundant clauses*. In contrast to the irredundant clauses φ, these clauses can be removed from consideration.
- σ denotes a sequence of literal-clause pairs $l{:}C$ with $l \in C$ that are required for solution reconstruction, as explained in detail in Sect. 6.1.

For some intuition on why we separate φ and ρ, note that *learned clauses*, i.e., clauses added through conflict analysis in CDCL solvers, are maintained separately from the clauses in the input formula, and can be forgotten (i.e., removed) since in pure CDCL they are entailed by the input formula. However, in the more generic context of inprocessing SAT solving captured by our framework, clauses in ρ may not be entailed by the original formula F. This is discussed in detail in Sect. 7 using clause addition as an example. In addition, for elimination techniques (such as BCE, variable elimination, and their variants) only the clauses in φ need to be considered when checking redundancy. Nevertheless, the clauses in ρ can be used for e.g. unit propagation.

5.1 Rules of Inprocessing

Our abstract framework for inprocessing SAT solving is based on four rules: LEARN, FORGET, STRENGTHEN, and WEAKEN, presented in Fig. 3. These rules characterize the set of legal next states $\varphi' \,[\, \rho' \,]\, \sigma'$ of a given current state $\varphi \,[\, \rho \,]\, \sigma$ in the form

$$\frac{\varphi \,[\, \rho \,]\, \sigma}{\varphi' \,[\, \rho' \,]\, \sigma'}.$$

Given a CNF formula F, a state $\varphi_k \,[\, \rho_k \,]\, \sigma_k$ is reachable from the state $F \,[\, \emptyset \,]\, \langle\rangle$ iff there is a sequence $\langle \varphi_0 \,[\, \rho_0 \,]\, \sigma_0, \ldots, \varphi_k \,[\, \rho_k \,]\, \sigma_k \rangle$ such that (i) $\varphi_0 = F$, $\rho_0 = \emptyset$, and $\sigma_0 = \langle\rangle$, and (ii) for each $i = 1, \ldots, k$, one of the rules in Fig. 3 allows the transition from $\varphi_{i-1} \,[\, \rho_{i-1} \,]\, \sigma_{i-1}$ to $\varphi_i \,[\, \rho_i \,]\, \sigma_i$. This sequence is called a derivation of $\varphi_k \wedge \rho_k$ from F.

The inprocessing rules are *correct* in the sense that they preserve satisfiability, i.e., starting from the state $F \,[\, \emptyset \,]\, \langle\rangle$, the following invariant holds for all $i = 1, \ldots, k$:

Formulas φ_i and $(\varphi_i \wedge \rho_i)$ are *both* satisfiability-equivalent to F.

The intuition behind these rules is as follows.

LEARN Allows for introducing (*learning*) a new clause C to the current redundant formula ρ. In the generic setting, the precondition $\sharp$ is that $\varphi \wedge \rho$ and $\varphi \wedge \rho \wedge C$ are satisfiability-equivalent.

FORGET Allows for *forgetting* a clause C from the current set of redundant clauses ρ.

STRENGTHEN Allows for *strengthening* φ by moving a clause C in the redundant formula $\rho \wedge C$ to φ.

WEAKEN Allows for *weakening* φ by moving a clause C in the current irredundant formula $\varphi \wedge C$ to ρ. In the generic setting, the precondition $\flat$ is that φ and $\varphi \wedge C$ are satisfiability-equivalent. (The literal l is related to instantiations of the rule based on specific redundancy properties, as further explained in Sect. 6 and Sect. 7.)

Notice that for unsatisfiable CNF formulas the generic precondition $\sharp$ allows for learning the empty clause to φ in a single step. Similarly, for satisfiable CNF formulas the

$$\frac{\varphi\,[\,\rho\,]\,\sigma}{\varphi\,[\,\rho \wedge C\,]\,\sigma}\ \sharp \qquad \frac{\varphi\,[\,\rho \wedge C\,]\,\sigma}{\varphi\,[\,\rho\,]\,\sigma} \qquad \frac{\varphi\,[\,\rho \wedge C\,]\,\sigma}{\varphi \wedge C\,[\,\rho\,]\,\sigma} \qquad \frac{\varphi \wedge C\,[\,\rho\,]\,\sigma}{\varphi\,[\,\rho \wedge C\,]\,\sigma, l{:}C}\ \flat$$

LEARN FORGET STRENGTHEN WEAKEN

Fig. 3. Inprocessing rules

generic precondition $\flat$ allows for weakening φ by moving *all* clauses in φ to ρ. However, in practice mostly polynomial-time checkable redundancy properties are of interest. Such properties are further discussed in Sect. 6 and Sect. 7.

Proposition 5. *The inprocessing rules in Fig. 3 are sound and complete in that:*

(i) *If F is unsatisfiable, then there is a derivation of an unsatisfiable $\varphi_k \wedge \rho_k$, where $k \geq 0$, from F using the rules (completeness).*

(ii) *If there is a derivation of an unsatisfiable $\varphi_k \wedge \rho_k$, where $k \geq 0$, from F using the rules, then F is unsatisfiable (soundness).*

Proof. (i) Since F is unsatisfiable, the LEARN rule can be used for learning the trivially unsatisfiable empty clause. (ii) We observe the following for any $i = 1, \ldots, k$. If LEARN is applied to enter state $\varphi_i\,[\,\rho_i\,]\,\sigma_i$ from $\varphi_{i-1}\,[\,\rho_{i-1}\,]\,\sigma_{i-1}$, by the precondition $\sharp$, φ_{i-1} and $\varphi_i \wedge \rho_i$ are satisfiability-equivalent. If STRENGTHEN or WEAKEN is applied, we have $\varphi_{i-1} \wedge \rho_{i-1} = \varphi_i \wedge \rho_i$. If FORGET is applied, we have $\varphi_{i-1} \wedge \rho_{i-1} \models \varphi_i \wedge \rho_i$. The claim then follows by induction on $i = k, \ldots, 1$. □

One could question whether the precondition $\sharp$ of LEARN, i.e., $\varphi \wedge \rho$ and $\varphi \wedge \rho \wedge C$ are satisfiability-equivalent, could be weakened to "φ and $\varphi \wedge C$ are satisfiability-equivalent". In other words, must the redundant clauses in ρ be taken into account for LEARN? To observe that ρ must indeed be included in $\sharp$, consider the CNF formula consisting of the single clause (a). From the initial state $a\,[\emptyset]\,\langle\rangle$ we obtain $\emptyset\,[a]\,\langle\rangle$ through WEAKEN. In case ρ were ignored in $\sharp$, it would then be possible to apply LEARN and derive $\emptyset\,[a \wedge \neg a]\,\langle\rangle$. However, this would violate the invariant of preserving satisfiability, since $a \wedge \neg a$ is unsatisfiable.

6 Instantiating the Rules Based on RAT

In contrast to the very generic preconditions $\sharp$ and $\flat$ under which the inprocessing rules were defined in the previous section, in practical SAT solving redundant clauses are learned and forgotten based on polynomial-time computable redundancy properties. In this section we give an instantiation of the inprocessing rules based on the polynomial-time computable property RAT. RAT is of special interest to us since, as will be shown in Sect. 7, known SAT solving techniques, including preprocessing, inprocessing, clause learning, and resolution, can be captured even when restricting the inprocessing rules using RAT. Moreover, under this property, a model of the original formula can be reconstructed in linear-time based on any model of any derivable φ_k using σ_k. This is important from the practical perspective due to the fact that in many applications a satisfying assignment for the original input formula F is required.

Preconditions Based on RAT. The preconditions of the inprocessing rules based on the property RAT are the following for a given state $\varphi_i\,[\,\rho_i\,]\,\sigma_i$.

LEARN: $\sharp$ is "C has RAT w.r.t. $\varphi_i \wedge \rho_i$".

Notice that LEARN under this precondition does not preserve logical equivalence. For example, consider the formula $F = (a \vee b)$. The LEARN rule can change $(a \vee b)\,[\emptyset]\,\langle\rangle$ into $(a \vee b)\,[C]\,\langle\rangle$, with $C = (\neg a \vee \neg b)$, since C has RAT on $\neg a$ w.r.t. F. The truth assignment $\tau = \{a = 1, b = 1\}$ satisfies F but does not satisfy $F \wedge C$.

WEAKEN: $\flat$ is "C has RAT on l w.r.t. φ_i".

Through weakening φ_i by moving a clause $C \in \varphi_i$ to ρ_{i+1}, the new φ_{i+1} may have more models than φ_i, since RAT does not preserve logical equivalence.

6.1 Solution Reconstruction

When the WEAKEN rule is used for a transition from a state $\varphi_i\,[\,\rho_i\,]\,\sigma_i$ to a state $\varphi_{i+1}\,[\,\rho_{i+1}\,]\,\sigma_{i+1}$, the set of models of φ_{i+1} can be a proper superset of the set of models of φ_i. For the practically relevant aspect of mapping any model of φ_{i+1} back to a model of φ_i, a literal pair $l{:}C$, where C is the clause moved from φ_i to ρ_{i+1}, is concatenated to the solution reconstruction stack σ_{i+1}. This is important when the redundancy property used does not guarantee preserving logical equivalence. More concretely, this is required if C has RAT but not e.g. AT.

For certain polynomial-time checkable redundancy properties, σ can be used for mapping models back to models of the original formula in linear time, as explained next. We describe a generic *model reconstruction algorithm* that can be applied in conjunction with the inprocessing rules in case the preconditions $\sharp$ and $\flat$ of LEARN and WEAKEN are restricted to RAT. In particular, for any CNF formula F and state $\varphi\,[\,\rho\,]\,\sigma$ that is reachable from $F\,[\,\emptyset\,]\,\langle\rangle$ using the inprocessing rules, given a model τ of φ, the reconstruction algorithm (Fig. 4) outputs a model of F solely based on σ and τ.

Reconstruction (literal-clause pair sequence σ, model τ of φ)
    while σ is not empty **do**
        remove the last literal-clause pair $l{:}C$ from σ
        if C is not satisfied by τ **then** $\tau := (\tau \setminus \{l = 0\}) \cup \{l = 1\}$
    return τ

Fig. 4. Pseudo-code of the model reconstruction algorithm

While the reconstruction algorithm may leave some variables unassigned in the output assignment (model of F), such variables can be arbitrarily assigned afterwards for establishing a full model of F.

Example 2. Consider the state $\varphi_i\,[\,\rho_i\,]\,\sigma_i$ with $\varphi_i = (a \vee b) \wedge (\neg a \vee \neg b)$, $\rho_i = \emptyset$ and $\sigma_i = \langle\rangle$. Apply WEAKEN to reach $\varphi_{i+1}\,[\,\rho_{i+1}\,]\,\sigma_{i+1}$, where $\varphi_{i+1} = (a \vee b)$, $\rho_{i+1} = (\neg a \vee \neg b)$, and $\sigma_{i+1} = \langle \neg a{:}(\neg a \vee \neg b)\rangle$. The assignment $\tau = \{a = 1, b = 1\}$ satisfies φ_{i+1} but not φ_i. The model reconstruction procedure will transform τ into $\{a = 0, b = 1\}$ which satisfies φ_i.

Proposition 6. *Given any CNF formula F, if a state $\varphi\,[\,\rho\,]\,\sigma$ is derivable from F using the inprocessing rules under preconditions based on* RAT*, then, given any model τ of φ,* Reconstruction(σ_i, τ) *returns a model of F.*

Proof. Follows from the proof of Proposition 1. Assume that $l{:}C$ is the last element in σ_i, τ is the current truth assignment, and that WEAKEN was applied to move C from φ_{i-1} to ρ_i based on the fact that C has RAT on l w.r.t. φ_{i-1}. By the proof of Proposition 1, there are at least two literals that are satisfied by τ in every clause containing $\neg l$ in $\varphi_{i-1} \setminus \{C\}$. Hence, in case $\tau(l) = 0$, we can flip this assignment to $\tau(l) = 1$. □

Interestingly, due to the generality of the inprocessing rules—as explained in the next section—this reconstruction algorithm covers model reconstruction for various simplification techniques that do not preserve logical equivalence, including specific reconstruction algorithms proposed for different cause elimination techniques [18,11] and variable elimination [18], and combinations thereof with other important techniques such as equivalence reasoning [24,3].

7 Capturing SAT Solving Techniques with the Inprocessing Rules

In this section we show how various existing inference techniques—including both known techniques and novel ideas—can be expressed as simple combinations of the LEARN, FORGET, STRENGTHEN, and WEAKEN rules. One should notice, however, that the inprocessing rules can be shown to naturally capture further inprocessing techniques. However, due to the page limit we are unable to discuss further techniques within this version of the paper. We also give examples of how incorrect variants of these techniques can be detected.

Clause elimination procedures based on redundancy property $\mathcal{P}$ can be expressed as deriving $\varphi\,[\,\rho\,]\,\sigma$ from $\varphi \wedge C\,[\,\rho\,]\,\sigma$ in a single step with the precondition that C has the property $\mathcal{P}$ w.r.t. φ. One step of clause elimination is simulated by two application steps of the inprocessing rules: 1. apply WEAKEN to move a redundant clause from φ to ρ; 2. apply FORGET to remove C from ρ. As explained in Sect. 6, the generic inprocessing rules can be instantiated using RAT as the redundancy property of the preconditions $\sharp$ and $\flat$. Since RAT covers all of the other clause redundancy properties discussed in Sect. 3 and 4 (such as blocked clauses, hidden tautologies, etc; also recall Fig. 2), it follows that all of the clause elimination procedures based on these properties are captured by our inprocessing rules, even when restricting the precondition to RAT.

As an example of incorrect clause elimination, consider the idea of eliminating C if it has the property $\mathcal{P}$ w.r.t. $\varphi \wedge \rho$ (and not w.r.t. just φ), allowing weakening φ based on ρ, i.e., also in case a clause in ρ subsumes C. This would allow using e.g. redundant learned clauses in ρ, which can be forgotten later on, for weakening φ. To see that this variant is incorrect consider $\varphi_i\,[\,\rho_i\,]\,\sigma_i$ where $\varphi_i = (a \vee \neg b) \wedge (\neg a \vee b) \wedge (\neg a \vee \neg b) \wedge (a \vee b \vee c) \wedge (a \vee b \vee \neg c)$ and $\rho_i = \emptyset$. Note that φ_i is unsatisfiable. The clause $(a \vee b)$ has AT w.r.t. φ_i, since $\text{ALA}(\varphi_i, (a \vee b))$ contains all literals, and hence applying LEARN gives $\varphi_{i+1} = \varphi_i$ and $\rho_{i+1} = (a \vee b)$. Now, $(a \vee b) \in \rho_{i+1}$ subsumes $(a \vee b \vee c) \in \varphi_{i+1}$, and incorrectly applying WEAKEN would give $\varphi_{i+2} = \varphi_{i+1} \setminus (a \vee b \vee c)$ and $\rho_{i+2} = \rho_{i+1} \wedge (a \vee b \vee c)$. However, φ_{i+2} is satisfiable, and the satisfiability-equivalence invariant is broken since $\varphi_{i+2} \wedge \rho_{i+2}$ is unsatisfiable. As a consequence, *it is not correct to use the clauses in* ρ *to eliminate an irrredundant clause* (such as hidden or asymmetric tautologies, blocked clauses, etc.), *unless the clauses, based on which the eliminated clause is redundant, are added to* φ *or are already part of* φ.

Pure Literal Elimination is an additional well-known clause elimination procedure: derive $\varphi\,[\,\rho\,]\,\sigma$ from $\varphi \wedge C\,[\,\rho\,]\,\sigma$ given that C contains a *pure literal* l (such that $\neg l$ does not appear in φ). It is easy to observe that this rule is also covered by our inprocessing rule: Any clause in φ that contains a pure literal l has RT (and thus RAT) on l w.r.t. φ. Notice that due to the WEAKEN precondition, only the irredundant clauses φ need to be considered, i.e., redundant (e.g., learned) clauses can still contain $\neg l$.

Clause addition procedures based on redundancy property $\mathcal{P}$ can be expressed as deriving $\varphi\,[\,\rho \wedge C\,]\,\sigma$ from $\varphi\,[\,\rho\,]\,\sigma$ in a single step with the precondition that C has the property $\mathcal{P}$ w.r.t. $\varphi \wedge \rho$. One step of clause addition is simulated by applying LEARN to add C to ρ. Similarly to clause elimination, the generic inprocessing rules can be instantiated using RAT as the redundancy property of the precondition $\sharp$. Again, since RAT covers all of the other clause redundancy properties discussed in Sect. 3 and 4, it follows that all of the clause addition procedures based on these properties are captured by the generic inprocessing rules.

Notice that some clause addition procedures do not preserve logical equivalence (recall Fig. 2), and hence can restrict the set of models of $\varphi \wedge \rho$. For such procedures, the inprocessing rules can be applied for checking correctness. As an example, consider *blocked clause addition* (BCA): for adding a clause C to ρ, it is required that C is blocked w.r.t. $\varphi \wedge \rho$. If C is only blocked w.r.t. φ, then BCA is not sound. Consider the formula $\varphi_0 = (a \vee \neg b) \wedge (\neg a \vee b) \wedge (a \vee c) \wedge (\neg c \vee b) \wedge (\neg a \vee \neg c)$. Notice that $(\neg a \vee \neg c)$ has RT (is blocked) on $\neg c$ w.r.t. φ_0. Hence $(\neg a \vee \neg c)$ can be moved from φ_0 to be part of ρ_1 by applying the WEAKEN rule: $\varphi_1 = \varphi_0 \setminus \{(\neg a \vee \neg c)\}$, $\rho_1 = \rho_0 \cup \{(\neg a \vee \neg c)\}$, and $\sigma_1 = \sigma_0 \cup \{\neg c{:}(\neg a \vee \neg c)\}$. Now the clause $(c \vee \neg b)$ is a RT on c w.r.t. φ_1, but not w.r.t. $\varphi_1 \wedge \rho_1$. Adding $(c \vee \neg b)$ to ρ to get $\rho_2 = \rho_1 \cup \{(c \vee \neg b)\}$ and $\varphi_2 = \varphi_1$ makes $\varphi_2 \wedge \rho_2$ unsatisfiable.

This brings us to an interesting observation of the framework. Continuing the above, if $(\neg a \vee \neg c)$ was removed (FORGET) after moving it to ρ (so $\rho_2 = \rho_1 \setminus \{(\neg a \vee \neg c)\}$, $\varphi_2 = \varphi_1$, and $\sigma_2 = \sigma_1$), then adding $(c \vee \neg b)$ to ρ via LEARN would be allowed ($\rho_3 = \rho_2 \setminus \{(\neg c \vee \neg b)\}$, $\varphi_3 = \varphi_2$, and $\sigma_3 = \sigma_2$) since $(c \vee \neg b)$ has RT on c w.r.t. $\varphi_3 \wedge \rho_3$. Now $\varphi_3 \wedge \rho_3 \wedge \mathrm{CNF}(\sigma_3)$ is unsatisfiable, where $\mathrm{CNF}(\sigma_3)$ is the conjunction of clauses in σ_3. Yet this does not cause a problem. The reconstruction method ensures that for every assignment satisfying φ a model of the original formula F can be constructed. Thus it also holds for assignments that satisfy $\varphi \wedge \rho$. *This illustrates that* LEARN *may add clauses to* ρ *that are not entailed by the clauses in the original formula.*

Clause Learning based on conflict graphs, which is central in modern CDCL solvers, can be simulated by the inprocessing rules. Since any conflict clause based on a conflict graph is derivable by trivial resolution from the current clause database [25], the inprocessing rules can simulate clause learning by simulating the steps of the resolution derivation, as explained next.

Resolution can also be simulated by the inprocessing rules in a straightforward way: For any φ, $(C \vee D)$ is an AT w.r.t. $\varphi \wedge (C \vee x) \wedge (D \vee \neg x)$, and thus $(C \vee D)$ can be learned by applying LEARN. This implies that *all* resolution-based simplification techniques can also be simulated. An example is *Hyper Binary Resolution* (HBR) [3]: Given a clause of the form $(l \vee l_1 \cdots \vee l_k)$ and k binary clauses of the form $(l' \vee \neg l_i)$, where $1 \leq i \leq k$, the hyper binary resolution rule allows to infer the *hyper binary resolvent* $(l \vee l')$ in one step. In essence, HBR simply encapsulates a sequence of specifically related resolution steps into one step.

Variable Elimination (VE) can also be simulated by our inprocessing rules. When applied in a bounded setting [7], VE is currently one of the most effective preprocessing techniques applied in SAT solvers. Variable elimination as a general version of VE for inprocessing can be characterized as the rule

$$\frac{\varphi \wedge \varphi_x \wedge \varphi_{\neg x} \,[\, \rho \wedge \rho_x \wedge \rho_{\neg x} \,]\, \sigma}{\varphi \wedge\ \varphi_x \otimes_x \varphi_{\neg x} \,[\, \rho \,]\, \sigma,\ x{:}\varphi_x, \neg x{:}\varphi_{\neg x}},$$

where F_l denotes the clauses in a CNF formula F that contain literal l, and $F_l \otimes_l F_{\neg l}$ is the lifting of the resolution operator to sets of clauses. Essentially, VE eliminates a variable x by producing all possible resolvents w.r.t. x, and removes at the same time all clauses containing x. Although not discussed in earlier work, our characterization takes into account the common practice that resolvents due to redundant clauses in ρ do not need to be produced.

To see that our inprocessing rules simulate VE, first apply LEARN to add the resolvents $\varphi_x \otimes \varphi_{\neg x}$ to ρ (all resolvents have AT w.r.t. φ). Second, apply STRENGTHEN to move the resolvents from ρ to φ. Now all clauses in φ_x have RS on x w.r.t. φ, and all clauses in $\varphi_{\neg x}$ have RS on $\neg x$ w.r.t. φ, and hence WEAKEN can be applied for making the clauses in φ_x and $\varphi_{\neg x}$ redundant, after which they can be removed using FORGET.

Notice that two variants of VE are distinguished [7]. The first, VE by *clause distribution* adds all the clauses of $\varphi_x \otimes \varphi_{\neg x}$ to φ. The second, VE by *substitution* adds only a subset of $\varphi_x \otimes \varphi_{\neg x}$ to φ in a satisfiability-preserving way. As a consequence, the latter variant may reduce the amount of unit propagations in the resulting formula compared to the former. However, under the inprocessing rules, the clauses produced by *clause distribution* but not by *substitution* can alternatively be added to ρ instead of φ, so that these clauses can be used subsequently for unit propagation but can still be considered redundant and thus be ignored in consecutive VE steps.

Partial Variable Elimination, as described below, is a novel variant of VE, which can also be naturally expressed via our inprocessing rules. Given a variable x and two subsets of clauses $S_x \subset \varphi_x$ and $S_{\neg x} \subset \varphi_{\neg x}$, if there are non-empty S_x and $S_{\neg x}$ such that all resolvents of $S_x \otimes (\varphi_{\neg x} \setminus S_{\neg x})$ and $S_{\neg x} \otimes (\varphi_x \setminus S_x)$ are tautologies, then we can apply VE *partially* by replacing $S_x \wedge S_{\neg x}$ in φ by $S_x \otimes S_{\neg x}$. We refer to this as *Partial Variable Elimination* (PVE). In practice, the VE rule is bounded by applying it only when the number of clauses is not increased. It is actually possible that PVE on x decreases the number of clauses, e.g., if $|S_x| = 1$ or $|S_{\neg x}| = 1$, while VE on x would increase the number of clauses. The correctness of PVE is immediate by the inprocessing rules, using a similar argument as in the case of VE.

Extended Resolution can also be simulated. This shows that LEARN, although perhaps not evident by its simple definition, is extremely powerful even when restricting the precondition to RAT only.

For a given CNF formula F, the *extension rule* [23] allows for iteratively adding definitions of the form $x \equiv a \wedge b$ (i.e. the CNF formula $(x \vee \neg a \vee \neg b) \wedge (\neg x \vee a) \wedge (\neg x \vee b)$) to F, where x is a new variable and a, b are literals in the current formula. The resulting formula $F \wedge E$ then consists of the original formula F and the *extension* E, the conjunction of the clauses iteratively added to F using the extension rule. In *Extended Resolution* [23] one can first apply the *extension rule* to add a conjunction of clauses (an *extension*) E to a CNF formula F, before using the resolution rule to construct a resolution proof of $F \wedge E$. This proof system is extremely powerful: surpassing the power of Resolution, it can even polynomially simulate extended Frege systems.

However, it is easy to observe that the LEARN rule simulates the extension rule: the clause $(x \vee \neg a \vee \neg b)$ has RAT on x w.r.t. $\varphi \wedge \rho$ and can thus be added to ρ by applying LEARN. The clauses $(\neg x \vee a)$ and $(\neg x \vee b)$ have RAT on $\neg x$ w.r.t. $\varphi \wedge (x \vee \neg a \vee \neg b) \wedge \rho$.

From a practical perspective, it follows that our inprocessing framework captures also the deduction applied in the recently proposed extensions of CDCL solvers that apply the Extension rule in a restricted fashion [26,27].

Finally, we would like to point out that the inprocessing rules capture various additional techniques that have proven important in practice. While we are unable (due to the page limit) to provide a more in-depth account of these techniques and how they are simulated by the inprocessing rules, such techniques include (as examples) *self-subsumption* (which has proven important both when combined with variable elimination [7] and when applied during search [28,29]), equivalence reasoning [24,3], including e.g. equivalent literal substitution, and also more recent techniques that can be defined for removing and adding literals from/to clauses (such as *hidden literal elimination* [13]).

8 Conclusion

Guaranteeing correctness of new inference techniques developed and implemented in state-of-the-art SAT solvers is becoming increasingly non-trivial as complex combinations of inference techniques are implemented within the solvers. We presented an abstract framework that captures the inference of inprocessing SAT solvers via four clean inference rules, providing a unified generic view to inprocessing, and furthermore captures sound solution reconstruction in a unified way. In addition to providing an in-depth understanding of the inferences underlying inprocessing solvers, we believe that this framework opens up possibilities for developing novel inprocessing and learning techniques that may lift the performance of SAT solvers even further.

References

1. Marques-Silva, J.P., Sakallah, K.A.: GRASP: a search algorithm for propositional satisfiability. IEEE Trans. Computers 48(5), 506–521 (1999)
2. Moskewicz, M.W., Madigan, C.F., Zhao, Y., Zhang, L., Malik, S.: Chaff: engineering an efficient SAT solver. In: Proc. DAC, pp. 530–535. ACM (2001)

3. Bacchus, F.: Enhancing Davis Putnam with extended binary clause reasoning. In: Proc. AAAI, pp. 613–619. AAAI Press (2002)
4. Bacchus, F., Winter, J.: Effective Preprocessing with Hyper-Resolution and Equality Reduction. In: Giunchiglia, E., Tacchella, A. (eds.) SAT 2003. LNCS, vol. 2919, pp. 341–355. Springer, Heidelberg (2004)
5. Subbarayan, S., Pradhan, D.K.: NiVER: Non-increasing Variable Elimination Resolution for Preprocessing SAT Instances. In: H. Hoos, H., Mitchell, D.G. (eds.) SAT 2004. LNCS, vol. 3542, pp. 276–291. Springer, Heidelberg (2005)
6. Gershman, R., Strichman, O.: Cost-Effective Hyper-Resolution for Preprocessing CNF Formulas. In: Bacchus, F., Walsh, T. (eds.) SAT 2005. LNCS, vol. 3569, pp. 423–429. Springer, Heidelberg (2005)
7. Eén, N., Biere, A.: Effective Preprocessing in SAT Through Variable and Clause Elimination. In: Bacchus, F., Walsh, T. (eds.) SAT 2005. LNCS, vol. 3569, pp. 61–75. Springer, Heidelberg (2005)
8. Jin, H., Somenzi, F.: An incremental algorithm to check satisfiability for bounded model checking. Electronic Notes in Theoretical Computer Science 119(2), 51–65 (2005)
9. Han, H., Somenzi, F.: Alembic: An efficient algorithm for CNF preprocessing. In: Proc. DAC, pp. 582–587. IEEE (2007)
10. Piette, C., Hamadi, Y., Saïs, L.: Vivifying propositional clausal formulae. In: Proc. ECAI, pp. 525–529. IOS Press (2008)
11. Heule, M.J.H., Järvisalo, M., Biere, A.: Clause Elimination Procedures for CNF Formulas. In: Fermüller, C.G., Voronkov, A. (eds.) LPAR-17. LNCS, vol. 6397, pp. 357–371. Springer, Heidelberg (2010)
12. Järvisalo, M., Biere, A., Heule, M.J.H.: Simulating circuit-level simplifications on CNF. Journal of Automated Reasoning (2012); OnlineFirst 2011
13. Heule, M.J.H., Järvisalo, M., Biere, A.: Efficient CNF Simplification Based on Binary Implication Graphs. In: Sakallah, K.A., Simon, L. (eds.) SAT 2011. LNCS, vol. 6695, pp. 201–215. Springer, Heidelberg (2011)
14. Heule, M.J.H., Järvisalo, M., Biere, A.: Covered clause elimination. In: LPAR-17 Short Papers (2010), http://arxiv.org/abs/1011.5202
15. Biere, A.: P{re,i}coSAT@SC 2009. In: SAT 2009 Competitive Event Booklet (2009)
16. Soos, M.: CryptoMiniSat 2.5.0, SAT Race 2010 solver description (2010)
17. Biere, A.: Lingeling, Plingeling, PicoSAT and PrecoSAT at SAT Race 2010. FMV Technical Report 10/1, Johannes Kepler University, Linz, Austria (2010)
18. Järvisalo, M., Biere, A.: Reconstructing Solutions after Blocked Clause Elimination. In: Strichman, O., Szeider, S. (eds.) SAT 2010. LNCS, vol. 6175, pp. 340–345. Springer, Heidelberg (2010)
19. Nieuwenhuis, R., Oliveras, A., Tinelli, C.: Solving SAT and SAT modulo theories: From an abstract Davis-Putnam-Logemann-Loveland procedure to DPLL(T). Journal of the ACM 53(6), 937–977 (2006)
20. Nieuwenhuis, R., Oliveras, A.: On SAT Modulo Theories and Optimization Problems. In: Biere, A., Gomes, C.P. (eds.) SAT 2006. LNCS, vol. 4121, pp. 156–169. Springer, Heidelberg (2006)
21. Larrosa, J., Nieuwenhuis, R., Oliveras, A., Rodríguez-Carbonell, E.: A framework for certified boolean branch-and-bound optimization. Journal of Automated Reasoning 46(1) (2011)
22. Andersson, G., Bjesse, P., Cook, B., Hanna, Z.: A proof engine approach to solving combinational design automation problems. In: Proc. DAC, pp. 725–730. ACM (2002)
23. Tseitin, G.S.: On the complexity of derivation in propositional calculus. In: Automation of Reasoning 2, pp. 466–483. Springer (1983)
24. Li, C.M.: Integrating equivalency reasoning into Davis-Putnam procedure. In: Proc. AAAI, pp. 291–296. AAAI Press (2000)

25. Beame, P., Kautz, H.A., Sabharwal, A.: Towards understanding and harnessing the potential of clause learning. J. Artif. Intell. Res. 22, 319–351 (2004)
26. Audemard, G., Katsirelos, G., Simon, L.: A restriction of extended resolution for clause learning SAT solvers. In: Proc. AAAI. AAAI Press (2010)
27. Huang, J.: Extended clause learning. Artificial Intelligence 174(15), 1277–1284 (2010)
28. Han, H., Somenzi, F.: On-the-Fly Clause Improvement. In: Kullmann, O. (ed.) SAT 2009. LNCS, vol. 5584, pp. 209–222. Springer, Heidelberg (2009)
29. Hamadi, Y., Jabbour, S., Saïs, L.: Learning for dynamic subsumption. In: Proc. ICTAI, pp. 328–335. IEEE (2009)

Logical Difference Computation with CEX2.5

Boris Konev, Michel Ludwig, and Frank Wolter

Department of Computer Science, University of Liverpool, United Kingdom
{konev,mludwig,wolter}@liverpool.ac.uk

Abstract. We present a new version of the CEX versioning tool for ontologies. CEX detects logical differences between acyclic terminologies in the lightweight description logic $\mathcal{EL}$ with role inclusions and domain and range restrictions. Depending on the application, CEX outputs differences between terminologies that capture derived concept inclusions, answers to instance queries, and answers to conjunctive queries. Experiments with versions of the NCI ontology are conducted to evaluate the performance of CEX and compare the three types of differences.

Keywords: Description Logics, Ontology Versioning, Logical Difference.

1 Introduction

In life sciences, healthcare, and other knowledge intensive areas, large scale terminologies are employed to provide a common vocabulary for a domain of interest together with descriptions of the meaning of terms built from the vocabulary and relationships between them. Two examples are the medical terminology SNOMED CT which contains more than 300 000 term definitions [6] and the National Cancer Institute ontology (NCI) consisting of more than 60 000 axioms [4]. Terminologies of this size and complexity cannot be developed and maintained without adequate automated versioning support. As a consequence, the development of ontology versioning tools and theoretical foundations for versioning have become a popular and an important research problem [5, 7, 9, 14–16].

In this paper we give an update on the CEX versioning tool which is the only purely logic-based tool for ontology versioning. The first version of CEX was presented in [10] and was able to compute a logical difference between acyclic $\mathcal{EL}$ terminologies that captures the different concept inclusions that follow from the two terminologies. More precisely, for any two acyclic $\mathcal{EL}$ terminologies and any signature Σ relevant for the comparison between the two terminologies, CEX computed a finite representation of the different concept inclusions over Σ that follow from one terminology but not the other. Recently, ontology based data access has become a major application of ontologies in general, and of $\mathcal{EL}$ terminologies in particular [12, 13, 17]. In this case, it is not sufficient to compare the derived concept inclusions of terminologies, but answers to instance queries or even conjunctive queries should be considered as well. Thus, we have extended CEX so as to cover three distinct types of logical differences: differences w.r.t. concept inclusions, answers to instance queries, and answers to conjunctive queries. Moreover, CEX now admits role inclusions and range and domain restrictions,

B. Gramlich, D. Miller, and U. Sattler (Eds.): IJCAR 2012, LNAI 7364, pp. 371–377, 2012.

and so acyclic $\mathcal{ELH}^r$ terminologies rather than only acyclic $\mathcal{EL}$ terminologies can be compared. The algorithms and theory behind CEX are presented in [11]. In contrast to the update presented here, the version of CEX discussed in [11] cannot compute differences w.r.t. conjunctive queries. In this paper, we therefore focus on experiments that show how moving from concept and instance queries to conjunctive queries influences the performance of CEX and the number of differences detected between distinct versions of NCI.

2 Preliminaries

An $\mathcal{ELH}^r$*-terminology* $\mathcal{T}$ is a finite set of role inclusions $r \sqsubseteq s$ and concept inclusions and equations of the form $A \sqsubseteq C$, $A \equiv C$, $\mathsf{ran}(r) \sqsubseteq C$, and $\exists r.\top \sqsubseteq C$ such that no concept name occurs more than once on the left-hand side, where A is a *concept name*, r, s are *role names*, $\mathsf{ran}(r)$ refers to the *range* of the role r and C, D are $\mathcal{EL}$*-concepts*, that is, expressions of the form $C := A \mid \top \mid C \sqcap D \mid \exists r.C$. (Complete definitions can be found in [11], see also [1] where $\mathcal{ELH}^r$ was introduced.) $\mathcal{T}$ is *acyclic* if no defined concept is used (directly or indirectly) in its definition. Instance data are represented by *ABox assertions* of the form $A(a)$ and $r(a,b)$, where a, b are *individual names*, A is a concept name and r is a role name. An *ABox* $\mathcal{A}$ is a non-empty finite set of ABox-assertions. The semantics of $\mathcal{ELH}^r$ can be given by interpreting terminologies and ABoxes as first-order (FO) sentences where concepts are formulas with one free variable, roles are binary predicates, and individual names are constants. For example, the inclusion $A \sqsubseteq \exists rB$ can be interpreted as $\forall x(A(x) \Rightarrow \exists y(r(x,y) \land B(y)))$. We use $\mathcal{T} \models \varphi$, or $(\mathcal{T}, \mathcal{A}) \models \varphi$, to denote that φ follows from $\mathcal{T}$, or $\mathcal{T} \cup \mathcal{A}$, respectively, in FO. An *instance query* α is of the form $r(a,b)$ or $C(a)$ with C an $\mathcal{EL}$-concept. α is *atomic* if it only contains one concept or role name. A *conjunctive query (CQ)* is a FO-formula $q(\boldsymbol{x}) = \exists \boldsymbol{y} \psi(\boldsymbol{x}, \boldsymbol{y})$, where ψ is constructed from atoms $A(t)$ and $r(t,t')$ using conjunction and t, t' range over individual names and individual variables from the sequences of variables $\boldsymbol{x}, \boldsymbol{y}$. $q(\boldsymbol{x})$ is *atomic* if ψ only contains one atom. A *signature* Σ is a finite set of concept and role names, and a Σ-concept (Σ-query, etc.) is a concept (query, etc.) that only uses concept and role names from Σ.

3 The CEX2.5 System

CEX2.5[1] takes as input two acyclic $\mathcal{ELH}^r$ terminologies $\mathcal{T}_1$, $\mathcal{T}_2$ and a signature Σ and analyses the following three types of logical difference:

- the Σ*-concept difference* between $\mathcal{T}_1$ and $\mathcal{T}_2$ is the set $\mathsf{cDiff}_\Sigma(\mathcal{T}_1, \mathcal{T}_2)$ of all Σ-role and Σ-concept inclusions α in $\mathcal{ELH}^r$ such that $\mathcal{T}_1 \models \alpha$ and $\mathcal{T}_2 \not\models \alpha$;
- the Σ*-instance difference* between $\mathcal{T}_1$ and $\mathcal{T}_2$ is the set $\mathsf{iDiff}_\Sigma(\mathcal{T}_1, \mathcal{T}_2)$ of pairs of the form $(\mathcal{A}, \alpha)$, where $\mathcal{A}$ is a Σ-ABox and α a Σ-instance query such that $(\mathcal{T}_1, \mathcal{A}) \models \alpha$ and $(\mathcal{T}_2, \mathcal{A}) \not\models \alpha$; and

[1] Available under an open-source license at `http://www.csc.liv.ac.uk/~michel/software/cex2/`

- the Σ*-query difference* between $\mathcal{T}_1$ and $\mathcal{T}_2$ is the set $\mathsf{qDiff}_\Sigma(\mathcal{T}_1, \mathcal{T}_2)$ of pairs $(\mathcal{A}, q(\boldsymbol{a}))$, where $\mathcal{A}$ is a Σ-ABox, $q(\boldsymbol{x})$ a Σ-CQ, and $\boldsymbol{a}$ a tuple of individual names in $\mathcal{A}$ such that $(\mathcal{T}_1, \mathcal{A}) \models q(\boldsymbol{a})$ and $(\mathcal{T}_2, \mathcal{A}) \not\models q(\boldsymbol{a})$.

If for one of these types of logical difference, the Σ-difference between $\mathcal{T}_i$ and $\mathcal{T}_j$ is empty for $\{i, j\} = \{1, 2\}$, then the two terminologies can be regarded as equivalent and replaced by each other in applications that use Σ-symbols only and for which the considered type of difference is appropriate. Notice that, for all three types of logical difference, if the Σ-difference between terminologies is not empty, then it is infinite. We distinguish between two modes in which CEX presents an approximation of this infinite Σ-difference to the user. First, it is shown in [11] that within every member of the Σ-difference, one can find an "elementary" difference which is either a role inclusion or

- for $\mathsf{cDiff}_\Sigma(\mathcal{T}_1, \mathcal{T}_2)$: a concept inclusion $C \sqsubseteq D$ in which either C is a concept name or an expression of the form $\mathsf{ran}(r)$ or $\exists r.\top$; or D is a concept name;
- for $\mathsf{iDiff}_\Sigma(\mathcal{T}_1, \mathcal{T}_2)$: a pair $(\mathcal{A}, \alpha)$ in which either $\mathcal{A}$ is a singleton ABox or α an atomic instance query;
- for $\mathsf{qDiff}_\Sigma(\mathcal{T}_1, \mathcal{T}_2)$: a pair $(\mathcal{A}, \alpha)$ in which either $\mathcal{A}$ is a singleton ABox or α an atomic CQ.

We call C and $\mathcal{A}$ the left-hand side of such an elementary difference and D and α its right-hand side. To abstract away from individuals/variables, the concept or role name of the atomic (or singleton) left or right-hand side of such an elementary difference is termed a Σ*-difference witness*. One can show that every Σ-concept difference witness is a Σ-instance difference witness is a Σ-query difference witness. Moreover, every left-hand side Σ-instance difference witness is a left-hand side Σ-concept difference witness, and every right-hand side Σ-query difference witness is a right-hand side Σ-instance difference witness.

Example 1. Consider the following terminologies $\mathcal{T}_1$ and $\mathcal{T}_2$

$$\begin{array}{llll} \mathcal{T}_1: & A \equiv \exists r.(A_1 \sqcap B_2) & \qquad \mathcal{T}_2: & A \sqsubseteq \exists r.(A_1 \sqcap B_2) \\ & A_2 \sqsubseteq B_2 & & A_2 \sqsubseteq B_2 \\ & E \sqsubseteq \exists s.F & & E \sqsubseteq \exists r_1.\top \sqcap \exists r_2.\top \\ & s \sqsubseteq r_1,\ s \sqsubseteq r_2 & & \end{array}$$

and signature $\Sigma = \{A, A_1, A_2, E, r, r_1, r_2\}$. Then

1) A is the only Σ-concept difference witness (and it is a right-hand side witness): it is a Σ-concept difference witness since the inclusion $\exists r.(A_1 \sqcap A_2) \sqsubseteq A$ is an elementary difference (observe that $\mathcal{T}_1 \models \exists r.(A_1 \sqcap A_2) \sqsubseteq A$ but $\mathcal{T}_2 \not\models \exists r.(A_1 \sqcap A_2) \sqsubseteq A$). No other Σ-concept difference witness exists since one can show that all elementary members of $\mathsf{cDiff}_\Sigma(\mathcal{T}_1, \mathcal{T}_2)$ have A on its right-hand side.

2) Similarly, A is the only Σ-instance difference witness (and it is again a right-hand side witness): an elementary difference is given by the pair $(\mathcal{A}_1, q_1)$, where $\mathcal{A}_1 = \{r(a, b), A_1(b), A_2(b)\}$ and $q_1 = A(a)$. No elementary Σ-instance difference without the atom A on its right-hand side exists.

3) The Σ-query difference witnesses are given by A and E (and A is a right-hand side witness while E is a left-hand side witness): in this case one can show that all elementary Σ-query differences either have the query $A(a)$ for some a on the right-hand side or the ABox $\{E(a)\}$ for some a on the left-hand side. Examples of such differences are the two pairs $(\mathcal{A}_1, q_1)$, $(\mathcal{A}_2, q_2)$ where $(\mathcal{A}_1, q_1)$ is as above, and $\mathcal{A}_2 = \{E(a)\}$ and $q_2 = \exists x(r_1(a,x) \wedge r_2(a,x))$. For the same terminologies and $\Sigma = \{E, F\}$, one can see that there are neither concept nor instance difference witnesses; however, as $(\mathcal{T}_1, \{E(a)\}) \models \exists x.F(x)$ but $(\mathcal{T}_2, \{E(a)\}) \not\models \exists x.F(x)$, there is a Σ-query difference between the terminologies and E is its (left-hand side) witness.

Note that the set of Σ-difference witnesses is uniquely determined by $\mathcal{T}_1, \mathcal{T}_2$ and Σ and gives a rather abstract description of the Σ-difference. This set is empty iff no Σ-difference exists and can be computed in polynomial time, for all three types of queries [11]. In its basic mode, CEX2.5 computes the set of all Σ-concept, instance and query witnesses and presents them (together with the information whether they are left or right-hand side witnesses) to the user. For a more detailed analysis of the Σ-difference between the two input terminologies $\mathcal{T}_1$ and $\mathcal{T}_2$, in its advanced mode CEX2.5 can also compute *examples* of elementary members of $\mathsf{cDiff}_\Sigma(\mathcal{T}_1, \mathcal{T}_2)$, $\mathsf{iDiff}_\Sigma(\mathcal{T}_1, \mathcal{T}_2)$, and $\mathsf{qDiff}_\Sigma(\mathcal{T}_1, \mathcal{T}_2)$ which illustrate why certain concept names are concept, instance, or query difference witnesses.

4 Experimental Results

In [11], we have conducted a detailed experimental evaluation of the performance of CEX2.5 in the concept and instance difference case. In this report we, therefore, focus on the CQ case and (a) compare the performance of CEX2.5 for the CQ case with its performance for the concept and instance case, and (b) compare the number of difference witnesses detected in the CQ case with the number of difference witnesses detected in the concept/instance case. The CEX2.5 system is implemented in OCaml, and it uses the reasoner CB [8] internally as classification engine. The experiments were conducted on PCs equipped with an Intel Core i5-2500 CPU and 4 GiB of main memory.

First, CEX2.5 is used to compare 71 consecutive acyclic $\mathcal{ELH}^r$-versions of the NCI Thesaurus.[2] For any two consecutive versions NCI_n and NCI_{n+1} within the considered range, we computed all instance and query difference witnesses together with corresponding examples for $\mathcal{T}_1 = \mathrm{NCI}_{n+1}$ and $\mathcal{T}_2 = \mathrm{NCI}_n$ on signatures $\Sigma = \mathsf{sig}(\mathrm{NCI}_n) \cap \mathsf{sig}(\mathrm{NCI}_{n+1})$. The results are given in Table 1, where only those comparisons are reproduced for which there are query difference witnesses which are not instance difference witnesses. The first two columns give the NCI versions, $|\mathsf{qRhs}_\Sigma(\cdot,\cdot)|$ is the number of right-hand side Σ-query difference witnesses (which always coincides with the number $|\mathsf{iRhs}_\Sigma(\cdot,\cdot)|$ of right-hand side Σ-instance difference witnesses). $|\mathsf{qLhs}_\Sigma(\cdot,\cdot)|$ and $|\mathsf{iLhs}_\Sigma(\cdot,\cdot)|$ are the number of left-hand side

[2] Full versions are available from `http://evs.nci.nih.gov/ftp1/NCI_Thesaurus/`. We refer the reader to [5] for additional information on NCI versions and note that the full versions contain inclusions that are not in acyclic $\mathcal{ELH}^r$.

Table 1. Detailed Results for Comparisons Between Consecutive $\mathcal{ELH}^r$-versions of the NCI Thesaurus Leading to Additional Conjunctive Query Differences

$\mathcal{T}_1$	$\mathcal{T}_2$	$\lvert\mathsf{qRhs}_\Sigma(\mathcal{T}_1,\mathcal{T}_2)\rvert = \lvert\mathsf{iRhs}_\Sigma(\mathcal{T}_1,\mathcal{T}_2)\rvert$	$\lvert\mathsf{qLhs}_\Sigma(\mathcal{T}_1,\mathcal{T}_2)\rvert$	$\lvert\mathsf{iLhs}_\Sigma(\mathcal{T}_1,\mathcal{T}_2)\rvert$	Time (s) (query)	Time (s) (instance)
03.12e	03.12a	49	1747	289	177.78	14.85
04.03n	04.02h	431	8277	5494	14.02	13.74
04.06i	04.05f	99	1147	1080	48.31	39.50
05.05d	05.03d	1007	2683	747	513.78	17.17
06.01c	05.12f	798	2066	2053	449.8	22.39
07.01d	06.12d	814	290	222	41.19	40.99

query and instance difference witnesses, respectively. One can see that in some cases there are significantly more query difference witnesses than instance difference witnesses. In fact, one of the conclusions one can draw from this table is that the instance difference between terminologies is not necessarily a good approximation of the query difference. Consequently, when terminologies are used to access instance data using CQs, a comparison at the concept or instance level cannot always replace an analysis tailored for CQs. Secondly, one can see that the time required to compute query witnesses can be significantly longer than the time necessary for detecting instance witnesses; we will comment on the reasons below.

To analyse the impact of the size of signatures on the running time of CEX2.5 and on the number of difference witnesses, in our second experiment CEX2.5 is used to compute concept, instance and query difference witnesses together with corresponding examples for $\mathcal{T}_1 = \mathrm{NCI}_{06.12d}$ and $\mathcal{T}_2 = \mathrm{NCI}_{07.01d}$ on randomly-generated signatures $\Sigma \subseteq \mathsf{sig}(\mathcal{T}_1) \cap \mathsf{sig}(\mathcal{T}_2)$. The signatures were composed of a varying number of concept names and 60 randomly-selected roles. For each considered sample size of concept names we generated 10 random signatures. The computation times and the number of difference witnesses that were detected on average for each sample size are depicted in Fig. 1. First note that in the concept and instance difference case the computation times on average never differed by more than one second and the same number of difference witnesses were computed. As in the previous experiment, on average there are significantly more query difference witnesses than concept and instance difference witnesses. Moreover, in contrast to the concept and instance difference case, the computation time in the query difference case increased with the number of concept names present in the considered signatures. Less than 287 MiB of memory were required in each of the comparisons involving NCI versions.

To evaluate the performance of CEX2.5 on very large terminologies, we compare three consecutive versions of SNOMED CT (January 2009, July 2009, and January 2010). We used CEX2.5 to compute instance and query difference witnesses with and without examples on the shared signature between two consecutive versions. All three versions of SNOMED CT considered have the same role names which are, therefore, also in the shared signature. In contrast to the experiments for NCI, in this case the set of query difference witnesses turned out to coincide with the set of instance difference witnesses and the computation times almost coincide: on average, 683 seconds for the instance witnesses and 672 seconds for the CQ witnesses.

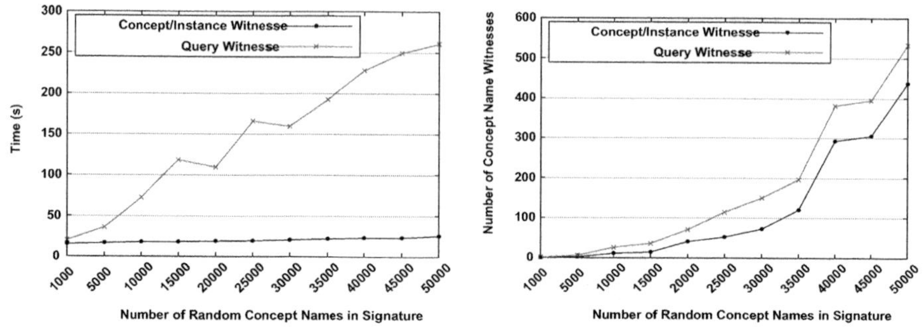

Fig. 1. Computation Time Required and Number of Difference Witnesses Detected between two Consecutive NCI Versions on Random Signatures

The running time rose to 1028 seconds and, respectively, 1006 seconds when examples were additionally computed. On average 2.84 GiB and, respectively, 2.92 GiB of memory were required for the computation.

Finally, in the experiments above, 687 813 examples of elementary differences between terminologies were computed. The average length (i.e., the number of occurrences of concept and role names) of an example was 5.98 with a maximal length of 98. It follows that in most cases the examples generated by CEX2.5 are sufficiently small to be analysed by a human user (note that, in theory, in the worst case minimal examples are of exponential size [11]).

We close with a discussion as to why in the NCI experiments computing left-hand side Σ-query difference witnesses takes longer than computing left-hand side Σ-instance difference witnesses (and why this is not the case for SNOMED CT). To check whether $A \in \Sigma$ is such a witness for $\mathcal{T}_1, \mathcal{T}_2$ both algorithms check whether there is a certain *Σ-simulation* between the minimal models $\mathcal{I}_{\mathcal{T}_1,\{A(a)\}}$ and $\mathcal{I}_{\mathcal{T}_2,\{A(a)\}}$ for the knowledge bases $(\mathcal{T}_1, \{A(a)\})$ and $(\mathcal{T}_2, \{A(a)\})$ [11,12]. The difference between the two cases is that for the instance difference witnesses a "standard" Σ-simulation between the node for a in $\mathcal{I}_{\mathcal{T}_1,\{A(a)\}}$ and the node for a in $\mathcal{I}_{\mathcal{T}_2,\{A(a)\}}$ is sufficient, whereas for the query difference the simulation has to, in addition, respect *intersections between Σ-roles* and has to be *global* (every node in $\mathcal{I}_{\mathcal{T}_1,\{A(a)\}}$ has to be simulated). The second condition is costly since it implies that one has to consider all nodes of $\mathcal{I}_{\mathcal{T}_1,\{A(a)\}}$ and find simulating nodes in $\mathcal{I}_{\mathcal{T}_2,\{A(a)\}}$ rather than consider nodes reachable from the node for a via Σ-paths only. In general, it therefore appears to be unavoidable that computation times for CQ are longer than for concept and instance queries. The SNOMED CT experiment is different: in this case Σ contains all role names in both terminologies and so any simulation of the node for a is a global simulation already.

We note that because of their importance in model checking and abstraction, a large variety of highly optimized algorithms computing simulations between Kripke models have been developed (e.g. [2,3]). In our implementation, however, we do *not* first construct the (potentially very large) minimal models and then check for Σ-simulation, but we check for Σ-simulation on-the-fly making heavy use of the condition that $\mathcal{T}_1$ and $\mathcal{T}_2$ are acyclic terminologies.

References

1. Baader, F., Brandt, S., Lutz, C.: Pushing the $\mathcal{EL}$ envelope further. In: Proceedings of OWLED 2009. CEUR Workshop Proceedings, vol. 529. CEUR-WS.org (2008)
2. Crafa, S., Ranzato, F., Tapparo, F.: Saving space in a time efficient simulation algorithm. Fundamenta Informaticae 108(1-2), 23–42 (2011)
3. van Glabbeek, R.J., Ploeger, B.: Correcting a Space-Efficient Simulation Algorithm. In: Gupta, A., Malik, S. (eds.) CAV 2008. LNCS, vol. 5123, pp. 517–529. Springer, Heidelberg (2008)
4. Golbeck, J., Fragaso, G., Hartel, F., Hendler, J., Oberhaler, J., Parsia, B.: The national cancer institute's thesaurus and ontology. Journal of Web Semantics 1(1), 75–80 (2003)
5. Gonçalves, R.S., Parsia, B., Sattler, U.: Analysing multiple versions of an ontology: A study of the NCI thesaurus. In: Proceedings of DL 2011. CEUR Workshop Proceedings, vol. 745. CEUR-WS.org (2011)
6. IHTSDO: SNOMED Clinical Terms User Guide. The International Health Terminology Standards Development Organisation (IHTSDO) (2008)
7. Jiménez-Ruiz, E., Grau, B.C., Horrocks, I., Llavori, R.B.: Supporting concurrent ontology development: Framework, algorithms and tool. Data & Knowledge Engineering 70(1), 146–164 (2011)
8. Kazakov, Y.: Consequence-driven reasoning for Horn SHIQ ontologies. In: Proceedings of IJCAI 2009, pp. 2040–2045 (2009)
9. Klein, M., Fensel, D., Kiryakov, A., Ognyanov, D.: Ontology Versioning and Change Detection on the Web. In: Gómez-Pérez, A., Benjamins, V.R. (eds.) EKAW 2002. LNCS (LNAI), vol. 2473, pp. 197–259. Springer, Heidelberg (2002)
10. Konev, B., Walther, D., Wolter, F.: The Logical Difference Problem for Description Logic Terminologies. In: Armando, A., Baumgartner, P., Dowek, G. (eds.) IJCAR 2008. LNCS (LNAI), vol. 5195, pp. 259–274. Springer, Heidelberg (2008)
11. Konev, B., Ludwig, M., Walther, D., Wolter, F.: The logical diff for the lightweight description logic $\mathcal{EL}$ (submitted), `http://www.csc.liv.ac.uk/~frank/publ/publ.html`
12. Lutz, C., Toman, D., Wolter, F.: Conjunctive query answering in the description logic $\mathcal{EL}$ using a relational database system. In: Proceedings of IJCAI 2009, pp. 2070–2075. AAAI Press, Menlo Park (2009)
13. Mei, J., Liu, S., Xie, G., Kalyanpur, A., Fokoue, A., Ni, Y., Li, H., Pan, Y.: A Practical Approach for Scalable Conjunctive Query Answering on Acyclic $\mathcal{EL}^+$ Knowledge Base. In: Bernstein, A., Karger, D.R., Heath, T., Feigenbaum, L., Maynard, D., Motta, E., Thirunarayan, K. (eds.) ISWC 2009. LNCS, vol. 5823, pp. 408–423. Springer, Heidelberg (2009)
14. Noy, N.F., Musen, M.A.: PromptDiff: A fixed-point algorithm for comparing ontology versions. In: Proceedings of AAAI 2002, pp. 744–750. AAAI Press, Menlo Park (2002)
15. Oliver, D.E., Shahar, Y., Shortliffe, E.H., Musen, M.A.: Representation of change in controlled medical terminologies. Artificial Intelligence in Medicine 15(1), 53–76 (1999)
16. Palma, R., Haase, P., Corcho, O., Gómez-Pérez, A.: Change representation for OWL 2 ontologies. In: Proceedings of OWLED 2009. CEUR Workshop Proceedings, vol. 529. CEUR-WS.org (2009)
17. Poggi, A., Lembo, D., Calvanese, D., Giacomo, G.D., Lenzerini, M., Rosati, R.: Linking data to ontologies. Journal of Data Semantics 10, 133–173 (2008)

Overview and Evaluation of Premise Selection Techniques for Large Theory Mathematics

Daniel Kühlwein, Twan van Laarhoven, Evgeni Tsivtsivadze, Josef Urban, and Tom Heskes*

Radboud University Nijmegen

Abstract. In this paper, an overview of state-of-the-art techniques for premise selection in large theory mathematics is provided, and new premise selection techniques are introduced. Several evaluation metrics are introduced, compared and their appropriateness is discussed in the context of automated reasoning in large theory mathematics. The methods are evaluated on the MPTP2078 benchmark, a subset of the Mizar library, and a 10% improvement is obtained over the best method so far.

1 Introduction: Formal Mathematics and Its AI Methods

In recent years, more and more mathematics is becoming available in a computer-understandable form [12]. A number of large formalization projects are progressing [11,13,15], formal mathematics is considered by influential mathematicians,[1] and new approaches and proof assistants are discussed and developed by interested newcomers [3,8,22,39].

As this happens, the users and developers of formal mathematics are increasingly faced with the problem of searching for relevant formal knowledge, analogous to the search problems started since the early days of the Internet. Web search has led to a large body of research of robust and scalable non-semantic methods in fields like information retrieval, machine learning, and data mining. On the other hand, formal mathematics has been traditionally focusing on exhaustive and precise deductive search methods, typically used on small, carefully manually pre-arranged search space. In nutshell, the difference between the former and the latter methods is that the former focus on heuristically finding knowledge that could be most relevant for solving semantically underspecified (typically natural language) queries, while the latter methods try to find a precise answer and a proof for a conjecture that is expressed with full semantic precision. The former methods are largely *inductive* and *data-driven* [29]: the "solutions" are unconfirmed suggestions derived from heuristics and previous evidence, and essential parts of the algorithms are typically obtained by learning from large corpora. The latter methods have so far been largely *deductive*

* The authors were supported by the NWO projects "MathWiki a Web-based Collaborative Authoring Environment for Formal Proofs" and "Learning2Reason".

[1] `http://gowers.wordpress.com/2008/07/28/more-quasi-automatic-theorem-proving`

B. Gramlich, D. Miller, and U. Sattler (Eds.): IJCAR 2012, LNAI 7364, pp. 378–392, 2012.

and *theory-driven*: the solutions are deduced in a logically correct way, and the algorithms are to a large extent specified by their programmers without inducing major parts of the programs from large datasets.

There are a number of interesting ways how existing deductive methods can be improved in the presence of previous knowledge. Some of them are mentioned below. An early research done by the Munich group [10] has already produced a number of ideas and advanced implementations, like for example the pattern-based proof guidance in E prover [26]. It seems that the recently appeared large corpora of formal knowledge allow AI combinations of inductive and deductive thinking (e.g., the MaLARea metasystem [37]) that can hardly be tried in other AI domains that lack precise semantics and the notion of formal proof.

On the other hand, the first and lasting necessity obvious since 2003, when the first large formal mathematical corpora became available to Automated Theorem Provers (ATPs), has been good selection of relevant premises for a given new conjecture. It has been shown that proper design and choice of knowledge selection heuristics can change the overall success of large-theory ATP techniques by tens of percents [1]. Such large improvements provide an incentive for further research, evaluations, and benchmarks developing the field of knowledge-based automated reasoning.

This paper develops this field in several ways. First, in Section 2 an overview of state-of-the-art techniques for premise selection in large-theory mathematics is provided, focusing on premise ranking. In Section 3 we present several relevant machine learning metrics developed for feasible training and evaluation of ranking algorithms. The premise selection methods are evaluated on the MPTP2078 benchmark in section 4, using the machine learning metrics as well as several different ATPs. The learning and ATP evaluation methods are compared, and the relevance of the machine learning metrics based on human-proof data is discussed, together with the performance of different methods and ATPs. Based on the findings, use of ensemble methods for aggregating premise selectors is proposed and initially tested in Section 5, and shown to further raise the overall ATP performance by 10% in comparison to the best method so far. Section 6 concludes and proposes directions for further research.

2 Premise Selection Algorithms

2.1 Premise Selection Setting

The typical setting for the task of premise selection is a large developed library of formally encoded mathematical knowledge, over which mathematicians attempt to prove new lemmas and theorems [5, 32, 36]. The actual mathematical corpora suitable for ATP techniques are only a fraction of all mathematics (e.g. about 50000 lemmas and theorems in the Mizar library) and started to appear only recently, but they already provide a corpus on which different heuristic methods can be defined, trained, and evaluated. Premise selection can be useful as a standalone service for the formalizers (suggesting relevant lemmas), or in conjunction with ATP methods that can attempt to find a proof from the relevant premises.

2.2 Learning-Based Ranking Algorithms

Learning-based ranking algorithms have a training and a testing phase and typically represent the data as points in pre-selected feature spaces. In the training phase the algorithm tries to fit one (or several) prediction functions to the data it is given. The result of the training is the best fitting prediction function which can then be used in the testing phase for evaluations.

In the typical setting presented above, the algorithms would train on all existing proofs in the library and be tested on the new theorem the mathematician wants to prove. We compare three different algorithms.

SNoW: SNoW (Sparse Network of Winnows) [6] is an implementation of (among others) the naive Bayes algorithm that has already been successfully used for premise selection (see e.g. [1, 32, 33]).

Naive Bayes is a statistical learning method based on Bayes' theorem with a strong (or naive) independence assumption. Given a new conjecture c and a premise p, SNoW computes the probability of p being needed to prove c, based on the previous use of p in proving conjectures that are similar to c. The similarity is in our case typically expressed using symbols and terms of the formulas. The independence assumption says that the (non-)occurrence of a symbol/term is not related to the (non-)occurrence of every other symbol/term.

MOR-CG: MOR-CG (Multi-Output Ranking Conjugate Gradient) is a kernel-based learning algorithm (see [29]) that is a new variation of our MOR algorithm described in [1]. The difference between MOR and MOR-CG are that MOR-CG uses a linear kernel instead of a Gaussian. Furthermore, MOR-CG uses conjugate-gradient descent to speed up the time needed for training.

Kernel-based algorithms do not aim to model probabilities, but instead try to minimize the expected loss of the prediction functions on the training data. For each premise p MOR-CG tries to find a function C_p such that for each conjecture c, $C_p(c) = 1$ iff p was used in the proof of c. Given a new conjecture c, we can evaluate the learned prediction functions C_p on c. The higher the value $C_p(c)$ the more relevant p is to prove c.

BiLi: BiLi (Bi-Linear) is a new algorithm that is based on a bilinear model of premise selection, similar to the work of Chu and Park [7]. Like MOR-CG, BiLi aims to minimize the expected loss. The difference lies in the kind of prediction functions they produce. In MOR-CG the prediction functions only take the features[2] of the conjecture into account. In BiLi, the prediction functions use the features of both the conjectures and the premises.[3] The bilinear model learns a weight for each combination of a conjecture feature together with a premise feature. Together, this weighted combination determines whether or not a premise is relevant to the conjecture.

---

[2] In our experiments each feature indicates the presence or absence of a certain symbol or term in a formula.

[3] This makes BiLi a bit closer to methods like SInE that symbolically compare conjectures with premises.

When the number of features becomes large, fitting a bilinear model becomes computationally more challenging. Therefore, in BiLi the number of features is first reduced to 100, using random projections [4]. To combat the noise introduced by these random projections, this procedure is repeated 20 times, and the averaged predictions are used for ranking the premises.

2.3 Other Algorithms Used in the Evaluation

SInE: The SInE (SUMO Inference Engine) is a heuristic state-of-the-art premise selection algorithm by Kryštof Hoder [14], recently also implemented in the E prover.[4] The basic idea is to use global frequencies of symbols in a problem to define their global *generality*, and build a relation linking each symbol S with all formulas F in which S is has the lowest global generality among the symbols of F. In common-sense ontologies, such formulas typically *define* the symbols linked to them, which is the reason for calling this relation a *D-relation.* Premise selection for a conjecture is then done by recursively following the D-relation, starting with the conjecture's symbols. For the experiments described here the E implementation of SInE has been used, because it can be instructed to select exactly N most relevant premises. This is compatible with the way how other premise rankers are used here, and it allows to compare the premise rankings produced by different algorithms for increasing values of N.[5]

Aprils: The Automated Prophesier of Relevance Incorporating Latent Semantics (APRILS) [24] is a signature-based premise selection method that employs Latent Semantic Analysis (LSA) [9] to define symbol and premise similarity. Latent semantics is a machine learning method that has been successfully used for example in the Netflix Prize,[6] and in web search. Its principle is to automatically derive "semantic" equivalence classes of words (like *car, vehicle, automobile*) from their co-occurrences in documents, and to work with such equivalence classes instead of the original words. In APRILS, formulas define the symbol co-occurrence, each formula is characterized as a vector over the symbols' equivalence classes, and the premise relevance is its dot product with the conjecture.

2.4 Techniques Not Included in the Evaluation

As a part of the overview, we also list important or interesting algorithms used for ATP knowledge selection that for various reasons do not fit the evaluation done here. Because of space contraints we refer readers to [34] for their discussion.

- The default premise selection heuristic used by the Isabelle/Sledgehammer export [17]. This is an Isabelle-specific symbol-based technique similar to SInE that would need to be evaluated on Isabelle data.

[4] `http://www.mpi-inf.mpg.de/departments/rg1/conferences/deduction10/slides/stephan-schulz.pdf`

[5] The exact parameters used for producing the E-SInE rankings are at `https://raw.github.com/JUrban/MPTP2/master/MaLARea/script/filter1`.

[6] `http://www.netflixprize.com`

- Goal directed ATP calculi including the *Conjecture Symbol Weight* clause selection heuristics in E prover [27] giving lower weights to symbols contained in the conjecture, the Set of Support (SoS) strategy in resolution/superposition provers, and tableau calculi like leanCoP [19] that are in practice goal-oriented.
- Model-based premise selection, as done by Pudlák's semantic axiom selection system for large theories [21], by the SRASS metasystem [30], and in a different setting by the MaLARea [37] metasystem.
- MaLARea [37] is a large-theory metasystem that loops between deductive proof and model finding (using ATPs and finite model finders), and learning premise-selection (currently using SNoW or MOR-CG) from the proofs and models to attack the conjectures that still remain to be proved.
- Abstract proof trace guidance implemented in the E prover by Stephan Schulz for his PhD [26]. Proofs are abstracted into clause patterns collected into a common knowledge base, which is loaded when a new problem is solved, and used for guiding clause selection. This is also similar to the *hints* technique in Prover9 [16].
- The MaLeCoP system [38] where the clause relevance is learned from all closed tableau branches, and the tableau extension steps are guided by a trained machine learner that takes as input features a suitable encoding of the literals on the current tableau branch.

3 Machine Learning Evaluation Metrics

Given a database of proofs, there are several possible ways to evaluate how good a premise selection algorithm is without running an ATP. Such evaluation metrics are used to estimate the best parameters (e.g. regularization, tolerance, step size) of an algorithm. We want to use the evaluation metrics that are the best indicators for the final ATP performance. The input for each metric is a ranking of the premises for a conjecture together with the information which premises where used to prove the conjecture (according to the training data).

Recall. *Recall@n* is a value between 0 and 1 and denotes the fraction of used premises that are among the top n highest ranked premises.

$$\text{Recall@}n = \frac{|\{\text{used premises}\} \cap \{n \text{ highest ranked premises}\}|}{|\{\text{used premises}\}|}$$

Recall@n is always less than *Recall@*$(n+1)$. As n increases, *Recall@n* will eventually converge to 1. Our intuition is that the better the algorithm, the faster its *Recall@n* converges to 1.

AUC. The AUC (Area under the ROC Curve) is the probability that, given a randomly drawn used premise and a randomly drawn unused premise, the used premise is ranked higher than the unused premise. Values closer to 1 show better performance.

Let $x_1, .., x_n$ be the ranks of the used premises and $y_1, .., y_m$ be the ranks of the unused premises. Then, the AUC is defined as

$$\text{AUC} = \frac{\sum_i^n \sum_j^m 1_{x_i > y_j}}{mn}$$

where $1_{x_i > y_j} = 1$ iff $x_i > y_j$ and zero otherwise.

100%Recall. *100%Recall* denotes the minimum n such that Recall@$n = 1$.

$$100\%\text{Recall} = \min\{n \mid \text{Recall@}n = 1\}$$

In other words *100%Recall* tells us how many premises (starting from the highest ranked one) we need to give to the ATP to ensure that all necessary premises are included.

4 Evaluation

4.1 Evaluation Data

The premise selection methods are evaluated on the large (*chainy*) problems from the MPTP2078 benchmark[7] [1]. These are 2078 related large-theory problems (conjectures) and 4494 formulas (conjectures and premises) in total, extracted from the Mizar Mathematical Library (MML). The MPTP2078 benchmark was developed to supersede the older and smaller MPTP Challenge benchmark (developed in 2006), while keeping the number of problems manageable for experimenting. Larger evaluations are possible,[8] but not convenient when testing a large number of systems with many different settings. MPTP2078 seems sufficiently large to test various hypotheses and find significant differences.

MPTP2078 also contains (in the smaller, *bushy* problems) for each conjecture the information about the premises used in the MML proof. This can be used to train and evaluate machine learning algorithms using a chronological order emulating the growth of MML. For each conjecture, the algorithms are allowed to train on all MML proofs that were done up to that conjecture.[9] For each of the 2078 problems, the algorithms predict a ranking of the premises.

4.2 Machine Learning Evaluation – Comparison of Predictions with Known Proofs

We first compare the algorithms introduced in section 2 using the machine learning evaluation metrics introduced in section 3. All evaluations are based on the training data, the human-written formal proofs from the MML. They do not take the possibility of alternative proofs into account.[10]

[7] Available at `http://wiki.mizar.org/twiki/bin/view/Mizar/MpTP2078`.

[8] See [2,35] for recent evaluations spanning the whole MML.

[9] This in particular means that the algorithms do not train on the data they were asked to predict.

[10] This could be improved in the future by adding alternative proofs, as discussed in section 6.

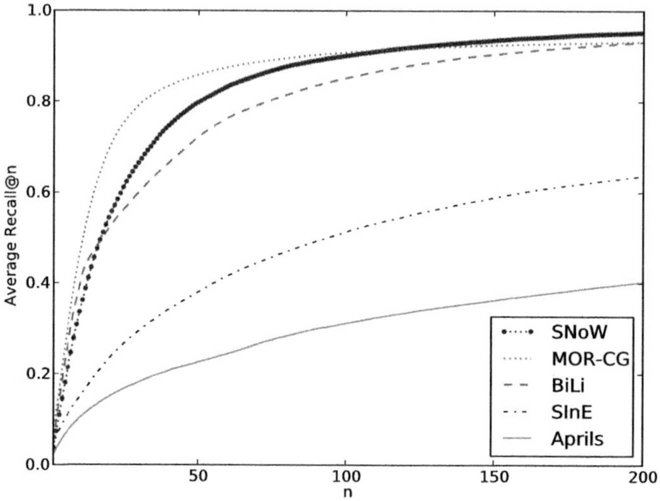

Fig. 1. Recall comparison of the premise selection algorithms

Recall. Figure 1 compares the average Recall@n of MOR-CG, BiLi, SNoW, SInE and Aprils for the top 200 premises over all 2078 problems. Higher values denote better performance. The graph shows that MOR-CG performs best, and Aprils worst. Note that here is a sharp distinction between the learning algorithms, which use the MML proofs and eventually reach a very similar recall, and the heuristic-based algorithms Aprils and SInE.

AUC. The average AUC of the premise selection algorithms is reported in table 1. Higher values mean better performance, i.e. a higher chance that a used premise is higher ranked than a unused premise. SNoW (97%) and BiLi (96%) have the best AUC scores with MOR-CG taking the third spot with an AUC of 88%. Aprils and SInE are considerably worse with 64% and 42% respectively. The standard deviation is very low with around 2% for all algorithms.

Table 1. AUC comparison of the premise selection algorithms

Algorithm	Avg. AUC	Std.
SNoW	0.9713	0.0216
BiLi	0.9615	0.0215
MOR-CG	0.8806	0.0206
Aprils	0.6443	0.0176
SInE	0.4212	0.0142

100%Recall. The comparison of the 100%Recall measure values can be seen in figure 2. For the first 115 premises, MOR-CG is the best algorithm. From then on, MOR-CG hardly increases and SNoW takes the lead. Eventually, BiLi almost catches up with MOR-CG. Again we can see a big gap between the performance of the learning and the heuristic algorithms with SInE and Aprils not even reaching 400 problems with 100%Recall.

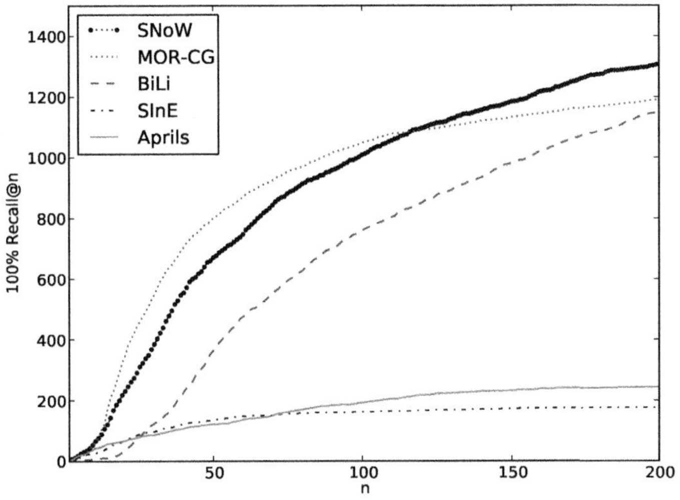

Fig. 2. 100%Recall comparison of the premise selection algorithms

Discussion. In all three evaluation metrics there is a clear difference between the performance of the learning-based algorithms SNoW, MOR-CG and BiLi and the heuristic-based algorithms SInE and Aprils. If the machine-learning metrics on the MML proofs are a good indicator for the ATP performance then there should be a corresponding performance difference in the number of problems solved. We investigate this in the following section.

4.3 ATP Evaluation

Vampire. In the first experiment we combined the rankings obtained from the algorithms introduced in section 2 with version 0.6 of the ATP Vampire[11] [23]. For each MPTP2078 problem (containing on average 1976.5 premises), we created 20 new problems, containing the 10, 20, ..., 200 highest ranked premises. The results can be seen in figure 3.

[11] All ATPs are run with 5s time limit on an Intel Xeon E5520 2.27GHz server with 24GB RAM and 8MB CPU cache. Each problem is always assigned one CPU. We use Vampire as our default ATP because of its good preformance in the CASC competitions, and because of its good performance on MML reported in [35].

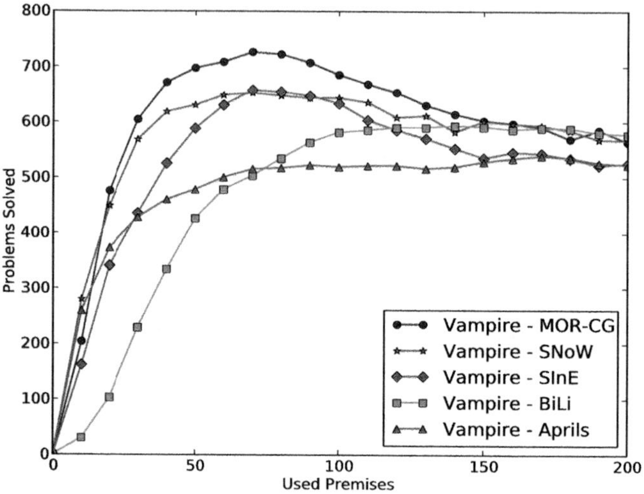

Fig. 3. Problems solved – Vampire

Apart from the first 10-premise batch and the three last batches, MOR-CG always solves the highest number of problems with a maximum of 726 problems with the top 70 premises. SNoW solves less problems in the beginning, but catches up in the end. BiLi solves very few problems in the beginning, but gets better as more premises are given and eventually is as good as SNoW and MOR-CG. The surprising fact (given the machine learning performance) is that SInE performs very well, on par with SNoW in the range of 60-100 premises. This indicates that SInE finds proofs that are very different from the human proofs. Furthermore, it is worth noting that most algorithms have their peak at around 70-80 premises. It seems that after that, the effect of increased premise recall is beaten by the effect of the growing ATP search space.

E, SPASS and Z3. We also compared the top three algorithms, MOR-CG, SNoW and SInE, with three other ATPs: E [27] (version 1.4), SPASS [40] (version 3.7) and Z3 [18] (version 3.2). The results can be seen in figure 4, 5, 6 respectively. In all three experiments, MOR-CG gives the best results. Looking at the number of problems solved by E we see that SNoW and SInE solve about the same number of problems when more than 50 premises are given. In the SPASS evaluation, SInE performs better than SNoW after the initial 60 premises. The results for Z3 are clearer, with (apart from the first run with the top 10 premises) MOR-CG always solving more problems than SNoW, and SNoW solving more problems than SInE. It is worth noting that independent of the learning algorithm, SPASS solves the fewest problems and Z3 the most, and that (at least up to the limit of 200 premises used) Z3 is hardly affected by having too many premises in the problems.

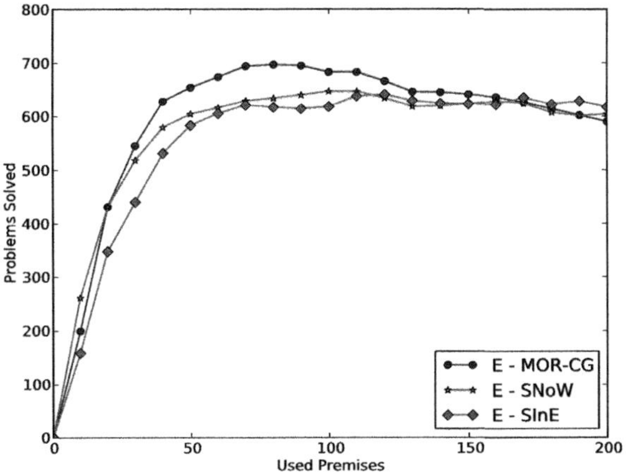

Fig. 4. Problems solved – E

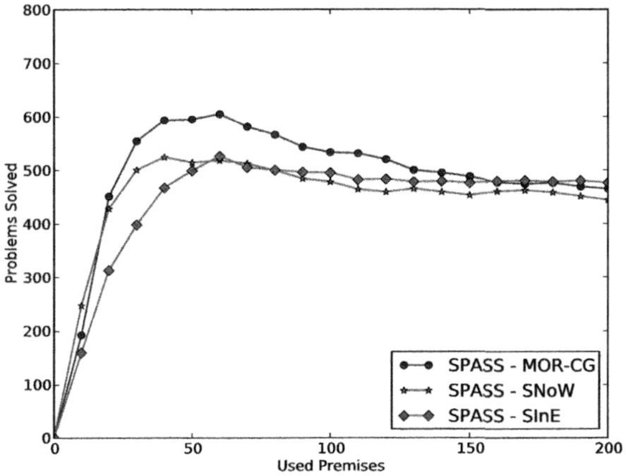

Fig. 5. Problems solved – SPASS

Discussion. The ATP evaluation shows that a good ML evaluation performance does not necessarily imply a good ATP performance and vice versa. E.g. SInE performs better than expected, and BiLi worse. A plausible explanation for this is that the human-written proofs that are the basis of the learning algorithms are not the best possible guidelines for ATP proofs, because there are a number of good alternative proofs: the total number of problems proved with Vampire by the union of all prediction methods is 1197, which is more (in 5s) than the 1105 problems that Vampire can prove in 10s when using only the premises used exactly in the human-written proofs. One possible way how to test this hypothesis (to a certain extent at least) would be to train the learning algorithms on all the ATP proofs that are found, and test whether the ML evaluation performance closer correlates with the ATP evaluation performance.

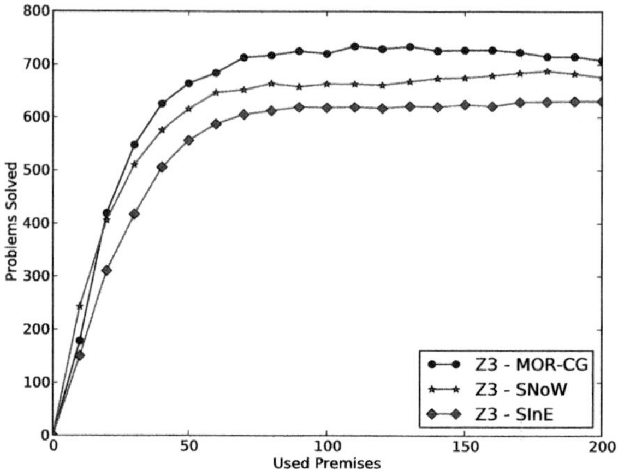

Fig. 6. Problems solved – Z3

The most successful 10s combination, solving 939 problems, is to run Z3 with the 130 best premises selected by MOR-CG, together with Vampire using the 70 best premises selected by SInE. It is also worth noting that when we consider all provers and all methods, 1415 problems can be solved.

It seems the heuristic and the learning based premise selection methods give rise to different proofs. In the next section, we try to exploit this by considering combinations of ranking algorithms.

5 Combining Premise Rankers

There is clear evidence about alternative proofs being feasible from alternative predictions. This should not be too surprising, because the premises are organized into a large derivation graph, and there are many explicit (and also quite likely many yet-undiscovered) semantic dependencies among them.

The evaluated premise selection algorithms are based on different ideas of similarity, relevance, and functional approximation spaces and norms in them. This also means that they can be better or worse in capturing different aspects of the premise selection problem (whose optimal solution is obviously undecidable in general, and intractable even if we impose some finiteness limits).

An interesting machine learning technique to try in this setting is the combination of different predictors. There has been a large amount of machine learning research in this area, done under different names. *Ensembles* is one of the most frequent, a recent overview of ensemble based systems is given in [20], while for example [28] deals with the specific task of aggregating rankers.

As a final experiment that opens the premise selection field to the application of advanced ranking-aggregation methods, we have performed an initial simple evaluation of combining two very different premise ranking methods: MOR-CG and SInE. The aggregation is done by simple weighted linear combination, i.e., the final ranking is obtained via weighted linear combination of the predicted individual rankings. We test a limited grid of weights, in the interval of [0, 1] with a step value of 0.25, i.e., apart from the original MOR-CG and SInE rankings we get three more weighted aggregate rankings as follows: $0.25 * \text{CG} + 0.75 * \text{SInE}$, $0.5 * \text{CG} + 0.5 * \text{SInE}$, and $0.75 * \text{CG} + 0.25 * \text{SInE}$. The following Figure 7 shows their ATP evaluation.

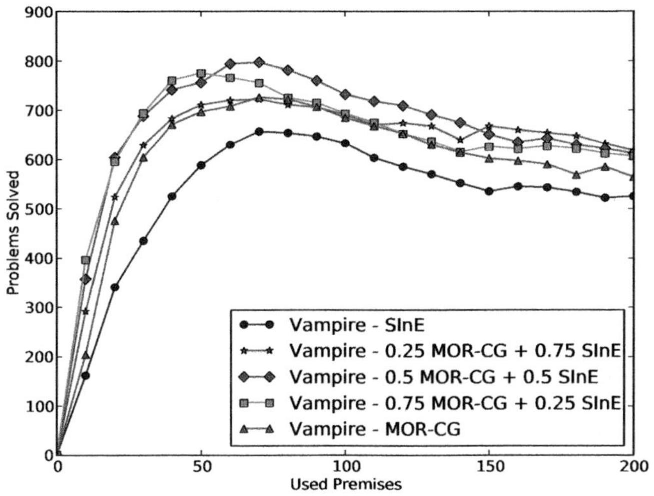

Fig. 7. Combining CG and SInE: Problems solved

The machine learning evaluation (done as before against the data extracted from the human proofs) is not surprising, and the graphs (which we omit due to space constraints) look like linear combinations of the corresponding figures for MOR-CG and SInE. The ATP evaluation (only Vampire was used) is a very different case. For example the equally weighted combination of MOR-CG and SInE solves over 604 problems when using only the top 20 ranked premises. The corresponding values for standalone MOR-CG resp. SInE are 476, resp. 341, i.e., they are improved by 27%, resp. 77%. The equally weighted combination solves 797 when using the top 70 premises, which is a 10% improvement over the best result of all methods (726 problems solved by MOR-CG when using the top 70 premises). Note that unlike the external combination mentioned above, this is done only in 5 seconds, with only one ATP, one premise selector, and one threshold.

6 Conclusion and Future Work

Heuristic and inductive methods seem indispensable for strong automated reasoning in large formal mathematics, and significant improvements can be achieved by their proper design, use and combination with precise deductive methods.

Knowing previous proofs and learning from them turns out to be important not just to mathematicians, but also for automated reasoning in large theories.

The possibility of the ultimate semantic (ATP) evaluation of the proposed premise rankings adds interesting "combined AI" aspects to the standard machine learning methods. Without expressive semantics the methods just try to predict the human proofs as closely as possible. Proposing alternative (and sometimes simpler) proofs is discouraged by the standard machine learning evaluation metrics. This produces interesting questions to AI researchers: given the explicit derivation graph of a large theory, and the precise semantics allowing this graph to grow further, what are good methods and metrics for (reasonably fast) training of premise selection methods? One pragmatic answer that we can give is to develop a growing database of ATP and human proofs, and other results (like counter-models), e.g., in a similar way as in the MaLARea metasystem, and use this growing database for training instead of just the human proofs, testing (and caching for further use) the ATP validity of new predictions on-demand.

We have evaluated practically all reasonably fast state-of-the-art premise selection techniques, tried some new ones, and currently experiment with more. This has produced a large amount of data on the most suitable (most orthogonal) combinations of premise selection systems, numbers of premises used, ATPs used, and ATP (currently E prover) strategies used. We further use machine learning in the spirit of the E-MaLeS system to determine optimal (either parallel or lower time-limit) combinations of these. These results are not included here due to space constraints.[12]

There is a trade-off between the precision and speed of the methods, and an interesting future work is to use fast methods as pre-selectors for more expensive methods. This is related to the problems of automated clustering the large theories, that can also be useful by itself for organizing and searching the large formal repositories. Clustering on a finer level is also one of the methods that could be used to further improve premise selection. It is quite likely that there are clusters of theorems that have the same logic power (their conjunctions are equal in the Lindenbaum algebra), and about the same strength when used with ATPs (the same conjecture can be proved from them in a similar number of steps). The current premise selection methods will likely recommend all such equivalent sets, which is blocking other (possibly necessary) premises, so heuristic identification of such (nearly) equivalent sets seems important. Including more semantics (for example evaluation in an evolving set of models as in MaLARea) in the learning and selection process could be one way how to achieve this.

We would like to make our strongest methods useful to as many formal mathematicians as possible. Some of them (like SNoW) have been used for MML and MPTP since 2003, but algorithms like MOR-CG and aggregated rankers are not deployed yet. We also hope to evaluate and deploy the algorithms at least for the Isabelle/Sledgehammer framework in near future.

[12] The fact that Z3 solves largely orthogonal sets of problems to Vampire is probably well known by now. Hence our focus on the differences between the premise selection methods.

References

1. Alama, J., Heskes, T., Kühlwein, D., Tsivtsivadze, E., Urban, J.: Premise Selection for Mathematics by Corpus Analysis and Kernel Methods. CoRR abs/1108.3446 (2011)
2. Alama, J., Kühlwein, D., Urban, J.: Automated and Human Proofs in General Mathematics: An Initial Comparison. In: Bjørner, N., Voronkov, A. (eds.) LPAR-18 2012. LNCS, vol. 7180, pp. 37–45. Springer, Heidelberg (2012)
3. Bem, J.: The Zermelo proof checker, `http://zermelo.org/`
4. Bingham, E., Mannila, H.: Random projection in dimensionality reduction: applications to image and text data. In: KDD, pp. 245–250 (2001)
5. Blanchette, J.C., Bulwahn, L., Nipkow, T.: Automatic Proof and Disproof in Isabelle/HOL. In: Tinelli, C., Sofronie-Stokkermans, V. (eds.) FroCos 2011. LNCS, vol. 6989, pp. 12–27. Springer, Heidelberg (2011)
6. Carlson, A., Cumby, C., Rosen, J., Roth, D.: The SNoW Learning Architecture. Tech. Rep. UIUCDCS-R-99-2101, UIUC Computer Science Department (May 1999), `http://cogcomp.cs.illinois.edu/papers/CCRR99.pdf`
7. Chu, W., Park, S.T.: Personalized recommendation on dynamic content using predictive bilinear models. In: Quemada, J., León, G., Maarek, Y.S., Nejdl, W. (eds.) WWW, pp. 691–700. ACM (2009)
8. Cramer, M., Fisseni, B., Koepke, P., Kühlwein, D., Schröder, B., Veldman, J.: The Naproche Project Controlled Natural Language Proof Checking of Mathematical Texts. In: Fuchs, N.E. (ed.) CNL 2009. LNCS, vol. 5972, pp. 170–186. Springer, Heidelberg (2010)
9. Deerwester, S.C., Dumais, S.T., Landauer, T.K., Furnas, G.W., Harshman, R.A.: Indexing by Latent Semantic Analysis. JASIS 41(6), 391–407 (1990)
10. Denzinger, J., Fuchs, M., Goller, C., Schulz, S.: Learning from Previous Proof Experience. Technical Report AR99-4, Institut für Informatik, Technische Universität München (1999)
11. Gonthier, G.: Advances in the Formalization of the Odd Order Theorem. In: van Eekelen, M., Geuvers, H., Schmaltz, J., Wiedijk, F. (eds.) ITP 2011. LNCS, vol. 6898, p. 2. Springer, Heidelberg (2011)
12. Hales, T.C.: Mathematics in the age of the Turing machine. Lecture Notes in Logic (to appear, 2012), `http://www.math.pitt.edu/~thales/papers/turing.pdf`
13. Hales, T.C., Harrison, J., McLaughlin, S., Nipkow, T., Obua, S., Zumkeller, R.: A revision of the proof of the Kepler Conjecture. Discrete & Computational Geometry 44(1), 1–34 (2010)
14. Hoder, K., Voronkov, A.: Sine Qua Non for Large Theory Reasoning. In: Bjørner, N., Sofronie-Stokkermans, V. (eds.) CADE 2011. LNCS, vol. 6803, pp. 299–314. Springer, Heidelberg (2011)
15. Klein, G., Andronick, J., Elphinstone, K., Heiser, G., Cock, D., Derrin, P., Elkaduwe, D., Engelhardt, K., Kolanski, R., Norrish, M., Sewell, T., Tuch, H., Winwood, S.: seL4: formal verification of an operating-system kernel. Commun. ACM 53(6), 107–115 (2010)
16. McCune, W.: Prover9 and Mace4 (2005-2010), `http://www.cs.unm.edu/~mccune/prover9/`
17. Meng, J., Paulson, L.C.: Lightweight relevance filtering for machine-generated resolution problems. J. Applied Logic 7(1), 41–57 (2009)
18. de Moura, L., Bjørner, N.: Z3: An Efficient SMT Solver. In: Ramakrishnan, C.R., Rehof, J. (eds.) TACAS 2008. LNCS, vol. 4963, pp. 337–340. Springer, Heidelberg (2008)

19. Otten, J., Bibel, W.: leanCoP: lean connection-based theorem proving. J. Symb. Comput. 36(1-2), 139–161 (2003)
20. Polikar, R.: Ensemble based systems in decision making. IEEE Circuits and Systems Magazine 6(3), 21–45 (2006)
21. Pudlak, P.: Semantic Selection of Premisses for Automated Theorem Proving. In: Sutcliffe et al. [31]
22. Reichelt, S.: The HLM proof assistant, http://hlm.sourceforge.net/
23. Riazanov, A., Voronkov, A.: The design and implementation of VAMPIRE. AI Commun. 15(2-3), 91–110 (2002)
24. Roederer, A., Puzis, Y., Sutcliffe, G.: Divvy: An ATP Meta-system Based on Axiom Relevance Ordering. In: Schmidt [25], pp. 157–162
25. Schmidt, R.A. (ed.): CADE-22. LNCS, vol. 5663. Springer, Heidelberg (2009)
26. Schulz, S.: Learning search control knowledge for equational deduction. DISKI, vol. 230. Infix Akademische Verlagsgesellschaft (2000)
27. Schulz, S.: E - A Brainiac Theorem Prover. AI Commun. 15(2-3), 111–126 (2002)
28. Sculley, D.: Rank aggregation for similar items. In: SDM. SIAM (2007)
29. Shawe-Taylor, J., Cristianini, N.: Kernel Methods for Pattern Analysis. Cambridge University Press, New York (2004)
30. Sutcliffe, G., Puzis, Y.: SRASS - A Semantic Relevance Axiom Selection System. In: Pfenning, F. (ed.) CADE 2007. LNCS (LNAI), vol. 4603, pp. 295–310. Springer, Heidelberg (2007)
31. Sutcliffe, G., Urban, J., Schulz, S. (eds.): Proceedings of the CADE-21 Workshop on Empirically Successful Automated Reasoning in Large Theories, Bremen, Germany, July 17. CEUR Workshop Proceedings, vol. 257. CEUR-WS.org (2007)
32. Urban, J.: MizarMode—an integrated proof assistance tool for the Mizar way of formalizing mathematics. Journal of Applied Logic 4(4), 414–427 (2006)
33. Urban, J.: MaLARea: a metasystem for automated reasoning in large theories. In: Sutcliffe et al. [31]
34. Urban, J.: An Overview of Methods for Large-Theory Automated Theorem Proving (Invited Paper). In: Höfner, P., McIver, A., Struth, G. (eds.) ATE Workshop. CEUR Workshop Proceedings, vol. 760, pp. 3–8. CEUR-WS.org (2011)
35. Urban, J., Hoder, K., Voronkov, A.: Evaluation of Automated Theorem Proving on the Mizar Mathematical Library. In: Fukuda, K., van der Hoeven, J., Joswig, M., Takayama, N. (eds.) ICMS 2010. LNCS, vol. 6327, pp. 155–166. Springer, Heidelberg (2010)
36. Urban, J., Rudnicki, P., Sutcliffe, G.: ATP and presentation service for Mizar formalizations. CoRR abs/1109.0616 (2011)
37. Urban, J., Sutcliffe, G., Pudlák, P., Vyskočil, J.: MaLARea SG1 - Machine Learner for Automated Reasoning with Semantic Guidance. In: Armando, A., Baumgartner, P., Dowek, G. (eds.) IJCAR 2008. LNCS (LNAI), vol. 5195, pp. 441–456. Springer, Heidelberg (2008)
38. Urban, J., Vyskočil, J., Štěpánek, P.: MaLeCoP Machine Learning Connection Prover. In: Brünnler, K., Metcalfe, G. (eds.) TABLEAUX 2011. LNCS, vol. 6793, pp. 263–277. Springer, Heidelberg (2011)
39. Voevodsky, V.: Univalent foundations project (2010) (manuscript), http://www.math.ias.edu/~vladimir/Site3/Univalent_Foundations.html
40. Weidenbach, C., Dimova, D., Fietzke, A., Kumar, R., Suda, M., Wischnewski, P.: SPASS Version 3.5. In: Schmidt [25], pp. 140–145

Branching Time? Pruning Time!

Markus Latte[1,*] and Martin Lange[2]

[1] Department of Computer Science, University of Munich, Germany
[2] School of Electrical Engineering and Computer Science, University of Kassel, Germany

Abstract. The full branching time logic CTL* is a well-known specification logic for reactive systems. Its satisfiability and model checking problems are well understood. However, it is still lacking a satisfactory sound and complete axiomatisation. The only proof system known for CTL* is Reynolds' which comes with an intricate and long completeness proof and, most of all, uses rules that do not possess the subformula property.

In this paper we consider a large fragment of CTL* which is characterised by disallowing certain nestings of temporal operators inside universal path quantifiers. This subsumes CTL$^+$ for instance. We present infinite satisfiability games for this fragment. Winning strategies for one of the players represent infinite tree models for satisfiable formulas. These can be pruned into finite trees using fixpoint strengthening and some simple combinatorial machinery such that the results represent proofs in a Hilbert-style axiom system for this fragment. Completeness of this axiomatisation is a simple consequence of soundness of the satisfiability games.

1 Introduction

Temporal logics originate from the philosophical tense logics [14] and are now important specification languages in computer science where they are being used to abstractly describe and verify the behaviour of reactive systems [11]. One of their most prominent examples is the full branching-time temporal logic CTL* [2]. Its satisfiability and model checking problem—i.e. the algorithmic nature of the logic—is well understood by now [18,3]. This cannot necessarily be said about the proof-theoretic nature of CTL*: despite CTL*'s long lifespan no clean and simple sound and complete axiomatisation has been found for it so far.

Various axiomatisations for simpler temporal logics like LTL for instance have been known for a long time, and others have been found since [7,5,10,9]. The same can be said about CTL [1,13,9,8]. These two logics enjoy the property that their formulas are modularly composed of a finite number of temporal operators and an axiomatisation can describe the handling of each of them. In CTL* though, the arbitrary mixture and nesting of path formulas leads to an essentially infinite number of temporal operators of which formulas are composed. A proper axiomatisation would have to capture their nature in a finite number of formula schemes. So far, the only successful attempt at presenting a sound and complete axiomatisation is Reynolds' [15]. However, it features an unsatisfactory system because of a rather intricate and difficult completeness proof. Most of

* Supported by the DFG Graduiertenkolleg 1480 (PUMA).

B. Gramlich, D. Miller, and U. Sattler (Eds.): IJCAR 2012, LNAI 7364, pp. 393–407, 2012.

all though, it contains rules which do not have the subformula property, for instance the rule (AA). The subformula property bounds the search space within a proof search. In particular for this rule, there is no (obvious) upper bound on the possible premises in terms of the conclusion because the rule introduces new atoms in its premise. The problem of finding a simple axiomatisation for CTL* with a neat completeness proof is therefore still open.

In this paper we make a step forward in this direction. We consider a large fragment of CTL* by restricting the nesting of temporal operators under a universal path quantifier. We call this fragment CTL$^\sharp$. Syntactically, it supersedes CTL and even CTL$^+$, and the satisfiability problem for CTL$^\sharp$ is therefore already 2EXPTIME-complete [6] which indicates that it is not an easy fragment. It also exceeds the expressive power of these two logics; for instance, it it possible to express that a path satisfying some fairness constraint exists.

Note that this fragment is not closed under complements in the sense that the negation of a formula with restrictions concerning *universal* path quantifiers has—in positive normal form—restrictions on *existential* path quantifiers. This is not a major problem. It simply means that one has to regard the context in which the logic is being used. We consider CTL$^\sharp$ in the context of satisfiability; in the context of validity one has the dual restrictions on existential path quantifiers.

We present a calculus of infinite games characterising the satisfiability problem for CTL$^\sharp$. We then employ combinatorial arguments and logical principles by which winning strategies in these games, which are infinite trees, can be made finite and loop-free. These are then used to derive a complete Hilbert-style axiomatisation for CTL$^\sharp$.

This is, as far as we know, the first attempt at approaching the CTL* axiomatisation problem on the syntactic route "from below". Other attempts are syntactic and "from above" in the sense that they consider a superlogic, for instance CTL* with past operators [16] which apparently makes things easier, or semantic in the sense that they redefine the class of structures over which the logic is interpreted [17].

The rest of the paper is organised as follows. Section "Branching Time" introduces the mentioned temporal logics; section "Playing Time" presents the satisfiability games; in section "Pruning Time" we show how to transform infinite winning strategies into finite ones; and section "Proving Time" presents the axiomatisation.

2 Branching Time

Transition Systems and Paths. A *transition system* is a tuple $\mathcal{T} = (\mathcal{S}, \longrightarrow, L)$ where $\mathcal{S}$ is a set of states, $\longrightarrow \subseteq \mathcal{S} \times \mathcal{S}$ a transition relation and $L : \mathcal{P} \to 2^{\mathcal{S}}$ a function that assigns to each q in some non-empty set $\mathcal{P}$ of atomic propositions the set of states $L(q)$ in which q holds. Here we assume the transition relation to be *total*: for all $s \in \mathcal{S}$ there is a $t \in \mathcal{S}$ such that $s \longrightarrow t$. A *path* is an infinite sequence $\pi = s_0, s_1, \ldots \in \mathcal{S}^\omega$ such that $s_i \longrightarrow s_{i+1}$ for all $i \in \mathbb{N}$. With π^k we denote the *k-th suffix* of π, namely the path $s_k, s_{k+1}, \ldots$.

CTL*. Formulas of the branching time temporal logic CTL* over $\mathcal{P}$ in positive normal form are given by the following grammar.

$$\varphi ::= q \mid \neg q \mid \varphi \vee \varphi \mid \varphi \wedge \varphi \mid \mathrm{E}\alpha \mid \mathrm{A}\alpha$$
$$\alpha ::= \varphi \mid \alpha \vee \alpha \mid \alpha \wedge \alpha \mid \mathrm{X}\alpha \mid \alpha \mathrm{U} \alpha \mid \alpha \mathrm{R} \alpha$$

where $q \in \mathcal{P}$. The formulas q and $\neg q$ are called *literals*. The constructors E and A are called *path quantifiers*; X, U and R are called *temporal operators*. Formulas derived as φ are called *state formulas*, those that are derived from α are called *path formulas*. The latter only occur as genuine subformulas of the former, i.e. a CTL* formula is one that is derived from φ in this grammar. Throughout this paper we will adopt the convention that small letters of the end of Greek alphabet, like φ, ψ, denote state formulas and small ones from the beginning, like α, β, denote path formulas. Note that every state formula is also a path formula.

We also use the standard abbreviations from temporal logic: $\mathrm{F}\alpha := \mathtt{tt}\,\mathrm{U}\,\alpha$ and $\mathrm{G}\alpha := \mathtt{ff}\,\mathrm{R}\,\alpha$ where $\mathtt{tt} := q \vee \neg q$ and $\mathtt{ff} := q \wedge \neg q$ for some $q \in \mathcal{P}$.

The *closure* of a formula ϑ is the least set $\mathsf{Cl}(\vartheta)$ that contains ϑ and satisfies the following.

- If $Q\alpha \in \mathsf{Cl}(\vartheta)$ for some $Q \in \{\mathrm{E}, \mathrm{A}\}$ then $\alpha \in \mathsf{Cl}(\vartheta)$.
- If $\alpha \wedge \beta \in \mathsf{Cl}(\vartheta)$ or $\alpha \vee \beta \in \mathsf{Cl}(\vartheta)$ then $\{\alpha, \beta\} \subseteq \mathsf{Cl}(\vartheta)$.
- If $\mathrm{X}\psi \in \mathsf{Cl}(\vartheta)$ then $\psi \in \mathsf{Cl}(\vartheta)$.
- If $\varphi \circ \psi \in \mathsf{Cl}(\vartheta)$ then $\{\psi, \varphi, \mathrm{X}(\varphi \circ \psi)\} \subseteq \mathsf{Cl}(\vartheta)$, for all $\circ \in \{\mathrm{U}, \mathrm{R}\}$.

Thus, the closure is essentially the set of all subformulas with the exception of the fixpoint-operators which we also include with a prefixed X-operator. Note that the size of the closure of some φ is linear in its syntactic length. We therefore use it as a measure for the size of a formula: $|\varphi| := |\mathsf{Cl}(\varphi)|$.

Formulas of CTL* are interpreted over states and paths of a transition system $\mathcal{T} = (\mathcal{S}, \longrightarrow, L)$, reflecting the two types of formulas.

$$
\begin{array}{ll}
\mathcal{T}, s \models q & \text{iff } q \in L(s) \\
\mathcal{T}, s \models \neg q & \text{iff } q \notin L(s) \\
\mathcal{T}, s \models \varphi \vee \psi & \text{iff } \mathcal{T}, s \models \varphi \text{ or } \mathcal{T}, s \models \psi \\
\mathcal{T}, s \models \varphi \wedge \psi & \text{iff } \mathcal{T}, s \models \varphi \text{ and } \mathcal{T}, s \models \psi \\
\mathcal{T}, s \models \mathrm{E}\alpha & \text{iff there is a path } \pi = s, s', \ldots \text{ with } \mathcal{T}, \pi \models \alpha \\
\mathcal{T}, s \models \mathrm{A}\alpha & \text{iff for all paths } \pi = s, s', \ldots \text{ we have } \mathcal{T}, \pi \models \alpha \\
\mathcal{T}, \pi \models \varphi & \text{iff } \mathcal{T}, s \models \varphi \quad \text{when } \varphi \text{ is a state formula and } \pi = s, s', \ldots \\
\mathcal{T}, \pi \models \alpha \vee \beta & \text{iff } \mathcal{T}, \pi \models \alpha \text{ or } \mathcal{T}, \pi \models \beta \\
\mathcal{T}, \pi \models \alpha \wedge \beta & \text{iff } \mathcal{T}, \pi \models \alpha \text{ and } \mathcal{T}, \pi \models \beta \\
\mathcal{T}, \pi \models \mathrm{X}\alpha & \text{iff } \mathcal{T}, \pi^1 \models \alpha \\
\mathcal{T}, \pi \models \alpha \mathrm{U} \beta & \text{iff there is a } k \in \mathbb{N} \text{ with } \mathcal{T}, \pi^k \models \beta \text{ and for all } j < k : \mathcal{T}, \pi^j \models \alpha \\
\mathcal{T}, \pi \models \alpha \mathrm{R} \beta & \text{iff for all } k \in \mathbb{N} : \mathcal{T}, \pi^k \models \beta \text{ or there is } j < k : \mathcal{T}, \pi^j \models \alpha
\end{array}
$$

Two (state) formulas are equivalent, written $\varphi \equiv \psi$, iff for all $\mathcal{T}$ and all states s we have: $\mathcal{T}, s \models \varphi$ iff $\mathcal{T}, s \models \psi$. A state formula φ is valid, written $\models \varphi$, iff $\mathcal{T}, s \models \varphi$ for any transition system $\mathcal{T}$ and any of its states s. Equally, a path formula α is valid, also written $\models \alpha$, if $\mathcal{T}, \pi \models \alpha$ for any transition system $\mathcal{T}$ and any path π in it. Note that $\models \alpha$ iff $\models \mathtt{A}\alpha$.

Finally, we introduce the *dual* $\neg\vartheta$ of a formula ϑ as follows: q and $\neg q$ are dual to each other; the usual deMorgan law's apply; path quantifiers are dual to each other as in $\neg\mathtt{E}\alpha := \mathtt{A}\neg\alpha$ and vice-versa; the next-operator is self-dual as in $\neg\mathtt{X}\varphi := \mathtt{X}\neg\varphi$; and the temporal fixpoint operators are dual to each other as in $\neg(\varphi\mathtt{U}\psi) := (\neg\varphi)\mathtt{R}(\neg\psi)$ and vice-versa. With negation around we can introduce implication $\alpha \to \beta$ as $\neg\alpha \vee \beta$ as usual. The next section explains why we have introduced formulas in positive normal form and avoided the use of negation as a first-class operators.

CTL$^\sharp$. The fragment CTL$^\sharp$ is obtained from CTL* in positive normal form by disallowing certain nestings of temporal operators inside a universal path quantifier: the arguments to an *until* formula in there must be state formulas. Formally, the syntax of CTL$^\sharp$ formulas is given by the following grammar.

$$\begin{aligned}
\varphi &::= q \mid \neg q \mid \varphi \vee \varphi \mid \varphi \wedge \varphi \mid \mathtt{E}\alpha \mid \mathtt{A}\beta \\
\alpha &::= \varphi \mid \alpha \vee \alpha \mid \alpha \wedge \alpha \mid \mathtt{X}\alpha \mid \alpha\mathtt{U}\alpha \mid \alpha\mathtt{R}\alpha \\
\beta &::= \varphi \mid \beta \vee \beta \mid \beta \wedge \beta \mid \mathtt{X}\beta \mid \varphi\mathtt{U}\varphi \mid \beta\mathtt{R}\beta
\end{aligned}$$

All the concepts introduced above for CTL* like the closure and the semantics clearly carry over to CTL$^\sharp$ as well. However, note that the dual $\neg\vartheta$ of a CTL$^\sharp$ formula ϑ need not be a CTL$^\sharp$ formula itself. A syntax for the *dual* of CTL$^\sharp$ is obtained from the one above by switching E and A. We set $\neg\text{CTL}^\sharp := \{\varphi \mid \neg\varphi \in \text{CTL}^\sharp\}$ as a subset of CTL*.

It is easy to see that many of the standard and simple types of properties like *safety*, *liveness*, *fairness*, etc. are expressible in some form or the other in CTL$^\sharp$, for instance through $\mathtt{AG}\, q_{safe}$, $\mathtt{AGEF}\, q_{live}$, $\mathtt{AGF}\, q_{fair}$. Also, it is possible to express a standard requirement for schedulers, namely that all requests need to be served at a later point: $\mathtt{AG}(q_{request} \to \mathtt{F}\, q_{serve})$.

However, it is for example not possible to say that all paths that are fair w.r.t. some predicate α satisfy some property β. This would be $\mathtt{A}(\mathtt{GF}\alpha \to \beta)$ which would be $\mathtt{A}(\mathtt{FG}\neg\alpha \vee \beta)$ in positive normal form and thus contain an R-formulas in an argument of a U-formula inside a universal path quantifier.

A prominent example of a CTL* formula which essentially bears much of the difficulty of finding a complete axiomatisation is the *limit closure formula*

$$\varphi_{\mathrm{LC}} \;:=\; q \wedge \mathtt{AG}\big(q \to \mathtt{EX}(q\,\mathtt{U}\,p)\big) \to \mathtt{EG}(q\,\mathtt{U}\,p)\,.$$

It is a valid CTL* formula, hence, its dual $\neg\varphi_{\mathrm{LC}}$ is unsatisfiable which is

$$q \wedge \mathtt{AG}\big(\neg q \vee \mathtt{EX}(q\,\mathtt{U}\,p)\big) \wedge \mathtt{AF}(\neg q\,\mathtt{R}\,\neg p)$$

in positive normal form. This is not a CTL$^\sharp$ formula though because the last conjunct is universally path quantified and contains a U-formula (of the abbreviated form F) which itself contains an R-formula in one of its arguments.

$$(\wedge)\ \frac{\varphi, \psi, \Phi}{\varphi \wedge \psi, \Phi} \qquad (\vee)\ \frac{\varphi_i, \Phi}{\varphi_0 \vee \varphi_1, \Phi}\ \exists, i \in \{0,1\} \qquad (\mathtt{ESt})\ \frac{\varphi, \mathtt{E}\Pi, \Phi}{\mathtt{E}(\varphi, \Pi), \Phi}$$

$$(\mathtt{E}\wedge)\ \frac{\mathtt{E}(\alpha, \beta, \Pi), \Phi}{\mathtt{E}(\alpha \wedge \beta, \Pi), \Phi} \qquad (\mathtt{Ett})\ \frac{\Phi}{\mathtt{E}\emptyset, \Phi} \qquad (\mathtt{EU})\ \frac{\mathtt{E}(\beta, \Pi), \Phi \ \mid\ \mathtt{E}(\alpha, \mathtt{X}(\alpha \mathtt{U} \beta), \Pi), \Phi}{\mathtt{E}(\alpha \mathtt{U} \beta, \Pi), \Phi}\ \exists$$

$$(\mathtt{E}\vee)\ \frac{\mathtt{E}(\alpha_i, \Pi), \Phi}{\mathtt{E}(\alpha_0 \vee \alpha_1, \Pi), \Phi}\ \exists, i \in \{0,1\} \qquad (\mathtt{ER})\ \frac{\mathtt{E}(\alpha, \beta, \Pi), \Phi \ \mid\ \mathtt{E}(\beta, \mathtt{X}(\alpha \mathtt{R} \beta), \Pi), \Phi}{\mathtt{E}(\alpha \mathtt{R} \beta, \Pi), \Phi}\ \exists$$

$$(\mathtt{A}\wedge)\ \frac{\mathtt{A}(\alpha, \Sigma), \mathtt{A}(\beta, \Sigma), \Phi}{\mathtt{A}(\alpha \wedge \beta, \Sigma), \Phi} \qquad (\mathtt{A}\vee)\ \frac{\mathtt{A}(\alpha, \beta, \Sigma), \Phi}{\mathtt{A}(\alpha \vee \beta, \Sigma), \Phi} \qquad (\mathtt{ASt})\ \frac{\varphi, \Phi \ \mid\ \mathtt{A}\Sigma, \Phi}{\mathtt{A}(\varphi, \Sigma), \Phi}\ \exists$$

$$(\mathtt{AU})\ \frac{\psi, \Phi \ \mid\ \varphi, \mathtt{A}(\mathtt{X}(\varphi \mathtt{U} \psi), \Sigma), \Phi \ \mid\ \mathtt{A}\Sigma, \Phi}{\mathtt{A}(\varphi \mathtt{U} \psi, \Sigma), \Phi}\ \exists, \text{if } \Sigma \text{ not U-pure}$$

$$(\mathtt{A}\vec{\mathtt{U}})\ \frac{\psi_j, \Phi \ \mid\ \{\varphi_i \mid i \in I\}, \mathtt{A}(\{\mathtt{X}(\varphi_i \mathtt{U} \psi_i) \mid i \in I\}), \Phi}{\mathtt{A}(\varphi_1 \mathtt{U} \psi_1, \ldots, \varphi_n \mathtt{U} \psi_n), \Phi}\ \exists, j \in [n], I \subseteq [n]$$

$$(\mathtt{AR})\ \frac{\mathtt{A}(\beta, \Sigma), \mathtt{A}(\alpha, \mathtt{X}(\alpha \mathtt{R} \beta), \Sigma), \Phi}{\mathtt{A}(\alpha \mathtt{R} \beta, \Sigma), \Phi}$$

$$(\mathtt{X}_0)\ \frac{\mathtt{A}\Sigma_1, \ldots, \mathtt{A}\Sigma_m}{\mathtt{AX}\Sigma_1, \ldots, \mathtt{AX}\Sigma_m, \Lambda} \qquad (\mathtt{X}_1)\ \frac{\mathtt{E}\Pi_i, \mathtt{A}\Sigma_1, \ldots, \mathtt{A}\Sigma_m}{\mathtt{EX}\Pi_1, \ldots, \mathtt{EX}\Pi_n, \mathtt{AX}\Sigma_1, \ldots, \mathtt{AX}\Sigma_m, \Lambda}\ \forall, i \in [n]$$

Fig. 1. Rules for the CTL$^\sharp$ satisfiability game

3 Playing Time

Configurations, Rules, Plays, and Winning Conditions. We present a game-theoretic characterisation of the satisfiability problem for CTL$^\sharp$. The game $\mathcal{G}(\vartheta)$ is played by two players $\exists$ and $\forall$ who want to show that ϑ is satisfiable, resp. is unsatisfiable. We fix a state formula $\vartheta \in$ CTL$^\sharp$ for the rest of this section.

A *block* is an element of $\{\mathtt{E}, \mathtt{A}\} \times 2^{\mathsf{Cl}(\vartheta)}$ written $\mathtt{E}\Pi$ or $\mathtt{A}\Sigma$. They represent the state formulas $\mathtt{E} \bigwedge \Pi$ and $\mathtt{A} \bigvee \Sigma$. Conversely, we identify a state formula $Q\alpha$ with the block $Q\{\alpha\}$ for $Q \in \{\mathtt{A}, \mathtt{E}\}$. A *configuration* is a set of state formulas and of blocks, for instance $\varphi_1, \ldots, \varphi_l, \mathtt{E}\Pi_1, \ldots, \mathtt{E}\Pi_n, \mathtt{A}\Sigma_1, \ldots, \Sigma_m$. The intended formula of such a configuration is the conjunction of its elements.

A formula set Σ is called U-pure if it consists of formulas of the form $\varphi_1 \mathtt{U} \varphi_2$ only. We write $\mathtt{X}\Sigma$ to denote the set $\{\mathtt{X}\psi \mid \psi \in \Sigma\}$ and equally for $\mathtt{X}\Pi$. A configuration Φ is *propositionally inconsistent* if there is a proposition q s.t. $\{q, \neg q\} \subseteq \Phi$.

The game $\mathcal{G}(\vartheta)$ starts in the initial configuration ϑ and proceeds according to the rules presented in Fig. 1. We write $[n]$ to denote $\{1, \ldots, n\}$. There are a few important comments to regard when reading the rules.

- They are to be read bottom-up, i.e. if the current configuration in a play is an instance of the pattern below then the player annotated to the right of a rule chooses one of the configurations on top to continue with. The respective player can choose from the alternatives which are separated by "|". Some rules are deterministic, i.e. no player is making a choice. The configuration on the top of a rule is called *premise* and that on the bottom *conclusion*.

- Formulas denoted $\varphi, \psi, \varphi_1, \ldots$ are state formulas according to the syntax of CTL*; formula denoted $\alpha, \beta, \alpha_1, \ldots$ are path (and therefore possibly also state) formulas. Λ always stands for a set of literals, Φ denotes an arbitrary set of blocks and state formulas, and Σ and Π denote a set of path formulas.
- As we identify the state formula $\mathrm{E}\alpha$ with the block $\mathtt{E}\{\alpha\}$—for instance—the rules $(\wedge)$ and $(\vee)$ can generate blocks.
- Although configurations and blocks are sets in the main, they are written as lists. However, a notation like $\mathtt{A}(\alpha \wedge \beta, \Sigma), \Phi$ implicitly states that $\alpha \wedge \beta \notin \Sigma$ and $\mathtt{A}(\alpha \wedge \beta, \Sigma) \notin \Phi$. Otherwise, a rule application could be repeated ad infinitum and hinders progress.

Note that in certain configurations several rules may apply, even rules for both players to perform choices. We therefore assume an arbitrary but fixed ordering on the rules which determines uniquely the rule that applies to a configuration. The exact ordering is irrelevant for the theory developed here.

In an application of a rule, the formula and the block which get transformed are called *principal formula* and *block*, respectively. Examples are the formula $\alpha \wedge \beta$ and the block $\mathtt{A}(\alpha \wedge \beta, \Sigma)$ in the instance of the rule $(\mathtt{A}\wedge)$ as shown in Fig. 1. In the rule $(\mathtt{A}\vec{\mathtt{U}})$ *all* formulas in the principal block are principal. The unaffected blocks and formulas are called *side formulas* taking blocks for formulas. Continuing the example, these are the formulas in Σ and in Φ.

A *play* is a possibly infinite sequence of configurations starting in the initial one and resulting from successive rule applications. Note that in every play, the intended formula of a configuration is in CTL$^\sharp$. Before we can define the winner of a play we need a technical definition capturing the unfulfilledness of least fixed point constructs.

Definition 1. A *component* of a configuration C is a state formula in C, an $\mathtt{A}$-block in C or a single formula inside an E-block contained in C. Let $\Phi_0, \Phi_1, \ldots$ be an infinite play. The rules induce a connection relation on components of adjacent configurations in this play, obtained from the game rules in a straightforward way. A component C in Φ_i is connected to a component C' in Φ_{i+1}, written $\langle \Phi_i, C\rangle \rightsquigarrow \langle \Phi_{i+1}, C'\rangle$, if either

- C is not principal and $C = C'$, or
- C is principal in this rule application and gets transformed into C'.

Example 2. To illustrate the second item consider an instance of the rule $(\mathtt{ASt})$ as shown Fig. 1 for $\varphi = \mathrm{E}\alpha$. For the left alternative, $\mathrm{E}\alpha$ becomes part of the configuration both as a state formula and as a block $\mathtt{E}\{\alpha\}$. Therefore, $\langle \cdot, \mathtt{A}(\mathrm{E}\alpha, \Sigma)\rangle \rightsquigarrow \langle \cdot, \mathrm{E}\alpha\rangle$ and, if α is a state formula, $\langle \cdot, \mathtt{A}(\mathrm{E}\alpha, \Sigma)\rangle \rightsquigarrow \langle \cdot, \alpha\rangle$ hold. For the other alternative, we have $\langle \cdot, \mathtt{A}(\mathrm{E}\alpha, \Sigma)\rangle \rightsquigarrow \langle \cdot, \mathtt{A}\Sigma\rangle$. On the other hand, a $\mathtt{U}$- and an $\mathtt{R}$-formula can grow by unfolding these fixed points. For example, the instance of the rule $(\mathtt{EU})$ in Fig. 1 yields $\langle \cdot, \alpha\mathtt{U}\beta\rangle \rightsquigarrow \langle \cdot, \mathtt{X}(\alpha\mathtt{U}\beta)\rangle$.

The following lemma is not hard to see. Note that only the unfolding rules for $\mathtt{U}$- and $\mathtt{R}$-formulas create in some sense larger configurations, but they introduce an $\mathtt{X}$-operator which has to be dealt with before the respective formula can be unfolded again.

Lemma 3. *Every infinite play contains infinitely many applications of rules* $(\mathtt{X}_0)$ *or* $(\mathtt{X}_1)$.

Definition 4. A *thread* in $\Phi_0, \Phi_1, \ldots$ is a sequence $C_0, C_1, \ldots$ of components such that $\langle \Phi_i, C_i \rangle \rightsquigarrow \langle \Phi_{i+1}, C_{i+1} \rangle$ for all $i \in \mathbb{N}$. It is called a *bad thread* if either

- there is a $\varphi \mathtt{U} \psi \in \mathsf{Cl}(\vartheta)$ s.t. $C_i = \varphi \mathtt{U} \psi$ for infinitely many i, or
- there is a block $\mathtt{A}\Sigma$ with Σ being U-pure, s.t. $C_i = \mathtt{A}\Sigma$ for infinitely many i.

In the first case we also speak of a bad E-thread, in the second case of a bad A-thread.

Hence, a play contains a *bad thread* if there is either a U-formula inside some E-blocks that regenerates itself infinitely often via the unfolding in rule (EU), or there is an A-block which contains no R-formula that regenerates itself in a similar way along this play. Player $\forall$ *wins a play* $\pi = \Phi_0, \Phi_1, \ldots$ if

($\forall$-1) there is an $n \in \mathbb{N}$ s.t. Φ_n is propositionally inconsistent, or
($\forall$-2) there is an $n \in \mathbb{N}$ s.t. $\mathtt{A}\emptyset \in \Phi_n$, or
($\forall$-3) π contains a bad thread.

In all other cases, player $\exists$ wins the play.

Determinacy. An important game-theoretic concept is *determinacy* meaning that for every game exactly one of the players has a winning strategy. The games presented here are determined. The proof is relatively simple by appealing to known determinacy results about games in general. We only need to identify the winning plays as being of a certain type, namely being recognisable by a co-Büchi automaton.

Lemma 5. *For a bad thread* $C_0, C_1, \ldots$ *in* $\Phi_0, \Phi_1, \ldots$ *either*

- *there is a* $k \in \mathbb{N}$ *and* $\varphi \mathtt{U} \psi \in \mathsf{Cl}(\vartheta)$ *s.t.* $C_i \in \{\varphi \mathtt{U} \psi, \mathtt{X}(\varphi \mathtt{U} \psi)\}$ *for all* $i \geq k$*, or*
- *there is a* $k \in \mathbb{N}$ *and a* U*-pure set* $\Sigma \subseteq \mathsf{Cl}(\vartheta)$ *s.t.* $C_i \in \{\mathtt{A}\Sigma, \mathtt{A}(\mathtt{X}\Sigma)\}$ *for all* $i \geq k$.

Proof. The case distinction follows Definition 4. For k we take one of the infinitely many values i mentioned in that definition. Finally, the game rules entail the properties along the corresponding suffix. □

Theorem 6. *The* CTL$^\sharp$ *satisfiability games are determined, i.e. for every* ϑ*, either* $\exists$ *or* $\forall$ *has a winning strategy for the game* $\mathcal{G}(\vartheta)$.

Proof. Following Lemma 5 and 3, the winning conditions can be represented as a co-Büchi condition and are therefore in the Borel hierarchy. The result then follows immediately from Martin's Theorem [12]. □

Soundness and Completeness. Due to lack of space we only sketch how one can prove that the games correctly characterise satisfiability in CTL$^\sharp$. It is possible to do this via explicit constructions of a model from a winning strategy for player $\exists$, etc. Instead, we appeal to a very similar system that is known to correctly characterise satisfiability for CTL* [4]. However, the syntactical restrictions of CTL* considered here supersede the distinction between traces and threads as exploited in [4].

Theorem 7. *Player* $\exists$ *has a winning strategy for the game* $\mathcal{G}(\vartheta)$ *iff* ϑ *is satisfiable.*

The games here differ from the system in [4] in the rules for universally path quantified blocks and in the winning conditions. There, the rules are simply dual to the ones for existentially path quantified blocks; in detail: rule (AU) and ($\vec{\text{AU}}$) here replace the dual version of (EU). Furthermore, in the system for full CTL*, a bad thread of A-blocks must not contain any infinitely regenerating R-formula. In order to prove soundness and completeness of the games for CTL$^\sharp$ it suffices to see that the CTL$^\sharp$ games essentially behave like the CTL* games when applied to a CTL$^\sharp$ formula. Now note that the defining property of being a CTL$^\sharp$ formula is having no genuine path formulas as arguments to a U-formula inside an A-block. State formulas, though, get removed from A-blocks with rule (ASt). This justifies rule (AU). Furthermore, if a sequence of connected A-blocks through a play contains no regenerating R-formula then the set of formulas in those blocks must eventually become U-pure because no U-formula in there can spawn off anything that remains in this set. Then the unfolding of U-formulas in a U-pure set can be synchronised which justifies rule ($\vec{\text{AU}}$). Finally, a bad thread in the sense of Def. 4 is a bad trace in the system for CTL*, and vice versa.

4 Pruning Time

By Thm. 6 and 7, $\neg\vartheta$ is a tautology iff player $\forall$ has a winning strategy in the game $\mathcal{G}(\vartheta)$. In the next section we will present an axiomatisation for CTL$^\sharp$, and in this section we develop the necessary tools in order to prove completeness thereof. We consider $\forall$'s winning strategy as a tree, namely the tree of all plays which conform to this strategy. I.e. at every position in which player $\exists$ makes a choice or a deterministic rule applies all successors are preserved in the tree. At position in which player $\forall$ makes a choice only the choice prescribed by the strategy is preserved in the tree. Clearly, such a tree is in general infinite. We turn it into a finite tree which essentially is a finite derivation for $\neg\vartheta$ in the axiom system of the next section.

As this axiomatisation should also be sound it clearly does not suffice to truncate the tree at arbitrary positions. Instead, the resulting finite tree should satisfy the following properties: (1) leaves should be unsatisfiable; and (2) unsatisfiability should be preserved in the direction towards the root. This is enough to yield completeness since it constructs a proof for an unsatisfiable $\neg\vartheta$, i.e. a valid ϑ.

The principles used to achieve (1) and (2) are the following. At nodes which are inconsistent or contain A$\emptyset$, it suffices to simply truncated the tree. This is possible on all plays that player $\forall$ wins with his winning conditions ($\forall$-1) or ($\forall$-2). The remaining paths in the tree contain bad threads. We use the principle of fixpoint strengthening in order to preserve satisfiability but disable infinite unfoldings of U-operators. In essence, this principle forbids the unfolding of a (set of) U-formulas in a certain context for the second time. Instead, the node becomes inconsistent and can be truncated as well.

An additional difficulty is the fact that this has to be done in the tree as a whole rather than on each branch separately—even though bad threads are properties of branches. Note that two branches with bad threads may have a common prefix but these threads may differ on that prefix. This is basically handled by a scheduling mechanism which strengthens least fixpoint formulas one-by-one.

$$(\mathtt{EU})^{\dagger}\ \frac{}{\mathtt{E}(\varphi\mathtt{U}_{\Pi\cup\Phi}\psi,\Pi),\Phi} \qquad (\mathtt{EU})^{*}\ \frac{\mathtt{E}(\beta,\Pi),\Phi \ \mid\ \mathtt{E}(\alpha,\mathtt{X}(\alpha\mathtt{U}_{\Gamma}\beta),\Pi),\Phi}{\mathtt{E}(\alpha\mathtt{U}_{\Gamma}\beta,\Pi),\Phi}$$

$$(\mathtt{EU})^{\natural}\ \frac{\mathtt{E}(\varphi\mathtt{U}\psi,\Pi),\Phi}{\mathtt{E}(\varphi\mathtt{U}_{\Gamma}\psi,\Pi),\Phi} \qquad (\mathtt{EU})^{\flat}\ \frac{\mathtt{E}(\beta,\Pi),\Phi \ \mid\ \mathtt{E}(\alpha,\mathtt{X}(\alpha\mathtt{U}_{\Pi\cup\Phi}\beta),\Pi),\Phi}{\mathtt{E}(\alpha\mathtt{U}\beta,\Pi),\Phi}$$

$$(\mathtt{A}\vec{\mathtt{U}})^{\dagger}\ \frac{}{\mathtt{A}(\varphi_1\mathtt{U}_{\Phi}\psi_1,\ldots,\varphi_n\mathtt{U}_{\Phi}\psi_n),\Phi} \qquad (\mathtt{A}\vec{\mathtt{U}})^{\natural}\ \frac{\mathtt{A}(\varphi_1\mathtt{U}\psi_1,\ldots,\varphi_n\mathtt{U}\psi_n),\Phi}{\mathtt{A}(\varphi_1\mathtt{U}_{\Gamma}\psi_1,\ldots,\varphi_n\mathtt{U}_{\Gamma}\psi_n),\Phi}$$

$$(\mathtt{A}\vec{\mathtt{U}})^{*}\ \frac{\psi_j,\Phi \ \mid\ \varphi_1,\ldots,\varphi_n,\mathtt{A}(\mathtt{X}(\varphi_1\mathtt{U}_{\Gamma}\psi_1),\ldots,\mathtt{X}(\varphi_n\mathtt{U}_{\Gamma}\psi_n)),\Phi}{\mathtt{A}(\varphi_1\mathtt{U}_{\Gamma}\psi_1,\ldots,\varphi_n\mathtt{U}_{\Gamma}\psi_n),\Phi}\ \exists j\in[n]$$

$$(\mathtt{A}\vec{\mathtt{U}})^{\flat}\ \frac{\psi_j,\Phi \ \mid\ \varphi_1,\ldots,\varphi_n,\mathtt{A}(\mathtt{X}(\varphi_1\mathtt{U}_{\Phi}\psi_1),\ldots,\mathtt{X}(\varphi_n\mathtt{U}_{\Phi}\psi_n)),\Phi}{\mathtt{A}(\varphi_1\mathtt{U}\psi_1,\ldots,\varphi_n\mathtt{U}\psi_n),\Phi}\ \exists j\in[n]$$

Fig. 2. Rules for annotated U-formulas in E-blocks

4.1 Annotations and Their Rules

For a set of formulas $\Gamma \subseteq \mathsf{Cl}(\vartheta)$ define $\varphi\mathtt{U}_{\Gamma}\psi := (\varphi \wedge \neg \bigwedge \Gamma)\mathtt{U}(\psi \wedge \neg \bigwedge \Gamma)$. The set Γ is called an *annotation* to the formula $\varphi\mathtt{U}\psi$. Note that annotating formulas can take us out of the CTL$^{\sharp}$ fragment. This, however, is just an observation and has no negative effect since its semantics is well-defined as a CTL*-formula.

The annotation is used to remember a set of side formulas. Informally, once the annotated formula occurs in a configuration with the same side formulas again, we like to truncate the play right after this repetition.

On an infinite play the U-formulas are eventually handled by the rules (EU) and $(\mathtt{A}\vec{\mathtt{U}})$ for $I = [n]$ only. Both rules are extended to operate on annotated formulas, introducing four new rules for each occurrence of U-formulas inside E- or A-quantifiers: one rule to create an annotation, one to keep it through the usual unfolding, one to erase it for following different branches with different bad threads, and one to terminate the play. These new rules are shown in Fig. 2.

The annotation in rule $(\mathtt{A}\vec{\mathtt{U}})^{\flat}$ is placed on all formulas in the block simultaneously. None of the rules can change or remove the annotation of a formula in such a block without affecting the other formulas. Hence, we actually annotate the whole block rather than each formula. However, for simplicity we focus on the annotation of formulas.

Lemma 8. *The conclusion in rule* $(\mathtt{EU})^{\dagger}$ *is unsatisfiable, and for the rules* $(\mathtt{EU})^{\natural}$, $(\mathtt{EU})^{*}$ *and* $(\mathtt{EU})^{\flat}$ *we have that if all premises are unsatisfiable then so is the conclusion.*

Proof. We detail the proof for the most difficult rule only, that is for $(\mathtt{EU})^{\flat}$. Assume that there is a transition system $\mathcal{T}$ and a path π such that

$$\mathcal{T},\pi \models \alpha\mathtt{U}\beta \ \wedge\ \bigwedge\Pi \ \wedge\ \bigwedge\Phi. \tag{1}$$

Let $k \in \mathbb{N}$ be such that $\mathcal{T},\pi^k \models \beta$, and $\mathcal{T},\pi^i \models \alpha \wedge \neg\beta$ for all $i < k$. Among all such paths satisfying (1), we choose a path π with a minimal k-value. Suppose that none of the premises is satisfiable. Thus, we have in particular

$$\mathcal{T}, \pi \not\models (\beta \vee (\alpha \wedge \mathtt{X}(\alpha \mathtt{U}_{\Pi \cup \Phi} \beta))) \wedge \bigwedge \Pi \wedge \bigwedge \Phi.$$

Then there is a $0 < \ell \leq k$ such that $\mathcal{T}, \pi^\ell \not\models \neg \bigwedge(\Pi \cup \Phi)$. Additionally, (1) yields $\mathcal{T}, \pi^\ell \models \alpha \mathtt{U} \beta$. Therefore, Eq. (1) holds for π^ℓ instead of π. But the k-value of π^ℓ is $k - \ell < k$. Thus, this is a contradiction to the minimality of k. □

Lemma 9. *The conclusion in rule* $(\mathtt{A}\vec{\mathtt{U}})^\dagger$ *is unsatisfiable, and for the rules* $(\mathtt{A}\vec{\mathtt{U}})^\natural$, $(\mathtt{A}\vec{\mathtt{U}})^*$ *and* $(\mathtt{A}\vec{\mathtt{U}})^\flat$ *we have that if all premises are unsatisfiable then so is the conclusion.*

Proof. Consider the rule $(\mathtt{A}\vec{\mathtt{U}})^\flat$. Assume that

$$\psi_j \wedge \bigwedge \Phi \text{ is unsatisfiable for all } j \in [n], \tag{2}$$

and that

$$\mathcal{T}, s \models \mathtt{A}\left(\bigvee\nolimits_{i \in [n]} \varphi_i \mathtt{U} \psi_i\right) \wedge \bigwedge \Phi \tag{3}$$

holds for a transition system $\mathcal{T}$ and a state s in $\mathcal{T}$.

For any such pair $(\mathcal{T}, s)$ satisfying (3) we associate a tree with unordered children. The tree is the unwinding of $\mathcal{T}$ which begins at s and ends at a node t whenever there is an $i \in [n]$ such that $\mathcal{T}, t \models \psi_i$ and $\mathcal{T}, r \models \varphi_i$ for any node r along the path from s, including, to t, excluding. Since an associated tree does not show any infinite path—nevertheless the tree might be infinite—the strict subtree-order is well-founded on associated trees.

Now, let $\mathcal{T}$ and s be such that they satisfy (3) and that their associated tree is a minimal one w.r.t. the strict subtree-order among all associated trees which originate from pairs which satisfy (3). By (2) the associated tree cannot consist of its root only. For the sake of contradiction, assume that the pair of $\mathcal{T}$ and s does not model the right premise of $(\mathtt{A}\vec{\mathtt{U}})^\flat$. Then there is a state r different from s, which corresponds to a node in the associated tree, such that $\mathcal{T}, r \not\models \neg \bigwedge \Phi$. However, the subtree at r is the associate tree of $\mathcal{T}$ and r. But this situation contradicts the choice of $\mathcal{T}$ and s.

The argument for the other rules are simpler. □

Remark 10. Although the rules in Fig. 2 are sound w.r.t. unsatisfiability they are *not* invertible in general. Consider the rule $(\mathtt{EU})^\natural$: the configuration $\mathtt{E}(p\mathtt{U}_{\{p\}}(p \wedge q))$ is unsatisfiable whereas $\mathtt{E}(p\mathtt{U}(p \wedge q))$ is satisfiable. Therefore, these rules are unsuitable for an incorporation into the game defined in Sect. 3 in the first place.

4.2 Truncating Infinite Trees

The following constructions consider (labelled) trees. To simplify the presentation we introduce some notations. For a tree T and nodes u, v in T we write $u <_T v$ iff u is a proper ancestor of v, $u \leq_T v$ iff $u <_T v$ or $u = v$, $T|_v$ for the subtree located at v such that v is the root of $T|_v$, and u is a *child* of v (v is *parent* of u, resp.) iff $v <_T u$ and $v <_T w <_T u$ for no node w in T. A path in T is a (finite or infinite) sequence such that any node in the sequence is a parent of the succeeding node if the latter exists. A branch is a path with begins at the root. Since $<_T$ is well-founded—but not a linear order in general—, $\min_T(V)$ denotes the set of minimal nodes w.r.t. $<_T$ in a set V of nodes.

For the remaining section assume that $\neg\vartheta$ is a tautology. Hence, player $\forall$ has a winning strategy for the game $\mathcal{G}(\vartheta)$, cf. Thm. 6 and 7. From now on, formulas and sets of formulas are assumed to be in or subsets of $\mathsf{Cl}(\vartheta)$.

Definition 11. We say that a tree T is a *(ϑ-)tree which follows a set of rules R* iff each node is labelled with a configuration for ϑ, the root is labelled with ϑ, and for each node v one of the following items holds.

- v is a leaf and the node is propositionally inconsistent or contains $\mathtt{A}\emptyset$.
- v has exactly one child, say w, such that v is the conclusion and w the premise of the same instance of the rule $(\mathtt{X}_0)$ or $(\mathtt{X}_1)$.
- For a rule in $R \setminus \{(\mathtt{X}_0), (\mathtt{X}_1)\}$ and for a principal block and formula(s), the set of children is the set of possible successor configurations which player $\exists$ can choose with this rule, principal block and formula(s). For instance, a node for the rule $(\mathtt{A}\vec{\mathtt{U}})$ has exactly $n+2^n$ children if n is the number of $\mathtt{U}$-formulas in the principal $\mathtt{A}$-block: The left hand of "|" admits n possibilities and the right hand 2^n.

Note that in such a tree any inner node uniquely determines the rule which was used in the justification for the second and the last item.

Lemma 12. *There is a ϑ-tree which follows the rules in Fig. 1.*

Proof. The winning strategy of player $\forall$ is taken for a tree. However, not all possible moves of player $\exists$ need to be considered. For the last item in Def. 11 it suffices to consider just one rule, principal block and formula(s) on $\exists$'s turn. Moreover, on every branch the first node is changed into a leaf iff the node is propositionally inconsistent or contains $\mathtt{A}\emptyset$. The obtained tree follows the said rules. □

We fix such a tree and call it T_ϑ. The label of a node v is written as $\ell(v)$. We may take the node for its label. As each branch in T_ϑ forms a game in the sense of Sect. 3—at least a prefix of a game for which the winner is already determined—, we may use the game-theoretic notations for the tree as well. For instance, every infinite branch in T_ϑ contains a bad thread.

To handle repetitions in an infinite branch of T_ϑ we set

$$\begin{aligned}\mathsf{Rep}(\vartheta) \;&:=\; \{\Sigma \mid \Sigma \in 2^{\mathsf{Cl}(\vartheta)} \text{ and } \Sigma \text{ is } \mathtt{U}\text{-pure}\}\\ &\;\cup\; \{\varphi \mid \varphi \in \mathsf{Cl}(\vartheta) \text{ and } \varphi \text{ is a } \mathtt{U}\text{-formula}\} \times 2^{\mathsf{Cl}(\vartheta)}.\end{aligned}$$

Definition 13. Let $\rho \in \mathsf{Rep}(\vartheta)$. A node v in the tree T_ϑ is a *repeated node for ρ* iff there is a path $v_1, \ldots, v_K$ in T_ϑ for some $K > 1$ such that $v_1 = v$, $\ell(v_1) = \ell(v_K)$ and one of the following items holds.

- $\rho \in 2^{\mathsf{Cl}(\vartheta)}$, v is the conclusion of an instance of $(\mathtt{A}\vec{\mathtt{U}})$, $\mathtt{A}\rho$ is principal for that application, and there is a sequence $\Sigma_1, \ldots, \Sigma_K$ of sets of formulas such that $\Sigma_1 = \Sigma_K = \rho$ and $\langle \ell(v_i), \mathtt{A}(\Sigma_i)\rangle \rightsquigarrow \langle \ell(v_{i+1}), \mathtt{A}(\Sigma_{i+1})\rangle$ for all $i \in [K-1]$.

- $\rho = (\varphi, \Pi)$ for $\{\varphi\} \cup \Pi \subseteq \mathsf{Cl}(\vartheta)$, v is the conclusion of an instance of $(\mathtt{EU})$, φ and $\mathtt{E}(\varphi, \Pi)$ are principals for that instance, and there are a sequence $\varphi_1, \ldots, \varphi_K$ of formulas and a sequence $\Pi_1, \ldots, \Pi_K$ of sets of formulas such that $\varphi_1 = \varphi_K = \varphi$, $\Pi_1 = \Pi_K = \Pi$, and $\langle \ell(v_i), \mathtt{E}(\varphi_i, \Pi_i)\rangle \rightsquigarrow \langle \ell(v_{i+1}), \mathtt{E}(\varphi_{i+1}, \Pi_{i+1})\rangle$ for all $i \in [K-1]$.

A node v_K is called *repeating node for v and ρ.*

Lemma 14. *Let $(v_i)_{i\in\mathbb{N}}$ be an infinite branch in T_ϑ. If the play $(\ell(v_i))_{i\in\mathbb{N}}$ contains a bad thread then there is a $\rho \in \mathsf{Rep}(\vartheta)$ such that for every i there are $j^-, j^+ \in \mathbb{N}$ with $i < j^- < j^+$, v_{j^-} is a repeated node for ρ, and v_{j^+} is their repeating node.*

Proof. Let $(C_i)_{i\in\mathbb{N}}$ be a bad thread in $(\ell(v_i))_{i\in\mathbb{N}}$. Lem. 5 yields two possibilities. We consider the first case only—the other is very similar. So, let k, φ, ψ as written in that alternative. Let $i \in \mathbb{N}$ be given. At least one of the rules $(\mathtt{X}_0)$ and $(\mathtt{X}_1)$ is applied infinitely often, cf. Lem. 3. Therefore and as only the rule $(\mathtt{EU})$ can modify a $\mathtt{U}$-formula, the rule $(\mathtt{EU})$ is also applied infinitely often with $\varphi\mathtt{U}\psi$ as principal formula. Because the amount of different instances of the rule $(\mathtt{EU})$ is finite, there are j^- and j^+ such that $\max(k, i) < j^- < j^+$, v_{j^-} is a repeated node for $\varphi\mathtt{U}\psi$, and v_{j^+} is their repeating node. Indeed, $(C_{i+j^- -1})_{i\in[j^+ - j^- +1]}$ is the φ-sequence as required in Def. 13. To this end, the ith element sequence of the Π-sequence is the set of side formula in the $\mathtt{E}$-block which hosts $C_{i+j^- -1}$. □

The truncation of T_ϑ to a finite tree is realized by the operation $\cdot \Downarrow_\vartheta \cdot$. To this end, the operation considers the elements of $\mathsf{Rep}(\vartheta)$ in some order. So, let $(\mathsf{Rep}_i)_{i\in[|\mathsf{Rep}(\vartheta)|]}$ be an enumeration of $\mathsf{Rep}(\vartheta)$.

Definition 15 **($\cdot \Downarrow_\vartheta \cdot$).** For a tree S and $i \in \mathbb{N}$, the application $S \Downarrow_\vartheta i$ returns a tree. If S is finite or $i > |\mathsf{Rep}(\vartheta)|$, $S \Downarrow_\vartheta i := S$. Otherwise, let

$$V^- := \min_S\{v \mid v \text{ is a repeated node for } \mathsf{Rep}_i \text{ in } S\},$$

$$V^+_{v^-} := \min_S\{v' \mid v' \text{ is a repeating node for } v^- \text{ and } \mathsf{Rep}_i \text{ in } S\}, \text{ and}$$

$$V^\perp := \min_S\{v \mid v \text{ and } v^+ \text{ are } \leq_S\text{-incomparable for all } v^- \in V^- \text{ and } v^+ \in V^+_{v^-}\}$$

where $v^- \in V^-$. For every $v \in V^\perp$ the operation replaces $S|_v$ by $(S|_v) \Downarrow_\vartheta (i+1)$. And for every $v \in V^-$ the operation annotates formulas and truncates subtree in S_v depending on Rep_i. If Rep_i is a set of $\mathtt{U}$-pure formulas $S \Downarrow_\vartheta i$ does the following. By definition, v is the conclusion of the rule $(\mathtt{A}\vec{\mathtt{U}})$ such that Rep_i is principal. The said rule got replaced by an instance of the rule $(\mathtt{A}\vec{\mathtt{U}})^\flat$. As this rule annotates formulas the operation proceeds away from the root as long as the current node is a proper ancestor of an $v' \in V^+_v$. Along this traversal, the rules got adjusted to the annotation. In particular, if the rule is $(\mathtt{A}\vec{\mathtt{U}})$ and the annotated formulas are principal, the rule got replaced by $(\mathtt{A}\vec{\mathtt{U}})^*$. When the traversal reaches an $v' \in V^+_v$ the node is replaced by the rule $(\mathtt{A}\vec{\mathtt{U}})^\dagger$. And as soon as the current node is not an ancestor of any $v' \in V^+_v$, the operation inserts the rule $(\mathtt{A}\vec{\mathtt{U}})^\natural$ to got rid of the annotation and skips the remaining subtree. The procedure for the other case—that is, Rep_i is a pair—is similar but replaces instances of the rule $(\mathtt{EU})$ by $(\mathtt{EU})^\flat$, $(\mathtt{EU})^*$, $(\mathtt{EU})^\natural$ and $(\mathtt{EU})^\dagger$. Since the set V^- is defined in terms of $\min_S$, the adjustments for the annotations do not interfere for different elements in V^-. This completes the description of $\cdot \Downarrow_\vartheta \cdot$.

Theorem 16. *$T_\vartheta \Downarrow_\vartheta 1$ is a finite ϑ-tree which follows the rules in Fig. 1 and 2.*

Proof. Each call $S \Downarrow_\vartheta i$ returns a tree such that its root and S's root share the same label. Therefore by construction, the tree $T_\vartheta \Downarrow_\vartheta 1$ follows the rules. If $\mathsf{Rep}(\vartheta) = \emptyset$ then $T_\vartheta \Downarrow_\vartheta 1 = T_\vartheta$ and by Lem. 14 the tree does not contain any bad thread. Since player $\forall$ wins $\mathcal{G}(\vartheta)$, T_ϑ must be finite. Now, suppose $\mathsf{Rep}(\vartheta) \neq \emptyset$. For the sake of contradiction, assume that $T_\vartheta \Downarrow_\vartheta 1$ is infinite. Since the tree is finitely branching, König's Lemma yields an infinite branch. The execution of $T_\vartheta \Downarrow_\vartheta 1$ leads to at most $|\mathsf{Rep}(\vartheta)| + 1$ nested invocations at a time. For each branch and each invocation the operation $\cdot \Downarrow_\vartheta \cdot$ inserts at most one node, namely the premise to the rule $(\mathtt{A}\vec{\mathtt{U}})^\natural$ or $(\mathtt{EU})^\natural$. Therefore, T_ϑ has an infinite branch. For simplicity we shall neither count nor name these additional nodes—hence the infinite branches in T_ϑ and in $T_\vartheta \Downarrow_\vartheta 1$ are identically equal. By definition, this branch contains a bad thread. Lem. 14 names a $\rho \in \mathsf{Rep}(\vartheta)$. Let i be such that $\mathsf{Rep}_i = \rho$. Since the branch is infinite, there was an invocation $T_\vartheta|_v \Downarrow_\vartheta i$ such that v lies on the infinite branch. Additionally, the same application of Lem. 14 yields two nodes v^- and v^+ on the infinite branch such that $v <_{T_\vartheta} v^- <_{T_\vartheta} v^+$, v^- is a repeated node for ρ, and v^+ is their repeating node. Among all such pairs (v^-, v^+) we minimize v^- and then v^+. Therefore, in the invocation of $T_\vartheta|_v \Downarrow_\vartheta i$ we have that $v^- \in V^-$ and $v^+ \in V^+_{v^-}$. The algorithm truncates the tree at v^+. This is a contradiction to the assumption that the infinite path in T_ϑ passes v^+. □

5 Proving Time

We present a Hilbert-style axiomatisation of $\neg\text{CTL}^\sharp$ and prove it to be sound and complete. An axiomatisation is a set of axioms and a set of rules. The axioms and rules may contain formula variables. A *proof* for some formula φ is a finite sequence $\varphi_0, \ldots, \varphi_n$ s.t. $\varphi = \varphi_n$ and for all $i = 0, \ldots, n$ we have: φ_i is either an instance of an axiom or follows from some $\varphi_0, \ldots, \varphi_{i-1}$ via an instantiation of one of the rules. We write $\vdash \varphi$ to denote that φ is provable.

The axiomatisation is derived from the satisfiability game rules in Fig. 1 and the amended rules in Fig. 2 in the following way. Take a rule in which player $\exists$ makes a choice among premisses $P_1, \ldots, P_n$ from a conclusion C. Then $\neg C$ should be provable from $\neg P_1, \ldots, \neg P_n$. The axioms and rules are presented in Fig. 3. All formula variables α, β and γ range over arbitrary CTL*-formula and φ, χ, and ψ over CTL*-state formula. The axiom (Ax-1) can be made finite using a textbook-like axiomatisation of propositional logic where path-quantified formulas are taken for propositions in the purely propositional logic.

Using the previous sections, soundness and completeness of this axiomatisation is relatively easy to establish.

Theorem 17 (Soundness). *For all $\varphi \in$ CTL*: if $\vdash \varphi$ then $\models \varphi$.*

Proof. By induction on the length of a proof. One easily establishes that all axioms are valid and all the rules preserves validity. In particular, the rules (Ru-3) and (Ru-4) are

(Ax-1) All substitution instances of propositional tautologies

(Ax-2) $\vdash \mathtt{E}(\alpha \vee \beta) \leftrightarrow \mathtt{E}\alpha \vee \mathtt{E}\beta$ (Ax-3) $\vdash \mathtt{E}(\varphi \wedge \alpha) \leftrightarrow \varphi \wedge \mathtt{E}\alpha$

(Ax-4) $\vdash \alpha \mathtt{U} \beta \leftrightarrow \beta \vee (\alpha \wedge \mathtt{X}(\alpha \mathtt{U} \beta))$ (Ax-5) $\vdash (\alpha \rightarrow \beta) \rightarrow (\mathtt{E}\alpha \rightarrow \mathtt{E}\beta)$

(Ax-6) $\vdash \mathtt{X}(\alpha \vee \beta) \leftrightarrow \mathtt{X}\alpha \vee \mathtt{X}\beta$ (Ax-7) $\vdash \mathtt{Ett}$

(Ax-8) $\vdash \mathtt{EXtt}$ (Ax-9) $\vdash \mathtt{EXE}\alpha \leftrightarrow \mathtt{EX}\alpha$

(Ax-10) $\vdash (\alpha \wedge \gamma)\mathtt{U}(\beta \wedge \gamma) \rightarrow \alpha \mathtt{U} \beta$ (Ru-1) If $\vdash \alpha \rightarrow \beta$ and $\vdash \alpha$ then $\vdash \beta$

(Ru-2) If $\vdash \mathtt{E}\alpha \rightarrow \mathtt{E}\beta$ then $\vdash \mathtt{EX}\alpha \rightarrow \mathtt{EX}\beta$

(Ru-3) If $\vdash \mathtt{A}((\gamma \vee \beta) \wedge (\gamma \vee \alpha \vee \mathtt{X}((\alpha \vee \neg\gamma)\mathtt{R}(\beta \vee \neg\gamma))))$ then $\vdash \mathtt{A}(\gamma \vee (\alpha \mathtt{R} \beta))$

(Ru-4) $\begin{cases} \text{If} \vdash \chi \rightarrow (\bigwedge_{j\in[n]} \psi_j \wedge (\bigvee_{j\in[n]} \varphi_j \vee \mathtt{E}(\bigwedge_{j\in[n]} \mathtt{X}((\varphi_j \vee \chi)\mathtt{R}(\psi_j \vee \chi))))) \\ \text{then} \vdash \chi \rightarrow \mathtt{E}(\bigwedge_{j\in[n]} \varphi_j \mathtt{R} \psi_j) \end{cases}$

Fig. 3. Axioms and rules of the $\neg$CTL$^\sharp$ axiomatisation

the negations of the rules $(\mathtt{EU})^\flat$ and $(\mathtt{A}\vec{\mathtt{U}})^\flat$ in the main. Indeed, the proofs of Lem. 8 and 9 also hold for arbitrary formulas in CTL* as long as the configuration is a state formula. □

Lemma 18. – *If Φ is propositionally inconsistent or contains $\mathtt{A}\emptyset$ then $\vdash \neg \bigwedge \Phi$.*

- *If Φ' is a premise and Φ the conclusion of $(\mathtt{X}_0)$ or $(\mathtt{X}_1)$, then $\vdash \neg \bigwedge \Phi'$ implies $\vdash \neg \bigwedge \Phi$.*
- *Let R be a rule in Fig. 1 or 2 apart from $(\mathtt{X}_0)$ and $(\mathtt{X}_1)$. For a fixed principal block and fixed principal formula(s), let $\Phi_1, \ldots \Phi_n$ be all possible premises to a conclusion Φ for the rule R. If $\vdash \neg \bigwedge \Phi_i$ for all $i \in [n]$ then we have $\vdash \neg \bigwedge \Phi$.*

Proof. The argument is mainly straightforward. For the rules of the satisfiability game it suffices to show that the conclusion implies the disjunction of the premises. Exceptions are $(\mathtt{X}_0)$, $(\mathtt{X}_1)$, $(\mathtt{EU})^\flat$ and $(\mathtt{A}\vec{\mathtt{U}})^\flat$. These rules can be proven sound by (Ru-2), (Ru-3) and (Ru-4). In some cases it is necessary to move side formulas first into and later out of the principal block. □

Theorem 19 (Completeness). *For all $\varphi \in \neg$CTL$^\sharp$: if $\models \varphi$ then $\vdash \varphi$.*

Proof. Suppose $\models \varphi$, i.e. $\neg\varphi$ is unsatisfiable. According to Thm. 7, player $\forall$ has a winning strategy for the game $\mathcal{G}(\neg\varphi)$. Thanks to Thm. 16 there is a $\neg\varphi$-tree T which follows the rules of Fig. 1 and 2 and whose root is $\neg\varphi$. By induction on the tree we can construct a proof for φ using Lemma 18. □

Altogether, the axiomatisation is sound and complete for $\neg$CTL$^\sharp$, and hence also for its fragments CTL and CTL$^+$.

6 Conclusion

A task for further work is obviously to extend these techniques to even larger fragments, let alone CTL* itself. It remains to be seen whether the syntactic restriction in CTL$^\sharp$, namely the fact that U-formulas inside of A-quantifiers must have state formulas as arguments, can be relaxed. If these are path formulas to some extend then a connected sequence of A-blocks which does not contain a regenerating R-formula need not become U-pure eventually. The problem is then to find a logical principle which is sound w.r.t. unsatisfiability and which allows one to strengthen such non-U-pure formula sets in order to become unsatisfiable after too many unwindings.

Another problem in this context is the fact that Lemma 8 cannot be made to work for universally path quantified formulas, i.e. by replacing each E in the corresponding rules with A. The reader is invited to attempt to prove the resulting statement. The problem essentially is that path formulas may hold somewhere in a path starting in a state but not in other paths starting in the same state. This, however, is true for state formulas.

References

1. Emerson, E.A., Halpern, J.Y.: Decision procedures and expressiveness in the temporal logic of branching time. Journal of Computer and System Sciences 30, 1–24 (1985)
2. Emerson, E.A., Halpern, J.Y.: "Sometimes" and "not never" revisited: On branching versus linear time temporal logic. Journal of the ACM 33(1), 151–178 (1986)
3. Emerson, E.A., Jutla, C.S.: The complexity of tree automata and logics of programs. SIAM Journal on Computing 29(1), 132–158 (2000)
4. Friedmann, O., Latte, M., Lange, M.: A Decision Procedure for CTL* Based on Tableaux and Automata. In: Giesl, J., Hähnle, R. (eds.) IJCAR 2010. LNCS, vol. 6173, pp. 331–345. Springer, Heidelberg (2010)
5. Gabbay, D., Pnueli, A., Shelah, S., Stavi, J.: The temporal analysis of fairness. In: Proc. 7th Symp. on Principles of Programming Languages, POPL 1980, pp. 163–173. ACM (1980)
6. Johannsen, J., Lange, M.: CTL^+ is Complete for Double Exponential Time. In: Baeten, J.C.M., Lenstra, J.K., Parrow, J., Woeginger, G.J. (eds.) ICALP 2003. LNCS, vol. 2719, pp. 767–775. Springer, Heidelberg (2003)
7. Kröger, F.: Temporal Logic of Programs. Springer (1987)
8. Kröger, F., Merz, S.: Temporal Logic and State Systems. Texts in Theoretical Computer Science. Springer (2008)
9. Lange, M., Stirling, C.: Focus games for satisfiability and completeness of temporal logic. In: Proc. 16th Symp. on Logic in Computer Science, LICS 2001, Boston, MA, USA. IEEE Computer Society Press (2001)
10. Lichtenstein, O., Pnueli, A.: Propositional temporal logics: Decidability and completeness. Logic Journal of the IGPL 8(1), 55–85 (2000)
11. Manna, Z., Pnueli, A.: The Temporal Logic of Reactive and Concurrent Systems Specification. Springer (1992)
12. Martin, D.A.: Borel determinacy. Ann. Math. 102, 363–371 (1975)
13. Penczek, W.: Branching time and partial order in temporal logics. In: Bolc, L., Szałas, A. (eds.) Time and Logic – A Computational Approach, pp. 179–228. UCL Press, London (1995)
14. Prior, A.N.: Time and modality. Oxford University Press, Oxford (1957)
15. Reynolds, M.: An axiomatization of full computation tree logic. Journal of Symbolic Logic 66(3), 1011–1057 (2001)
16. Reynolds, M.: An axiomatization of PCTL*. Information and Computation 201(1), 72–119 (2005)
17. Reynolds, M.: A tableau for bundled CTL*. Journal of Logic and Computation 17(1), 117–132 (2007)
18. Sistla, A.P., Clarke, E.M.: The complexity of propositional linear temporal logics. Journal of the Association for Computing Machinery 32(3), 733–749 (1985)

New Algorithms for Unification Modulo One-Sided Distributivity and Its Variants

Andrew M. Marshall and Paliath Narendran

University at Albany–SUNY
College of Computing and Information
Computer Science Department
{marshall,dran}@cs.albany.edu

Abstract. An algorithm for unification modulo *one-sided* distributivity is an early result by Tiden and Arnborg [14]. Unfortunately the algorithm presented in the paper, although correct, has recently been shown not to be polynomial time bounded as claimed [11]. In addition, for some instances, there exist most general unifiers that are exponentially large with respect to the input size. In this paper we first present a new polynomial time algorithm that solves the *decision* problem for a non-trivial subcase, based on a typed theory, of unification modulo one-sided distributivity. Next we present a new polynomial algorithm that solves the *decision* problem for unification modulo one-sided distributivity. A construction, employing string compression, is used to achieve the polynomial bound.

1 Introduction

Equational unification has long been a core component of automated deduction and more recently has found application in symbolic cryptographic protocol analysis. In particular, the algorithm for unification modulo a *one-sided* distributivity axiom

$$X \times (Y + Z) = X \times Y + X \times Z$$

is an early result by Tiden and Arnborg [14]. More recently this theory has been of interest in protocol analysis due to the fact that many cryptographic operators satisfy this property. Unfortunately the algorithm presented in the paper, although correct, has recently been shown not to be polynomial time bounded as claimed [11]. In addition, for some instances, there exists most general unifiers (mgus) that are exponentially large with respect to the input size. In this paper we examine the *decision* problem for one-sided distributivity. More formally we consider the decision problem for *elementary* unification modulo this theory, where the terms can only contain symbols in the signature of the theory and variables. This is the theory considered by Tiden and Arnborg [14]. We first present a new polynomial time algorithm which solves the *decision* problem for a non-trivial subcase, based on a typed theory. This subcase happens to be sufficient to express the negative complexity result in [11]. Next we present a new polynomial algorithm which solves the *decision* problem for unification modulo one-sided distributivity. We employ string compression through the use of

B. Gramlich, D. Miller, and U. Sattler (Eds.): IJCAR 2012, LNAI 7364, pp. 408–422, 2012.

straight line programs, which allows us to achieve the polynomial bound. Some pioneering work on using compression in unification has been done in [4] and [7]. Due to space restrictions some proof details, figures, and a minor appendix have been left out of this paper. For the interested reader these details can be found in the technical report [9].

2 Preliminaries

We use the standard notation of equational unification [2] and term rewriting systems [1]. Let X, Y, W and Z denote variables. A set of equations S is said to be in *standard form* over a signature F if and only if every equation in S is of the form $X =^? t$ where X is a variable and t, a term over F, is one of the following: (a) a variable different from X, (b) a constant, or (c) a term of depth 1 that contains no constants. We say S is *in standard form* if and only if it is in standard form over the entire signature. It is not generally difficult to decompose equations of a given problem into simpler standard forms.

A set of equations is said to be in *dag-solved form* (or *d-solved form*) if and only if they can be arranged as a list $X_1 =^? t_1, \ldots, X_n =^? t_n$ where (a) each left-hand side X_i is a distinct variable, and (b) $\forall 1 \leq i \leq j \leq n$: X_i does not occur in t_j ([5]).

Definition 1. *A straight-line program (SLP)*[1] *is a context-free grammar, $G = (\Sigma, N, P)$. Where Σ is the set of terminal symbols (these will correspond to a set of "label" variables in this paper), N is a set of nonterminal symbols and P as set of grammar productions. P contains only two types of productions: $N_i \to a$ and $N_i \to N_j N_k$ with $i > j, k$, where N_i, N_j, N_k are nonterminals and a is a terminal. The SLP generates exactly one string corresponding to the top nonterminal.*

The size of a SLP can be defined in several ways. We use the following definitions from [6]. For any terminal, a, define the $depth(a) = 0$ and for any nonterminal N_i

$$depth(N_i) = max\,\{depth(N_{j_i}) + 1 \mid N_i \to N_{j_1} N_{J_2}\}$$

We can define the depth of the SLP as the depth of its top nonterminal. The *size* of a SLP, S, is defined to be the number of productions and is denoted as $\|S\|$. We denote the length of a string produced by a SLP S by $|S|$. Note that the lengths are represented in binary.

3 Typed System and Single Homomorphism

We present a typed system interpretation of one-way distributive unification. We begin with the simplest non-trivial subcase, the case of a *single homomorphism*. This is non-trivial because the exponential complexity result in [11] holds in this

[1] These are also known as Singleton Context-Free Grammars in [7].

case as well. Consider a 'type' system based on two types τ_1 and τ_2. We let all left multiplication variables be of type τ_1 and all right variables of type τ_2. Thus

$$\times : \tau_1 * \tau_2 \to \tau_2,$$
$$+ : \tau_2 * \tau_2 \to \tau_2,$$

Therefore, if there is only a single variable of type τ_1 in the input equations then we can consider the multiplication operation as a homomorphism h over $+$. This is the single homomorphism case, it restricts the number of valid terms from the general case but it is sufficient for encoding the exponential example in [11] and it yields a much simplified decision algorithm.

Definition 2. *We define the following relations (X, Y and Z are variables):*
$X \succ_h Y$ *if* $X = h(Y)$ $\quad$ $X \succ_{l_+} Y$ *if* $X = Y + Z$.
$X \succ_{r_+} Z$ *if* $X = Y + Z$ $\quad$ $X \succ_a Z$ *if* $X = Y + Z$ *or* $X = Z + Y$.

We denote the transitive closure of a relation R as R^+. For a unification problem S in standard form we construct the following two graphs (similar to the graphs used in [14]).

Definition 3. *A* path labeled dependency graph *($\mathcal{LD}$) is a directed graph such that the nodes in the graph correspond to variables of type τ_2. We form two kinds of edges:*
(i) Lateral *edges, where for each equation of the form $X =^? T \times Y$, we have an edge from node X to node Y labeled with a* label variable, *T of type τ_1. In the single homomorphism case $T = h$ and the labels are numbers corresponding to the number of homomorphisms. Thus, for single edges corresponding to a single homomorphism the label is* 1. *For paths corresponding to multiple homomorphisms the label is m, where m is the number of homomorphisms.*
(ii) Downward *edges, where for each equation of the form $X =^? X_1 + X_2$, we have directed edges from node X to node X_1 and from node X to node X_2.*

Definition 4. *The* path labeled propagation graph *($\mathcal{LP}$) is a directed simple graph. Its vertices are the equivalence classes of the symmetric, reflexive, and transitive closure of the relation defined by $\succ_h$ on the $\mathcal{LD}$ graph for the same system. Edges exist between equivalence classes $[X]$ and $[Y]$ if there exists variables $U \in [X]$ and $V \in [Y]$ such that $U \succ_a V$.*

These graphs, mainly the $\mathcal{LD}$, will be the primary data structure and will be processed in a similar fashion to the original algorithm [14] but using a modified set of graph *saturation* rules. The $\mathcal{LP}$ graph is included due to a specific type of non-unifiable (as was proven in [14]) system. An example of this is the following set of equations. $\{X =^? V + Y,\ X =^? h(Y)\}$. It can be seen that this will cause infinite application of rule (vii). This problem was solved in [14] by using the *propagation graph*. This is due to the fact that this type of non-unifiable systems will cause cycles in the propagation graph. Therefore, each time the algorithm updates the $\mathcal{LD}$ graph it also updates the $\mathcal{LP}$ graph and checks for cycles. Likewise, if cycles are found the algorithm terminates with failure. We now introduce a set of saturation/inference rules. Rule (0) acts on the system S and rules (i) through (vii) act on the $\mathcal{LD}$ graph of S. Rule (0) is simple variable replacement. Rules (i) - (iii) are cancellation rules that follow directly from [14]. Rule (vi) is a failure rule that corresponds to occur-check type errors. Rules (iv), (v), (vii) are path completion rules. Rule (vii) is the same path propagation rule from the Tiden and Arnborg algorithm, justified by the axioms of the system; see [14]. *But, in rule (vii) we do not create the new variables W_1 and W_2 unless W has no children variables related along a $\succ_a$ edge.* We denote a path and its label π between nodes X and Y by $X \xrightarrow{\pi} Y$. For the single homomorphism case paths are simply the composition of many occurrences of the homomorphism. Therefore, a path π is essentially h^j for some $j \in \mathbb{N}$ and the path length, $|\pi|$, is just j. During saturation we derive *path constraints* of the form $\pi_1 =^? \pi_2$ or $\pi_1 \prec^? \pi_2$. For the single homomorphism case, because there is just one homomorphism, $\pi_1 =^? \pi_2$ is simply a check if the lengths are equal, i.e., if $|\pi_1| = |\pi_2|$. For the prefix check $\pi_1 \prec^? \pi_2$, in the single homomorphism case we only need to check if the length of π_1 is less then π_2, i.e., $|\pi_1| < |\pi_2|$. It is important to note that path lengths are kept in binary representation. This compression is significant as it allows us to avoid exponential growth in the path lengths. In addition to path constraints we will need to perform several *path computations*, specifically we need to *concatenate paths* and *compute path suffixes*. These operations can be accomplished, in the single homomorphism case, by simple addition and subtraction.

(0) $$\frac{S \uplus \{U =^? V\}}{\{U \mapsto V\}(S) \cup \{U =^? V\}} \quad \text{if } U \text{ occurs in } S$$

(i) $$\frac{X \xrightarrow{\eta} Y, \quad X \xrightarrow{\pi} Z, \quad |\eta| = |\pi|}{X \xrightarrow{\pi} Z, \quad Y =^? Z}$$

(ii) $$\frac{X \xrightarrow{\pi} Y, \quad Z \xrightarrow{\pi} Y}{X =^? Z, \quad Z \xrightarrow{\pi} Y}$$

(iii) $$\frac{U =^? U_1 + U_2,\ U =^? U_3 + U_4,}{U =^? U_1 + U_2,\ U_3 = U_1,\ U_4 = U_2}$$

(iv) $$\frac{X \xrightarrow{h^j} Y, \quad X \xrightarrow{h^i} Z, \quad j < i}{Y \xrightarrow{h^{i-j}} Z, \quad X \xrightarrow{h^i} Z}$$

(v) $$\frac{X \xrightarrow{h^i} Y,\ Y \xrightarrow{h^j} Z}{X \xrightarrow{h^{i+j}} Z, \quad Y \xrightarrow{h^j} Z}$$

(vi) $$\frac{S}{FAIL} \quad \text{if } \mathcal{LP} \text{ or } \mathcal{LD} \text{ are cyclic}$$

(vii) $$\frac{U \xrightarrow{\eta} W,\ U =^? U_1 + U_2,}{U \xrightarrow{\eta} W,\ W =^? W_1 + W_2,\ U_1 \xrightarrow{\eta} W_1,\ U_2 \xrightarrow{\eta} W_2}$$

We will need to apply the rules in a specific order to ensure we maintain a polynomial time bound. Let $\sim_h$ stand for the reflexive, symmetric and transitive closure of $\succ_h$. Thus $\sim_h$ defines a set of equivalence classes over a set of variables. Denote these classes as $[Y]_h$. We can note that the $\mathcal{LP}$ graph has exactly these classes as its nodes. We can define a strict partial ordering $>_h$ on the $\sim_h$-equivalence classes based on $\succ_a$. That is, $[X]_h >_h [Y]_h$ if and only if there exist $K_1 \in [X]_h$ and $K_2 \in [Y]_h$ such that $K_1 \succ_a K_2$.

Given these rules the decision algorithm for the single homomorphism subcase is presented in Algorithm 1. We next discuss the correctness and complexity of Algorithm 1, most of these results will follow directly from [14].

Algorithm 1. Unification modulo a Single Homomorphism

(Input: A system of equations in standard form)
(1: Generate data structures) Generate the graphs, $\mathcal{LD}$ and $\mathcal{LP}$.
(2: Clean up the system) Exhaustively apply the rules (0), (i), (ii), (iii) and (iv).
(3: Error checking) Apply graph cycle checking to the two graphs (i.e., rule (vi)). If a cycle is found *stop* with failure.
(4: Process equivalence class) Select an equivalence class based on the strict partial ordering $>_h$. That is, we select the largest element of $>_h$ that has not yet been processed. Thus, if we select the class $[X]_h$ then there does not exist a class $[Y]_h$ such that $[Y]_h$ has not been processed and $[Y]_h >_h^+ [X]_h$. Clearly, if $>_h$ is not a strict partial ordering then there is a cycle in the $\mathcal{LP}$ graph.

First we apply rule (v) — this is done by starting with the sink node of the path and working back to the start node of the path. Once rule (v) has been exhaustively applied we apply rule (vii) if applicable. (Note that after rule (v) the class will be a one-level tree thus when rule (vii) is applied W must be the sink)
(5: Check if Complete) If no inference rules can be applied and no cycles exist, then exit with success, else return to Step 2.

3.1 Correctness

Correctness of the inference rules can be assured due to the correctness proof of the algorithm presented in [14] and the following lemmas.

Lemma 1. *Soundness of rules (i) through (vii) are direct consequences of Lemmas (5) through (7) of [14] and variable replacement.*

Proof. Rules (i) through (iv) and (vi) through (vii) follow from Lemmas (5) through (7) of [14] and rule (v) follows from simple variable replacement.

It remains to be shown that if the algorithm terminates without failure then the system is indeed unifiable.

Lemma 2. *Given a system S in standard form if no failure errors occur the Algorithm 1 transforms S, through its $\mathcal{LD}$ graph representation, into dag-solved form.*

We need now to show that if the system is not unifiable that the algorithm correctly reports that as a failure. Directly from [14] we get the following two results.

Lemma 3. *Cycles in the $\mathcal{LP}$ graph for a system S in standard form imply that S is not unifiable.*

Lemma 4. *Cycles in the $\mathcal{LD}$ graph for a system S in standard form imply that S is not unifiable.*

Lemma 3 is due to the fact that a cycle in $\mathcal{LP}$ corresponds to a system requiring an infinite unifier, see [14]. Lemma 4 corresponds to occur check type errors (caught by rule (vi), graph cycle checking) and thus implies non-unifiability.

Theorem 1. *Algorithm 1 is correct.*

Proof. Follows from Lemma 1 to Lemma 4.

3.2 Complexity

First we get the following result from the cancellative nature of the rules (i) through (iii).

Lemma 5. *Given a $\mathcal{LD}$ graph rules (0)-(iii) can only be applied a polynomial number of times with respect to the initial set of nodes in the graph.*

In addition, we get the following clear result.

Lemma 6. *Given a $\mathcal{LD}$ graph rule (iv) can only be applied a polynomial number of times with respect to the initial set of nodes in the graph.*

Lemma 7. *Each equivalence class formed by closure along $\succ_h$-related nodes has a unique sink.*

Lemma 8. *Processing an equivalence class (Step 4) takes polynomial time with respect to the number of variables in the class.*

Proof. Applying (v) exhaustively starting from the sink is therefore bounded linearly by the number of nodes in the class. In addition, the number of applications of (vii) is also bounded by the number of nodes in the class.

Lemma 9. *The number of $\sim_h$-equivalence classes for a system S in standard form can never increase.*

Proof. New variables created by rule (vii) don't create new equivalence classes as they are added to pre-existing classes.

Lemma 10. *A maximum of 2 new nodes can be added to an equivalence class from any equivalence class one level higher, by $\succ_h$.*

In addition by rule (vii) we get the following.

Lemma 11. *During processing the number of lateral edges added to a $\sim_h$-equivalence class from a higher, by $\succ_h$, $\sim_h$-equivalence class cannot exceed the number of nodes in the higher equivalence class.*

Combining the above results we get the following.

Theorem 2. *The running time of Algorithm 1 is polynomial with respect to the initial set of equations.*

4 General Algorithm

We now consider the general problem, with no type system. As in the single homomorphism case we interpret the equations of a unification problem as graphs.

Definition 5. *we define the following additional relations:*
$X \succ_{r_*} Y$ *if* $X = Z \times Y$ $X \succ_m Y$ *if* $X \succ_{l_*} Y$ *or* $X \succ_{r_*} Y$ $X \succ_{l_*} Z$ *if* $X = Z \times Y$

We make the following additions/modifications to the $\mathcal{LD}$ graph.

Definition 6. *Downward edges remain the same, we make the following modifications to the lateral edges and add one additional type of edge.*

(1) Lateral *Edges, where for each equation of the form* $X =^? Z \times Y$, *we have an edge from node* X *to node* Y *labeled with the top nonterminal of a SLP generating the* label variable, Z. *Label variables are kept as straight line programs, where the terminals corresponds to the label variable. Each label variable,* Z, *is given a unique single production SLP. Therefore, lateral edge and path labels correspond to the top nonterminal of the SLP generating the label variables corresponding to those edges. We denote a path and its label,* π, *of one or more lateral edges between nodes* X *and* Y *by* $X \xrightarrow{\pi} Y$. *For the general case paths are the composition of any number of the label variables. Therefore, a path* π *is notation for a path* $X_{i_1} \cdot X_{i_2} \cdot \ldots \cdot X_{i_m}$ *for some* $m \in \mathbb{N}$ *and is kept altogether compressed as a SLP.*

(2) Relation *Edges, where for each node* X *in the graph such that there exists a path* $X \xrightarrow{\pi} Y$ *and for each terminal/label variable* K_i *in the SLP* π, *we have a single edge from* X *to the* node K_i *in the graph. These edges will only be used for cycle checking and could even be generated just before the graph is checked for cycles in the algorithm.*

The initial $\mathcal{LD}$ graph will be built from an initial unification problem, S, in standard form. That initial graph will not have any composite paths labeled by a *SLP* with more then one production. The "composite" paths will be added later by the algorithm. In addition, when constructing the $\mathcal{LD}$ graph each variable X from the set of label variables is given a unique *SLP*. For example, a label variable X would be given a *SLP* $\pi_X \to X$ and all lateral edges formed by an equation with X as the label variable would be labeled by π_X. This implies that different lateral edges can have the same edge label and improves efficiency. Example, in the $\mathcal{LD}$ graph of Figure 1 the edges $X \to Y$ and $L_2 \to L_3$ have the same *SLP* label because they used the same label variable in the equations $X =^? Z_1 \times Y$ and $L_2 =^? Z_1 \times L_3$. The *Relation* edges are simply used to track the relation between a node and the variables labeling a path from that node. This information is needed as variables can both be label variables and nodes in the graph. Let us consider an example $\mathcal{LD}$ graph for the following system of equations (r denotes relation edges).

$$X =^? X_1 + X_2,\ X =^? \pi_1 \times Y,\ Y =^? \pi_2 \times L,\ L =^? L_1 + L_2,\ L_1 =^? \pi_5 \times K,\ L_2 =^? \pi_1 \times L_3$$

where the *SLP*s are: $\pi_1 \to Z_1$, $\pi_2 \to Z_2$ and $\pi_5 \to \pi_3\pi_4$, $\pi_3 \to \pi_2\pi_2$, $\pi_4 \to \pi_1\pi_1$.

The corresponding $\mathcal{LD}$ graph is given in Figure 1.

We make the following modifications to the $\mathcal{LP}$ graph.

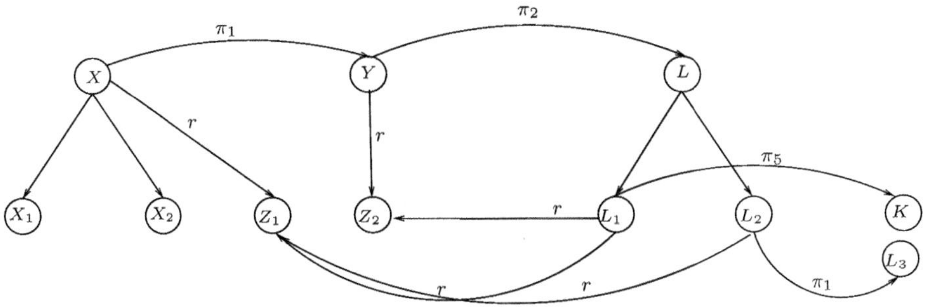

Fig. 1. LD graph example

Definition 7. *Vertices are now the equivalence classes of the symmetric, reflexive, and transitive closure of the relation defined by $\succ_{r_*}$ on the new $\mathcal{LD}$ graph. Edges between equivalence classes are still based on the $\succ_a$ relation.*

The $\mathcal{LP}$ graph is included due to a specific type of non-unifiable (as was proven in [14]) system. This problem was solved in [14] by using the *propagation graph*, which is equivalent to the $\mathcal{LP}$ graph. This is due to the fact that this type of non-unifiable systems will cause cycles in the propagation graph. One consequence of this new graph interpretation is that the label variable paths can grow exponentially in length with respect to the initial set of label variables. This can be seen using the same example used to prove the exponential result in [11]. If the algorithm presented below (without compression) is applied to the system in [11], we get label paths of exponential length. The growth is due to the path string being copied and then doubled at each consecutive level. Although this doubling of the string leads to the exponential growth it also requires the re-use of the string and this suggests the use of string compression. Therefore, we keep each of these paths compressed in the form of straight line programs or $SLPs$.

The algorithm will work by "saturating" the graphs. A set of transformation rules is used to either convert the graph into a solved form or detect a cycle in the graph. The first case implies unifiability and the second non-unifiability. During saturation we derive *path constraints* of the form $\pi_1 =^? \pi_2$ or $\pi_1 \prec^? \pi_2$. The constraint $\pi_1 \prec^? \pi_2$, is a prefix check (i.e., whether the string produced by the SLP π_1 is a prefix of the string produced by the SLP π_2) and $\pi_1 =^? \pi_2$, similarly, is an equality check. In addition to path constraints we will need to perform several *path computations:* specifically we need to *concatenate paths*, *compute path suffixes* and find a single pair of *mismatched terminals* in two unequal SLP produced strings, all without decompressing the $SLPs$.

4.1 Algorithm Presentation

We first present a set of inference rules for a system S in standard form that act on the $\mathcal{LD}$ graph for S. The rules are applied to the graph $\mathcal{LD}$ and as that graph is updated the $\mathcal{LP}$ is updated.

(0) $$\frac{S \uplus \{U =^? V\}}{\{U \mapsto V\}(S) \cup \{U =^? V\}} \quad \text{if } U \text{ occurs in } S$$

(i) $$\frac{U =^? U_1 + U_2,\ U =^? U_3 + U_4,}{U =^? U_1 + U_2,\ U_3 = U_1,\ U_4 = U_2}$$

(ii) $$\frac{X \xrightarrow{\pi} Y, \quad Z \xrightarrow{\eta} Y \quad \eta = \pi}{X =^? Z, \quad X \xrightarrow{\pi} Y}$$

(iii) $$\frac{X \xrightarrow{\eta} Y, \quad X \xrightarrow{\pi} Z, \quad \eta = \pi}{Y =^? Z, \quad X \xrightarrow{\pi} Z}$$

(iv) $$\frac{X \xrightarrow{\eta} Y, \quad X \xrightarrow{\pi} Z, \quad \eta \prec \pi}{Y \xrightarrow{\eta^{-1}\pi} Z, \quad X \xrightarrow{\pi} Z}$$

(v) $$\frac{X \xrightarrow{\eta} Y, \quad X \xrightarrow{\pi} Y, \quad \eta \neq \pi}{\eta =^? \pi}$$

(vi) $$\frac{X \xrightarrow{\eta} Y, \quad X \xrightarrow{\pi} Z, \quad \eta \not\prec \pi}{\eta \prec^? \pi} \quad \text{if } |\eta| < |\pi|$$

(vii) $$\frac{X \xrightarrow{\eta} Y, \quad X \xrightarrow{\pi} Y, \quad \eta = \pi}{X \xrightarrow{\eta} Y}$$

(viii) $$\frac{U \xrightarrow{\eta} W,\ U =^? U_1 + U_2,\ W =^? W_1 + W_2}{W =^? W_1 + W_2,\ U_1 \xrightarrow{\eta} W_1,\ U_2 \xrightarrow{\eta} W_2,\ U \xrightarrow{\eta} W}$$

(ix) $$\frac{U \xrightarrow{\eta} W,\ U =^? U_1 + U_2,}{W =^? W_1 + W_2,\ U_1 \xrightarrow{\eta} W_1,\ U_2 \xrightarrow{\eta} W_2,\ U \xrightarrow{\eta} W}$$

(x) $$\frac{S}{FAIL} \quad \text{if } \mathcal{LP} \text{ or } \mathcal{LD} \text{ are cyclic}$$

(xi) $$\frac{X \xrightarrow{\pi} Y,\ Y \xrightarrow{\eta} Z}{X \xrightarrow{\sigma=(\pi\eta)} Z, \quad Y \xrightarrow{\eta} Z}$$

Rule (0) is simply a variable replacement rule but it has a special action on label variables: *if a label variable is equated to a non-label variable then the non-label variable is replaced by the label variable.* This rule acts directly on the system S by doing variable replacement whenever there is an equation between two variables. Note, whenever a rule creates an equation of the form $X =^? Y$, those two nodes in the graph are equated and rule (0) applies that equation to the set S. Rules (i) through (xi) act on the $\mathcal{LD}$ graph. Rule (i) is due to the cancellative nature of the $+$ operator ([14]). Rules (ii), (iii) and (iv) are due to the cancellative nature of $\times$ ([14]). Rules (v) and (vi) check the path constraints and attempt to find label variables that have to be equated. These rules and rules (iv) and (xi) are explained in more detail in Section 4.4. Rules (ix) and (viii) directly correspond to the splitting rule of [14] and are a direct consequence of the distributive axiom. These two versions are just modifications to work in the modified graph setting. The difference between the two rules is that rule (ix) creates new variables as rule (viii) does not. Rule (x) is a failure rule, which roughly corresponds to detecting a cycle in the dependency graph in the Tidén-Arnborg algorithm. Redundant edges are removed by rule (vii). Finally, rule (xi) is a path completion rule, justified by the soundness of variable replacement. This rule is also responsible for building the new $SLPs$, σ, with more then one production. More details on rule (xi) are given in Section 4.4. Let $\sim_r$ stand for the reflexive, symmetric and transitive closure of $\succ_{r_*}$. Thus $\sim_r$ defines a set of equivalence classes over a set of variables. Denote these classes as $[Y]_r$. We can note that the $\mathcal{LP}$ graph has exactly these classes as its nodes. We can define a strict partial ordering $\gg_r$ on the $\sim_r$-equivalence classes based on $\succ_a$. That is, $[X]_r \gg_r [Y]_r$ if and only if there exist $K_1 \in [X]_r$ and $K_2 \in [Y]_r$ such that $K_1 \succ_a K_2$. We now give Algorithm 2 with the details.

Algorithm 2. One-sided Distributive Unification

(Input: A system of equations in standard form)

(1: Generate data structures) Generate the 2 graphs, $\mathcal{LD}$ and $\mathcal{LP}$. Make a note of the initial label variables in S; denote this set as $\mathcal{V}$.

(2: Clean up the system) Exhaustively apply the following composite rule: $(0 + i + ii + iii + vii)$

(3: Error checking) Apply graph cycle checking to the graphs (i.e., rule (x)). If a cycle is found *stop* with failure. If the graphs have no cycles and are in dag-solved form, *exit* with success.

(4: Process equivalence class) Select an equivalence class based on the strict partial ordering $\gg_r$. That is, we select the largest element of $\gg_r$ that has not yet been processed. Thus, if we select the class $[X]_r$ then there does not exist a class $[Y]_r$ such that $[Y]_r$ has not been processed and $[Y]_r \gg_r^+ [X]_r$.
Process the selected class using the following composite rule:
$(v + vi)(iv)^!(xi)^!(viii)(ix)^!$
Rule (ix) is applied by starting with the sink node of the path and working back to the start node of the path. In addition, if rule (v) or rule (vi) is applied label variables will be equated therefore we goto step (5).

(5: Checking) If any of the variables in $\mathcal{V}$ are equated go back to Step 1 else go back to Step 2. That is, if label variables are equated go back to step 1.

Let r_1 and r_2 denote inference rules. Then, $r_1^!$ indicates *exhaustive* application of the rule r_1. Therefore, the composite rule $r_1^! r_2$ means to apply r_1 until it cannot be applied any more and then try to apply r_2. Note that even if r_1 cannot be applied the rule $r_1^! r_2$ can still be used if r_2 can be applied. Thus $r_1^!$ does not indicate that r_1 *must* be applied but rather that if r_1 can be applied we do so exhaustively. $r_1 + r_2$ indicates choice: apply rule r_1 *or* rule r_2. Therefore, the last composite rule implies that rule (ix) has the lowest priority and that rule (viii) is only applied once in the processing of a single equivalence class.

We also keep and update SLP string length information as saturation of the graph proceeds. As the lengths of the strings produced by any SLP grow or decrease as a result of application of rules (i)-(xi), this information is updated immediately by one of the following rules, i.e., an application of (xi) causes an application of (d). Let l and m be string lengths (kept in binary representation). The $SLPs$ are built bottom up as are the strings they generate allowing us to keep track of that information. The following set of rules explain how the string length information is modified by the corresponding graph saturation rules. This information is needed for some of the SLP algorithms.

(d) $$\frac{X \xrightarrow{l} Y, \qquad Y \xrightarrow{m} Z}{X \xrightarrow{l+m} Z}$$

(e) $$\frac{U \xrightarrow{l} W,\ U =^? U_1 + U_2,\ W =^? W_1 + W_2}{U_1 \xrightarrow{l} W_1,\ U_2 \xrightarrow{l} W_2}$$

(f) $$\frac{X \xrightarrow{l} Y, \qquad X \xrightarrow{m} Z, \qquad l < m}{Y \xrightarrow{m-l} Z}$$

4.2 Label Variables

We need several results about the label variables and their interaction with the new variables. Let $\mathcal{V}_0$ denote the set of initial label variables for a system S and $\mathcal{V}$ the set of label variables at any point during the application of Algorithm 2 on S. Let $\mathcal{Z}$ denote the set of fresh variables created by rule (ix).

Lemma 12. *During application of Algorithm 2, $\mathcal{V} \cap \mathcal{Z} = \emptyset$.*

Proof. By the definition it is not possible to apply rule (0) such that a newly created variable is made a label variable.

Lemma 13. *During application of Algorithm 2 on a system S in standard form, $|\mathcal{V}| \leq |\mathcal{V}_0|$*

Proof. This follows directly from rule (0) and Lemma 12.

Thus we can safely assume that there will never be an equation of the form $X =^? Z \times Y$, where Z is new. But we could have a relation $Z \succ_{l_*} X$ where X is some pre-existing variable. This implies that there exists a variable K and an equation of the form $K =^? L \times Z$. This leads to the observation (assuming the modified variable replacement method) that we only need to apply rule (ix) to the last variable in a path of $\succ_{r_*}$-related variables.

4.3 Soundness and Completeness

We recall several results already proven by Tidén and Arnborg on one-sided distributive unification.

Theorem 3. *Tidén and Arnborg [14]*

1. *(Lemma 5) A set of equations $\{U =^? X_1 \circ X_2,\ U =^? T_1 \circ T_2\}$ has precisely the same solutions as the set of equations $\{U =^? X_1 \circ X_2,\ X_1 =^? T_1,\ X_2 =^? T_2\}$, where $\circ$ is $\times$ or $+$.*
2. *(Lemma 6) Every unifier for the set of equations $\{U =^? V \times W,\ W =^? W_1 + W_2,\ X =^? V \times W_1,\ Y =^? V \times W_2\}$ is a unifier for the set of equations $\{U =^? V \times W,\ U =^? X + Y\}$.*

Lemma 14. *Rules (0) through (ix) and rule (xi) are sound.*

Proof. Rule (xi) and rule (0) follows from the soundness of variable replacement. Rule (i) follows from Theorem 3 part 1. Now consider rules (ii) and (iii), since $\pi = \eta$, these rules follow from $|\pi|$ applications of Theorem 3 part 1. The same holds for rule (iv) except η is a prefix. Rules (v) and (vi) also follow from Theorem 3 part 1. Rule (vii) is simply removing duplicates. Finally, the soundness of rules (ix) and (viii) follows from Theorem 3 part 2.

Lemma 15. *If Algorithm 2 exits with success on a system S in standard form, then S is unifiable.*

Proof. The result follows from the fact that the set of inference rules transforms S into a dag-solved form.

Lemma 16. *If Algorithm 2 terminates with failure on a system S, then S is not unifiable.*

4.4 Graph and SLP Operations

The $\mathcal{LD}$ graph is updated by the algorithm as the inference rules operate on it. The $\mathcal{LP}$ is built from the $\mathcal{LD}$ and thus can be updated after updating the $\mathcal{LD}$ graph. We note that the $\mathcal{LD}$ and $\mathcal{LP}$ graphs can use standard cycle checking algorithms.

Lemma 17. *The $\mathcal{LP}$ and $\mathcal{LD}$ graphs for a system S in standard form can be checked for cycles in polynomial time with respect to the size of S.*

The $SLPs$ are formed by first encoding the label variables as $SLPs$. Each unique label variable get a unique SLP For example, when creating the $\mathcal{LD}$ graph for two equations $X =^? Y * Z$ and $K =^? Y * L$ only one SLP is created, $\pi_Y \rightarrow Y$, and two edges are labeled by that SLP, i.e., by the top nonterminal π_Y. Then, larger/additional SLPs are formed, bottom up, by the inference rules (xi) and (iv). In addition, *we only keep a single copy of each unique SLP.* This implies we only keep the set of all productions. Then, when creating a new larger SLP we need only create the new top production.
Rule (xi): Rule (xi) forms a new SLP by concatenating two existing $SLPs$. Rule (ix) is applied by starting with the sink node of the path and working back to the start node of the path. This simply ensures a minimal number of applications of rule (xi). To concatenate two $SLPs$, $I = (\Sigma, N_I, P_I)$ and $J = (\Sigma, N_I, P_J)$,[2] we create a new SLP, $K = (\Sigma, N_K, P_K)$. Let π_I and π_J be the top nonterminals of I and J respectively. Then, $N_K = N_I \cup N_J \cup \{\pi_K\}$ and $P_K = P_I \cup P_J \cup \{\pi_K \rightarrow \pi_I \pi_J\}$. This is a simplified version, with just two $SLPs$, of the method presented in [7]. Therefore, the following result easily follows.

Lemma 18. *Let $I = (\Sigma, N_I, P_I)$ and $J = (\Sigma, N_I, P_J)$ be two $SLPs$. Then there exists a SLP $K = (\Sigma, N_K, P_K)$ that generates the concatenation of the two strings generated by I and J such that $\|K\| = |P_I \cup P_J| + 1$ and $depth(K) \leq max(depth(I), depth(J)) + 1$.*

[2] Because all the $SLPs$ use the set of label variables as the set of terminals, Σ is the same for all $SLPs$.

Additional algorithms for doing concatenation can be found in [6, 13].

Rules (ii) and (iii): These two rules require that we can decide if two compressed strings are equal, $\pi_1 =^? \pi_2$. The area of fully compressed pattern matching is an active area and there are many algorithms that will solve this problem, in polynomial time with respect to the size of the SLP. We cite the following, non-exhaustive, list of papers for excellent algorithms; [8, 10, 13, 12].

Rule (iv): We can partially order the nodes in each equivalence class based on the lateral edges, i.e., based on the $\succ_{r_*}$ relation. Rule (iv) is applied based on this partial ordering, starting from the source nodes and working down to the sinks. Therefore we do not apply rule (iv) to a node X if rule (iv) can be applied to a node Y such that there is a lateral path from Y to X. Rule (iv) requires that we can decide if one SLP π_1 is a prefix of an SLP π_2, $\pi_1 \prec^? \pi_2$, in polynomial time with respect to π_2. This problem has been solved in [8]. We also need to extract the suffix in compressed form, $\pi_3 = {\pi_1}^{-1}\pi_2$. Because we build the SLPs bottom up and keep the length information, a simple polynomial-time recursive algorithm can accomplish this. We could also use the general method for prefixes and suffixes developed in [7]. Therefore, we get the following result.

Lemma 19. *Let $I = (\Sigma, N_I, P_I)$ and $J = (\Sigma, N_I, P_J)$ be two SLPs such that the string generated by J is a prefix of the string generated by I. Then, there exists a SLP $K = (\Sigma, N_K, P_K)$ that generates the suffix of the string generated by I after removing the prefix string generated J such that $\|K\| \leq \|I\| + depth(I)$ and $depth(K) \leq depth(I)$.*

Rules (v) and (vi): These rules essentially handle the situation where two label paths should be equal, or one a prefix of the other, but are found not to be. We then need to check if they can be made equal. We accomplish this by finding at least one pair, (X, Y), of terminals (label variables) in the corresponding SLPs such that these terminals form a mismatch, $X \neq Y$. One pair will do for each application of the rule because by the cancellative nature of $\times$ all mismatched pairs of terminals *must* be equated. In [3] a nice polynomial algorithm for finding the first mismatch is developed. A mismatch can also be found using the algorithms in [8, 10]. The way these rules work in Algorithm 1 is that if in the $\mathcal{LD}$ graph one of the rules is satisfied a pair of label variables will be found and equated (through the use of rule (0)). This will cause the set of label variables, $\mathcal{V}$, to be reduce and thus the number of label variables in the system S to be reduced. The algorithm then returns to step 1 and rebuilds a new $\mathcal{LD}$ graph from the newly modified system, S.

4.5 Complexity

Lemma 20. *After processing any equivalence class, the number of sinks for that class is at most one and every non-sink node in the class has exactly one outgoing edge.*

Proof. If there is no sink in the class, then this implies a cycle and thus a non-unifiable system. In addition, there must be at least one source node. It can be seen that rules (ii), (iii), (iv), (v) and (vi) ensure that all the nodes in the class have at most a single outgoing edge.

Lemma 21. *The maximum number of new variables added to the system S is equal to twice the number of equivalence classes.*

Lemma 22. *Let $[X]_r$ be a $\sim_r$-equivalence class. Assume there exist K $\sim_r$-equivalence classes one level above $[X]_r$ by the $>_r$ ordering. Let us denote the K classes as $C_1, C_2, \ldots, C_K$ and assume that each class C_i contains n_i variables, such that $N_K = \sum_{i=1}^{K} n_i$. Then the total number of lateral edges added to $[X]_r$ by the K higher classes is $\leq 2 * N_K$.*

Lemma 23. *The maximum number of lateral edges added to any $\sim_r$-equivalence class of a simple system S is $O(N+M)$, where N is the initial number of variables in S and M is the number of equivalence classes.*

Lemma 24. *The number of $\sim_r$ equivalence classes never increases.*

Proof. Rule (ix) is the only rule that creates new variables but these variables are contained in pre-existing equivalence classes.

Lemma 25. *The number of inference rule applications used during a single application of step (2) of Algorithm 2 is polynomially bounded by the number of edges in the $\mathcal{LD}$-graph at the start of step (2).*

Proof. Clearly rules (i) - (iii) and (vii) are bounded by the number of edges.

Lemma 26. *The number of inference rule applications used to process a single equivalence class (step (4) of Algorithm 2) is polynomially bounded with respect to the number of lateral edges and nodes in that class.*

Finally, we need to bound the size of the $SLPs$.

Lemma 27. *The largest, in size, SLP constructed by Algorithm 2 on any unification problem S is polynomially bounded by the initial number of lateral edges in S.*

Theorem 4. *The worst-case running time of Algorithm 2 is a polynomial in the initial number of variables N and the initial number of $\mathcal{LD}$ graph edges L.*

Proof-Sketch: Let us consider each step in Algorithm 2.
Step 1: By Lemma 23 and Lemma 21, building the graphs can be done in polynomial time. Denote this polynomial as P_1. Step 2: Combining Lemma 27 with the cancellative rules, (i)-(iii), we get that Step (2) is bounded by a polynomial in N and L. Denote this polynomial as P_2. Step 3: By Lemma 17 cycle checking the graphs is polynomially bounded by the size of the graphs. Denote this polynomial as P_3. Step 4: By Lemma 26, Lemma 27, Lemma 23 and Lemma 21, the time to process each class is polynomially bounded in L and N. Denote this polynomial as P_4. Step 5: If M be the number of equivalence classes and V the number of initial label variables, by Lemma 13, V does not increase and

by Lemma 24, M does not increase. Therefore, we apply steps (2) through (5) M times, once for each equivalence class. We can reset (applying step (1)) and apply them again for each class but this only happens when we equate two label variables. Thus this cannot happen more than V times. Finally, we can see that the running time of Algorithm 2 is bounded by the following polynomial: $\mathcal{P} = V(P_1 + M(P_2 + P_3 + P_4))$.

Acknowledgements. We wish to thank the referees for their comments and suggestions.

References

[1] Baader, F., Nipkow, T.: Term rewriting and all that. Cambridge University Press, New York (1998)

[2] Baader, F., Snyder, W.: Unification theory. In: Alan Robinson, J., Voronkov, A. (eds.) Handbook of Automated Reasoning, pp. 445–532. Elsevier, MIT Press (2001)

[3] Gascón, A., Godoy, G., Schmidt-Schauß, M.: Unification with Singleton Tree Grammars. In: Treinen, R. (ed.) RTA 2009. LNCS, vol. 5595, pp. 365–379. Springer, Heidelberg (2009)

[4] Gascòn, A., Godoy, G., Schmidt-Schauß, M.: Unification and matching on compressed terms. ACM TOCL 12(4) (2011)

[5] Jouannaud, J.-P., Kirchner, C.: Solving equations in abstract algebras: A rule-based survey of unification. In: Computational Logic - Essays in Honor of Alan Robinson, pp. 257–321 (1991)

[6] Levy, J., Schmidt-Schauß, M., Villaret, M.: Monadic Second-Order Unification Is NP-Complete. In: van Oostrom, V. (ed.) RTA 2004. LNCS, vol. 3091, pp. 55–69. Springer, Heidelberg (2004)

[7] Levy, J., Schmidt-Schauß, M., Villaret, M.: The complexity of monadic second-order unification. SIAM J. Computation 38(3), 1113–1140 (2008)

[8] Lifshits, Y.: Processing Compressed Texts: A Tractability Border. In: Ma, B., Zhang, K. (eds.) CPM 2007. LNCS, vol. 4580, pp. 228–240. Springer, Heidelberg (2007)

[9] Marshall, A.M., Narendran, P.: New algorithms for unification modulo one-sided distributivity and its variants. Technical Report SUNYA-CS-12-02, University at Albany-SUNY (2012), `http://www.cs.albany.edu/~ncstrl/treports/Data/README.html`

[10] Miyazaki, M., Shinohara, A., Takeda, M.: An Improved Pattern Matching Algorithm for Strings in Terms of Straight-line Programs. In: Hein, J., Apostolico, A. (eds.) CPM 1997. LNCS, vol. 1264, pp. 1–11. Springer, Heidelberg (1997)

[11] Narendran, P., Marshall, A.M., Mahapatra, B.: On the complexity of the Tiden-Arnborg algorithm for unification modulo one-sided distributivity. In: UNIF 24. EPTCS, vol. 42, pp. 54–63 (2010)

[12] Plandowski, W.: Testing Equivalence of Morphisms on Context-free Languages. In: van Leeuwen, J. (ed.) ESA 1994. LNCS, vol. 855, pp. 460–470. Springer, Heidelberg (1994)

[13] Rytter, W.: Application of Lempel-Ziv factorization to the approximation of grammar-based compression. Theoretical Computer Science 302, 211–222 (2003)

[14] Tidén, E., Arnborg, S.: Unification problems with one-sided distributivity. J. Symb. Comput. 3(1/2), 183–202 (1987)

Reachability Analysis of Program Variables

Đurica Nikolić[1,2] and Fausto Spoto[1]

[1] Dipartimento di Informatica, University of Verona
[2] Microsoft Research - University of Trento Centre for Computational and Systems Biology
{durica.nikolic,fausto.spoto}@univr.it

Abstract. A variable v reaches a variable w if there is a path from the memory location bound to v to the one bound to w. This information is important for improving the precision of other static analyses, such as side-effects, field initialization, cyclicity and path-length, as well as of more complex analyses built upon them, such as nullness and termination. We present a provably correct constraint-based reachability analysis for Java bytecode. Our constraint is a graph whose nodes are program points and whose arcs propagate reachability information according to the semantics of bytecodes. The analysis has been implemented in the Julia static analyzer. Experiments that we performed on non-trivial Java and Android programs show a gain in precision due to a reachability information, whose presence also reduces the cost of nullness and termination analyses.

1 Introduction

Static analysis of computer programs allows us to statically gather information about their run-time behavior, making it possible to prove that these programs do not perform illegal operations (such as division by zero or dereference of `null`), do not give rise to erroneous executions (such as infinite loops) or do not divulge information (such as security authorizations or GPS position) in an incorrect way.

Dynamic allocation of objects is heavily used in real life programs. These objects are instantiated on demand, their number is not statically known and they can reference other objects (through *fields*). Such references can be updated at run-time. In this paper we present, formalize and implement a provably correct abstraction of the run-time, dynamically allocated memory, that we call *reachability*. We say that a variable v *reaches* a variable w if w holds an object reachable from v, by following (different objects') fields from the object held in the location bound to v. For instance, after an assignment `v.next.next = w`, we can state that v reaches w. Reachability is distinct from *sharing* i.e., being able to reach a shared object. For instance, after the statement `v.next = w.next`, we can state that v and w share. If v reaches w then v and w share, but the converse might not hold. Hence reachability is more *precise*, i.e., it induces a finer, more concrete abstraction of the computational states than sharing analysis. Our analysis is constraint-based: constraints are built from the syntax of the program and their solution is a correct approximation of reachability. A companion paper [14] includes full definitions and proofs.

B. Gramlich, D. Miller, and U. Sattler (Eds.): IJCAR 2012, LNAI 7364, pp. 423–438, 2012.

Reachability has been applied to several static analyses:

Side-Effects Analysis: Side-effects analysis tracks (among other things) which parameters p of a method might be affected by its execution in the sense that the method might update a field of an object reachable from p. Namely, if the method performs an assignment a.f=b, this affects p only if p reaches a. If we used sharing rather than reachability information, that would lead to a loss of precision, since it might be the case that p and a share but the assignment modifies an object unreachable from p.

Field Initialization Analysis: It is often the case that a field is initialized by all of the constructors of its defining class *before being read* by these constructors. Spotting this frequent situation is important for many analyses, including nullness [15,22]. Hence, we want to know whether a field read operation a=expression.f inside a constructor can actually read field f of the this object, being initialized by the constructor. This happens only if this reaches expression. Again, sharing would be less precise here.

Cyclicity Analysis: An assignment a.f=b might make a *cyclical* (i.e., point to a cyclical data structure), but only if b reaches a. Originally, this analysis was built upon sharing information [16], but analysis of reachable variables helps here.

Path-Length Analysis: Path-length is a data structure measure used in termination analysis [23]. It is the maximum number of pointer dereferences that can be followed from a program variable. An assignment a.f=b can only modify the path-length of the program variables that share with a, according to the original definition of path-length [23]. Reachability analysis improves this approximation, since the path-length of a program variable v is actually modified only if v *reaches* a.

These analyses, among others, are implemented in our Julia tool (http://www.juliasoft.com). They are building blocks of larger *tools*, such as a nullness and a termination checker. The former spots where a program might throw a null-pointer exception at run-time; the latter if method calls might diverge. A tool performs its supporting analyses (the *building blocks*) in distinct threads, parallel on multi-core hardware.

Our experiments show that reachability improves side-effects, field initialization and nullness analysis of non-trivial Java and Android programs. However there is no improvement for cyclicity, path-length and termination analysis of the same programs, but only of sample programs from the international termination competition. That is because termination often depends on loops over integer counters rather than on recursion over data structures, as is the case in those samples (probably unusual and artificial). An unexpected effect of reachability is, however, an increase in the speed of both tools.

Reachability analysis belongs to the group of *pointer analyses*, that support other static analyses. Plenty of papers consider them: [9] surveys more than 75 papers. Different properties of pointers give rise to different kinds of pointer analyses: *alias*, *sharing*, *points-to* and *shape* analyses. Possible (definitive) alias analysis discovers the pairs of variables that might (must) point to the same memory location. If two variables are alias, they are also reachable from each other, but the opposite might not hold. Sharing analysis [21] determines whether two variables might ever reach the same object at run-time. Reachability entails sharing, but the opposite, in general, does not hold. Points-to analysis [20,10,11,17,8] computes the objects that a pointer variable might refer to at run-time. Usually, points-to analysis performs a conservative approximation of the heap, which is then used to compute points-to information for the whole program. In [20], points-to graphs are precise approximations of the run-time heap memory and can be used to over-approximate the reachability information. Points-to information is much more concrete than our reachability information. Shape analysis determines heap *shape invariants* [18,19,3,7]. These analyses are quite concrete and capture aliasing and points-to information, as well as other properties such as cyclicity or acyclicity. These are often encoded as first-order logic formulae and theorem provers are used to determine their validity. Reachability can, of course, be abstracted from these very precise approximations of the memory, but we wanted here an analysis that uses the most abstract (i.e., the simplest) domain able to express reachability between variables.

There is also another notion of reachability [13], slightly different from ours. The *reachability predicate* determines whether a memory location reaches another one, usually along *one* particular *field* of *one* particular *data structure*, while our definition of reachable locations deals with *arbitrary fields* of *arbitrary data structures*. That predicate is used in [6,1,4] for abstraction of programs, as one particular case of predicate abstraction [2].

2 Operational Semantics

We present here a formal operational semantics of Java bytecode, inspired by the standard informal semantics [12]. The same semantics is used in [22], while [23] uses its denotational form. Java bytecode is the form of instructions executed by the Java Virtual Machine (JVM). Our formalization is at bytecode level for several reasons: there is a small number of bytecode instructions, compared to varieties of source statements; bytecode lacks complexities such as inner classes; our implementation of reachability analysis is at bytecode level, bringing formalism, implementation and proofs closer.

For simplicity, we assume that the only primitive type is int and that reference types are *classes* containing *instance fields* and *instant methods* only. Our implementation handles all Java types and bytecodes, as well as classes with static fields and methods. We analyze bytecode preprocessed into a control flow graph, i.e., a directed graph of *basic blocks*, with no jumps inside the blocks. [ins / *rest*] $\rightarrow b_1 \cdots \rightarrow b_m$ denotes a block of code starting at instruction ins, possibly followed by more bytecodes *rest* and linked to m subsequent blocks $b_1, \ldots, b_m$. Exception handlers start with a catch. A conditional, virtual method call, or selection of an exception handler becomes a block with many subsequent blocks, starting with a *filtering* bytecode such as exception_is K for exception handlers.

Example 1. Fig. 2 shows the basic blocks of the constructor in Fig. 1. There is a branch at the call to the constructor of `java.lang.Object`, that might throw an exception (like every call). If this happens, the exception is first caught and then re-thrown to the caller of the constructor. Otherwise, the execution continues with 2 blocks storing the formal parameters (locals 1 and 2) into the fields of `this` (local 0) and then returns. □

Bytecodes operate on *variables*, which encompass both stack elements and local variables. A standard algorithm [12] infers their static types.

```java
public class ListStudent {
  public Student head;
  public ListStudent tail;

  public ListStudent(Student head,
                     ListStudent tail) {
    this.head = head;
    this.tail = tail;
  }
}
```

Fig. 1. Our running example

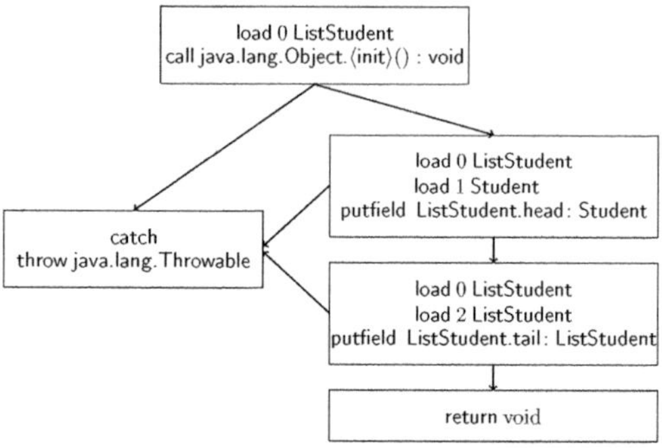

Fig. 2. Representation of the constructor from Fig. 1

Definition 1 (Classes). *The set of* classes $\mathbb{K}$ *of a program is partially ordered w.r.t. the* subclass relation $\leq$*:* $\mathsf{t}\leq\mathsf{t}'$ *if* t *(respectively* t'*) is a subclass (respectively superclass) of* t' *(respectively* t*). Every class has at most one* direct superclass *and an arbitrary number of direct subclasses. A* type *is an element of* $\mathbb{T}=\{\mathsf{int}\}\cup\mathbb{K}$*, ordered by the extension of* $\leq$ *with* $\mathsf{int}\leq\mathsf{int}$*. A class* $\kappa\in\mathbb{K}$ *has* fields $\kappa.f\!:\!\mathsf{t}$ *(field* f *of type* $\mathsf{t}\in\mathbb{T}$ *defined in* κ*), where* κ *and* t *are often omitted. We let* $\mathbb{F}(\kappa)=\{\kappa'.f\!:\!\mathsf{t}' \mid \kappa\leq\kappa'\}$ *be the fields defined in* κ *or in any of its superclasses. A class* κ *has* methods $\kappa.m(\vec{\mathsf{t}})\!:\mathsf{t}$ *(method* m*, defined in* κ*, with arguments of type* $\vec{\mathsf{t}}$*, returning a value of type* $\mathsf{t}\in\mathbb{T}\cup\{\mathsf{void}\}$*), where* κ*,* $\vec{\mathsf{t}}$*, and* t *are often omitted. Constructors are methods named* init *that return* void*.*

Definition 2 (Type environment). *Let* V *be the set of* variables *from* $L=\{l_0,\ldots,l_m\}$ *(*local *variables) and* $S=\{s_0,\ldots,s_n\}$ *(*stack *variables). A* type environment *is a function* $\tau: V\rightarrow\mathbb{T}$*. Its* domain *is written as* $\mathsf{dom}(\tau)$*. The set of all type environments is* $\mathcal{T}$*.*

Definition 3 (State). *A* value *is an element of* $\mathbb{Z}\cup\mathbb{L}\cup\{\mathtt{null}\}$*, where* $\mathbb{L}$ *is an infinite set of* memory locations*. A* state *over* $\tau\in\mathcal{T}$ *is a pair* $\langle\langle l \parallel s\rangle,\mu\rangle$ *where* l *is an array of values for the local variables in* $\mathsf{dom}(\tau)$*,* s *is a stack of values for the stack variables in* $\mathsf{dom}(\tau)$*, which grows leftwards, and* μ *is a* memory*, or* heap*, that binds locations to* objects*. The empty stack is denoted by* ε*. We often use another representation for a state:* $\langle\rho,\mu\rangle$*, where an* environment ρ *maps each* $l_k\in L$ *to its value* $l[k]$ *and each* $s_k\in S$ *to its value* $s[k]$*. An object* o *has class* $o.\kappa$ *(is an* instance *of* $o.\kappa$*) and has an internal environment* $o.\phi$ *that maps every field* $\kappa'.f:\mathsf{t}'\in\mathbb{F}(o.\kappa)$ *into its value* $(o.\phi)(\kappa'.f:\mathsf{t}')$*. A value* v has type t *in* $\langle\rho,\mu\rangle$ *if:* $v\in\mathbb{Z}$ *and* $\mathsf{t}=\mathsf{int}$*, or* $v=\mathtt{null}$ *and* $\mathsf{t}\in\mathbb{K}$*, or* $v\in\mathbb{L}$*,* $\mathsf{t}\in\mathbb{K}$ *and* $\mu(v).\kappa\leq\mathsf{t}$*. In a state* $\langle\rho,\mu\rangle$ *over* τ*, we require that* $\rho(v)$ *has type* $\tau(v)$ *for any* $v\in\mathsf{dom}(\tau)$ *and* $(o.\phi)(\kappa'.f:\mathsf{t}')$ *has type* t' *for every* $o\in\mathsf{rng}(\mu)$ *(range* μ*) and every* $\kappa'.f\!:\!\mathsf{t}'\in\mathbb{F}(o.\kappa)$*. The set of states is* Ξ*. We write* Ξ_τ *when we want to fix the type environment* τ*.*

Example 2. Let $\tau=[l_1\mapsto\mathsf{ListStudent}; l_2\mapsto\mathsf{int}; l_3\mapsto\mathsf{Student}; l_4\mapsto\mathsf{ListStudent}]\in\mathcal{T}$ and consider the state $\sigma=\langle\rho,\mu\rangle\in\Sigma_\tau$ shown in Fig. 3. The environment ρ maps variables l_1, l_2, l_3 and l_4 to values ℓ_2, 2, ℓ_3 and ℓ_4, respectively; the memory μ maps locations ℓ_2 and ℓ_4 to objects o_2 and o_4 of class ListStudent and location ℓ_3 to object o_3 of class Student. Objects are shown as boxes with a class tag and an internal environment mapping fields to values. For instance, fields head and tail of o_4 contain ℓ_3 and ℓ_2, respectively. □

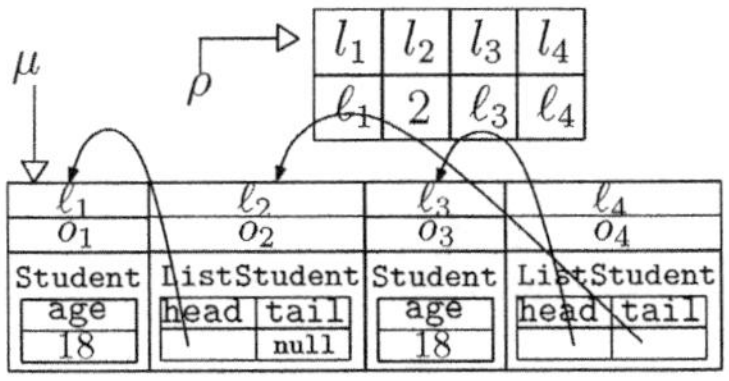

Fig. 3. A JVM state $\sigma = \langle \rho, \mu \rangle$

We assume that states are well-typed, i.e., variables hold values consistent with their static types. Since the JVM supports exceptions, we distinguish between *normal* states Ξ and *exceptional* states $\underline{\Xi}$, which arise *immediately after* bytecode instructions throwing an exception and have a stack of height 1 containing a location bound to the thrown exception. When we denote a state by σ, we do not specify if it is normal or exceptional. If we want to stress that, we write $\langle\langle l \parallel s\rangle, \mu\rangle$ or $\underline{\langle\langle l \parallel s\rangle, \mu\rangle}$.

The semantics of an instruction ins is a partial map $ins : \Sigma_\tau \to \Sigma_{\tau'}$ from *initial* to *final* states. The number and type of local variables and stack elements at its start are specified by τ. The formal semantics is given in [14]. We discuss it informally below.

Basic Instructions. const v pushes $v \in \mathbb{Z}$ on the top of the stack. Like any other bytecode except catch, it is defined only when the JVM is in a normal state. The latter starts the exceptional handlers from an exceptional state and is, therefore, undefined on a normal state. dupt duplicates the top of the stack, of type t. load k t pushes on the stack the value of local variable number k, l_k, which must exist and have type t. Conversely, store k t pops the top of the stack of type t and writes it in local variable l_k; it might potentially enlarge the set of local variables. In our formalization, conditional bytecodes are used in complementary pairs (such as ifne t and ifeq t), at a conditional branch. For instance, ifeq t checks whether the top of the stack, of type t, is 0 when t = int or `null` when t $\in \mathbb{K}$. Otherwise, its semantics is undefined.

Object-Manipulating Instructions. These bytecode instructions create or access objects in memory. new κ pushes on the stack a reference to a new object o of class κ, whose fields are initialized to a default value: `null` for reference fields, and 0 for integer fields [12]. getfield $\kappa.f$: t reads the field $\kappa.f$: t of a receiver object r popped from the stack, of type κ. putfield $\kappa.f$: t writes the top of the stack, of type t, inside field $\kappa.f$: t of the object pointed to by the underlying value r, of type κ.

Exception-Handling Instructions. throw κ throws the top of the stack, of type $\kappa \le$ Throwable. catch starts an exception handler: it takes an exceptional state and transforms it into a normal state at the beginning of the handler. After catch, exception_is$K$ selects an appropriate handler depending on the run-time class of the exception.

Method Call and Return. We use an activation stack of states. Methods can be redefined in object-oriented code, so a call instruction has the form call $m_1 \dots m_k$, enumerating an over-approximation of the set of possible run-time targets [14].

3 Reachability

In this section we formalize our notion of *reachability* between two program variables.

Definition 4 (Locations reachable from a variable). *Let* $\tau \in \mathcal{T}$. *The set of* locations reachable from a variable $a \in \mathsf{dom}(\tau)$ in a state $\sigma = \langle \rho, \mu \rangle \in \Sigma_\tau$ *is* $\mathsf{L}_\sigma(a) = \bigcup_{i \ge 0} \mathsf{L}^i_\sigma(a)$,

$\mathsf{L}^0_\sigma(l_1) = \{\ell_2\}$ $\mathsf{L}^1_\sigma(l_1) = \mathsf{L}_\sigma(l_1) = \{\ell_1, \ell_2\}$
$\mathsf{L}^0_\sigma(l_2) = \mathsf{L}_\sigma(l_2) = \varnothing$
$\mathsf{L}^0_\sigma(l_3) = \mathsf{L}_\sigma(l_3) = \{\ell_3\}$
$\mathsf{L}^0_\sigma(l_4) = \{\ell_4\}$ $\mathsf{L}^1_\sigma(l_4) = \{\ell_2, \ell_3, \ell_4\}$ $\mathsf{L}^2_\sigma(l_4) = \mathsf{L}_\sigma(l_4) = \{\ell_1, \ell_2, \ell_3, \ell_4\}$

$\mathsf{T}^0(\mathsf{Object})$	$= \mathsf{T}(\mathsf{Object})$ $= \{\mathsf{Object}, \mathsf{Student}, \mathsf{ListStudent}\}$
$\mathsf{T}^0(\mathsf{Student})$ $\mathsf{T}^1(\mathsf{Student})$	$= \{\mathsf{Object}, \mathsf{Student}\}$ $= \mathsf{T}(\mathsf{Student})$ $= \{\mathsf{int}, \mathsf{Object}, \mathsf{Student}\}$
$\mathsf{T}^0(\mathsf{ListStudent})$ $\mathsf{T}^1(\mathsf{ListStudent})$ $\mathsf{T}^2(\mathsf{ListStudent})$	$= \{\mathsf{ListStudent}, \mathsf{Object}\}$ $= \{\mathsf{ListStudent}, \mathsf{Object}, \mathsf{Student}\}$ $= \mathsf{T}(\mathsf{ListStudent})$ $= \{\mathsf{int}, \mathsf{ListStudent}, \mathsf{Object}, \mathsf{Student}\}$

Fig. 4. Example of computation of reachable locations and types

where $\mathsf{L}^i_\sigma(a)$ *are the locations reachable from* a *in at most* i *steps:* $\mathsf{L}^i_\sigma(a) = \{\rho(a)\} \cap \mathbb{L}$ *if* $i = 0$, *and* $\mathsf{L}^i_\sigma(a) = \mathsf{L}^{i-1}_\sigma(a) \cup \bigcup_{\ell \in \mathsf{L}^{i-1}_\sigma(a)} (\mathsf{rng}(\mu(\ell).\phi) \cap \mathbb{L})$ *if* $i > 0$.

Definition 5 (Reachability between variables). *Let* $\tau \in \mathcal{T}$, $\sigma = \langle \rho, \mu \rangle \in \Sigma_\tau$ *and variables* $a, b \in \mathsf{dom}(\tau)$. *We say that* b *is* reachable *from* a *in* σ *or, equivalently, that* a *reaches* b *in* σ, *denoted as* $a \leadsto^\sigma b$, *iff* $\rho(b) \in \mathsf{L}_\sigma(a)$.

We also introduce a notion of static reachability between types.

Definition 6 (Reachability between types). *Let* $\mathsf{t} \in \mathbb{T}$. *The set of types* compatible *with* t *is* $\mathsf{compatible}(\mathsf{t}) = \{\mathsf{t}' \mid \mathsf{t} \le \mathsf{t}' \text{ or } \mathsf{t}' \le \mathsf{t}\}$. *The* set of types reachable from t *is* $\mathsf{T}(\mathsf{t}) = \bigcup_{i \ge 0} \mathsf{T}^i(\mathsf{t})$, *where* $\mathsf{T}^i(\mathsf{t})$ *are the types reachable from* t *in at most* i *steps:* $\mathsf{T}^i(\mathsf{t}) = \mathsf{compatible}(\mathsf{t})$ *if* $i = 0$, *and* $\mathsf{T}^i(\mathsf{t}) = \mathsf{T}^{i-1}(\mathsf{t}) \cup \bigcup_{\kappa \in \mathsf{T}^{i-1}(\mathsf{t}) \cap \mathbb{K},\ \kappa'.f:\mathsf{t}' \in \mathbb{F}(\kappa)} \mathsf{compatible}(\mathsf{t}')$ *if* $i > 0$. *We say that* $\mathsf{t}' \in \mathbb{T}$ *is reachable from* t *if* $\mathsf{t}' \in \mathsf{T}(\mathsf{t})$, *and we denote it as* $\mathsf{t} \leadsto \mathsf{t}'$.

Example 3. Consider $\sigma \in \Sigma_\tau$ from Ex. 2. On the left of Fig. 4 we give, for each $l_i \in \mathsf{dom}(\tau)$ and $j \ge 0$, the set of reachable locations from l_i in σ in at most j steps until the fixpoint is reached. Hence, $l_1 \leadsto^\sigma l_1$, $l_1 \leadsto^\sigma l_2$, $l_3 \leadsto^\sigma l_3$, $l_4 \leadsto^\sigma l_1$, $l_4 \leadsto^\sigma l_2$, $l_4 \leadsto^\sigma l_3$, $l_4 \leadsto^\sigma l_4$. Assume that class Student contains only one field, of type int. ListStudent and Student are subclasses of Object. Fig. 4 reports on the right the types reachable from these three classes: ListStudent$\leadsto$Student, Object$\leadsto$Student, Student$\leadsto$Object, Object$\leadsto$Student, etc. □

Reachability between types can be used to conservatively approximate possible pairs of variables that might reach each other.

Lemma 1. *Let* $\tau \in \mathcal{T}$, $\sigma \in \Sigma_\tau$ *and* $a, b \in \mathsf{dom}(\tau)$. *If* $a \leadsto^\sigma b$, *then* $\tau(a) \leadsto \tau(b)$.

Example 4. Since $l_4 \leadsto^\sigma l_3$ (Ex. 3), by Lemma 1, also $\tau(l_4) \leadsto \tau(l_3)$ holds. In fact, Ex. 3 shows that $\tau(l_4) = \mathsf{ListStudent} \leadsto \mathsf{Student} = \tau(l_3)$. □

4 Reachability Analysis

We define here an abstract interpretation of the concrete semantics of Section 2 w.r.t. the property of reachability between variables (Definition 5). This will be an actual algorithm for interprocedural, whole-program reachability analysis. We follow here the abstract interpretation approach [5], that allows us to define a static analysis from the formal specifications of the property of interest and the semantics of the language.

The concrete semantics works over concrete states (Definition 3), that our abstract interpretation abstracts into ordered pairs of variables.

Definition 7 (Concrete and Abstract Domain). *Given a type environment* $\tau \in \mathcal{T}$, *we define the* concrete domain *over* τ *as* $\mathsf{C}_\tau = \langle \wp(\Sigma_\tau), \subseteq \rangle$ *and the* abstract domain *over* τ *as the powerset of the set of ordered pairs of variables* $\mathsf{A}_\tau = \langle \wp(\mathsf{dom}(\tau) \times \mathsf{dom}(\tau)), \subseteq \rangle$. *For every* $v, w \in \mathsf{dom}(\tau)$, *we write* $v \rightsquigarrow w$ *to denote the ordered pair* $\langle v, w \rangle$.

An abstract element $R \in \mathsf{A}_\tau$ represents those concrete states whose reachability information is over-approximated by the pairs of variables in R (*possible* reachability).

Definition 8 (Concretization map). *For every* $\tau \in \mathcal{T}$, *we define the* concretization map $\gamma_\tau : \mathsf{A}_\tau \to \mathsf{C}_\tau$ *as* $\gamma_\tau = \lambda R.\{\sigma \in \Sigma_\tau \mid \forall a, b \in \mathsf{dom}(\tau). a \rightsquigarrow^\sigma b \Rightarrow a \rightsquigarrow b \in R\}$.

Both C_τ and A_τ are complete lattices. Moreover, we proved γ_τ co-additive, and therefore it is the concretization map of a Galois connection [5] and A_τ is actually an abstract domain, in the sense of abstract interpretation.

Our analysis is constraint-based: we build an *abstract constraint graph* from the source code of a Java bytecode program. There is a node for each bytecode b in the program, containing an element of A_τ, where τ is the static type information at the beginning of b. An arc linking the nodes corresponding to two bytecodes b_1 and b_2 propagates the reachability information from b_1 to b_2. Here, the exact meaning of *propagates* depends on b_1, since each bytecode has different effects on reachability.

Definition 9 (ACG). *Let* P *be the program under analysis (i.e., a control flow graph of basic blocks for each method or constructor). The* abstract constraint graph *(ACG) of* P *is a directed graph* $\langle V, E \rangle$ *(nodes, arcs) where:*

- V *contains a node* $\boxed{\mathsf{ins}}$, *for every bytecode instruction* ins *of* P;
- V *contains nodes* $\boxed{\mathsf{exit@}m}$ *and* $\boxed{\mathsf{exception@}m}$ *for each method or constructor* m *in* P, *and these nodes correspond to the normal and exceptional end of* m;
- E *contains directed (multi-)arcs with one or two sources and always one sink;*
- *for every arc in* E, *there is a* propagation rule, *i.e., a function over* A, *from the reachability information at its source(s) to the reachability information at its sink.*

The arcs in E *are built from* P *as follows. We assume that* τ *and* τ' *are the static type information at and immediately after the execution of a bytecode* ins, *respectively. Moreover, we assume that* τ *contains* j *stack elements and* i *local variables. In the following we discuss different types of arcs.*

Sequential Arcs. *If* ins *is a bytecode in* P, *distinct from* call, *immediately followed by a bytecode* ins′, *distinct from* catch, *then a simple arc is built from* $\boxed{\mathsf{ins}}$ *to* $\boxed{\mathsf{ins}'}$, *with one of the propagation rules #1-#7 in Fig. 5.*

Final Arcs. *For each* return t *and* throw κ *occurring in a method or in a constructor* m *of* P, *there are simple arcs from* $\boxed{\mathsf{return\ t}}$ *to* $\boxed{\mathsf{exit@}m}$ *and from* $\boxed{\mathsf{throw}\ \kappa}$ *to* $\boxed{\mathsf{exception@}m}$ *respectively, with one of the propagation rules #8-#10 in Fig. 5.*

#1	dup t	$\lambda R.R \cup R[s_{j-1} \mapsto s_j] \cup \{s_{j-1}\rightsquigarrow s_j, s_j \rightsquigarrow s_{j-1} \mid s_{j-1}\rightsquigarrow s_{j-1} \in R\}$
#2	new κ	$\lambda R.R \cup \{s_j \rightsquigarrow s_j\}$
#3	load k t	$\lambda R.R \cup R[l_k \mapsto s_j] \cup \{l_k \rightsquigarrow s_j, s_j \rightsquigarrow l_k \mid l_k \rightsquigarrow l_k \in R\}$
#4	store k t	$\lambda R.\{(a\rightsquigarrow b)[s_{j-1} \mapsto l_k] \mid a\rightsquigarrow b \in R \land a, b \neq l_k\}$
#5	getfield f:t	$\lambda R.\{a\rightsquigarrow b \in R \mid a, b \neq s_{j-1}\} \cup \{s_{j-1}\rightsquigarrow b \in R \mid \mathsf{t}\rightsquigarrow \tau(b)\} \cup$ $\{a\rightsquigarrow s_{j-1} \mid a \in \mathsf{dom}(\tau) \land \tau(a)\rightsquigarrow \mathsf{t} \land [a \text{ and } s_{j-1} \text{ might share at getfield } f\text{:t}]\}$
#6	putfield f:t	$\lambda R.\{a\rightsquigarrow b \in R \mid a, b \notin \{s_{j-1}, s_{j-2}\}\} \cup$ $\{a\rightsquigarrow b \mid a, b \notin \{s_{j-1}, s_{j-2}\} \land a\rightsquigarrow s_{j-2} \in R \land s_{j-1}\rightsquigarrow b \in R\}$
#7	const v, catch, ifne t, ifeq t	$\lambda R.\{a\rightsquigarrow b \in R \mid a, b \in \mathsf{dom}(\tau')\}$
#8	return void	$\lambda R.\{a\rightsquigarrow b \in R \mid a, b \notin \{s_0, \dots, s_{j-1}\}\}$
#9	return t	$\lambda R.\{(a\rightsquigarrow b)[s_{j-1} \mapsto s_0] \mid a\rightsquigarrow b \in R \land a, b \notin \{s_0, \dots, s_{j-2}\}\}$
#10	throw κ	$\lambda R.\{(a\rightsquigarrow b)[s_{j-1} \mapsto s_0] \mid a\rightsquigarrow b \in R \land a, b \notin \{s_0, \dots, s_{j-2}\}\} \cup \{s_0\rightsquigarrow s_0\}$
#11	throw κ	$\lambda R.\{(a\rightsquigarrow b)[s_{j-1} \mapsto s_0] \mid a\rightsquigarrow b \in R \land a, b \notin \{s_0, \dots, s_{j-2}\}\} \cup \{s_0\rightsquigarrow s_0\}$
#12	call $m_1 \dots m_k$	$\lambda R.\{a\rightsquigarrow b \in R \mid a, b \notin \{s_0, \dots, s_{j-1}\}\} \cup \{s_0\rightsquigarrow s_0\}$ $\cup\{a\rightsquigarrow s_0 \mid a \in \{l_0, \dots, l_{i-1}\} \land \tau(a)\rightsquigarrow \mathtt{Throwable}\}$ $\cup\{s_0\rightsquigarrow a \mid a \in \{l_0, \dots, l_{i-1}\} \land \mathtt{Throwable}\rightsquigarrow \tau(a)\}$
#13	new κ, getfield f:t, putfield f:t	$\lambda R.\{a\rightsquigarrow b \mid a\rightsquigarrow b \in R \land a, b \notin \{s_0, \dots, s_{j-1}\}\} \cup \{s_0\rightsquigarrow s_0\}$
#14	call $m_1 \dots m_k$	$\lambda R.\left\{(a\rightsquigarrow b)\begin{bmatrix} s_{j-\pi}\mapsto l_0 \\ \dots \\ s_{j-1}\mapsto l_{\pi-1}\end{bmatrix} \middle\vert\ a\rightsquigarrow b \in R \land a, b \in \{s_{j-\pi}, \dots, s_{j-1}\}\right\}$

Fig. 5. Propagation rules of simple arcs

Exceptional Arcs. *For each* ins *throwing an exception, immediately followed by a* catch, *a arc is built from* $\boxed{\text{ins}}$ *to* $\boxed{\text{catch}}$, *with one of the propagation rules #11 – #13 in Fig. 5.*

Parameter Passing Arcs. *For each* $\text{ins}_c = \text{call } m_1 \dots m_k$ *to a method with* π *parameters (including* this*), we build a simple arc from* $\boxed{\text{ins}_c}$ *to the node corresponding to the first bytecode of* m_w *with the propagation rule #14 in Fig. 5, for each* $1 \leq w \leq k$.

Return Value Arcs. *For each* $\text{ins}_c = \text{call } m_1 \dots m_k$ *to a method with* π *parameters (including* this*) returning a value of type* $\mathsf{t} \in \mathbb{K}$ *and each subsequent bytecode* ins′ *distinct from* catch, *we build a multi-arc from* $\boxed{\text{ins}_c}$ *and* $\boxed{\text{exit@}m_w}$ *(2 sources, in that order) to* $\boxed{\text{ins}'}$ *with the propagation rule #15 defined in Fig. 6, for each* $1 \leq w \leq k$.

Side-Effects Arcs. *For each* $\text{ins}_c = \text{call } m_1 \dots m_k$ *to a method with* π *parameters (including* this*) and each subsequent bytecode* ins′ , *we build a multi-arc from* $\boxed{\text{ins}_c}$ *and* $\boxed{\text{exit@}m_w}$ *(2 sources, in that order) to* $\boxed{\text{ins}'}$, *where* ins′ *is not a* catch, *or from* $\boxed{\text{ins}_c}$ *and* $\boxed{\text{exception@}m_w}$ *(2 sources, in that order) to* $\boxed{\text{catch}}$, *for each* $1 \leq w \leq k$. *The propagation rule #16 is given in Fig. 6, where* $\max = j - \pi$ *if* ins′ *is not a* catch *and* $\max = 0$ *otherwise.*

The **sequential arcs** link an instruction to its immediate successors. For instance, the arc #1, starting from a node corresponding to a dup t, states that the reachability approximation at that node can be found at its successor's node as well ($\lambda R.R$). On the other hand, since s_j, the new topmost stack element (new top), is an alias of s_{j-1}, the former topmost stack element (old top), it is clear that every variable reaching s_{j-1} (or, respectively, that is reachable from s_{j-1}) also reaches s_j (respectively, is reachable from s_j): $\lambda R.R \cup R[s_{j-1} \mapsto s_j]$. For the same reason, we must assume that, if s_{j-1} reaches itself (i.e., if the old top was not `null`) then, immediately after the dupt, s_j might reach

Rule	Propagation rule
#15	$\lambda R_1.\lambda R_2.\{s_{j-\pi}{\rightsquigarrow}s_{j-\pi} \mid s_0{\rightsquigarrow}s_0 \in R_2\}$ $\cup \{ a{\rightsquigarrow}s_{j-\pi} \mid$ 1. $a \in \mathsf{dom}(\tau') \setminus \{s_{j-\pi}\} \wedge$ 2. $\tau'(a){\rightsquigarrow}\mathsf{t} \wedge$ 3. $\exists j - \pi \le p < j$ s.t. a might share with s_p at call $m_1 \dots m_k \wedge$ 4. if a is definitely alias of s_p at call $m_1 \dots m_k$ and no store $l_{p-j+\pi}$ occurs in m_w, then $l_{p-j+\pi}{\rightsquigarrow}s_0 \in R_2 \}$ $\cup \{ s_{j-\pi}{\rightsquigarrow}b \mid$ 1. $b \in \mathsf{dom}(\tau') \setminus \{s_{j-\pi}\} \wedge$ 2. $\mathsf{t}{\rightsquigarrow}\tau'(b) \wedge$ 3. $\exists j - \pi \le p < j$ s.t. $s_p{\rightsquigarrow}b \in R_1 \wedge$ 4. if b is definitely alias of s_p at call $m_1 \dots m_k$ and no store $l_{p-j+\pi}$ occurs in m_w, then $s_0{\rightsquigarrow}l_{p-j+\pi} \in R_2 \}$
#16	$\lambda R_1.\lambda R_2.$ $\{ a{\rightsquigarrow}b \mid$ $[a{\rightsquigarrow}b \in R_1 \wedge a, b \in \{l_0, \dots, l_{i-1}, s_0, \dots, s_{\max-1}\}] \vee$ $[$ 1. $a, b \in \{l_0, \dots, l_{i-1}, s_0, \dots, s_{\max-1}\} \wedge$ 2. $\tau'(a){\rightsquigarrow}\tau'(b) \wedge$ 3. $\exists j - \pi \le p_a < j$ s.t. a might share with s_{p_a} at call $m_1 \dots m_k \wedge$ 4. $\exists j - \pi \le p_b < j$ s.t. $p_b{\rightsquigarrow}b \in R_1 \wedge$ 5. if $\exists j - \pi \le q_a < j$ s.t. a is definitely alias of s_{q_a} at call $m_1 \dots m_k$ and if $\exists j - \pi \le q_b < j$ s.t. b is definitely alias of s_{q_b} at call $m_1 \dots m_k$ and no store $l_{q_a-j+\pi}$ nor store $l_{q_b-j+\pi}$ occurs in m_i, then $l_{q_a-j+\pi}{\rightsquigarrow}l_{q_b-j+\pi} \in R_2$ $]\}$

Fig. 6. Propagation rules of mulit-arcs

s_{j-1} and vice versa, which leads to rule #1. Rule #5 is more interesting: getfield f: t replaces the old top of the stack, s_{j-1}, with the value of its field f. Hence all reachability pairs that do not consider s_{j-1} are still valid after the execution of the getfield f: t: $\lambda R.\{a{\rightsquigarrow}b \in R \mid a, b \neq s_{j-1}\}$. But we have to consider which variable b might be reached from the field ($s_{j-1}{\rightsquigarrow}b$) and which variable a might reach the field ($a{\rightsquigarrow}s_{j-1}$). For b, we observe that if the field reaches b, then also its containing object (i.e., the old top of the stack) had to reach b before the getfield f:t (i.e., $s_{j-1}{\rightsquigarrow}b \in R$); for better precision we consider only those pairs of variables that satisfy type reachability requirement, i.e., $\mathsf{t}{\rightsquigarrow}\tau(b)$. For a, we rely on a pessimistic (but conservative) assumption: every variable a might reach the field after the getfield$f$: t, as long as the field has a reference type such that $\tau(a){\rightsquigarrow}\mathsf{t}$ and as long as a shares with the top of the stack before the instruction. Rule #6 states that a reachability pair at a putfield f:t instruction remains valid just after that instruction, provided that it did not deal with the topmost two values of the stack s_{j-1} and s_{j-2}, that disappear. Moreover, since this instruction writes s_{j-1} in a field of s_{j-2}, it might introduce reachability from a to b, when a reaches the receiver s_{j-2} and the value s_{j-1} reaches b before the putfield f:t.

The **final arcs** feed nodes [exit@m] and [exception@m] for each method or constructor m. The former contains all states at the end of a normal execution of m; the latter contains those at the end of an exceptional execution of m. Hence [exit@m] is the sink of an arc from every return t in m. The propagation rule states that the stack is emptied at the end of execution of m (#8) or only one element survives, the return value (#9). Similarly, [exception@m] is the sink node of every throwκ instruction that has no exception handler in m (i.e., it has no successors in m). Rule #10 states that all stack elements, but the topmost one s_{j-1}, disappear. The latter is renamed into the exception object s_0, and is always non-`null` (thus, $s_0{\rightsquigarrow}s_0$). We observe that only a throw κ is allowed to throw an exception to the caller since, in our representation of the code as basic blocks, all other instructions that might throw an exception are always linked to an exception handler, possibly minimal (as the two putfield in Fig. 2).

The **exceptional arcs** link every instruction that might throw an exception to the catch at the beginning of their exception handler(s). Rules #10 and #11 are identical, but the latter is applied when throw κ has a successor. Rule #12 states a pessimistic assumption about the exceptional states after a method call: the reachability pairs before the call can survive as long as they do not deal with stack elements. The thrown object s_0 is non-null (thus, $s_0 \rightsquigarrow s_0$) and conservatively assumed to reach and be reached from every local variable a, as long as the static types allow it.

The **parameter passing arcs** connect each method call to the beginning of a method m_w that it might call. Rule #14 renames the actual parameters of m_w, i.e., $s_{j-\pi}, \ldots, s_{j-1}$, into its formal parameters, i.e., $l_0, \ldots, l_{\pi-1}$.

There exists a **return value multi-arc** for each target m_w of a call. Rule #15 considers R_1 and R_2, approximations at the node corresponding to the call and at node $\boxed{\text{exit@}m_w}$. It builds the reachability pairs related to the returned value $s_{j-\pi}$, in the caller. Namely, $s_{j-\pi}$ reaches itself if the return value in the callee (held in the only stack element s_0 at its end) reaches itself. Moreover, a variable a of the caller might reach that returned value ($a \rightsquigarrow s_{j-1}$) if it exists after the call and it is not $s_{j-\pi}$ itself (condition 1); if the static types allow it (condition 2); if a shares with at least one actual parameter s_p (condition 3); moreover, if a is a definite alias of the actual parameter s_p whose corresponding formal parameter $l_{p-j+\pi}$ is never re-assigned inside the callee m_w, then it must also be the case that $l_{p-j+\pi}$ reaches the returned value s_0 (condition 4). Variables b that might be reachable from the returned value $s_{j-\pi}$ are determined in a symmetrical way. It is worth noting that the result of the call can reach a variable b only if b is reachable from at least one actual parameter s_p of the call at call-time ($s_p \rightsquigarrow b \in R_1$).

The **side-effects multi-arcs** enrich the reachability information already known at call-time with some additional pairs of variables whose presence is due to the side-effects of the call. Rule #16 adds a new pair $a \rightsquigarrow b$ if it satisfies the following conditions: a and b must exist after the call and must not be the returned value nor the exception thrown by m_w (condition 1); the static types of a and b must allow their reachability (condition 2); moreover, a must share with at least one actual parameter of the call and b must be reachable from at least one actual parameter of the call (conditions 3 and 4, respectively); finally, if a and b are definite aliases of two actual parameters q_a and q_b of the call whose corresponding formal parameters $l_{q_a-j+\pi}$ and $l_{q_b-j+\pi}$ are not re-assigned inside m_w, then $l_{q_a-j+\pi}$ must reach $l_{q_b-j+\pi}$ at the end of m_w (condition 5).

Propagation rules #15 and #16 use possible sharing and definite aliasing between program variables. If these data are missing, one can always assume the worst, least precise hypothesis. In our experiments (Section 5) reachability analysis is performed inside the nullness and termination tools of Julia, that already perform definite aliasing and possible sharing analyses, so they have no additional cost. The precision of the analysis would benefit from a possible inlining of frequently used methods, so that their calling contexts are not merged into one. However, this is not implemented in Julia.

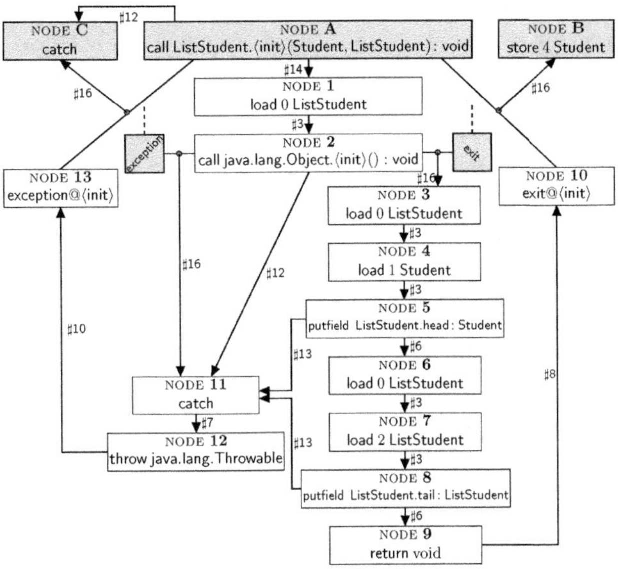

Fig. 7. The ACG for the constructor in Fig. 2

An ACG is *solved* by finding a reachability approximation at each node, consistent with the propagation rules of the arcs. Since these propagation rules are monotonic, a minimal solution exists and can be computed through a fixpoint calculation. This solution is the *reachability analysis* of the program, and has been proven sound [14].

Theorem 1 (Soundness). *Let* ins *and* $\sigma \in \Sigma_\tau$ *be a bytecode instruction and a state reached by an execution of the* main *method of a program, and let* $R_{\mathsf{ins}} \in \mathsf{A}_\tau$ *be the reachability approximation computed by our analysis at* $\boxed{\mathsf{ins}}$. *Then,* $\sigma \in \gamma_\tau(R_{\mathsf{ins}})$.

Example 5. Fig. 7 shows the ACG built for the constructor in Fig. 2. It also shows, in grey, three nodes of a caller of this constructor (nodes A, B and C) and two nodes of the callee of call `java.lang.Object.⟨init⟩() : void`, to exemplify the arcs related to method call and return. Arcs are decorated with the number of their associated propagation rule. Note that the graph for the whole program includes other nodes and arcs. Suppose that at node A, which invokes the constructor, there are four stack elements and four local variables and that we know, from previous static analyses, that a correct possible sharing information is $\mathsf{share}_A = \{\langle s_0, s_1\rangle, \langle l_3, s_2\rangle, \langle l_1, s_3\rangle\}$ (only these pairs of variables might share), while a correct definite aliasing information is $\mathsf{alias}_A = \{\langle s_0, s_1\rangle, \langle l_3, s_2\rangle\}$ (those pairs of variables must be alias, but there might be others). Moreover, suppose that this call occurs in a context with reachability information $S_A = \{l_1 \rightsquigarrow l_1, l_3 \rightsquigarrow l_3, l_1 \rightsquigarrow s_3, l_3 \rightsquigarrow s_2, s_2 \rightsquigarrow l_3, s_0 \rightsquigarrow s_0, s_0 \rightsquigarrow s_1, s_1 \rightsquigarrow s_0, s_1 \rightsquigarrow s_1, s_2 \rightsquigarrow s_2, s_3 \rightsquigarrow s_3\}$. The constructor stores the locations held in its parameters s_2 and s_3 into the fields `head` and `tail` of the newly created object, whose location is, in turn, held in s_0 and s_1. Moreover, s_2 and l_3 are definite aliases at node A, hence we expect that, after any non-exceptional execution of the call (node B), l_3 is reachable from s_0. Node A is linked to node 1 through an arc with propagation rule #14, whose application on S_A gives an approximation of the reachability information at node 1, $S_1 = \{l_0 \rightsquigarrow l_0, l_1 \rightsquigarrow l_1, l_2 \rightsquigarrow l_2\}$. Similarly, we determine the approximations of the reachability information of the other nodes. For instance, $S_2 = \{l_0 \rightsquigarrow l_0, l_1 \rightsquigarrow l_1, l_2 \rightsquigarrow l_2, l_0 \rightsquigarrow s_0, s_0 \rightsquigarrow l_0, s_0 \rightsquigarrow s_0\}$, $S_3 = S_1$, etc. In particular, $S_{10} = \{l_0 \rightsquigarrow l_0, l_0 \rightsquigarrow l_1, l_0 \rightsquigarrow l_2, l_1 \rightsquigarrow l_1, l_2 \rightsquigarrow l_2\}$ and there is a side-effect arc from nodes A and 10 to node B, whose propagation rule #16 applied to S_A and S_{10} gives $S_B = \{l_1 \rightsquigarrow l_1, l_1 \rightsquigarrow s_0, l_1 \rightsquigarrow l_3, l_3 \rightsquigarrow l_3, s_0 \rightsquigarrow l_3, s_0 \rightsquigarrow s_0\}$. As expected, $s_0 \rightsquigarrow l_3 \in S_B$. □

5 Experiments

We have implemented our reachability analysis inside the Julia analyzer for Java and Android (`http://www.juliasoft.com`). Our first aim was to evaluate the cost of the reachability analysis itself and verify whether it actually improves the precision of side-effects, field initialization and cyclicity, as hinted in Section 1. The second aim was to verify if the extra reachability information improves the precision of the nullness and termination checking tools available in Julia, that use side-effects, field initialization, cyclicity and path-length as (some of their) supporting analyses. We do not have any measure of precision for path-length analysis, so we do not evaluate its improvements directly but only as a component of the termination checking tool. To reach these goals, we have analyzed some Java and Android programs, with reachability analysis turned off and then on. Most of these samples are Android applications: Mileage, OpenSudoku, Solitaire and TiltMazes[1]; ChimeTimer, Dazzle, OnWatch and Tricorder[2]; TxWthr[3]. There are also some Java programs: JFlex is a lexical analyzers generator[4]; Plume is a library by Michael D. Ernst[5]; Nti is a non-termination analyzer by Étienne Payet[6]; Lisimplex is a numerical simplex implementation by Ricardo Gobbo[7]. The others are sample programs taken from the Android 3.1 distribution by Google.

Fig. 8 reports time and precision of reachability analysis on a Linux quad-core Intel Xeon machine running at 2.66GHz, with 8 gigabytes of RAM. Times are always below 41 seconds. Average precision is 45.07% which means that, given two variables v and w of reference type at a given program point, in more than half of the cases the analysis proves that v does not reach w. A smaller percentage, here, means better precision. Fig. 8 shows that reachability analysis improves the precision of the side-effects analysis and has positive effects on field initialization as well. Instead, cyclicity analysis seems unaffected. Sharing analysis is always used in these experiments, both when we use reachability information and when we do not compute it. Thus, this figure shows the importance of having also reachability information instead of just sharing information.

Fig. 9 presents our experiments with the nullness and termination tools of Julia and reports their runtime, including reachability analysis. In 8 cases over 24, the extra reachability information improves the precision of the nullness checking tool. But

[1] `http://f-droid.org/repository/browse/`
[2] `http://moonblink.googlecode.com/svn/trunk/`
[3] `http://typoweather.googlecode.com/svn/trunk/`
[4] `http://jflex.de`
[5] `http://code.google.com/p/plume-lib`
[6] `http://personnel.univ-reunion.fr/epayet/Research/NTI/NTI.html`
[7] `http://sourceforge.net/projects/lisimplex`

program	language	source lines	analyzed lines	reach. analysis		prec. of side-effects analysis		prec. of field initial. analysis		prec. of cyclicity analysis	
				time	prec	without reach.	with reach.	without reach.	with reach.	without reach.	with reach.
BluetoothChat	Android	616	84415	21.26	56.01%	645.99	540.23	2185	2325	12.85%	12.85%
ChimeTimer	Android	1090	89565	23.39	47.04%	730.68	618.08	2348	2486	13.54%	13.54%
Dazzle	Android	1791	77828	24.23	46.99%	309.89	225.96	2417	2447	22.19%	22.19%
GestureBuilder	Android	502	84346	23.90	64.11%	667.70	557.52	2162	2282	16.57%	16.57%
Home	Android	870	87413	18.54	55.78%	693.80	584.01	2274	2415	10.89%	10.89%
HoneycombGallery	Android	948	71558	16.71	23.84%	333.25	242.32	2131	2175	33.33%	33.33%
JFlex	Java	7681	40779	7.19	39.59%	357.59	243.89	1092	1146	33.67%	33.67%
JetBoy	Android	839	65174	16.37	64.54%	281.48	198.71	2173	2202	11.79%	11.79%
Lisimplex	Java	768	49303	16.26	47.98%	637.69	347.96	1356	1433	14.13%	14.13%
LunarLander	Android	538	57675	14.92	66.40%	270.87	191.07	1880	1911	18.11%	18.11%
Mileage	Android	5877	104009	32.12	43.73%	959.30	804.98	2636	2794	25.45%	25.45%
NotePad	Android	705	73742	17.96	36.59%	293.57	218.17	2108	2139	37.50%	37.50%
Nti	Java	2372	13486	2.44	47.90%	24.11	13.51	465	467	32.59%	32.59%
OnWatch	Android	6295	112423	29.59	41.00%	1299.51	796.89	3232	3399	32.59%	32.59%
OpenSudoku	Android	5877	90810	40.68	44.81%	440.36	344.92	2622	2660	22.57%	22.57%
Plume	Java	8586	43637	17.75	24.17%	186.31	126.71	1316	1335	57.11%	57.11%
Real3D	Android	1228	74350	17.81	43.55%	497.94	400.73	2093	2189	36.43%	36.43%
SampleSyncAdapter	Android	978	65971	18.48	34.59%	328.80	235.68	2111	2142	42.77%	42.77%
SoftKeyboard	Android	703	58088	10.96	51.90%	174.01	116.96	2112	2131	11.21%	11.21%
Solitaire	Android	3905	62065	18.67	32.23%	243.19	166.57	1957	1982	50.06%	50.06%
TicTacToe	Android	607	59160	13.40	58.56%	228.27	154.35	1919	1943	20.73%	20.73%
TiltMazes	Android	1853	89653	21.14	15.66%	650.45	562.57	2313	2454	71.57%	71.57%
Tricorder	Android	5317	98389	26.69	46.39%	783.59	663.23	2806	2942	33.91%	33.91%
TxWthr	Android	2024	74537	16.97	48.33%	309.24	229.79	2220	2258	15.39%	15.39%
average precision					**45.07%**	**472.81**	**361.86 (-23.47%)**	**2080.33**	**2152.37 (+3.46%)**	**26.84%**	**26.84% (+0.00%)**

Fig. 8. Cost and precision of reachability analysis, and its effects on the precision of side-effects, field initialization and cyclicity analyses. *Source lines* counts non-comment non-blank lines of codes. *Analyzed lines* includes the portion of `java.*`, `javax.*` and `android.*` libraries analyzed with each program and is a more faithful measure of the analyzed codebase. Times are in seconds. For reachability analysis, precision is the ratio of pairs of variables $\langle v, w \rangle$ s.t. the analysis concludes that v might reach w, over the total number of pairs of variables of reference type: the lower the ratio, the higher the precision (the reatio never reaches 0% in practice, since real-life programs contain reachability). For side-effects analysis, precision is the average number of fields modified or read by a method or constructor: the lower the numbers, the better the precision. For field initialization analysis, precision is the number of fields of reference type proven to be always initialized before being read, in all constructors of their defining class: the higher the numbers, the better the precision. For cyclicity analysis, precision is the average number of variables of reference type proven to hold a non-cyclical data structure; the higher the numbers, the better the precision

program	null. without reach.			null. with reach.			term. without reach.			term. with reach.		
	time	ws	prec	time	ws	prec	time	ws	prec	time	ws	prec
BluetoothChat	368.43	**22*****	93.65%	301.31	**19*****	94.23%	158.96	2	33.33%	141.78	2	33.33%
ChimeTimer	343.01	4	98.36%	360.28	4	98.36%	178.87	1	83.33%	183.81	1	83.33%
Dazzle	223.16	26	97.99%	220.78	26	97.99%	120.34	0	100.00%	126.07	0	100.00%
GestureBuilder	261.25	16	92.37%	288.51	16	92.37%	153.33	0	100.00%	151.83	0	100.00%
Home	314.66	27	94.27%	312.55	27	94.27%	166.98	8	38.46%	163.39	8	38.46%
HoneycombGallery	177.32	12	97.79%	179.90	12	97.79%	105.96	0	100.00%	101.47	0	100.00%
JFlex	87.06	71	97.03%	86.10	71	97.03%	300.84	66	53.52%	321.03	66	53.52%
JetBoy	138.99	20**	97.42%	140.64	20**	97.42%	85.91	3	57.14%	85.38	3	57.14%
Lisimplex	251.09	20**	96.94%	202.76	20**	96.94%	160.07	9	70.97%	153.36	9	70.97%
LunarLander	118.75	4	99.30%	121.25	4	99.30%	72.49	3*	0.00%	68.41	3*	0.00%
Mileage	503.90	**102**	97.40%	501.02	**95**	97.67%	387.68	12	69.23%	381.99	12	69.23%
NotePad	194.52	**18**	96.50%	199.19	**17**	96.50%	103.64	0	100.00%	101.49	0	100.00%
Nti	14.06	12	98.93%	16.15	12	98.93%	43.70	70	36.94%	43.53	70	36.94%
OnWatch	898.36	**74**	97.91%	518.55	**65**	98.18%	385.00	6	86.96%	371.32	6	86.96%
OpenSudoku	284.30	124*	95.93%	286.72	124*	95.93%	458.01	6	90.32%	467.34	6	90.32%
Plume	106.67	**59**	98.82%	116.75	**58**	98.83%	208.81	86	60.00%	187.92	86	60.00%
Real3D	203.62	19*	98.14%	195.76	19*	98.14%	116.42	2	60.00%	112.22	2	60.00%
SampleSyncAdapter	156.31	3	99.51%	152.45	3	99.51%	91.90	2	60.00%	89.61	2	60.00%
SoftKeyboard	104.21	**14**	95.78%	103.83	**13**	95.94%	70.45	0	100.00%	67.96	0	100.00%
Solitaire	153.51	63	92.59%	147.54	63	92.59%	207.09	11	86.08%	203.92	11	86.08%
TicTacToe	115.38	0	100.00%	118.27	0	100.00%	79.69	1	85.71%	78.02	1	85.71%
TiltMazes	281.43	**18**	98.20%	276.54	**14**	98.83%	188.56	1	88.89%	174.63	1	88.89%
Tricorder	415.17	**54**	98.29%	407.51	**52**	98.41%	252.25	12	80.33%	257.36	12	80.33%
TxWthr	200.16	48	97.85%	191.88	48	97.85%	109.76	6	70.00%	105.08	6	70.00%
sum of the times	**5915.32**			**5456.24 (-7.77%)**			**4206.71**			**4138.92 (-1.62%)**		
sum of the warnings	**830**			**802 (-3.38%)**			**307**			**307 (+0.00%)**		

Fig. 9. Our experiments with the nullness and termination tools of Julia. Times are in seconds. For nullness analysis, *ws* counts the warnings issued by Julia (possible dereference of null, possibly passing null to a library method) and *prec* reports its precision, as the ratio of the dereferences proved safe over their total number (100% is the maximal precision). For termination analysis, *ws* counts the warnings issued by Julia (constructors or methods possibly diverging) and *prec* reports its precision, as the ratio of the constructors or methods proved to terminate over the total number of constructors or methods containing loops or recursive (100% is the maximal precision). Asterisks stand for actual bugs in the programs. Boldface highlights the cases where reachability improves the precision of the tools

this never happens for termination, consistently with the fact that cyclicity is not improved (Fig. 8). This is because the methods of the programs that we have analyzed terminate since they perform loops over numerical counters or iterators. There is no complex case of recursion over data structures dynamically allocated in memory (lists or trees) where cyclicity would help. To investigate further the case of termination analysis, we have applied Julia to the set of (very tiny) programs used in the international termination competition that is performed every year. Those programs, although small and often unrealistic, are nevertheless interesting since the proof of their termination often requires non-trivial arguments, also related to objects dynamically allocated in memory. Over a total of 164 test programs, the reachability information allows Julia to prove the termination of six more tests: LinkedList, List, ListDuplicate, PartitionList, Test5 and Test6, by supporting a more precise cyclicity and path-length analysis.

For both nullness and termination checking, the presence of reachability analysis actually reduces the total runtime of the tools. This is because reachability helps subsequent analyses, in particular side-effects analysis, and prevents them from generating too much spurious information. For instance, side-effects analysis computes much smaller sets of affected fields per method (Fig. 8, compare the 7th and the 8th columns).

6 Conclusion

We have introduced, formalized and implemented a provably sound (see [14] for proofs) constraint-based reachability analysis for Java bytecode. Its implementation inside the Julia static analyzer is able to scale to programs containing 100k lines of code. Our experiments show that the reachability analysis improves the precision and efficiency of the side-effects, field initialization and nullness analyses, already performed by Julia.

Our constraint-based approach has been used to develop aliasing and sharing analyses of our tool (never published and with completely different propagation rules). We plan to use it in the future to formalize and prove correct other static analyses as well.

References

1. Balaban, I., Pnueli, A., Zuck, L.D.: Shape Analysis by Predicate Abstraction. In: Cousot, R. (ed.) VMCAI 2005. LNCS, vol. 3385, pp. 164–180. Springer, Heidelberg (2005)
2. Ball, T., Millstein, T., Rajamani, S.K.: Polymorphic Predicate Abstraction. ACM Trans. on Programming Languages and Systems 27, 314–343 (2005)
3. Calcagno, C., Distefano, D., O'Hearn, P., Yang, H.: Compositional Shape Analysis by Means of Bi-Abduction. In: Proc. of the 36th POPL, pp. 289–300. ACM, New York (2009)
4. Chatterjee, S., Lahiri, S., Qadeer, S., Rakamaric, Z.: A Low-Level Memory Model and an Accompanying Reachability Predicate. STTT 11(2), 105–116 (2009)
5. Cousot, P., Cousot, R.: Abstract Interpretation: A Unified Lattice Model for Static Analysis of Programs by Construction or Approximation of Fixpoints. In: Proceedings of the 4th POPL, pp. 238–252. ACM (1977)
6. Dams, D.R., Namjoshi, K.S.: Shape Analysis through Predicate Abstraction and Model Checking. In: Zuck, L.D., Attie, P.C., Cortesi, A., Mukhopadhyay, S. (eds.) VMCAI 2003. LNCS, vol. 2575, pp. 310–323. Springer, Heidelberg (2002)

7. Distefano, D., O'Hearn, P.W., Yang, H.: A Local Shape Analysis Based on Separation Logic. In: Hermanns, H. (ed.) TACAS 2006. LNCS, vol. 3920, pp. 287–302. Springer, Heidelberg (2006)
8. Hardekopf, B.C.: Pointer Analysis: Building a Foundation for Effective Program Analysis. Ph.D. thesis, University of Texas at Austin, Austin, TX, USA (2009)
9. Hind, M.: Pointer Analysis: Haven't We Solved This Problem Yet? In: Proceedings of PASTE 2001, pp. 54–61. ACM, New York (2001)
10. Lhoták, O.: Program Analysis Using Binary Decision Diagrams. Ph.D. thesis, McGill University (2006)
11. Lhoták, O., Chung, K.C.A.: Points-to Analysis with Efficient Strong Updates. In: Proceedings of the 38th POPL, pp. 3–16. ACM (2011)
12. Lindholm, T., Yellin, F.: The Java^{TM} Virtual Machine Specification, 2nd edn. Addison-Wesley (1999)
13. Nelson, G.: Verifying Reachability Invariants of Linked Structures. In: Proc. of the 10th POPL, pp. 38–47 (1983)
14. Nikolić, D., Spoto, F.: Reachability Analysis of Program Variables, `http://profs.sci.univr.it/~nikolic/download/IJCAR2012/IJCAR2012Ext.pdf`
15. Papi, M.M., Ali, M., Correa, T.L., Perkins, J.H., Ernst, M.D.: Practical Pluggable Types for Java. In: Proceedings of the ISSTA 2008, pp. 201–212. ACM, Seattle (2008)
16. Rossignoli, S., Spoto, F.: Detecting Non-cyclicity by Abstract Compilation into Boolean Functions. In: Emerson, E.A., Namjoshi, K.S. (eds.) VMCAI 2006. LNCS, vol. 3855, pp. 95–110. Springer, Heidelberg (2005)
17. Rountev, A., Milanova, A., Ryder, B.G.: Points-to Analysis for Java Using Annotated Constraints. In: Proceedings of the 16th OOPSLA, pp. 43–55. ACM (2001)
18. Sagiv, M., Reps, T., Wilhelm, R.: Solving Shape-Analysis Problems in Languages with Destructive Updating. ACM Trans. on Programming Languages and Systems 20, 1–50 (1998)
19. Sagiv, M., Reps, T., Wilhelm, R.: Parametric Shape Analysis via 3-Valued Logic. ACM Trans. Program. Lang. Syst. 24, 217–298 (2002)
20. Salcianu, A.D.: Pointer Analysis for Java Programs: Novel Techniques and Applications. Ph.D. thesis, Massachusetts Institute of Technology, Cambridge, MA, USA (2006)
21. Secci, S., Spoto, F.: Pair-Sharing Analysis of Object-Oriented Programs. In: Hankin, C., Siveroni, I. (eds.) SAS 2005. LNCS, vol. 3672, pp. 320–335. Springer, Heidelberg (2005)
22. Spoto, F., Ernst, M.D.: Inference of Field Initialization. In: Proceedings of the 33rd ICSE, pp. 231–240. ACM, Waikiki (2011)
23. Spoto, F., Mesnard, F., Payet, E.: A Termination Analyzer for Java Bytecode Based on Path-Length. ACM Trans. on Programming Languages and Systems 32(3), 1–70 (2010)

Playing Hybrid Games with KeYmaera*

Jan-David Quesel[1] and André Platzer[2]

[1] University of Oldenburg, Department of Computing Science, Germany
quesel@informatik.uni-oldenburg.de
[2] Carnegie Mellon University, Computer Science Department, Pittsburgh, PA, USA
aplatzer@cs.cmu.edu

Abstract. We propose a new logic, called *differential dynamic game logic* (dDG$\mathcal{L}$), that adds several game constructs on top of differential dynamic logic (d$\mathcal{L}$) so that it can be used for hybrid games. The logic dDG$\mathcal{L}$ is a conservative extension of d$\mathcal{L}$, which we exploit for our implementation of dDG$\mathcal{L}$ in the theorem prover KeYmaera. We provide rules for extending the d$\mathcal{L}$ sequent proof calculus to handle the dDG$\mathcal{L}$ constructs by identifying analogs to operators of d$\mathcal{L}$. We have implemented dDG$\mathcal{L}$ in an extension of KeYmaera and verified a case study in which a robot satisfies a joint safety and liveness objective in a factory automation scenario, in which the factory may perform interfering actions independently.

Keywords: differential dynamic logic, hybrid games, sequent calculus, theorem proving, logics for hybrid systems, factory automation.

1 Introduction

One relevant question when analyzing complex physical systems is whether one component is able to meet a given safety requirement no matter what its environment does. Consider an autonomous robot moving around in a robotic factory environment. Global decision planning is infeasible, so the robot has limited knowledge about what the other elements of the factory will decide to do. If there is any probabilistic information about the decisions of agents, stochastic system models can be used for verification [9]. Otherwise, the question can be considered as a game between the component and its environment. The mathematical model for interacting discrete control and continuous evolutions is called hybrid system [8]. The game theoretic extension is called hybrid games.

Hybrid games [12,14,4,1,15] have two types of actions: discrete jumps, which update the value of a variable instantaneously, and continuous evolutions along solutions of differential equations. Time only passes for the latter action. Hence, hybrid games are a natural extension of timed games [5], which only support

* This material is based upon work supported by the German Research Council (DFG) in SFB/TR 14 AVACS, the National Science Foundation under NSF CAREER Award CNS-1054246 and NSF EXPEDITION CNS-0926181, grant no. CNS-0931985, and the Army Research Office under Award No. W911NF-09-1-0273.

B. Gramlich, D. Miller, and U. Sattler (Eds.): IJCAR 2012, LNAI 7364, pp. 439–453, 2012.

clocks with differential equation $x' = 1$ and only allow variables to be reset to 0 and not assigned arbitrarily. Fairly restricted classes of hybrid games have been shown to be decidable (see e.g. [4,1,15]), but the general case is undecidable. Tomlin et al. [14] study hybrid games for controller synthesis. They give a numerical algorithm for computing controllable predecessors and thus checking if there is a controller that drives the system into a safe state.

Our approach to hybrid games is based on logic and built on top of differential dynamic logic (d$\mathcal{L}$) [7,8], which is a dynamic logic [3] for hybrid systems instead of the conventional discrete programs that dynamic logic has originally been invented for. The logic d$\mathcal{L}$ has modal formulas $[\alpha]\phi$ and $\langle\alpha\rangle\phi$ for each hybrid system α. The d$\mathcal{L}$ formula $[\alpha]\phi$ expresses that all states reachable by following hybrid system α satisfy ϕ and $\langle\alpha\rangle\phi$ expresses that at least one state reachable by α satisfies ϕ, where ϕ is an arbitrary d$\mathcal{L}$ formula. The logic d$\mathcal{L}$ is closed under all operators of first-order logic and nesting of modalities.

With these operators, we can express simple games in d$\mathcal{L}$ [8]. For example, when F is a hybrid system describing a factory and R a hybrid system describing a robot, then a formula of the form $[F]\langle R\rangle \mathit{safe}$ can be used to express that, for all behaviors of a factory F, the robot R can choose at least one behavior ensuring safety (represented by some d$\mathcal{L}$ formula *safe*). This is a simple game expressible in d$\mathcal{L}$, but it stops after one round of interactions by the factory player and the robot player. In order to say that the robot is still safe if it reacts appropriately after the factory changed its mind in response to the robot's first choice, we can use the formula $[F]\langle R\rangle(\mathit{safe} \wedge [F]\langle R\rangle \mathit{safe})$. We can do so for any given number of rounds of interactions of F and R, but we typically want to say that the system will be safe for *any* number of interactions of F and R, not just for 2.

In this paper, we propose a logic that can state those properties using several game constructs on top of d$\mathcal{L}$, including repetition operators $(G)^{[*]}$ and $(G)^{\langle*\rangle}$ to say that game G repeats. The difference between both operators is which player decides how often to repeat the game. They decide how often to repeat before the game starts. For example, the dDG$\mathcal{L}$ formula $([F]\langle R\rangle)^{[*]}\mathit{safe}$ expresses that, no matter how often the player responsible for $(\cdot)^{[*]}$ decides to repeat the game $[F]\langle R\rangle$, the state resulting from those alternating choices by F and R is safe.

In order to prove such properties, we lift the induction principles of d$\mathcal{L}$ to dDG$\mathcal{L}$. A dDG$\mathcal{L}$ formula $(G)^{[*]}\phi$ behaves in some ways like the d$\mathcal{L}$ or dDG$\mathcal{L}$ formula $[\alpha^*]\phi$ (where α^* is the hybrid system that repeats α). In both cases, we consider all possible numbers of iterations, because we do not know how often it will be repeated. The dDG$\mathcal{L}$ formula $(G)^{\langle*\rangle}\phi$ has similarities to the d$\mathcal{L}$ formula $\langle\alpha^*\rangle\phi$, since in both cases, we can choose some number of repetitions. Yet, G is a hybrid game, whereas α is a hybrid system. Nevertheless, we show that the induction principles of invariants and variants lift from d$\mathcal{L}$ to dDG$\mathcal{L}$.

We prove that dDG$\mathcal{L}$ is a conservative extension of d$\mathcal{L}$ and, thus, our theorem prover KeYmaera [11] for d$\mathcal{L}$ can be extended such that it can be used to prove hybrid games expressible in dDG$\mathcal{L}$. We develop a proof calculus for the specifics of dDG$\mathcal{L}$ and implement it in KeYmaera. We develop and verify a case study in which a mobile robot satisfies a joint safety and liveness objective in a factory automation scenario, in which the factory may perform interfering actions.

2 Hybrid Programs and dDG$\mathcal{L}$

Syntax. We use the *hybrid program* (HP) notation for hybrid systems. We sketch the syntax of these programs as defined in [7,8]. The syntax of HPs is shown together with an informal semantics in Tab. 1. The basic terms (called θ in the table) are either rational numbers, real-valued variables or arithmetic expressions (with operators $+, -, \cdot, /$) built from those.

Discrete jumps are modeled by $x := \theta$. Their effect is to assign the value of the term θ to the variable x. Continuous evolutions, on the other hand, are modeled by $x' = \theta \,\&\, \chi$. Here, the variables evolve along the solution of the differential equation (x' denotes the derivative of x w.r.t. to time), without leaving the evolution domain characterized by the formula χ. If there is no evolution domain restriction, i.e., $\chi \equiv \mathit{true}$, we just write $x' = \theta$. Note that $x' = \theta$ can be a system of differential equations.

To test conditions on the program flow the test action $?\chi$ is used. If the formula χ holds in the current state, the action has no effect. Otherwise, it aborts the program execution and the execution is discarded. The nondeterministic choice $\alpha \cup \beta$ expresses alternatives in the behavior of the hybrid system. The sequential composition $\alpha; \beta$ expresses that β starts after α finishes. Nondeterministic repetition α^* says that HP α repeats an arbitrary number of times. These operations can be combined to form any other classical control structure [8].

The assignment $x := *$ nondeterministically assigns a real value to x, thereby expressing unbounded nondeterminism. This nondeterminism can be restricted by combining basic programs. For instance, the idiom $x := *; ?\phi$ assigns any value to x such that the formula ϕ holds.

Based on this program notation for the behavior of hybrid systems, we separately define *hybrid games*. The idea behind our notion of hybrid games is to use operators somewhat similar to those of hybrid programs, but for games on top of full hybrid systems. The particular hybrid games that we consider here are two-player games produced by the following grammar (α is a HP):

$$G ::= [\alpha] \mid \langle\alpha\rangle \mid (G_1 \cap G_2) \mid (G_1 \cup G_2) \mid (G_1 G_2) \mid (G)^{[*]} \mid (G)^{\langle * \rangle}$$

Table 1. Statements of hybrid programs (χ is a first-order formula, α, β are HPs)

Statement	Effect
$\alpha;\ \beta$	sequential composition, performing first α and then β afterwards
$\alpha \cup \beta$	nondeterministic choice, following either α or β
α^*	nondeterministic repetition, repeating α some $n \geq 0$ times
$x_1 := \theta_1, .., x_n := \theta_n$	simultaneously assign θ_i to variables x_i by a discrete assignment
$x := *$	nondeterministic assignment of an arbitrary real number to x
$(x_1' = \theta_1, \ldots, x_n' = \theta_n \,\&\, \chi)$	continuous evolution of x_i along the differential equation system $x_i' = \theta_i$, restricted to evolution domain χ
$?\chi$	test if formula χ holds at current state, abort otherwise

By $\mathcal{G}$, we denote the set of all such hybrid games. The intuition behind these games is as follows. The game is played by two players, which we call *Verifier* and *Falsifier* who play by the following rules: In the game $[\alpha]$ Falsifier resolves the nondeterminism whereas in the game $\langle\alpha\rangle$ Verifier is allowed to do so. Observe that our notion of hybrid games is built *on top of full hybrid systems*, that is, every hybrid program α is, by way of $[\alpha]$ or $\langle\alpha\rangle$, directly a hybrid game. The game (G_1G_2) is the sequential composition of games, where game G_2 is played right after game G_1 has finished. In a game $(G_1 \cap G_2)$ Falsifier may decide whether the game proceeds with G_1 or with G_2. In the game $(G_1 \cup G_2)$ this choice is made by Verifier. Repetitive game playing is possible using the iteration constructs $(G)^{[*]}$ and $(G)^{\langle * \rangle}$, where, for the first one, Falsifier decides how many iterations are played and, for the latter one, Verifier makes the choice. Note that the choice on the number of iterations has to be made by advance notice. That is, the player responsible for controlling the iteration decides how often G is repeated when the game starts and announces it to the other player.

Winning conditions for the games are formulated in dDG$\mathcal{L}$ as postconditions of games. A *strategy* for a player determines how to resolve the nondeterminism under his control based on the result of the game played so far. The nondeterminisms inside a hybrid system are resolved by choosing which real values to assign to x when executing $x := *$ statements, which branch to follow for choices $\cup$, the number of loop iterations, and how long to follow continuous flows.

The dDG$\mathcal{L}$-*formulas* are first-order formulas over the reals extended by hybrid games. They are defined by the following grammar (θ_i are terms, x is a variable, $\sim \in \{<, \leq, =, \geq, >, \neq\}$, ϕ and ψ are formulas, and G is a hybrid game):

$$\phi ::= \theta_1 \sim \theta_2 \mid \neg\phi \mid \phi \wedge \psi \mid \phi \vee \psi \mid \phi \rightarrow \psi \mid \phi \leftrightarrow \psi \mid \forall x\, \phi \mid \exists x\, \phi \mid G\, \phi$$

A dDG$\mathcal{L}$ formula $G\,\phi$ is valid if Verifier has a strategy to ensure that ϕ holds after playing the game G. Therefore, the goal of Verifier is to make ϕ true while that of Falsifier is complementary, i.e., to make ϕ false. Note that the formula ϕ itself might contain another game.

Consider the dDG$\mathcal{L}$ formula $([\alpha])^{\langle * \rangle}\phi$ which expresses that there is a number of repetitions n, such that the formula ϕ holds after n repetitions of α. Note that this dDG$\mathcal{L}$ formulas is not equivalent to $[\alpha^*]\,\phi$, which would demand that it holds for all possible numbers of executions of α. It is also not equivalent to $\langle\alpha^*\rangle\,\phi$ as this would give control to Verifier over the (possibly unbounded) nondeterminism during the executions of α. A similar observation can be made for $(\langle\alpha\rangle)^{[*]}\phi$ which says that the program α is always able to ensure ϕ by appropriate choices of the nondeterminisms in α. Combining the repetition operator and the choice operators we can express properties like $(\langle\beta\rangle\,[\alpha] \cup [\alpha]\,\langle\beta\rangle)^{[*]}\phi$. This formula means that ϕ holds after any number of iterations (as Falsifier has control over the number of iterations) while Verifier can control (by $\cup$) for each iteration if he wants to move first according to β or Falsifier has to move first according to α.

Semantics. Next, we define the semantics of dDG$\mathcal{L}$. For a set of variables V, denote by $Sta(V)$ the set of states, i.e., all mappings of type $V \rightarrow \mathbb{R}$. Let $val(\nu, \theta)$ denote the valuation of a term θ in a state ν.

Definition 1 (Transitions of hybrid programs). *The* transition relation, $\rho(\alpha)$, of a HP α, *specifies which state ω is reachable from a state ν by operations of the hybrid system α and is defined as follows*

1. $(\nu, \omega) \in \rho(x_1 := \theta_1, \ldots, x_n := \theta_n)$ *iff* $\nu[x_1 \mapsto val(\nu, \theta_1)] \ldots [x_n \mapsto val(\nu, \theta_n)]$ *equals state ω. Particularly, the value of other variables $z \notin \{x_1, \ldots, x_n\}$ remains constant, i.e., $val(\nu, z) = val(\omega, z)$.*
2. $(\nu, \omega) \in \rho(x := *)$ *iff state ω is identical to ν except for the value of x, which can be any real number (could be identical to the previous valuation of x).*
3. $(\nu, \omega) \in \rho(x_1' = \theta_1, \ldots, x_n' = \theta_n \,\&\, \chi)$ *iff there is a continuous function $f : [0, r] \to Sta(V)$ from $f(0) = \nu$ to $f(r) = \omega$, which solves the system of differential equations, i.e., for all $i \in [1, n]$, $val(f(\zeta), x_i)$ has a derivative of value $val(f(\zeta), \theta_i)$ at each time $\zeta \in (0, r)$. Other variables remain constant: $val(f(\zeta), y) = val(\nu, y)$ for $y \neq x_i$, for all $i \in [1, n]$ and $\zeta \in [0, r]$. And the evolution domain χ is respected: $val(f(\zeta), \chi) = true$ for each $\zeta \in [0, r]$.*
4. $\rho(?\chi) = \{(\nu, \nu) \ : \ \nu \models \chi\}$ *where $\nu \models \chi$ is defined as in first-order logic.*
5. $\rho(\alpha \cup \beta) = \rho(\alpha) \cup \rho(\beta)$
6. $\rho(\alpha; \beta) = \rho(\alpha) \circ \rho(\beta) = \{(\nu, \omega) : (\nu, z) \in \rho(\alpha), (z, \omega) \in \rho(\beta)$ *for some state* $z\}$
7. $(\nu, \omega) \in \rho(\alpha^*)$ *iff there are $n \in \mathbb{N}$ and $\nu = \nu_0, \ldots, \nu_n = \omega$ with $(\nu_i, \nu_{i+1}) \in \rho(\alpha)$ for all $0 \leq i < n$.*

The semantics of formulas of dDG$\mathcal{L}$ is defined as follows.

Definition 2 (Interpretation of dDG$\mathcal{L}$ formulas). *The interpretation $\models$ of a* dDG$\mathcal{L}$ *formula w.r.t. state ν uses the standard meaning of first-order logic:*

1. $\nu \models \theta_1 \sim \theta_2$ *iff* $val(\nu, \theta_1) \sim val(\nu, \theta_2)$ *for* $\sim \in \{=, \leq, <, \geq, >\}$
2. $\nu \models \phi \wedge \psi$ *iff* $\nu \models \phi$ *and* $\nu \models \psi$*, accordingly for* $\neg, \vee, \rightarrow, \leftrightarrow$
3. $\nu \models \forall x\, \phi$ *iff* $\omega \models \phi$ *for all ω that agree with ν except for the value of x*
4. $\nu \models \exists x\, \phi$ *iff* $\omega \models \phi$ *for some ω that agrees with ν except for the value of x*

Statements about hybrid games G and programs α have the following semantics

5. $\nu \models [\alpha]\phi$ *iff* $\omega \models \phi$ *for all ω with* $(\nu, \omega) \in \rho(\alpha)$,
6. $\nu \models \langle\alpha\rangle\phi$ *iff* $\omega \models \phi$ *for some ω with* $(\nu, \omega) \in \rho(\alpha)$,
7. $\nu \models (G_1 \cup G_2)\phi$ *iff* $\nu \models G_1\phi$ *or* $\nu \models G_2\phi$,
8. $\nu \models (G_1 \cap G_2)\phi$ *iff* $\nu \models G_1\phi$ *and* $\nu \models G_2\phi$,
9. $\nu \models (G_1 G_2)\phi$ *iff* $\nu \models G_1(G_2\phi)$,
10. $\nu \models (G)^{[*]}\phi$ *iff* $\nu \models (G^n)\phi$ *holds for all* $n \in \mathbb{N}$,
11. $\nu \models (G)^{\langle * \rangle}\phi$ *iff* $\nu \models (G^n)\phi$ *holds for some* $n \in \mathbb{N}$,

where G^n denotes the n-times sequential composition of G and $G^0\phi \equiv \phi$. A formula ϕ is valid (denoted by $\models \phi$) iff $\nu \models \phi$ holds for all states $\nu \in Sta(V)$.

These definitions are abstract. They do not refer to how games are played. Therefore, we now provide a structural operational semantics for games, formally define the notions of play and strategy, and then prove that the existence of a winning strategy for Verifier coincides with the notion of satisfaction in Def. 2. For a game

$$\text{(F1)}\ \frac{(\nu,\omega)\in\rho(\alpha)}{[\alpha]@\nu \to *@\omega}\quad \text{(F2)}\ \frac{\rho(\alpha)=\emptyset}{[\alpha]@\nu \to \top@\nu}\quad \text{(F3)}\ \frac{G@\nu \to G'@\omega}{G\cap H@\nu \to G'@\omega}$$

$$\text{(F4)}\ \frac{G\cap H@\nu \to G'@\omega}{H\cap G@\nu \to G'@\omega}\quad \text{(F5)}\ \frac{n\in\mathbb{N}}{(G)^{[*]}@\nu \to G^n@\nu}$$

$$\text{(V1)}\ \frac{(\nu,\omega)\in\rho(\alpha)}{\langle\alpha\rangle@\nu \to *@\omega}\quad \text{(V2)}\ \frac{\rho(\alpha)=\emptyset}{\langle\alpha\rangle@\nu \to \bot@\nu}\quad \text{(V3)}\ \frac{G@\nu \to G'@\omega}{G\cup H@\nu \to G'@\omega}$$

$$\text{(V4)}\ \frac{G\cup H@\nu \to G'@\omega}{H\cup G@\nu \to G'@\omega}\quad \text{(V5)}\ \frac{n\in\mathbb{N}}{(G)^{\langle*\rangle}@\nu \to G^n@\nu}$$

$$\text{(S1)}\ \frac{G@\nu \to *@\omega}{(G\ H)@\nu \to H@\omega}\quad \text{(S2)}\ \frac{G@\nu \to \bot@\omega}{(G\ H)@\nu \to \bot@\omega}\quad \text{(S3)}\ \frac{G@\nu \to \top@\omega}{(G\ H)@\nu \to \top@\omega}$$

Fig. 1. Structural Operational Semantics of Hybrid Games (Verifier can only control V and S rules and Falsifier can only control F and S rules)

G and a state ν, we use $G@\nu$ to denote that the game is in the position where starting from state ν the game will follow the transitions of G.

The operational semantics for the games is structured into three types of actions: those controllable by Falsifier (prefixed with F), those controllable by Verifier (prefixed with V), and those for modelling sequential composition (prefixed with S). We add special games $*$, $\top$, and $\bot$ to denote the possible outcomes of a game. In the latter two cases either of the players was unable to make another move. The game terminates in $*$ after the players played all their actions.

Definition 3 (Structural Operational Semantics of Games). *For a game G its operational semantics $[\![G]\!]$ is given by the rules defined in Fig. 1. Here, G^n denotes the n-times sequential composition of G. The semantics provides a relation between game positions, i.e. $[\![G]\!] \subseteq \mathcal{G} \times Sta(V) \times \mathcal{G} \times Sta(V)$.*

Following the rules of the structural operational semantics defines a transition system that is possibly uncountably branching, due to the non-determinism in hybrid programs, e.g. in choosing evolution times. Observe that each path is of finite length, because the number of iterations is chosen non-deterministically but a priori. Note this semantics does not yet define who decides which options to follow. In particular, the structural operational semantics of $\cap$ and $\cup$ is still the same and that of $(\cdot)^{[*]}$ and $(\cdot)^{\langle*\rangle}$ is still the same, but they will differ as soon as we define which player gets to choose. The Verifier can choose V rules and the Falsifier can choose F rules. The S rules are determined anyway.

To determine whether a strategy is compatible with a game, we define the closure of games under a subgame relation. This closure gives the set of all game positions (ignoring the state of system variables) that can occur in a play.

Definition 4 (Closure under Subgame). *For a game G its closure under subgame, $cl(G)$, is defined inductively as:*

- $cl([\alpha]) = \{[\alpha]\}$ *and* $cl(\langle\alpha\rangle) = \{\langle\alpha\rangle\}$

- $cl(G_1G_2) = \{G_1G_2\} \cup cl(G_1) \cup cl(G_2)$
- $cl(G_1 \cup G_2) = \{G_1 \cup G_2\} \cup cl(G_1) \cup cl(G_2)$
- $cl(G_1 \cap G_2) = \{G_1 \cap G_2\} \cup cl(G_1) \cup cl(G_2)$
- $cl((G)^{[*]}) = \{(G)^{[*]}\} \cup \bigcup_{n\in\mathbb{N}} cl(G^n)$ *and* $cl((G)^{\langle*\rangle}) = \{(G)^{\langle*\rangle}\} \cup \bigcup_{n\in\mathbb{N}} cl(G^n)$

Definition 5 (Strategy). *A strategy* $s : \mathcal{G} \times Sta(V) \to (\mathcal{G} \cup \{*, \top, \bot\}) \times Sta(V)$ *is a mapping between game positions. A strategy s is called compatible with a game G if its actions are allowed, i.e.,* $((g@\nu) \to s(g@\nu)) \in [\![G]\!]$ *for all* $g \in cl(G)$ *and for all* $\nu \in Sta(V)$.

Using this notion of strategy, we now formalize the rules of the game by determining which player gets to choose from the actions of the operational semantics.

Definition 6 (Play). *Given a game* $G \in \mathcal{G}$, *a state* $\nu \in Sta(V)$, *and two compatible strategies (one for Falsifier f and one for Verifier v), a play* $p_{f,v}(G@\nu)$ *is defined by the following algorithm:*

while $G \notin \{*, \bot, \top\}$ do
  Match the form of G:
    Case $[\alpha]$, $G_1 \cap G_2$, or $(G_1)^{[*]}$ then $G@\nu := f(G@\nu)$ // *Falsifier chooses*
    Case $\langle\alpha\rangle, G_1 \cup G_2$, or $(G_1)^{\langle*\rangle}$ then $G@\nu := v(G@\nu)$ // *Verifier chooses*
    Case G_1G_2 then do
      $G@\nu := p_{f,v}(G_1@\nu)$ // *play* G_1
      If $G = *$ then $G := G_2$ // *if* G_1 *terminated with* $*$ *move to* G_2
    od
od // *the result is* $G@\nu$ *with* $G \in \{*, \bot, \top\}$

Definition 7 (Winning). *Given a game G and a* dDG$\mathcal{L}$ *formula* ϕ *as winning condition. For an initial state* ν, *the game G is won by Verifier iff G ends in a position* $H@\omega$ *where either* $H = *$ *and* $\omega \models \phi$, *or* $H = \top$. *Otherwise, Falsifier wins (i.e. the game is zero-sum). For a* dDG$\mathcal{L}$*-formula* ϕ *a strategy s is called winning in a game G if, by applying this strategy, Falsifier (resp. Verifier) wins every play of G regardless of which strategy Verifier (resp. Falsifier) follows.*

Lemma 1. *For a state* ν, $\nu \models G\phi$ *iff Verifier has a strategy in the game G with winning condition* ϕ *started in position* $G@\nu$ *such that he wins the game.*

The proof of Lemma 1 can be found in [13].

Corollary 1. *The formula* $G\phi$ *is valid iff Verifier has a winning strategy in the game G for the winning condition* ϕ.

A crucial point for the design of dDG$\mathcal{L}$ is that we want it to be conservative with respect to differential dynamic logic in the sense that all d$\mathcal{L}$ formulas are dDG$\mathcal{L}$ formulas and that any d$\mathcal{L}$ formula is valid in the semantics of dDG$\mathcal{L}$ if and only it was valid in the original semantics for d$\mathcal{L}$. This allows us to transfer soundness results for proof calculus rules from d$\mathcal{L}$ to dDG$\mathcal{L}$ and extend our theorem prover KeYmaera with additional proof rules for handling the extra dDG$\mathcal{L}$ constructs.

Theorem 1 (Conservative Extension). *Differential dynamic game logic is a conservative extension of differential dynamic logic (A proof is in [13]).*

3 Proof Rules for dDGℒ

In this section, we present a sound but incomplete proof calculus for dDGℒ. It is incomplete, because hybrid systems are not semidecidable [8]. The calculus symbolically executes the hybrid games and hybrid programs. Thereby the dDGℒ calculus reduces properties of hybrid games to dℒ properties of hybrid programs, which it, in turn, reduces to validity questions of formulas in first-order logic over the reals like the dℒ proof calculus does [7].

Substitutions are defined only on first-order formulas. Therefore, state changes performed by assignments and continuous evolutions are kept as a simultaneous assignment $\mathcal{J}$ of the form $x_1 := \theta_1, \ldots, x_n := \theta_n$. When the formula was successfully transformed into a first-order one, these jump sets are applied as substitutions. See [7,8] for more details on this matter. We denote by $\forall^G \phi$ the universal closure of the formula ϕ w.r.t. the variables changed within the game G. The sequent $\Gamma \vdash \Delta$ is an abbreviation for $\bigwedge_{\phi \in \Gamma} \phi \to \bigvee_{\psi \in \Delta} \psi$. Proof rules are applied from the desired conclusion (goal below bar) to the resulting premises (above bar) that need to be proved instead.

Definition 8 (Rules [8]). *Calculus rules are defined from the rule schemata presented in Fig. 2 using the following definitions:*

1. *If*

$$\frac{\phi_1 \vdash \psi_1 \quad \ldots \quad \phi_n \vdash \psi_n}{\phi_0 \vdash \psi_0}$$

is an instance of a rule schema in Fig. 2, then

$$\frac{\Gamma, \langle \mathcal{J} \rangle \phi_1 \vdash \langle \mathcal{J} \rangle \psi_1, \Delta \quad \ldots \quad \Gamma, \langle \mathcal{J} \rangle \phi_n \vdash \langle \mathcal{J} \rangle \psi_n, \Delta}{\Gamma, \langle \mathcal{J} \rangle \phi_0 \vdash \langle \mathcal{J} \rangle \psi_0, \Delta}$$

can be applied as a proof rule, where Γ, Δ are arbitrary (possible empty) finite sets of context formulas and $\mathcal{J}$ is a (possibly empty) discrete jump set.

2. *Symmetric schemata*

$$\frac{\phi}{\psi}$$

can be applied on either side of the sequent as

$$\frac{\Gamma, \langle \mathcal{J} \rangle \phi \vdash \Delta}{\Gamma, \langle \mathcal{J} \rangle \psi \vdash \Delta} \quad \text{or as} \quad \frac{\Gamma \vdash \langle \mathcal{J} \rangle \phi, \Delta}{\Gamma \vdash \langle \mathcal{J} \rangle \psi, \Delta}$$

Again they do not alter the context. Additionally, we use the abbreviation $\langle\!|\alpha|\!\rangle$ if the rule is independent of the player controlling the action α.

3. *The existential quantifier elimination rule applies to all goals containing variable X at once: If $\phi_1 \vdash \psi_1, \ldots, \phi_n \vdash \psi_n$ is the list of all open goals (i.e., goals that have not been proved yet) of the proof that contain the free variable X, then the following instance can be applied as a proof rule:*

$$\frac{\vdash QE(\exists X \bigwedge_i (\phi_i \vdash \psi_i))}{\phi_1 \vdash \psi_1 \quad \ldots \quad \phi_n \vdash \psi_n}$$

$$\text{(D1)}\ \frac{\phi \wedge \psi}{\langle ?\phi\rangle\, \psi} \quad \text{(D3)}\ \frac{\langle\!|\alpha|\!\rangle\, \langle\!|\beta|\!\rangle\, \phi}{\langle\!|\alpha;\beta|\!\rangle\, \phi} \quad \text{(D5)}\ \frac{[\alpha]\,\phi \wedge [\beta]\,\phi}{[\alpha \cup \beta]\,\phi} \quad \text{(D7)}\ \frac{\phi \wedge [\alpha;\alpha^*]\,\phi}{[\alpha^*]\,\phi}$$

$$\text{(D2)}\ \frac{\phi \rightarrow \psi}{[?\phi]\,\psi} \quad \text{(D4)}\ \frac{\langle\alpha\rangle\,\phi \vee \langle\beta\rangle\,\phi}{\langle\alpha\cup\beta\rangle\,\phi} \quad \text{(D6)}\ \frac{\phi \vee \langle\alpha;\alpha^*\rangle\,\phi}{\langle\alpha^*\rangle\,\phi} \quad \text{(D8)}\ \frac{\phi^{\theta_1\ldots\theta_n}_{x_1\ldots x_n}}{\langle\!|x_1 := \theta_1, \ldots, x_n := \theta_n|\!\rangle\,\phi}$$

$$\text{(D9)}\ \frac{\forall t[x := t]\phi}{[x := *]\,\phi} \quad \text{(D11)}\ \frac{\exists t \geq 0\, \forall 0 \leq \tilde{t} \leq t\, \langle x := y_v(\tilde{t})\rangle\, \chi \rightarrow \langle x := y_v(t)\rangle\,\phi}{\langle x' = \theta \& \chi\rangle\,\phi}$$

$$\text{(D10)}\ \frac{\exists t[x := t]\phi}{\langle x := *\rangle\,\phi} \quad \text{(D12)}\ \frac{\forall t \geq 0\, \forall 0 \leq \tilde{t} \leq t\, [x := y_v(\tilde{t})]\, \chi \rightarrow [x := y_v(t)]\,\phi}{[x' = \theta \& \chi]\,\phi}$$

$$\text{(D13)}\ \frac{\vdash \forall^{[\alpha]}(\phi \rightarrow [\alpha]\,\phi)}{\phi \vdash [\alpha^*]\,\phi} \quad \text{(D14)}\ \frac{\vdash \forall^{\langle\alpha\rangle}\forall n > 0(\varphi(n) \rightarrow \langle\alpha\rangle\,\varphi(n-1))}{\exists n\varphi(n) \vdash \langle\alpha^*\rangle\,\exists n(n \leq 0 \wedge \varphi(n))}$$

$$\text{(D15)}\ \frac{\vdash \phi(s(X_1,\ldots,X_n))}{\vdash \forall x\phi(x)} \quad \text{(D17)}\ \frac{\vdash \phi(X)}{\vdash \exists x\phi(x)}$$

$$\text{(D16)}\ \frac{\phi(s(X_1,\ldots,X_n)) \vdash}{\exists x\phi(x) \vdash} \quad \text{(D18)}\ \frac{\vdash \mathrm{QE}(\exists X \bigwedge_i(\phi_i \vdash \psi_i))}{\phi_1 \vdash \psi_1 \quad \ldots \quad \phi_n \vdash \psi_n}$$

$$\text{(D19)}\ \frac{\phi(X) \vdash}{\forall x\phi(x) \vdash} \quad \text{(D20)}\ \frac{\vdash \mathrm{QE}(\forall X\phi(X) \vdash \psi(X))}{\phi(s(X_1,\ldots,X_n)) \vdash \psi(s(X_1,\ldots,X_n))}$$

$$\text{(G1)}\ \frac{G_1\phi \vee G_2\phi}{(G_1 \cup G_2)\phi} \quad \text{(G3)}\ \frac{G_1(G_2\phi)}{(G_1G_2)\phi} \quad \text{(G5)}\ \frac{\phi \vee G\,(G)^{\langle*\rangle}\phi}{(G)^{\langle*\rangle}\phi}$$

$$\text{(G2)}\ \frac{G_1\phi \wedge G_2\phi}{(G_1 \cap G_2)\phi} \quad \text{(G4)}\ \frac{\phi \wedge G\,(G)^{[*]}\phi}{(G)^{[*]}\phi} \quad \text{(G6)}\ \frac{\vdash \forall^G(\phi \rightarrow \psi)}{G\phi \vdash G\psi}$$

$$\text{(G7)}\ \frac{\vdash \forall^G(\phi \rightarrow G\phi)}{\phi \vdash (G)^{[*]}\phi} \quad \text{(G8)}\ \frac{\vdash \forall^G\forall n > 0(\varphi(n) \rightarrow G\,(\varphi(n-1)))}{\exists n\varphi(n) \vdash (G)^{\langle*\rangle}\exists n(n \leq 0 \wedge \varphi(n))}$$

- t and $\tilde{t}$ are fresh logical variables, y_v is the solution of the symbolic initial value problem ($\dot{x} = \theta, x(0) = v$).
- Logical variable n does neither occur in α nor G.
- $\phi^{\theta_1\ldots\theta_n}_{x_1\ldots x_n}$ denotes the formula where each x_i is substituted by θ_i simultaneously. The assignment in rule D8 must be admissible [8], otherwise it is added to the jump context $\langle\mathcal{J}\rangle$.
- X is a new logical variable.
- QE: quantifier elimination procedure (can only be applied to first-order formulas).
- For D18 $\phi_i \vdash \psi_i$ are the only branches where X occurs as a free logical variable.

Fig. 2. Free-variable proof calculus for dDG$\mathcal{L}$

Figure 2 shows a proof calculus for dDG$\mathcal{L}$. Together with rules for dealing with propositional logic (including a cut rule), the calculus rules D1-D20 form the original proof calculus for d$\mathcal{L}$ [7,8]. To handle the new game constructs appearing in dDG$\mathcal{L}$ the rules G1-G8 are used. The rules D1-D12 are equivalences for transforming and decomposing hybrid programs. The rules D13 (resp. D14) allow reasoning about loops using induction (resp. proving convergence).

For handling first-order quantifiers, we use rules D15-D20 from d$\mathcal{L}$ [7]. They perform Skolemization [8] to allow for removing the modalities within the formulas using other rules. After the modalities are dealt with, the quantifiers are reintroduced and quantifier elimination (QE) is performed.

The rules G1 and G2 are equivalences to reason about the choice operations on games. They are the game-equivalents of D5 and D4. Rule G3 transforms sequential compositions such that they can be handled by the original rules of the d$\mathcal{L}$ calculus. It takes the form of D3. The rules G4-G5 allow for unwinding of the game loops. Rule G7 follows the pattern of D13, but allows induction over game loops that are under Falsifier's control. If Verifier can establish that a formula ϕ holds after any run of game G that started in an arbitrary state satisfying ϕ, then, by induction, ϕ holds for an arbitrary number of plays. Rule G8 follows the pattern of D14 and can be used to show properties of game loops that are under Verifier's control. We can be sure that there is a number of iterations after which the postcondition $\varphi(n)$ holds for some $n \leq 0$ if G can be controlled by Verifier such that the state converges w.r.t. $\varphi(n)$. Here, the existence of some n such that $\varphi(n)$ holds serves as an induction anchor. As for each play started in an arbitrary state where $n > 0$ and $\varphi(n)$ holds, Verifier can assure that after playing the game G the formula $\varphi(n-1)$ holds, thus the game can be forced to eventually reach a state where $n \leq 0$ and $\varphi(n)$ holds. Note that n must not occur in the game G as otherwise it would be bound by the game instead of the quantifier prefix in the postcondition and thus falsify our induction. Additionally, the generalization rule G6 can be used to strengthen postconditions. This rule can, for example, be used to add induction anchors and use cases to the rules G7 and G8.

The purpose of the calculus is to provide a framework for deriving valid dDG$\mathcal{L}$ formulas syntactically. A calculus is sound iff all formulas derived by applying the calculus rules are indeed valid.

Definition 9 (Soundness). *A calculus rule* $\frac{\phi_1,\ldots,\phi_n}{\psi_1,\ldots,\psi_n}$ *is sound iff validity of the premises* $\phi_1 \wedge \cdots \wedge \phi_n$ *implies the validity of the conclusions* $\psi_1 \wedge \cdots \wedge \psi_n$.

Theorem 2. *The* dDG$\mathcal{L}$ *calculus rules presented in Fig. 2 are sound.*

The soundness proofs for the rules D1-D20 in [8] are valid for dDG$\mathcal{L}$ as well, because dDG$\mathcal{L}$ is a conservative extension of d$\mathcal{L}$ (see Theorem 1).

The soundness of the rules G1 and G2 is obvious from the semantics of the operators $\cup$ and $\cap$ on games. For the rule G3 the soundness follows directly from the definition of the sequential composition. The soundness of the unwinding rules G4 (resp. G5) is a direct consequence of the semantics of $(G)^{[*]}$ (resp. $(G)^{\langle * \rangle}$). For proving soundness of rule G6 a similar pattern to that in [8] can be applied. The game G can only change the variables that occur in G. Therefore, if $\phi \rightarrow \psi$ and $G\phi$ holds independent of how the variables occurring in G are evaluated, ψ also holds after playing G.

Soundness of the induction rule G7 and the convergence rule G8 can be shown by induction over the number of executions of the loop, in analogy to the soundness proofs for D13 and D14. The proof of Theorem 2 can be found in [13].

4 Case Study: Robotic Factory Automation

To demonstrate the applicability of our approach we model a factory automation scenario in which an autonomous robot moves in an automatic factory. For scalability reasons, central coordination and planning become infeasible, so the factory is set up as a collection of autonomous agents pursuing goals that may not be known globally. The robot has a secondary objective of reaching certain target positions, but its primary objective is to stay safe, i.e., neither leave the factory site nor bump into its surrounding wall, which could damage the robot.

Model. We model a robot with position (x, y), velocity $\boldsymbol{v} = (v_x, v_y)$, and acceleration $\boldsymbol{a} = (a_x, a_y)$ on a 2 dimensional rectangular factory ground (Fig. 3). There are two conveyor belts. One pointing in x-direction and one pointing in y-direction. The factory may independently decide to activate the conveyor belts, in which case they increase the velocity of the robot. The robot may decide to move in any direction. Therefore, it can decelerate to try to compensate for this increased speed. The goal of the robot is to avoid crashing into any wall and avoid other machines using the belt.

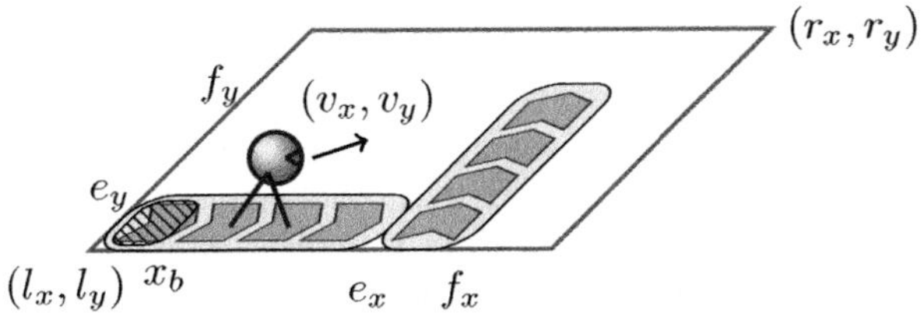

Fig. 3. Sketch of the robotic factory automation site

A sketch of the factory site is provided in Fig. 3. One conveyor belt is of y-width e_y between positions l_x and e_x and moves in x-direction if activated. Between e_x and f_x there is a belt of y-width f_y moving in y-direction. The shaded region in Fig. 3 indicates a region that has to be cleared within ε time units after the system was started, because other robotic elements of the factory may occupy this space then and not watch out for our robot. For simplicity, the robot is initially located at the lower left end (l_x, l_y) of the factory site. The conveyor belt in x-direction has a maximal velocity of c_x and that in y-direction of c_y. The conveyor belts accelerate very quickly, so we simply consider them to accelerate instantaneously. Thus, upon activation, their effect is to increase the velocity of the robot by a discrete assignment instantaneously if the robot is currently located on the conveyor belt that got activated. The robot itself can accelerate with any acceleration of absolute value at most $A = 2$ and that acceleration can be applied in x-direction (acceleration a_x) or y-direction (acceleration a_y) or both. The robot can activate a brake that will slow it down. The difference between braking and just accelerating in the opposite direction is that braking does not allow changes of the sign of the velocity but instead stops at velocity 0.

Specification. As the robot is a moving object and cannot come to a standstill instantaneously, certain conditions have to be satisfied to allow safe operation. Therefore, we assume the following conditions on the scenario. We require that the point x_b can be reached by accelerating for at most time ε, the x-belt moves to the right (if activated), and after passing the belts there is enough space to brake from the velocity we reach by accelerating for four cycles (each of duration $\leq \varepsilon$) and possible extra velocity gained when a conveyor belt activates:

$$x_b < \frac{1}{2}A\varepsilon^2 \wedge c_x > 0 \wedge (c_x + 4A\varepsilon)^2 \leq 2A(r_x - f_x) \tag{1}$$

For the y-direction we assume

$$c_y > 0 \wedge c_y^2 \leq 2A(r_y - l_y) \ , \tag{2}$$

i.e., the y-belt moves upwards (if activated) and there is enough space for the robot to compensate for the effects of the conveyor belt by braking long enough without having to leave the factory ground.

Even though these constraints limit the possible scenarios, we haven proven that a strategy for the robot exists such that it meets its objectives. Figure 4 shows the hybrid game describing the robotic factory scenario. The game is structured as follows. First the environment, i.e., Falsifier, chooses a number of iterations. In each iteration, the environment may choose to activate one of the conveyor belts if the robot is on it. Afterwards, the robot (i.e. Verifier) chooses his accelerations in x and y-direction. The clock t_s is reset to measure the cycle time (i.e. $t_s \leq \varepsilon$), then the robot chooses if it wants to brake or possibly to drive backwards (w.r.t. to its current direction). The time for the continuous evolution is then chosen by the environment within the cycle time constraint and possibly the zero crossing of one of the velocities. Thus, accelerating for ε time units can take many iterations of the loop as ε only provides an upper bound on the cycle time. Further, note that for the braking case if the velocity in a direction is 0 then that acceleration is set to 0 as well to avoid time deadlocks. Also the robot has to ensure that his choices for the acceleration are compatible with the current velocities: for a velocity v and an acceleration a, if the robot wants to brake, i.e. reduce the velocity to zero, then the product va has to be non-positive.

The winning condition for the robot is to stay safe, i.e.,:

$$l_x \leq x \leq r_x \wedge l_y \leq y \leq r_y \wedge (t \geq \varepsilon \rightarrow (x \geq x_b \vee y \geq e_y))$$

The robot must stay within the rectangle of the points (l_x, l_y) and (r_x, r_y) but has to leave the rectangle (l_x, l_y) and (x_b, e_y) after ε time units. The latter requirement models that uncooperative robotic elements might enter that region. Note that the number of iterations is chosen when the game starts not when specifying the system. Sensor and communication delays are not modelled explicitly here. Since they are beyond control for the robot, the number of iterations and the evolution durations are chosen by the factory environment. How long the robot needs to work in the factory is also decided by the factory, so the robot needs to guarantee safety for all times. However, whether the robot actually has a strategy is quite subtle. Simple strategies like always accelerating, or always braking are bound to fail and accelerating for exactly ε time units is not possible as the environment determines the actual cycle time and might not allow changing the acceleration at that exact point in time. The robot has to navigate the factory very carefully, react to changes in the conveyor belt activation as needed, and robustly adapt to the number of control loop repetitions and (possibly erratic) cycle durations chosen by the factory environment.

Verification. We consider an instance of the case study that is parametric w.r.t. ε, c_x, c_y, and x_b, but we fix $l_x = l_y = 0$, $r_x = r_y = 10$, $e_x = 2$, $e_y = 1$, $f_x = 3$,

$$\Big([?true \cup (?(x < e_x \wedge y < e_y \wedge \text{eff}_1 = 1);\ v_x := v_x + c_x;\ \text{eff}_1 := 0)$$
$$\cup (?(e_x \le x \wedge y \le f_y \wedge \text{eff}_2 = 1);\ v_y := v_y + c_y;\ \text{eff}_2 := 0)\ ]$$
$$\langle\, a_x := *;\ ?(-A \le a_x \le A);\ a_y := *;\ ?(-A \le a_y \le A);\ t_s := 0\,\rangle$$
$$([x' = v_x, y' = v_y, v_x' = a_x, v_y' = a_y, t' = 1, t_s' = 1 \& t_s \le \varepsilon\,]$$
$$\cup(\langle ?a_x v_x \le 0 \wedge a_y v_y \le 0;\ \text{if } v_x = 0 \text{ then } a_x := 0 \text{ fi};\ \text{if } v_y = 0 \text{ then } a_y := 0 \text{ fi}\rangle$$
$$[x' = v_x, y' = v_y, v_x' = a_x, v_y' = a_y, t' = 1, t_s' = 1 \& t_s \le \varepsilon \wedge a_x v_x \le 0 \wedge a_y v_y \le 0]))\Big)^{[*]}$$

Fig. 4. Description of game for robotic factory automation scenario (RF)

$f_y = 10$. We have verified the following propositions using KeYmaera [11], to which we added dDG$\mathcal{L}$ proof rules. To establish the desired property, we first show that the robot can stay within the factory site whatever the factory does.

Proposition 1. *The following* dDG$\mathcal{L}$ *formula is valid, i.e., there is a strategy for Verifier in the game depicted in Fig. 4 that achieves the postcondition:*

$$(x = y = 0 \wedge (1) \wedge (2)) \to (RF)(l_x \le x \le r_x \wedge l_y \le y \le r_y)$$

When proving this property, we focus on the case where the robot is not driving towards the lower left corner; see Fig. 3.

Again allowing for arbitrary movement in x-direction, we analyze, for a projection to the x-axis, a more complex postcondition, where the robot has to leave the shaded region but stay inside the factory site.

Proposition 2. *The following* dDG$\mathcal{L}$ *formula is valid, i.e., there is a strategy for Verifier in the game in Fig. 4 projected to the x-axis (denoted $RF|_x$) that achieves the postcondition:*

$$(x = y = 0 \wedge (1)) \to (RF|_x)(l_x \le x \le r_x \wedge (t \ge \varepsilon \to (x \ge x_b)))$$

In the proof of Proposition 2 we prove the following inductive invariant:

$$\begin{aligned}
&\text{eff}_1 \in \{0,1\} \wedge x \ge l_x \wedge v_x \ge 0 \wedge (t \ge \varepsilon \to x \ge x_b) \wedge (v_x + c_x \text{eff}_1)^2 \le 2A(r_x - x)\\
&\wedge \Big(x < x_b \to t \le \varepsilon \wedge (x_b - x \le \frac{1}{2}A\varepsilon^2 - \frac{1}{2}At^2\\
&\wedge (\text{eff}_1 = 1 \to v_x = At) \wedge (\text{eff}_1 = 0 \to v_x = At + c_x)\\
&\wedge r_x - x \ge \frac{(v_x + \text{eff}_1 c_x)^2}{2A} + A(2\varepsilon - t)^2 + 2(2\varepsilon - t)(v_x + \text{eff}_1 c_x))\Big)
\end{aligned} \tag{3}$$

The invariant says that enough space remains to brake before reaching the right end of the factory ground. Additionally, if the point x_b has not yet been passed then the time is not up and the distance to the right wall is bounded by the space the robot can cover by accelerating ε time units and the distance it could already have covered within the current runtime. Further, the distance to the far right side is large enough to accelerate for another $2\varepsilon - t$ time units and brake afterwards without hitting the wall. The 2ε time units are necessary as Falsifier chooses how long to evolve and the robot may accelerate for ε time units before it can react again. Therefore, the robot may have to accelerate when clock t is almost ε and then may not react again within the next ε time units.

The KeYmaera proof [13] for Proposition 1 has 2471 proof steps on 742 branches, with 159 interactive steps. The proof for Proposition 2 has 375079 proof steps on 10641 branches (1673 interactive steps). The interactive steps provide the invariant and simplify the resulting arithmetic. Note that Proposition 1 is significantly simpler than Proposition 2, because there is a simpler strategy that ensures safety (Proposition 1), whereas the dDG$\mathcal{L}$ formula in Proposition 2 is only valid when the robot follows a subtle strategy to leave the shaded region quickly enough without picking up too much speed to get itself into trouble when conveyor belts decide to activate. Specifically, Proposition 2 needs the much more complicated invariant (3). Also, the a priori restriction (and thus strategy choice) to the case where the robot is driving in the direction towards larger x and larger y values reduces the proof for Proposition 1 significantly.

As every strategy witnessing Proposition 2 is compatible with some strategy witnessing Proposition 1, we claim that the robot meets its requirements.

5 Related Work

Our approach to hybrid games has some resemblance to Parikh's propositional game logic (GL) [6] for propositional games. But dDG$\mathcal{L}$ is a conservative extension of d$\mathcal{L}$ [7] with hybrid programs as hybrid system models. We refer to [10] for an identification of the fundamental commonalities and differences of GL versus d$\mathcal{L}$. Axiom K and Gödel's generalization rule stop to hold for games [10], but are used in KeYmaera, which had been designed for hybrid systems not games. It is, thus, crucial that K and Gödel's generalization are still sound for dDG$\mathcal{L}$. Unlike GL, dDG$\mathcal{L}$ has an advance notice semantics, i.e., the players announce the number of repetitions of a loop when it starts.

Vladimerou et. al. [15] extended o-minimal hybrid games [1] to STORMED hybrid games and proved decidability of optimal reachability. These hybrid games are based on STORMED hybrid systems, which require that all system actions point towards a common direction. Unfortunately, neither STORMED hybrid games nor their special case of o-minimal hybrid games are expressive enough for our needs. Our factory automation scenario is not STORMED, e.g., because some actions decrease the velocity (when the robot brakes) and some trajectories increase it (when the conveyor belt activates).

Tomlin et. al. [14] present an algorithm to compute maximal controlled invariants for hybrid games with continuous inputs. The class of games they consider is more general than ours as they allow inputs to differential equations to be controlled by both players, thereby added a differential game component. However, the general class of games they consider is so large, the algorithm presented is semi-decidable only for certain classes of systems, e.g., systems specified as timed or linear hybrid automata, or o-minimal hybrid systems. They further present numerical techniques to compute approximations of their reach set computation operators. However, these sometimes give unsound results. Additionally, it only works for differentiable value functions. Extending these ideas, Gao et. al. [2] present a different technique for the same approach. The drawback is that the players can neither force discrete transitions to happen nor influence which location is reached by a discrete transition.

In contrast to these automata-based approaches to hybrid games we do not consider concurrent choices of actions. However, it is, for instance, possible to model voting for the next evolution time, provided the players can announce their choices in given orders. A precedence for player actions is often times encoded into the semantics of hybrid game automata, e.g. controller actions have precedence over environment actions in [15], whereas d$\mathcal{DGL}$ offers more flexibility in modeling these syntactically for the particular needs of the application.

References

1. Bouyer, P., Brihaye, T., Chevalier, F.: O-minimal hybrid reachability games. Logical Methods in Computer Science 6(1) (2009)
2. Gao, Y., Lygeros, J., Quincampoix, M.: On the Reachability Problem for Uncertain Hybrid Systems. IEEE Transactions on Automatic Control 52(9) (September 2007)
3. Harel, D.: First-Order Dynamic Logic. LNCS, vol. 68. Springer, Heidelberg (1979)
4. Henzinger, T.A., Horowitz, B., Majumdar, R.: Rectangular Hybrid Games. In: Baeten, J.C.M., Mauw, S. (eds.) CONCUR 1999. LNCS, vol. 1664, pp. 320–335. Springer, Heidelberg (1999)
5. Maler, O., Pnueli, A., Sifakis, J.: On the Synthesis of Discrete Controllers for Timed Systems (An Extended Abstract). In: Mayr, E.W., Puech, C. (eds.) STACS 1995. LNCS, vol. 900, pp. 229–242. Springer, Heidelberg (1995)
6. Parikh, R.: The logic of games and its applications. In: Annals of Discrete Mathematics, pp. 111–140. Elsevier (1985)
7. Platzer, A.: Differential dynamic logic for hybrid systems. J. Autom. Reas. 41(2), 143–189 (2008)
8. Platzer, A.: Logical Analysis of Hybrid Systems: Proving Theorems for Complex Dynamics. Springer, Heidelberg (2010)
9. Platzer, A.: Stochastic Differential Dynamic Logic for Stochastic Hybrid Programs. In: Bjørner, N., Sofronie-Stokkermans, V. (eds.) CADE 2011. LNCS, vol. 6803, pp. 446–460. Springer, Heidelberg (2011)
10. Platzer, A.: Differential game logic for hybrid games. Tech. Rep. CMU-CS-12-105, School of Computer Science, Carnegie Mellon University, Pittsburgh (March 2012)
11. Platzer, A., Quesel, J.-D.: KeYmaera: A Hybrid Theorem Prover for Hybrid Systems (System Description). In: Armando, A., Baumgartner, P., Dowek, G. (eds.) IJCAR 2008. LNCS (LNAI), vol. 5195, pp. 171–178. Springer, Heidelberg (2008)
12. Quesel, J.-D., Fränzle, M., Damm, W.: Crossing the Bridge between Similar Games. In: Fahrenberg, U., Tripakis, S. (eds.) FORMATS 2011. LNCS, vol. 6919, pp. 160–176. Springer, Heidelberg (2011)
13. Quesel, J.D., Platzer, A.: Playing Hybrid Games with KeYmaera. Tech. Rep. 84, SFB/TR 14 AVACS (April 2012), `http://www.avacs.org` ISSN: 1860–9821
14. Tomlin, C., Lygeros, J., Sastry, S.: A Game Theoretic Approach to Controller Design for Hybrid Systems. Proceedings of IEEE 88, 949–969 (2000)
15. Vladimerou, V., Prabhakar, P., Viswanathan, M., Dullerud, G.: Specifications for decidable hybrid games. Theoretical Computer Science 412(48), 6770–6785 (2011)

The QMLTP Problem Library for First-Order Modal Logics

Thomas Raths* and Jens Otten

Institut für Informatik, University of Potsdam
August-Bebel-Str. 89, 14482 Potsdam-Babelsberg, Germany
{traths,jeotten}@cs.uni-potsdam.de

Abstract. The Quantified Modal Logic Theorem Proving (QMLTP) library provides a platform for testing and evaluating automated theorem proving (ATP) systems for first-order modal logics. The main purpose of the library is to stimulate the development of new modal ATP systems and to put their comparison onto a firm basis. Version 1.1 of the QMLTP library includes 600 problems represented in a standardized extended TPTP syntax. Status and difficulty rating for all problems were determined by running comprehensive tests with existing modal ATP systems. In the presented version 1.1 of the library the modal logics K, D, T, S4 and S5 with constant, cumulative and varying domains are considered. Furthermore, a small number of problems for multi-modal logic are included as well.

1 Introduction

Problem libraries are essential tools when developing and testing *automated theorem proving* (ATP) systems for various logics. Popular examples are the TPTP library [23] for classical logic and the ILTP library [18] for intuitionistic logic. These libraries help many developers to test and improve their ATP system and, hence, have stimulated the development of more efficient ATP systems.

Modal logics extend classical logic with the modalities "it is necessarily true that" and "it is possibly true that" represented by the unary operators $\Box$ and $\Diamond$, respectively. The (Kripke) semantics of modal logics is defined by a set of worlds constituting classical logic interpretations, and a binary accessibility relation on this set. $\Box F$ or $\Diamond F$ are true in a world w, if F is true in *all* worlds accessible from w or *some* world accessible from w, respectively. First-order or *quantified* modal logics extend propositional modal logics by *domains* specifying sets of objects that are associated with each world, and the standard universal and existential quantifiers [4, 10].

First-order modal logics allow a natural and compact knowledge representation. The subtle combination of the modal operators and first-order logic enables specifications on epistemic, dynamic and temporal aspects, and on infinite sets of objects. For this reason, first-order modal logics have applications, e.g., in planning, natural language processing, program verification, querying knowledge bases, and modeling communication. In these applications, modalities are used to represent incomplete knowledge, programs,

* This work is partly funded by the German Science Foundation DFG under reference number KR858/9-1.

B. Gramlich, D. Miller, and U. Sattler (Eds.): IJCAR 2012, LNAI 7364, pp. 454–461, 2012.

or to contrast different sources of information. These applications would benefit from a higher degree of automation. Consequently there is a real need for efficient ATP systems for first-order modal logic whose development is still in its infancy.

The *Quantified Modal Logic Theorem Proving (QMLTP) library* provides a comprehensive set of standardized problems in first-order modal logic and, thus, constitutes a basis for testing and evaluating the performance of ATP systems for modal logics. The main purpose of the QMLTP library is to stimulate the development of new calculi and ATP systems for first-order modal logics. It puts the testing and evaluation of ATP systems for first-order modal logics on a firm basis, makes meaningful system evaluations and comparisons possible, and allows to measure practical progress in the field.

There already exist a few benchmark problems and methods for some *propositional* modal logics, e.g. there are some scalable problem classes [1] and procedures that generate formulas randomly in a normal form [16]. For *first-order* modal logics, there are only small collections of formulas available, e.g., a small set of formulas used for testing the ATP system GQML-Prover [25]. Version 5.1.0 of the TPTP library also contains some modal problems, mostly from textbooks, formulated in a typed higher-order language. All these existing sets of problems are included in the QMLTP library.

This paper introduces release v1.1 of the QMLTP library. It describes how to obtain and use the QMLTP library, provides details about the contents of the library, and information about status, difficulty rating and syntax of the problems.

2 Obtaining and Using the Library

The QMLTP library is available online at `http://www.iltp.de/qmltp/`. The main file containing the modal problems is structured into three subdirectories:

`Documents`	–	contains papers, statistic files, and other documents.
`Problems`	–	contains a directory for each domain with problem files.
`TPTP2X`	–	contains the tptp2X tool and the format files.

There are a few important conditions that should be observed when presenting results of modal ATP systems based on the QMLTP library; see [23]. The release number of the QMLTP library and the version of the tested ATP system including all settings must be documented. Each problem should be referred to by its unique name and no part of the problems may be modified. No reordering of axioms, hypotheses and/or conjectures is allowed. Only the syntax of problems may be changed, e.g., by using the tptp2X tool (see Section 3.4). The header information (see Section 3.3) of each problem may not be exploited by an ATP system.

It is a good practice to make at least the executable of an ATP system available whenever performance results or statistics based on the QMLTP library are presented. This makes the verification and validation of performance data possible.

3 Contents of the QMLTP Library

Figure 1 provides a summary of the contents of release v1.1 of the QMLTP library.

Table 1. Overall statistics of the QMLTP library v1.1

Number of problem domains	11	
Number of problems	600	(100%)
Number of first-order problems	421	(70%)
Number of propositional problems	179	(30%)
Number of uni-modal problems	580	(97%)
Number of multi-modal problems	20	(3%)

3.1 The QMLTP Domain Structure

The 600 problems of the library are divided into problem classes or *domains*. These domains are APM, GAL, GLC, GNL, GSE, GSV, GSY, MML, NLP, SET, and SYM.[1]

1. APM – *applications mixed.*
   10 problems from planning, querying databases, natural language processing and communication, and software verification [5, 7, 8, 19, 21, 22].
2. GAL/GLC/GNL/GSE/GSV/GSY – *Gödel's embedding.*
   245 problems are generated by using Gödel's embedding of intuitionistic logic into the modal logic S4 [13]. The original problems are from the following domains of the TPTP library: ALG (general algebra), LCL (logic calculi), NLP (natural language processing), SET (set theory), SWV (software verification), SYN (syntactic).
3. MML – *multi-modal logic.*
   20 problems in a multi-modal logic syntax from various textbooks and applications, e.g., security protocols and dialog systems [8, 21, 22].
4. NLP/SET – *classical logic.*
   80 problems from the NLP and SET domains of the TPTP library [23]; these allow comparisons with classical ATP systems.
5. SYM – *syntactic modal.*
   175 problems from various textbooks [9–12, 17, 20, 25] and 70 problems from the TANCS-2000 system competition for modal ATP systems [14].

3.2 Modal Problem Status and Difficulty Rating

As already done in the TPTP library, each problem is assigned a status and a rating.

The *rating* determines the difficulty of a problem with respect to current state-of-the-art ATP systems. It is the fraction of state-of-the-art systems which are *not* able to solve a problem within a given time limit. For example a rating of 0.3 indicates that 30% of the state-of-the-art systems do *not* solve the problem; a problem with rating of 1.0 cannot be solved by any state-of-the-art system. A *state-of-the-art* system is an ATP system whose set of solved problems is not subsumed by any other ATP system.

The *status* is either `Theorem`, `Non-Theorem` or `Unsolved`; see [18]. For problems with status `Theorem` (i.e. valid) or `Non-Theorem` (i.e. invalid) at least one of the considered ATP systems has found a proof or counter model, respectively. Problems with

[1] The domains MML, NLP, and SET were added in v1.1 of the QMLTP library.

Table 2. QMLTP library v1.1: Status and rating summary for all uni-modal problems

Logic	Domain Condition	Modal Status: Theorem	Modal Status: Non-Theorem	Modal Status: Unsolved	Modal Rating: 0.00	Modal Rating: 0.01–0.49	Modal Rating: 0.50–0.99	Modal Rating: 1.00	∑
K	varying	105	134	341	72	0	167	341	580
	cumulative	136	204	240	54	42	244	240	580
	constant	161	159	260	54	50	216	260	580
D	varying	180	243	157	65	46	312	157	580
	cumulative	200	257	123	53	74	330	123	580
	constant	219	242	119	55	90	316	119	580
T	varying	228	173	179	102	49	250	179	580
	cumulative	255	158	167	73	94	246	167	580
	constant	274	144	162	74	110	234	162	580
S4	varying	278	146	156	125	53	246	156	580
	cumulative	342	120	118	91	106	265	118	580
	constant	358	106	116	88	121	255	116	580
S5	varying	362	115	103	156	62	259	103	580
	cumulative	441	64	75	109	179	217	75	580
	constant	443	64	73	110	179	218	73	580

`Unsolved` status have not been solved by any ATP system. No inconsistencies between the output of the ATP systems were found. The status is specified with respect to a particular modal logic and a particular *domain condition*.

When determining the modal status, the standard semantics of first-order modal logics are considered; see e.g. [10]. Term designation is assumed to be rigid, i.e., terms denote the same object in each world, and terms are local, i.e., any ground term denotes an existing object in every world. For release v1.1 of the QMLTP library all rating and status information is with respect to the first-order modal logics K, D, T, S4, or S5 with constant, cumulative or varying domain condition [10, 25].

To determine the modal rating and status of the problems, all existing ATP systems for first-order modal logic were used. These are LEO-II 1.2.6-M1.0, Satallax 2.2-M1.0, MleanSeP 1.2, MleanTAP 1.3, f2p-MSPASS 3.0 and MleanCoP 1.2. Not all systems support all modal logics or domain conditions. LEO-II [3] and Satallax [6] are ATP systems for typed higher-order logic. To deal with modal logic, both ATP systems use an embedding of quantified modal logic into simple type theory [2].[2] LEO-II uses an extensional higher-order resolution calculus. Satallax employs a complete ground tableau calculus for higher-order logic. MleanSeP and MleanTAP are compact ATP systems for several first-order modal logics.[3] MleanSeP is a compact implementation of the standard modal sequent calculi and performs an analytic proof search. MleanTAP implements an analytic free-variable prefixed tableau calculus and uses an additional prefix unification. f2p-MSPASS uses a non-clausal instance-based method

[2] LEO-II and Satallax were used as these are the two higher-order ATP system that solved the highest number of problems at CASC-23 and CASC-J5 [24].

[3] Available at `http://www.leancop.de/mleansep/` and `http://www.leancop.de/mleantap/`

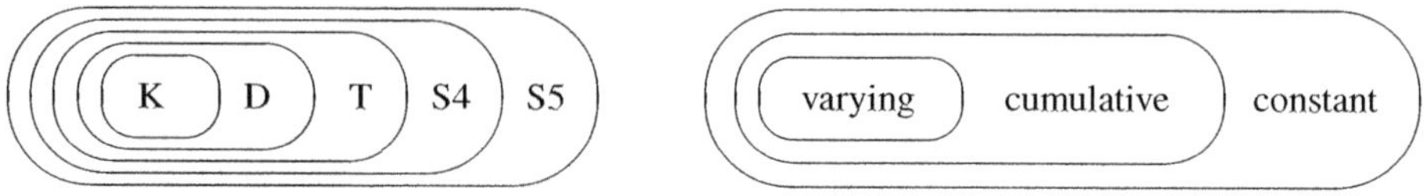

Fig. 1. The theoremhood relationship between different modal logics and domain conditions

and the MSPASS 3.0 system for propositional modal logic. MleanCoP implements a connection calculus for first-order modal logic extended by prefixes and a prefix unification algorithm [15]. Table 2 shows statistics about the status and rating of the problems in the QMLTP library for all uni-modal logics under consideration. The 20 problems for multi-modal logic contain 14 theorems and 6 unsolved problems. Figure 1 shows the theoremhood relationship, which is also reflected in Table 2.

3.3 Naming, Syntax and Presentation

Similar to the TPTP library, each problem is given an unambiguous name. The problem *name* has the form `DDD.NNN+V[.SSS].p` consisting of the mnemonic `DDD` of its domain, the number `NNN` of the problem, its version number `V`, and an optional parameter `SSS` indicating the size of the instance. For example `SYM001+1.p` is (the first version of) the first problem in the domain SYM.

For the *syntax* of the problems the Prolog syntax of the TPTP library [23] is extended by the modal operators. The two Prolog atoms "`#box`" and "`#dia`" are used for representing $\Box$ and $\Diamond$, respectively. The formulas $\Box F$ and $\Diamond F$ are then represented by "`#box:F`" and "`#dia:F`", respectively (see Figure 2). For multi-modal logic the modal operators $\Box_i$ and $\Diamond_i$ are represented by "`#box(i)`" and "`#dia(i)`", respectively, in which the index `i` is a Prolog atom. Furthermore, for multi-modal logic the `set_logic` command of the new TPTP *process instruction language* is used to specify the semantics of the used modal operators. For example,

```
tpi(1,set_logic,modal([(a,[s4,constant]),(b,[d,constant])])).
```

determines the specific semantics of the multi-modal operators $\Box_a$, $\Diamond_a$, $\Box_b$, and $\Diamond_b$.

A header with useful information is added to the *presentation* of each problem. It is adapted from the TPTP library and includes information about the file name, the problem description, the modal status and the modal difficulty rating. An example file of a first-order modal problem is shown in Figure 2.

3.4 Tools and Prover Database

The TPTP library provides the tptp2X tool for transforming and converting the syntax of TPTP problem files. This tool can be used for the QMLTP library as well. *Format files* for all existing modal ATP systems are included in the library. They are used together with the tptp2X tool to convert the problems in the QMLTP library into the input syntax of existing modal ATP systems. The prover database of the library provides information about published modal ATP systems. For each system some basic information is

```
%------------------------------------------------------------------------------
% File       : SYM001+1 : QMLTP v1.1
% Domain     : Syntactic (modal)
% Problem    : Barcan scheme instance. (Ted Sider's qml wwf 1)
% Version    : Especial.
% English    : if for all x necessarily f(x), then it is necessary that for
%              all x f(x)
%
% Refs       : [Sid09] T. Sider. Logic for Philosophy. Oxford, 2009.
%            : [Brc46] [1] R. C. Barcan. A functional calculus of first
%              order based on strict implication. Journal of Symbolic Logic
%              11:1-16, 1946.
% Source     : [Sid09]
% Names      : instance of the Barcan formula
%
% Status     :        varying       cumulative    constant
%                K    Non-Theorem   Non-Theorem   Theorem        v1.1
%                D    Non-Theorem   Non-Theorem   Theorem        v1.1
%                T    Non-Theorem   Non-Theorem   Theorem        v1.1
%                S4   Non-Theorem   Non-Theorem   Theorem        v1.1
%                S5   Non-Theorem   Theorem       Theorem        v1.1
%
% Rating     :        varying       cumulative    constant
%                K    0.50          0.75          0.25           v1.1
%                D    0.75          0.83          0.17           v1.1
%                T    0.50          0.67          0.17           v1.1
%                S4   0.50          0.67          0.17           v1.1
%                S5   0.50          0.20          0.20           v1.1
%
%  term conditions for all terms:  designation: rigid,  extension: local
%
% Comments :
%------------------------------------------------------------------------------
qmf(con,conjecture,
(( ! [X] : (#box : ( f(X) ) ) ) => (#box : ( ! [X] : ( f(X) ) )))).
%------------------------------------------------------------------------------
```

Fig. 2. Problem file SYM001+1

provided, e.g., author, web page, short description, references, and test runs on two example problems. A summary and a detailed list of the performance results of the modal ATP system on the problems in the QMLTP library are included as well.

4 Conclusion

Extensive testing is an integral part of any software development. Logical problem libraries provide a platform for testing ATP systems. Hence, they are crucial for the development of correct and efficient ATP systems. Despite the fact that modal logics are considered as one of the most important non-classical logics, the implementation of ATP systems for (first-order) modal logic is still in its infancy.

The QMLTP library provides a comprehensive set of problems for testing ATP systems for first-order modal logic. Version 1.1 includes 600 problems with almost 9.000 status and rating information. It will make meaningful systems evaluations and comparisons possible and help to ensure that published results reflect the actual performance of an ATP system. Experiences with existing libraries have shown that they stimulate the development of novel, more efficient calculi and implementations. The availability of modal ATP systems that are sufficiently efficient will promote their employment within real applications. This will generate more modal problems from actual applications, which in turn will be included in the QMLTP library (provided that they will be submitted). The few multi-modal problems in the current release have already stimulated the implementation of the first ATP systems for first-order multi-modal logic. Future versions of the library will include more problems for multi-modal logic.

Like other problem libraries the QMLTP library is an ongoing project. All interested users are invited to submit new (first-order) modal problems and new modal ATP systems to the QMLTP library.

Acknowledgements. The authors would like to thank all people who have contributed to the QMLTP library so far. In particular, Geoff Sutcliffe and Christoph Benzmüller for many helpful discussions and suggestions.

References

1. Balsiger, P., Heuerding, A., Schwendimann, S.: A Benchmark Method for the Propositional Modal Logics K, KT, S4. Journal of Automated Reasoning 24, 297–317 (2000)
2. Benzmüller, C.E., Paulson, L.C: Quantified Multimodal Logics in Simple Type Theory. Seki Report SR-2009-02, Saarland University (2009) ISSN 1437-4447
3. Benzmüller, C.E., Paulson, L.C., Theiss, F., Fietzke, A.: LEO-II - A Cooperative Automatic Theorem Prover for Classical Higher-Order Logic (System Description). In: Armando, A., Baumgartner, P., Dowek, G. (eds.) IJCAR 2008. LNCS (LNAI), vol. 5195, pp. 162–170. Springer, Heidelberg (2008)
4. Blackburn, P., van Bentham, J., Wolter, F.: Handbook of Modal Logic. Elsevier, Amsterdam (2006)
5. Boeva, V., Ekenberg, L.: A Transition Logic for Schemata Conflicts. Data & Knowledge Engineering 51(3), 277–294 (2004)
6. Brown, C.E.: Reducing Higher-Order Theorem Proving to a Sequence of SAT Problems. In: Bjørner, N., Sofronie-Stokkermans, V. (eds.) CADE 2011. LNCS, vol. 6803, pp. 147–161. Springer, Heidelberg (2011)
7. Calvanese, D., De Giacomo, G., Lembo, D., Lenzerini, M., Rosati, R.: EQL-Lite: Effective First-Order Query Processing in Description Logics. In: Veloso, M.M. (ed.) IJCAI 2007, pp. 274–279 (2007)
8. Fariñas del Cerro, L., Herzig, A., Longin, D., Rifi, O.: Belief Reconstruction in Cooperative Dialogues. In: Giunchiglia, F. (ed.) AIMSA 1998. LNCS (LNAI), vol. 1480, pp. 254–266. Springer, Heidelberg (1998)
9. Fitting, M.: Types, Tableaus, and Goedel's God. Kluwer, Amsterdam (2002)
10. Fitting, M., Mendelsohn, R.L.: First-Order Modal Logic. Kluwer, Amsterdam (1998)
11. Forbes, G.: Modern Logic. A Text in Elementary Symbolic Logic. OUP, Oxford (1994)
12. Girle, R.: Modal Logics and Philosophy. Acumen Publ. (2000)
13. Gödel, K.: An Interpretation of the Intuitionistic Sentential Logic. In: Hintikka, J. (ed.) The Philosophy of Mathematics, pp. 128–129. Oxford University Press, Oxford (1969)
14. Massacci, F., Donini, F.M.: Design and Results of TANCS-2000 Non-classical (Modal) Systems Comparison. In: Dyckhoff, R. (ed.) TABLEAUX 2000. LNCS (LNAI), vol. 1847, pp. 50–56. Springer, Heidelberg (2000)
15. Otten, J.: Implementing Connection Calculi for First-order Modal Logics. In: 9th International Workshop on the Implementation of Logics (IWIL 2012), Merida, Venezuela (2012)

16. Patel-Schneider, P.F., Sebastiani, R.: A New General Method to Generate Random Modal Formulae for Testing Decision Procedures. Journal of Articial Intelligence Research 18, 351–389 (2003)
17. Popcorn, S.: First Steps in Modal Logic. Cambridge University Press, Cambridge (1994)
18. Raths, T., Otten, J., Kreitz, C.: The ILTP Problem Library for Intuitionistic Logic. Journal of Automated Reasoning 38(1-3), 261–271 (2007)
19. Reiter, R.: What Should a Database Know? Journal of Logic Programming 14(1-2), 127–153 (1992)
20. Sider, T.: Logic for Philosophy. Oxford University Press, Oxford (2009)
21. Stone, M.: Abductive Planning With Sensing. In: AAAI 1998, Menlo Park, CA, pp. 631–636 (1998)
22. Stone, M.: Towards a Computational Account of Knowledge, Action and Inference in Instructions. Journal of Language and Computation 1, 231–246 (2000)
23. Sutcliffe, G.: The TPTP Problem Library and Associated Infrastructure: The FOF and CNF Parts, v3.5.0. Journal of Automated Reasoning 43(4), 337–362 (2009)
24. Sutcliffe, G.: The 5th IJCAR automated theorem proving system competition – CASC-J5. AI Communications 24(1), 75–89 (2011)
25. Thion, V., Cerrito, S., Cialdea Mayer, M.: A General Theorem Prover for Quantified Modal Logics. In: Egly, U., Fermüller, C. (eds.) TABLEAUX 2002. LNCS (LNAI), vol. 2381, pp. 266–280. Springer, Heidelberg (2002)

Correctness of Program Transformations as a Termination Problem*

Conrad Rau, David Sabel, and Manfred Schmidt-Schauß

Goethe-University Frankfurt am Main, Germany
{rau,sabel,schauss}@ki.informatik.uni-frankfurt.de

Abstract. The diagram-based method to prove correctness of program transformations includes the computation of (critical) overlappings between the analyzed program transformation and the (standard) reduction rules which result in so-called forking diagrams. Such diagrams can be seen as rewrite rules on reduction sequences which abstract away the expressions and allow additional expressive power, like transitive closures of reductions. In this paper we clarify the meaning of forking diagrams using interpretations as infinite term rewriting systems. We then show that the termination problem of forking diagrams as rewrite rules can be encoded into the termination problem for conditional integer term rewriting systems, which can be solved by automated termination provers. Since the forking diagrams can be computed automatically, the results of this paper are a big step towards a fully automatic prover for the correctness of program transformations.

1 Introduction

This work is motivated from proving correctness of program transformations in program calculi that model core languages of functional programming languages. For instance, Haskell [13] is modeled by the calculus *LR* [21], Concurrent Haskell [14] is modeled by the calculus CHF [19], and Alice ML[1] is modeled by the calculus $\lambda(\mathsf{fut})$ [12,11]. A *program transformation* transforms one program into another one. It is correct if the semantics of the program is unchanged, i.e. the programs before and after the transformation are semantically equivalent. *Correctness* of program transformations plays an important role in several fields of computer science: Optimizations applied while compiling programs are program transformations and their correctness thus ensures correct compilation. For software verification programs are transformed or simplified to show properties of programs, of course these transformations must be correct. In code refactoring programs are redesigned, but the semantics of the programs must not be changed, i.e. the transformations must be correct.

As semantics (or equality) of programs we choose *contextual equivalence* [10,15], since it is a natural notion of program equivalence which can directly

* This work was supported by the DFG under grant SCHM 986/9-1.

[1] `http://www.ps.uni-saarland.de/alice/`

B. Gramlich, D. Miller, and U. Sattler (Eds.): IJCAR 2012, LNAI 7364, pp. 462–476, 2012.

be defined on top of the operational semantics. Two programs are contextually equivalent if their termination behavior is indistinguishable if they are used as subprograms in any surrounding larger program (which are called the *contexts*, denoted by C). For deterministic and expressive programming languages it is sufficient to observe whether the program's execution terminates successfully, since there are enough contexts to discriminate obviously different programs.

Proving two expressions to be contextually equivalent starting from the definition is inconvenient, since all program contexts must be considered. Several methods and theoretical tools have been developed to ease the proofs, however, depending on properties of the program calculus. In this paper we concentrate on the so-called *diagram-based method* to prove correctness of program transformations, which was successfully used for several calculi, e.g., [7,9,11,21,18,19]. Diagram uses that are similar to ours also appear in [1]. Related work on diagram methods is [22], who aim at meaning preservation and make a distinction between standard reduction and transformation. Also [8] propose the use of diagrams to prove meaning preservation during compilation.

The diagram method, as we use it, is syntactic in nature, where the steps can roughly be described as follows: Let $\xrightarrow{T}$ be a program transformation, i.e. a binary relation on expressions. First a set of overlappings between the standard reduction of the calculus and the transformation $\xrightarrow{T}$ is computed, resulting in a so-called (finite) *complete set of forking diagrams* for $\xrightarrow{T}$. The second task is to show that for all expressions e_1, e_2, and contexts C such that $C[e_1] \xrightarrow{T} C[e_2]$: The program $C[e_1]$ converges, if and only if, the program $C[e_2]$ converges. Starting with a successful reduction sequence (evaluation) for $C[e_1]$ (or $C[e_2]$, resp.), we construct a successful reduction sequence for $C[e_2]$ ($C[e_1]$, resp.) by an induction, where the forking diagrams are used like a (non-deterministic) rewriting system and the normal form is the desired evaluation.

Our current research goal is to automate the manual proofs in the diagram method. We already proposed an extended unification algorithms which performs the computation of the forking diagrams for the call-by-need lambda calculus and for the above mentioned calculus LR [16,17]. We will show that the missing part of the correctness proof, i.e. using the diagrams and induction, can be performed by showing (innermost) termination of a term rewriting system that can be constructed from the forking diagrams. The termination proof can then be automated using termination provers like AProVE [5], TTT2 [6], and CiME [2].

In this paper we rigorously analyze the use of forking diagrams as rewriting problems on reduction sequences. The goal is twofold: to encode the induction proofs as a termination proof of TRSs and also to clarify the intermediate steps thereby showing in a general way that the encoding method is sound. The forking diagrams are denoted by an expressive language, also permitting transitive closure. They only speak about the arrows (perhaps labeled) of a reduction, and completely abstract away the expressions. To show that the encoding is correct, we provide a link to the concrete reductions on expressions, which requires two levels of abstractions. Finally, we will show that the termination problem can be expressed (or encoded) by extended term rewriting systems, which are conditional integer term rewrite systems (ITRS) (see e.g., [4]). Since AProvE can only show innermost termination of ITRS, our encodings are carefully designed to require innermost termination only. We applied these encodings to the diagrams of the calculus *LR* and the manually computed forking diagrams in [21] and used AProVE to show termination (and thus correctness) of several program transformations automatically.

Structure of the Paper. In Sect. 2 we introduce the notions of a program calculus, contextual equivalence, and correct program transformations. In Sect. 3 we explain the diagram-based method, and introduce several abstractions for those diagrams and the corresponding rewriting systems. Our main result is obtained in Theorem 3.27 showing that correctness of program transformations can be encoded as a termination problem. In Sect. 4 we apply our techniques to transformations of the calculus *LR* and show the step-wise encoding of the diagrams of two transformations into an integer term rewriting system for which AProVE can automatically prove termination. Finally, we conclude in Sect. 5.

2 Calculi and Program Transformations

In this section we introduce the notion of a program calculus, contextual equivalence, and correctness of program transformations.

Definition 2.1. *A* program calculus *is a tuple* $(\mathcal{E}, \mathcal{C}, \xrightarrow{sr}, \mathcal{A}, \mathcal{L})$ *where* $\mathcal{E}$ *is the set of* expressions, $\mathcal{C}$ *is the set of* contexts, *where* $C \in \mathcal{C}$ *is a function from* $\mathcal{E}$ *into* $\mathcal{E}$, $\xrightarrow{sr} \subseteq \mathcal{E} \times \mathcal{E} \times \mathcal{L}$ *is a reduction relation,* $\mathcal{A} \subseteq \mathcal{E}$ *is a set of* answers, *and* $\mathcal{L}$ *is a finite set of labels. We assume that there is a context* $[\cdot] \in \mathcal{C}$, *such that* $[e] = e$ *for all* $e \in \mathcal{C}$, *and that* $\mathcal{C}$ *is a monoid with* $[\cdot]$ *as unit, such that* $(C_1C_2)[e] = C_1[C_2[e]]$. *We write* $\xrightarrow{sr,l} \subseteq \xrightarrow{sr}$ *for reductions with label* $l \in \mathcal{L}$.

The contexts $\mathcal{C}$ consist of all expressions of $\mathcal{E}$ where one subexpression is replaced by the context hole. The reduction $\xrightarrow{sr}$ is a small step reduction as the standard reduction of the calculus, where the labels distinguish different kinds of reductions. We do not require that answers $a \in \mathcal{A}$ are $\xrightarrow{sr}$-irreducible. The converse relation is always written by reversing the arrows.

Example 2.2. The program calculus $LR = (\mathcal{E}, \mathcal{C}, \xrightarrow{sr}, \mathcal{A}, \mathcal{L})$ [21] is an extended call-by-need lambda calculus where expressions $\mathcal{E}$ comprise abstractions, applications, data-constructors, case-expressions, `letrec` for recursive shared bindings, and `seq` for strict evaluation. $\mathcal{C}$ is the set of contexts. The standard reduction $\xrightarrow{sr}$ of *LR* is called *normal order reduction* denoted by $\xrightarrow{n}$ and the answers are so-called *weak head normal forms.* The set of labels $\mathcal{L}$ are the names of the standard reductions, e.g. *seq*, *lbeta*, and *llet*.

The *evaluation* of a program expression $e \in \mathcal{E}$ is a sequence of standard reduction steps to some answer $a \in \mathcal{A}$, i.e. $e \xrightarrow{sr,*} a$, where $\xrightarrow{sr,*}$ denotes the reflexive-transitive closure of $\xrightarrow{sr}$. If such an evaluation exists, then we write $e\Downarrow$ and say e *converges*, otherwise we write $e\Uparrow$ and say e *diverges*. The semantics of expressions is given by contextual equivalence:

Definition 2.3. *For a program calculus* $(\mathcal{E}, \mathcal{C}, \xrightarrow{sr}, \mathcal{A}, \mathcal{L})$ contextual preorder $\leq_c$ *and* contextual equivalence $\sim_c$ *on expressions* $e, e' \in \mathcal{E}$ *are defined as follows:*

$$e \leq_c e' \iff \forall C \in \mathcal{C} : C[e]\Downarrow \implies C[e']\Downarrow \quad \text{and} \quad e \sim_c e' \iff e \leq_c e' \wedge e' \leq_c e.$$

Definition 2.4. *A* program transformation $\xrightarrow{T} \subseteq (\mathcal{E} \times \mathcal{E})$ *is a binary relation on expressions, and* $\xrightarrow{T}$ *is called* correct *iff* $\xrightarrow{T} \subseteq \sim_c$

Definition 2.5. *A relation* $R \subseteq \mathcal{E} \times \mathcal{E}$ *is called* convergence-preserving *iff* $(e, e') \in R \wedge e\Downarrow \implies e'\Downarrow$*. If* R *and its inverse relation* R^- *are convergence-preserving then we say* R *is* convergence-equivalent.

Often, the correctness proof for $\xrightarrow{T}$ is done by applying the diagram method to a modified transformation $\xrightarrow{T'}$, and then using theorems of the calculus.

Definition 2.6. *A program transformation* $\xrightarrow{T'}$ *is* CP-sufficient *for a program transformation* $\xrightarrow{T}$ *iff convergence preservation of* $\xrightarrow{T'}$ *implies* $\xrightarrow{T} \subseteq \leq_c$.

For a transformation $\xrightarrow{T}$, let $\xrightarrow{\mathcal{C}(T)} := \{(C[e], C[e']) \mid e \xrightarrow{T} e', C \in \mathcal{C}\}$. Then the transformation $\xrightarrow{\mathcal{C}(T)}$ is CP-sufficient for $\xrightarrow{T}$. However, this still requires to inspect all contexts $C \in \mathcal{C}$ for proving correctness of transformation $\xrightarrow{T}$. In many calculi a so-called context lemma holds (see e.g. [3,20]) which shows that the relation $\xrightarrow{\mathcal{R}(T)} := \{(R[e], R[e']) \mid e \xrightarrow{T} e', R \in \mathcal{R}\}$ is CP-sufficient for $\xrightarrow{T}$, where $\mathcal{R} \subset \mathcal{C}$ are so-called reduction contexts. The following corollary describes the method for proving correctness. It follows directly from Definitions 2.3 and 2.6.

Corollary 2.7. *If* $\xrightarrow{T'}$ *is CP-sufficient for* $\xrightarrow{T}$, $\xleftarrow{T''}$ *is CP-sufficient for* $\xleftarrow{T}$*, and* $\xrightarrow{T'}$ *and* $\xleftarrow{T''}$ *are both convergence-preserving, then* $\xrightarrow{T}$ *is correct.*

Proving convergence preservation of a transformation $\xrightarrow{T}$ requires as a base case to inspect what happens if an answer $a \in \mathcal{A}$ is transformed by $\xrightarrow{T}$.

Definition 2.8. *A* program transformation $\xrightarrow{T}$ *is called* answer-preserving (weakly answer-preserving)*, if* $a \in \mathcal{A}$ *and* $a \xrightarrow{T} e$ *imply* $e \in \mathcal{A}$ *(*$e\Downarrow$*, respectively).*

3 Proving Correctness of Program Transformations

Throughout this section we assume that a program calculus $(\mathcal{E}, \mathcal{C}, \xrightarrow{sr}, \mathcal{A}, \mathcal{L})$ is given. In this section we explain our diagram-based method to prove convergence preservation of a transformation $\xrightarrow{T}$. For showing that $e_0\Downarrow$ is implied by $e_1 \xrightarrow{T} e_0$ and $e_1\Downarrow$, we start with a sequence of reductions $a \xleftarrow{sr,l_n} e_n \xleftarrow{sr,l_{n-1}} \ldots \xleftarrow{sr,l_1} e_1 \xrightarrow{T} e_0$ where $a \in \mathcal{A}$ and *rewrite* this sequence resulting in a sequence $a' \xleftarrow{sr,l'_m} e'_m \xleftarrow{sr,l'_{m-1}} \ldots \xleftarrow{sr,l'_0} e_0$ (with $a' \in \mathcal{A}$) validating that $e_0\Downarrow$ holds. If this is possible for all $e_1 \xrightarrow{T} e_0$ then convergence preservation of $\xrightarrow{T}$ is proven.

Definition 3.1. *Let* $\{\xrightarrow{T_1}, \ldots, \xrightarrow{T_k}\}$ *be a set of program transformations. Then a* concrete reduction sequence (RS) *is a string of elements in* $\xleftarrow{sr} \cup \bigcup_{1 \le i \le k}(\xRightarrow{T_i} \times \{i\}) \cup \{(a, a, \xrightarrow{id}) \mid a \in \mathcal{A}\}$ *with the restrictions that* $(a, a, \xrightarrow{id})$ *can only be the leftmost element, and that two subsequent elements* $(e_1, e_2, d)(e_3, e_4, d')$ *are only permitted if* $e_2 = e_3$. *We write* $e_1 \xLeftarrow{sr,l} e_2$ *for* (e_1, e_2, l), $e_1 \xRightarrow{T_i} e_2$ *for* (e_1, e_2, i), *and* $a \xrightarrow{id} a$ *for* $(a, a, \xrightarrow{id})$. *An RS is a* converging concrete reduction sequence (cRS) *if its leftmost reduction is of the form* $a \xrightarrow{id} a$. *Let* cRS *be the set of all cRSs.*

We write RSs like reduction sequences, e.g. $e_1 \xLeftarrow{sr,l} e_2\ e_2 \xRightarrow{T} e_3$ is written as $e_1 \xLeftarrow{sr,l} e_2 \xRightarrow{T} e_3$. A *rewrite rule* on RSs is a rule $S_1 \rightsquigarrow S_2$ where S_1, S_2 are RSs.

Definition 3.2. *Let* D *be a set of rewrite rules on RSs. Then the pair* $(\mathsf{cRS}, \xrightarrow{D})$ *is a string rewrite system, called a* concrete rewrite system on RSs *(CRSRS).*

3.1 Abstract Reduction Sequences

For reasoning we use abstract reduction sequences (ARS), which abstract away concrete expressions, and where abstract symbols represent the reductions and transformations, and a special constant A represents answers. To distinguish concrete and abstract reductions we use solid lines on the abstract level (i.e. $\xrightarrow{sr}$ instead of $\xRightarrow{sr}$), in contrast to doubly lined-arrows on the concrete level.

We also provide an interpretation of ARSs which maps them into concrete sequences. Note that there may be ARSs without a corresponding RS. We define two variants of abstract reduction sequences, those that must start with an answer and a more general variant which may start with any expression.

Definition 3.3. *An* abstract reduction sequence *(ARS) is a finite sequence* $I_n \ldots I_1$, *and a* converging ARS *(cARS) is a finite sequence* $AI_n \ldots I_1$ *where* A *is a constant representing any answer,* $n \ge 0$. *The symbol* I_j *may either be the symbol* $\xleftarrow{sr,l}$ *with* $l \in \mathcal{L}$ *representing a (labeled) standard reduction, the symbol* $\xleftarrow{sr,x}$ *where* x *is a variable,* $\xleftarrow{sr,\tau}$ *with* $\tau \notin \mathcal{L}$ *where* τ *represents a union of labels, or the symbol* $\xrightarrow{T_i}$ *representing transformation* T_i. *Any symbol can also be extended by* $+$ *representing the transitive closure of the corresponding reduction or transformation. Symbols that have* sr *as a part are called* sr-symbols, *and other symbols are called* transformation-symbols.

An ARS or cARS that does not contain variables is called ground, *and a ground ARS or cARS is called* simple *if there is no occurrence of* $+$. *An ARS or cARS that does not contain* $\xleftarrow{sr,\tau}$*-symbols is called* τ*-free.*

Definition 3.4. *Let* S *be a simple ARS (or* S *be a simple cARS, resp.) and* $M \subseteq \mathcal{L}$ *be a set of labels. The* interpretation w.r.t. M *is the set* $\mathcal{I}_M(S)$ *of RSs (cRSs, resp.) defined recursively by the following cases, where* S_1, S_2 *are non-empty sequences,* ϵ *denotes the empty sequence, and* $e_1 \bowtie e_2$ *means a RS that starts with expression* e_1 *and ends with expression* e_2.

$$\mathcal{I}_M(\epsilon) := \emptyset \qquad \mathcal{I}_M(A) := \{a \xrightarrow{id} a \mid a \in \mathcal{A}\}$$
$$\mathcal{I}_M(\xleftarrow{sr,l}) := \{e_1 \xLeftarrow{sr,l} e_2 \mid e_2 \xRightarrow{sr,l} e_1\} \qquad \mathcal{I}_M(\xrightarrow{T_i}) := \{e_1 \xRightarrow{T_i} e_2 \mid e_1 \xRightarrow{T_i} e_2\}$$
$$\mathcal{I}_M(\xleftarrow{sr,\tau}) := \{e_1 \xLeftarrow{sr,l} e_2 \mid e_2 \xRightarrow{sr,l} e_1, l \in M\}$$
$$\mathcal{I}_M(S_1S_2) := \{e_1 \bowtie e_2 \bowtie' e_3 \mid e_1 \bowtie e_2 \in \mathcal{I}_M(S_1), e_2 \bowtie' e_3 \in \mathcal{I}_M(S_2)\}$$

3.2 Rewriting by Forking and Answer Diagrams

ARSs are used in the so-called forking diagrams [7,21], which represent the overlappings of transformation steps with standard reductions on the abstract level.

General forking diagrams are a *finite* representation of all overlappings between a transformation step and a standard reduction step, and are suitable for automated encoding. They may contain + for transitive closure and label-variables. For clarifying their meaning we introduce simple forking diagrams (without label-variables and transitive closures, but with τ).

Definition 3.5. *A* general forking diagram *for a transformation* $\xRightarrow{T}$ *is a rewrite rule* $S_L \leadsto S_R$ *where* S_L, S_R *are* τ*-free ARSs and:*

- S_L *is of the form* $I_n \dots I_1 \xrightarrow{T}$ *with* $n \geq 0$ *and all* I_i *are sr-symbols.*
- S_R *is of the form* $J_m \dots J_1 I'_{m'} \dots I'_1$ *with* $m, m' \geq 0$ *where all* J_i *are transformation-symbols and* I'_i *are sr-symbols.*

A simple forking diagram $S_L \leadsto S_R$ *is defined like a forking diagram where* S_L *and* S_R *are simple ARSs (which are not necessarily* τ*-free).*

$\xleftarrow{sr,l_n} \dots \xleftarrow{sr,l_1} \xrightarrow{T} \leadsto \xrightarrow{T_1} \dots \xrightarrow{T_{m'}} \xleftarrow{sr,l'_m} \dots \xleftarrow{sr,l'_1}$ is a forking diagram as shown to the right, where the left hand side of the rule (the solid arrows) form a fork and the right hand side (the dashed arrows) join the fork.

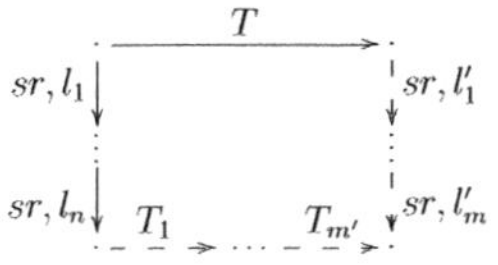

Example 3.6. The transformation $\xRightarrow{llet}$ of the calculus LR (Example 2.2) is defined by two rules, where Env is an environment of the form $y_1 = e_1, \dots, y_n = e_n$:

`letrec` Env_1 `in(letrec` Env_2 `in` e`)` $\xRightarrow{llet}$ `letrec` Env_1, Env_2 `in` e

`letrec` $Env_1, y =$ `(letrec` Env_2 `in` e'`)` `in` $e \xRightarrow{llet}$ `letrec` $Env_1, Env_2, y = e'$ `in` e

A (complete) set of general forking diagrams for the transformation $\xRightarrow{iS,llet}$ (which is CP-sufficient for the transformation $\xRightarrow{llet}$) consists of five diagrams:

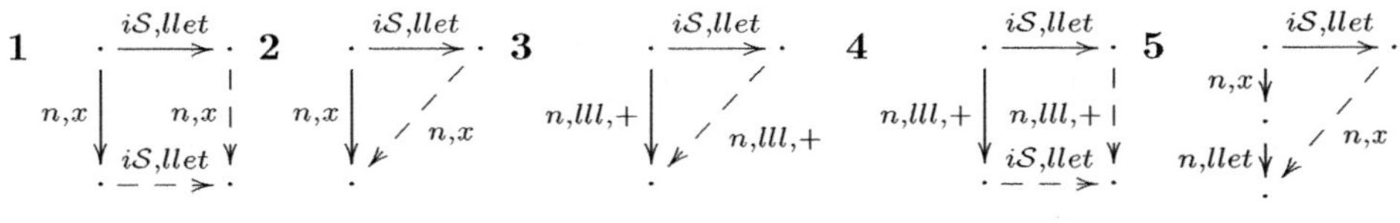

Definition 3.7. *Let* D *be a set of rewrite rules of the form* $S_L \leadsto S_R$ *where* S_L, S_R *are simple ARSs. Let* $\mathsf{cARS}(D)$ *be the set of simple cARSs that can be built by the symbols occurring in* D*. Then the string rewriting system* $(\mathsf{cARS}(D), \xrightarrow{D})$ *is called a* simple rewrite system on abstract reduction sequences *(SRSARS).*

On the concrete level, the interpretation of a simple forking diagram is a set of rewrite rules on (concrete) reduction sequences:

Definition 3.8. *The interpretation $\mathcal{I}_M(S_L \rightsquigarrow S_R)$ of a simple forking diagram $S_L \rightsquigarrow S_R$ w.r.t. a set of labels $M \subseteq \mathcal{L}$ is defined as*

$$\mathcal{I}_M(S_L \rightsquigarrow S_R) := \{e_1 \bowtie e_2 \rightsquigarrow e_1 \bowtie' e_2 \mid e_1 \bowtie e_2 \in \mathcal{I}_M(S_L), e_1 \bowtie' e_2 \in \mathcal{I}_M(S_R)\}$$

We will also interpret general forking diagrams as sets of simple forking diagrams (and thus also as rewrite rules on RSs using $\mathcal{I}_M$). We first introduce the notion of a variable interpretation which assigns concrete labels or τ to symbols $\xleftarrow{sr,x}$ and the notion of an expansion which unfolds the symbols containing a + for the transitive closure of a reduction or transformation.

Definition 3.9. *A* simple expansion *Exp_k (where $k \geq 1$) expands symbols as follows: $Exp_k(\xrightarrow{T_i,+}) = \underbrace{\xrightarrow{T_i} \dots \xrightarrow{T_i}}_{k \text{ times}}$ and $Exp_k(\xleftarrow{sr,l,+}) = \underbrace{\xleftarrow{sr,l} \dots \xleftarrow{sr,l}}_{k \text{ times}}$ where $l \in \mathcal{L} \cup \{\tau\}$, and $Exp_k(I) = I$ otherwise. For a simple ARS (or cARS) $S = I_n \dots I_1$ we define $Exp_\pi(S) := Exp_{\pi(1)}(I_n) \dots Exp_{\pi(n)}(I_1)$ where $\pi : \mathbb{N} \to \mathbb{N}$, and Exp_π denotes the* expansion *for π. For a set of labels $M \subseteq \mathcal{L}$ a* variable interpretation *$V_{\neg M}$ maps any $\xleftarrow{sr,x}$-symbol to a symbol $\xleftarrow{sr,l}$ where $l = V_{\neg M}(x) \in (\mathcal{L} \setminus M) \cup \{\tau\}$.*

General forking diagrams are interpreted as a set of simple forking diagrams:

Definition 3.10. *For a general forking diagram $S_L \rightsquigarrow S_R$ and $M \subseteq \mathcal{L}$ the translation $\mathcal{J}_M(S_L \rightsquigarrow S_R)$ is a set of simple forking diagrams $\mathcal{J}_M(S_L \rightsquigarrow S_R) :=$*

$$\left\{ Exp_\pi(V_{\neg M}(S_L)) \rightsquigarrow Exp_{\pi'}(V_{\neg M}(S_R)) \;\middle|\; \begin{array}{l} V_{\neg M} \text{ is a variable interpretation for} \\ M,\ Exp_\pi \text{ and } Exp_{\pi'} \text{ are expansions} \end{array} \right\}$$

We also use $\mathcal{J}_M$ for sets of forking diagrams, where the resulting sets are joined.

Example 3.11. Let D be the third diagram from Example 3.6. For $\mathcal{L} = \{lll, llet, seq, \dots\}$ and $M = \mathcal{L} \setminus \{lll, llet\}$ the translation $\mathcal{J}_M(D)$ is $\{\xleftarrow{n,lll}\xrightarrow{iS,llet} \rightsquigarrow \xleftarrow{n,lll}, \xleftarrow{n,lll}\xleftarrow{n,lll}\xrightarrow{iS,llet} \rightsquigarrow \xleftarrow{n,lll}, \xleftarrow{n,lll}\xrightarrow{iS,llet} \rightsquigarrow \xleftarrow{n,lll}\xleftarrow{n,lll}, \dots\}$. If D is the second diagram from Example 3.6 then $\mathcal{J}_M(D) = \{\xleftarrow{n,lll}\xrightarrow{iS,llet} \rightsquigarrow \xleftarrow{n,lll}, \xleftarrow{n,llet}\xrightarrow{iS,llet} \rightsquigarrow \xleftarrow{n,llet}, \xleftarrow{n,\tau}\xrightarrow{iS,llet} \rightsquigarrow \xleftarrow{n,\tau}\}$.

With $DF(\xRightarrow{T})$ we denote a set of forking diagrams for a transformation $\xRightarrow{T}$.

Definition 3.12. *A set of forking diagrams $DF(\xRightarrow{T})$ for transformation $\xRightarrow{T}$ is called* complete *for a set of labels $M \subseteq \mathcal{L}$, if any concrete reduction sequence of the form $a \xLeftarrow{sr,l_n} e_n \xLeftarrow{sr,l_{n-1}} \dots \xLeftarrow{sr,l_1} e_1 \xRightarrow{T} e_0$ where $a \in \mathcal{A}$, $n > 0$, and $l_i \in \mathcal{L}$ is rewritable by the CRSRS* (cRS, $\xrightarrow{\mathcal{I}_M(\mathcal{J}_M(DF(\xRightarrow{T})))}$).

In ARSs the label τ is used to represent standard reductions which are not explicitly mentioned in the diagrams, i.e. $\xleftarrow{sr,\tau}$ is interpreted as $\bigcup_{i=1}^{m} \xLeftarrow{sr,l_i}$ where $l_1, \dots, l_m$ are the labels of $\mathcal{L}$ that do not occur in the general forking diagram.

Now that forking diagrams and their semantics are defined there are two further tasks: (i) We have also to deal with reductions $a \stackrel{T}{\Rightarrow} \ldots$, and (ii) diagrams may use several transformations $\stackrel{T_i}{\Rightarrow}$ in S_R. Thus for (i) we introduce answer diagrams, and for (ii) we will join forking diagrams of a set of transformations.

Definition 3.13. *An* answer diagram *for transformation* $\stackrel{T}{\Rightarrow}$ *is a rewrite rule of the form* $A \stackrel{T}{\rightharpoonup} \rightsquigarrow S$ *where* S *is a* τ*-free cARS. A* simple answer diagram *is defined analogously where* τ*-labels in* S *are allowed, but* S *is a simple cARS.*

The interpretation of a simple answer diagram w.r.t. a set $M \subseteq \mathcal{L}$ *is*

$$\mathcal{I}_M(A \stackrel{T}{\rightharpoonup} \rightsquigarrow S) := \{a_1 \bowtie e \rightsquigarrow a_2 \bowtie' e \mid a_1 \bowtie e \in \mathcal{I}_M(A \stackrel{T}{\rightharpoonup}), a_2 \bowtie' e \in \mathcal{I}_M(S)\}$$

We extend $\mathcal{I}_M$ *to sets of simple answer diagrams joining the resulting sets.*

For an answer diagram the set of simple answer diagrams w.r.t. a set of labels $M \subseteq \mathcal{L}$ *is computed by the function* $\mathcal{J}_M$ *which is defined as follows:*

$$\mathcal{J}_M(A \stackrel{T}{\rightharpoonup} \rightsquigarrow S) = \left\{ A \stackrel{T}{\rightharpoonup} \rightsquigarrow S' \;\middle|\; \begin{array}{l} S' \in Exp_\pi(V_{\neg M}(S)),\ Exp_\pi \text{ is an expansion,} \\ V_{\neg M} \text{ is a variable interpretation for } M \end{array} \right\}$$

We extend $\mathcal{J}_M$ *to sets of answer diagrams such that the resulting sets are joined.*

A set of answer diagrams $DA(\stackrel{T}{\Rightarrow})$ *for transformation* $\stackrel{T}{\Rightarrow}$ *is* complete w.r.t. a set of labels M *iff the set* $\mathcal{I}_M(\mathcal{J}_M(DA(\stackrel{T}{\Rightarrow})))$ *contains a rewrite rule with left hand side matching any possible cRS* $a \stackrel{id}{=} a \stackrel{T}{\Rightarrow} e$ *for* $a \in \mathcal{A}$ *and* $e \in \mathcal{E}$*.*

Note that for an answer-preserving transformation $\stackrel{T}{\Rightarrow}$ a complete set of answer diagrams is $\{A \stackrel{T}{\rightharpoonup} \rightsquigarrow A\}$ and for a weakly answer-preserving transformation a complete set of answer diagrams can be constructed such that all answer diagrams are of the form $A \stackrel{T}{\rightharpoonup} \rightsquigarrow AI_m \ldots I_1$ where every I_j is an *sr*-symbol.

Definition 3.14. *Let* $DA(\stackrel{T_1}{\Longrightarrow}), \ldots, DA(\stackrel{T_n}{\Longrightarrow})$ *be sets of answer diagrams and* $DF(\stackrel{T_1}{\Longrightarrow}), \ldots, DF(\stackrel{T_n}{\Longrightarrow})$ *be sets of general forking diagrams. Let* M *be all labels of* $\mathcal{L}$ *that do not occur in any of the diagrams. The union* D *of these sets of diagrams is called* complete *for transformations* $\stackrel{T_1}{\Longrightarrow}, \ldots, \stackrel{T_n}{\Longrightarrow}$ *iff every set* $DA(\stackrel{T_i}{\Longrightarrow})$ *is complete for* $\stackrel{T_i}{\Rightarrow}$ *w.r.t.* M*, every set* $DF(\stackrel{T_i}{\Longrightarrow})$ *is complete for* $\stackrel{T_i}{\Rightarrow}$ *w.r.t.* M*, and the only transformations occurring in the diagrams are* $\stackrel{T_1}{\longrightarrow}, \ldots, \stackrel{T_n}{\longrightarrow}$*.*

We write $\mathcal{I}(D)$ instead of $\mathcal{I}_M(D)$ ($\mathcal{I}(\mathcal{J}(D))$ instead of $\mathcal{I}_M(\mathcal{J}_M(D))$, resp.) for a complete set D, since the set M is fixed by the completeness definition.

3.3 Proving Convergence Preservation

Definition 3.15. *A string rewriting system* $(O, \rightharpoonup)$ *is* leftmost terminating, *iff it is terminating w.r.t. the* leftmost rewriting relation $\rightharpoonup_l$*:*
$S_1 I_n \ldots I_1 S_2 \rightharpoonup_l S_1 S_R S_2$ *iff* $I_n \ldots I_1 \rightharpoonup S_R$ *and* $S_1 I_n \ldots I_2$ *is* $\rightharpoonup$*-irreducible.*

Proposition 3.16. *Let* $D = \bigcup_{i=1}^{n} DA(\xrightarrow{T_i}) \cup \bigcup_{i=1}^{n} DF(\xrightarrow{T_i})$ *be complete for* $\xrightarrow{T_1},\ldots,\xrightarrow{T_n}$. *If the SRSARS* $(\mathsf{cARS}(\mathcal{J}(D)), \xrightarrow{\mathcal{J}(D)})$ *is (leftmost) terminating then the CRSRS* $(\mathsf{cRS}, \xrightarrow{\mathcal{I}(\mathcal{J}(D))})$ *is (leftmost) terminating.*

Proof. For termination, the claim holds, since a nonterminating rewriting sequence for the CRSRS $(\mathsf{cRS}, \xrightarrow{\mathcal{I}(\mathcal{J}(D))})$ can easily be transfered into a nonterminating rewriting sequence of the SRSARS $(\mathsf{cARS}(\mathcal{J}(D)), \xrightarrow{\mathcal{J}(D)})$. For leftmost termination the claim also holds, since completeness of D implies that always the leftmost transformation step is rewritten by the CRSRS, and there is a corresponding rewrite rule in $\mathcal{J}(D)$ which must be leftmost, since all left hand sides of rules in $\mathcal{J}(D)$ are of the form $\xleftarrow{sr,l_n} \ldots \xleftarrow{sr,l_1} \xrightarrow{T_j}$.

Proposition 3.17. *Let* $D = \bigcup_{i=1}^{n} DA(\xrightarrow{T_i}) \cup \bigcup_{i=1}^{n} DF(\xrightarrow{T_i})$ *be complete for* $\xrightarrow{T_1},\ldots,\xrightarrow{T_n}$. *Let the CRSRS* $(\mathsf{cRS}, \xrightarrow{\mathcal{I}(\mathcal{J}(D))})$ *be terminating (leftmost terminating, resp.). Then the transformations* $\xrightarrow{T_1},\ldots,\xrightarrow{T_n}$ *are convergence-preserving.*

Proof. Let $e_1 \xrightarrow{T_i} e_0$ where $e_1\Downarrow$. Let $a \xleftarrow{id} a \xleftarrow{sr,l_n} e_n \xleftarrow{sr,l_{n-1}} \ldots \xleftarrow{srl_1} e_1$ be a cRS witnessing $e_1\Downarrow$. We compute a normal form of the cRS $a \xleftarrow{id} a \xleftarrow{sr,l_n} e_n \xleftarrow{sr,l_{n-1}} \ldots \xleftarrow{srl_1} e_1 \xrightarrow{T_i} e_0$ using leftmost rewriting of the CRSRS $(\mathsf{cRS}, \xrightarrow{\mathcal{I}(\mathcal{J}(D))})$. We only have to argue that this normal form is of the form $a' \xleftarrow{id} a' \xleftarrow{sr,l'_m} e'_m \ldots \xleftarrow{sr,l'_m} e_0$, which implies $e_0\Downarrow$. The definition of forking and answer diagrams and the completeness conditions imply that any rewrite step $\xrightarrow{\mathcal{I}(\mathcal{J}(D))}$ transforms a cRS into a cRS where the contained reductions are $\xleftarrow{sr,l}$-reductions and $\xrightarrow{T_j}$-transformations. Completeness of the diagrams ensures that the reduction sequence is modifiable by $\xrightarrow{\mathcal{I}(\mathcal{J}(D))}$ as long as $\xrightarrow{T_j}$-transformations are contained in the sequences. □

In general the other direction does not hold, since the SRSARS may be nonterminating, while the CRSRS is terminating. Propositions 3.16 and 3.17 imply:

Theorem 3.18. *Let* $D = \bigcup_{i=1}^{n} DA(\xrightarrow{T_i}) \cup \bigcup_{i=1}^{n} DF(\xrightarrow{T_i})$ *be complete for* $\xrightarrow{T_1},\ldots,\xrightarrow{T_n}$, *and let the SRSARS* $(\mathsf{cARS}(\mathcal{J}(D)), \xrightarrow{\mathcal{J}(D)})$ *be (leftmost) terminating. Then the transformations* $\xrightarrow{T_1},\ldots,\xrightarrow{T_n}$ *are convergence-preserving.*

3.4 A Rewriting System with Finitely Many Rules

A naive approach that encodes general diagrams with transitive closures adds rules $\xrightarrow{T,+} \rightsquigarrow \xrightarrow{T}\xrightarrow{T,+}$ and $\xrightarrow{T,+} \rightsquigarrow \xrightarrow{T}$ for any symbol $\xrightarrow{T,+}$. However, this is useless, since it it leads to nontermination. Hence, we provide another encoding which is suitable for automation. It translates +-symbols as nondeterministic rules using natural numbers to avoid nontermination. The translation is a little bit complex, since it has to respect the leftmost rewriting and it treats +-symbols in left hand sides and right hand sides differently.

Definition 3.19. *A (converging, resp.)* abstract reduction sequence with natural numbers *(NARS) (or cNARS, resp.) is a sequence $I_n \dots I_1$ $(AI_n \dots I_1$, resp.) where A represents any answer and each I_j is a symbol of the form $\xleftarrow{sr,l}$, $\xrightarrow{T_i}$, $\langle w \rangle, \langle w, k \rangle, \langle w, \overline{k}+1 \rangle$ where $l \in \mathcal{L} \cup \{\tau\}$, where $w \in W$ for a set of names W with $W \cap (\mathcal{L} \cup \{\tau\}) = \emptyset$, and k is either a natural number ($k \in \mathbb{N}$) or a* number variable, *i.e. a variable that may only be instantiated by natural numbers, and $\overline{k}$ is always a number variable. A NARS (cNARS, resp.) is called* ground *iff it does not contain number variables.*

Definition 3.20. *A* number substitution σ *assigns a natural number to any number variable. The extension of σ to NARS-symbols is the identity except for the cases $\sigma(\langle w, k \rangle) = \langle w, \sigma(k) \rangle$, $\sigma(\langle w, k+1 \rangle) = \langle w, k' \rangle$, where k is a number variable, and $k' = \sigma(k) + 1 \in \mathbb{N}$.*

We now define rewriting on ground cNARSs.

Definition 3.21. *Let D be a set of rules of the form $S_L \rightsquigarrow S_R$ where S_L, S_R are NARSs. Let $\mathsf{gcNARS}(D)$ be the set of all ground cNARS that can be built by instantiating the symbols occurring in D by any number substitution σ. The rewriting system $(\mathsf{gcNARS}(D), \xrightarrow{D})$ is called an* encoded rewriting system on abstract reduction systems *(ERSARS) where $\xrightarrow{D}$ is defined by: If $S = S'S'_L S''$, $S_L \rightsquigarrow S_R \in D$, σ is a number substitution with $\sigma(S_L) = S'_L$, then $S \xrightarrow{D} S'\sigma(S_R)S''$.*

Definition 3.22. *For a general forking or answer diagram $S_L \rightsquigarrow S_R$ and $M \subseteq \mathcal{L}$ the translation $\mathcal{V}_M$ is a finite set of rewrite rules over ground ARSs:*

$\mathcal{V}_M(S_L \rightsquigarrow S_R) := \bigcup \{V_{\neg M}(S_L) \rightsquigarrow V_{\neg M}(S_R) \mid V_{\neg M}$ *is a variable interpretation*$\}$

Given a set $D = \bigcup_i \{S_{i,L} \rightsquigarrow S_{i,R}\}$ of general forking and general answer diagrams and $M \subseteq \mathcal{L}$ the translation $\mathcal{K}_M(D)$ is defined as follows:

First all (usual) variables are interpreted, resulting in the set $D' := \bigcup_i \{\mathcal{V}_M(S_{i,L} \rightsquigarrow S_{i,R})\}$. For every rule $S_L \rightsquigarrow S_R \in D'$ the set $\mathcal{K}_M(D)$ contains a rule $\mathcal{K}_L(S_L) \to \mathcal{K}_R(S_R)$ perhaps together with some further rules.

Construction of $\mathcal{K}_L(S_L)$: *Let $S_L = I_n \dots I_1 \xrightarrow{T_i}$ where I_j is either $\xleftarrow{sr,l_j}$, or $\xleftarrow{sr,l_j,+}$, or (for $j = 1$) $I_j = A$. Let $K_j := \langle w_j \rangle$ if $I_j = \xleftarrow{sr,l_j,+}$ and $K_j := I_j$ otherwise, where $w_j \in W$ are fresh names (chosen fresh for any new rule). Then we set $\mathcal{K}_L(S_L) := K_n \dots K_1 \xrightarrow{T_i}$. For any I_j which is of the form $\xleftarrow{sr,l_j,+}$ we add two so-called* contraction rules*: $\xleftarrow{sr,l_j} K_{j-1} \dots K_1 \xrightarrow{T_i} \rightsquigarrow K_j K_{j-1} \dots K_1 \xrightarrow{T_i}$ and $\xleftarrow{sr,l_j} K_j K_{j-1} \dots K_1 \xrightarrow{T_i} \rightsquigarrow K_j K_{j-1} \dots K_1 \xrightarrow{T_i}$.*

Construction of $\mathcal{K}_R(S_R)$**:** *Let $S_R = I_n \dots I_1$. If none of the I_j contains a $+$ then the translation is $\mathcal{K}_R(S_R) := I_n \dots I_1$. Otherwise there is at least one $+$. Let $L_j := \xrightarrow{T_i}$ if $I_j = \xrightarrow{T_i,+}$, $L_j := \xleftarrow{sr,l_j}$ if $I_j = \xleftarrow{sr,l_j,+}$ and $L_j := I_j$ otherwise.*

Let $w'_j \in W$ (for $j \in \{1, \dots, n\}$) be fresh names. Let I_a be the rightmost I_j that contains a $+$, then we set $\mathcal{K}_R(S_R) := \langle w'_a, k \rangle L_{a-1} \dots L_1$ where k is a number variable. For all I_j with $I_j = \xrightarrow{T_i,+}$ or $I_j = \xleftarrow{sr,l_j,+}$ we additionally add so-called *expansion rules* $\langle w'_j, k+1\rangle \rightsquigarrow \langle w'_j, k\rangle L_j$ *and* $\langle w'_j, 1\rangle \rightsquigarrow L_n \dots L_j$. *If there exists* $m > j$ *where* I_m *contains a* $+$, *then for the smallest such* m *we also add the expansion rule* $\langle w'_j, k+1\rangle \rightsquigarrow \langle w'_m, k\rangle L_{m-1} \dots L_j$.

For complete sets of diagrams, M is the set of labels that do not occur in any of the diagrams. In this case we omit the index M in $\mathcal{K}_M$.

The symbols $\langle w_i\rangle$ and $\langle w'_j, k\rangle$ together with the additional rules are used to interpret the transitive closure symbols on the left and the right hand side of rules in forking and answer diagrams. It is easy to verify that any rewriting sequence using only the contraction rules must be finite, and also that any rewriting sequence using only the expansions rules is also finite.

Example 3.23. Let D be the set consisting of the third diagram from Example 3.6. For $\mathcal{L} = \{lll, llet, seq, \dots\}, M = \mathcal{L} \setminus \{lll, llet\}$ the translation $\mathcal{K}_M(D)$ is: $\{\langle w\rangle \xrightarrow{iS,llet} \rightsquigarrow \langle w', k\rangle,\ \xleftarrow{n,lll} \xrightarrow{iS,llet} \rightsquigarrow \langle w\rangle \xrightarrow{iS,llet},\ \xleftarrow{n,lll} \langle w\rangle \xrightarrow{iS,llet} \rightsquigarrow \langle w\rangle \xrightarrow{iS,llet},$ $\langle w', k+1\rangle \rightsquigarrow \langle w', k\rangle \xleftarrow{n,lll},\ \langle w', 1\rangle \rightsquigarrow \xleftarrow{n,lll}\}$

Lemma 3.24. *Let* $D = \bigcup_{i=1}^{n} DA(\xRightarrow{T_i}) \cup \bigcup_{i=1}^{n} DF(\xRightarrow{T_i})$ *be complete for* $\xRightarrow{T_1}, \dots, \xRightarrow{T_n}$ *and* $S_L \rightsquigarrow S_R \in \mathcal{J}(D)$. *Then* $S_L \xrightarrow{\mathcal{K}(D),*}_l S_R$.

Proof. Since $S_L \rightsquigarrow S_R \in \mathcal{J}(D)$ there are expansions $Exp_\pi, Exp_{\pi'}$ and a variable interpretation $V_{\neg M}$ such that $S_L = Exp_\pi(V_{\neg M}(S'_L))$ and $S_R = Exp_{\pi'}(V_{\neg M}(S'_R))$ where $S'_L \rightsquigarrow S'_R \in D$. Let $S'_L = I'_n \dots I'_1 \xrightarrow{T_i}$, i.e. $S_L = Exp_{\pi(1)}(V_{\neg M}(I'_n)) \dots Exp_{\pi(n)}(V_{\neg M}(I'_1)) \xrightarrow{T_i}$. Let $K_j := \langle w_j\rangle$ if $V_{\neg M}(I'_j)$ contains a $+$ and $K_j := V_{\neg M}(I'_j)$ otherwise. Using the contraction rules introduced by $\mathcal{K}(D)$ we rewrite $S_L = Exp_{\pi(1)}(V_{\neg M}(I'_n)) \dots Exp_{\pi(n)}(V_{\neg M}(I'_1)) \xrightarrow{T_i} \xrightarrow{\mathcal{K}(D),*}_l$ $Exp_{\pi(1)}(V_{\neg M}(I'_n)) \dots K_1 \xrightarrow{T_i} \xrightarrow{\mathcal{K}(D),*}_l \dots \xrightarrow{\mathcal{K}(D),*}_l K_n \dots K_1 \xrightarrow{T_i} = \mathcal{K}(V_{\neg_M}(S'_L))$. All these steps are leftmost, since the rightmost symbol $\xrightarrow{T_i}$ is always part of the redex and always kept. Now we apply the rule $\mathcal{K}(V_{\neg_M}(S'_L)) \xrightarrow{\mathcal{K}(D)}_l \mathcal{K}(V_{\neg_M}(S'_R))$ (which is again leftmost) and have to show that $\mathcal{K}(V_{\neg_M}(S'_R))$ can be rewritten into S_R by leftmost rewriting using $\xrightarrow{\mathcal{K}(D)}$.

If S_R does not contain a $+$-symbol, then this is obvious. Suppose that S_R contains at least one $+$-symbol. Let $S'_R = J'_m \dots J'_1$, i.e. $S_R = Exp_{\pi'(1)}(V_{\neg M}(J'_m)) \dots Exp_{\pi'(m)}(V_{\neg M}(J'_1))$ and let $L_j = \xrightarrow{T_i}$ if $V_{\neg M}(J'_j) = \xrightarrow{T_i,+}$, $L_j = \xleftarrow{sr,l}$ if $V_{\neg M}(J'_j) = \xleftarrow{sr,l,+}$, and $L_j = V_{\neg M}(J'_j)$ otherwise. Moreover let us assume that $J'_{a_r}, \dots J'_{a_1}$ are the symbols that contain a $+$, such that for $i, j \in \{1, \dots, r\}$ with $i \neq j$ we have $a_i < a_j$. Then $S_R = Q_m \dots Q_1$ where every Q_j consists of k_j repetitions of L_j, i.e. it is a string of the form $L_j \dots L_j$. Let $s := \sum_{i \in \{a_1, \dots, a_k\}} k_i$. We choose this number s during the rewriting step $\mathcal{K}(V_{\neg_M}(S'_L)) \xrightarrow{\mathcal{K}(D)}_l \mathcal{K}(V_{\neg_M}(S'_R))$, and then iteratively build the string S_R using the expansion rules introduced by $\mathcal{K}(D)$: $\mathcal{K}(V_{\neg_M}(S'_L)) \xrightarrow{\mathcal{K}(D)}_l \mathcal{K}(V_{\neg_M}(S'_R)) =$

$\langle w'_{a_1}, s\rangle Q_{a_1-1}\cdots Q_1 \xrightarrow{\mathcal{K}(D),*}_l \langle w'_{a_2}, s - k_{a_1}\rangle Q_{a_2-1}\cdots Q_1 \xrightarrow{\mathcal{K}(D),*}_l \cdots \xrightarrow{\mathcal{K}(D),*}_l$ $\langle w'_{a_r}, k_{a_r}\rangle Q_{a_r-1}\cdots Q_1 \xrightarrow{\mathcal{K}(D),*}_l Q_m \cdots Q_1 = S_R$. All steps are leftmost, since always the leftmost symbol is reduced (which is of the form $\langle w'_j, s'\rangle$). □

Proposition 3.25. *Let* $D = \bigcup_{i=1}^n DA(\xRightarrow{T_i}) \cup \bigcup_{i=1}^n DF(\xRightarrow{T_i})$ *be complete for* $\xRightarrow{T_1}, \ldots, \xRightarrow{T_n}$. *Then leftmost termination of the ERSARS* $(\mathsf{gcNARS}(\mathcal{K}(D)), \xrightarrow{\mathcal{K}(D)})$ *implies leftmost termination of the SRSARS* $(\mathsf{cARS}(\mathcal{J}(D)), \xrightarrow{\mathcal{J}(D)})$.

Proof. We show that a leftmost diverging rewriting sequence of the SRSARS can be transformed to a leftmost diverging rewriting sequence of the ERSARS. Assume there is a diverging reduction. We consider a single step $S_1 S_L S_2 \xrightarrow{\mathcal{J}(D)}_l S_1 S_R S_2$ from this diverging reduction. Then S_L must be of the form $I_1 \ldots I_n \xrightarrow{T_i}$ where all I_k are sr-symbols or A. If S_1 does not contain a transformation-symbol, then Lemma 3.24 implies that $S_1 S_L S_2 \xrightarrow{\mathcal{K}(D),*}_l S_1 S_R S_2$: The rewriting step must be leftmost, since at the beginning a transformation step is required on the end of the redex, and since the rewriting generates S_R from right to left.

We now consider the case that S_1 contains other transformation symbols, w.l.o.g. let $S_1 = S_3 \xrightarrow{T_j} S_4$ such that S_4 does not contain transformation-symbols. Then perhaps there are some leftmost rewriting steps possible inside $S_3 \xrightarrow{T_j}$ using $\xrightarrow{\mathcal{K}(D),*}_l$: These can only be steps using the contraction rules. Since contraction rules cannot remove the rightmost transformation-symbol in the redex and since they are terminating, the following rewriting sequence is possible $S_3 \xrightarrow{T_j} S_4 S_L S_2 \xrightarrow{\mathcal{K}(D),*}_l S'_3 \xrightarrow{T_j} S_4 S_L S_2 \xrightarrow{\mathcal{K}(D),*}_l S'_3 \xrightarrow{T_j} S_4 S_R S_2$. Any rewriting sequence of $\xrightarrow{\mathcal{K}(D)}_l$ now cannot modify the prefix $S'_3 \xrightarrow{T_j}$. Moreover, all rewriting steps of $\xrightarrow{\mathcal{J}(D)}_l$ starting with $S_3 \xrightarrow{T_j} S_4 S_R S_2$ also do not modify the prefix $S_3 \xrightarrow{T_j}$ and thus it does not make a difference if we replace S_3 by S'_3. □

Proposition 3.25 and Theorem 3.18 imply:

Theorem 3.26. *Let* $D = \bigcup_{i=1}^n DA(\xRightarrow{T_i}) \cup \bigcup_{i=1}^n DF(\xRightarrow{T_i})$ *be complete for* $\xRightarrow{T_1}, \ldots, \xRightarrow{T_n}$. *Then leftmost termination of the ERSARS* $(\mathsf{gcNARS}(\mathcal{K}(D)), \xrightarrow{\mathcal{K}(D)})$ *implies that all transformations* $\xRightarrow{T_i}$ *are convergence preserving.*

Theorem 3.27. *Let* $\xRightarrow{T_1}$ *be CP-sufficient for* $\xRightarrow{T}$ *and let* $\xLeftarrow{T'_1}$ *be CP-sufficient for* $\xLeftarrow{T}$ *and let* $D = \bigcup_{i=1}^n DA(\xRightarrow{T_i}) \cup \bigcup_{i=1}^n DF(\xRightarrow{T_i})$ *be complete for* $\xRightarrow{T_1}, \ldots, \xRightarrow{T_n}$, $D' = \bigcup_{i=1}^m DA(\xLeftarrow{T'_i}) \cup \bigcup_{i=1}^m DF(\xLeftarrow{T'_i})$ *be complete for* $\xLeftarrow{T'_1}, \ldots, \xLeftarrow{T'_m}$, *such that both ERSARSs* $(\mathsf{gcNARS}(\mathcal{K}_M(D)), \xrightarrow{\mathcal{K}_M(D)})$ *and* $(\mathsf{gcNARS}(\mathcal{K}_M(D')), \xrightarrow{\mathcal{K}_M(D')})$ *are (leftmost) terminating. Then* $\xRightarrow{T}$ *is a correct program transformation.*

Proof. Theorem 3.26 shows that $\xRightarrow{T_1}$ and $\xLeftarrow{T'_1}$ are convergence preserving and thus CP-sufficiency shows that $\xRightarrow{T}$ is a correct program transformation.

4 Encoding ARSs and Sets of Diagrams as ITRSs

For the automation of correctness proofs we left open how to check for leftmost termination of an ERSARS derived by complete sets of forking and answer diagrams according to Theorems 3.27.

If the diagrams do not contain transitive closures, then the ERSARS is also an SRSARS with finitely many rules. In this case the SRSARS can be encoded as a term rewriting system: A step $\xrightarrow{T_i}$ is encoded as 1-ary function symbol *ti*, a step $\xleftarrow{sr,l_i}$ is encoded as a 1-ary function symbol *srli*, and the answer token A is encoded as a constant A. The string rewriting rules are translated into term rewriting rules, where left and right hand sides are both encoded from *right to left*, e.g. the rule $\xleftarrow{sr,l_1}\xrightarrow{T_1} \rightsquigarrow \xrightarrow{T_2}\xleftarrow{sr,l_2}$ is encoded as the term rewriting rule $t1(srl1(X)) \to srl2(t1(X))$ where X is a variable. It is easy to verify that leftmost termination of the SRSARS is implied by innermost termination of the TRS. We illustrate this encoding by an example from [21] for the calculus *LR* (see Example 2.2). We consider the transformation $\xrightarrow{seq}$, which is used for sequentialization, and reduces an expression $\texttt{seq}\ e_1\ e_2$ to e_2 if e_1 is a value or bound to a value. The complete set of general forking diagrams $DF(\xrightarrow{iS,seq})$ for the transformation $\xrightarrow{iS,seq}$, which is CP-sufficient for the transformation $\xrightarrow{seq}$, is:

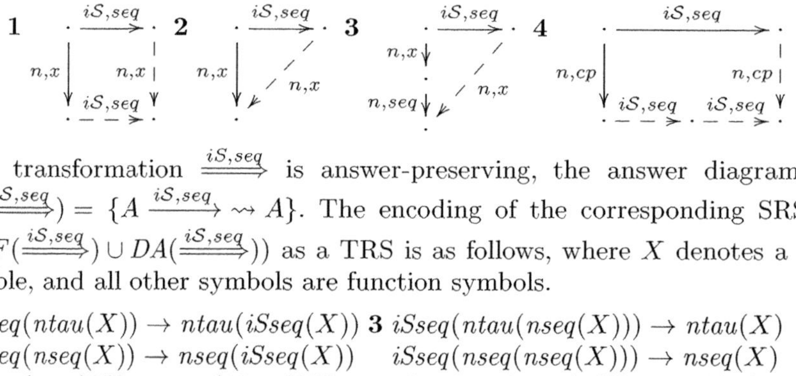

Since transformation $\xrightarrow{iS,seq}$ is answer-preserving, the answer diagrams are $DA(\xrightarrow{iS,seq}) = \{A \xrightarrow{iS,seq} \rightsquigarrow A\}$. The encoding of the corresponding SRSARS $\mathcal{J}(DF(\xrightarrow{iS,seq}) \cup DA(\xrightarrow{iS,seq}))$ as a TRS is as follows, where X denotes a term-variable, and all other symbols are function symbols.

1 $iSseq(ntau(X)) \to ntau(iSseq(X))$
$iSseq(nseq(X)) \to nseq(iSseq(X))$
$iSseq(ncp(X)) \to ncp(iSseq(X))$

2 $iSseq(ntau(X)) \to ntau(X)$
$iSseq(nseq(X)) \to nseq(X)$
$iSseq(ncp(X)) \to ncp(X)$

3 $iSseq(ntau(nseq(X))) \to ntau(X)$
$iSseq(nseq(nseq(X))) \to nseq(X)$
$iSseq(ncp(nseq(X))) \to ncp(X)$

4 $iSseq(ncp(X)) \to ncp(iSseq(iSseq(X)))$

Answer diagram: $iSseq(A) \to A$

For $\xrightarrow{iS,seq}$, and also for its inverse $\xleftarrow{iS,seq}$ innermost termination of the encoded complete diagram sets could be automatically shown via AProVE. Hence by Theorem 3.18 we can conclude correctness of the transformation.

If transitive closures occur on right hand sides of the diagrams, then an encoding into a usual TRS is not possible, since the corresponding rule in the ERSARS introduces a free number variable (which is then used for the expansion of the transitive closures). However, *conditional integer term rewriting systems* (ITRSs) is a formalism that fits for encoding ERSARS, since they allow free variables on right hand sides of rules which may only be instantiated by normal forms during rewriting. Also integers as well as conditions including arithmetic operations, comparison of integers, and Boolean connectives are already present in ITRSs (see e.g., [4]). Moreover, innermost termination of ITRSs can also be treated by the automated termination prover AProVE [5,4]. Since innermost termination of the encoded ITRSs then implies leftmost termination of an ERSARS we can use AProVE to show correctness of program transformations. The translation is as before, where the introduced names w_i in contraction rules are encoded as 1-ary function symbols, the names w'_i in expansion rules are encoded as 2-ary function symbols, natural numbers are represented by integers, and number variables are represented by variables together with constraints. Example 3.6 shows a complete set of forking diagrams for the transformation $\xrightarrow{iS,llet}$, and Example 3.23 shows the encoding of the third diagram as an ERSARS. An encoding of these rules as an ITRS is as follows where X, K are variables and all other symbols are function symbols:

$$iSllet(w(X)) \to v(K, X) \quad \text{with } K > 0$$
$$iSllet(w(nlll(X))) \to iSllet(w(X)) \qquad iSllet(nlll(X)) \to iSllet(w(X))$$
$$v(K, X) \to nlll(v(K-1, X)) \quad \text{if } K > 1 \qquad v(1, X) \to nlll(X)$$

The first rule encodes the diagram, the other rules are contraction rules (using the function symbol w), and expansion rules (using the function symbol v). The first constraint $K > 0$ ensures that a positive integer is chosen, and the constraint $K > 1$ ensures that K is a positive integer after rewriting. Innermost termination of the ITRS-encoded complete sets of forking and answer diagrams for $\xrightarrow{iS,llet}$ and for $\xleftarrow{iS,llet}$ can be checked using AProVE. This implies leftmost termination of the corresponding ERSARSs and thus by Theorem 3.27 correctness of the transformation *llet* is shown automatically.

We encoded complete sets of diagrams for several program transformations from [21] and they could all be shown as innermost terminating using AProvE. The encoded diagrams and the termination proofs can be found on our website[2].

5 Conclusion

Future work is to connect the automated termination prover with the diagram calculator of [16,17] and thus to complete the tool for automated correctness proofs of program transformations. Another direction is to check more sets of diagrams which may require more sophisticated encoding techniques.

Acknowledgments. We thank Carsten Fuhs for pointing us to ITRS and his support on AProVE and the anonymous reviewers for their valuable comments.

References

1. Barendregt, H.P.: The Lambda Calculus. Its Syntax and Semantics. North-Holland, Amsterdam (1984)
2. Contejean, E., Courtieu, P., Forest, J., Pons, O., Urbain, X.: Automated certified proofs with CiME3. In: Schmidt-Schauß, M. (ed.) RTA 22. LIPIcs, vol. 10, pp. 21–30. Schloss Dagstuhl - Leibniz-Zentrum fuer Informatik (2011)

[2] `http://www.ki.cs.uni-frankfurt.de/research/dfg-diagram/auto-induct/`

3. Ford, J., Mason, I.A.: Formal foundations of operational semantics. Higher Order Symbol. Comput. 16(3), 161–202 (2003)
4. Fuhs, C., Giesl, J., Plücker, M., Schneider-Kamp, P., Falke, S.: Proving Termination of Integer Term Rewriting. In: Treinen, R. (ed.) RTA 2009. LNCS, vol. 5595, pp. 32–47. Springer, Heidelberg (2009)
5. Giesl, J., Schneider-Kamp, P., Thiemann, R.: AProVE 1.2: Automatic Termination Proofs in the Dependency Pair Framework. In: Furbach, U., Shankar, N. (eds.) IJCAR 2006. LNCS (LNAI), vol. 4130, pp. 281–286. Springer, Heidelberg (2006)
6. Korp, M., Sternagel, C., Zankl, H., Middeldorp, A.: Tyrolean Termination Tool 2. In: Treinen, R. (ed.) RTA 2009. LNCS, vol. 5595, pp. 295–304. Springer, Heidelberg (2009)
7. Kutzner, A., Schmidt-Schauß, M.: A nondeterministic call-by-need lambda calculus. In: Felleisen, M., Hudak, P., Queinnec, C. (eds.) 3rd ICFP, pp. 324–335. ACM (1998)
8. Machkasova, E., Turbak, F.A.: A Calculus for Link-Time Compilation. In: Smolka, G. (ed.) ESOP 2000. LNCS, vol. 1782, pp. 260–274. Springer, Heidelberg (2000)
9. Mann, M.: Congruence of bisimulation in a non-deterministic call-by-need lambda calculus. Electron. Notes Theor. Comput. Sci. 128(1), 81–101 (2005)
10. Morris, J.H.: Lambda-Calculus Models of Programming Languages. PhD thesis. MIT (1968)
11. Niehren, J., Sabel, D., Schmidt-Schauß, M., Schwinghammer, J.: Observational semantics for a concurrent lambda calculus with reference cells and futures. Electron. Notes Theor. Comput. Sci. 173, 313–337 (2007)
12. Niehren, J., Schwinghammer, J., Smolka, G.: A concurrent lambda calculus with futures. Theoret. Comput. Sci. 364(3), 338–356 (2006)
13. Peyton Jones, S.: Haskell 98 language and libraries: the Revised Report. Cambridge University Press (2003)
14. Peyton Jones, S., Gordon, A., Finne, S.: Concurrent Haskell. In: Steele, G. (ed.) 23th POPL, pp. 295–308. ACM (1996)
15. Plotkin, G.D.: Call-by-name, call-by-value, and the lambda-calculus. Theoret. Comput. Sci. 1, 125–159 (1975)
16. Rau, C., Schmidt-Schauß, M.: Towards correctness of program transformations through unification and critical pair computation. In: Fernandez, M. (ed.) 24th UNIF. EPTCS, vol. 42, pp. 39–54 (2010)
17. Rau, C., Schmidt-Schauß, M.: A unification algorithm to compute overlaps in a call-by-need lambda-calculus with variable-binding chains. In: Baader, F., Morawska, B., Otop, J. (eds.) 25th UNIF, pp. 35–41 (2011)
18. Sabel, D., Schmidt-Schauß, M.: A call-by-need lambda-calculus with locally bottom-avoiding choice: Context lemma and correctness of transformations. Math. Structures Comput. Sci. 18(03), 501–553 (2008)
19. Sabel, D., Schmidt-Schauß, M.: A contextual semantics for Concurrent Haskell with futures. In: Hanus, M. (ed.) 13th PPDP, pp. 101–112. ACM (2011)
20. Schmidt-Schauß, M., Sabel, D.: On generic context lemmas for higher-order calculi with sharing. Theoret. Comput. Sci. 411(11-13), 1521–1541 (2010)
21. Schmidt-Schauß, M., Schütz, M., Sabel, D.: Safety of Nöcker's strictness analysis. J. Funct. Programming 18(04), 503–551 (2008)
22. Wells, J.B., Plump, D., Kamareddine, F.: Diagrams for Meaning Preservation. In: Nieuwenhuis, R. (ed.) RTA 2003. LNCS, vol. 2706, pp. 88–106. Springer, Heidelberg (2003)

Fingerprint Indexing for Paramodulation and Rewriting

Stephan Schulz

Institut für Informatik, Technische Universität München,
D-80290 München, Germany
schulz@eprover.org

Abstract. Indexing is critical for the performance of first-order theorem provers. We introduce *fingerprint indexing*, a non-perfect indexing technique that is based on short, constant length vectors of samples of term positions ("fingerprints") organized in a trie. Fingerprint indexing supports matching and unification as retrieval relations. The algorithms are simple, the indices are small, and performance is very good in practice.

We demonstrate the performance of the index both in relative and absolute terms using large-scale profiling.

1 Introduction

Saturating theorem provers like Vampire [3] and E [5] are among the most powerful ATP systems for first-order logic. These systems work in a refutational setting. The state of the proof search is represented by a set of clauses. It is manipulated using inference rules. The most important inferences (resolution and paramodulation/superposition) and simplifications (rewriting and subsumption) use two or more premises—usually a main premise and one or more additional side premises. This requires the system to find potential inference partners for a given clause in the potentially large set of clauses representing the search state.

The performance can be improved if inference partners are not found by sequential search, but via an *index*. An index, in this context, is a data structure with associated algorithms that allows the efficient retrieval of terms or clauses from the indexed set that are in a given *retrieval relation* with a query.

E has featured *(perfect) discrimination tree indexing* [2] for forward rewriting since version 0.1. It added *feature vector indexing* [6] for subsumption in version 0.8. In this paper, we present *fingerprint indexing*, a new, non-perfect term indexing technique that can be seen as a generalization of top symbol hashing. It shares the basic structure of feature vector indexing (indexed objects are represented by finite-length vectors organized in a trie), and combines it with ideas from coordinate and path indexing [7,2,1] (values in the index vectors represent the occurrence of symbols at certain positions in terms). The index can be used for retrieving candidate terms unifiable with a query term, matching a query term, or being matched by a query term. The index data structure has very low memory use, and all operations for maintenance and candidate retrieval are fast in practice. Variants of fingerprint indexing have been incorporated into E for backwards simplification and superposition, with generally positive results.

B. Gramlich, D. Miller, and U. Sattler (Eds.): IJCAR 2012, LNAI 7364, pp. 477–483, 2012.

2 Background

We use standard terminology for first-order logic. A signature consists of a finite set F of function symbols with associated arities. We write $f|_n$ to indicate that f has arity $n \in \mathbb{N}_0$. We assume an enumerable set V of *variables* disjoint from F, typically denoted by $x, y, z, x_0, \ldots$, or by upper-case X0,X1 if represented in TPTP syntax. Terms, subterms, literals and clauses are defined as usual.

A *substitution* is a mapping $\sigma : V \to \mathit{Term}(F, V)$ with the property that $\mathit{Dom}(\sigma) = \{x \in V \mid \sigma(x) \neq x\}$ is finite. It can be extended to terms, atoms, literals and clauses. A *matcher* from a term s to another term t is a substitution σ such that $\sigma(s) \equiv t$. A *unifier* of two terms s and t a substitution σ such that $\sigma(s) \equiv \sigma(t)$. If s and t are unifiable, a *most general unifier (mgu)* for them exists, and is unique up to variable renaming.

A *(potential) position* in a term is a sequence $p \in \mathbb{N}^*$ over natural numbers. We use ϵ to denote the empty position. The set of positions in a term, $\mathrm{pos}(t)$ is defined as follows: If $t \equiv x \in V$, then $\mathrm{pos}(t) = \{\epsilon\}$. Otherwise $t \equiv f(t_1, \ldots, t_n)$. In this case $\mathrm{pos}\,(t) = \{\epsilon\} \cup \{i.p \mid 1 \leq i \leq n, p \in \mathrm{pos}(t_i)\}$. The subterm of t at position $p \in \mathrm{pos}(t)$ is defined recursively: if $p = \epsilon$, then $t|_p = t$. Otherwise, $p \equiv i.p'$ and $t \equiv f(t_1, \ldots, t_n)$. In that case, $t_p = t_i|_{p'}$. The top symbol of $x \in V$ is $\mathrm{top}(x) = x$ and the top symbol of $f(t_1, \ldots, t_n)$ is $\mathrm{top}(f(t_1, \ldots, t_n)) = f$.

Positions can be extended to literals (selecting a term in a literal) and clauses (selecting a term in a literal in a clause) easily if we assume an arbitrary, but fixed ordering of terms in literals and literals in clauses.

Modern saturating calculi are instantiated with a *term ordering*. This ordering is lifted to literals and clauses. Generating inferences can be restricted to (subterms of) maximal terms of maximal literals. Simplification allows the replacement of clauses with equivalent smaller clauses.

3 Fingerprint Indexing

We will now introduce *fingerprints* of terms, and show that the compatibility of the respective fingerprints is a required condition for the existence of a unifier (or matcher) between two terms. The basic idea is that the application of a substitution never removes an existing position from a term, nor will it change an existing function symbol in a term.

Consider a potential position p and a term t (as a running example, assume $t = g(f(x, a))$. Then the following cases are possible:

1. p is a position in t and $t|_p$ is a variable (e.g. $p = 1.1$)
2. p is a position in t and $t|_p$ is a non-variable term (e.g. $p = 1.2$ or $p = \epsilon$)
3. p is not a position in t, but there exists an instance $\sigma(t)$ with $p \in \mathrm{pos}(\sigma(t))$ (e.g. $p = 1.1.1.2$ with $\sigma = \{x \mapsto f(a, b))\}$
4. p is not a position in t or any of its instances (e.g. $p = 2.1$)

These four cases have different implications for unification. Two terms which, at the same position, have different function symbols, cannot be unified. Stated positively, if we search for terms unifiable with a query term t, and $\text{top}(t|_p) = f$, we only need to consider terms s where $\text{top}(s|_p)$ can potentially become f.

To formalize this, consider the following definition: Let $F' = F \uplus \{\mathbf{A}, \mathbf{B}, \mathbf{N}\}$ (the set of *fingerprint feature values* for F). The *general fingerprint feature function* is a function gfpf : $Term(F, V) \times \mathbb{N}^* \to F'$ defined by:

$$\text{gfpf}(t,p) = \begin{cases} \mathbf{A} & \text{if } p \in \text{pos}(t),\ t|_p \in V \\ \text{top}(t|_p) & \text{if } p \in \text{pos}(t),\ t|_p \notin V \\ \mathbf{B} & \text{if } p = q.r,\ q \in \text{pos}(t) \text{ and } t|_q \in V \text{ for some } q \\ \mathbf{N} & \text{otherwise} \end{cases}$$

A *fingerprint feature function* is a function fpf : $Term(F, V) \to F'$ defined by $\text{fpf}(t) = \text{gfpf}(t,p)$ for a fixed $p \in \mathbb{N}^*$.

Now assume two terms, s and t, and a fingerprint feature function fpf. Assume $u = \text{fpf}(s)$ and $v = \text{fpf}(t)$. The values u and v are *compatible for unification* if they are marked with a **Y** in the *Unification* table of Figure 1. They are *compatible for matching from s onto t*, if they are marked with a **Y** in the *Matching* table in Figure 1. It is easy to show by case distinction that compatibility of the fingerprint feature values is a necessary condition for unification or matching.

Unification

	f_1	f_2	**A**	**B**	**N**
f_1	**Y**	**N**	**Y**	**Y**	**N**
f_2	**N**	**Y**	**Y**	**Y**	**N**
A	**Y**	**Y**	**Y**	**Y**	**N**
B	**Y**	**Y**	**Y**	**Y**	**Y**
N	**N**	**N**	**N**	**Y**	**Y**

Matching

	f_1	f_2	**A**	**B**	**N**
f_1	**Y**	**N**	**N**	**N**	**N**
f_2	**N**	**Y**	**N**	**N**	**N**
A	**Y**	**Y**	**Y**	**N**	**N**
B	**Y**	**Y**	**Y**	**Y**	**Y**
N	**N**	**N**	**N**	**N**	**Y**

Fig. 1. Fingerprint feature compatibility for unification and matching (down onto across). f_1 and f_2 are arbitrary but distinct.

Now assume $n \in \mathbb{N}$. A *fingerprint function* is a function fp : $Term(F, V) \to (F')^n$ with the property that $\pi_n^i \circ \text{fp}$ (the projection onto the ith element of the result) is a fingerprint feature function for all $i \in \{1, \ldots, n\}$. A *fingerprint* is the result of the application of a fingerprint function to a term, i.e. a vector of n elements over F'. We will in the following assume a fixed fingerprint function fp.

Two fingerprints for s and t are unification-compatible (or compatible for matching from s onto t) if they are component-wise so compatible.

Theorem 1. *Assume an arbitrary fingerprint function* fp. *If* $\text{fp}(t_1)$ *and* $\text{fp}(t_2)$ *are not unification compatible, then* t_1 *and* t_2 *are not unifiable. If* $\text{fp}(t_1)$ *and* $\text{fp}(t_2)$ *are not compatible for matching* t_1 *onto* t_2*, then* t_1 *does not match* t_2.

A fingerprint function defines an equivalence on $Term(F, V)$, and we can use the fingerprints to organize any set of terms into disjoint subsets, each sharing a fingerprint. If we want to find terms in a given relation (unifiable or matchable)

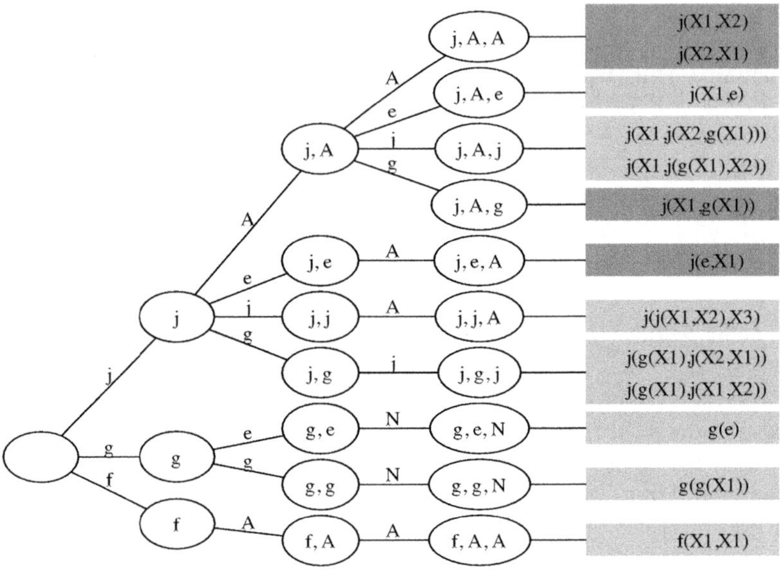

Fig. 2. Example fingerprint index

to a query term, we only need to consider terms from those subsets for which the query relation holds on the fingerprints.

However, we do not need to linearly compare fingerprints to find unification or matching candidates. A *fingerprint index* is a constant-depth *trie* over fingerprints that associates the indexed term sets with the leaves of the trie.

As an example, consider $F = \{j|_2, f|_2, g|_1, a|_0, b|_0, e|_0\}$ and fp : $Term(F, V) \rightarrow (F')^3$ defined by $\text{fp}(t) = \langle \text{gfpf}(t, \epsilon), \text{gfpf}(t, 1), \text{gfpf}(t, 2)\rangle$.

Figure 2 shows a fingerprint index for fp. When we query the index for terms unifiable with $t = j(e, g(X)))$, we first compute $\text{fp}(t) = \langle j, e, g\rangle$. At each node in the tree we follow all branches labeled with feature values unification-compatible with the corresponding fingerprint value of the query. At the root, only the branch labelled with j has to be considered. At the next node, branches e and **A** are compatible. Finally, three leaves (marked in darker gray), with a total of 4 terms, are unification-compatible with the query. In this case, all 4 terms found actually are unifiable with the query.

Even fairly short fingerprints are sufficient to achieve good performance of the index. Computing these small fingerprints is computationally cheap, and so is insertion and removal of fingerprints from the trie.

Since each term has a unique fingerprint, it is stored at exactly one leaf. To find all retrieval candidates, we traverse the trie recursively, collecting candidates from all leaves. Since all terms at a leaf are compatible with all fingerprints leading to it, and since all terms are represented at most once in the index, we only need to form the union of the candidate sets at all matching leaves. This is a major advantage compared to coordinate indexing, where it is necessary to compute the intersection of candidate sets for each coordinate. The same applies to path indexing, where e.g. Vampire goes to great lengths to optimize this bottleneck [4]. Moreover, since each term has a single fingerprint and is represented only once in the trie, there are at most as many fingerprints in an index as there are indexed terms. Thus, memory consumption of the index scales at worst linearly with the number of indexed terms.

4 Implementation

We have implemented fingerprint indexing in our theorem prover E to speed up superposition and backwards rewriting. For this purpose, we have added three global indices to E, called the *backwards-rewriting index*, the *paramodulation-from index*, and the *paramodulation-into* index.

The *backwards-rewriting index* contains all (potentially rewritable) subterms of processed clauses. Each term is associated with the set of all processed clauses it occurs in. Given a new unit clause $l \simeq r$, we find all rewritable clauses by finding all leaves compatible with the fingerprint of l, try to match l onto each of the terms t stored at the leaf, and, in the case of success, verify if $\sigma(l) > \sigma(r)$. If and only if this is the case, all clauses associated with the term are rewritable with the new unit clause (and hence are removed from the set of processed clauses). Note that in this implementation the (potentially expensive) ordering check only has to be made once for every t, not once per occurrence of t.

For the paramodulation indices, we use a somewhat more complex structure. The FP-Trie indexes sets of terms with the same fingerprint. For each term, we store a set of clauses in which this term occurs (at a position potentially compatible with the superposition restrictions). Finally, with each of these clauses, we store the positions in which the term occurs. The paramodulation-into index is organized analogously.

5 Experimental Results

To measure the performance of fingerprint indexing, we performed a series of experiments. All tests use problems from the set of 15386 untyped first-order problems in the TPTP problem library [8], Version 5.2.0. The full test data and the version of the prover used for the test runs are archived at `http://www.eprover.eu/E-eu/FPIndexing.html`. All tests were run with a time limit of 300 seconds on 2.4 GHz Intel Xeon CPUs under the Linux 2.6.18-164.el5 SMP Kernel in 64 bit mode.

We have instrumented the prover by adding profiling code to measure the time spent in parts of the program without the overhead of a standard profiler. For quantitative analysis of the run times, we only use cases where the proof search followed very similar lines for the indexed and non-indexed case (as evidenced by clause counts). This resulted in 5824 problems used for comparison.

Table 1. CPU times for different parts of the proof process (in seconds)

Index	Run time	Sat time	PM time	PMI time	MGU time	BR time	BRI time
NoIdx	16062.392	14078.300	8980.320	0.000	2545.080	2280.250	0.000
FP0	16644.127	14835.130	9904.120	26.380	4360.330	1846.280	41.440
FP0FP	9581.606	8211.010	3633.590	27.950	1322.530	1071.210	42.030
FP1	7006.758	6145.870	1816.100	25.710	450.760	379.570	40.150
FP2	6200.043	5556.330	1345.440	28.900	199.600	104.340	43.300
FP3D	6107.780	5463.240	1266.820	31.410	150.880	91.430	46.040
FP4M	6050.617	5423.820	1197.720	33.640	109.870	64.740	49.620
FP5M	6088.364	5455.180	1203.240	38.250	107.860	65.630	53.520
FP6M	6000.177	5385.810	1181.710	38.240	99.110	39.010	55.660
FP7	6022.196	5404.150	1179.250	41.880	95.880	38.400	57.610
FP8X2	6066.482	5429.390	1193.820	56.430	88.580	37.710	77.400
NPDT	6082.246	5434.760	1184.750	64.910	83.110	33.200	79.910

We include results for a number of different versions: NoIdx (no indexing), FP0 (pseudo-fingerprint of lengths 0), FP0FP (pseudo-fingerprint emulating optimizations in the unindexed version), FP1 (sampling at ϵ, equivalent to top-symbol hashing, FP2 (ϵ, 1), FP3D (ϵ, 1, 1.1), FP4M (ϵ, 1, 2, 1.1), FP5M ($\epsilon, 1, 2, 3, 1.1$), FP6M ($\epsilon, 1, 2, 3, 1.1, 1.2$), FP7 ($\epsilon$, 1, 2, 1.1, 1.2, 2.1, 2.2), FP8X2 (ϵ, 1, 2, 3, 4, 1.1, 1.2, 1.3, 2.1, 2.2, 2.3, 3.1, 3.2, 3.3, 1.1.1, 2.1.1), NPDT (non-perfect discrimination trees).

Table 1 shows the result of the time measurements, summed over all 5824 problems. "Run time" is the total run time of the prover. "Sat time" is spent in the main saturation loop, "PM time" is for paramodulation/superposition, "PMI time" is for paramodulation index maintenance. "MGU time" is the time for unification. "BR time" and "BRI time" are the times used for backwards-rewriting and backward-rewrite index maintenance.

Comparing the unindexed version with the FP6M index, total run time decreases by more than 60%. The time spent for unification itself has been reduced by a factor of about 25. The total time for unification-related code (i.e. index maintenance and unification) amounts to less than 2.5% of the total run time.

Comparing the times for FP6M with the times for discrimination tree indexing, we see that overall fingerprint indexing outperforms discrimination tree indexing, if not by a large margin. Time for actual unification is slightly lower for discrimination tree indexing, but index maintenance is slightly more expensive.

We see an even stronger improvement for backwards rewriting. The time for the operation itself drops more than 58-fold. Time for index maintenance is of the same order of magnitude. Taking index maintenance into account, the total time for backward rewriting improved by a factor of about 25.

Figure 3(a) shows a scatter plot of run times for NoIdx and FP6M. Please note the double logarithmic scale. For the vast majority of problems, the indexed version is significantly, and often dramatically, faster, while there are no problems for which the conventional version is more than marginally faster. Figure 3(b) compares FP6M and discrimination tree indexing. For most problems, performance is nearly identical.

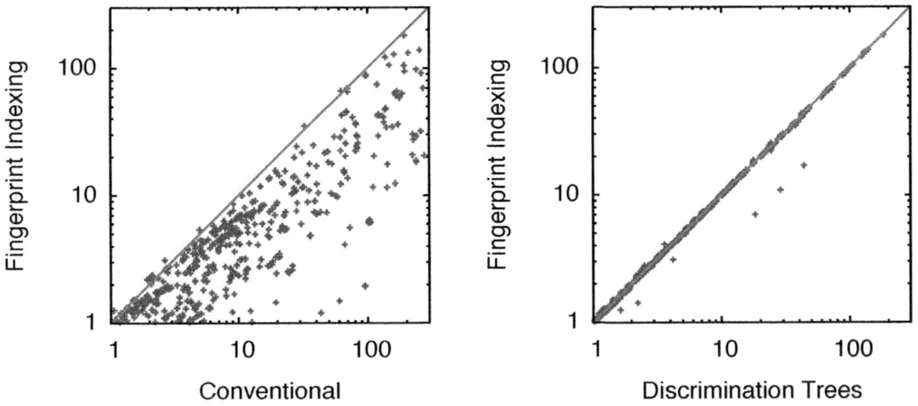

Fig. 3. Scatter plots of run times (in seconds) for FP6M over (a) non-indexed and (b) non-perfect discrimination tree implementations

6 Conclusion

In this paper, we have introduced *fingerprint indexing*, a lightweight indexing technique that is easy to implement, has a small memory footprint, and shows excellent performance in practice.

In the future, we will further investigate the influence of different fingerprint functions, and evaluate if further gains can be made by automatically generating a good fingerprint function based on the signature.

Acknowledgements. I thank the University of Miami's Center for Computational Science HPC team for making their cluster available for the experimental evaluation.

References

1. Graf, P.: Term Indexing. LNCS (LNAI), vol. 1053. Springer, Heidelberg (1996)
2. McCune, W.: Experiments with Discrimination-Tree Indexing and Path Indexing for Term Retrieval. Journal of Automated Reasoning 9(2), 147–167 (1992)
3. Riazanov, A., Voronkov, A.: The Design and Implementation of VAMPIRE. Journal of AI Communications 15(2/3), 91–110 (2002)
4. Riazanov, A., Voronkov, A.: Efficient Instance Retrieval with Standard and Relational Path Indexing. In: Baader, F. (ed.) CADE 2003. LNCS (LNAI), vol. 2741, pp. 380–396. Springer, Heidelberg (2003)
5. Schulz, S.: E – A Brainiac Theorem Prover. Journal of AI Communications 15(2/3), 111–126 (2002)
6. Schulz, S.: Simple and Efficient Clause Subsumption with Feature Vector Indexing. In: Proc. of the IJCAR 2004 Workshop on Empirically Successful First-Order Theorem Proving, Cork, Ireland (2004)
7. Stickel, M.E.: The Path-Indexing Method for Indexing Terms. Technical Note 473, AI Center, SRI International, Menlo Park, California, USA (October 1989)
8. Sutcliffe, G.: The TPTP Problem Library and Associated Infrastructure: The FOF and CNF Parts, v3.5.0. Journal of Automated Reasoning 43(4), 337–362 (2009)

Optimization in SMT with $\mathcal{LA}(\mathbb{Q})$ Cost Functions*

Roberto Sebastiani and Silvia Tomasi

DISI, University of Trento, Italy

Abstract. In the contexts of automated reasoning and formal verification, important *decision* problems are effectively encoded into Satisfiability Modulo Theories (SMT). In the last decade efficient SMT solvers have been developed for several theories of practical interest (e.g., linear arithmetic, arrays, bit-vectors). Surprisingly, very little work has been done to extend SMT to deal with *optimization* problems; in particular, we are not aware of any work on SMT solvers able to produce solutions which minimize cost functions over *arithmetical* variables. This is unfortunate, since some problems of interest require this functionality.

In this paper we start filling this gap. We present and discuss two general procedures for leveraging SMT to handle the minimization of $\mathcal{LA}(\mathbb{Q})$ cost functions, combining SMT with standard minimization techniques. We have implemented the procedures within the MathSAT SMT solver. Due to the absence of competitors in AR and SMT domains, we have experimentally evaluated our implementation against state-of-the-art tools for the domain of *linear generalized disjunctive programming (LGDP)*, which is closest in spirit to our domain, on sets of problems which have been previously proposed as benchmarks for the latter tools. The results show that our tool is very competitive with, and often outperforms, these tools on these problems, clearly demonstrating the potential of the approach.

1 Introduction

In the contexts of automated reasoning (AR) and formal verification (FV), important *decision* problems are effectively encoded into and solved as Satisfiability Modulo Theories (SMT) problems. In the last decade efficient SMT solvers have been developed, that combine the power of modern conflict-driven clause-learning (CDCL) SAT solvers with dedicated decision procedures ($\mathcal{T}$-Solvers) for several first-order theories of practical interest like, e.g., those of linear arithmetic over the rationals ($\mathcal{LA}(\mathbb{Q})$) or the integers ($\mathcal{LA}(\mathbb{Z})$), of arrays ($\mathcal{AR}$), of bit-vectors ($\mathcal{BV}$), and their combinations. (See [11] for an overview.)

Many SMT-encodable problems of interest, however, may require also the capability of finding models that are *optimal* wrt. some cost function over continuous arithmetical variables. [1] E.g., in (SMT-based) *planning with resources* [33] a plan for achieving a certain goal must be found which not only fulfills some resource constraints (e.g.

* R. Sebastiani is supported by Semiconductor Research Corporation under GRC Custom Research Project 2009-TJ-1880 WOLFLING and GRC Research Project 2012-TJ-2266 WOLF.

[1] Although we refer to quantifier-free formulas, as it is frequent practice in SAT and SMT, with a little abuse of terminology we often call "Boolean variables" the propositional atoms and we call "variables" the Skolem constants x_i in $\mathcal{LA}(\mathbb{Q})$-atoms like, e.g., "$3x_1 - 2x_2 + x_3 \leq 3$".

B. Gramlich, D. Miller, and U. Sattler (Eds.): IJCAR 2012, LNAI 7364, pp. 484–498, 2012.

on time, gasoline consumption, ...) but that also minimizes the usage of some such resource; in SMT-based *model checking with timed or hybrid systems* (e.g. [9]) you may want to find executions which minimize some parameter (e.g. elapsed time), or which minimize/maximize the value of some constant parameter (e.g., a clock time-out value) while fulfilling/violating some property (e.g., minimize the closure time interval of a rail-crossing while preserving safety). This also involves, as particular subcases, problems which are traditionally addressed as *linear disjunctive programming (LDP)* [10] or *linear generalized disjunctive programming (LGDP)* [25,28], or as SAT/SMT with Pseudo-Boolean (PB) constraints and Weighted Max-SAT/SMT problems [26,19,24,15,7]. Notice that the two latter problems can be easily encoded into each other.

Surprisingly, very little work has been done to extend SMT to deal with optimization problems [24,15,7]; in particular, to the best of our knowledge, all such works aim at minimizing cost functions over *Boolean* variables (i.e., SMT with PB cost functions or MAX-SMT), whilst we are not aware of any work on SMT solvers able to produce solutions which minimize cost functions over *arithmetical* variables. Notice that the former can be easily encoded into the latter, but not vice versa (see §2).

In this paper we start filling this gap. We present two general procedures for adding to SMT($\mathcal{LA}(\mathbb{Q}) \cup \mathcal{T}$) the functionality of finding models minimizing some $\mathcal{LA}(\mathbb{Q})$ cost variable —$\mathcal{T}$ being some (possibly empty) stably-infinite theory s.t. $\mathcal{T}$ and $\mathcal{LA}(\mathbb{Q})$ are signature-disjoint. These two procedures combine standard SMT and minimization techniques: the first, called *offline*, is much simpler to implement, since it uses an incremental SMT solver as a black-box, whilst the second, called *inline*, is more sophisticate and efficient, but it requires modifying the code of the SMT solver. (This distinction is important, since the source code of most SMT solvers is not publicly available.)

We have implemented these procedures within the MATHSAT5 SMT solver [5]. Due to the absence of competitors from AR and SMT domains, we have experimentally evaluated our implementation against state-of-the-art tools for the domain of LGDP, which is closest in spirit to our domain, on sets of problems which have been previously proposed as benchmarks for the latter tools. (Notice that LGDP is limited to plain $\mathcal{LA}(\mathbb{Q})$, so that, e.g., it cannot handle combination of theories like $\mathcal{LA}(\mathbb{Q}) \cup \mathcal{T}$.) The results show that our tool is very competitive with, and often outperforms, these tools on these problems, clearly demonstrating the potential of the approach.

Related Work. The idea of optimization in SMT was first introduced by Nieuwenhuis & Oliveras [24], who presented a very-general logical framework of "SMT with progressively stronger theories" (e.g., where the theory is progressively strengthened by every new approximation of the minimum cost), and present implementations for Max-SAT/SMT based on this framework. Cimatti et al. [15] introduced the notion of "Theory of Costs" $\mathcal{C}$ to handle PB cost functions and constraints by an ad-hoc and independent "$\mathcal{C}$-solver" in the standard lazy SMT schema, and implemented a variant of MathSAT tool able to handle SAT/SMT with PB constraints and to minimize PB cost functions. The SMT solver YICES [7] also implements Max-SAT/SMT, but we are not aware of any document describing the procedures used there.

Mixed Integer Linear Programming (MILP) is an extension of Linear Programming (LP) involving both discrete and continuous variables. A large variety of techniques and tools for MILP are available, mostly based on efficient combinations of LP, *branch-and-bound* search mechanism and *cutting-plane* methods (see e.g. [20]). SAT techniques have also been incorporated into these procedures for MILP (see [8]).

Linear Disjunctive Programming (LDP) problems are LP problems where linear constraints are connected by conjunctions and disjunctions [10]. Closest to our domain, *Linear Generalized Disjunctive Programming (LGDP)*, is a generalization of LDP which has been proposed in [25] as an alternative model to the MILP problem. Unlike MILP, which is based entirely on algebraic equations and inequalities, the LGDP model allows for combining algebraic and logical equations with Boolean propositions through Boolean operations, providing a much more natural representation of discrete decisions. Current approaches successfully address LGDP by reformulating and solving it as a MILP problem [25,32,27,28]; these reformulations focus on efficiently encoding disjunctions and logic propositions into MILP, so as to be fed to an efficient MILP solver like CPLEX.

Content. The rest of the paper is organized as follows: in §2 we define the problem addressed, and show how it generalizes many known optimization problems; in §3 we present our novel procedures; in §4 we present an experimental evaluation; in §5 we briefly conclude and highlight directions for future work.

2 Optimization in SMT($\mathcal{LA}(\mathbb{Q}) \cup \mathcal{T}$)

We assume the reader is familiar with the main concepts of Boolean and first-order logic. Let $\mathcal{T}$ be some stably infinite theory with equality s.t. $\mathcal{LA}(\mathbb{Q})$ and $\mathcal{T}$ are signature-disjoint, as in [23]. ($\mathcal{T}$ can be itself a combination of theories.) We call an *Optimization Modulo* $\mathcal{LA}(\mathbb{Q}) \cup \mathcal{T}$ *problem, OMT*($\mathcal{LA}(\mathbb{Q}) \cup \mathcal{T}$), a pair $\langle \varphi, \mathsf{cost} \rangle$ such that φ is a SMT($\mathcal{LA}(\mathbb{Q}) \cup \mathcal{T}$) formula and cost is a $\mathcal{LA}(\mathbb{Q})$ variable occurring in φ, representing the cost to be minimized. The problem consists in finding a model $\mathcal{M}$ for φ (if any) whose value of cost is minimum. We call an *Optimization Modulo* $\mathcal{LA}(\mathbb{Q})$ *problem (OMT*($\mathcal{LA}(\mathbb{Q})$)) an SMT($\mathcal{LA}(\mathbb{Q}) \cup \mathcal{T}$) problem where $\mathcal{T}$ is empty. If φ is in the form $\varphi' \wedge (\mathsf{cost} < c)$ [resp. $\varphi' \wedge \neg(\mathsf{cost} < c)$] for some value $c \in \mathbb{Q}$, then we call c an *upper bound* [resp. *lower bound*] for cost. If ub [resp lb] is the minimum upper bound [resp. the maximum lower bound] for φ, we also call the interval [lb, ub[the *range* of cost.

These definitions capture many interesting optimizations problems. First, it is straightforward to encode LP, LDP and LGDP into OMT($\mathcal{LA}(\mathbb{Q})$) (see [29] for details).

Pseudo-Boolean (PB) constraints (see [26]) in the form $(\sum_i \mathbf{a}_i X^i \leq b)$, s.t. X^i are Boolean atoms and $\mathbf{a}_i$ constant values in $\mathbb{Q}$, and cost functions $\mathsf{cost} = \sum_i \mathbf{a}_i X^i$, are encoded into OMT($\mathcal{LA}(\mathbb{Q})$) by rewriting each PB-term $\sum_i \mathbf{a}_i X^i$ into the $\mathcal{LA}(\mathbb{Q})$-term $\sum_i \mathbf{x}_i$, $\mathbf{x}$ being an array of fresh $\mathcal{LA}(\mathbb{Q})$ variables, and by conjoining to φ the formula:

$$\bigwedge_i((\neg X^i \vee (\mathbf{x}_i = \mathbf{a}_i)) \wedge (X^i \vee (\mathbf{x}_i = 0))). \quad (1)$$

Moreover, since Max-SAT (see [19]) [resp. Max-SMT (see [24,15,7])] can be encoded into SAT [resp. SMT] with PB constraints (see e.g. [24,15]), then optimization problems for SAT with PB constraints and Max-SAT can be encoded into OMT($\mathcal{LA}(\mathbb{Q})$), whilst those for SMT($\mathcal{T}$) with PB constraints and Max-SMT can be encoded into OMT($\mathcal{LA}(\mathbb{Q}) \cup \mathcal{T}$) (assuming $\mathcal{T}$ matches the definition above).

We remark the deep difference between OMT($\mathcal{LA}(\mathbb{Q})$)/OMT($\mathcal{LA}(\mathbb{Q}) \cup \mathcal{T}$) and the problem of SAT/SMT with PB constraints and cost functions (or Max-SAT/SMT) addressed in [24,15]. With the latter problem, the cost is a deterministic consequence of a truth assignment to the atoms of the formula, so that the search has only a Boolean component, consisting in finding the cheapest truth assignment. With OMT($\mathcal{LA}(\mathbb{Q})$)/ OMT($\mathcal{LA}(\mathbb{Q}) \cup \mathcal{T}$), instead, for every satisfying assignment μ it is also necessary to find the minimum-cost $\mathcal{LA}(\mathbb{Q})$-model for μ, so that the search has both a Boolean and a $\mathcal{LA}(\mathbb{Q})$-component.

3 Procedures for OMT($\mathcal{LA}(\mathbb{Q})$) and OMT($\mathcal{LA}(\mathbb{Q}) \cup \mathcal{T}$)

It may be noticed that very naive OMT($\mathcal{LA}(\mathbb{Q})$) or OMT($\mathcal{LA}(\mathbb{Q}) \cup \mathcal{T}$) procedures could be straightforwardly implemented by performing a sequence of calls to an SMT solver on formulas like $\varphi \wedge (\mathsf{cost} \geq \mathsf{l}_i) \wedge (\mathsf{cost} < \mathsf{u}_i)$, each time restricting the range $[\mathsf{l}_i, \mathsf{u}_i[$ according to a linear-search or binary-search schema. With the former schema, however, the SMT solver would repeatedly generate the same $\mathcal{LA}(\mathbb{Q})$-satisfiable truth assignment, each time finding a cheaper model for it. With the latter schema the efficiency should improve; however, an initial lower-bound should be necessarily required as input (which is not the case, e.g., of the problems in §4.2.)

In this section we present more sophisticate procedures, based on the combination of SMT and minimization techniques. We first present and discuss an *offline* schema (§3.1) and an *inline* (§3.2) schema for an OMT($\mathcal{LA}(\mathbb{Q})$) procedure; then we show how to extend them to the OMT($\mathcal{LA}(\mathbb{Q}) \cup \mathcal{T}$) case (§3.3).

In what follows we assume the reader is familiar with the basics about CDCL SAT solvers and lazy SMT solvers. A detailed background section on that is available on the extended version of this paper [29]; for a much more detailed description, we refer the reader, e.g., to [22,11] respectively.

3.1 An Offline Schema for OMT($\mathcal{LA}(\mathbb{Q})$)

The general schema for the offline OMT($\mathcal{LA}(\mathbb{Q})$) procedure is displayed in Algorithm 1. It takes as input an instance of the OMT($\mathcal{LA}(\mathbb{Q})$) problem, plus optionally values for lb and ub (which are implicitly considered to be $-\infty$ and $+\infty$ if not present), and returns the model $\mathcal{M}$ of minimum cost and its cost u (the value ub if φ is $\mathcal{LA}(\mathbb{Q})$-inconsistent). We represent φ as a set of clauses, which may be pushed or popped from the input formula-stack of an incremental SMT solver.

First, the variables l, u (defining the current range) are initialized to lb and ub respectively, the atom PIV to $\top$, and $\mathcal{M}$ is initialized to be an empty model. Then the procedure adds to φ the bound constraints, if present, which restrict the search within the range $[\mathsf{l}, \mathsf{u}[$ (row 2). [2] The solution space is then explored iteratively (rows 3-26), reducing at each loop the current range $[\mathsf{l}, \mathsf{u}[$ to explore, until the range is empty. Then $\langle \mathcal{M}, \mathsf{u} \rangle$ is returned —$\langle \emptyset, \mathsf{ub} \rangle$ if there is no solution in $[\mathsf{lb}, \mathsf{ub}[$— $\mathcal{M}$ being the model of minimum cost u. Each loop may work in either *linear-search* or *binary-search* mode,

[2] Of course literals like $\neg(\mathsf{cost} < -\infty)$ and $(\mathsf{cost} < +\infty)$ are not added.

Algorithm 1. Offline OMT($\mathcal{LA}(\mathbb{Q})$) Procedure based on Mixed Linear/Binary Search.

Require: $\langle\varphi, \mathsf{cost}, \mathsf{lb}, \mathsf{ub}\rangle$ {ub can be $+\infty$, lb can be $-\infty$}
1: $\mathsf{l} \leftarrow \mathsf{lb}$; $\mathsf{u} \leftarrow \mathsf{ub}$; $\mathsf{PIV} \leftarrow \top$; $\mathcal{M} \leftarrow \emptyset$
2: $\varphi \leftarrow \varphi \cup \{\neg(\mathsf{cost} < \mathsf{l}), (\mathsf{cost} < \mathsf{u})\}$
3: **while** ($\mathsf{l} < \mathsf{u}$) **do**
4:     **if** (BinSearchMode()) **then** {Binary-search Mode}
5:         $\mathsf{pivot} \leftarrow \mathsf{ComputePivot}(\mathsf{l}, \mathsf{u})$
6:         $\mathsf{PIV} \leftarrow (\mathsf{cost} < \mathsf{pivot})$
7:         $\varphi \leftarrow \varphi \cup \{\mathsf{PIV}\}$
8:         $\langle\mathsf{res}, \mu\rangle \leftarrow \mathsf{SMT.IncrementalSolve}(\varphi)$
9:         $\eta \leftarrow \mathsf{SMT.ExtractUnsatCore}(\varphi)$
10:     **else** {Linear-search Mode}
11:         $\langle\mathsf{res}, \mu\rangle \leftarrow \mathsf{SMT.IncrementalSolve}(\varphi)$
12:         $\eta \leftarrow \emptyset$
13:     **end if**
14:     **if** ($\mathsf{res} = \text{SAT}$) **then**
15:         $\langle\mathcal{M}, \mathsf{u}\rangle \leftarrow \mathsf{Minimize}(\mathsf{cost}, \mu)$
16:         $\varphi \leftarrow \varphi \cup \{(\mathsf{cost} < \mathsf{u})\}$
17:     **else** {res = UNSAT }
18:         **if** ($\mathsf{PIV} \not\in \eta$) **then**
19:             $\mathsf{l} \leftarrow \mathsf{u}$
20:         **else**
21:             $\mathsf{l} \leftarrow \mathsf{pivot}$
22:             $\varphi \leftarrow \varphi \setminus \{\mathsf{PIV}\}$
23:             $\varphi \leftarrow \varphi \cup \{\neg\mathsf{PIV}\}$
24:         **end if**
25:     **end if**
26: **end while**
27: **return** $\langle\mathcal{M}, \mathsf{u}\rangle$

driven by the heuristic BinSearchMode(). Notice that if $\mathsf{u} = +\infty$ or $\mathsf{l} = -\infty$, then BinSearchMode() returns false.

In **linear-search mode**, steps 4-9 and 21-23 are not executed. First, an incremental SMT($\mathcal{LA}(\mathbb{Q})$) solver is invoked on φ (row 11). (Notice that, given the incrementality of the solver, every operation in the form "$\varphi \leftarrow \varphi \cup \{\phi_i\}$" [resp. $\varphi \leftarrow \varphi \setminus \{\phi_i\}$] is implemented as a "push" [resp. "pop"] operation on the stack representation of φ; it is also very important to recall that during the SMT call φ is updated with the clauses which are learned during the SMT search.) η is set to be empty, which forces condition 18 to hold. If φ is $\mathcal{LA}(\mathbb{Q})$-satisfiable, then it is returned res =SAT and a $\mathcal{LA}(\mathbb{Q})$-satisfiable truth assignment μ for φ. Thus Minimize is invoked on (the subset of $\mathcal{LA}(\mathbb{Q})$-literals of) μ, returning the model $\mathcal{M}$ for μ of minimum cost u ($-\infty$ iff the problem in unbounded). The current solution u becomes the new upper bound, thus the $\mathcal{LA}(\mathbb{Q})$-atom $(\mathsf{cost} < \mathsf{u})$ is added to φ (row 16). Notice that if the problem is unbounded, then for some μ Minimize will return $-\infty$, forcing condition 3 to be false and the whole process to stop. If φ is $\mathcal{LA}(\mathbb{Q})$-unsatisfiable, then no model in the current cost range $[\mathsf{l}, \mathsf{u}[$ can be found; hence the flag l is set to u, forcing the end of the loop.

In **binary-search mode** at the beginning of the loop (steps 4-9), the value pivot $\in$ $]\mathsf{l}, \mathsf{u}[$ is computed by the heuristic function ComputePivot (in the simplest form, pivot is $(\mathsf{l}+\mathsf{u})/2$), the (possibly new) atom PIV $\stackrel{\text{def}}{=}$ (cost $<$ pivot) is pushed into the formula stack, so that to temporarily restrict the cost range to $[\mathsf{l}, \mathsf{pivot}[$; then the incremental SMT solver is invoked on φ, this time activating the feature SMT.ExtractUnsatCore, which returns also the subset η of formulas in (the formula stack of) φ which caused the unsatisfiability of φ. This exploits techniques similar to unsat-core extraction [21]. (In practice, it suffices to say if PIV $\in \eta$.) If φ is $\mathcal{LA}(\mathbb{Q})$-satisfiable, then the procedure behaves as in linear-search mode. If instead φ is $\mathcal{LA}(\mathbb{Q})$-unsatisfiable, we look at η and distinguish two subcases. If PIV does not occur in η, this means that $\varphi \setminus \{\mathsf{PIV}\}$ is $\mathcal{LA}(\mathbb{Q})$-inconsistent, i.e. there is no model in the whole cost range $[\mathsf{l}, \mathsf{u}[$. Then the procedure behaves as in linear-search mode, forcing the end of the loop. Otherwise, we can only conclude that there is no model in the cost range $[\mathsf{l}, \mathsf{pivot}[$, so that we still need exploring the cost range $[\mathsf{pivot}, \mathsf{u}[$. Thus l is set to pivot, PIV is popped from φ and its negation is pushed into φ. Then the search proceeds, investigating the cost range $[\mathsf{pivot}, \mathsf{u}[$.

We notice an important fact: if BinSearchMode() always returned true, then Algorithm 1 would not necessarily terminate. In fact, an SMT solver invoked on φ may return a set η containing PIV even if $\varphi \setminus$ PIV is $\mathcal{LA}(\mathbb{Q})$-inconsistent. Thus, e.g., the procedure might got stuck into a infinite loop, each time halving the cost range right-bound (e.g., $[-1, 0[$, $[-1/2, 0[$, $[-1/4, 0[$,..). To cope with this fact, however, it suffices that BinSearchMode() returns false infinitely often, forcing then a "linear-search" call which finally detects the inconsistency. (In our implementation, we have empirically experienced the best performance with one linear-search loop after every binary-search one, because satisfiable calls are typically much cheaper than unsatisfiable ones.)

Under such hypothesis, it is straightforward to see the following facts: (i) Algorithm 1 terminates, in both modes, because there are only a finite number of candidate truth assignments μ to be enumerated, and steps 15-16 guarantee that the same assignment μ will never be returned twice by the SMT solver; (ii) it returns a model of minimum cost, because it explores the whole search space of candidate truth assignments, and for every suitable assignment μ Minimize finds the minimum-cost model for μ; (iii) it requires polynomial space, under the assumption that the underlying CDCL SAT solver adopts a polynomial-size clause discharging strategy (which is typically the case of SMT solvers, including MATHSAT).

In a nutshell, Minimize is a simple extension of the simplex-based $\mathcal{LA}(\mathbb{Q})$-Solver of [16] which is invoked after one solution is found, minimizing it by standard Simplex techniques. We recall that the algorithm in [16] can handle strict inequalities. Thus, if the input problem contains strict inequalities, then Minimize temporarily treats them as non-strict ones and finds the minimum-cost solution with standard Simplex techniques. If such minimum-cost solution $\mathbf{x}$ of cost min lays only on non-strict inequalities, then $\mathbf{x}$ is a solution; otherwise, for some $\delta > 0$ and for every cost $c \in]\mathsf{min}, \mathsf{min} + \delta]$ there exists a solution of cost c. (If needed, such solution is computed using the techniques for handling strict inequalities described in [16].) Thus the value min is returned, tagged as a non-strict minimum, so that the constraint (cost $\leq$ min) rather than (cost $<$ min) is added to φ.

Discussion. We remark a few facts about this procedure.

First, if Algorithm 1 is interrupted (e.g., by a timeout device), then u can be returned, representing the best approximation of the minimum cost found so far.

Second, the incrementality of the SMT solver plays an essential role here, since at every call SMT.IncrementalSolve resumes the status of the search of the end of the previous call, only with tighter cost range constraints. (Notice that at each call here the solver can reuse all previously-learned clauses.) To this extent, one can see the whole process as only one SMT process, which is interrupted and resumed each time a new model is found, in which cost range constraints are progressively tightened.

Third, we notice that in Algorithm 1 all the literals constraining the cost range (i.e., $\neg(\mathsf{cost} < \mathsf{l})$, $(\mathsf{cost} < \mathsf{u})$) are always added to φ as unit clauses; thus inside SMT.IncrementalSolve these literals are immediately unit-propagated, becoming part of each truth assignment μ from the very beginning of its construction. (We recall that the SMT solver invokes incrementally $\mathcal{LA}(\mathbb{Q})$-Solver also while building an assignment μ (*early pruning calls* [11].)) As soon as novel $\mathcal{LA}(\mathbb{Q})$-literals are added to μ which prevent it from having a $\mathcal{LA}(\mathbb{Q})$-model of cost in $[\mathsf{l}, \mathsf{u}[$, the $\mathcal{LA}(\mathbb{Q})$-solver invoked on μ by early-pruning calls returns UNSAT and the $\mathcal{LA}(\mathbb{Q})$-lemma $\neg\eta$ describing the conflict $\eta \subseteq \mu$, triggering theory-backjumping and -learning. To this extent, SMT.IncrementalSolve implicitly plays a form of *branch & bound*: (i) decide a new literal l and propagate the literals which derive from l ("branch") and (ii) backtrack as soon as the current branch can no more be expanded into models in the current cost range ("bound").

Fourth, in binary-search mode, the range-partition strategy may be even more aggressive than that of standard binary search, because the minimum cost u returned in row 15 can be significantly smaller than pivot, so that the cost range is more than halved.

Finally, unlike with other domains (e.g., search in a sorted array) the binary-search strategy here is not "obviously faster" than the linear-search one, because the unsatisfiable calls to SMT.IncrementalSolve are typically much more expensive than the satisfiable ones, because they must explore the whole Boolean search space rather than only a portion of it (although with a higher pruning power, due to the stronger constraint induced by the presence of pivot). Thus, we have a tradeoff between a typically much-smaller number of calls plus a stronger pruning power in binary search versus an average much smaller cost of the calls in linear search. To this extent, it is possible to use dynamic/adaptive strategies for ComputePivot (see [30]).

3.2 An Inline Schema for OMT($\mathcal{LA}(\mathbb{Q})$)

With the inline schema, the whole optimization procedure is pushed inside the SMT solver by embedding the range-minimization loop inside the CDCL Boolean-search loop of the standard lazy SMT schema. The SMT solver, which is thus called only once, is modified as follows.

Initialization. The variables $\mathsf{lb}, \mathsf{ub}, \mathsf{l}, \mathsf{u}, \mathsf{PIV}, \mathsf{pivot}, \mathcal{M}$ are brought inside the SMT solver, and are initialized as in Algorithm 1, steps 1-2.

Range Updating and Pivoting. Every time the search of the CDCL SAT solver gets back to decision level 0, the range $[\mathsf{l}, \mathsf{u}[$ is updated s.t. u [resp. l] is assigned the lowest [resp. highest] value u_i [resp. l_i] such that the atom $(\mathsf{cost} < \mathsf{u}_i)$ [resp. $\neg(\mathsf{cost} < \mathsf{u}_i)$] is currently assigned at level 0. (If $\mathsf{u} \leq \mathsf{l}$, or two literals $l, \neg l$ are both assigned at level 0, then the procedure terminates, returning the current value of u.) Then BinSearchMode() is invoked: if it returns true, then ComputePivot computes pivot $\in]\mathsf{l}, \mathsf{u}[$, and the (possibly new) atom $\mathsf{PIV} \stackrel{\text{def}}{=} (\mathsf{cost} < \mathsf{pivot})$ is decided to be true (level 1) by the SAT solver. This mimics steps 4-7 in Algorithm 1, temporarily restricting the cost range to $[\mathsf{l}, \mathsf{pivot}[$.

Decreasing the Upper Bound. When an assignment μ propositionally satisfying φ is generated which is found $\mathcal{LA}(\mathbb{Q})$-consistent by $\mathcal{LA}(\mathbb{Q})$-Solver, μ is also fed to Minimize, returning the minimum cost min of μ; then the unit clause $(\mathsf{cost} < \mathsf{min})$ is learned and fed to the backjumping mechanism, which forces the SAT solver to backjump to level 0, then unit-propagating $(\mathsf{cost} < \mathsf{min})$. This case mirrors steps 14-16 in Algorithm 1, permanently restricting the cost range to $[\mathsf{l}, \mathsf{min}[$. Minimize is embedded within $\mathcal{LA}(\mathbb{Q})$-Solver, so that it is called incrementally after it, without restarting its search from scratch.

As a result of these modifications, we also have the following typical scenario.

Increasing the Lower Bound. In binary-search mode, when a conflict occurs s.t. the conflict analysis of the SAT solver produces a conflict clause in the form $\neg\mathsf{PIV} \vee \neg\eta'$ s.t. all literals in η' are assigned true at level 0 (i.e., $\varphi \wedge \mathsf{PIV}$ is $\mathcal{LA}(\mathbb{Q})$-inconsistent), then the SAT solver backtracks to level 0, unit-propagating $\neg\mathsf{PIV}$. This case mirrors steps 21-23 in Algorithm 1, permanently restricting the cost range to $[\mathsf{pivot}, \mathsf{u}[$.

Although the modified SMT solver mimics to some extent the behaviour of Algorithm 1, the "control" of the range-restriction process is handled by the standard SMT search. To this extent, notice that also other situations may allow for restricting the cost range: e.g., if $\varphi \wedge \neg(\mathsf{cost} < \mathsf{l}) \wedge (\mathsf{cost} < \mathsf{u}) \models (\mathsf{cost} \bowtie \mathsf{m})$ for some atom $(\mathsf{cost} \bowtie \mathsf{m})$ occurring in φ s.t. $\mathsf{m} \in [\mathsf{l}, \mathsf{u}[$ and $\bowtie \in \{\leq, <, \geq, >\}$, then the SMT solver may backjump to decision level 0 and propagate $(\mathsf{cost} \bowtie \mathsf{m})$, further restricting the cost range.

The same considerations about the offline procedure in §3.1 hold for the inline version. The efficiency of the inline procedure can be further improved as follows.

First, in binary-search mode, when a truth assignment μ with a novel minimum min is found, not only $(\mathsf{cost} < \mathsf{min})$ but also $\mathsf{PIV} \stackrel{\text{def}}{=} (\mathsf{cost} < \mathsf{pivot})$ is learned as unit clause. Although redundant from the logical perspective because $\mathsf{min} < \mathsf{pivot}$, the unit clause PIV allows the SAT solver for reusing all the clauses in the form $\neg\mathsf{PIV} \vee C$ which have been learned when investigating the cost range $[\mathsf{l}, \mathsf{pivot}[$. (In Algorithm 1 this is done implicitly, since PIV is not popped from φ before step 16.) Moreover, the $\mathcal{LA}(\mathbb{Q})$-inconsistent assignment $\mu \wedge (\mathsf{cost} < \mathsf{min})$ may be fed to $\mathcal{LA}(\mathbb{Q})$-Solver and the negation of the returned conflict $\neg\eta \vee \neg(\mathsf{cost} < \mathsf{min})$ s.t. $\eta \subseteq \mu$, can be learned, which prevents the SAT solver from generating any assignment containing η.

Second, in binary-search mode, if the $\mathcal{LA}(\mathbb{Q})$-Solver returns a conflict set $\eta \cup \{\mathsf{PIV}\}$, then it is further asked to find the maximum value max s.t. $\eta \cup \{(\mathsf{cost} < \mathsf{max})\}$ is also $\mathcal{LA}(\mathbb{Q})$-inconsistent. (This is done with a simple modification of the algorithm in [16].) If $\mathsf{max} \geq \mathsf{u}$, then the clause $C^* \stackrel{\text{def}}{=} \neg\eta \vee \neg(\mathsf{cost} < \mathsf{u})$ is used do drive backjumping and learning instead of $C \stackrel{\text{def}}{=} \neg\eta \vee \neg\mathsf{PIV}$. Since $(\mathsf{cost} < \mathsf{u})$ is permanently assigned at level 0, the dependency of the conflict from PIV is removed. Eventually, instead of using C

to drive backjumping to level 0 and propagating $\neg$PIV, the SMT solver may use C^*, then forcing the procedure to stop.

3.3 Extensions to OMT($\mathcal{LA}(\mathbb{Q}) \cup \mathcal{T}$)

The procedures of §3.1 and §3.2 extend to the OMT($\mathcal{LA}(\mathbb{Q}) \cup \mathcal{T}$) case straightforwardly as follows. We assume that the underlying SMT solver handles $\mathcal{LA}(\mathbb{Q}) \cup \mathcal{T}$, and that φ is a $\mathcal{LA}(\mathbb{Q}) \cup \mathcal{T}$ formula (which for simplicity and wlog we assume to be pure [23]).

Algorithm 1 is modified as follows. First, SMT.IncrementalSolve in step 8 or 11 is asked to return also a $\mathcal{LA}(\mathbb{Q}) \cup \mathcal{T}$-model $\mathcal{M}$. Then Minimize is invoked on the pair $\langle \mathsf{cost}, \mu_{\mathcal{LA}(\mathbb{Q})} \cup \mu_{ei} \rangle$, s.t. $\mu_{\mathcal{LA}(\mathbb{Q})}$ is the truth assignment over the $\mathcal{LA}(\mathbb{Q})$-atoms in φ returned by the solver, and μ_{ei} is the set of equalities $(x_i = x_j)$ and strict inequalities $(x_i < x_j)$ on the shared variables x_i which are true in $\mathcal{M}$. (The equalities and strict inequalities obtained from the others by the transitivity of $=, <$ can be omitted.)

The implementation of an inline OMT($\mathcal{LA}(\mathbb{Q}) \cup \mathcal{T}$) procedures comes nearly for free if the SMT solver handles $\mathcal{LA}(\mathbb{Q}) \cup \mathcal{T}$-solving by *Delayed Theory Combination* [13], with the strategy of case-splitting automatically disequalities $\neg(x_i = x_j)$ into the two inequalities $(x_i < x_j)$ and $(x_j < x_i)$, which is implemented in MATHSAT. If so the solver enumerates truth assignments in the form $\mu' \stackrel{\text{def}}{=} \mu_{\mathcal{LA}(\mathbb{Q})} \cup \mu_{eid} \cup \mu_{\mathcal{T}}$, where (i) μ' propositionally satisfies φ, (ii) μ_{eid} is a set of interface equalities $(x_i = x_j)$ and disequalities $\neg(x_i = x_j)$, containing also one inequality in $\{(x_i < x_j), (x_j < x_i)\}$ for every $\neg(x_i = x_j) \in \mu_{eid}$; then $\mu'_{\mathcal{LA}(\mathbb{Q})} \stackrel{\text{def}}{=} \mu_{\mathcal{LA}(\mathbb{Q})} \cup \mu_{ei}$ and $\mu'_{\mathcal{T}} \stackrel{\text{def}}{=} \mu_{\mathcal{T}} \cup \mu_{ed}$ are passed to the $\mathcal{LA}(\mathbb{Q})$-Solver and $\mathcal{T}$-Solver respectively, μ_{ei} and μ_{ed} being obtained from μ_{eid} by dropping the disequalities and inequalities respectively. [3]

If this is the case, it suffices to apply Minimize to $\mu'_{\mathcal{LA}(\mathbb{Q})}$, then learn $(\mathsf{cost} < \mathsf{min})$ and use it for backjumping, as in §3.2.

For lack of space we omit here a detailed justification that the above procedures compute OMT($\mathcal{LA}(\mathbb{Q}) \cup \mathcal{T}$), which is presented in the extended paper [29]. In short, they correspond to apply the techniques of §3.1, §3.2 to look for minimum-cost $\mathcal{T}$-satisfiable and $\mathcal{LA}(\mathbb{Q})$-satisfiable truth-assignments for the $\mathcal{LA}(\mathbb{Q}) \cup \mathcal{T}$ formula $\varphi' \stackrel{\text{def}}{=} \varphi \wedge \bigwedge_{x_i, x_j \in Shared(\varphi)} ((x_i = x_j) \vee (x_i < x_j) \vee (x_j < x_i))$, which is equivalent to φ, each time passing to $\mathcal{T}$-solver, $\mathcal{LA}(\mathbb{Q})$-Solver and Minimize only the relevant literals.

4 Experimental Evaluation

We have implemented both the OMT($\mathcal{LA}(\mathbb{Q})$) procedures and the inline OMT($\mathcal{LA}(\mathbb{Q}) \cup \mathcal{T}$) procedures of §3 on top of MATHSAT [5] (thus we refer to them as OPT-MATHSAT). We consider four different configurations of OPT-MATHSAT,

[3] In [13] $\mu' \stackrel{\text{def}}{=} \mu_{\mathcal{LA}(\mathbb{Q})} \cup \mu_{ed} \cup \mu_{\mathcal{T}}$, μ_{ed} being a truth assignment over the interface equalities, and as such a set of equalities and disequalities. However, since typically a SMT($\mathcal{LA}(\mathbb{Q})$) solver handles disequalities $\neg(x_i = x_j)$ by case-splitting them into $(x_i < x_j) \vee (x_j < x_i)$, the assignment considers also one of the two strict inequalities, which is ignored by the $\mathcal{T}$-Solver and is passed to the $\mathcal{LA}(\mathbb{Q})$-Solver instead of the corresponding disequality.

depending on the approach (offline vs. inline, denoted by "-OF" and "-IN") and the search schema (linear vs. binary, denoted by "-LIN" and "-BIN"). [4]

Due to the absence of competitors on OMT($\mathcal{LA}(\mathbb{Q}) \cup \mathcal{T}$), we evaluate the performance of our four configurations of OPT-MATHSAT by comparing them against GAMS v23.7.1 [14] on OMT($\mathcal{LA}(\mathbb{Q})$) problems. GAMS provides two reformulation tools, LOGMIP v2.0 [4] and JAMS [3] (a new version of the EMP solver [2]), both of them allow to reformulate LGDP models by using either big-M (BM) or convex-hull (CH) methods [25,28]. We use CPLEX v12.2 [18] (through an OSI/CPLEX link) to solve the reformulated MILP models. All the tools were executed using default options, as indicated to us by the authors [31].

Notice that OPT-MATHSAT uses *infinite precision arithmetic* whilst, to the best of our knowledge, the GAMS tools implement standard *floating-point arithmetic*.

All tests were executed on 2.66 GHz Xeon machines with 4GB RAM running Linux, using a timeout of 600 seconds. The correctness of the minimum costs min found by OPT-MATHSAT have been cross-checked by another SMT solver, YICES [7], by detecting the inconsistency within the bounds of $\varphi \wedge (\mathsf{cost} < \mathsf{min})$ and the consistency of $\varphi \wedge (\mathsf{cost} = \mathsf{min})$ (if min is non-strict), or of $\varphi \wedge (\mathsf{cost} \leq \mathsf{min})$ and $\varphi \wedge (\mathsf{cost} = \mathsf{min} + \epsilon)$ (if min is strict), ϵ being some very small value. All tools agreed on the final results, apart from tiny rounding errors, [5] and, much more importantly, from some noteworthy exceptions on the smt-lib problems (see §4.2).

In order to make the experiments reproducible, the full-size plots, a Linux binary of OPT-MATHSAT, the problems, and the results are available at [1].[6]

4.1 Comparison on LGDB Problems

We first performed our comparison over two distinct benchmarks, strip-packing and zero-wait job-shop scheduling problems, which have been previously proposed as benchmarks for LOGMIP and JAMS by their authors [32,27,28]. We have adopted the encoding of the problems into LGDP given by the authors. [7]

The Strip-Packing Problem. Given a set N of rectangles of different length L_j and height H_j, $j \in 1,..,N$, and a strip of fixed width W but unlimited length, the *strip-packing* problem aims at minimizing the length L of the filled part of the strip while filling the strip with all rectangles, without any overlap and any rotation. We considered the LGDP model provided by [27] and a corresponding OMT($\mathcal{LA}(\mathbb{Q})$) encoding.

We randomly generated benchmarks according to a fixed width W of the strip and a fixed number of rectangles N. For each rectangle $j \in N$, length L_j and height H_j

[4] Here "-LIN" means that BinSearchMode() always returns false, whilst "-BIN" denotes the mixed linear-binary strategy described in §3.1 to ensure termination.

[5] GAMS +CPLEX often gives some errors $\leq 10^{-5}$, which we believe are due to the printing floating-point format: (e.g. "`3.091250e+00`"); notice that OPT-MATHSAT uses infinite-precision arithmetic, returning values like, e.g. "`7728125177/2500000000`".

[6] We cannot distribute the GAMS tools since they are subject to licencing restrictions. See [14].

[7] Examples are available at `http://www.logmip.ceride.gov.ar/newer.html` and at `http://www.gams.com/modlib/modlib.htm`

Procedure	Strip-packing											
	$W = \sqrt{N}/2$						$W = 1$					
	$N = 9$		$N = 12$		$N = 15$		$N = 9$		$N = 12$		$N = 15$	
	#s.	time	#s.	time	#s.	time	#s.	time	#s.	time	#s.	time
OPT-MATHSAT-LIN-OF	100	51	100	600	93	7862	100	588	90	4555	18	1733
OPT-MATHSAT-LIN-IN	100	32	100	449	96	8057	100	578	91	4855	22	3216
OPT-MATHSAT-BIN-OF	100	48	100	641	90	8712	100	641	88	4385	19	2251
OPT-MATHSAT-BIN-IN	100	32	100	458	96	9706	100	554	92	5892	21	3257
JAMS(BM)+CPLEX	100	381	66	8631	12	1411	50	161	88	4344	46	5978
JAMS(CH)+CPLEX	98	3414	23	1011	0	0	50	887	62	7784	14	3034
LOGMIP(BM)+CPLEX	100	239	78	10266	12	1170	100	164	91	3850	51	6619
LOGMIP(CH)+CPLEX	100	3004	27	2481	1	437	100	2032	70	7406	17	3860

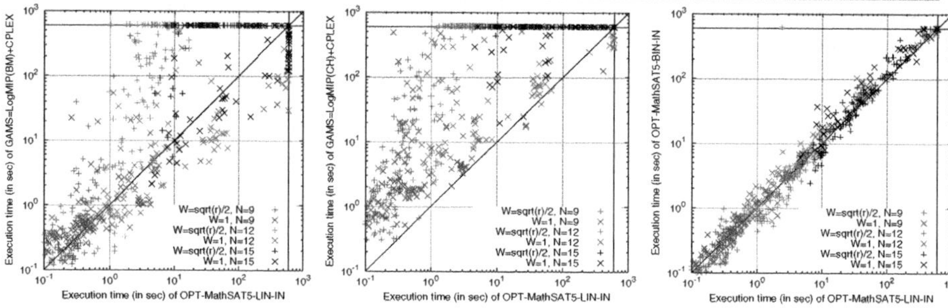

Fig. 1. Table: results (# of solved instances, cumulative time in seconds for solved instances) for OPT-MATHSAT and GAMS (using LOGMIP and JAMS) on 100 random instances each of the strip-packing problem for N rectangles, where $N = 9, 12, 15$, and width $W = \sqrt{N}/2$, 1. Scatter-plots: comparison of the best configuration of OPT-MATHSAT (OPT-MATHSAT-LIN-IN) against LOGMIP(BM)+CPLEX (left), LOGMIP(CH)+CPLEX (center) and OPT-MATHSAT-BIN-IN (right).

are selected in the interval $]0, 1]$ uniformly at random. The upper bound ub is computed with the same heuristic used by [27], which sorts the rectangles in non-increasing order of width and fills the strip by placing each rectangles in the bottom-left corner, and the lower bound lb is set to zero. We generated 100 samples each for 9, 10 and 11 rectangles and for two values of the width $\sqrt{N}/2$ and 1[8].

The table of Figure 1 shows the number of solved instances and their cumulative execution time for different configurations of OPT-MATHSAT and GAMS on the randomly-generated formulas. The scatter-plots of Figure 1 compare the best-performing version of OPT-MATHSAT, OPT-MATHSAT-LIN-IN, against LOGMIP with BM and CH reformulation (left and center respectively); the figure also compares the two inline versions OPT-MATHSAT-LIN-IN and OPT-MATHSAT-BIN-IN (right).

[8] Notice that with $W = \sqrt{N}/2$ the filled strip looks approximatively like a square, whilst $W = 1$ is the average of two 2 rectangles.

Procedure	Job-shop					
	$I = 9, J = 8$		$I = 10, J = 8$		$I = 11, J = 8$	
	#s.	time	#s.	time	#s.	time
OPT-MATHSAT-LIN-OF	97	360	97	1749	92	9287
OPT-MATHSAT-LIN-IN	97	314	97	1436	93	7232
OPT-MATHSAT-BIN-OF	97	619	97	3337	85	13286
OPT-MATHSAT-BIN-IN	97	412	97	1984	93	9166
JAMS(BM)+CPLEX	100	263	100	1068	100	4458
JAMS(CH)+CPLEX	83	22820	6	2533	0	0
LOGMIP(BM)+CPLEX	100	259	100	1066	100	4390
LOGMIP(CH)+CPLEX	86	23663	6	2541	0	0

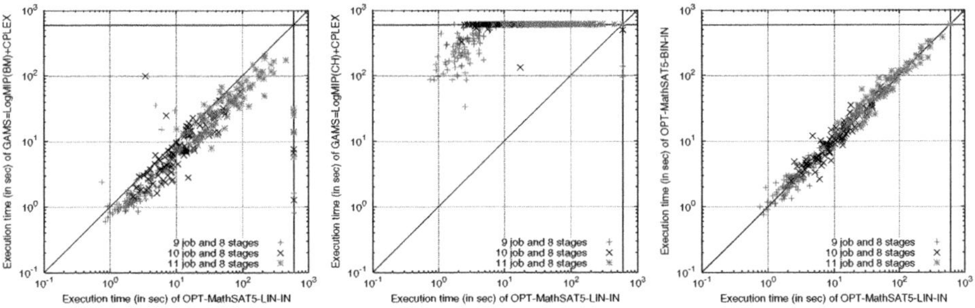

Fig. 2. Table: results (# of solved instances, cumulative time in seconds for solved instances) for OPT-MATHSAT and GAMS on 100 random samples each of the job-shop problem, for $J = 8$ stages and $I = 9, 10, 11$ jobs. Scatter-plots: comparison of the best configuration of OPT-MATHSAT (OPT-MATHSAT-LIN-IN) against LOGMIP(BM)+CPLEX (left), LOGMIP(CH)+CPLEX (center) and OPT-MATHSAT-BIN-IN (right).

The Zero-Wait Jobshop Problem. Consider the scenario where there is a set I of jobs which must be scheduled sequentially on a set J of consecutive stages with zero-wait transfer between them. Each job $i \in I$ has a start time s_i and a processing time t_{ij} in the stage $j \in J_i$, J_i being the set of stages of job i. The goal of the *zero-wait job-shop scheduling* problem is to minimize the makespan, that is the total length of the schedule. In our experiments, we used the LGDP model used in [27] and a corresponding OMT($\mathcal{LA}(\mathbb{Q})$) encoding.

We randomly generated benchmarks according to a fixed number of jobs I and a fixed number of stages J. For each job $i \in I$, start time s_i and processing time t_{ij} of every job are selected in the interval $]0, 1]$ uniformly at random. We consider a set of 100 samples each for 9, 10 and 11 jobs and 8 stages. We set no value for ub and $\mathsf{lb} = 0$.

The table of Figure 2 shows the number of solved instances and their cumulative execution time for different configurations of OPT-MATHSAT and GAMS on the randomly-generated formulas. The scatter-plots of Figure 2 compare the best-performing version of OPT-MATHSAT, OPT-MATHSAT-LIN-IN, against LOGMIP with BM and CH reformulation (left and center respectively); the figure also compares the two inline versions OPT-MATHSAT-LIN-IN and OPT-MATHSAT-BIN-IN (right).

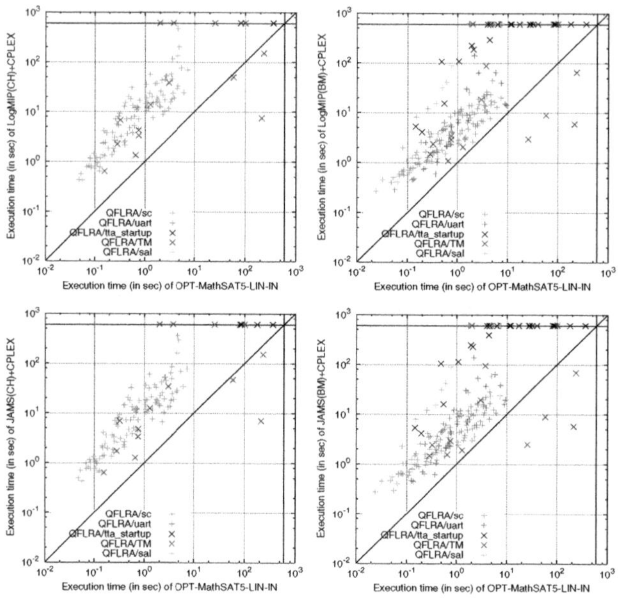

Fig. 3. Scatter-plots of the pairwise comparisons on the smt-lib $\mathcal{LA}(\mathbb{Q})$ satisfiable instances between OPT-MATHSAT-BIN-IN and the two versions of LOGMIP (up) and JAMS. (down).

Discussion. The results in Figures 1 and 2 suggest some considerations.

Comparing the different version of OPT-MATHSAT, overall the -LIN options seems to perform a little better than and -BIN options (although gaps are not dramatic): in fact, OPT-MATHSAT-LIN-OF performs most often a little better than OPT-MATHSAT-BIN-OF, OPT-MATHSAT-BIN-IN performances are slightly better than to those of OPT-MATHSAT-LIN-IN. We notice that the -IN options behave uniformly better than the -OF options.

Comparing the different versions of the GAMS tools, we see that (i) on strip-packing instances LOGMIP reformulations lead to better performance than JAMS reformulations, (ii) on job-shop instances they produce substantially identical results. For both reformulation tools, the "BM" versions uniformly outperform the "CH" ones.

Comparing the different versions of OPT-MATHSAT against all the GAMS tools, we notice that (i) on strip-packing problems all versions of OPT-MATHSAT most often outperform all GAMS tools, (ii) on job-shop problems OPT-MATHSAT outperforms the "CH" versions whilst it is beaten by the "BM" ones.

4.2 Comparison on SMT-LIB Problems

We compare OPT-MATHSAT against GAMS also on the satisfiable $\mathcal{LA}(\mathbb{Q})$-formulas (QF_LRA) in the SMT-LIB [6]. They are divided into six categories: `sc`, `uart`, `sal`, `TM`, `tta_startup`, and `miplib`. [9] Since we have no information on lower bounds

[9] Notice that other SMT-LIB categories like `spider_benchmarks` and `clock_synchro` do not contain satisfiable instances and are thus not reported here.

on these problems, we use the linear-search version OPT-MATHSAT-LIN-IN. Since we have no control on the origin of each problem and on the name and meaning of the variables, we selected iteratively one variable at random as cost variable, dropping it if the resulting minimum was $-\infty$. This forced us to eliminate a few instances, in particular all `miplib` ones.

We first noticed that some results for GAMS have some problem (see Table 1 in [29]). Using the default options, on $\approx$ 60 samples over 193, both GAMS tools with the CH option returned "unfeasible" (inconsistent), whilst the BM ones, when they did not timeout, returned the same minimum values as OPT-MATHSAT. (We recall that all OPT-MATHSAT results were cross-checked, and that the four GAMS tool were fed with the same files.) Moreover, on four `sal` instances the two GAMS tools with BM options returned a wrong minimum value "0", with "CH" they returned "unfeasible", whilst OPT-MATHSAT returned the minimum value "2"; by modifying a couple of parameters from their default value, namely "`eps`" and "`bigM Mvalue`", the results become unfeasible also with BM options. (We conjecture that these problems may be caused, at least in part, by the fact that GAMS tools use floating-point rather than infinite-precision arithmetic; nevertheless, this issue may deserve further investigation.)

After eliminating all flawed instances, the results appear as displayed in Figure 3. OPT-MATHSAT solved all problems within the timeout, whilst GAMS did not solve many samples. Moreover, with the exception of 3-4 samples, OPT-MATHSAT always outperforms the GAMS tool, often by more than one order magnitude.

5 Conclusions and Future Work

This research opens the possibility for several interesting future directions. A short-term goal is to improve the efficiency and applicability of OPT-MATHSAT: we plan to (i) investigate and implement novel mixed linear/binary-search strategies and heuristics (ii) extend the experimentation to novel sets of problems, possibly investigating ad-hoc customizations. A middle-term goal is to extend the approach to $\mathcal{LA}(\mathbb{Z})$ or mixed $\mathcal{LA}(\mathbb{Q}) \cup \mathcal{LA}(\mathbb{Z})$, by exploiting the solvers which are already present in MATHSAT [17]. A much longer-term goal is to investigate the feasibility of extending the technique to deal with non-linear constraints, possibly using MINLP tools as $\mathcal{T}$-Solver/Minimize.

References

1. `http://disi.unitn.it/~rseba/ijcar12/ijcar12-tarball.tar.gz`
2. EMP, `http://www.gams.com/dd/docs/solvers/emp.pdf`
3. JAMS, `http://www.gams.com/`
4. LogMIP v2.0, `http://www.logmip.ceride.gov.ar/index.html`
5. MathSAT 5, `http://mathsat.fbk.eu/`
6. SMT-LIB, `http://www.smtlib.org/`
7. Yices, `http://yices.csl.sri.com/`
8. Achterberg, T., Berthold, T., Koch, T., Wolter, K.: Constraint Integer Programming: A New Approach to Integrate CP and MIP. In: Trick, M.A. (ed.) CPAIOR 2008. LNCS, vol. 5015, pp. 6–20. Springer, Heidelberg (2008)
9. Audemard, G., Cimatti, A., Korniłowicz, A., Sebastiani, R.: SAT-Based Bounded Model Checking for Timed Systems. In: Proc. FORTE 2002. LNCS, vol. 2529. Springer (2002)

10. Balas, E.: Disjunctive programming: Properties of the convex hull of feasible points. Discrete Applied Mathematics 89(1-3), 3–44 (1998)
11. Barrett, C., Sebastiani, R., Seshia, S.A., Tinelli, C.: Satisfiability Modulo Theories. In: Biere et al [12], ch. 26, pp. 825–885 (February 2009)
12. In: Biere, A., Heule, M.J.H., van Maaren, H., Walsh, T. (eds.) Handbook of Satisfiability. Frontiers in Artificial Intelligence and Applications, vol. 185, IOS Press (February 2009)
13. Bozzano, M., Bruttomesso, R., Cimatti, A., Junttila, T.A., Ranise, S., van Rossum, P., Sebastiani, R.: Efficient Theory Combination via Boolean Search. Information and Computation 204(10) (2006)
14. Brooke, A., Kendrick, D., Meeraus, A., Raman, R.: GAMS - A User's Guide. In: GAMS Development Corporation, Washington, DC, USA (2011)
15. Cimatti, A., Franzén, A., Griggio, A., Sebastiani, R., Stenico, C.: Satisfiability Modulo the Theory of Costs: Foundations and Applications. In: Esparza, J., Majumdar, R. (eds.) TACAS 2010. LNCS, vol. 6015, pp. 99–113. Springer, Heidelberg (2010)
16. Dutertre, B., de Moura, L.: A Fast Linear-Arithmetic Solver for DPLL(T). In: Ball, T., Jones, R.B. (eds.) CAV 2006. LNCS, vol. 4144, pp. 81–94. Springer, Heidelberg (2006)
17. Griggio, A.: A Practical Approach to Satisfiability Modulo Linear Integer Arithmetic. Journal on Satisfiability, Boolean Modeling and Computation - JSAT 8, 1–27 (2012)
18. IBM. IBM ILOG CPLEX Optimizer (2010), `http://www-01.ibm.com/software/integration/optimization/cplex-optimizer/`
19. Li, C.M., Manyà, F.: MaxSAT, Hard and Soft Constraints. In: Biere et al. [12], ch.19, pp. 613–631 (February 2009)
20. Lodi, A.: Mixed Integer Programming Computation. In: 50 Years of Integer Programming 1958-2008, pp. 619–645. Springer (2009)
21. Lynce, I., Marques-Silva, J.: On Computing Minimum Unsatisfiable Cores. In: SAT (2004)
22. Marques-Silva, J.P., Lynce, I., Malik, S.: Conflict-Driven Clause Learning SAT Solvers. In: Biere et al [12], ch.4, pp. 131–153 (February 2009)
23. Nelson, C.G., Oppen, D.C.: Simplification by cooperating decision procedures. TOPLAS 1(2), 245–257 (1979)
24. Nieuwenhuis, R., Oliveras, A.: On SAT Modulo Theories and Optimization Problems. In: Biere, A., Gomes, C.P. (eds.) SAT 2006. LNCS, vol. 4121, pp. 156–169. Springer, Heidelberg (2006)
25. Raman, R., Grossmann, I.: Modelling and computational techniques for logic based integer programming. Computers and Chemical Engineering 18(7), 563–578 (1994)
26. Roussel, O., Manquinho, V.: Pseudo-Boolean and Cardinality Constraints. In: Biere et al [12], ch. 22, pp. 695–733 (February 2009)
27. Sawaya, N.W., Grossmann, I.E.: A cutting plane method for solving linear generalized disjunctive programming problems. Comput. Chem. Eng. 29(9), 1891–1913 (2005)
28. Sawaya, N.W., Grossmann, I.E.: A hierarchy of relaxations for linear generalized disjunctive programming. European Journal of Operational Research 216(1), 70–82 (2012)
29. Sebastiani, R., Tomasi, S.: Optimization in SMT with LA(Q) Cost Functions. Technical Report DISI-12-003, DISI, University of Trento (January 2012), `http://disi.unitn.it/~rseba/ijcar12/DISI-12-003.pdf`
30. Sellmann, M., Kadioglu, S.: Dichotomic Search Protocols for Constrained Optimization. In: Stuckey, P.J. (ed.) CP 2008. LNCS, vol. 5202, pp. 251–265. Springer, Heidelberg (2008)
31. Vecchietti, A.: Personal communication (2011)
32. Vecchietti, A., Grossmann, I.: Computational experience with logmip solving linear and nonlinear disjunctive programming problems. In: Proc. of FOCAPD, pp. 587–590 (2004)
33. Wolfman, S., Weld, D.: The LPSAT Engine & its Application to Resource Planning. In: Proc. IJCAI (1999)

Synthesis for Unbounded Bit-Vector Arithmetic

Andrej Spielmann and Viktor Kuncak

School of Computer and Communication Sciences (I&C)
École Polytechnique Fédérale de Lausanne (EPFL), Switzerland

Abstract. We propose to describe computations using QFPAbit, a language of quantifier-free linear arithmetic on unbounded integers with bitvector operations. We describe an algorithm that, given a QFPAbit formula with input and output variables denoting integers, generates an efficient function from a sequence of inputs to a sequence of outputs, whenever such function on integers exists. The starting point for our method is a polynomial-time translation mapping a QFPAbit formula into the sequential circuit that checks the correctness of the input/output relation. From such a circuit, our synthesis algorithm produces solved circuits from inputs to outputs that are no more than singly exponential in size of the original formula. In addition to the general synthesis algorithm, we present techniques that ensure that, for example, multiplication and division with large constants do not lead to an exponential blowup, addressing a practical problem with a previous approach that used the MONA tool to generate the specification automata.

1 Introduction

Over the past decades, a number of decision procedures has been developed and integrated into satisfiability modulo theory (SMT) solvers. Among the primary uses of this technology so far has been verification and error finding. Recently, researchers started using this technology for software synthesis [11]. In the line of work on complete functional synthesis, researchers proposed to generalize decision procedures for infinite domains to *synthesis procedures* [7].

The basic idea is to describe fragments of code using formulas in a decidable logic. Such a formula specifies a relation between inputs and outputs. A synthesis procedure then compiles this formula into a program that maps inputs into outputs, and whose behavior corresponds to invoking a decision procedure on that particular constraint. The resulting program is guaranteed to satisfy the specification. Synthesis procedures have been described for, e.g., parameterized Presburger arithmetic [7], using a constructive version of quantifier elimination.

For domains such as integer arithmetic, automata-based methods can have a number of advantages compared to quantifier elimination, including the ability to support operations on unbounded bitvectors. Motivated by these observations, in related previous work [4] researchers considered synthesis of specifications expressed in weak monadic second-order logic of one successor (WS1S), which is equivalent to Presburger arithmetic with bitwise logical operators. In contrast to automata-based approaches to reactive synthesis [1, 2, 5, 8], this approach uses automata to encode relations on integers,

B. Gramlich, D. Miller, and U. Sattler (Eds.): IJCAR 2012, LNAI 7364, pp. 499–513, 2012.

which means that the causality restriction of Church's synthesis problem does not apply. The synthesized function for this problem cannot always be given as a one-pass finite-state transducer. The approach [4] synthesizes a two pass transducer, where the first pass generates a sequence that abstracts the tree of possible executions, whereas the second pass processes this sequence backwards to choose an acceptable sequence of outputs. The previous implementation of this approach used the MONA tool [6] to transform the given specification formula into an automaton accepting a sequence of bits of combined input/output vectors. This implementation therefore suffered from the explicit-state representation used by MONA. The most striking problem is multiplication by constants, where a subformula $x = c * y$ leads to circuits of size proportional to c and thus exponential in the binary representation of c.

To overcome the difficulties with explicit-state representation used in the implementation of [4], in this paper we investigate an approach that directly uses circuit representations for both specifications and implementations. To avoid the non-elementary worst-case complexity [12] of transforming WS1S formulas to automata, we use as our specification language quantifier-free Presburger arithmetic with bitvector operations [9], denoted QFPAbit. We describe a polynomial-time transformation between sequential circuits and QFPAbit. We then present an algorithm for transforming sequential circuit representations of input/output relations into systems of sequential circuits that map inputs into outputs. The worst-case complexity of our translation is bounded by a singly exponential function of the specification circuit size. Building on this general result, we identify optimizations that exploit the structure of specifications to reduce the potential for exponential explosion. Our prototype implementation confirms the improved asymptotic behavior of this synthesis approach, and is available for download from `http://lara.epfl.ch/w/cisy`. Additional details of our constructions are available in the technical report [10].

2 Preliminaries

2.1 Quantifier-Free Presburger Arithmetic with Bit-Vector Logical Operators

Presburger Arithmetic with Bit-vector Logical Operators is the structure of integers with addition and bit-vector logical operations acting on the binary two's complement representation of the integers. Let V be a finite set of variables. Let $c \in \mathbb{Z}$, $x \in V$, and $\%$ ranges over $=, \neq, <, \leq, >, \geq$. The following is the grammar of QFPAbitterms and formulas.

$$T := c \mid x \mid T + T \mid cT \mid \bar{\neg}T \mid T \bar{\wedge} T \mid T \bar{\vee} T$$
$$F := T \,\%\, T \mid \neg F \mid F \wedge F \mid F \vee F \mid F \rightarrow F \mid F \leftrightarrow F$$

Variables range over the set of integers $\mathbb{Z}$. The bitvector logical operators act on the two's complement encoding of numbers [9]:

$$\langle x_k, \ldots x_0 \rangle_{\mathbb{Z}} = -2^k x_k + \Sigma_{i=0}^{k-1} 2^i x_i.$$

A property of this encoding is that replicating the most significant bit does not change the value. This justifies our definition of the bit-vector operators because for any two numbers we can always find encodings that have the same length. By *the* two's complement encoding of a number, we mean its shortest possible encoding. Given a QFPAbit formula F over the set of variables $V = \{x_1, ...x_n\}$, we say that a valuation $val : V \to \mathbb{Z}$ satisfies F if F is true when each occurrence of a variable x_i evaluates to $val(x_i)$. We say that F is satisfiable if there exists a valuation that satisfies F. Note that the identity $-x = \bar{\neg}\, x + 1$ holds for all x.

We can use QFPAbit formulae to define languages over $\Sigma = \{0,1\}^n$. Let F be a QFPAbit formula over the variables $V = \{x_1, ...x_n\}$. Let $w \in \Sigma^+$ be a word of length m. By $w(j)$ denote the j-th letter of w, indexing from 0, so that the initial letter is denoted $w(0)$. Each $w(j)$ is a vector of dimension n, let $w_i(j)$ denote the the i-th coordinate of $w(j)$. Define a valuation $val_w : V \to \mathbb{Z}$ by $val_w(x_i) = \langle w_i(m-1), ...w_i(0)\rangle_{\mathbb{Z}}$. Thus, in the matrix whose columns are the letters of w, the i-th row represents the encoding of $val_w(x_i)$ with the most significant bit coming first. The language defined by the formula is $L(F) = \{w \in \Sigma^+ | val_w \text{ satisfies } F\}$.

2.2 Sequential Circuits

A combinational boolean circuit K is a pair (G, σ) where G is a finite directed acyclic graph and $\sigma : U \to \{AND, OR, NOT\}$ is a labeling function such that U is the set of vertices of G whose in-degree is greater than zero. We require that whenever $\sigma(x) = NOT$ then x has in-degree of one.

We call the vertices in U the *gates*. We denote the vertices of in-degree zero I and call them *inputs*; we denote the vertices of out-degree zero O, and call them *outputs*.

Given a boolean valuation $i : I \to \{true, false\}$, we define a valuation v on all vertices of G as follows:

$$v(x) = \begin{cases} i(x), & \text{if } x \in I \\ \neg v(\gamma(x)), & \text{if } x \in U \wedge \sigma(x) = NOT \\ \bigwedge_{y \in \Gamma(x)} v(y), & \text{if } x \in U \wedge \sigma(x) = AND \\ \bigvee_{y \in \Gamma(x)} v(y), & \text{if } x \in U \wedge \sigma(x) = OR \end{cases}$$

where $\gamma(x)$ denotes the single neighbor of x connected to it by an edge directed towards x and $\Gamma(x)$ denotes the set of all neighbors of x connected to it by edges directed towards x. We call the values of v on O the output values of K for input i.

The values of the outputs of a *combinational* boolean circuit, defined above, depend only on a single set of inputs and can be represented in a truth table. We next review (clocked) *sequential* circuits, which are equivalent to deterministic finite automata but compactly represent the set of states and the transition function.

A clocked sequential circuit (or SC, for short) is a tuple $(K, M, \mathsf{store}, \mathsf{load}, \mathsf{init})$ where

- K is a combinational boolean circuit with inputs and outputs I and O;
- M is a set of D-type flip-flops;
- $\mathsf{store} : M \to O$;
- $\mathsf{load} : M \to I$;
- $\mathsf{init} : M \to \{true, false\}$.

The load and store functions describe how the data input of each flip-flop is connected to a unique output of K and how the Q-output of each flip-flop is connected to a unique input of K. Such a backward-connected output-input pair will be denoted as a *state variable*. We call the inputs of K that are not in the image of load the *input variables* and call the outputs of K that are not in the image of store the *output variables*.

The SC works in clock pulses. It takes as input a stream that for every clock pulse contains values for all input variables, and produces as output a stream that for every clock pulse contains values of all the output variables, computed by K. In every clock pulse, K is provided with input values and it computes output values. The values for K's inputs corresponding to state variables are loaded from the flip-flops and the values for its inputs corresponding to input variables are provided by the input stream. Some of the values of K's outputs are stored in the flip-flops for the use in the next clock cycle, as determined by store.

The values stored in the flip-flops at the beginning of the first clock cycle are called initial values of the state variables and they are given by init.

Notice that a circuit with n input variables and m output variables can be viewed as a machine that, given a word from $(\{0,1\}^n)^+$, produces a word of the same length in $(\{0,1\}^m)^+$.

We can also use a SC to recognize a language.

Definition 1. *Let C be a SC with one output variable o and n input variables. We say that C accepts the word $w \in \{0,1\}^n$ if the value of o in the last cycle is* 1 *when the circuit is given w as input, one letter at each clock cycle.*

The language of C is $L(C) = \{w \in \{0,1\}^n | C$ accepts $w\}$.

Some of the standard finite state machine operations can be efficiently performed on the sequential circuit representations. Given a SC C with input variables $v_1, ...v_n$, state variables $q_1, ...q_n$ and output o, and a SC C' that uses the same input variables $v_1, ...v_n$ and has state variables $q'_1, ...q'_n$ and output o', we can construct a circuit $\neg C$ by simply appending a NOT gate at o and making the output of the NOT gate the output of $\neg C$. Similarly, we can construct circuits $C \wedge C'$ and $C \vee C'$ by connecting the outputs of C and C' to an appropriate logical gate, whose output will become the output of the composite circuit. It can easily be seen that 1) $L(\neg C) = (\{0,1\}^n)^+ \backslash L(C)$; 2) $L(C \wedge C') = L(C) \cap L(C')$; 3) $L(C \vee C') = L(C) \cup L(C')$.

3 Translations between QFPAbit and Sequential Circuits

This section establishes correspondence between QFPAbit and sequential circuits by providing translations in both directions that maintain a close correspondence between the accepted languages.

3.1 Reduction from QFPAbit to Sequential Circuits

Since we have already shown how to construct boolean combinations of sequential acceptor circuits, it is enough to find a set of basic QFPAbit formulae out of which all QFPAbit formulae can be built using logical connectives, and then show how these basic formulae can be translated to SCs.

Definition 2. *Let $w \in \Sigma^+$ with $\Sigma = \{0,1\}^n$ as usually. Suppose*

$$w = \begin{pmatrix} w_1(0) \\ \vdots \\ w_n(0) \end{pmatrix} \begin{pmatrix} w_1(1) \\ \vdots \\ w_n(1) \end{pmatrix} \cdots \begin{pmatrix} w_1(m) \\ \vdots \\ w_n(m) \end{pmatrix}$$

Let $S \subseteq \{1,...n\}$ be non-empty. We define the projection of w onto the coordinates S to be the string $w^S = w^S(0)...w^S(m)$, where $w^S(i)$ is the column vector $(w_j(i))_{j \in S} \in \{0,1\}^{|S|}$. For a language $L \subset \Sigma^+$, we define the projection of L onto the coordinates S to be the language $L^S = \{w^S | w \in L\}$. Note that L^S is a language over the alphabet $\{0,1\}^{|S|}$.

Every QFPAbit formula is a boolean combination of atomic formulae of the form $T_1 \% T_2$ where T_1 and T_2 are terms and $\% \in \{=, \neq, <, \leq, >, \geq\}$. We will now show how to transform any formula F into a new one where the atoms will be of a more restricted form. The new formula will have more variables than F, but when projected onto the variables occurring in F their languages will be the same. We apply the following sequence of transformations:

1. Replace all atomic relations by equalities and strict "less-than" inequalities using the fact that $T_1 < T_2$ if and only if $T_1 + (-1)T_2 < 0$.
2. Remove all instances of multiplication by constants other than -1 and powers of two by exploiting the fact that any term of the form cT is equal to a sum of terms of the form $2^k T$ corresponding to c's two's complement encoding.
3. Remove all instances of multiplication by -1 by replacing every sub-term of the form $(-1)T$ by $\bar{\neg} T + 1$. This equivalence follows easily from the definition of the two's complement encoding.
4. Move all additions to separate conjuncts on the highest level of the formula by replacing every occurence of $T_1 + T_2$ by a fresh variable s and adding conjuncts $s = x + y$, $x = T_1$ and $y = T_2$ to the formula, where x and y are also fresh variables.
5. Move all multiplications by a constant 2^k, which are the only multiplications now left in the formula, to conjuncts on the highest level of the formula by replacing every occurence of $2^k T$ by a fresh variable x and adding $x = 2^k y$ and $y = T$ as conjuncts to the formula, where y is another fresh variable.
6. Replace every additive occurrence of an integer constant c inside a larger term by a fresh variable y_c and add a conjunct $y_c = c$ to the formula.

Let us call the formula that we obtain G. It has size that is polynomial in the size of F and and it consists only of atoms of the following five forms: (i) $T < 0$; (ii) $T_1 = T_2$; (iii) $y = c$; (iv) $x = 2^k t$; (v) $s = x + y$, where x, y, s and t are variables, c is an integer constant and T, T_1, T_2 are terms that contain exclusively variables and bit-vector logical operators.

It is easy to construct SCs for atoms of each of these four forms. For details of these constructions along with circuit diagrams, see our technical report [10]. The general flavor of these circuits is that they compare streams of binary digits. The most complicated case is (iv), where the circuit compares a binary stream to a version of itself shifted by a constant number of bits. Each of the sub-circuits for cases (i),(ii) and (v) has only a constant number of state variables. In case (iii), the number of state variables is proportional to the logarithm of the constant c and in case (iv) it is proportional to k.

Finally, we compose the partial specification circuits by boolean operations to find a SC for G. The correctness of this synthesis procedure is expressed in the following theorem.

Theorem 1. *Let C_F be the circuit obtained from a QFPAbit formula F using the above synthesis procedure. Let V be the set of variables occuring in F. Then*

$$L(C_F)^V = L(F).$$

Moreover, both the the number of gates of C_F and the running time of the synthesis procedure are polynomial in the number of symbols of F. The number of input variables of C is the same as that of F and the number of C's state variables is proportional to the number of symbols of F.

3.2 Reduction from Sequential Circuits to QFPAbit

Let C be a sequential circuit with an underlying combinational circuit $K = (G, \sigma)$, n input variables $\{v_1, ...v_n\}$, m state variables $\{q_1, ...q_m\}$ and output variables $\{o_1, ...o_l\}$. Let $I : \{q_1, ...q_m\} \to \{0, 1\}$ be the initial assignment of values to the state variables. Let U be the set of all gates of K other than those corresponding to the output variables and state variables of C. We will pretend that the elements of U can be used as identifiers for QFPAbit variables and construct a QFPAbit formula with variables $\{v_1, ...v_n, q_1, ...q_m, o_1, ...o_l\} \cup U$, such that for every satisfying assignment, the two's complement encodings of the values of the variables describes the evolution of the values of the corresponding variables and gates in a run of C. Although the QFPAbit variables have the same names as the variables and gates of the circuit, it should be clear from the context which ones do we mean.

We will refer to the values of the gates and inputs of the automaton in the k-th clock cycle by $q_1(k), ...q_m(k)$,$v_1(k), ...v_n(k)$, $o_1(k)...o_l(k)$ and $x(k)$ for all $x \in U$. In the cycle when the inputs are $q_1(i), ...q_m(i), v_1(i), ...v_n(i)$, the values of all the gates in U will be $x(k)$, the output variables will be $o_1(i), ...o_l(i)$ and the outputs corresponding to state variables at that cycle will be denoted $q_1(i+1), ...q_m(i+1)$, because they serve as inputs for the next cycle. We start the numbering of clock cycles from 0.

We will be abusing notation slightly by writing $\sigma(v)(x_1, ..., x_k)$ for some gate v and boolean values $x_1, ..., x_k$ to mean the application of the boolean function represented by $\sigma(v)$ to $x_1, ..., x_k$. Then for all $j \in \{1, ..., m\}$, $k \in \{1, ..., l\}$, $x \in U$ and all $i \in \{0, ..., N-1\}$ where N is the length of the input word, the run of C on that input word is characterized by the following four equations:

$$q_j(0) = I(q_j) \tag{1}$$

$$x(i) = \sigma(x)(\Gamma(x)(i)) \tag{2}$$

$$o_k(i) = \sigma(o_k)(\Gamma(o_k)(i)) \tag{3}$$

$$q_j(i+1) = \sigma(q_j)(\Gamma(q_j)(i)) \tag{4}$$

where, just like in our definition of a combinational circuit, $\Gamma(v)$ denotes the part of the neighborhood of a gate v connected to it with incoming edges, and $\Gamma(v)(i)$ denotes a vector of values of these nodes in clock cycle i.

We next build a QFPAbit formula for which every satisfying evaluation is such that **the reverse** of the bit-sequences of the values it assigns to the variables conform to the above conditions. Since C treats all numbers as starting with the most significant bit, in our QFPAbit representation this will be reversed and hence $x(0)$, $q_j(0)$ and $o_k(0)$ will refer to the least significant bits of the encoding of the values of the variables.

For any gate v of K, let $\bar{\sigma}_v$ be the formula obtained by applying the bit-vector logical operator corresponding to $\sigma(v)$ to the variables in $\Gamma(v)$. Then the following formula can be used to describe the evolution of the digits of q_j:

$$q_j = 2\sigma_{qj}\bar{\vee}I(q_j)$$

The justification is as follows. Taking the bitwise disjunction of a number with 1 or 0 preserves all the digits except the least significant one, which is set to 1 or 0 respectively. Multiplication by 2 induces a shift to the left of the two's complement encoding of a number. Hence the above formula establishes that every bit of q_j is equal to the next bit of σ_{qj} except for the first (least significant) one, which is equal to $I(q_j)$. This ensures that equations (1) and (4) are satisfied.

Similarly, the formulas $o_j = \bar{\sigma}_{oj}$ and $x = \bar{\sigma}_x$ assert that the reverse binary encodings of o_j and x, for some $x \in U$, correspond to their values in the run of C on the given input as described by equations (3) and (2).

Since the most significant digit in a two's complement encoding can be replicated without changing the value of the represented number, QFPAbit formulas have the property that the last letter of a word in a formula's language can be repeated arbitrarily many times to obtain another word inside the language. In the underlying circuit, this would translate to a "blindness" towards the repetition of the initial input letter, which is a property that not all circuits have. In general we cannot find a formula whose language contains exactly those words whose reverse encodes a run of the circuit.

The way to treat this problem is to construct a formula that contains a clause saying "the variables are only simulating the circuit for a finite number of steps and then are allowed to deviate". That way we obtain a formula for which to every possible satisfying evaluation corresponds a word describing the run of the circuit. However, each such valuation will also represent an infinite number of longer incorrect descriptions of a run of the circuit.

For succintness, let $\Delta_{qj} \equiv 2\bar{\sigma}_{qj}\bar{\vee}I(q_j)$. Let y be a fresh variable and consider the formula

$$F_C \equiv 1 + ((y-1)\bar{\vee}y) = 2y \wedge y > 1 \wedge \left[\bigwedge_{j=1}^{m}(q_j\bar{\wedge}(y-1)) = (\Delta_{qj}\bar{\wedge}(y-1))\right]$$
$$\wedge \left[\bigwedge_{j=1}^{l}(o_j\bar{\wedge}(y-1)) = (\bar{\sigma}_{oj}\bar{\wedge}(y-1))\right] \wedge \left[\bigwedge_{x\in U}(x\bar{\wedge}(y-1)) = (\bar{\sigma}_x\bar{\wedge}(y-1))\right].$$

The subformula $1 + ((y-1)\bar{\vee}y) = 2y \wedge y > 1$ asserts that y is a power of two, say $y = 2^k$, and that k is at least 1. Therefore the two's complement encoding of $(y-1)$ is $\langle 0, ..., 0, 1, ..., 1\rangle_{\mathbb{Z}}$ with an arbitrary number of zeros and exactly k ones. So the clauses of the form $(T_1\bar{\wedge}(y-1)) = (T_2\bar{\wedge}(y-1))$ assert that the k least significant digits of T_1 and T_2 are the same. The rest of the digits can be arbitrary.

Theorem 2. *For any given satisfying valuation of F_C, $y = 2^k$ for some k. The first k bits, presented in the reverse order, of the bit-sequences corresponding to two's complement encodings of $q_1, ...q_m$ and $o_1...o_l$ describe the evolution of the values of those variables throughout the first k clock cycles of the run of C on the input word given by the reverse two's complement encodings of $v_1, ...v_n$, as specified by the equations (1)-(4).*

Now we show how this translates to acceptor circuits defining a language:

Theorem 3. *Suppose C is a SC with one output o and let $F'_C \equiv F_C \wedge o < 0$. Then $L(F'_C) \neq \emptyset$ if and only if $L(C) \neq \emptyset$*

Proof. *The clause $o < 0$ is true if and only if the first digit of o is one. It follows from the above discussion of F_C that for every word w of length k providing encoding of values for $v_1, ..., v_n$, there exist infinitely many satisfying evaluations for F_C under which $y = 2^k$ and the reverse encoding of the values of the variables describes the run of C on the reverse of w. Now suppose that C accepts w. This happens if and only if in the last, k-th, clock cycle the value of the output bit is one. But this is if and only if the first digit of the value of o in F'_C is one. Therefore the described evaluations satisfy F'_C if and only if C accepts w. This means that the language of C is non-empty if and only if F'_C is satisfiable.* ■

To summarize, we have described polynomial-size translations between QFPAbit and sequential circuits going both ways. For every QFPAbit formula we can construct a sequential circuit recognizing the same language. For every sequential circuit we can construct a QFPAbit formula that contains variables representing inputs, outputs and state variables of the circuit, and it is satisfied only by valuations that assign these variables values whose binary encoding in reverse describes an initial portion of the evolution of the circuit's variables during a run. If the circuit has only one output then it is an acceptor circuit and in this case we can construct a QFPAbit formula which is satisfiable if and only if the language of the SC is non-empty. Moreover, the formula will accept a language such that for every word w in this language, an initial part of w projected onto the input variables and reversed is a word in the language of the SC.

4 From Specification Circuits to Transducer Circuits

Given a specification written as a QFPAbit formula, we have shown how to build a specification circuit of a size linear in the size of the formula. Provided that the variables of the formula, and thus the inputs of the automaton are partitioned into two groups, $\bar{i}$ and $\bar{o}$, interpreted as the inputs and the outputs of the synthesized function, we will now show how to construct a set of circuits that will work as a transducer, i.e. given a word from the "$\bar{i}$-projection" of the language, produce an output word from the "$\bar{o}$-projection" of the language such that together they satisfy the specification, if such an output word exists. The structure of our algorithm is similar to the one presented in [4]. Our use of the word "transducer" does not refer to the traditional notion of Finite State Transducers, but to a more complicated machine with the following main features. Our transducer reads the whole input twice. The first time from the beginning to the end to generate the exhaustive run of the projection of the specification circuit onto the input variables, and the second time backwards, determining concrete states and output letters within the exhaustive run. In the meantime it uses an amount of memory proportional to the length of the input. This allows us to express functions for which it is not possible to determine the output before reading the entire input, which is needed to obtain complete synthesis for QFPAbit.

In contrast to [4], we will be using sequential circuits instead of automata. This more concrete implementation allows us to perform an optimization that will ensure that the presence of large integer constants in the formula does not necessarily cause a blow-up in the size of the transducer proportional to the value of that constant, as was the case with the previous approach. Moreover, even if a state-space expansion does occur, the size of our circuits is guaranteed to be singly-exponential in the size of the specification formula. No such bound on the size of the automata was provided in [4].

In Section 4.2 we study two more optimization techniques - how to exploit the circumstance when the specification formula is either a conjunction or a disjunction of sub-formulas to build the transducer as a composition of smaller transducers.

Definition 3. *Given a (non-)deterministic automaton* $A = (\Sigma_V, Q, init, F, T)$ *over variables* V *and a set* $I \subset V$*, the projection of* A *to* I*, denoted by* A^I*, is the non-deterministic automaton* $(\Sigma_I, Q, init, F, T_I)$ *with* $T_I = \{(q, \sigma_I, q') \in Q \times \Sigma_I \times Q | \exists \sigma \in \Sigma_V.(q, \sigma, q') \in T \wedge \sigma^I = \sigma_I\}$.

Since it is natural to view a sequential circuit as a DFA, we also allow ourselves to talk about projections of sequential circuits.

Definition 4. *The exhaustive run* ρ *of an automaton* $A = (\Sigma, Q, init, F, T)$ *on a word* $w \in \Sigma^*$ *is a sequence of sets of states* $S_1, ...S_{|w|+1}$ *such that (i)* $S_1 = init$ *and (ii) for all* $1 \leq |w|, S_{i+1} = \{q' \in Q | \exists q \in S_i.(q, w_i, q') \in T\}$.

Suppose the specification circuit is a sequential circuit C with input variables $\bar{i} \cup \bar{o}$, state variables $\bar{q}$ and one output variable determining the acceptance. Here by each of $\bar{i}, \bar{o}$ and $\bar{q}$ we actually mean vectors of variables wide n, l and m bits respectively. We will also be using $\bar{i}, \bar{o}$ and $\bar{q}$ to denote the sets of individual variables comprising each of the vectors.

We now partition the state variables as follows. We let $\bar{s}$ be the largest set of state variables such that the value of each of them in the $(N+1)$-st clock cycle depends only on the values of $\bar{i}$ and the state variables inside $\bar{s}$ in the N-th clock cycle. In particular, they do not depend on the values of $\bar{o}$. We denote all the other state variables as $\bar{r}$ and we will assume that $\bar{r}$ is a vector of width m_1 and $\bar{s}$ is of width m_2.

The set $\bar{s}$ can be determined by exploring the graph of dependencies amongst the variables of $\bar{q}$ and $\bar{o}$. We can determine whether a formula $\varphi(x)$, for example one defining the value of a q_j in the next clock cycle, depends on a variable x, which it contains, by using a SAT-solver to check whether the formula $\varphi(true) \leftrightarrow \varphi(false)$ is valid.

We will now describe the operation of our transducer, which consists of three circuits that we call C', ϕ and τ. Circuit ϕ is a combinational circuit and the other two are sequential. Their roles are analogous to those of the deterministic automaton A' and functions ϕ and τ in [4]. Our specification circuit C fulfills the responsibility of the specification automaton A used in [4].

C' performs two tasks. First, it runs the part of C that computes the sequence of values of $\bar{s}$ as C consumes $\bar{i}$. In parallel with this, C' also simulates the exhaustive run of the projection of C onto the input variables $\bar{i}$. So running C' with the sequence of values for $\bar{i}$ as input will generate a sequence of values for $\bar{s}$ together with a sequence of sets of possible values for the rest of the state variables, which are $\bar{r}$. We will store this trace in a memory from which it can later be read in the reversed order.

This separation of sets $\bar{s}$ and $\bar{r}$ is one of the main improvements in our approach over previous work. It takes advantage of the simple idea that when projecting a deterministic automaton onto a subset of its input variables, it is possible that the transitions within a subset of the states of the automaton remain deterministic even with the restricted alphabet, and hence that part of the automaton does not need to go through an exponential expansion due to the projection. This optimization applies in particular in the case when the specification formula contains division of a term that is completely determined by $\bar{i}$-variables by a power of 2. An intuitive explanation is the following. The specification circuit for the formula $x = 2^k t$ verifies whether the encoding of x is a copy of the encoding of t shifted to the left by k bits. Therefore it needs k state variables to remember the past k bits of x. The values of these k state variables are independent of t and hence if x is an $\bar{i}$-variable, which means that we are performing division, then these k state variables will belong to $\bar{s}$ and they will not participate in the state-space explosion of C'. On the other hand, this optimization does not apply if x is an $\bar{o}$-variable, i.e. when we are performing multiplication.

The purpose of ϕ is to find inside the last stored set of possible states for $\bar{r}$ one which is, combined with the last stored value of $\bar{s}$, an accepting state of C.

Eventually, we run τ, which reconstructs a whole accepting run of C by tracing backwards through the stored exhaustive run of its projection onto the input variable set $\bar{i}$, using the accepting state determined by ϕ as a starting point. During this backward run it constructs a sequence of $\bar{o}$ letters that is the final output of the transducer.

4.1 Implementation of C', ϕ and τ as Circuits

For C', consider the circuit in the figure in Appendix A, which has state variables $R_1, ...R_{2^{m_1}}$ and $\bar{s}$, and no outputs.

Let C_1 and C_2 denote the sub-circuits of C for computing $\bar{r}$ and $\bar{s}$ respectively. In the figure, we denote the corresponding combinational circuits behind these SCs by K_1 and K_2. We let $C_{\bar{i}}$ be the projection automaton obtained from C_1 by projecting it onto the $\bar{i}$-variables. The intended meaning of the state variables $R_1, ...R_{2^{m_1}}$ of C' is that R_k is set to true if and only if at that point the non-deterministic automaton $C_{\bar{i}}$ could be in the state number k. Since there are exactly 2^{m_1} possible states of $C_{\bar{i}}$, we can make some arbitrary assignment of the possible states of $C_{\bar{i}}$ to the R_k's. Initially, A' is in a state where all variables R_k are 0 except for one, corresponding to the initial state of $C_{\bar{i}}$. The initial value of $\bar{s}$ is also determined by the given initial state of C.

The $\bar{r}_i$ and $\bar{o}_j$ denoted in italics represent constant bit-vectors given as input to each of the 2^{ml} copies of C_1. The indexes are assigned so that $\bar{r}_j$ is the assignment of state variables of $C_{\bar{i}}$ corresponding to the state which is represented by R_j. Hence each of the C_2-subcircuits produces an outcome $\bar{r}$-state for a given combination of a previous state and values for the $\bar{o}$-variables.

Each of the AND-like-gates with an R_k inscription is understood to have negations at an appropriate combination of its inputs, so that it returns true if and only if its input $\bar{r}$ represents the $\bar{r}$-state corresponding to R_k and also the incoming signal from the state variable R_j is true. This last condition has the effect of considering the output only of those sub-circuits for which the input state $\bar{r}_j$ is actually one of the possible states in the exhausting run of $C_{\bar{i}}$ at the moment.

The last layer of ordinary OR-gates just has the effect that if any of the possible combinations of an active previous state and an $\bar{o}$-letter produces the state corresponding to R_k then R_k is set to one in the next cycle. The main idea of this circuit is that for every state of C that is possible at the present clock cycle, it tries every possible $\bar{o}$-letter to produce the set of all possible states in the next clock cycle.

Now assume that the sequence of states this circuit goes through while reading an input word is saved in a memory from where it can readily be read in the reverse order. Recall that ϕ is supposed to find an accepting state of C amongst the possible states encoded in the last state of C' - that is, in the combination of the "exhaustive state" of $C_{\bar{i}}$ encoded by $R_1, ...R_{2^{m_1}}$ and the deterministic part of the state, $\bar{s}$. A slight divergence between deterministic automata used in [4] and our variant of sequential circuits is that whether the circuit accepts depends not only on the current value of its state variables but also on the value of all its inputs - the circuit accepts simply when it outputs a 1. To account for this, our ϕ circuit has to choose both a state from amongst the states possible in the penultimate clock cycle of the run of C', and a suitable $\bar{o}$-letter, such that the resulting state is accepting. If such state and $\bar{o}$-letter do not exist, the user is notified that for the given sequence of values for the $\bar{i}$-variables there exists no satisfying sequence of values for the $\bar{o}$-variables. The implementation of ϕ is a circuit very similar to that for C', also containing 2^{m_1+l} copies of K_1. However, since it only needs to be run for one clock cycle, it is a combinational circuit rather than a sequential one.

Finally, we use a very similar circuit for the function τ. In each clock cycle, it takes as input a transition $\langle S', \bar{i}, S\rangle$ of C' and a state $\bar{q} \in S$ and generates a state $\bar{q}' \in S'$ and an output symbol $\bar{o}$ such that there is a valid transition in C from $\bar{q}$ to $\bar{q}'$ while reading the letter obtained by combining $\bar{i}$ with $\bar{o}$. This is again implemented by guessing combinations of an appropriate $\bar{o}$-letter and $\bar{r}$-state, so τ consists of 2^{m_1+l} copies of K_1 and some servicing circuitry. The output of τ and also the final output of the transducer is the sequence of $\bar{o}$ letters. Notice that it comes in the reverse order, respective to $\bar{i}$.

4.2 Constructing Transducer as a Composition of Transducers for Sub-formulas

Suppose that the specification formula F on its highest level is a disjunction of sub-formulas $\varphi_1, ...\varphi_k$. Then we can build a transducer for each of them separately and run them in parallel. If for a given input any of the transducers finds an output satisfying the sub-formula corresponding to that transducer, say φ_i, then this output can be taken to be the global output. If φ_i mentions only a subset of the output variables then values for the remaining ones can be picked arbitrarily.

If the specification formula, on the other hand, is a conjunction of sub-formulas, then we also have to mind dependencies between the variables.

Definition 5. *We say that a* QFPAbit *formula ψ over variables V uniquely determines a set of variables $\bar{x}$ as a function of a set of variables $\bar{y}$, if for any partial valuation $val_{\bar{y}} : \bar{y} \to \mathbb{Z}$ that only assigns values to the variables of $\bar{y}$, the set of satisfying valuations of ψ that extend val is non-empty and all of them give all the variables in $\bar{o}$ the same values.*

If the specification formula F is a conjunction of sub-formulas $\varphi_1, ...\varphi_k$, we can apply the following reasoning. Suppose that there exists $\bar{o}' \subseteq \bar{o}$ such that some φ_j uniquely determines the values of $\bar{o}'$ as a function of $\bar{i}$. Now suppose that $val : \bar{i} \cup \bar{o} \to \mathbb{Z}$ is a satisfying valuation for F. Then, in particular, it is a satisfiing valuation for φ_j and it assigns $\bar{o}'$ the same values as *any* satisfying valuation of φ_j that gives the $\bar{i}$'s the same values as val.

This means that we can build an independent transducer for φ_j and use its output to fix the values of $\bar{o}'$ in F, allowing us to build a smaller transducer for the rest of the variables. Notice that the values that the transducer for φ_j computes for those variables that have not been proven to be uniquely determined by $\bar{i}$ must be ignored, because their values need not be a part of a satisfying valuation for the rest of F.

In practice, we can use this fact to construct a sequence of transducers with increasing number of $\bar{i}$ variables and decreasing number of $\bar{o}$ variables. We scan through the list of conjuncts of F and whenever we find one, say φ_j, in which some subset of $\bar{o}$ variables is uniquely determined by the $\bar{i}$ variables, we build a transducer for it, reclassify the uniquely determined $\bar{o}$-variables to $\bar{i}$-variables in F and repeat the process, wiring the appropriate outputs of the transducer for φ_j to become the inputs of the next transducer. If it turns out that in a particular conjunct, all the occurring $\bar{o}$-variables are uniquely determined, this whole conjunct can be removed from F.

Notice that for regularly occuring conjuncts of a standard form, like for example equality assertions involving standard arithmetical operations, we will not have to invoke the general transducer-synthesis method described at the beginning of this section. Instead, we can use potentially more efficient pre-computed circuits loaded from a library. This can, for example, be applied in the case when the conjunct asserts that an $\bar{o}$-variable is a constant multiple of a term that is uniquely determined by the $\bar{i}$ variables.

The length of the resulting sequence of transducers is at most quadratic in the number of $\bar{o}$ variables, which can be seen by inspecting the running time of the trivial algorithm that loops through the cojuncts in an arbitrary fixed order and halts when during an iteration examining all the conjuncts it can not reclassify any new $\bar{o}$-variables to $\bar{i}$-variables.

Obviously, this optimization is useful only if the specification formula F is in fact a conjunction containing conjuncts that do have the property of uniquely determining some of the $\bar{o}$-variables as a function of the $\bar{i}$ variables. As discussed in Section 3.1, before building the specification circuit we first pre-process the input formula, so that the formula that is eventually used for building the circuit is

$$G \equiv F' \wedge \varphi_1 \wedge ... \wedge \varphi_n$$

where each of the φ_i has one of the following forms: (i) $x = 2^k t$; (ii) $x = c$; (iii) $s = x + y$; (iv) $T_1 = T_2$, where x, y, t, s are variables, c is an integer constant and T_1, T_2 are terms built out of variables and bit-vector logical operations. F' is a boolean combination of atoms of similar forms, but at the present time we do not have methods for investigating variable dependencies in non-atomic formulas.

On the other hand, for each of the φ_j's we can exactly determine which $\bar{o}$-variables are uniquely determined by the $\bar{\imath}$-variables. In case (i), if at least one of the variables present is an $\bar{\imath}$-variable then the other is determied. In case (i), variable x is determined, and in case (iii), if at least two of the variables are $\bar{\imath}$-variables then the last one is determined. In case (iv), since T_1 and T_2 contain only variables and bit-vector logical operations, the equality holds exactly if the propositional formulas corresponding to T_1 and T_2 evaluate to the same boolean value in every clock cycle. Therefore it is enough to investigate which $\bar{o}$ variables are uniquely determined by the $\bar{\imath}$ variables in the propositional formula $\hat{T}_1 \leftrightarrow \hat{T}_2$, where $\hat{T}_1$ and $\hat{T}_2$ are propositional formulas obtained from T_1 and T_2 by replacing the bit-vector logical operators by standard boolean operators and treating the QFPAbit variables as propositional variables.

Example. We demonstrate the usefulness of this optimization technique on an example. Let us forget for a moment that our language contains an out-of-the-box plus operator and suppose we would like to synthesize a function for performing addition and outputting the sequence of carry bits at the same time. It can be specified in QFPAbit as follows.

$$(s = x\bar{\oplus}y\bar{\oplus}c) \land (c = 2((x\bar{\land}y)\bar{\lor}(x\bar{\land}c)\bar{\lor}(y\bar{\land}c)))$$

where x and y are designated as inputs and s and c are outputs representing the sum and the sequence of carry bits respectively. Clearly, the right-hand conjunct determines c uniquely, given values for x and y. Our prototype implementation is able to detect this and builds a transducer which is a composition of two parts - one for the right-hand conjunct, which computes the value of c given values for x and y, and one for the left-hand conjunct that computes the value of s given values for x, y and c. Due to this factorisation, the total number of gates in all the circuits involved is $7.2\times$ smaller than when we enforce the building of a single monolithic transducer for the whole formula.

To conclude the discussion of this optimization technique, let us look closer at how it applies to those φ_j's that are of form $x = 2^k t$. Because of the way how these conjuncts originate during the pre-processing of the specification formula, often both x and t are output variables. If after inspecting some other conjuncts we manage to specify one of them as an input variable, the other is immediately determined by it and we will be able to remove this conjunct from the formula and construct an efficient transducer for it. We can summarize this in the following lemma.

Lemma 1. *Suppose that the original formula, before pre-processing, contains multiplication by a constant c in a context of the form $T_1[cT] = T_2$ such that either all the $\bar{o}$-variables occuring in T are uniquely determined by the $\bar{\imath}$-variables, or the $\bar{o}$-variables of T occur nowhere else in T_1 and T_2 and the value of a fresh variable x is uniquely determined in the formula $T_1[x] = T_2$. Then the total size of all the circuits of the transducer obtained by the procedure described in this section will be proportional to the logarithm of c.*

5 Conclusion

We have presented a synthesis procedure that starts from QFPAbit description of an input/output relation, generates a sequential circuit of a polynomial size, and then transforms this circuit into a synthesized system of sequential circuits that maps a sequence of inputs into a sequence of outputs.

The described synthesis procedure improves the previous work by two independent optimizations. We have built a prototype implementation that allowed us to show on examples that these techniques work and are important.

Acknowledgements. The idea of replacing synthesis from WS1S with synthesis from QFPAbit as well as a polynomial translation from QFPAbit into circuits originated in a discussion between Barbara Jobstmann and Viktor Kuncak in October 2010. Barbara Jobstmann was also suggesting decomposing specifications and performing synthesis modularly. We thank Aarti Gupta for pointing to the related work in her PhD thesis [3] as well as Sharad Malik and Paolo Ienne for useful discussions.

References

1. Bloem, R., Jobstmann, B., Piterman, N., Pnueli, A., Sa'ar, Y.: Synthesis of reactive(1) designs. J. Comput. Syst. Sci. 78(3), 911–938 (2012)
2. Buchi, J., Landweber, L.: Solving sequential conditions by finite-state strategies. Transactions of the American Mathematical Society 138(295-311), 5 (1969)
3. Gupta, A.: Inductive Boolean Function Manipulation: A Hardware Verification Methodology for Automatic Induction. PhD thesis, CMU (1994)
4. Hamza, J., Jobstmann, B., Kuncak, V.: Synthesis for regular specifications over unbounded domains. In: Formal Methods in Computer-Aided Design (FMCAD), pp. 101–109. IEEE (2010)
5. Jobstmann, B., Bloem, R.: Optimizations for LTL synthesis. In: FMCAD (2006)
6. Klarlund, N., Møller, A., Schwartzbach, M.I.: MONA Implementation Secrets. In: Yu, S., Păun, A. (eds.) CIAA 2000. LNCS, vol. 2088, pp. 182–194. Springer, Heidelberg (2001)
7. Kuncak, V., Mayer, M., Piskac, R., Suter, P.: Complete functional synthesis. In: ACM SIGPLAN Conf. Programming Language Design and Implementation, PLDI (2010)
8. Rabin, M.: Automata on infinite objects and Church's problem. Regional Conference Series in Mathematics, vol. 13. American Mathematical Society (1972)
9. Schuele, T., Schneider, K.: Verification of Data Paths Using Unbounded Integers: Automata Strike Back. In: Bin, E., Ziv, A., Ur, S. (eds.) HVC 2006. LNCS, vol. 4383, pp. 65–80. Springer, Heidelberg (2007)
10. Spielmann, A., Kuncak, V.: On synthesis for unbounded bit-vector arithmetic. Technical Report EPFL-REPORT-174801, EPFL (2012)
11. Srivastava, S., Gulwani, S., Foster, J.S.: From program verification to program synthesis. In: POPL (2010)
12. Stockmeyer, L., Meyer, A.R.: Cosmological lower bound on the circuit complexity of a small problem in logic. J. ACM 49(6), 753–784 (2002)

A Schema of Circuit C'

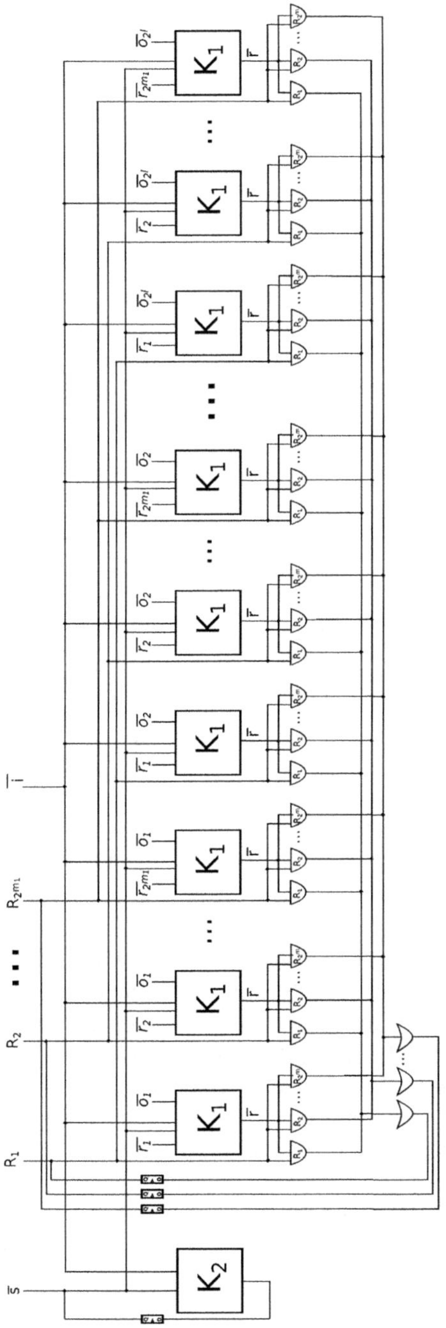

Extended Caching, Backjumping and Merging for Expressive Description Logics

Andreas Steigmiller[1], Thorsten Liebig[2], and Birte Glimm[1]

[1] Ulm University, Ulm, Germany
firstname.lastname@uni-ulm.de
[2] derivo GmbH, Ulm, Germany
liebig@derivo.de

Abstract. With this contribution we push the boundary of some known optimisations such as caching to the very expressive Description Logic $\mathcal{SROIQ}$. The developed method is based on a sophisticated dependency management and a precise unsatisfiability caching technique, which further enables better informed tableau backtracking and more efficient pruning. Additionally, we optimise the handling of cardinality restrictions, by introducing a strategy called pool-based merging.

We empirically evaluate the proposed optimisations within the novel reasoning system Konclude and show that the proposed optimisations indeed result in significant performance improvements.

1 Motivation

Tableau algorithms are dominantly used in sound and complete reasoning systems, which are able to deal with ontologies specified in the OWL 2 DL ontology language [16]. Such algorithms are usually specified in terms of Description Logics (DLs) [1], which provide the formal basis for OWL, e.g., OWL 2 is based on the DL $\mathcal{SROIQ}$ [11].

To our knowledge, all competitive systems for reasoning with $\mathcal{SROIQ}$ knowledge bases such as FaCT++ [19], HermiT,[1] jFact,[2] or Pellet [17] use a variant of the tableau method – a refutation-based calculus that systematically tries to construct an abstraction of a model for a given query by exhaustive application of so called tableau rules.

Due to the wide range of modelling constructs supported by expressive DLs, the typically used tableau algorithms have a very high worst-case complexity. Developing optimisations to nevertheless allow for highly efficient implementations is, therefore, a long-standing research area in DLs (see, e.g., [13,20]). A very effective and widely implemented optimisation technique is "caching", where one caches, for a set of concepts, whether they are known to be, or can safely be assumed to be, satisfiable or unsatisfiable [4]. If the set of concepts appears again in a model abstraction, then a cache-lookup allows for skipping further applications of tableau rules. Caching even allows for implementing worst-case optimal decision procedures for $\mathcal{ALC}$ [6].

Unfortunately, with increasing expressivity some of the widely used optimisations become unsound. For instance, naively caching the satisfiability status of interim

[1] http://www.hermit-reasoner.com
[2] http://jfact.sourceforge.net/

B. Gramlich, D. Miller, and U. Sattler (Eds.): IJCAR 2012, LNAI 7364, pp. 514–529, 2012.

results easily causes unsoundness in the presence of inverse roles due to their possible interactions with universal restrictions [1, Chapter 9]. On the other hand, for features such as cardinality restrictions there are nearly no optimisations yet. An attempt to use algebraic methods [9,5], i.e., by combining a tableau calculus with a procedure to solve systems of linear (in)equations, performs well, but requires significant changes to the calculus and has not (yet) been extended to very expressive DLs such as $\mathcal{SROIQ}$.

Our contribution in this paper is two-fold. We push the boundary of known optimisations, most notably caching, to the expressive DL $\mathcal{SROIQ}$. The developed method is based on a sophisticated dependency management and a precise unsatisfiability caching technique, which further enables better informed tableau backtracking and more efficient pruning (Section 3). In addition we optimise the handling of cardinality restrictions, by introducing a strategy called *pool-based merging* (Section 4). Our techniques are grounded in the widely implemented tableau calculus for $\mathcal{SROIQ}$ [11], which makes it easy to transfer our results into existing tableau implementations. The presented optimisations are implemented within a novel reasoning system, called Konclude [15]. Our empirical evaluation shows that the proposed optimisations result in significant performance improvements (Section 5).

2 Preliminaries

Model construction calculi, such as tableau, decide the consistency of a knowledge base $\mathcal{K}$ by trying to construct an abstraction of a model for $\mathcal{K}$, a so-called "completion graph". A completion graph G is a tuple $(V, E, \mathcal{L}, \dot{\neq})$, where each node $x \in V$ represents one or more individuals, and is labelled with a set of concepts, $\mathcal{L}(x)$, which the individuals represented by x are instances of; each edge $\langle x, y\rangle$ represents one or more pairs of individuals, and is labelled with a set of roles, $\mathcal{L}(\langle x, y\rangle)$, which the pairs of individuals represented by $\langle x, y\rangle$ are instances of. The relation $\dot{\neq}$ records inequalities, which must hold between nodes, e.g., due to at-least cardinality restrictions.

The algorithm works by initialising the graph with one node for each Abox individual/nominal in the input KB, and using a set of expansion rules to syntactically decompose concepts in node labels. Each such rule application can add new concepts to node labels and/or new nodes and edges to the completion graph, thereby explicating the structure of a model. The rules are repeatedly applied until either the graph is fully expanded (no more rules are applicable), in which case the graph can be used to construct a model that is a *witness* to the consistency of $\mathcal{K}$, or an obvious contradiction (called a *clash*) is discovered (e.g., both C and $\neg C$ in a node label), proving that the completion graph does not correspond to a model. The input knowledge base $\mathcal{K}$ is *consistent* if the rules (some of which are non-deterministic) can be applied such that they build a fully expanded, clash free completion graph. A cycle detection technique called *blocking* ensures the termination of the algorithm.

2.1 Dependency Tracking

Dependency tracking keeps track of all dependencies that cause the existence of concepts in node labels, roles in edge labels as well as accompanying constrains such as inequalities that must hold between nodes. Dependencies are associated with so-called *facts*, defined as follows:

Definition 1 (Fact). *We say that* G *contains a* concept fact $C(x)$ *if* $x \in V$ *and* $C \in \mathcal{L}(x)$, G *contains a* role fact $r(x, y)$ *if* $\langle x, y\rangle \in E$ *and* $r \in \mathcal{L}(\langle x, y\rangle)$, *and* G *contains an* inequality fact $x \dot{\neq} y$ *if* $x, y \in V$ *and* $(x, y) \in \dot{\neq}$. *We denote the set of all (concept, role, or inequality) facts in* G *as* Facts_G.

Dependencies now relate facts in a completion graph to the facts that caused their existence. Additionally, we annotate these relations with a running index, called dependency number, and a branching tag to track non-deterministic expansions:

Definition 2 (Dependency). *Let* d *be a pair in* $\mathsf{Facts}_G \times \mathsf{Facts}_G$. *A* dependency *is of the form* $d^{n,b}$ *with* $n \in \mathbf{N}_0$ *a* dependency number *and* $b \in \mathbf{N}_0$ *a* branching tag.

We inductively define the dependencies for G, *written* Dep_G. *If* G *is an initial completion graph, then* $\mathsf{Dep}_G = \emptyset$. *Let* R *be a tableau rule applicable to a completion graph* G *with* $\{c_0, \ldots, c_k\}$ *a minimal set of facts in* G *that satisfy the preconditions of* R. *If* $\mathsf{Dep}_G = \emptyset$, *then* $n_m = b_m = 0$, *otherwise, let* $n_m = max\{n \mid d^{n,b} \in \mathsf{Dep}_G\}$ *and* $b_m = max\{b \mid d^{n,b} \in \mathsf{Dep}_G\}$. *If* R *is non-deterministic, then* $b_R = 1 + b_m$, *otherwise* $b_R = 0$. *Let* G' *be the completion graph obtained from* G *by applying* R *and let* $c'_0, \ldots, c'_\ell$ *be the newly added facts in* G', *then*

$$\mathsf{Dep}_{G'} = \mathsf{Dep}_G \cup \{(c'_j, c_i)^{n,b} \mid 0 \leq i \leq k, 0 \leq j \leq \ell, n = n_m + 1 + (j * k) + i,\\ b = max\{\{b_R\} \cup \{b' \mid (c_i, c)^{n',b'} \in \mathsf{Dep}_G\}\}\}.$$

The branching tag indicates which facts were added non-deterministically:

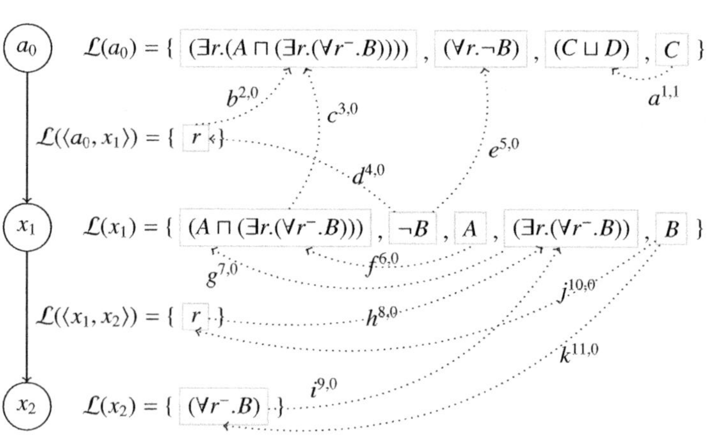

Fig. 1. Tracked dependencies for all facts in the generated completion graph

Definition 3 (Non-deterministic Dependency). *For $d^{n,b} \in Dep_G$ with $d = (c_1, c_2)$, let $D_d = \{(c_2, c_3)^{n',b'} \mid (c_2, c_3)^{n',b'} \in Dep_G\}$. The dependency $d^{n,b}$ is a* non-deterministic dependency *in G if $b > 0$ and either $D_d = \emptyset$ or $max\{b' \mid (c, c')^{n',b'} \in D_d\} < b$.*

Figure 1 illustrates a completion graph obtained in the course of testing the consistency of a knowledge base with three concept assertions:

$$a_0 : (\exists r.(A \sqcap (\exists r.(\forall r^-.B)))) \qquad a_0 : (\forall r.\neg B) \qquad a_0 : (C \sqcup D).$$

Thus, the completion graph is initialised with the node a_0, which has the three concepts in its label. Initially, the set of dependencies is empty. For the concepts and roles added by the application of tableau rules, the dependencies are shown with dotted lines, labelled with the dependency. The dependency number increases with every new dependency. The branching tag is only non-zero for the non-deterministic addition of C to the label of a_0 in order to satisfy the disjunction $(C \sqcup D)$. Note the presence of a clash due to B and $\neg B$ in the label of x_1.

3 Extended Caching and Backtracking

In the following we introduce improvements to caching and backjumping by presenting a more informed dependency directed backtracking strategy that also allows for extracting precise unsatisfiability cache entries.

3.1 Dependency Directed Backtracking

Dependency directed backtracking is an optimisation that can effectively prune irrelevant alternatives of non-deterministic branching decisions. If branching points are not involved in clashes, it will not be necessary to compute any more alternatives of these branching points, because the other alternatives cannot eliminate the cause of the clash. To identify involved non-deterministic branching points, all facts in a completion graph are labelled with information about the branching points they depend on. Thus, the united information of all clashed facts can be used to identify involved branching points. A typical realisation of dependency directed backtracking is backjumping [1,20], where the dependent branching points are collected in the dependency sets for all facts.

3.2 Unsatisfiability Caching

Another widely used technique to increase the performance of a tableau implementation is caching, which comes in two flavours: satisfiability and unsatisfiability caching. For the former, one caches sets of concepts, e.g., from node labels, that are known to be satisfiable. In contrast, for an unsatisfiability cache, we cache sets of concepts that are unsatisfiable. For such a cached set, any *superset* is also unsatisfiable. Thus, one is interested in caching a minimal, unsatisfiable set of concepts. Although the caching of satisfiable and unsatisfiable sets of concepts is often considered together, we focus here on the unsatisfiability caching problem since the two problems are quite different in nature and already the required data structure for an efficient cache retrieval can differ significantly.

Definition 4 (Unsatisfiability Cache). *Let $\mathcal{K}$ be a knowledge base, $\mathsf{Con}_{\mathcal{K}}$ the set of (sub-)concepts that occur in $\mathcal{K}$. An* unsatisfiability cache $\mathsf{UC}_{\mathcal{K}}$ *for $\mathcal{K}$ is a subset of $2^{\mathsf{Con}_{\mathcal{K}}}$ such that each* cache entry $S \in \mathsf{UC}_{\mathcal{K}}$ *is an unsatisfiable set of concepts. An* unsatisfiability retrieval *for $\mathsf{UC}_{\mathcal{K}}$ and a completion graph G for $\mathcal{K}$ takes a set of concepts $S \subseteq \mathsf{Con}_{\mathcal{K}}$ from a node label of G as input. If $\mathsf{UC}_{\mathcal{K}}$ contains a set $S_{\bot} \subseteq S$, then $S_{\bot}$ is returned; otherwise, the empty set is returned.*

Deciding when we can safely create a cache entry rapidly becomes difficult with increasing expressivity of the used DL. Already with blocking on tableau-based systems for the DL $\mathcal{ALC}$ care has to be taken to not generate invalid cache entries [7]. There are some approaches for caching with inverse roles [2,3,6], where possible propagations over inverse roles from descendant nodes are taken into account. The difficulty increases further in the presence of nominals and, to the best of our knowledge, the problem of caching with inverses and nominals has not yet been addressed in the literature. In this setting, it is difficult to determine, for a node x with a clash in its label, which nodes (apart from x) are also labelled with unsatisfiable sets of concepts. Without nominals and inverse roles, we can determine the ancestor y of x with the last non-deterministic expansion and consider the labels of all nodes from x up to y as unsatisfiable. With inverse roles, a non-deterministic rule application on a descendant node of x can be involved in the creation of the clash, whereby the node labels that can be cached as unsatisfiable become limited.

In order to demonstrate the difficulties with inverse roles, let us assume that the example in Figure 1 is extended such that $((\forall r^{-}.B) \sqcup E) \in \mathcal{L}(x_2)$ and that $(\forall r^{-}.B) \in \mathcal{L}(x_2)$ results from the non-deterministic expansion of the disjunction. For the resulting clash in $\mathcal{L}(x_1)$, it is not longer sufficient to consider only non-deterministic expansions on ancestor nodes. The label of x_2 cannot be cached because some facts $(\neg B)$ involved in the clash are located on different nodes (x_1). Furthermore, if trying the disjunct E also leads to a clash, the disjunction $((\forall r^{-}.B) \sqcup E)$ in $\mathcal{L}(x_2)$ is unsatisfiable in the context of *this* completion graph. Nevertheless, a cache entry cannot be generated because (at least) the first disjunct involves facts of an ancestor node. In order to also handle inverse roles, it would, therefore, be necessary to remember all nodes or at least the minimum node depth involved in the clashes of all alternatives. In the presence of nominals, it further becomes necessary to precisely manage the exact causes of clashes, e.g., via tracking the dependencies as presented in Section 2.1. If such a technique is missing, often the only option is to deactivate caching completely [17,20].

Since node labels can have many concepts that are not involved in any clashes, the precise extraction of a small set of concepts that are in this combination unsatisfiable would yield better entries for the unsatisfiability cache. With an appropriate subset retrieval potentially more similar also unsatisfiable node labels can be found within the cache. We call this technique *precise caching*. Although techniques to realise efficient subset retrieval exist [10], unsatisfiability caches based on this idea are only implemented in very few DL reasoners [8]. Furthermore, the often used backjumping only allows the identification of all branching points involved in a clash, but there is no information about how the clash is exactly caused. As a result, only complete node labels can be saved in the unsatisfiability cache. We refer to this often used form of caching combined with only an equality cache retrieval as *label caching*.

For precise caching, the selection of an as small as possible but still unsatisfiable subset of a label as cache entry should be adjusted to the cache retrieval strategy, i.e., the strategy of when the cache is queried in the tableau algorithm. Going back to the example in Figure 1, for the node x_1 the set $\{\neg B, (\exists r.(\forall r^-.B))\}$ could be inserted into the cache as well as $\{\neg B, (A \sqcap (\exists r.(\forall r^-.B)))\}$. The number of cache entries should, however, be kept small, because the performance of the retrieval decreases with an increasing number of entries. Thus, the insertion of concepts for which the rule application is cheap (e.g., concept conjunction) should be avoided. Concepts that require the application of non-deterministic or generating rules are more suitable, because the extra effort of querying the unsatisfiability cache before the rule application can be worth the effort. Optimising cache retrievals for incremental changes further helps to efficiently handle multiple retrievals for the same node with identical or slightly extended concept labels.

The creation of new unsatisfiability cache entries based on dependency tracking can be done during backtracing, which is also coupled with the dependency directed backtracking as described next. Basically all facts involved in a clash are backtraced to collect the facts that cause the clash within one node, whereby then an unsatisfiability cache entry can be created.

3.3 Dependency Backtracing

The dependency tracking defined in Section 2.1 completely retains all necessary information to exactly trace back the cause of the clash. Thus, this *backtracing* is qualified to identify all involved non-deterministic branching points for the dependency directed backtracking and also to identify small unsatisfiable sets of concepts that can be used to create new unsatisfiability cache entries.

Algorithm 1 performs the backtracing of facts and their tracked dependencies in the presence of inverse roles and nominals. If all facts and their dependencies are collected on the same node while backtracing, an unsatisfiability cache entry with these facts can be generated, assuming all facts are concept facts. As long as no nominal or Abox individual occurs in the backtracing, the unsatisfiability cache entries can also be generated while all concept facts have the same node depth. Thus, an important task of the backtracing algorithm is to hold as many facts as possible within the same node depth to allow for the generation of many cache entries. To realise the backtracing, we introduce the following data structure:

Definition 5 (Fact Dependency Node Tuple). *A* fact dependency node tuple for *G is a triple $\langle c, d^{n,b}, x\rangle$ with $c \in \mathsf{Facts}_G, d^{n,b} \in \mathsf{Dep}_G$ and $x \in V$. Abbreviatory we also write $\langle C, d^{n,b}, x\rangle$ if c is the concept fact $C(x)$.*

If a clash is discovered in the completion graph, a set of fact dependency node tuples is generated for the backtracing. Each tuple consists of a fact involved in the clash, an associated dependency and the node where the clash occurred. The algorithm gets this set T of tuples as input and incrementally traces the facts back from the node with the clash to nodes with depth 0 (Abox individuals or root nodes).

In each loop round (line 3) some tuples of T are exchanged with tuples, whose facts are the cause of the exchanged one. To identify which tuple has to be traced back first, the current minimum node depth (line 4) and the maximum branching tag (line 5) are

---

Algorithm 1. Backtracing Algorithm

Require: A set of fact dependency node tuples T obtained from clashes

 1: **procedure** DEPENDENCYBACKTRACING(T)
 2:     $pendingUnsatCaching \leftarrow false$
 3:     **loop**
 4:         $min_D \leftarrow$ MINIMUMNODEDEPTH(T)
 5:         $max_B \leftarrow$ MAXIMUMBRANCHINGTAG(T)
 6:         $A \leftarrow \{t \in T \mid$ NODEDEPTH(t)$> min_D \wedge$ HASDETERMINISTICDEPENDENCY(t)$\}$
 7:         $C \leftarrow \emptyset$
 8:         **if** $A \neq \emptyset$ **then**
 9:             $pendingUnsatCaching \leftarrow true$
10:             **for all** $t \in A$ **do**
11:                 $T \leftarrow (T \setminus t) \cup$ GETCAUSETUPLESBYDEPENDENCY(t)
12:             **end for**
13:         **else**
14:             $B \leftarrow \{t \in T \mid$ NODEDEPTH(t)$> min_D \wedge$ BRANCHINGTAG(t)$= max_B\}$
15:             **if** $B = \emptyset$ **then**
16:                 **if** $pendingUnsatCaching = true$ **then**
17:                     $pendingUnsatCaching \leftarrow$ TRYCREATEUNSATCACHEENTRY(T)
18:                 **end if**
19:                 **if** HASNODEPENDENCY(t) **for all** $t \in T$ **then**
20:                     $pendingUnsatCaching \leftarrow$ TRYCREATEUNSATCACHEENTRY(T)
21:                     **return**
22:                 **end if**
23:                 $C \leftarrow \{t \in T \mid$ BRANCHINGTAG(t)$= max_B\}$
24:             **end if**
25:             $t \leftarrow$ ANYELEMENT($B \cup C$)
26:             **if** HASDETERMINISTICDEPENDENCY(t) **then**
27:                 $T \leftarrow (T \setminus t) \cup$ GETCAUSETUPLESBYDEPENDENCY(t)
28:             **else**
29:                 $b \leftarrow$ GETNONDETERMINISTICBRANCHINGPOINT(t)
30:                 **if** ALLALTERNATIVESOFNONDETBRANCHINGPOINTPROCESSED(b) **then**
31:                     $T \leftarrow T \cup$ LOADTUPLESFROMNONDETBRANCHINGPOINT(b)
32:                     $T \leftarrow (T \setminus t) \cup$ GETCAUSETUPLESBYDEPENDENCY(t)
33:                     $T \leftarrow$ FORCETUPLESBEFOREBRANCHINGPOINT(T, b)
34:                     $pendingUnsatCaching \leftarrow$ TRYCREATEUNSATCACHEENTRY(T)
35:                 **else**
36:                     $T \leftarrow$ FORCETUPLESBEFOREBRANCHINGPOINT(T, b)
37:                     SAVETUPLESTONONDETBRANCHINGPOINT(T, b)
38:                     JUMPBACKTO(max_B)
39:                     **return**
40:                 **end if**
41:             **end if**
42:         **end if**
43:     **end loop**
44: **end procedure**

---

extracted from the tuples of T. All tuples, whose facts are located on a deeper node and whose dependencies are deterministic, are collected in the set A. Such tuples will be directly traced back until their facts reach the current minimum node depth (line 10-12). If there are no more tuples on deeper nodes with deterministic dependencies, i.e., $A = \emptyset$, the remaining tuples from deeper nodes with non-deterministic dependencies and the current branching tag are copied into B (line 14) in the next round. If B is not empty, one of these tuples (line 25) and the corresponding non-deterministic branching point (line 29) are processed. The backtracing is only continued, if all alternatives of the branching point are computed as unsatisfiable. In this case, all tuples, saved from the backtracing of other unsatisfiable alternatives, are added to T (line 31). Moreover, for c the fact in t, t can be replaced with tuples for the fact on which c non-deterministically depends (line 32).

For a possible unsatisfiability cache entry all remaining tuples, which also depend on the non-deterministic branching point, have to be traced back until there are no tuples with facts of some alternatives of this branching point left (line 33). An unsatisfiability cache entry is only generated (line 34), if all facts in T are concept facts for the same node or on the same node depth.

Unprocessed alternatives of a non-deterministic branching point have to be computed before the backtracing can be continued. It is, therefore, ensured that tuples do not consist of facts and dependencies from this alternative, which also allows for releasing memory (line 36). The tuples are saved to the branching point (line 37) and the algorithm jumps back to an unprocessed alternative (line 38).

If B is also empty, but there are still dependencies to previous facts, some tuples based on the current branching tag have to remain on the current minimum node depth. These tuples are collected in the set C (line 23) and are processed separately one per loop round, similar to the tuples of B, because the minimum node depth or maximum branching tag may change. The tuples of C can have deterministic dependencies, which are processed like the tuples of A (line 27). If all tuples have no more dependencies to previous facts, the algorithm terminates (line 21).

Besides the creation of unsatisfiability cache entries after non-deterministic dependencies (line 34), cache entries may also be generated when switching from a deeper node to the current minimum node depth in the backtracing (line 9 and 17) or when the backtracing finishes (line 20). The function that tries to create new unsatisfiability cache entries (line 17, 20, and 34) returns a Boolean flag that indicates whether the attempt has failed, so that the attempt can be repeated later.

For an example, we consider the clash $\{\neg B, B\}$ in the completion graph of Figure 1. The initial set of tuples for the backtracing is T_1 (see Figure 2). Thus, the minimum node depth for T_1 is 1 and the maximum branching tag is 0. Because there are no tuples on a deeper node, the sets A and B are empty for T_1. Since all clashed facts are generated deterministically, the dependencies of the tuples have the current maximum branching tag 0 and are all collected into the set C. The backtracing continues with one tuple t from C, say $t = \langle B, k^{11,0}, x_1 \rangle$. The dependency k of t relates to the fact $(\forall r^{-}.B)(x_2)$, which is a part of the cause and replaces the backtraced tuple t in T_1. The resulting set T_2 is used in the next loop round. The minimum node depth and the maximum branching tag remain unchanged, but the new tuple has a deeper node depth and is traced back with a higher

$$T_1 = \{\langle \neg B, d^{4,0}, x_1\rangle, \langle \neg B, e^{5,0}, x_1\rangle, \langle B, j^{10,0}, x_1\rangle, \langle B, k^{11,0}, x_1\rangle\}$$
$$\downarrow$$
$$T_2 = \{\langle \neg B, d^{4,0}, x_1\rangle, \langle \neg B, e^{5,0}, x_1\rangle, \langle B, j^{10,0}, x_1\rangle, \langle (\forall r^-.B), i^{9,0}, x_2\rangle\}$$
$$\downarrow$$
$$T_3 = \{\langle \neg B, d^{4,0}, x_1\rangle, \langle \neg B, e^{5,0}, x_1\rangle, \langle B, j^{10,0}, x_1\rangle, \langle (\exists r.(\forall r^-.B)), g^{7,0}, x_1\rangle\}$$
$$\downarrow$$
$$T_4 = \{\langle \neg B, d^{4,0}, x_1\rangle, \langle \neg B, e^{5,0}, x_1\rangle, \langle r(x_1, x_2), h^{8,0}, x_1\rangle, \langle (\exists r.(\forall r^-.B)), g^{7,0}, x_1\rangle\}$$
$$\downarrow$$
$$T_5 = \{\langle \neg B, d^{4,0}, x_1\rangle, \langle \neg B, e^{5,0}, x_1\rangle, \langle (\exists r.(\forall r^-.B)), g^{7,0}, x_1\rangle\}$$
$$\downarrow$$
$$T_6 = \{\langle \neg B, d^{4,0}, x_1\rangle, \langle (\forall r.\neg B), -, a_0\rangle, \langle (\exists r.(\forall r^-.B)), g^{7,0}, x_1\rangle\}$$
$$\downarrow$$
$$T_7 = \{\langle r(a_0, x_1), b^{2,0}, x_1\rangle, \langle (\forall r.\neg B), -, a_0\rangle, \langle (A \sqcap (\exists r.(\forall r^-.B))), c^{3,0}, x_1\rangle\}$$
$$\downarrow$$
$$T_8 = \{\langle (\exists r.(A \sqcap (\exists r.(\forall r^-.B)))) -, a_0\rangle, \langle (\forall r.\neg B), -, a_0\rangle\}$$

Fig. 2. Backtracing sequence of tuples as triggered by the clash of Figure 1

priority to enable unsatisfiability caching again. Thus, $\langle (\forall r^-.B), i^{9,0}, x_2\rangle$ is added to the set A and then replaced by its cause, leading to T_3. Additionally, a pending creation of an unsatisfiability cache entry is noted, which is attempted in the third loop round since A and B are empty. The creation of a cache entry is, however, not yet sensible and deferred since T_3 still contains an atomic clash. Let $t = \langle B, j^{10,0}, x_1\rangle \in C$ be the tuple from T_3 that is traced back next. In the fourth round, the creation of a cache entry is attempted again, but fails because not all facts are concepts facts. The backtracing of $\langle r(x_1, x_2), h^{8,0}, x_1\rangle$ then leads to T_5. In the following round an unsatisfiability cache entry is successfully created for the set $\{\neg B, (\exists r.(\forall r^-.B))\}$. Assuming that now the tuple $\langle \neg B, e^{5,0}, x_1\rangle$ is traced back, we obtain T_6, which includes the node a_0. Thus, the minimum node depth changes from 1 to 0. Two more rounds are required until T_8 is reached. Since all remaining facts in T_8 are concept assertions, no further backtracing is possible and an additional cache entry is generated for the set $\{(\exists r.(A \sqcap (\exists r.(\forall r^-.B)))), (\forall r.\neg B)\}$.

If a tuple with a dependency to node a_0 had been traced back first, it would have been possible that the first unsatisfiability cache entry for the set $\{\neg B, (\exists r.(\forall r^-.B))\}$ was not generated. In general, it is not guaranteed that an unsatisfiability cache entry is generated for the node where the clash is discovered if there is no non-deterministic rule application and if the node is not a root node or an Abox individual. Furthermore, if there are facts that are not concept facts, these can be backtraced with higher priority, analogous to the elements of the set A, to make unsatisfiability cache entries possible again. To reduce the repeated backtracing of identical tuples in different rounds, an additional set can be used to store processed tuples for the alternative for which the backtracing is performed.

The backtracing can also be performed over nominal and Abox individual nodes. However, since Abox and absorbed nominal assertions such as $\{a\} \sqsubseteq C$ have no previous dependencies, this can lead to a distributed backtracing stuck on different nodes. In this case, no unsatisfiability cache entries are possible.

A less precise caching can lead to an adverse interaction with dependency directed backtracking. Consider the example of Figure 3, where the satisfiability of the combination of the concepts $(\exists r.(\exists s.(A \sqcap B)))$, $((C_1 \sqcap \forall r.C) \sqcup (C_2 \sqcap \forall r.C))$, and $((D_1 \sqcap \forall r.(\forall s.\neg A)) \sqcup (D_2 \sqcap \forall r.(\forall s.\neg A)))$ is tested. Note that, in order to keep the figure readable, we no longer

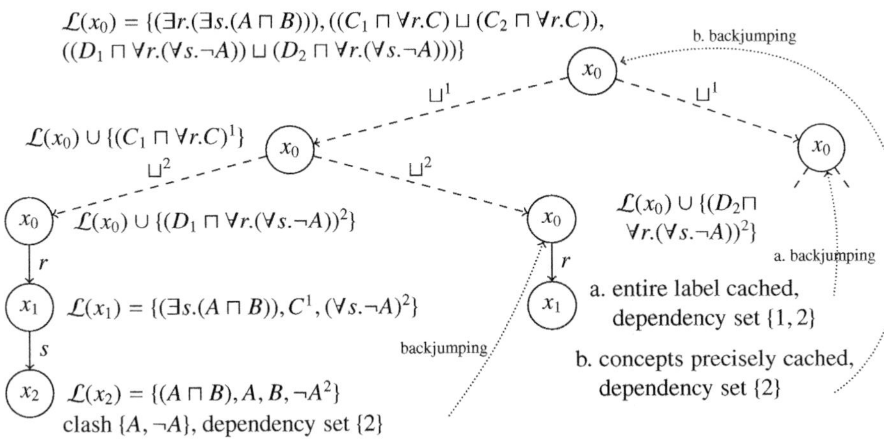

Fig. 3. More pruned alternatives due to dependency directed backtracking and precise caching (b.) in contrast to label caching (a.)

show complete dependencies, but only the branching points for non-deterministic decisions. First, the two disjunctions are processed. Assuming that the alternative with the disjuncts $(C_1 \sqcap \forall r.C)$ and $(D_1 \sqcap \forall r.(\forall s.\neg A))$ is considered first (shown on the left-hand side of Figure 3), an r-successor x_1 with label $\{(\exists s.(A \sqcap B)), C^1, (\forall s.\neg A)^2\}$ is generated. The branching points indicate which concepts depend on which non-deterministic decision. For example, C is in $\mathcal{L}(x_1)$ due to the disjunct $(C_1 \sqcap \forall r.C)$ of the first non-deterministic branching decision (illustrated in Figure 3 with the superscript 1). In the further generated s-successor x_2 a clash is discovered. For the only involved non-deterministic branching point 2, we have to compute the second alternative. Thus, an identical r-successor x_1 is generated again for which we can discover the unsatisfiability with a cache retrieval. If the entire label of x_1 was inserted to the cache, the dependent branching points of all concepts in the newly generated node x_1 would have to be considered for further dependency directed backtracking. Thus, the second alternative of the first branching decision also has to be evaluated (c.f. Figure 3, a.). In contrast, if the caching was more precise and only the combination of the concepts $(\exists s.(A \sqcap B))$ and $(\forall s.\neg A)$ was inserted into the unsatisfiability cache, the cache retrieval for the label of node x_1 would return the inserted subset. Thus, only the dependencies associated to the concepts of the subset could be used for further backjumping, whereby it would not be necessary to evaluate the remaining alternatives (c.f. Figure 3, b.).

4 Optimised Merging

At-most cardinality restrictions require the non-deterministic merging of role neighbours until the cardinality restriction is satisfied. Only for cardinalities of 1, merging is deterministic. The usual merging approach [11], which can still be found in several available reasoner implementations, employs a $\leq$-rule that shrinks the number of role neighbours by one with each rule application. Each such merging step gathers pairs of potentially mergeable neighbouring nodes. For each merging pair a branch is generated in which the merging of the pair is executed. Without optimisations, this approach leads to an inefficient implementation since for merging problems that require more than one merging step, several identical merging combinations have to be evaluated multiple times. Throughout this section, we consider the following example: a node in the completion graph has four r-neighbours w, x, y and z, which have to be merged into two nodes. The naive approach described above leads to eighteen non-deterministic alternatives: in the first of two necessary merging steps there are $\sum_{i=1}^{n-1} i$, i.e., six possible merging pairs. A second merging step is required to reduce the remaining three nodes to two. If the merging rule is applied again without any restrictions, each second merging step generates three more non-deterministic alternatives. However, only seven of these eighteen alternatives overall are really different. For example, the combination wxy, z, where the nodes w, x and y have been merged, can be generated by $merge(merge(w, x), y)$, $merge(merge(w, y), x)$ and $merge(merge(x, y), w)$.

The problem is very similar to the syntactic branching search [1], where unsatisfiable concepts of non-disjoint branches might have to be evaluated multiple times. The semantic branching technique is commonly used to avoid such redundant evaluations and in the merging context an analogous approach can be very beneficial.

In order to apply this technique, all nodes of previously tested merging pairs are set to be pairwise distinct. For example, when merging (w, x) in the first merging step leads to a clash, w and x are set to be distinct because this combination has been tested and should be avoided in future tests. In the second alternative, the nodes w and y are merged, which leads to $wy \dot{\neq} x$. As a result of the inequality, $merge(merge(w, y), x)$ is never tested in the second merging step (Figure 4). If also merging w and y fails, a further inequality $w \dot{\neq} y$ is added. Finally, for the last two alternatives of the first merging step the inequality constraints prevent further merging and show that these alternatives are unsatisfiable. Summing up, with the inequalities the total number of non-deterministic alternatives can be reduced to nine in this example. Unfortunately, similarly sophisticated merging techniques can hardly be found in current reasoners.

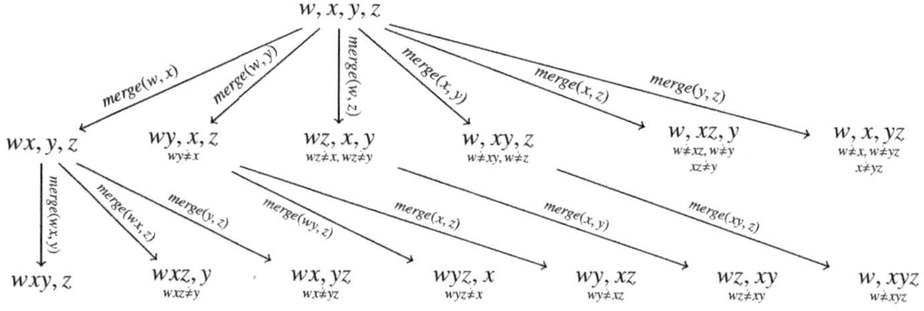

Fig. 4. Non-deterministic merging alternatives with added inequality information

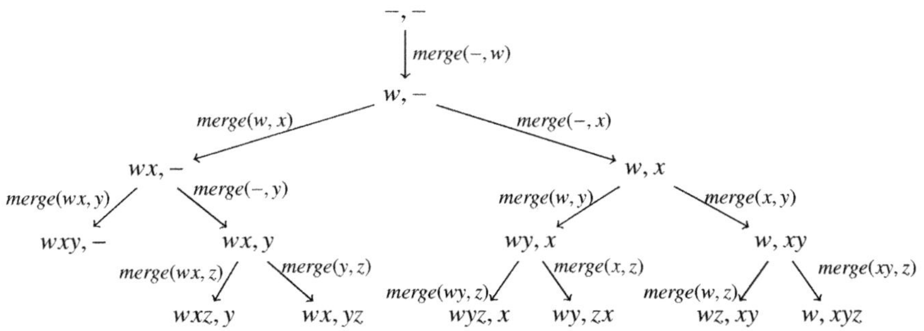

Fig. 5. Pool-based merging approach to avoid redundant evaluation of previous merging attempts

Apart from using the inequality information, the *pool-based merging* method that we propose also prevents the redundant evaluation of previously computed merging attempts. Furthermore it works very well in combination with dependency directed backtracking due to the thin and uniform branching tree.

Regarding the implementation of the pool-based merging method, the nodes that have to be merged are managed in a queue. Each merging step takes the next node from the queue and non-deterministically inserts this node into a so-called *pool*, where the number of pools corresponds to the required cardinality. All pools are considered as distinct and nodes within one pool are merged together. If there are several empty pools, we will only generate one alternative, where the node is inserted in one of these empty pools. If several empty pools were initialised with the same node, once again redundant merging combinations would have to be evaluated. For the example, the generated merging combinations due to the pool based merging procedure are illustrated in Figure 5. At the beginning, all nodes are in the queue and both pools are empty. In the first merging step the node w is taken from the queue and inserted to the first empty pool. In the second step the next node x is non-deterministically inserted into the first pool together with the node w or into another empty pool. This process continues until the cardinality restriction is satisfied. Note that z is not removed from the queue for the alternative shown on the left-hand side since the cardinality is already satisfied. If a clash occurs in an alternative, all relevant merging steps can be identified with the dependency directed backtracking. Different insertion alternatives are, therefore, only tested for nodes that are involved in the clashes. In the worst-case also the pool based merging is systematically testing all possible combinations, but the different generation of these alternatives prevents redundant evaluations. Other tableau expansions rules for $\mathcal{SROIQ}$, such as the *choose*- or the *NN*-rule, are not influenced by the merging method, consequently also qualified cardinality restrictions are supported in combination with the pool based merging.

5 Evaluation

Our Konclude reasoning system implements the enhanced optimisation techniques for $\mathcal{SROIQ}$ described above. In the following, we first compare different caching methods.

Furthermore, we benchmark our pool-based merging technique against the standard pair-based approach that is used in most other systems. A comparison of Konclude with other reasoners can be found in the accompanying technical report [18].

We evaluate dependency directed backtracking and unsatisfiability caching with the help of concept satisfiability tests from the well-known DL 98 benchmark suite [12] and spot tests regarding cardinality restrictions and merging first proposed in [14]. From the DL 98 suite we selected satisfiable and unsatisfiable test cases (with _n resp. _p postfixes) and omitted those for which unsatisfiability caching is irrelevant and tests that were too easy to serve as meaningful and reproducible sample.

With respect to caching, we distinguish between precise caching and label caching as described in Section 3.2. To recall, precise caching stores precise cache entries consisting of only those backtraced sets of concepts that are explicitly known to cause an unsatisfiability in combination with subset retrieval, while label caching stores and returns only entire node labels.

Independent of the caching method, we distinguish between unfiltered and relevant dependencies for further dependency backtracing after a cache hit. *Unfiltered dependency* denotes the backtracing technique that uses all the concept facts and their dependencies within a node label, for which the unsatisfiability has been found in the cache. In contrast, *relevant dependency* uses only those facts and dependencies of a node label for further backtracing that caused the unsatisfiability (as if the unsatisfiability would be found without caching).

Konclude natively maintains relevant dependencies and implements precise unsatisfiability caching based on hash data structures [10] in order to efficiently facilitate subset cache retrieval. Figure 6 shows the total number of processed non-deterministic alternatives for five configurations of caching precision and dependency handling for a selection of test cases solvable within one minute. Note that runtime is not a reasonable basis of comparison since all configuration variants of Figure 6 have been implemented (just for the purpose of evaluation) on top of the built-in and computationally more costly precise caching approach. System profiling information, however, strongly indicate that building and querying the precise unsatisfiability cache within Konclude is

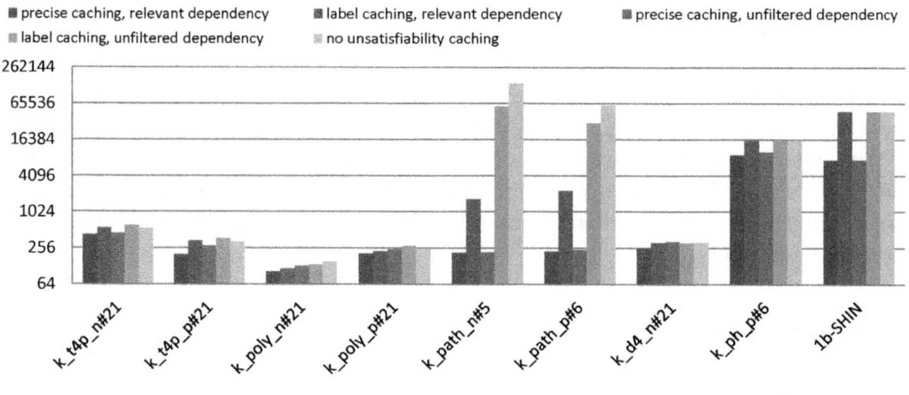

Fig. 6. Log scale comparison of processed alternatives for different caching methods

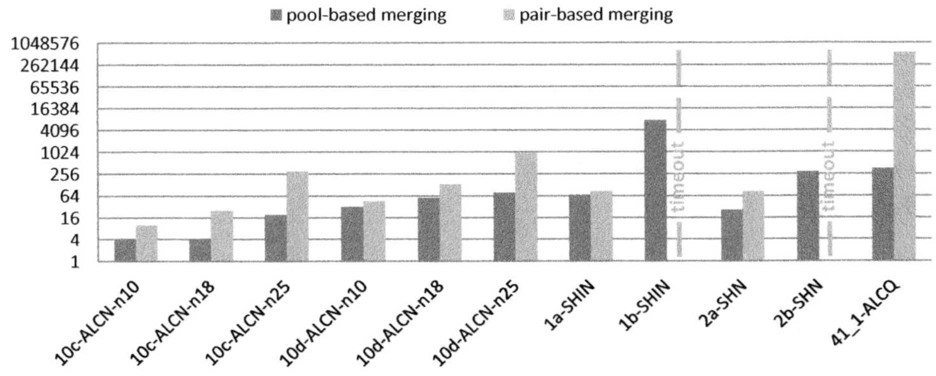

Fig. 7. Processed alternatives (on a logarithmic scale) for different merging methods

negligible in terms of execution time compared to the saved processing time for disregarded alternatives. However, we have experienced an increase of memory usage by a worst-case factor of two in case of dependency tracking in comparison to no dependency handling.

Figure 6 reveals that, amongst the tested configurations, precise caching provides the most effective pruning method. For some test cases it can reduce the number of nondeterministic alternatives by two orders of magnitude in comparison to label caching with unfiltered dependencies. Particularly the test cases k_path_n/p are practically solvable for Konclude only with precise caching at their largest available problem size (#21). The difference between relevant and unfiltered dependencies is less significant at least within our set of test cases.

Figure 7 compares pool-based with pair-based merging in terms of non-deterministic alternatives that have to be processed in order to solve selected test cases from [14]. In addition to the built-in pool-based merging we also added pair-based merging to our Konclude system. The test cases 10c and 10d are variants of the original test case 10a in terms of different problem sizes (10c) as well as more hidden contradictions nested within disjunctions (10d). The pool-based approach introduced in Sec. 4 clearly dominates the naive pair-based merging, especially when dealing with satisfiable problems (1b and 2b) and expressive DLs. Note that the test cases 1b and 2b are only solvable with pool-based merging within a one minute timeout. The required reasoning times strongly correlate to the number of processed alternatives for all test cases of Figure 7.

6 Conclusions

We have presented a range of optimisation techniques that can be used in conjunction with the very expressive DL $\mathcal{SROIQ}$. The presented dependency management allows for more informed backjumping, while also supporting the creation of precise cache unsatisfiability entries. In particular the precise caching approach can reduce the number of tested non-deterministic branches by up to two orders of magnitude compared to standard caching techniques. Regarding cardinality constraints, the presented pool-based merging technique also achieves a significant improvement and a number of test cases can only be solved with this optimisation within an acceptable time limit. Both techniques are well-suited for the integration into existing tableau implementations for $\mathcal{SROIQ}$ and play well with other commonly implemented optimisation techniques.

References

1. Baader, F., Calvanese, D., McGuinness, D., Nardi, D., Patel-Schneider, P. (eds.): The Description Logic Handbook: Theory, Implementation, and Applications, 2nd edn. Cambridge University Press (2007)
2. Ding, Y., Haarslev, V.: Tableau caching for description logics with inverse and transitive roles. In: Proc. 2006 Int. Workshop on Description Logics, pp. 143–149 (2006)
3. Ding, Y., Haarslev, V.: A procedure for description logic $\mathcal{ALCFI}$. In: Proc. 16th European Conf. on Automated Reasoning with Analytic Tableaux and Related Methods, TABLEAUX 2007 (2007)
4. Donini, F.M., Massacci, F.: EXPTIME tableaux for $\mathcal{ALC}$. J. of Artificial Intelligence 124(1), 87–138 (2000)
5. Faddoul, J., Farsinia, N., Haarslev, V., Möller, R.: A hybrid tableau algorithm for $\mathcal{ALCQ}$. In: Proc 18th European Conf. on Artificial Intelligence (ECAI 2008), pp. 725–726 (2008)
6. Goré, R., Widmann, F.: Sound Global State Caching for *ALC* with Inverse Roles. In: Giese, M., Waaler, A. (eds.) TABLEAUX 2009. LNCS, vol. 5607, pp. 205–219. Springer, Heidelberg (2009)
7. Haarslev, V., Möller, R.: Consistency Testing: The RACE Experience. In: Dyckhoff, R. (ed.) TABLEAUX 2000. LNCS (LNAI), vol. 1847, pp. 57–61. Springer, Heidelberg (2000)
8. Haarslev, V., Möller, R.: High performance reasoning with very large knowledge bases: A practical case study. In: Proc. 17th Int. Joint Conf. on Artificial Intelligence (IJCAI 2001), pp. 161–168. Morgan Kaufmann (2001)
9. Haarslev, V., Sebastiani, R., Vescovi, M.: Automated Reasoning in $\mathcal{ALCQ}$ via SMT. In: Bjørner, N., Sofronie-Stokkermans, V. (eds.) CADE 2011. LNCS, vol. 6803, pp. 283–298. Springer, Heidelberg (2011)
10. Hoffmann, J., Koehler, J.: A new method to index and query sets. In: Proc. 16th Int. Conf. on Artificial Intelligence (IJCAI 1999), pp. 462–467. Morgan Kaufmann (1999)
11. Horrocks, I., Kutz, O., Sattler, U.: The even more irresistible $\mathcal{SROIQ}$. In: Proc.10th Int. Conf. on Principles of Knowledge Representation and Reasoning (KR 2006), pp. 57–67. AAAI Press (2006)
12. Horrocks, I., Patel-Schneider, P.F.: DL systems comparison. In: Proc. 1998 Int. Workshop on Description Logics (DL 1998), vol. 11, pp. 55–57 (1998)
13. Horrocks, I., Patel-Schneider, P.F.: Optimizing description logic subsumption. J. of Logic and Computation 9(3), 267–293 (1999)
14. Liebig, T.: Reasoning with OWL – system support and insights –. Tech. Rep. TR-2006-04, Ulm University, Ulm, Germany (September 2006)
15. Liebig, T., Steigmiller, A., Noppens, O.: Scalability via parallelization of OWL reasoning. In: Proc. Workshop on New Forms of Reasoning for the Semantic Web: Scalable & Dynamic (NeFoRS 2010) (2010)
16. OWL Working Group, W.: OWL 2 Web Ontology Language: Document Overview. W3C Recommendation (October 27, 2009), `http://www.w3.org/TR/owl2-overview/`
17. Sirin, E., Parsia, B., Grau, B.C., Kalyanpur, A., Katz, Y.: Pellet: A practical OWL-DL reasoner. J. of Web Semantics 5(2), 51–53 (2007)

18. Steigmiller, A., Liebig, T., Glimm, B.: Extended caching, backjumping and merging for expressive description logics. Tech. Rep. TR-2012-01, Ulm University, Ulm, Germany (2012), http://www.uni-ulm.de/fileadmin/website_uni_ulm/iui/Ulmer_Informatik_Berichte/2012/UIB-2012-01.pdf
19. Tsarkov, D., Horrocks, I.: FaCT++ Description Logic Reasoner: System Description. In: Furbach, U., Shankar, N. (eds.) IJCAR 2006. LNCS (LNAI), vol. 4130, pp. 292–297. Springer, Heidelberg (2006)
20. Tsarkov, D., Horrocks, I., Patel-Schneider, P.F.: Optimizing terminological reasoning for expressive description logics. J. of Automated Reasoning 39, 277–316 (2007)

KBCV – Knuth-Bendix Completion Visualizer*

Thomas Sternagel[1,2] and Harald Zankl[2]

[1] Master Program in Computer Science
[2] Institute of Computer Science,
University of Innsbruck, Austria

Abstract. This paper describes a tool for Knuth-Bendix completion. In its interactive mode the user only has to select the orientation of equations into rewrite rules; all other computations (including necessary termination checks) are performed internally. Apart from the interactive mode, the tool also provides a fully automatic mode. Moreover, the generation of (dis)proofs in equational logic is supported. Finally, the tool outputs proofs in a certifiable format.

Keywords: term rewriting, completion, equational logic, automation.

1 Introduction

The *Knuth-Bendix Completion Visualizer* (KBCV) is an interactive/automatic tool for Knuth-Bendix completion and equational logic proofs. This paper describes KBCV version 1.7, which features a command-line and a graphical user interface as well as a Java-applet version. The tool is available under the *GNU Lesser General Public License 3* at

`http://cl-informatik.uibk.ac.at/software/kbcv`

Completion is a procedure which takes as input a finite set of equations E (and nowadays optionally a reduction order $>$) and attempts to construct a terminating and confluent term rewrite system (TRS) R which is equivalent to E, i.e., their equational theories coincide. In case the completion procedure succeeds, R represents a decision procedure for the word problem of E. Now two terms are equivalent with respect to E if and only if they reduce to the same normal form with respect to R.

The computation is done by generating a finite sequence of intermediate TRSs which constitute approximations of the equational theory of E. Following Bachmair and Dershowitz [2] the completion procedure can be modeled as an inference system like system $\mathcal{C}$ in Figure 1. The inference rules work on pairs (E, R) where E is a finite set of equations and R is a finite set of rewrite rules. The goal is to transform an initial pair $(E, \varnothing)$ into a pair $(\varnothing, R)$ such that R is terminating, confluent and equivalent to E. In our setting a completion procedure based on these rules may succeed (find R after finitely many steps), loop, or fail. In

* This research is supported by FWF P22467 and a grant of the Hypo Tirol Bank.

B. Gramlich, D. Miller, and U. Sattler (Eds.): IJCAR 2012, LNAI 7364, pp. 530–536, 2012.

$$\text{DEDUCE}\ \frac{(E, R)}{(E \cup \{s \approx t\}, R)}\ \text{if}\ s\ {}_R\!\leftarrow u \rightarrow_R t \qquad \text{ORIENT}\ \frac{(E \cup \{s \stackrel{.}{\approx} t\}, R)}{(E, R \cup \{s \rightarrow t\})}\ \text{if}\ s > t$$

$$\text{COMPOSE}\ \frac{(E, R \cup \{s \rightarrow t\})}{(E, R \cup \{s \rightarrow u\})}\ \text{if}\ t \rightarrow_R u \qquad \text{DELETE}\ \frac{(E \cup \{s \approx s\}, R)}{(E, R)}$$

$$\text{COLLAPSE}\ \frac{(E, R \cup \{s \rightarrow t\})}{(E \cup \{u = t\}, R)}\ \text{if}\ s \stackrel{\sqsupset}{\rightarrow}_R u \qquad \text{SIMPLIFY}\ \frac{(E \cup \{s \stackrel{.}{\approx} t\}, R)}{(E \cup \{u \stackrel{.}{\approx} t\}, R)}\ \text{if}\ s \rightarrow_R u$$

Fig. 1. Inference rules for completion with a fixed reduction order ($\mathcal{C}$)

Figure 1 a reduction order $>$ is provided as part of the input. We use $s \stackrel{\sqsupset}{\rightarrow}_R u$ to express that s is reduced by a rule $\ell \rightarrow r \in R$ such that ℓ cannot be reduced by another rule $s \rightarrow t \in R$. The notation $s \stackrel{.}{\approx} t$ denotes either of $s \approx t$ and $t \approx s$.

Writing $(E, R) \vdash_{\mathcal{C}} (E', R')$ to indicate that (E', R') is obtained from (E, R) by one of the inference rules of system $\mathcal{C}$ we define a *completion procedure*:

Definition 1. *A* completion procedure *is a program that accepts as input a finite set of equations E_0 (together with a reduction order $>$) and uses the inference rules of Figure 1 to construct a sequence*

$$(E_0, \varnothing) \vdash_{\mathcal{C}} (E_1, R_1) \vdash_{\mathcal{C}} (E_2, R_2) \vdash_{\mathcal{C}} (E_3, R_3) \vdash_{\mathcal{C}} \cdots$$

Such a sequence is called a run *of the completion procedure on input E_0 and $>$. A finite run $(E_0, \varnothing) \vdash^n_{\mathcal{C}} (\varnothing, R_n)$ is* successful *if R_n is locally confluent.*

The following result follows from [1, Theorem 7.2.8] specialized to finite runs.

Lemma 2. *Let $(E_0, \varnothing) \vdash^n_{\mathcal{C}} (\varnothing, R_n)$ be a successful run of completion. Then R_n is terminating, confluent, and equivalent to E_0.* □

In the sequel we assume familiarity with term rewriting, equational logic, and completion [1]. The remainder of this paper is organized as follows. In the next section the main features of KBCV are presented before Section 3 addresses implementation issues and experimental results. Section 4 concludes.

2 Features

KBCV offers two modes for completion, namely the Normal Mode (Section 2.1) and the Expert Mode (Section 2.2). In the GUI the user can change the mode via the menu entry *View* at any time. Irregardless of the chosen mode, termination checks are performed automatically, following the recent approach from [11]. By default, an incremental LPO is constructed and maintained by the tool but also external termination tools are supported (this option is not available in the applet version). For convenience KBCV stores a history that allows to step backwards (and forwards again) in interactive completion proofs. Apart from completion proofs, the tool can generate proofs in equational logic (Section 2.3) and produces output in a certifiable format.

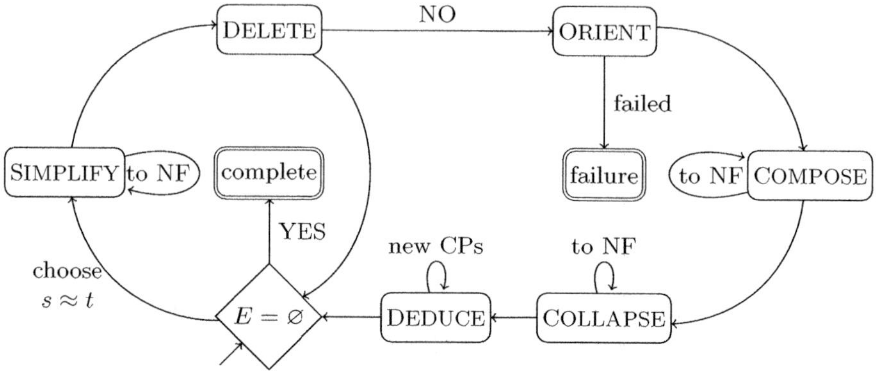

Fig. 2. Flow chart for the efficient completion procedure

2.1 Normal Mode

In *normal mode* the user can switch between *efficient* and *simple* completion. The efficient procedure executes all inference rules from Figure 1 in a fixed order, while the simple procedure considers a subset only.

Efficient Completion. The *efficient completion procedure* (following Huet [4], see Figure 2) takes a set of equations E as input and has three possible outcomes: It may terminate successfully, it may loop indefinitely, or it may fail because an equation could not be oriented into a rewrite rule.

While $E \neq \varnothing$ the user chooses an equation $s \approx t$ from E. The terms in this equation are simplified to normal form by using SIMPLIFY exhaustively. In the next step the equation is deleted if it was trivial and if so the next iteration of the loop starts. Otherwise (following the transition labeled NO) the user suggests the orientation of the equation into a rule and ORIENT performs the necessary termination check. Here the procedure might fail if the equation cannot be oriented (in either direction) with the used termination technique. But if the orientation succeeds the inferred rule is used to reduce the right-hand sides of (other) rules to normal form (COMPOSE) while COLLAPSE rewrites the left-hand sides of rules, which transforms rules into equations that go back to E. In this way the set of rules in R is kept as small as possible at all times. Afterwards DEDUCE is used to compute (all) critical pairs (between the new rule and the old rules and between the new rule and itself). If still $E \neq \varnothing$ the next iteration of the loop begins and otherwise the procedure terminates successfully yielding the terminating and confluent (complete) TRS R equivalent to the input system E.

Simple Completion. The simple procedure (following the *basic* completion procedure [1, Figure 7.1]) makes no use of COMPOSE and COLLAPSE, which means that the inference rule DEDUCE immediately follows ORIENT. Hence although correct, this procedure is not particularly efficient.

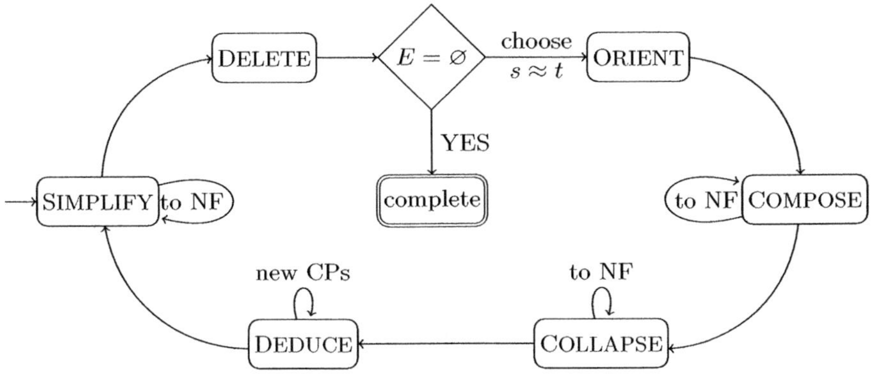

Fig. 3. Flow chart for the automatic mode

2.2 Expert Mode

Inference System. In the *expert mode* the user can select the equations and rewrite rules on which the desired inference rules from Figure 1 should be applied on. If no equations/rules are selected explicitly then all equations/rules are considered. For efficiency reasons DEDUCE does only add critical pairs emerging from overlaps that have not yet been considered. KBCV notifies the user if a complete R equivalent to the input E is obtained.

Automatic Mode. At any stage of the process the user can press the button *Completion* which triggers the automatic mode of KBCV where it applies the inference rules according to the loop in Figure 3. Pressing the button again (during the completion attempt) stops the automatic mode and shows the current state (of the selected thread, see below). It is also possible to specify an upper limit on the loops performed in Figure 3 (*Settings* → *Automatic Completion*). This is especially useful to step through a completion proof with limit 1.

In Figure 3 the rules SIMPLIFY and DELETE operate on all equations and are applied exhaustively. If $E = \varnothing$ then R is locally confluent (since the previous DEDUCE considered all remaining critical pairs) and the procedure successfully terminates. Note that in contrast to the completion procedure from Figure 2 the automatic mode postpones the choice of the equation $s \approx t$. Hence KBCV can choose an equation of minimal length *after* simplification (which is typically beneficial for the course of completion) for the rule ORIENT. To maximize power, KBCV executes two threads in parallel which have different behavior for ORIENT. The first thread prefers to orient equations from left-to-right and if this is not possible it tries a right-to-left orientation (the second thread behaves dually). If this also fails another equation is selected in the next turn. (Note that it is possible that some later equation can be oriented which then simplifies the problematic equation such that it can be oriented or deleted.) A thread fails if no equation in E can be oriented in the ORIENT step.

2.3 Equational Logic and Certification

Since KBCV stores how rules have been deduced from equations [9], in command-line mode the command `showh` lists how rules/equations have been derived and allows to trace back the completion steps that gave rise to a rule/equation. The same mechanism facilitates KBCV to automatically transform a join $s \to_R^* \cdot {}_R^*\!\leftarrow t$ with respect to the current system R (which need not be complete yet) into a conversion with respect to the input system E, i.e., $s \leftrightarrow_E^* t$, and further into equational proofs with respect to E (*File* → *Equational Proof*).

If E could be completed into a TRS R, the recent work in [9] allows KBCV to export proof certificates (*File* → *Export Equational Proof* and *File* → *Export Completion Proof*) in CPF, a certification proof format for rewriting.[1] These proof certificates can be certified by CeTA [10], i.e., checked by a trustable program generated from the theorem prover Isabelle. Apart from the input system E and the completed TRS R such a certificate must also contain a proof that E and R are equivalent, e.g., by giving an explicit conversion $\ell \leftrightarrow_E^* r$ for each $\ell \to r \in R$.

3 Implementation and Experiments

KBCV is implemented in Scala,[2] an object-functional programming language which compiles to Java Byte Code. For this reason KBCV is portable and runs on Windows and Linux machines. We have developed a term library in Scala (`scala-termlib`, available from KBCV's homepage) of approximately 1700 lines of code. KBCV builds upon this library and has an additional 4500 lines of code.

Besides the stand-alone version of KBCV there also is a Java-Applet version available online. The stand-alone version has three different modes: The text mode where one can interact with KBCV via the console, the graphic mode using a graphical user interface implemented in `java.swing`, and the hybrid mode where the text mode and the graphic mode are combined. In text mode typing `help` yields a list of all available commands, whereas in graphic (hybrid) mode or the Java-Applet you can select *Help* → *User Manual* to get a description of the user interface.

The stand-alone version of KBCV is able to call third party termination checkers whereas the Java-Applet version is limited to the internal LPO for termination proofs.

As input KBCV supports the XML-format for TRSs[3] and also a subset of the older TRS-format.[4] (Only one `VAR` and one `RULES` section are allowed in this order. No theory or strategy annotations are supported.) In both cases rules are interpreted as equations.

In addition KBCV supports another file format for the export and import of command logs to save and load user specific settings of KBCV. This format lists all executed commands within KBCV in a human readable form, like:

---

1 `http://cl-informatik.uibk.ac.at/software/cpf`
2 `http://www.scala-lang.org/`
3 `http://www.termination-portal.org/wiki/XTC_Format_Specification`
4 `http://www.lri.fr/~marche/tpdb/format.html`

Table 1. Experimental results on 115 systems

	LPO			termination tool		
	KBCV	MKBTT	MAXCOMP	KBCV	MKBTT	Slothrop
completed	85	70	86	86	81	71
LS94_P1	✓			✓		
SK90_3.26	✓			✓		
Slothrop_cge					✓	
Slothrop_equiv_proof_or					✓	
WS06_proofreduction					✓	

```
load ../examples/gene.trs
orient > 1
simplify
...
```

Saving the current command log is done via (*File* → *Export Command Log*) and loading works alike (*File* → *Load Command Log*). Command logs saved in the file `.kbcvinit` are loaded automatically on program startup.

Although the major attraction of KBCV clearly is its interactive mode, in the sequel experimental results demonstrate that its automatic mode can compete with state-of-the-art completion tools. To this end we extend [5, Table 1] with data for KBCV (considering 115 problems from the distribution of MKBTT).[5] Hence Table 1[6] compares KBCV with MKBTT [8], MAXCOMP [5], and Slothrop [11]. Within a time limit of 300 seconds, KBCV completes 85 systems using its internal LPO and succeeds on an additional system when calling the external termination tool $\mathsf{T_TT_2}$ [6]. Slothrop [11] was the first tool to construct reduction orders on the fly using external termination tools and obtains 71 completed systems. MKBTT [8] adopts this approach, but additionally features *multi-completion*, i.e., considering multiple reduction orderings at the same time. Finally, the strategy of MAXCOMP [5] is to handle all suitable candidate TRSs (terminating and maximal) at once. MAXCOMP can complete 86 systems with LPO but since the search for maximal TRSs is coupled with the search for the reduction order this approach does not support external termination tools. All tools together can complete 95 systems. The lower part of Table 1 shows those systems which only one tool could complete within the given time limit. Here KBCV completed two systems where all other tools failed.

All 86 completion proofs found by KBCV (Table 1) could be certified by CeTA [10] (see Section 2.3). Since recently, MKBTT can also provide proof certificates but currently neither MAXCOMP nor Slothrop support them.

[5] `http://cl-informatik.uibk.ac.at/software/mkbtt`

[6] KBCV data available from `http://cl-informatik.uibk.ac.at/software/kbcv/experiments/12ijcar`

4 Conclusion

In this paper we have presented KBCV, a tool that supports interactive completion proofs. Hence it is of particular interest for students and users that are exposed to the area of completion for the first time or want to follow a completion proof step by step. Its automatic mode can compete with modern completion tools (Slothrop, MKBTT, MAXCOMP) that use more advanced techniques for completion (completion with external termination tools, multi-completion, maximal-completion) but lack an interactive mode. Since KBCV records how rules have been derived, it can produce certifiable output of completion proofs and can construct (dis)proofs in equational logic.

Unfailing completion [3] is a variant of Knuth-Bendix completion, which sacrifices confluence for ground confluence. One possible direction for future work would be to integrate unfailing completion into KBCV. Another issue is to gain further efficiency by a smart design of the employed data structure [7].

Acknowledgments. We thank Christian Sternagel and the reviewers for helpful comments and suggestions concerning efficiency.

References

1. Baader, F., Nipkow, T.: Term Rewriting and *All That*. Cambridge University Press, New York (1999)
2. Bachmair, L., Dershowitz, N.: Equational inference, canonical proofs, and proof orderings. Journal of the ACM 41(2), 236–276 (1994)
3. Bachmair, L., Dershowitz, N., Plaisted, D.: Completion without failure. In: Resolution of Equations in Algebraic Structures. Rewriting Techniques, vol. 2, pp. 1–30 (1989)
4. Huet, G.P.: A complete proof of correctness of the Knuth-Bendix completion algorithm. J. Comput. Syst. Sci. 23(1), 11–21 (1981)
5. Klein, D., Hirokawa, N.: Maximal completion. In: Schmidt-Schauß, M. (ed.) RTA 2011. LIPIcs, vol. 10, pp. 71–80. Schloss Dagstuhl, Dagstuhl (2011)
6. Korp, M., Sternagel, C., Zankl, H., Middeldorp, A.: Tyrolean Termination Tool 2. In: Treinen, R. (ed.) RTA 2009. LNCS, vol. 5595, pp. 295–304. Springer, Heidelberg (2009)
7. Lescanne, P.: Completion Procedures as Transition Rules + Control. In: Díaz, J., Orejas, F. (eds.) TAPSOFT 1989. LNCS, vol. 351, pp. 28–41. Springer, Heidelberg (1989)
8. Sato, H., Winkler, S., Kurihara, M., Middeldorp, A.: Multi-completion with Termination Tools (System Description). In: Armando, A., Baumgartner, P., Dowek, G. (eds.) IJCAR 2008. LNCS (LNAI), vol. 5195, pp. 306–312. Springer, Heidelberg (2008)
9. Sternagel, T., Thiemann, R., Zankl, H., Sternagel, C.: Recording completion for finding and certifying proofs in equational logic. In: IWC (2012)
10. Thiemann, R., Sternagel, C.: Certification of Termination Proofs Using CeTA. In: Berghofer, S., Nipkow, T., Urban, C., Wenzel, M. (eds.) TPHOLs 2009. LNCS, vol. 5674, pp. 452–468. Springer, Heidelberg (2009)
11. Wehrman, I., Stump, A., Westbrook, E.: SLOTHROP: Knuth-Bendix Completion with a Modern Termination Checker. In: Pfenning, F. (ed.) RTA 2006. LNCS, vol. 4098, pp. 287–296. Springer, Heidelberg (2006)

A PLTL-Prover Based on Labelled Superposition with Partial Model Guidance

Martin Suda[1,2,3,⋆] and Christoph Weidenbach[1,⋆⋆]

[1] Max-Planck-Institut für Informatik, Saarbrücken, Germany
[2] Saarland University, Saarbrücken, Germany
[3] Charles University, Prague, Czech Republic

Abstract. Labelled superposition (LPSup) is a new calculus for PLTL. One of its distinguishing features, in comparison to other resolution-based approaches, is its ability to construct partial models on the fly. We use this observation to design a new decision procedure for the logic, where the models are effectively used to guide the search. On a representative set of benchmarks, our implementation is then shown to considerably advance the state of the art.

1 Introduction

Labelled superposition (LPSup) is a new calculus for Propositional Linear Temporal Logic (PLTL). In previous work [7] we have shown a saturation based approach to deciding PLTL with LPSup. Here we instead rely on the ability of LPSup to generate partial models on the fly and use a SAT solver to drive the search and select inferences. This typically leads to a fast discovery of models, but also drastically reduces the number of inferences that need to be performed before an instance can be shown unsatisfiable.

Our method doesn't work with PLTL formulas directly, but instead relies on a certain normal form, which we review in Sect. 2. Algorithms for deciding PLTL formulas are inherently complicated, because one needs to show the existence of an infinite path through the world structure. Our algorithm is described in two steps in Sect. 3. First we show how a certain modification of bounded model checking can be turned into a complete method for the reachability tasks. This is then used as a subroutine to decide whole PLTL. Although the ideas underlying our algorithm rely on the theory of [7] that cannot be repeated here in full due to lack of space, we provide the most important ideas to understand our approach. In the final section 4, we compare LS4, an implementation of our algorithm, to other existing PLTL-provers on a representative set of benchmarks. The results clearly indicate that partial model guidance considerably improves the state of the art of symbolic based approaches to PLTL satisfiability checking.

⋆ Supported by Microsoft Research through its PhD Scholarship Programme.
⋆⋆ Supported by the German Transregional Collaborative Research Center SFB/TR 14 AVACS.

B. Gramlich, D. Miller, and U. Sattler (Eds.): IJCAR 2012, LNAI 7364, pp. 537–543, 2012.

2 Preliminaries

The language of propositional formulas and clauses over a given signature $\Sigma = \{p, q, \ldots\}$ of propositional variables is defined in the usual way. By propositional *valuation*, or simply a *world*, we mean a mapping $W : \Sigma \to \{0, 1\}$. We write $W \models P$ if a propositional formula P is satisfied by W. The syntax of PLTL is an extension of the propositional one by temporal operators $\Box, \Diamond, \mathsf{U}, \ldots$ We do not detail the syntax here, due to lack of space, but will instead directly rely on so called *Separated Normal Form (SNF)* to which any PLTL formula can be translated by a satisfiability preserving transformation with at most linear increase in size [6]. The semantics of PLTL is based on a discrete linear model of time, where the structure of possible time points is isomorphic to $\mathbb{N}$: An *interpretation* is an infinite sequence $(W_i)_{i \in \mathbb{N}}$ of worlds. In order to talk about two neighboring worlds at once we introduce a primed copy of the basic signature: $\Sigma' = \{p', q', \ldots\}$. Primes can also be applied to formulas and valuation with the obvious meaning. Formulas over $\Sigma \cup \Sigma'$ can be evaluated over the respective joined valuation: When both W_1 and W_2 are valuations over Σ, we write $[W_1, W_2]$ as a shorthand for the mapping $W_1 \cup (W_2)' : (\Sigma \cup \Sigma') \to \{0, 1\}$.

The input of our method is a refinement of SNF based on the results of [3]:

Definition 1. *A* PLTL-specification S *is a quadruple* (Σ, I, T, G) *such that*

- Σ *is a finite propositional signature,*
- I *is a set of* initial *clauses* C_i *(over the signature* Σ*),*
- T *is a set of* step *clauses* $C_t \vee D'_t$ *(over joined signature* $\Sigma \cup \Sigma'$*),*
- G *is a set of* goal *clauses* C_g *(over the signature* Σ*).*[1]

An interpretation $(W_i)_{i \in \mathbb{N}}$ *is a* model *of* $S = (\Sigma, I, T, G)$ *if*

1. *for every* $C_i \in I$, $W_0 \models C_i$,
2. *for every* $i \in \mathbb{N}$ *and every* $C_t \vee D'_t \in T$, $[W_i, W_{i+1}] \models C_t \vee D'_t$,
3. *there is infinitely many indices* j *such that for every* $C_g \in G$, $W_j \models C_g$.

A PLTL-specification S is *satisfiable* if it has a model.

Our algorithm for deciding satisfiability of PLTL-specifications to be described next relies on incremental SAT solver technology as described in [5]. There each call to the SAT solver is parameterized by a set of unit assumptions. It either returns a model of all the clauses inserted to the solver that also satisfies the given assumptions, or the UNSAT result along with a subset of the given assumptions that were needed in the proof. Negation of literals from the returned subset can be seen as a new conflict clause that has just been shown semantically entailed by the clauses stored in the solver.

[1] The specification stands for the PLTL formula $(\bigwedge C_i) \wedge \Box (\bigwedge (C_t \vee \bigcirc D_t)) \wedge \Box\Diamond (\bigwedge C_g)$ and can be understood as a symbolic representation of a Büchi automaton recognizing the set of all models of the original input formula.

3 The Algorithm

In order to explain the basic mechanics of our algorithm we first focus on a simpler problem of reaching a goal world only once. That is, given a specification $S = (\Sigma, I, T, G)$, we first try to establish whether there is a finite sequence of worlds $W_0, W_1, \ldots, W_k$ such that $W_0 \models I$, $W_k \models G$ and $[W_i, W_{i+1}] \models T$ for every two neighboring worlds W_i and W_{i+1}. An algorithm for this problem known from verification is called Bounded Model Checking (BMC) [2], where one looks for such a sequence by successively trying increasing values of k (starting with $k = 0$) and employs a SAT solver to answer the respective satisfiability queries[2] until a model for the sequence is found. We modify this approach in order to gain more information from the individual runs with answer UNSAT and to be able to ensure termination in the case of an overall unsatisfiable input.

The idea is to use multiple instances of the solver, as many as there is worlds in the current sequence, and build the sequence progressively, from the beginning towards the end. Each individual *solver instance* contains variables of the joined signature ($\Sigma \cup \Sigma'$) and thus represents two neighboring worlds. However, only the primed part is actually used for SAT solving. As the search proceeds forward, the world model constructed over Σ'-variables in the solver instance i is transformed to a set of assumptions over the Σ-variables for the instance $(i + 1)$. If a world model cannot be completed in the instance $(i + 1)$ due to inconsistency (the current world sequence cannot be extended by one more step) the instance $(i+1)$ returns a *conflict clause* over its assumptions on Σ-variables, which is *propagated* back and added to the solver instance i as a clause over Σ'. Thus the instance i will now produce a different world model, a model which additionally satisfies the added conflict clause. The whole situation is depicted in Fig. 1. We can also see from there how the individual solver instances are initialized. The first contains only the clauses from I (and doesn't depend on assumptions), all the other instances contain clauses from T, and the last instance, additionally, the clauses from G.

One *round* of the algorithm (for a specific value of k) ends either by building an overall sequence of worlds of length $k + 1$, which is a reason for termination with result SAT, or by deriving an empty clause in the first solver instance. Standard BMC would then simply increase k and continue searching for longer sequences. We can do better than that. By analyzing the overall proof[3] of the empty clause, we may discover it doesn't depend on (has not been derived with the help of) I or G in which case we terminate and report overall UNSAT: the same proof will also work for larger values of k. Even if the proof depends on both I, G (and of course T), we can still perform the following check: Define *layer* j as the set of all clauses that depend on G and have been propagated to the solver instance that lies j steps before the last one. Formally, we set layer 0 to be

[2] For every fixed k the question becomes whether there exist a model over $\bigcup_{i=0}^{k} \Sigma^{(k)}$ of the formula $I^{(0)} \wedge \bigwedge_{i=0}^{k-1} T^{(i)} \wedge G^{(k)}$, where $T^{(i)}$ stands for T primed i times, etc.

[3] Proof recording is not needed on the SAT solver side. The described analysis can be implemented with the help of so called *marker literals* as explained, e.g., in [1].

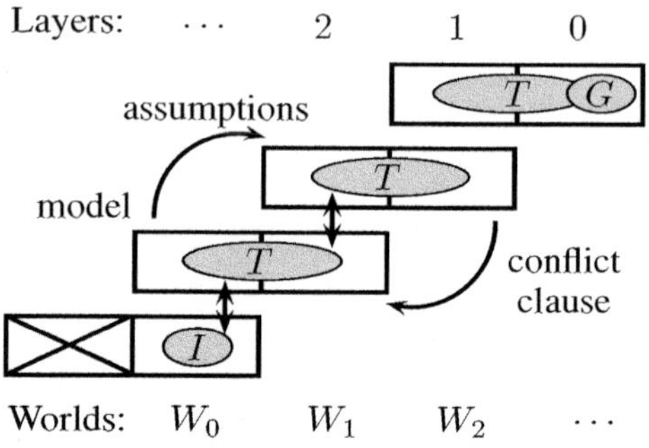

Fig. 1. The information exchanged between the individual solver instances. Completed world model is passed forward as a set of variable assumptions for the next instance. Failed run delivers a conflict clause over the variable assumptions so that the previous instance can be updated and SAT checking may find a different model. The very first solver instance doesn't depend on assumptions. When the last instance reports SAT, we have an overall model for the reachability task.

equal to G. Now, if there are indices $j_1 \neq j_2$ such that layers j_1 and j_2 are equal, we also terminate the algorithm with result UNSAT: we have just discovered a *layer repetition* in the proof, which means we know how we would derive empty clauses also in any of the rounds to come.[4] Note that the case of repeating layers is bound to occur, if not preceded by a different reason for termination, as there are only finitely many different sets of clauses over Σ. This shows the overall termination, and thus completeness, of our modification of BMC.

We now move to providing an algorithm for the general case, where a goal world is required to be shown reachable infinitely many times. In that algorithm we use the above described procedure as a basic building block. In fact, we call the configuration of solver instances as the one in Fig. 1 a *block*. The algorithm starts by building the first block exactly as described above. If this first step doesn't succeed in providing a sequence of worlds, leading from a world satisfying I to a one satisfying G, we terminate with result UNSAT. Otherwise we continue adding new blocks, but now the first solver instance of each new block no longer contains the clauses from I and is instead connected via the model/assumptions link described before to the world represented by the last solver instance of the previous block. This way we continue producing a sequence of blocks, each block being itself a sequence of solver instances (see Fig. 2), the whole thing representing a partial (unfinished) overall model of the given specification. As in the above procedure, each block grows from its initial length 1, and is only extended when necessary and just by one solver instance at a time.

For termination, we perform the following *model repetition check* to recognize a satisfiable specification. Each time a particular run of the SAT solver constructs

[4] Intuitively, the proof can be "cut" at the index j_1, and the part between j_1 and j_2 inserted arbitrarily many times, thus giving rise to proofs of arbitrary length.

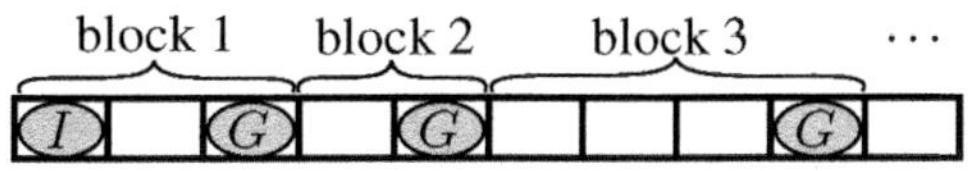

Fig. 2. The layout of blocks as the search proceeds forward. The copies of clauses form T that occupy the positions of every pair of neighboring worlds are not depicted.

a new world model, i.e. a new world in the sequence, we scan all the worlds of previous blocks, and if one of them is equal to the new one, we terminate and report SAT. Note that only the worlds of previous blocks are eligible for the test, because we need to ensure that at least one world satisfying the clauses from G lies between the two repeating ones. The particular infinite sequence formally required as a model of the specification in now easily seen to be represented by the world sequence constructed so far, where the segment of worlds between the repetition will be traversed infinitely often.

Recognizing unsatisfiable specifications is again based on proof analysis. Note that now more than one block (or more precisely the set of goal clauses thereof) may be involved in the derivation of the empty clause. Each time an empty clause is derived, we *extend* the latest block involved in the proof by one additional solver instance and discard any blocks further to the right of it (in the sense of Fig. 2). Then we resume the search. As before each block maintains a sequence of *layers* of clauses. This time layer j contains the clauses that depend on *the block's own copy of* G and have been propagated to a solver instance that lies j steps before its last one. Detecting layer repetition for the first block incurs termination with the result UNSAT as before. If we detect repetition in a block which is not the first one, we perform a so called *leap* inference: A particular repeating layer is selected (see [7] for the details) and its clauses are globally added to the set G. Then the current block is discarded and the search continues from the last solver instance of the previous block. By construction,[5] this last instance currently doesn't provide a model for the strengthened set G, which is a key observation for proving overall termination of the algorithm, because it implies that the leap inference can only be applied finitely many times.

4 Experimental Evaluation

We implemented our algorithm in C++ with Minisat [5] version 2.2 as the backend solver. Although a more efficient implementation with just one solver instance (over an extended signature and special decision heuristic) seems possible, we really use multiple instances of the solver as described before, because it allows us to use the solver in a blackbox manner. An additional abstraction

[5] The intuition behind the leap inference is the following: We have just discovered that none of the successor worlds of the lastly visited G-world is itself a G-world. Thus the lastly visited G-world doesn't have the vital property of lying on a loop in the state space and may be safely discarded from consideration.

layer over the solver has been developed that allows us to mark any clause by a set of block indices it depends on in a form of *marker literals* [1]. That is how we perform the proof analysis described in the previous section without an actual need of true proof recording on the SAT solver side. The standard `set` container is used to represent layers and a simple linear pass implements both the model repetition check and the layer repetition check. We found out that most of the overall running time is typically spent inside individual calls to Minisat and therefore didn't attempt any further optimizations of the checks. As an additional trick we adapted the variable and clause elimination preprocessing of Minisat [4] to be also used on our inputs. This is done only once, before the actual algorithm starts. Special care needs to be taken, because of the dependency between the variables in Σ and Σ'. Moreover, we still need to be able to separate the clauses after elimination into sets I, T and G, which can be achieved by a clever use of marker literals.

Table 1. Number of problems (SAT/UNSAT) solved by each prover – timelimit 60s

problem set	# problems	LS4	trp++	'satisfiable'	'model'
TRP-suite	22	2/20	2/19	2/13	2/13
HW-reach	465	38/55	3/30	0/0	0/0
HW-live	118	38/15	7/7	3/4	0/1

We compared our implementation[6], which we call LS4, with clausal temporal resolution prover trp++[7] version 2.1 and two tableaux-based decision procedures implemented in the PLTL module of the Logics Workbench[8] Version 1.1, namely the 'satisfiable' function and the 'model' function. All the tests were performed on our servers with 3.16 GHz Xeon CPU, 16 GB RAM, running Debian 5.0. We collected several benchmark sets from different sources. The TRP-suite consists of 22 problems available on the web page of trp++[7] in the TOY and FO subdirectories. Further, we translated into PLTL the benchmarks from Hardware Model Checking Competition (HWMCC) 2011[9]. We obtained a set of 465 problems, here denoted HW-reach, from the safety checking track, and 118 problems from the liveness track, HW-live. Note that the competition examples are natively described as circuits in the form of And-Inverter Graphs; these were translated into clause form by standard techniques. The results from these benchmarks are summarized in Table 1. For each prover we report the number of satisfiable and unsatisfiable problems solved in 60 seconds. For a second test we generated formulas from several scalable families[6] and in Table 2 we report for each family the maximal size a prover was able to solve in 60 seconds.

We can see that LS4 is the only system to solve all the problems in the TRP-suite in the given time limit. It also by far outperforms the other systems on

[6] `http://www.mpi-inf.mpg.de/~suda/ls4.html`
[7] `http://www.csc.liv.ac.uk/~konev/software/trp++/`
[8] `http://www.lwb.unibe.ch/`
[9] `http://fmv.jku.at/hwmcc11/`

Table 2. Maximal formula size (from the range) solved in 60s by the provers.

formula family	size range	LS4	trp++	'satisfiable'	'model'
C1	1-100	100	100	3	100
C2	2-20	19	20	3	2
bincnt_u	1-16	10	16	11	6
bincnt_s	1-16	10	11	11	7
binflip_u	2-10	10	5	6	3
binflip_a	2-10	10	5	6	4

the problems coming from verification.[10] The formula families let us see that guidance by a partial model is not always an advantage. For example, on the bincnt_u family, LS4 has to construct an exponentially long path before it starts deriving conflict clauses. Moreover, these need to be propagated back trough all the worlds of the path before the final contradiction is reached. On the other side, the binflip families are already more difficult for the saturation based prover trp++. For example, on binflip_u of size 5, trp++ generates 1494299 resolvents before deriving the empty clause (in 3.67s), while LS4 needs only 1891 calls to Minisat and derives 936 non-empty conflict clauses before reaching the same conclusion (and spends 0.01s on that). To sum up, our test results demonstrate that LS4 considerably advances the state of the art in PLTL satisfiability checking.

References

[1] Asín, R., Nieuwenhuis, R., Oliveras, A., Rodríguez-Carbonell, E.: Efficient Generation of Unsatisfiability Proofs and Cores in SAT. In: Cervesato, I., Veith, H., Voronkov, A. (eds.) LPAR 2008. LNCS (LNAI), vol. 5330, pp. 16–30. Springer, Heidelberg (2008)

[2] Biere, A., Cimatti, A., Clarke, E., Zhu, Y.: Symbolic Model Checking without BDDs. In: Cleaveland, W.R. (ed.) TACAS 1999. LNCS, vol. 1579, pp. 193–207. Springer, Heidelberg (1999)

[3] Degtyarev, A., Fisher, M., Konev, B.: A Simplified Clausal Resolution Procedure for Propositional Linear-Time Temporal Logic. In: Egly, U., Fermüller, C. (eds.) TABLEAUX 2002. LNCS (LNAI), vol. 2381, pp. 85–99. Springer, Heidelberg (2002)

[4] Eén, N., Biere, A.: Effective Preprocessing in SAT Through Variable and Clause Elimination. In: Bacchus, F., Walsh, T. (eds.) SAT 2005. LNCS, vol. 3569, pp. 61–75. Springer, Heidelberg (2005)

[5] Eén, N., Sörensson, N.: An Extensible SAT-solver. In: Giunchiglia, E., Tacchella, A. (eds.) SAT 2003. LNCS, vol. 2919, pp. 502–518. Springer, Heidelberg (2004)

[6] Fisher, M., Dixon, C., Peim, M.: Clausal temporal resolution. ACM Trans. Comput. Logic 2, 12–56 (2001)

[7] Suda, M., Weidenbach, C.: Labelled Superposition for PLTL. In: Bjørner, N., Voronkov, A. (eds.) LPAR-18. LNCS, vol. 7180, pp. 391–405. Springer, Heidelberg (2012)

[10] Note that a dedicated tool `suprove`, the winner of the safety checking track of HWMCC 2011, solved 395 problems in 900s. The winner of the liveness track, the tool `tip`, solved 77 problems within the same timelimit.

Stratification in Logics of Definitions

Alwen Tiu*

Research School of Computer Science
The Australian National University

Abstract. Proof systems for logics with recursive definitions typically impose a strict syntactic stratification on the body of a definition to ensure cut elimination and consistency of the logics, i.e., by forbidding any negative occurrences of the predicate being defined. Often such a restriction is too strong, as there are cases where such negative occurrences do not lead to inconsistency. Several logical frameworks based on logics of definitions have been used to mechanise reasoning about properties of operational semantics and type systems. However, some of the uses of these frameworks actually go beyond what is justified by their logical foundations, as they admit definitions which are not strictly stratified, e.g., in the formalisation of logical-relation type of arguments in typed λ-calculi. We consider here a more general notion of stratification, which allows one to admit some definitions that are not strictly stratified. We outline a novel technique to prove consistency and a partial cut elimination result, showing that every derivation can be transformed into a certain head normal form, by simulating its cut reductions in an infinitary proof system. We demonstrate this technique for a specific logic, but it can be extended to other richer logics.

1 Introduction

Proof systems for logics with definitions or fixed points typically impose a syntactic stratification condition on the body of a recursive definition to make sure that its corresponding fixed point operator has a fixed point. Often such a restriction is too strong; for example, the logics of definitions in [9,4,20,12] all impose a strict stratification where no negative occurrences of a recursive predicate is allowed in the body of its definition. This strict stratification rules out the obvious kinds of inconsistent definitions, such as one that defines an atom to be its negation, e.g., $p \triangleq p \supset \bot$. It is shown in [20] that admitting this definition of p breaks cut-elimination and leads to inconsistency, when the contraction rule is presence. However, this strict stratification also rules out some consistent definition, in the sense that it has a fixed point, such as $odd\ (s\ X) \triangleq odd\ X \supset \bot$, where s denotes the successor function on natural numbers. For a more interesting example of a consistent but not strictly stratified definition, consider the definition of a *reducibility set*, used in normalisation proofs for typed λ-calculi [22,8]:

$$red\ t\ (\alpha \to \beta) \triangleq \forall s. red\ s\ \alpha \supset red\ (t\ s)\ \beta$$

* Supported by the ARC Discovery Grant DP110103173.

B. Gramlich, D. Miller, and U. Sattler (Eds.): IJCAR 2012, LNAI 7364, pp. 544–558, 2012.

which says that a term t of type $\alpha \to \beta$ is reducible iff for every reducible term s of type α, $(t\ s)$ is a reducible term of type β.

Several frameworks based on logics of definitions have been designed for the task of specifying and reasoning about operational semantics [12,17,23,16,6,1]. The underlying logics for these frameworks all impose the strict stratification on definitions. However, some of the more interesting applications of these frameworks actually make use of non-strictly stratified definitions that are not supported by their logical foundations, e.g., the formalisation of reducibility proofs [7]; the definition of the satisfiability relations for a modal logic for process calculi [15]; and encodings of object logics [19]. In this paper, we extend the logical foundations so these uses can be justified formally. We show here how this can be done for an extension of an intuitionistic logic called $FO\lambda^{\Delta\mathbb{N}}$ [12], with a more general notion of stratification, which we call the logic LD. But the methods shown here should be extendable to richer logics of definitions.

The stratification condition in $FO\lambda^{\Delta\mathbb{N}}$ is defined by first assigning a natural number to each predicate symbol p, called its *level*, and denoted by $lvl(p)$. One then extends this definition of levels to formulas, requiring, among others, that the level of an atomic formula is the same as the level of the predicate symbol in the formula, and that the level of $A \supset B$ is the maximum of $lvl(A)+1$ and $lvl(B)$. Then a definition $H \triangleq B$ is stratified iff $lvl(H) \geq lvl(B)$. This immediately rules out the definition $p \triangleq p \supset \bot$ and the definition of *odd* above, as there can be no level assignments to predicates that can stratify these definitions. Assuming that all definitions are stratified this way, cut elimination can be proved [12], using a reducibility argument adapted from normalisation proofs by Tait [22] and Martin-Löf [11]. In the cut-elimination proof of $FO\lambda^{\Delta\mathbb{N}}$, the role of stratification is reflected in the definition of the reducibility sets, which are defined by induction on the levels of formulas. In extending the cut elimination proof for $FO\lambda^{\Delta\mathbb{N}}$, the difficulty is really in showing that the definition of reducibility sets is well-founded; once this is done, the technicality of the cut elimination proof itself can draw on a variety of proofs done for similar systems [12,23,24,1].

We propose here an extension to the definition of stratification in $FO\lambda^{\Delta\mathbb{N}}$ by using level assignments that take into account the arguments in atomic formulas. One problem with this level assignment is that in general we cannot compare the levels of non-ground formulas, e.g., *odd* x and *odd* y are not comparable without knowing the relation between x and y. Another complication is that $FO\lambda^{\Delta\mathbb{N}}$ (and related logics) incorporates a case analysis rule on definitions (called def_L) that may instantiate eigenvariables in a sequent. A consequence of this is that the notions of reducibility sets and levels of formulas have to be stable under arbitrary substitutions. It seems technically challenging to define a well-founded notion of reducibility using this extended stratification without propagating the dependencies between variables throughout the cut-elimination proofs, and complicating the already complicated proofs.

To avoid the complications with having to deal with variables in stratification, we prove cut-elimination indirectly via a ground version of LD, called LD^∞, where no eigenvariables are present in the derivations. This is achieved by utilising Schütte's ω-rule [21] for the introdution rule for $\forall$ on the right (and dually, $\exists$ on the left). We then show, in Section 5, that cut reductions in LD can be simulated by cut reductions in LD^∞. This would normally entail that cut reduction in LD terminates. However, due to a peculiar interaction between cut reductions and the case analysis rule def_L, one can only obtain a weaker result, that cuts can always be pushed up over other rules, but may not be completely eliminated. More specifically, we show that any applications of the cut rule can be pushed up above the case analysis rule, resulting in a derivation in what we call a head normal form. As we shall see in Section 5, the existence of a head normal form for every derivation entails a number of properties usually associated with cut elimination: syntactic consistency (no proofs of $\bot$), and the disjunction and the existential properties. The full cut elimination result, however, holds for a subset of LD defined by *positive formulas* and positive definitions (Section 6).

Despite the fact that we can prove only a limited form of cut-elimination, we argue that, in terms of applications in reasoning about computational models, we still retain much of the use of a full cut-elimination result. One of the most important applications of cut-elimination in a proof-theory-based logical framework is that it can be used as a tool to prove the adequacy of the encoding of a computation model in logic. Adequacy here means a (meta) property that the encoding captures its intended computational meaning. This is typically proved by exploiting the structures of cut-free proofs and the fact that cut-free provability is complete. In this sense, the existence of head normal form is not a strong enough guarantee that the adequacy of the encodings can always be proved. Fortunately, as we shall argue in Section 6, if one considers computational models specified via Horn clauses, there is a natural embedding into positive definitions in LD, for which we have full cut-elimination, so the usual adequacy proofs can still be carried out. Many interesting examples fall within this class, e.g., all examples done using a two-level approach to reasoning in Abella [5,7]. We shall show a couple of examples of positive definitions in Section 3. One of them is an embedding of classical logic, via Gentzen-Gödel's translation and Kolmogorov's translation. We also prove formally the equivalence between the two inside LD.

Due to space constraints, some proofs are omitted, but they can be found in the extended version of the paper.

2 The Logic LD

The logic LD is a small variant of the logic $FO\lambda^{\Delta\mathbb{N}}$ [12], with a modified notion of stratification of recursive definitions. The core logic itself is an intuitionistic first-order fragment of Church's simple type theory. Formulas and terms in LD are just simply typed λ-terms. We shall assume familiarity with simply typed λ-calculus, and recall only the basic notions in the following. We assume a countable set of typed variables, ranged over by x_τ, y_τ, etc. We shall omit the type index when the type information is not important or can be inferred from the context of discussion. Terms are constructed from typed variables, λ-abstraction and application. We consider only well-formed terms, typed using the usual typing rules for simply typed λ-calculus. A term t is *ground* if it contains no free

$$\frac{}{B \longrightarrow B}\ init \qquad \frac{B, B, \Gamma \longrightarrow C}{B, \Gamma \longrightarrow C}\ c_L \qquad \frac{\Gamma \longrightarrow C}{B, \Gamma \longrightarrow C}\ w_L$$

$$\frac{\Delta_1 \longrightarrow B_1 \quad \cdots \quad \Delta_n \longrightarrow B_n \quad B_1, \ldots, B_n, \Gamma \longrightarrow C}{\Delta_1, \ldots, \Delta_n, \Gamma \longrightarrow C}\ mc, \text{where } n \geq 0$$

$$\frac{}{\Gamma \longrightarrow \top}\ \top_R \qquad \frac{B_i, \Gamma \longrightarrow D}{B_1 \wedge B_2, \Gamma \longrightarrow D}\ \wedge_L, i \in \{1,2\} \qquad \frac{\Gamma \longrightarrow B \quad \Gamma \longrightarrow C}{\Gamma \longrightarrow B \wedge C}\ \wedge_R$$

$$\frac{}{\bot \longrightarrow B}\ \bot_L \qquad \frac{B, \Gamma \longrightarrow D \quad C, \Gamma \longrightarrow D}{B \vee C, \Gamma \longrightarrow D}\ \vee_L \qquad \frac{\Gamma \longrightarrow B_i}{\Gamma \longrightarrow B_1 \vee B_2}\ \vee_R, i \in \{1,2\}$$

$$\frac{\Gamma \longrightarrow B \quad C, \Gamma \longrightarrow D}{B \supset C, \Gamma \longrightarrow D}\ \supset_L \qquad \frac{B, \Gamma \longrightarrow C}{\Gamma \longrightarrow B \supset C}\ \supset_R$$

$$\frac{B[t/x], \Gamma \longrightarrow C}{\forall_\tau x.B, \Gamma \longrightarrow C}\ \forall_L \qquad \frac{\Gamma \longrightarrow B[y/x]}{\Gamma \longrightarrow \forall_\tau x.B}\ \forall_R \qquad \frac{B[y/x], \Gamma \longrightarrow C}{\exists_\tau x.B, \Gamma \longrightarrow C}\ \exists_L \qquad \frac{\Gamma \longrightarrow B[t/x]}{\Gamma \longrightarrow \exists_\tau x.B}\ \exists_R$$

Fig. 1. Core inference rules of LD

$$\frac{\Gamma \longrightarrow B\theta}{\Gamma \longrightarrow A}\ def_R, \mathrm{dfn}(A, H \triangleq B, id, \theta) \qquad \frac{\{\{\mathcal{S}_i\}_{i \in \mathcal{I}(H \triangleq B, \theta, \rho)} \mid \mathrm{dfn}(A, H \triangleq B, \theta, \rho)\}}{\Gamma, A \longrightarrow C}\ def_L$$

$$\frac{}{\Gamma \longrightarrow nat\ 0}\ nat_R \qquad \frac{\Gamma \longrightarrow nat\ t}{\Gamma \longrightarrow nat\ (s\ t)}\ nat_R \qquad \frac{\longrightarrow I\ 0 \quad I\ x \longrightarrow I\ (s\ x) \quad \Gamma, I\ t \longrightarrow C}{\Gamma, nat\ t \longrightarrow C}\ nat_L$$

Fig. 2. Introduction rules for recursive definitions and natural numbers

occurrences of variables. We write $s =_\lambda t$ when s and t are equal modulo $\beta\eta$ coversions. The set of free variables in t is denoted by $fv(t)$.

A type is either a *base type*, or a function type $\alpha \to \beta$, where α and β are types. We write $t : \tau$ to mean that the term t has type τ. The notion of a type here is used only to enforce well-formedness of syntactic expressions; their inhabitants are syntactic objects, i.e., λ-terms. So a function type such as $\alpha \to \beta$ is not inhabited by functions in the semantic sense, but by syntactic expressions (i.e., λ-abstractions). We assume that every base type has at least one constant of that type; hence all types are inhabited. This assumption is important to be able to simulate cut reductions of LD in its corresponding infinitary version (see Section 5). Following Church's notation, formulas are λ-terms of a base type o. Logical connectives are constants of the appropriate types: $\bot$ (false) and $\top$ (true) of type o; $\wedge$ (and), $\vee$ (or), $\supset$ (implies) of type $o \to o \to o$; and $\exists_\tau$ (exists), and $\forall$ (for all), of type $(\tau \to o) \to o$, where τ does not contain type o. When writing down formulas, we follow the conventional notation, so rather than writing $\wedge AB$ or $\forall_\tau\ (\lambda x.P)$, we simply write $A \wedge B$ and $\forall_\tau x.P$. We assume a collection of typed predicate symbols, ranged over by p, q, r. Atomic formulas are formulas of the form $p\,\vec{t}$ where p is a predicate symbol. We use capital letters to range over formulas, but reserve A to denote atomic formulas. A sequent is an expression of the form $\Gamma \longrightarrow C$, where Γ is a multiset of formulas and C

is a formula, and they are all in $\beta\eta$-normal form. We consider only variables of "first-order" types, in the sense that their types do not contain o.

We assume the usual notion of capture-avoiding substitutions for λ-calculus. Substitutions are ranged over by θ, ρ, etc. The domain of a substitution θ is denoted by $dom(\theta)$, and its range denoted by $ran(\theta)$. We use the notation $[t_1/x_1, \ldots, t_n/x_n]$ to enumerate substitutions. We assume that the range of every substitution is normalised. A substitution θ is a *ground substitution* if its range consists only of ground terms. We write $t\theta$ to denote the term resulting from applying the mapping θ to t. The identity substitution is denoted by id. Composition of substitutions, denoted by $\theta \circ \rho$, is defined as $(\theta \circ \rho)(x) = (x\rho)\theta$. The restriction of a susbtitution θ to a set of variables V is denoted by $\theta|_V$.

The introduction rules for the standard logical connectives for LD are the same as in $FO\lambda^{\Delta\mathbb{N}}$. These are given in Figure 1. The rule mc (for 'multi-cut') is a generalisation of the cut rule. The $\exists_L$ and $\forall_R$ rules have the usual side condition that y is not free in the conclusions of the rules. In $\exists_R$ and $\forall_L$, the term t has type τ. As in $FO\lambda^{\Delta\mathbb{N}}$, the core logic of LD is extended with a proof theoretic notion of definitions [12], and a principle of induction for natural numbers.

Definition 1 (Recursive definitions). *A* definition clause *in* LD *has the form* $\forall\vec{x}.[p\ \vec{t} \triangleq B]$ *where* $fv(\vec{t}) = \{\vec{x}\}$ *and* $fv(B) \subseteq \{\vec{x}\}$. *The atomic formula* $p\ \vec{t}$ *is called the* head *of the definition clause, and* B *is called the* body. *The symbol* $\triangleq$ *is not a logical connective; it is used only as a notation to separate the head and the body of a definition clause. A definition is a finite set of definition clauses.*

We do not assume that every predicate symbol has a definition clause associated with it. We shall allow some predicates to be undefined.

We assume that, in a definition clause $\forall\vec{x}.H \triangleq B$, the variables $\vec{x}$ are not used anywhere else in derivations or other definition clauses. We can think of $\vec{x}$ as reserve variables used only in definition clauses. So they never occur in the range of any susbtitutions, but they can occur in the domain for specific cases as we shall see later. Since we only have a finite number of clauses, this assumption is harmless, as we always have enough variables to be used in derivations and ranges of substitutions. With this convention, we shall often remove the quantifier $\vec{x}$ from the above definition clause, and simply write $H \triangleq B$.

For determinacy reasons, we shall restrict to definitions of the form $\forall\vec{x}.H \triangleq B$ where H is a *higher-order pattern* term [14,18], i.e., every occurrence of x in H is applied to only distinct bound variables. This restriction is immaterial as one can always rewrite any given definition to an equivalent pattern clause. That is, if there is a non-pattern clause $p\,\vec{t}\,u \triangleq B$ where $\vec{t}$ are pattern terms and u is not, then one redefines p as follows: $p\,\vec{t}\,x \triangleq B \wedge (x = u)$ where x is a new variable, and where the equality symbol is defined via the pattern definition $\forall y.(y = y) \triangleq \top$.

We assume that each ground atomic formula A is assigned a natural number, $lvl(A)$, called its *level*. We require that $lvl(p\vec{t}) = 0$ when p is an undefined predicate symbol or $p = nat$. When using LD, we need to be explicit about how such a level assignment to ground atomic formulas can be defined. In the following we generalise the notion of level to non-atomic ground formula. To define the level of a quantified formula, we take the least upper bound of the levels of all its instances; thus the level in this case can be a (countable) ordinal. The level of a non-atomic closed formula is defined recursively as follows:

- $lvl(\bot) = lvl(\top) = 0$.
- $lvl(A \wedge B) = lvl(A \vee B) = max(lvl(A), lvl(B))$.
- $lvl(A \supset B) = max(lvl(A) + 1, lvl(B))$.
- $lvl(\exists_\tau x.B) = lvl(\forall_\tau x.B) = \sup\{lvl(B[t/x]) \mid t \text{ is a ground term of type } \tau\}$.

Definition 2 (Stratification of definitions). *A definition clause* $\forall \vec{x}.[A \triangleq B]$ *is* stratified *if for every ground substitution* $\theta = [\vec{t}/\vec{x}]$, *we have* $lvl(A\theta) \geq lvl(B\theta)$.

We shall assume from now on that all definition clauses are stratified.

The relation $\text{dfn}(A, H \triangleq B, \theta, \rho)$ holds whenever $A\theta =_\lambda H\rho$ and $dom(\rho) = fv(H)$. The latter is not neccessary, but it simplifies some proofs. Because of the variable convention and the pattern restriction on definitions, we have:

Lemma 3. *If* $\text{dfn}(A, H \triangleq B, \delta, \theta)$ *and* $\text{dfn}(A, H \triangleq B, \delta, \rho)$, *then* $\theta = \rho$.

The introduction rules for a defined predicate are basically unfolding of its definitions (reading the rules bottom up). They are shown in Figure 2, i.e., the rules def_L and def_R. Recall that we do not assume all predicates are defined, so if a predicate is undefined, the introduction rules def_L and def_R are not applicable.

The right-introduction rule def_R matches an atom with a definition clause. The def_L rule is the dual of def_R: we consider all definition clause $H \triangleq B$ and *all* substitution θ and all substitution ρ such that $\text{dfn}(A, H \triangleq B, \theta, \rho)$. Each $\mathcal{I}(H \triangleq B, \theta, \rho)$ is a non-empty countable index set, and each $\mathcal{S}_i$, for $i \in \mathcal{I}(H \triangleq B, \theta, \rho)$, is an occurrence of the sequent $\Gamma\theta, B\rho \longrightarrow C\theta$. That is, for each (θ, ρ) such that $\text{dfn}(A, H \triangleq B, \theta, \rho)$ holds, there could be more than one (possibly infinitely many) occurrences of the sequent $\Gamma\theta, B\rho \longrightarrow C\theta$ in the premises of def_L. This is a generalisation of def_L in $FO\lambda^{\Delta\mathbb{N}}$, where the set $\mathcal{I}(H \triangleq B, \theta, \rho)$ is a singleton. It is easy to see that the above def_L and its $FO\lambda^{\Delta\mathbb{N}}$ version are interchangable without affecting provability. The need for the more general version of def_L is purely technical; it allows one to account for all possible ways of instantiating a derivation, so that cut reduction is stable under instantiations of eigenvariables.

For practical uses of the logic, it is enough to consider a finitary version of def_L using the notion of *complete sets of unifiers* (CSU) [10]. Instead of considering all possible θ and ρ such that $\text{dfn}(A, H \triangleq B, \theta, \rho)$, one instead consider all unifiers γ that are in the CSU of A and H. This version of def_L using CSU is shown to be equivalent to the version of def_L that uses dfn in [12].

If the premise set of def_L is empty, i.e., if A does not unify with the head of any definition clause, then the conclusion is proved. For example, given a definition such as $p\ (s\ x) \triangleq p\ x$, one can show that $p\ 0 \longrightarrow \bot$ by an application of def_L, because $p\ 0$ cannot possibly be unified with $p\ (s\ x)$.

We now extend LD with a predicate nat to encode the natural numbers. We assume a given type nt with two constructors: $0 : nt$ (zero) and $s : nt \rightarrow nt$ (successor function). The predicate nat has type $nt \rightarrow o$, and its introduction rules are given in Figure 2, i.e., the rules nat_R and nat_L. Note that the type nt only enforces syntactic well-formedness of terms, e.g., one cannot form a term

$s\ (\lambda.0)$ of type nt. The semantic notion of a set of natural numbers is encoded through the predicate nat, via its introduction rules. The rule nat_L embodies the induction principle over natural numbers. The term I here is an *inductive invariant*, which is a ground term of type $nt \to o$, and x is a new variable not appearing in the conclusion of the rule.

In terms of proof-theoretic strength, LD, like $FO\lambda^{\Delta\mathbb{N}}$, is at least as expressive as an intuitionistic version of Peano arithmetic (see [13] for a proof). As we do not allow structural induction rules to apply to arbitrary definitions, the kind of well-ordering that one can encode (using a definition) in LD and prove is restricted to at most those isomorphic to natural numbers.

We say that a derivation Π is a *premise derivation* of another derivation Π' if Π occurs as the derivation of one of the premises of the last rule in Π'. The *height* of a derivation Π, denoted by $ht(\Pi)$, is defined as:

$$ht(\Pi) = \sup\{ht(\Pi') \mid \Pi' \text{ is a premise derivation of } \Pi\} + 1.$$

Saturated Derivations. An important property that is needed in cut-elimination for LD is the stability of cut reduction under substitutions of eigenvariables. We now define a subclass of LD derivations, called saturated derivations, that satisfies this property. We then show that every LD derivation can be saturated.

To ease presentation, we shall introduce some notation to denote derivations. We shall abbreviate the following derivation

$$\frac{\begin{array}{c}\Pi_1\\ \Delta_1 \longrightarrow B_1\end{array} \quad \cdots \quad \begin{array}{c}\Pi_n\\ \Delta_n \longrightarrow B_n\end{array} \quad \begin{array}{c}\Pi\\ B_1, \dots, B_n, \Gamma \longrightarrow C\end{array}}{\Delta_1, \dots, \Delta_n, \Gamma \longrightarrow C}\ mc$$

as $mc(\Pi_1, \dots, \Pi_n, \Pi)$. Whenever we use such a notation, it is implicit that the right-hand side of the end sequent of Π_i is in the multiset of formulas in the left-hand side of the end sequent of Π. We use the notation: $def_L(\{\mathfrak{D}(H \triangleq D, \theta, \rho) \mid \mathrm{dfn}(A, H \triangleq D, \theta, \rho)\})$ to denote a derivation ending with def_L, with sets of premise derivations $\mathfrak{D}(H \triangleq D, \theta, \rho)$. Again, it is implicit that A is in the left-hand side of the end sequent of the derivation. The notation $def_R(\Pi, H \triangleq B)$ denotes a derivation ending with def_R, with premise derivation Π, where the right-hand side of the end sequent is matched against the definition clause $H \triangleq B$.

We identify derivations that differ only in the choice of eigenvariables that are not free in the end sequents. It is easy to see that one can always apply renaming to those variables to avoid clashes with variables in any given substitution θ. So in the definition of an application of a substitution θ to Π, we shall assume that eigenvariables that are not free in the end sequent are not affected by θ.

Definition 4. *Let Π be a derivation in* LD *and let θ be a substitution. We define a derivation $\Pi[\theta]$ in* LD *by induction on the height of Π as follows:*

1. *Suppose $\Pi = def_L(\{\mathfrak{D}(H \triangleq B, \rho, \rho') \mid \mathrm{dfn}(A, H \triangleq B, \rho, \rho')\})$. Then $\Pi[\theta] = def_L(\{\mathfrak{D}(H \triangleq B, \theta \circ \delta, \delta' \mid \mathrm{dfn}(A\theta, H \triangleq B, \delta, \delta')\})$. Note that $\mathrm{dfn}(A\theta, H \triangleq B, \delta, \delta')$ implies $\mathrm{dfn}(A, H \triangleq B, \theta \circ \delta, \delta')$ so all derivations in the set $\mathfrak{D}(H \triangleq B, \theta \circ \delta, \delta')$ are premise derivations of Π.*

2. *If Π ends with any other rule with premise derivations $\{\Pi_i\}_{i\in\mathcal{I}}$ for some index set $\mathcal{I}$, then $\Pi[\theta]$ ends with the same rule with premise derivations $\{\Pi_i[\theta]\}_{i\in\mathcal{I}}$.*

Lemma 5. *Π and $\Pi[id]$ are the same derivation.*

Lemma 6. *$(\Pi[\theta])[\rho]$ and $\Pi[\theta \circ \rho]$ are the same derivation.*

Substitution does not increase the height of a derivation. It might yield a smaller derivation because some branches in def_L may be pruned.

Lemma 7. *$ht(\Pi[\theta]) \leq ht(\Pi)$.*

Definition 8. *A derivation Π in* LD *is* saturated *if the following hold:*

- *Suppose $\Pi = def_L(\{\mathfrak{D}(H \triangleq B, \theta, \rho) \mid \mathrm{dfn}(A, H \triangleq B, \theta, \rho)\})$. Then Π is saturated if whenever $\mathrm{dfn}(A, H \triangleq B, \theta, \rho)$ and $\Psi \in \mathfrak{D}(H \triangleq B, \theta, \rho)$, then Ψ is saturated and for every δ, $\Psi[\delta] \in \mathfrak{D}(H \triangleq B, \theta \circ \delta, \rho')$ where $\rho' = (\rho \circ \delta)|_{fv(H)}$.*
- *If Π ends with any other rule, then it is saturated iff its premise derivations are all saturated.*

Intuitively, a saturated derivation is a derivation in which all instances of def_L are closed under substitution, i.e., if Π is a premise derivation of an instance of def_L, then $\Pi[\theta]$ is also in the premise set of the same def_L.

Lemma 9. *If Π is saturated then $\Pi[\sigma]$ is also saturated.*

Lemma 10. *If a sequent is derivable, then there exists a saturated derivation of the same sequent.*

Cut Reduction. We now define a set of reduction rules to eliminate applications of mc in LD derivations. The cut reduction rules for LD are minor variations of those in $FO\lambda^{\Delta\mathbb{N}}$ [12]. We show here the interesting cases where def_L is involved. Let $\Xi = mc(\Pi_1, \dots, \Pi_n, \Pi)$ be the redex to reduce, where each Π_i ends with the sequent $\Delta_i \longrightarrow B_i$ and Π ends with $B_1, \dots, B_n, \Gamma \longrightarrow C$.

- *Case def_R/def_L:* Suppose $\Pi_1 = def_R(\Pi_1', H \triangleq D)$, where $\mathrm{dfn}(B_1, H \triangleq D, id, \rho)$, and Π ends with a def_L applied to the cut formula B_1, i.e.,

$$\Pi = def_L(\{\mathfrak{D}(H' \triangleq D', \theta, \sigma) \mid \mathrm{dfn}(B_1, H' \triangleq D', \theta, \sigma)\}).$$

  For any $\Psi \in \mathfrak{D}(B_1, H \triangleq D, id, \delta)$, the derivation $mc(\Pi_1', \Pi_2, \dots, \Pi_n, \Psi)$ is a reduct of Ξ. Note that there could be infinitely many reducts for Ξ, if $\mathfrak{D}(H \triangleq D, id, \delta)$ is infinite.
- *Case $def_L/\circ_L$:* Π ends with a left rule $\circ_L$ other than the rule c_L acting on B_1 and $\Pi_1 = def_L(\{\mathfrak{D}(H \triangleq D, \theta, \rho) \mid \mathrm{dfn}(A, H \triangleq D, \theta, \rho)\})$ for some $A \in \Delta_1$. Let

$$\mathfrak{D}'(H \triangleq D, \theta, \rho) = \{mc(\Psi, \Pi_2[\theta], \dots, \Pi_n[\theta], \Pi[\theta]) \mid \Psi \in \mathfrak{D}(H \triangleq D, \theta, \rho)\}.$$

  Then Ξ reduces to $def_L(\{\mathfrak{D}'(H \triangleq D, \theta, \rho) \mid \mathrm{dfn}(A, H \triangleq D, \theta, \rho)\})$.

Given Lemma 9, it is not difficult to see that cut reduction preserves the property of a being a saturated derivation.

Lemma 11. *If Ξ is a saturated derivation in* LD *ending with mc, then every reduct of Ξ is also saturated.*

3 Examples

We consider a couple of simple examples of recursive definitions in LD. For more advanced examples of non-strictly stratified definitions, but that still can be made stratified according to Definition 2, see for example, strong normalisation proofs for typed λ-calculi in Abella [5,7].[1] All the proofs, except for the adequacy theorem, have been formalised in the theorem prover Abella [5]. Abella is an implementation of a richer logic, but the formalisation uses only inference rules permitted by LD. The adequacy theorem can be proved by induction on the structures of cut-free derivations in LD.

Odd numbers. Consider the following two mutually recursive definitions (the mutual recursion is just to make the example slightly more interesting):

$$\forall x.p\ (s\ x) \triangleq q\ x \supset \bot \qquad \forall y.q\ (s\ y) \triangleq p\ y \supset \bot$$

and the standard definition of odd numbers:

$$odd\ (s\ 0) \triangleq \top \qquad \forall x.odd\ (s\ (s\ x)) \triangleq odd\ x.$$

Notice that $p\ 0 \supset \bot$ is provable, because $p\ 0$ cannot be unified with the head $p\ (s\ x)$ of the definition clause of p. The same observation leads to provability of $q\ 0 \supset \bot$, which in turn implies provability of $p\ (s\ 0)$. Thus the 'base case' for the definition of p (and likewise, q) is implicitly covered by the fact that failure of unification of an atom on the left with its definitions turns it into a proof of its negation. This is a unique feature of logics of definitions that separates them from other fixed point logics. These clauses can be stratified by the following level assignment: $lvl(p\ t) = |t|$, $lvl(q\ t) = |t|$ and $lvl(odd\ t) = 1$, where $|t|$ is the size of the term t. We can prove in LD that these two definitions of odd numbers coincide: $\forall x.nat\ x \supset (p\ x \equiv odd\ x)$ where $\equiv$ here stands for logical equivalence.

Encoding classical logic. Let us introduce a type o' to denote object-logic formulas. We consider here the propositional classical logic, where the connectives are encoded as the following constants: $\hat{\wedge}, \hat{\vee}, \Rightarrow : o' \to o' \to o'$ and $\hat{\bot} : o'$. Atomic object-logic formulas are encoded using the constant $\mathrm{atm} : \iota \to o'$ for some base type ι. We shall need to do induction over the structures of formulas. As we do not have induction rules for recursive definitions, this has to be done indirectly by indexing the definition of object formulas with natural numbers, and do induction on natural numbers instead.

$$\begin{array}{ll}
\mathrm{fm}\ (s\ I)\ (A \hat{\wedge} B) \triangleq \mathrm{fm}\ I\ A \wedge \mathrm{fm}\ I\ B. & \mathrm{fm}\ I\ \hat{\bot} \triangleq \top. \\
\mathrm{fm}\ (s\ I)\ (A \hat{\vee} B) \triangleq \mathrm{fm}\ I\ A \wedge \mathrm{fm}\ I\ B. & \mathrm{fm}\ I\ (\mathrm{atm} X) \triangleq \top. \\
\mathrm{fm}\ (s\ I)\ (A \Rightarrow B) \triangleq \mathrm{fm}\ I\ A \wedge \mathrm{fm}\ I\ B. & \mathrm{form}\ A \triangleq \exists I.nat\ I \wedge \mathrm{fm}\ I\ A.
\end{array}$$

[1] Strictly speaking, these examples on λ-calculi are outside LD, as they require the ∇-quantifier [23]. But the techniques shown here can be extended to include ∇.

$$
\begin{array}{ll}
g\ (\text{atm}\ X) \triangleq \neg\neg(\text{atom}\ X) & k\ (\text{atm}\ X) \triangleq \neg\neg(\text{atom}\ X).\\
g\ \hat{\bot} \triangleq \bot. & k\ \hat{\bot} \triangleq \bot.\\
g\ (A \hat{\wedge} B) \triangleq g\ A \wedge g\ B. & k\ (A \hat{\wedge} B) \triangleq \neg\neg(k\ A \wedge k\ B).\\
g\ (A \Rightarrow B) \triangleq g\ A \supset g\ B. & k\ (A \Rightarrow B) \triangleq \neg\neg(k\ A \supset k\ B).\\
g\ (A \hat{\vee} B) \triangleq \neg(\neg(g\ A) \wedge \neg(g\ B)). & k\ (A \hat{\vee} B) \triangleq \neg\neg(k\ A \vee k\ B).
\end{array}
$$

Fig. 3. Gentzen-Gödel's translation (left), and Kolmogorov's translation (right)

There are at least two ways to encode classical logic into intuitionistic logic: one via Gentzen-Gödel's negative translation, and the other via Kolmogorov's double negation translation [25]. The definition clauses for both are given in Figure 3, where we abbreviate $F \supset \bot$ as $\neg F$. Note that atomic object level formulas are interpreted in the meta-logic using an undefined predicate atom.

Both definitions can be stratified by letting $lvl(g\ t) = lvl(k\ t) = |t|^2$ for every ground object-level formula t. For example, in the last clause we have:

$$lvl(\neg\neg(k\ A \vee k\ B)) = max(|A|^2, |B|^2) + 2 \leq (|A| + |B| + 1)^2 = lvl(A \hat{\vee} B).$$

Theorem 12 (Adequacy). *F is valid classicaly iff $(g\ F)$ is provable in* LD.

Theorem 13. *The formula $\forall F.$*form *$F \supset (g\ F \equiv k\ F)$ is provable in* LD.

4 A Ground Proof System with Recursive Definitions

We now define a ground version of LD, called LD^∞, where no eigenvariables occur in the derivations. This allows one to define a well-founded notion of reducible derivations which is crucial to the cut-elimination proof of $FO\lambda^{\Delta\mathbb{N}}$. Head normalisation of LD can then be proved by simulating its cut reductions in LD^∞. We shall not detail the cut elimination proof for LD^∞ here, as this is straightforward once the notion of reducibility can be defined. We shall review in this section the definitions of normalizability and reducibility as used in $FO\lambda^{\Delta\mathbb{N}}$. For more detailed explanations of these notions, the reader is referred to [11,12].

The rules of LD^∞ are identical to LD, except for $\forall_R$, $\exists_L$ and nat_L; these are replaced with their infinitary versions, given below, where $\mathcal{T}_\alpha$ is the (countable) set of ground terms of type α.

$$\frac{\{\Gamma \longrightarrow B[t/x]\}_{t \in \mathcal{T}_\tau}}{\Gamma \longrightarrow \forall_\tau x.B}\ \forall_R \qquad \frac{\{B[t/x], \Gamma \longrightarrow C\}_{t \in \mathcal{T}_\tau}}{\exists_\tau x.B, \Gamma \longrightarrow C}\ \exists_L$$

$$\frac{\longrightarrow D\ 0 \quad \{D\ n \longrightarrow D\ (s\ n)\}_{n \in \mathcal{T}_{nt}} \quad D\ t, \Gamma \longrightarrow C}{nat\ t, \Gamma \longrightarrow C}\ nat_L$$

The infinitary form of $\forall_R$ above is essentially Schütte's ω-rule [21]. As the $\forall_R$ rule is applicable to $\forall$ of type nt, strictly speaking there is no need for nat_L in LD^∞, as the $\forall_R$-rule is powerful enough to capture the induction scheme. We keep the rules to simplify the translation from LD to LD^∞. The cut reduction rules for

LD^∞ are a straightforward adaptation of those for LD; they are identical except for those involving the modified rules above. The definition of applications of substitutions to a derivation in Definition 4 also applies to LD^∞.

Definition 14. *The set of* normalizable *derivations is the smallest set that satisfies:*

1. *If a derivation Π ends with mc, then it is normalizable if for every θ, every reduct of $\Pi[\theta]$ is normalizable.*
2. *If a derivation ends with any rule other than mc, then it is normalizable if the premise derivations are normalizable.*

It is obvious that normalizability implies cut-elimination.

In the definition of normalizability, in the case where a derivation ends with *mc*, we require that *all* the reducts of all its instances are normalizable. This is different from $FO\lambda^{\Delta\mathbb{N}}$ where only one of its reducts (for every instance) is required to be normalizable. Thus our definition captures a notion strong normalization with respect to 'head' reduction (as we always reduce lowest instances of *mc*). Although McDowell and Miller's proof is claimed for weak normalization, the actual proof itself can be adapted for strong normalization without much change; see for example, variants of their proofs in [23,24].

The *level of a derivation Π* is the level of the formula on the right-hand side of the end sequent of Π. We define a family $\{\mathcal{R}_i\}_i$ of reducible sets of derivations, by (transfinite) induction on the levels of derivations. We say that a derivation is *reducible* if it is in $\mathcal{R}_i$ where i is the level of the derivation.

Definition 15. *The set $\mathcal{R}_i$ of reducible i-level derivations is the smallest set of derivations of level i satisfying:*

1. *If Π ends with mc, then $\Pi \in \mathcal{R}_i$ if for every θ, every reduct of $\Pi[\theta]$ is in $\mathcal{R}_i$.*
2. *If Π ends with $\supset_R$:* $\dfrac{\begin{array}{c}\Pi' \\ \Gamma, B \longrightarrow C\end{array}}{\Gamma \longrightarrow B \supset C}\supset_R$, *then it is in $\mathcal{R}_i$ if for every θ and every j-level reducible derivation Ψ of $\Delta \longrightarrow B$, where $j = lvl(B)$, we have $mc(\Psi, \Pi'[\theta]) \in R_k$ where $k = lvl(C)$.*
3. *If a derivation ends with $\supset_L$ or nat_L, then it is in $\mathcal{R}_i$ if the right premise derivation is in $\mathcal{R}_i$ and the other premises are normalizable.*
4. *If a derivation ends with any other rule, then it is in $\mathcal{R}_i$ if the premise derivations are reducible.*

As all definitions are stratified, it follows that the definition of reducibility above is well-founded. It can be shown that that reducibility implies normalizability, following the same proof as in $FO\lambda^{\Delta\mathbb{N}}$ [12]. The main cut-elimination proof consists of showing that every derivation is reducible.

Theorem 16. *Cut elimination holds for* LD^∞.

Corollary 17. *The logic* LD^∞ *is consistent, i.e., there is no proof of* $\bot$ *in* LD^∞.

5 Simulating Cut Reductions of LD in LD^∞

To make use of the cut elimination result for LD^∞ to prove partial cut elimination of LD, we need to prove some certain commutation results between cut reduction, substitution of eigenvariables in derivations and the translation from LD to LD^∞.

Definition 18. *Let Π be a derivation of a ground sequent $\Gamma \longrightarrow C$ in* LD. *We define a derivation $[\![\Pi]\!]$ of $\Gamma \longrightarrow C$ in LD^∞ by induction on $ht(\Pi)$ as follows:*

1. *If Π is as shown below left, then $[\![\Pi]\!]$ is as shown below right:*

$$\dfrac{\begin{matrix}\Pi' \\ \Gamma \longrightarrow B\end{matrix}}{\Gamma \longrightarrow \forall_\alpha x.B}\ \forall_R \qquad \dfrac{\left\{\begin{matrix}[\![\Pi'[t/x]]\!] \\ \Gamma \longrightarrow B[t/x]\end{matrix}\right\}_{t\in\mathcal{T}_\alpha}}{\Gamma\theta \longrightarrow \forall_\alpha x.B}\ \forall_R$$

2. *Π ends with $\exists_L$: this is dual to the previous case.*
3. *Π ends with nat_L:* $\dfrac{\begin{matrix}\Pi_0 \\ \longrightarrow D\ 0\end{matrix} \quad \begin{matrix}\Pi_s \\ D\ x \longrightarrow D\ (s\ x)\end{matrix} \quad \begin{matrix}\Pi' \\ \Gamma, D\ u \longrightarrow C\end{matrix}}{\Gamma, nat\ u \longrightarrow C}\ nat_L.$

   Then $[\![\Pi]\!]$ is

$$\dfrac{\begin{matrix}[\![\Pi_0]\!] \\ \longrightarrow D\ 0\end{matrix} \quad \left\{\begin{matrix}[\![\Pi_s[n/x]]\!] \\ D\ n \longrightarrow D\ (s\ n)\end{matrix}\right\}_{n\in\mathcal{T}_{nt}} \quad \begin{matrix}[\![\Pi']\!] \\ \Gamma, D\ u \longrightarrow C\end{matrix}}{\Gamma, nat\ u \longrightarrow C}\ nat_L\ .$$

4. *If Π ends with any other rule, with premise derivations $\{\Pi_i\}_{i\in\mathcal{I}}$ for some index set $\mathcal{I}$, then $[\![\Pi]\!]$ ends with the same rule with premise derivations $\{[\![\Pi_i]\!]\}_{i\in\mathcal{I}}$.*

Given a saturated derivation Π of $\Gamma \longrightarrow C$ in LD and a ground substitution θ, we say that θ is a *grounding substitution for Π* if $fv(\Gamma, C) \subseteq dom(\theta)$.

Lemma 19. *Let Ξ be a saturated derivation of $\Delta \longrightarrow C$ in* LD *ending with mc. Let θ be a grounding substitution for Ξ. If Ξ reduces to Ξ' in* LD *then $[\![\Xi[\theta]]\!]$ reduces to $[\![\Xi'[\theta]]\!]$ in LD^∞.*

Proof. Suppose $\Xi = mc(\Pi_1, \ldots, \Pi_n, \Pi)$ and it reduces to Ξ' in LD. For most cases of the cut reduction, it is straightforward to verify that $[\![\Xi[\theta]]\!]$ reduces to $[\![\Xi'[\theta]]\!]$ in LD^∞. We show here one case involving the reduction rule def_R/def_L.

Suppose Π_1 ends with def_R with premise derivation Π_1', where $\mathrm{dfn}(B_1, H \triangleq D, id, \rho)$, and $\Pi = def_L(\{\mathfrak{D}(H' \triangleq D', \sigma, \delta) \mid \mathrm{dfn}(B_1, H' \triangleq D', \sigma, \delta)\})$. Suppose $\Xi' = mc(\Pi_1', \Pi_2, \ldots, \Pi_n, \Psi)$ for some $\Psi \in \mathfrak{D}(H \triangleq D, id, \rho)$. In this case,

$$\Pi[\theta] = def_L(\{\mathfrak{D}'(H' \triangleq D', \theta \circ \sigma, \delta) \mid \mathrm{dfn}(B_1\theta, H' \triangleq D', \sigma, \delta)\}).$$

Because Π is saturated, it follows that $\Psi[\theta] \in \mathfrak{D}(H \triangleq D, \theta, \rho') = \mathfrak{D}'(H \triangleq D, id, \rho')$. Then we have $[\![\Pi_1[\theta]]\!] = def_R([\![\Pi_1'[\theta]]\!], H \triangleq D)$ where $\mathrm{dfn}(A\theta, H \triangleq D, id, \rho')$. Moreover, the following is a reduct of $[\![\Xi[\theta]]\!]$:

$$\Xi'' = mc([\![\Pi_1'[\theta]]\!], [\![\Pi_2[\theta]]\!], \ldots, [\![\Pi_n[\theta]]\!], [\![\Psi[\theta]]\!])$$

because $\Psi[\theta] \in \mathfrak{D}'(H \triangleq D, id, \rho')$. So we have

$$\begin{aligned}\Xi'' &= mc([\![\Pi_1'[\theta]]\!], [\![\Pi_2[\theta]]\!], \ldots, [\![\Pi_n[\theta]]\!], [\![\Psi[\theta]]\!]) \\ &= [\![mc(\Pi_1', \Pi_2, \ldots, \Pi_n, \Psi)[\theta]]\!] = [\![\Xi'[\theta]]\!].\end{aligned}$$

□

Definition 20. *A derivation Π in* LD *is said to be in* head normal form *if every instance of mc in Π appears above an instance of def_L.*

Head normalisation can be proved using Lemma 19; given an LD-derivation Π ending with mc, we apply a grounding substitution, say θ (which must exist because types are non-empty), to get $[\![\Pi[\theta]]\!]$ and simulate any sequence of reductions from Π with its image reductions in LD^∞ from $[\![\Pi[\theta]]\!]$. Note that as an instance of mc may be pushed up over a def_L instance in LD, and because a substitution may prune some branches of def_L, not all premises of Π are retained in $\Pi[\theta]$. As a result, we are not able to conclude that all cut instances above def_L instances in Π are eliminable.

Theorem 21 (Head normalisation). *Every derivation Π of a sequent in* LD *can be transformed into a derivation Π', in head normal form, of the same sequent.*

Theorem 22 (Disjunction property). *$B \vee C$ is derivable in* LD *if and only if either B is derivable or C is derivable in* LD.

Theorem 23 (Existential property). *$\exists_\tau x.B$ is derivable in* LD *if and only if there exists a term $t : \tau$ such that $B[t/x]$ is derivable in* LD.

6 Cut-Elimination for a Positive Fragment of LD

Let Σ be a set of defined predicate symbols, with two subsets (not necessarily distinct) Σ^+ of *positive predicates* and Σ^- of *negative predicates*, such that $\Sigma = \Sigma^- \cup \Sigma^+$. The sets of *$\Sigma$-positive* and *$\Sigma$-negative formulas* are defined via the grammars below, respectively:

$$\begin{aligned} P &::= p\,\vec{s} \mid nat\ u \mid r\,\vec{s} \mid P \wedge P \mid P \vee P \mid N \supset P \mid \exists x.P \\ N &::= q\,\vec{t} \mid r\,\vec{t} \mid N \wedge N \mid N \vee N \mid P \supset N \mid \forall x.N \end{aligned}$$

where $p \in \Sigma^+$, $q \in \Sigma^-$ and r is any undefined predicate symbol. The set Σ is *closed* if for every $p \in \Sigma^+$ (resp. Σ^-), and every definition clause $p\,\vec{t} \triangleq B$, B is a Σ-positive formula (resp. Σ-negative formula). A formula B is a *positive formula* (resp. *negative formula*) if there exists a closed Σ such that B is a Σ-positive (resp. Σ-negative) formula. Note that a formula B can be both positive and negative, e.g., when the formula and all definitions for the predicate symbols occuring in B are quantifier-free. A definition clause is *positive* if its body is a positive formula. A sequent $\Gamma \longrightarrow C$ is a *positive sequent* if every formula in Γ is negative and C is positive. Obviously, any cut-free derivation of a ground positive sequent does not use the introduction rules $\forall_R$, $\exists_L$ or nat_L, which means that provability for ground positive sequents in LD and LD$^\infty$ coincide. Therefore, for the positive fragment, cut-elimination for LD$^\infty$ implies cut-elimination for LD.

Theorem 24. *A ground positive sequent is derivable in* LD *if and only if it is cut-free derivable in* LD.

Positive formulas and positive definitions are already expressive enough to capture models of computations that can be encoded via logic programming. This follows from the fact that Horn clauses in logic programming can be encoded as positive definitions, and Horn goals can be encoded as positive formulas. That is, every Horn clause $\forall \vec{x}.B \supset H$ is encoded as the definition $\forall \vec{x}.H \triangleq B$. It is easy to see that the def_R rule simulates the backchaining rule in logic programming. So in proving the adequacy result of an encoding of a computation model encoded via Horn clauses, we can use the fact that cut-free proof search is complete.

7 Related and Future Work

There is a long series of works on logics of definitions, often extended with induction/co-induction proof rules [9,4,20,12,17,23,6,1]. All these works enforce a strict stratification that forbids negative occurrences of a recursive predicate in its definition. A different approach to proving cut elimination in the presence of non-strictly stratified definitions is recently being considered by Baelde and Nadathur [2]. Instead of considering a recursive definition as defining a fixed point, they view it as a rewrite rule (on propositions), and use the deduction modulo framework [3] to prove cut elimination. However, their introduction rules for atoms do not allow case analysis to be applied to recursive definitions.

On the application side, it would be interesting to use linear logic as the base logic, so that one could do the kind of embedding of different object logics in the meta logic as it is done in [19]. It would also be interesting to investigate whether the techniques shown here can be adapted to a type theoretic setting. Finally, it should be straightforward to extend the simulation technique as we show here to establish head normalisation for extensions of LD with the ∇ quantifier, e.g., the logic $\mathcal{G}$ [6] with the more general stratification conditions we consider here.

References

1. Baelde, D.: Least and greatest fixed points in linear logic. ACM Trans. Comput. Log. 13(1), 2 (2012)
2. Baelde, D., Nadathur, G.: Combining deduction modulo and logics of fixed point definitions. Accepted to LICS 2012 (2012)
3. Dowek, G., Werner, B.: Proof normalization modulo. Journal of Symbolic Logic 68(4), 1289–1316 (2003)

4. Eriksson, L.-H.: A Finitary Version of the Calculus of Partial Inductive Definitions. In: Eriksson, L.-H., Hallnäs, L., Schroeder-Heister, P. (eds.) ELP 1991. LNCS (LNAI), vol. 596, pp. 89–134. Springer, Heidelberg (1992)
5. Gacek, A.: The Abella Interactive Theorem Prover (System Description). In: Armando, A., Baumgartner, P., Dowek, G. (eds.) IJCAR 2008. LNCS (LNAI), vol. 5195, pp. 154–161. Springer, Heidelberg (2008)
6. Gacek, A., Miller, D., Nadathur, G.: Combining generic judgments with recursive definitions. In: LICS, pp. 33–44. IEEE Computer Society (2008)
7. Gacek, A., Miller, D., Nadathur, G.: Reasoning in Abella about structural operational semantics specifications. Electr. Notes Theor. Comput. Sci. 228, 85–100 (2009)
8. Girard, J.-Y., Taylor, P., Lafont, Y.: Proofs and Types. Cambridge University Press (1989)
9. Hallnäs, L.: Partial inductive definitions. Theor. Comput. Sci. 87(1), 115–142 (1991)
10. Huet, G.: A unification algorithm for typed λ-calculus. Theoretical Computer Science 1, 27–57 (1975)
11. Martin-Löf, P.: Hauptsatz for the intuitionistic theory of iterated inductive definitions. In: Proc. of the Second Scandinavian Logic Symposium. Studies in Logic and the Foundations of Mathematics, vol. 63, pp. 179–216. North-Holland (1971)
12. McDowell, R., Miller, D.: Cut-elimination for a logic with definitions and induction. Theoretical Computer Science 232, 91–119 (2000)
13. McDowell, R., Miller, D.: Reasoning with higher-order abstract syntax in a logical framework. ACM Trans. Comput. Log. 3(1), 80–136 (2002)
14. Miller, D.: A logic programming language with lambda-abstraction, function variables, and simple unification. J. Logic and Computation 1(4), 497–536 (1991)
15. Miller, D.: Encoding generic judgments: Preliminary results. Electr. Notes Theor. Comput. Sci. 58(1), 59–78 (2001)
16. Miller, D., Tiu, A.: A proof theory for generic judgments. ACM Trans. Comput. Logic 6(4), 749–783 (2005)
17. Momigliano, A., Tiu, A.: Induction and Co-induction in Sequent Calculus. In: Berardi, S., Coppo, M., Damiani, F. (eds.) TYPES 2003. LNCS, vol. 3085, pp. 293–308. Springer, Heidelberg (2004)
18. Nipkow, T.: Functional unification of higher-order patterns. In: Vardi, M. (ed.) LICS 1993, pp. 64–74. IEEE (June 1993)
19. Pimentel, E., Miller, D.: On the Specification of Sequent Systems. In: Sutcliffe, G., Voronkov, A. (eds.) LPAR 2005. LNCS (LNAI), vol. 3835, pp. 352–366. Springer, Heidelberg (2005)
20. Schroeder-Heister, P.: Rules of definitional reflection. In: Vardi, M. (ed.) LICS, pp. 222–232. IEEE Computer Society Press, IEEE (June 1993)
21. Schütte, K.: Proof theory. In: Grundlehren der mathematischen Wissenschaften, Springer (1977)
22. Tait, W.W.: Intensional interpretations of functionals of finite type I. Journal of Symbolic Logic 32(2), 198–212 (1967)
23. Tiu, A.: A Logical Framework for Reasoning about Logical Specifications. PhD thesis, Pennsylvania State University (May 2004)
24. Tiu, A., Momigliano, A.: Cut elimination for a logic with induction and co-induction. CoRR, abs/1009.6171 (2010)
25. Troelstra, A.S., Schwichtenberg, H.: Basic Proof Theory. Cambridge University Press (1996)

Diabelli: A Heterogeneous Proof System*

Matej Urbas and Mateja Jamnik

Computer Laboratory, University of Cambridge, UK
{Matej.Urbas,Mateja.Jamnik}@cl.cam.ac.uk

Abstract. We present Diabelli, a formal reasoning system that enables users to construct so-called heterogeneous proofs that intermix sentential formulae with diagrams.

1 Introduction

Despite the fact the people often prove theorems with a mixture of formal sentential methods as well as the use of more informal representations such as diagrams, mechanised reasoning tools still predominantly use sentential logic to construct proofs – diagrams have not yet made their way into traditional formal proof systems. In this paper, we do just that: we present a novel heterogeneous theorem prover Diabelli that allows users to seamlessly mix traditional logic with diagrams to construct completely formal heterogeneous proofs – Fig. 1 shows an example of such a proof. In particular, Diabelli connects a state-of-the-art sentential theorem prover Isabelle [1] with our new formal diagrammatic theorem prover for spider diagrams called Speedith [2]. The interactive user interface allows for displaying typical sentential proof steps as well as visual diagrammatic statements and inferences. The derived heterogeneous proof is certified to be (logically) correct. Our heterogeneous framework is designed to allow reasonably easy plugin of other external proof tools (sentential or diagrammatic), thus potentially widening the domain of problems that can be tackled heterogeneously.

The motivation for our work is not to produce shorter and faster proofs, but to provide a different perspective on formulae as well as to enable a flexible way of proving them. The users can switch representation and type of proof steps at any point, which gives them options and freedom in the way they construct the proof, tailored to their level of expertise and their cognitive preference. The integration in Diabelli's framework benefits both: diagrammatic reasoners gain proof search automation and expressiveness of sentential reasoners, while sentential reasoners gain access to another, diagrammatic view of the formula and its proof which might provide a better insight into the problem.

2 Heterogeneous Reasoning Components

The kind of problems that can be tackled in our Diabelli heterogeneous framework depends on the choice of sentential and diagrammatic reasoners, thus on the domains of Isabelle and Speedith. Isabelle, as a general purpose theorem prover, covers numerous different domains. Speedith's domain is monadic first-order logic with equality

* Supported by EPSRC Advanced Research Fellowship GR/R76783 (Jamnik), EPSRC Doctoral Training Grant and Computer Lab Premium Research Studentship (Urbas).

B. Gramlich, D. Miller, and U. Sattler (Eds.): IJCAR 2012, LNAI 7364, pp. 559–566, 2012.

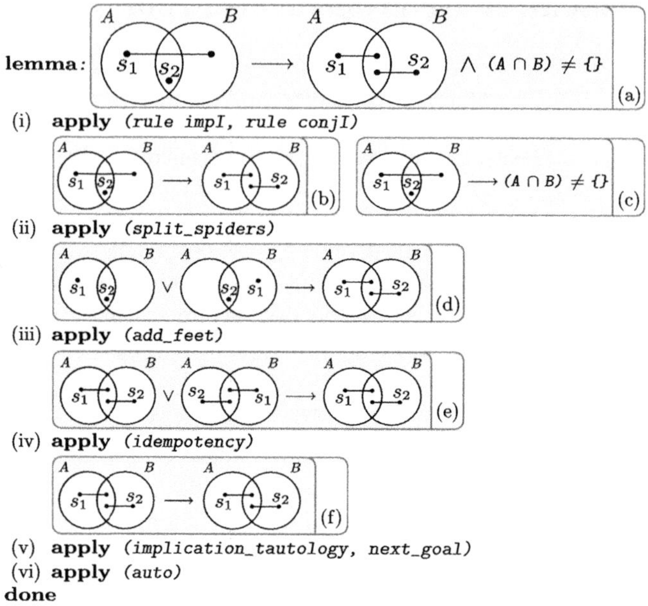

Fig. 1. A heterogeneous proof of a statement with sentential steps (i), (v), (vi) and diagrammatic steps (ii), (iii), (iv). (i) splits the lemma into a diagrammatic and a sentential sub-goal. (ii), (iii) and (iv) prove the diagrammatic sub-goal. (v) discharges the goal (f) and proceeds to the sentential goal in (c), which is discharged by *auto* in (vi) – a powerful proof tactic in Isabelle. All steps are verified by Isabelle's logical kernel.

(MFOLE). Thus, heterogeneously, in Diabelli system we can currently prove all theorems that contain subformulae of MFOLE provided they are expressed with spider diagrams. Even though this domain is decidable, so potentially Isabelle alone could sententially prove any theorem of MFOLE, our motivation lies elsewhere, namely in heterogeneous proofs and in showing the feasibility of formal heterogeneous reasoning, rather than in sentential proofs alone. Spider diagrams are a case study and a prototype diagrammatic language for the Diabelli framework. When Diabelli is extended with new theorem provers, the domain of problems covered will extend accordingly.

2.1 The Diagrammatic Reasoner

Spider Diagrams. The language of spider diagrams is used in Speedith, which constitutes the diagrammatic reasoning part of our heterogeneous system. It is equivalent to MFOLE, and is sound and complete (see [3] for details).

Spider diagrams are composed of *contours*, *zones*, *spiders*, *shading*, and *logical connectives*. *Contours* are closed curves which represent sets and assert relationships between them through relative spatial positioning. For example, the enclosure of one contour in another denotes the subset and superset relations. Fig. 1(c) shows a diagram containing two contours named with *labels* A and B. The set of all contour labels in a diagram d is denoted by $L(d)$.

A *zone* is an area lying within some (or no) collection of contours and lying outside the others. It denotes a set and is represented as a pair of finite, disjoint sets of contour labels (in, out), such that $in \cup out = L(d)$. A zone (in, out) lies within contours in and outside of contours out. The diagrams in Fig. 1(c) contains the zones $(\emptyset, \{A, B\})$, $(\{A\}, \{B\})$, $(\{B\}, \{A\})$, and $(\{A, B\}, \emptyset)$. The set of zones in a diagram is denoted by $Z(d)$. The zones in $Z(d)$ can be *shaded*, denoted by $ShZ(d)$. Shading indicates that the zone's only elements are the spiders, which places an upper bound on the cardinality of the set.

Spiders are connected acyclic graphs whose nodes (*feet*) lie in a set of zones. A spider asserts the existence of an element in the union of the zones where its feet reside. The set of all spiders in a diagram d is denoted by $S(d)$, and the function $\eta : S(d) \rightarrow \mathcal{P}(Z(d)) \setminus \emptyset$ returns the *habitat*, that is, the set of zones in which a spider's feet are placed. The diagrams in Fig. 1(c) contains two spiders s_1 and s_2, with s_1, for example, having the habitat $\{(\{A\}, \{B\}), (\{B\}, \{A\})\}$.

The diagrams considered so far are called *unitary* diagrams. Spider diagrams can be negated with the $\neg$ operator, and joined with binary logical connectives $\wedge$ (conjunction), $\vee$ (disjunction), $\Rightarrow$ (implication), and $\Leftrightarrow$ (equivalence) into *compound diagrams* (e.g., Fig. 1(b)). For a complete formal specification of the semantics of spider diagrams, see [2]. In Isabelle/HOL we formalise it with the `sd_sem` interpretation function (Sec. 3.1).

Diagrammatic Inference Rules. Spider diagrams are equipped with inference rules [2,3] that are all proved to be sound, hence proofs derived by using them are guaranteed to be correct. Step (iii) in Fig. 1 from diagrams in (d) to (e) shows an application of the diagrammatic inference rule *add feet* (which adds feet to an existing spider to assert that it could live in another region too). Speedith allows interactive application of this and a number of additional inference rules (see [2] for complete specification).

Speedith's Architecture. Speedith [2] is our implementation of an interactive diagrammatic theorem prover for spider diagrams. It has four main components:

1. abstract representation of spider-diagrammatic statements,
2. the reasoning kernel with proof infrastructure,
3. verification of diagrammatic proofs, including input and output system for importing and exporting formulae in many different formats, and
4. visualisation of spider-diagrammatic statements.

The abstract representation of spider diagrams is used internally in Speedith to represent all spider-diagrammatic formulae (see [2] for details). Speedith can be used as a standalone interactive proof assistant, but it can also be easily plugged into other systems via its extensible mechanism for import and export of spider diagrams. Currently, spider diagrams can be exported to Isabelle/HOL formulae or a textual format native to Speedith; and for import, MFOLE formulae need to be translated to Speedith's native textual format.

2.2 The Sentential Reasoner - Isabelle

For the sentential part of our heterogeneous reasoning framework, we chose Isabelle, which is a general purpose interactive proof assistant. This choice was arbitrary, any other highly expressive and interactive theorem prover could be used. In particular, Diabelli requires the reasoner to provide a way to interactively enter proof goals and proof steps. The reasoner should also be able to output its current proof goals and incorporate new proof steps from external tools. In the future we could add a requirement that the reasoner should support storage of arbitrary data – this could be utilized when a diagrammatic language is not translatable into a reasoner's format.

3 Integration of Diagrammatic and Sentential Reasoners

The Diabelli framework integrates diagrammatic and sentential provers on two levels. Firstly, it connects them via drivers that in case of Speedith and Isabelle contain a bidirectional translation procedures and a formal definition of the semantics of diagrams – this is presented next. Secondly, the interactive construction of heterogeneous proofs is facilitated through Diabelli's graphical user interface, which is presented in Sec. 4.

3.1 Interpretation of Spider Diagrams in Isabelle/HOL

To formally define the semantics of spider diagrams we specify a theory of spider diagrams in Isabelle/HOL. In particular, we provide a formalisation of the abstract representation and its interpretation, which verifies that our encoding is faithful to the formalisation in [3]. The main part of this theory is the function `sd_sem` (Def. 2) which translates the abstract representation of spider diagrams to Isabelle/HOL formulae. The function `sd_sem` interprets a data structure `SD` (Def. 1), which closely matches the abstract representation of spider diagrams.

Definition 1. *The* `SD` *data structure captures the abstract representation of spider diagrams and is defined in Isabelle/HOL as:*

```
datatype SD = NullSD
            | CompoundSD {operator: bool =>...=> bool, args: sd list}
            | UnitarySD {habitats: region list, sm_zones: zone set}
```

The unitary diagram contains a list of regions (spider habitats that η generates) and a set of zones. Regions are sets of zones, zones are pairs of in- and out-contours, and contours are native Isabelle/HOL sets. The data structure `SD` does not contain a list of spider names S – they are generated by the interpretation function `usd_sem` (see Def. 3).

Definition 2. *The* `sd_sem` *function takes as an argument a description of the spider diagram and produces a FOL formula that corresponds to the meaning of the given spider diagram. The function is defined in Isabelle/HOL as:*

```
fun sd_sem (spider_diagram : SD) =
    NullSD -> True
  | CompoundSD operator args -> apply operator (map sd_sem args)
  | UnitarySD habitats sm_zones -> usd_sem habitats sm_zones
```

The function `sd_sem` formally specifies the semantics of spider diagrams. The *null spider diagram* is interpreted as the logical truth constant. The *compound spider diagrams* are interpreted as a composition of a number of spider diagrams with `operator`. The interpretation of a *unitary spider diagram* is central to the specification and is defined by the function `usd_sem` in Def. 3.

Definition 3. *The* usd_sem *function interprets a unitary spider diagram which is given through a list of regions (i.e., habitats of all spiders) and a set of shaded zones. It produces a FOL formula that describes the meaning of the unitary spider diagram:*

```
fun usd_sem (habs: region list, sm_zones: zone set) =
    for each h in habs
            conjunct ∃s. s ∈ ⋃_{zone∈h} sd_zone_sem zone
            spiders ← spiders ∪ s
    ∧ distinct spiders
    ∧ ∀z ∈ sm_zones. sd_zone_sem z ⊆ spiders
```

where we use this shorthand: (**for each** x **in** [x1...xn] **conjunct** P(x)) $\equiv$ P(x1)∧...∧ P(xn). We define sd_zone_sem as:

```
fun sd_zone_sem (in, out) = [⋂_{c∈in} set_of c] \ [⋃_{c∈out} set_of c]
```

where in and out are the sets of in- and out-contours of the given zone. For every habitat h, usd_sem introduces a fresh existentially quantified variable s (a spider) and asserts that s lives in the region defined by h. Once all variables s_i are introduced, they are declared distinct and shaded zones are interpreted as sets that may only contain spiders.

Here is an example of how sd_sem interprets the spider diagram from Fig. 1(b):

```
(i)   sd_sem CompoundSD{operator →,
                 [UnitarySD{[[(A,B),(B,A)],[(AB,∅)],{}},
                  UnitarySD{[[(A,B),(AB,∅)],[(B,A),(AB,∅)]],{}}]}
(ii)  (usd_sem UnitarySD{[[(A,B),(B,A)],[(AB,∅)],{}}) →
      (usd_sem UnitarySD{[[(A,B),(AB,∅)],[(B,A),(AB,∅)]],{}})
(iii) (∃s1.s1∈(sd_zone_sem(A,B) ∪ sd_zone_sem(B,A)) ∧
       ∃s2.s2∈ sd_zone_sem(AB,∅) ∧ distinct[s1,s2]) →
      (∃s1.s1∈(sd_zone_sem(A,B) ∪ sd_zone_sem(AB,∅)) ∧
       ∃s2.s2∈(sd_zone_sem(B,A) ∪ sd_zone_sem(AB,∅)) ∧ distinct[s1,s2])
(iv)  (∃s1.s1∈(A\B ∪ B\A) ∧ ∃s2.s2∈(A∩B) ∧ distinct[s1,s2]) →
      (∃s1.s1∈(A\B ∪ A∩B) ∧ ∃s2.s2∈(B\A ∪ A∩B) ∧ distinct[s1,s2])
```

where step (i) is a call to the sd_sem function, which applies the operator and calls the function usd_sem in step (ii). usd_sem existentially quantifies variables with their regions in step (iii). Lastly, in step (iv), sd_zone_sem interprets zones as sets, which produces the final formula.

3.2 Translation of Isabelle/HOL to Spider Diagrams

Above, we showed how diagrams are translated to MFOLE expressions via sd_sem function. We now give a translation in the other direction: from MFOLE expressions in Isabelle/HOL to spider diagrams.

An algorithm for conversion from MFOLE formulae to spider diagrams exists, but it was shown to be intractable for practical applications [4]. Consequently, Diabelli currently translates formulae that are in a specific form, called SNF (*spider normal form*), which is based on the sd_sem function. Whilst SNF is a syntactic subset of MFOLE, it is important to note that it is able to express any spider-diagrammatic formula. An example of an SNF formula of a compound diagram was given in line (iv) of the translation example in Sec. 3.1 above.

The translation procedure recursively descends into Isabelle's formulae, which are internally represented as trees, and returns the abstract representation of the corresponding spider diagram by essentially reversing the *sd_sem* function. A future goal is to extend this translation with heuristics that would cover a wider range of formulae.

4 Architecture of Diabelli

The architecture of Diabelli heterogeneous framework with Isabelle and Speedith plugins is illustrated in Fig. 2. Diabelli utilizes a plugin system of the I3P framework [5]

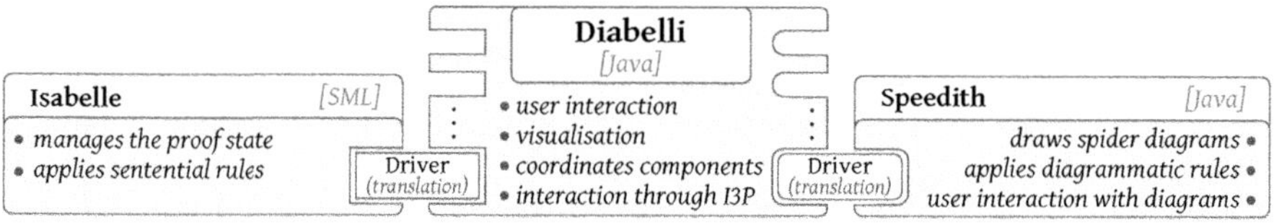

Fig. 2. The architecture of the Diabelli framework with Isabelle and Speedith

to connect the user interfaces of the sentential and diagrammatic provers. Fig. 3 shows Diabelli's graphical user interface. User commands are passed from I3P to Isabelle to

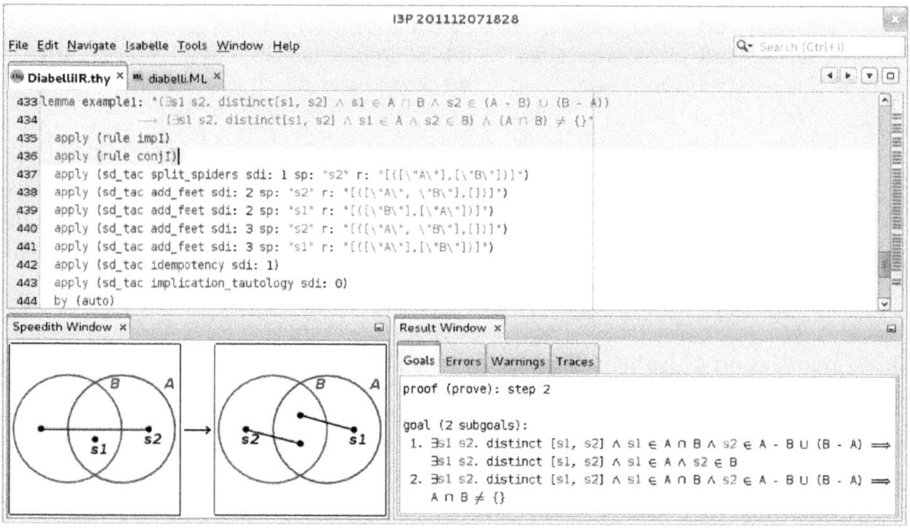

Fig. 3. A screenshot of a heterogeneous proof in Diabelli

execute them. The results are returned back to I3P, which displays them in a separate result window. The commands are read from a user-edited theory file, which may contain, for example, custom definitions, lemmas, and proof scripts. Diabelli's users may instruct I3P step by step to issue the sentential commands to Isabelle or the diagrammatic instructions to Speedith.

Diagrammatic instructions take the same textual form in the theory file as any other Isabelle tactic (see the `sd_tac` entries in Fig. 3). These instructions, however, are not manually written by the user, but are generated by Diabelli through the user's point-and-click interactions with the diagram (note the Speedith sub-window in Fig. 3).

Diabelli automatically presents every translatable sentential sub-goal as a spider diagram in the Speedith window. If the user chooses to apply a diagrammatic inference rule on (part of) it, then Speedith executes the rule and produces a transformed diagram, which it passes back to Diabelli to replace the old sub-goal.

Diabelli currently connects only one general purpose theorem prover with a single diagrammatic language and its logic. However, we designed the Diabelli framework to be easily extended with new diagrammatic or sentential languages and logical systems by leveraging on the extensibility of the I3P framework that can manage multiple provers at the same time. Formalising the requirements for Diabelli plugins remains part of our future work.

5 General Observations

Diabelli is a novel heterogeneous reasoning framework that is a proof of concept for the connection of more powerful provers. It demonstrates how sentential and diagrammatic theorem provers can be integrated into a single heterogeneous framework. We show that using these to interactively reason with mixed diagrammatic and sentential inference steps is feasible and formally verifiable – this is breaking new ground in mechanised reasoning. Diabelli provides an intuitive interface for people wanting to understand the nature of proof. It more closely models how humans solve problems than existing state-of-the-art proof tools, is adaptable and flexible to the needs of the user, and capitalises on the advantages of each individual proof system integrated into Diabelli.

Applying either or both, diagrammatic and sentential proof steps is seamless. The normal workflow of Isabelle is not modified by the diagrammatic subsystem, moreover, the diagrammatic steps may be applied whenever a sentential formula can be translated into the diagrammatic language.

Closest to Diabelli is the Openproof [6] framework which is the only other existing system that facilitates the construction of heterogeneous reasoning systems. However, it does not integrate existing reasoning systems, but rather provides a way of combining different representations and logics.

Diabelli is implemented in SML and Java; its sources are available from `https://gitorious.org/speedith`. With Diabelli, we can heterogeneously prove all theorems that contain subformulae of MFOLE expressed with spider diagrams – this is a significant range and depth of theorems.

Despite our focus on the language of spider diagrams, Diabelli introduces a way to extend its scope to other domains. It is designed as a plugable system for seamless integration of other diagrammatic and sentential theorem provers – we are currently developing drivers for other systems to demonstrate the scalability of the Diabelli framework. A future direction is to establish a formal and concise specification of the plugin interface required to add new systems that extend Diabelli's problem domain.

References

1. Paulson, L.C.: Isabelle - A Generic Theorem Prover. LNCS, vol. 828. Springer, Heidelberg (1994)
2. Urbas, M., Jamnik, M., Stapleton, G., Flower, J.: Speedith - a diagrammatic reasoner for spider diagrams. In: Diagrams. LNCS. Springer (2012)
3. Howse, J., Stapleton, G., Taylor, J.: Spider Diagrams. LMS JCM 8, 145–194 (2005)
4. Stapleton, G., Howse, J., Taylor, J., Thompson, S.J.: The expressiveness of spider diagrams. JLC 14(6), 857–880 (2004)
5. Gast, H.: Towards a Modular Extensible Isabelle Interface. In: 22nd TPHOLs, pp. 10–19. TU Muenchen, Institut fuer Informatik (2009)
6. Barker-Plummer, D., Etchemendy, J., Liu, A., Murray, M., Swoboda, N.: Openproof - A Flexible Framework for Heterogeneous Reasoning. In: Stapleton, G., Howse, J., Lee, J. (eds.) Diagrams 2008. LNCS (LNAI), vol. 5223, pp. 347–349. Springer, Heidelberg (2008)

Author Index

GPSR Compliance

The European Union's (EU) General Product Safety Regulation (GPSR) is a set of rules that requires consumer products to be safe and our obligations to ensure this.

If you have any concerns about our products, you can contact us on ProductSafety@springernature.com

In case Publisher is established outside the EU, the EU authorized representative is:

Springer Nature Customer Service Center GmbH
Europaplatz 3
69115 Heidelberg, Germany

Batch number: 09496765

Printed by Printforce, the Netherlands